From Kant to Hilbert:
A Source Book in the Foundations of Mathematics

Volume II

From Kant to Hilbert: A Source Book in the Foundations of Mathematics

Volume II

WILLIAM EWALD
The Law School
University of Pennsylvania

CLARENDON PRESS · OXFORD

Oxford University Press, Walton Street, Oxford, OX2 6DP

Oxford New York
Athens Auckland Bangkok Bombay
Calcutta Cape Town Dar es Salaam Delhi
Florence Hong Kong Istanbul Karachi
Kuala Lumpur Madras Madrid Melbourne
Mexico City Nairobi Paris Singapore
Taipei Tokyo Toronto

and associated companies in
Berlin Ibadan

Oxford is a trade mark of Oxford University Press

Published in the United States
by Oxford University Press Inc., New York

First published 1996
First published in paperback 1999

A catalogue record for this book is available from the British Library

Library of Congress Cataloging in Publication Data
From Kant to Hilbert: readings in the foundations of mathematics/William Ewald.
Includes bibliographical references and indexes

ISBN 0 19 850535 3 (Vol. 1)
ISBN 0 19 850536 1 (Vol. 2)
ISBN 0 19 850537 X (set)

I. Mathematics-Philosophy. I. Ewald, William Bragg, III 1954–
QA8.6.F77 1996 510'.1–dc20 96–1586

Printed in Great Britain by
Bookcraft (Bath) Ltd
Midsomer Norton, Avon.

To
C.R.E.
and
T.H.B.E

Von allen Hülfsmitteln, welche der menschliche Geist zur Erleichterung seines Lebens, d.h. der Arbeit, in welcher das Denken besteht, bis jetzt erschaffen hat, ist keines so folgenreich und so untrennbar mit seiner innersten Natur verbunden, wie der Begriff der *Zahl*. Die Arithmetik, deren einziger Gegenstand dieser Begriff ist, ist schon jetzt eine Wissenschaft von unermesslicher Ausdehnung und es ist keinem Zweifel unterworfen, dass ihrer ferneren Entwicklung gar keine Schranken gesetzt sind; ebenso unermesslich ist das Feld ihrer Anwendung, weil jeder denkende Mensch, auch wenn er dies nicht deutlich fühlt, ein Zahlenmensch, ein Arithmetiker ist.

– Richard Dedekind

CONTENTS

Volume I

Volume II

16
Georg Friedrich Bernhard Riemann (1826–1866)

The last few selections have dealt with algebra and its applications to logic, topics that were pursued largely by mathematicians in Britain and America. With Riemann we turn to the intellectual world of late nineteenth-century Germany, and to geometry. Previous selections—especially those from Kant, Lambert, Bolzano, and Gauss—have already broached the main traditional problems of the foundations of geometry. These are: the truth and provability of Euclid's axiom of parallels; the nature of an axiom system; the possibility of alternative geometries; the role of intuition (*Anschauung*) in mathematical thought; the relationship of geometry to the study of empirical space. Riemann's short paper 'On the hypotheses which underlie geometry' was to inaugurate a new era in the study of space, and was, in time, to have a profound influence on the way mathematicians viewed each of the traditional problems.

Riemann's paper was published posthumously in 1868, at about the same time that the work of Gauss, Bolyai, and Lobatchevsky on non-Euclidean geometry was becoming generally known.[a] The combined authority of Gauss and Riemann gave impetus to the study of non-Euclidean geometry, and the ideas in Riemann's paper were immediately taken up and explored by Clifford, Klein, von Helmholtz, Beltrami, Christoffel, and Lipschitz. Many years later, Klein recalled that

> The publication of ‖*Riemann 1868*‖ occurred just at the time when I was beginning to occupy myself independently with mathematical problems. So I still have vivid memories of the extraordinary impact Riemann's train of thought made on the young mathematicians of the day. Much seemed to us dark and difficult to understand, and yet of unfathomable depth, where the modern mathematician, who has already absorbed all these things into his mode of thought from the outset, only admires the clarity and fecundity of the exposition (*Klein 1926–7*, Vol. ii, p. 165).

[a] The first public mention of Gauss's ideas occurred in a memorial address by Sartorius von Waltershausen, *Gauss zum Gedächtnis*, in 1856. The address is reprinted in *Gauss 1863–1929* (Vol. viii, pp. 267–8). The publication of Gauss's correspondence with Schumacher, including the geometrical correspondence, began in 1862; but the full extent of Gauss's work was not known until the publication of Volume 8 of the *Werke* in 1900. The contributions of Lobatchevsky and Bolyai, published in the 1830s, were ignored until *Baltzer 1866–7* called attention to them. For a discussion of the evolution of Gauss's views on non-Euclidean geometry and their gradual emergence into public view after his death, see the Introductory Note to the selections from Gauss in Volume 1.

Riemann, the son of a Protestant minister, entered the University of Göttingen in 1846. He spent the years 1847–9 in Berlin, where he studied under Jacobi and Dirichlet. He returned to Göttingen in 1849, and in the 1850s had one of the most brilliantly productive decades in the history of mathematics. To mention some of the high points: in 1851 he earned his doctorate in Göttingen with a dissertation on complex function theory; in it, he introduced the idea of Riemann surfaces. In 1853 he submitted his *Habilitation* thesis—a landmark in the theory of Fourier series. In 1854 he delivered his *Habilitation* lecture on non-Euclidean geometry (*Riemann 1868*), a lecture which overturned the study of geometry, and which is reproduced below. In 1857 he published his seminal article on Abelian functions; in 1859, his paper on the Riemann zeta-function. In that year, Riemann succeeded Dirichlet as professor of mathematics at Göttingen; Dirichlet had himself succeeded Gauss in 1855. Among Riemann's colleagues and friends was Dedekind, who, with Heinrich Weber, was to edit his papers. (The second edition of Riemann's papers contains philosophical material from the *Nachlass*; this additional material was edited by Max Noether, the father of Emmy.)

The paper reproduced below on the foundations of geometry was Riemann's formal lecture to qualify him to teach in the Philosophical Faculty at Göttingen. It was read, in the presence of Gauss, on 10 June 1854. (Gauss, an old man, was to die the following year.) The story of the lecture is recounted by Dedekind in his biographical sketch of Riemann (*Riemann 1876*, pp. 515–17). On 23 December 1853 (according to Dedekind) Riemann wrote to his brother Wilhelm 'At the beginning of December I submitted my *Habilitation* thesis, and was required to propose three topics for a trial lecture, of which the faculty then chooses one. I had prepared the first two, and hoped that they would pick one of them; but Gauss chose the third, so now I am in a pinch and must prepare it.' Dedekind says that, 'Gauss, contrary to tradition, chose the third proposed topic rather than the first because he was curious to hear how such a difficult subject would be handled by so young a man; the lecture, which surpassed all his expectations, caused him the greatest astonishment, and on the way home from the faculty meeting he expressed to Wilhelm Weber, with uncharacteristic excitement, his highest praise for the depth of the thoughts Riemann had presented.' (Incidentally, the young Dedekind's own *Habilitation* lecture, 'On the introduction of new functions in mathematics', translated below, was read in the presence of Gauss ten days after Riemann's.)

Riemann's lecture distinguished between what are now called *topological* properties of space and its *metric* properties; this distinction enabled him to generalize traditional geometry in two ways. First, he subsumed the study of three-dimensional spaces under the study of n-dimensional manifolds; second, he observed that the same underlying n-dimensional manifold can admit different metrics, which then give rise to different geometries. So Euclidean space becomes the doubly special case of a three-dimensional topological manifold of constant curvature with a vanishing curvature tensor.

Riemann's paper falls into three parts, which treat the topological, metric, and physical properties of space respectively. In part one, Riemann begins by mentioning Gauss and Herbart as precursors. The influence of Gauss on his thought is obvious; that of Herbart, less so. Johann Friedrich Herbart (1776–1841), professor of philosophy at Königsberg and Göttingen, was a figure in whose ideas Riemann was deeply interested. Riemann's *Nachlass* in Göttingen contains many reading notes and discussions of Herbart's philosophical ideas; for a published sample, see the philosophical fragments reproduced as an appendix to *Riemann 1876*. But despite Riemann's mention of Herbart in the *Habilitation* lecture, it is not clear precisely in what manner Herbart influenced his geometrical investigations. (For a discussion of this problem, see *Scholz 1982*.)

Riemann next introduces the notions of a continuous manifold (I, §1) and of an *n*-dimensional manifold (I, §2). His treatment of these topics is intuitive and imprecise, perhaps because his lecture was intended for an audience of non-mathematicians. But it seems, both from this paper and from the technical results in the so-called *Pariserarbeit* (*1861*), that Riemann's conception of a manifold was essentially the modern one—i.e. he realized that manifolds are characterized by the fact that, locally, they resemble $\mathbf{R}^n$. (Riemann's discussion of *n*-manifolds probably was also influenced by the work of Jacobi (*1834*) and Grassmann (*1844*) in the theory of multidimensional algebras, even though Riemann does not mention their names.)

In Part Two, Riemann first takes up the problem of determining the length of a curve in a way that is independent of its position (II, §1). This part of his paper explicitly develops Gauss's ideas on differential geometry. Before Gauss, mathematicians had studied curved surfaces as figures embedded in three-dimensional Euclidean space. Gauss showed how the geometric properties of a surface could be obtained from the study of the intrinsic, local properties of the surface, without any reference to the surrounding space (*Gauss 1827*). It is not clear whether Gauss realized that the choice of a different metric would lead to a non-Euclidean geometry; nor, apparently, did he see that, because geometric properties could be intrinsically defined, a curved surface could be treated as a space in its own right. Riemann, however, drew both conclusions.

Riemann next defines the *curvature* of a manifold (generalizing Gauss's notion of the total curvature of a surface), and he discusses the special case of a manifold of constant curvature. Finally, in Part Three, he observes that, because a three-dimensional topological manifold can admit different metrics, the geometric properties of space cannot be determined by its topological properties alone, but must be determined by observation and experiment.

Although Dedekind praised Riemann's paper as a 'masterpiece of non-technical presentation' (*Riemann 1876*, p. 517), the work is sketchy, compressed, and chary with the mathematical details. But there exist several masterly commentaries that attempt to fill in the gaps. See in particular the edition of Riemann's paper edited by Hermann Weyl (*Riemann 1919*), or the notes by Heinrich Weber (heavily influenced by Dedekind) in *Riemann 1876*, pp. 384–99. For a modern analysis of the mathematics of Riemann's paper, see *Spivak 1970*, especially Chapter 4. For a historical description of the background to Riemann's work and its impact on the mathematicians of the last decades of the nineteenth century, see the account by Felix Klein (*Klein 1926–7*).

Although Riemann's *Habilitation* lecture, 'On the hypotheses which lie at the foundation of geometry', was delivered in 1854, it was not widely known until its publication in 1868, two years after Riemann's death; the lecture has been assigned the later date in order not to create misconceptions about the timing of Riemann's influence on the geometers of the nineteenth century. The translation is a revised version of the translation by William Kingdon Clifford, which originally appeared in *Nature* (*Clifford 1873*). The revisions are by William Ewald, and are chiefly designed to bring the terminology up to date. References to *Riemann 1868* should be to the part and section numbers (for example, II, §2) of the original German edition.

A. ON THE HYPOTHESES WHICH LIE AT THE FOUNDATION OF GEOMETRY (*RIEMANN 1868*)

PLAN OF THE INVESTIGATION

It is known that geometry assumes, as things given, both the notion of space and the first principles of constructions in space. It gives definitions of them which are merely nominal, while the true determinations appear in the form of axioms. The relation of these assumptions remains consequently in darkness; we neither perceive whether and how far their connection is necessary, nor, *a priori*, whether it is possible.

From Euclid to Legendre (to name the most famous of modern reforming geometers) this darkness was cleared up neither by mathematicians nor by such philosophers as concerned themselves with it. The reason of this is doubtless that the general notion of multiply extended magnitudes (in which space-magnitudes are included) remained entirely unworked. I have in the first place, therefore, set myself the task of constructing the notion of a multiply extended magnitude out of general notions of magnitude. It will follow from this that a multiply extended magnitude is capable of different metric relations, and consequently that space is only a particular case of a triply extended magnitude. But hence flows as a necessary consequence that the propositions of geometry cannot be derived from general notions of magnitude, but that the properties which distinguish space from other conceivable triply extended magnitudes are only to be deduced from experience. Thus arises the problem, to discover the simplest matters of fact from which the metric relations of space may be determined; a problem which from the nature of the case is not completely determinate, since there may be several systems of matters of fact which suffice to determine the metric relations of space—the most important system for our present purpose being that which Euclid has laid down as a foundation. These matters of fact are—like all matters of fact—not necessary, but only of empirical certainty; they are hypotheses. We may therefore investigate their probability, which within the limits of observation is of course very great, and inquire about the justice of their extension beyond the limits of observation, on the side both of the infinitely great and of the infinitely small.

I. Notion of an N-fold extended magnitude

In proceeding to attempt the solution of the first of these problems, the development of the notion of a multiply extended magnitude, I think I may the more claim indulgent criticism in that I am not practised in such undertakings of a philosophical nature where the difficulty lies more in the notions themselves than in the construction; and that, besides some very short hints on the matter given by Privy Councillor Gauss in his second memoir on Biquadratic Residues, in the *Göttingen Gelehrte Anzeige*, and in his Jubilee-book, and some philosophical researches of Herbart, I could make use of no previous labours.

§1. Notions of magnitude are only possible where there is an antecedent general notion which admits of different specializations ⟦Bestimmungsweisen⟧. According as there exists among these specializations a continuous path from one to another or not, they form a *continuous* or *discrete* manifold: the individual specializations are called in the first case points, in the second case elements, of the manifold. Notions whose specializations form a *discrete* manifold are so common that, at least in the cultivated languages, any things being given, it is always possible to find a notion in which they are included. (Hence mathematicians might unhesitatingly found the theory of discrete magnitudes upon the postulate that certain given things are to be regarded as equivalent.) On the other hand, so few and far between are the occasions for forming notions whose specializations make up a *continuous* manifold, that the only simple notions whose specializations form a multiply extended manifold are the positions of perceived objects and colours. More frequent occasions for the creation and development of these notions occur first in higher mathematics.

Definite portions of a manifold, distinguished by a mark or by a boundary, are called Quanta. Their comparison with regard to quantity is accomplished in the case of discrete magnitudes by counting, in the case of continuous magnitudes by measuring. Measure consists in the superposition of the magnitudes to be compared; it therefore requires a means of using one magnitude as the standard for another. In the absence of this, two magnitudes can only be compared when one is a part of the other; in which case also we can only determine the more or less and not the how much. The researches which can in this case be instituted about them form a general division of the science of magnitude in which magnitudes are regarded not as existing independently of position and not as expressible in terms of a unit, but as regions in a manifold. Such researches have become a necessity for many parts of mathematics, e.g., for the treatment of many-valued analytical functions; and the want of them is no doubt a chief cause why the celebrated theorem of Abel and the achievements of Lagrange, Pfaff, Jacobi for the general theory of differential equations, have so long remained unfruitful. Out of this general part of the science of extended magnitude in which nothing is assumed but what is contained in the notion of it, it will suffice for the present purpose to bring into prominence two points; the first of which relates to the construction of the notion of a multiply extended manifold, the second relates to the reduction of determinations of place in a given manifold to determinations of quantity, and will make clear the true character of an *n*-fold extension.

§2. If in the case of a notion whose specializations ‖Bestimmungsweisen‖ form a continuous manifold one passes from a certain specialization in a definite way to another, the specializations passed over form a simply extended manifold, whose true character is that in it a continuous progress from a point is possible only on two sides, forwards or backwards. If one now supposes that this manifold in its turn passes over into another entirely different, and again in a definite way, namely so that each point passes over into a definite point of the other, then all the specializations so obtained form a doubly extended manifold. In a similar manner one obtains a triply extended manifold, if one imagines a doubly extended one passing over in a definite way to another entirely different; and it is easy to see how this construction may be continued. If one regards the variable object instead of the determinable notion of it, this construction may be described as a composition of a variability of $n + 1$ dimensions out of a variability of n dimensions and a variability of one dimension.

§3. I shall now show how conversely one may resolve a variability whose region is given into a variability of one dimension and a variability of fewer dimensions. To this end let us suppose a variable piece of a manifold of one dimension—reckoned from a fixed origin, that the values of it may be comparable with one another—which has for every point of the given manifold a definite value, varying continuously with the point; or, in other words, let us take a continuous function of position within the given manifold, which, moreover, is not constant throughout any part of that manifold. Every system of points where the function has a constant value, forms then a continuous manifold of fewer dimensions than the given one. These manifolds pass over continuously into one another as the function changes; we may therefore assume that out of one of them the others proceed, and speaking generally this may occur in such a way that each point passes over into a definite point of the other; exceptional cases (the study of which is important) may here be left unconsidered. Hereby the determination of position in the given manifold is reduced to a determination of quantity and to a determination of position in a manifold of less dimensions. It is now easy to show that this manifold has $n - 1$ dimensions when the given manifold is n-fold extended. By repeating then this operation n times, the determination of position in an n-fold extended manifold is reduced to n determinations of quantity, and therefore the determination of position in a given manifold is reduced to a finite number of determinations of quantity when this is possible. There are manifolds in which the determination of position requires not a finite number, but either an endless series or a continuous manifold of determinations of quantity. Such manifolds are, for example, the possible determinations of a function for a given region, the possible shapes of a solid figure, &c.

II. Metric-relations of which a manifold of *n* dimensions is capable on the assumption that lines have a length independent of position, and consequently that every line may be measured by every other

Having constructed the notion of a manifold of n dimensions, and found that its true character consists in the property that the determination of position in it may be reduced to n determinations of magnitude, we come to the second of the problems proposed above, viz. the study of the metric relations of which such a manifold is capable, and of the conditions which suffice to determine them. These metric relations can only be studied in abstract notions of magnitude, and their dependence on one another can only be represented by formulæ. On certain assumptions, however, they are decomposable into relations which, taken separately, are capable of geometric representation; and thus it becomes possible to express geometrically the calculated results. In this way, to come to solid ground, we cannot, it is true, avoid abstract considerations in our formulæ, but at least the results of calculation may subsequently be presented in a geometric form. The foundations of these two parts of the question are established in the celebrated memoir of Gauss, *Disquisitiones generales circa superficies curvas*.

§1. Metric determinations require that quantity should be independent of position, which may happen in various ways. The hypothesis which first presents itself, and which I shall here develop, is that according to which the length of lines is independent of their position, and consequently every line is measurable by means of every other. Position-fixing being reduced to quantity-fixings, and the position of a point in the n-dimensioned manifold being consequently expressed by means of n variables $x_1, x_2, x_3, \ldots x_n$, the determination of a line comes to the giving of these quantities as functions of one variable. The problem consists then in establishing a mathematical expression for the length of a line, and to this end we must consider the quantities x as expressible in terms of certain units. I shall treat this problem only under certain restrictions, and I shall confine myself in the first place to lines in which the ratios of the increments dx of the respective variables vary continuously. We may then conceive these lines broken up into elements, within which the ratios of the quantities dx may be regarded as constant; and the problem is then reduced to establishing for each point a general expression for the linear element ds starting from that point, an expression which will thus contain the quantities x and the quantities dx. I shall suppose, secondly, that the length of the linear element, to the first order, is unaltered when all the points of this element undergo the same infinitesimal displacement, which implies at the same time that if all the quantities dx are increased in the same ratio, the linear element will vary also in the same ratio. On these suppositions, the linear element may be any homogeneous function of the first degree of the quantities dx, which is unchanged when we change the signs of all the dx, and in which the arbitrary constants are continuous functions of the quantities x. To find the simplest cases, I shall seek first an expression for manifolds of $n - 1$ dimensions which are everywhere equidistant from the origin of the linear element; that is, I shall seek a continuous function of position whose values distinguish them from one another. In going outwards from the origin, this must either increase in all directions or decrease in all directions; I assume that it increases in all directions, and therefore has a minimum at that point. If, then, the first and second differential coefficients of this function are finite, its first differential must vanish, and the second differential cannot become negative; I assume that it is always positive. This differential expression, then, of the second order remains constant when ds remains constant, and increases in the duplicate ratio when the dx, and therefore also ds, increase in the same ratio; it must therefore be ds^2 multiplied by a constant, and consequently ds is the square root of an always positive integral homogeneous function of the second order of the quantities dx, in which the coefficients are continuous functions of the quantities x. For Space, when the position of points is expressed by rectilinear co-ordinates, $ds = \sqrt{\Sigma(dx)^2}$; Space is therefore included in this simplest case. The next case in simplicity includes those manifolds in which the line-element may be expressed as the fourth root of a quartic differential expression. The investigation of this more general kind would require no really different principles, but would take considerable time and throw little new light on the theory of space, especially as the results cannot be geometrically expressed; I restrict myself, therefore, to those manifolds in which the line-element is expressed as the square root of a quadric differential expression. Such an expression we can transform into another similar one if we substitute for the n independent variables functions of n new independent variables. In this way, however, we cannot transform any expression into any other; since the expression contains $\frac{1}{2}n(n + 1)$ coefficients which are arbitrary functions of the independent variables; now by the introduction of new variables we can only satisfy n conditions, and therefore make no more than n of the coefficients equal to given quantities. The remaining $\frac{1}{2}n(n - 1)$ are then entirely determined by the nature of the continuum to be represented, and consequently $\frac{1}{2}n(n - 1)$ functions of positions are required for the determination of its measure-relations. Manifolds in which, as in the Plane and in Space, the line-element may be reduced to the form $\sqrt{\Sigma dx^2}$, are therefore only a particular case of the manifolds to be here investigated; they require a special name, and therefore these manifolds in which the square of the line-element may be expressed as the sum of the squares of complete differentials I will call *flat*. In order now to review the true varieties of all the continua which may be represented in the assumed form, it is necessary to get rid of difficulties arising from the mode of representation, which is accomplished by choosing the variables in accordance with a certain principle.

§2. For this purpose let us imagine that from any given point the system of shortest lines going out from it is constructed; the position of an arbitrary point may then be determined by the initial direction of the geodesic in which it lies, and by its distance measured along that line from the origin. It can therefore be expressed in terms of the ratios dx_0 of the quantities dx in this geodesic, and of the length s of this line. Let us introduce now instead of the dx_0 linear functions dx of them, such that the initial value of the square of the line-element shall equal the sum of the squares of these expressions, so that the independent variables are now the length s and the ratios of the quantities dx. Lastly, take instead of the dx quantities $x_1, x_2, x_3, \ldots x_n$ proportional to them, but such that the sum of their squares $= s^2$. When we introduce these quantities, the square of the line-element is Σdx^2 for infinitesimal values of the x, but the term of next order in it is equal to a homogeneous function of the second order of the $\frac{1}{2}n(n-1)$ quantities $(x_1\, dx_2 - x_2\, dx_1)$, $(x_1\, dx_3 - x_3\, dx_1) \ldots$, an infinitesimal, therefore, of the fourth order; so that we obtain a finite quantity on dividing this by the square of the infinitesimal triangle, whose vertices are $(0, 0, 0, \ldots)$, $(x_1, x_2, x_3, \ldots)$, $(dx_1, dx_2, dx_3, \ldots)$. This quantity retains the same value so long as the x and the dx are included in the same binary linear form, or so long as the two geodesics from 0 to x and from 0 to dx remain in the same surface-element; it depends therefore only on place and direction. It is obviously zero when the manifold represented is flat, i.e., when the squared line-element is reducible to Σdx^2, and may therefore be regarded as the measure of the deviation of the manifold from flatness at the given point in the given surface-direction. Multiplied by $-\frac{3}{4}$ it becomes equal to the quantity which Privy Councillor Gauss has called the total curvature of a surface. For the determination of the metric relations of a manifold capable of representation in the assumed form we found that $\frac{1}{2}n(n-1)$ place-functions were necessary; if, therefore, the curvature at each point in $\frac{1}{2}n(n-1)$ surface-directions is given, the metric relations of the continuum may be determined from them—provided there be no identical relations among these values, which in fact, to speak generally, is not the case. In this way the metric relations of a manifold in which the line-element is the square root of a quadric differential may be expressed in a manner wholly independent of the choice of independent variables. A method entirely similar may for this purpose be applied also to the manifold in which the line-element has a less simple expression, e.g., the fourth root of a quartic differential. In this case the line-element, generally speaking, is no longer reducible to the form of the square root of a sum of squares, and therefore the deviation from flatness in the squared line-element is an infinitesimal of the second order, while in those manifolds it was of the fourth order. This property of the last-named continua may thus be called flatness of the smallest parts. The most important property of these continua for our present purpose, for whose sake alone they are here investigated, is that the relations of the twofold ones may be geometrically represented by surfaces, and of the morefold ones may be reduced to those of the surfaces included in them; which now requires a short further discussion.

§3. In the idea of surfaces, together with the intrinsic metric relations in which only the length of lines on the surfaces is considered, there is always mixed up the position of points lying out of the surface. We may, however, abstract from external relations if we consider such deformations as leave unaltered the length of lines—i.e., if we regard the surface as bent in any way without stretching, and treat all surfaces so related to each other as equivalent. Thus, for example, any cylindrical or conical surface counts as equivalent to a plane, since it may be made out of one by mere bending, in which the intrinsic metric relations remain, and all theorems about a plane—therefore the whole of planimetry—retain their validity. On the other hand they count as essentially different from the sphere, which cannot be changed into a plane without stretching. According to our previous investigation the intrinsic metric relations of a twofold extent in which the line-element may be expressed as the square root of a quadric differential, which is the case with surfaces, are characterized by the total curvature. Now this quantity in the case of surfaces is capable of a visible interpretation, viz., it is the product of the two curvatures of the surface, or multiplied by the area of a small geodesic triangle, it is equal to the spherical excess of the same. The first definition assumes the proposition that the product of the two radii of curvature is unaltered by mere bending; the second, that in the same place the area of a small triangle is proportional to its spherical excess. To give an intelligible meaning to the curvature of an n-fold extension at a given point and in a given surface-direction through it, we must start from the fact that a geodesic proceeding from a point is entirely determined when its initial direction is given. According to this we obtain a determinate surface if we prolong all the geodesics proceeding from the given point and lying initially in the given surface-direction; this surface has at the given point a definite curvature, which is also the curvature of the n-fold continuum at the given point in the given surface-direction.

§4. Before we make the application to space, some considerations about flat manifolds in general are necessary; i.e., about those in which the square of the line-element is expressible as a sum of squares of complete differentials.

In a flat n-fold extension the total curvature is zero at all points in every direction; it is sufficient, however (according to the preceding investigation), for the determination of metric relations, to know that at each point the curvature is zero in $\frac{1}{2}n(n-1)$ independent surface-directions. Manifolds whose curvature is constantly zero may be treated as a special case of those whose curvature is constant. The common character of these continua whose curvature is constant may be also expressed thus, that figures may be moved in them without stretching. For clearly figures could not be arbitrarily shifted and turned round in them if the curvature at each point were not the same in all directions. On the other hand, however, the metric relations of the manifold are entirely determined by the curvature; they are therefore exactly the same in all directions at one point as at another, and consequently the same constructions can be made from it: whence it follows that in aggregates with constant curvature figures may have any arbitrary position given them. The metric relations of these manifolds depend only on the value of the curvature, and in relation to the analytic expression it may be remarked that if this value is denoted by α, the expression for the line-element may be written

$$\frac{1}{1 + \frac{1}{4}\alpha\Sigma x^2}\sqrt{\Sigma dx^2}.$$

§5. The theory of *surfaces* of constant curvature will serve for a geometric illustration. It is easy to see that surfaces whose curvature is positive may always be rolled on a sphere whose radius is unity divided by the square root of the curvature; but to review the entire manifold of these surfaces, let one of them have the form of a sphere and the rest the form of surfaces of revolution touching it at the equator. The surfaces with greater curvature than this sphere will then touch the sphere internally, and take a form like the outer portion (from the axis) of the surface of a ring; they may be rolled upon zones of spheres having less radii, but will go round more than once. The surfaces with less positive curvature are obtained from spheres of larger radii, by cutting out the lune bounded by two great half-circles and bringing the section-lines together. The surface with curvature zero will be a cylinder standing on the equator; the surfaces with negative curvature will touch the cylinder externally and be formed like the inner portion (towards the axis) of the surface of a ring. If we regard these surfaces as *locus in quo* for surface-regions moving in them, as Space is *locus in quo* for bodies, the surface-regions can be moved in all these surfaces without stretching. The surfaces with positive curvature can always be so formed that surface-regions may also be moved arbitrarily about upon them without *bending*, namely (they may be formed) into sphere-surfaces; but not those with negative curvature. Besides this independence of surface-regions from position there is in surfaces of zero curvature also an independence of *direction* from position, which in the former surfaces does not exist.

III. Application to Space

§1. By means of these inquiries into the determination of the metric relations of an n-fold extension the conditions may be declared which are necessary and sufficient to determine the metric properties of space, if we assume the independence of line-length from position and expressibility of the line-element as the square root of a quadric differential, that is to say, flatness in the smallest parts.

First, they may be expressed thus: that the curvature at each point is zero in three surface-directions; and thence the metric properties of space are determined if the sum of the angles of a triangle is always equal to two right angles.

Secondly, if we assume with Euclid not merely an existence of lines independent of position, but of bodies also, it follows that the curvature is everywhere constant; and then the sum of the angles is determined in all triangles when it is known in one.

Thirdly, one might, instead of taking the length of lines to be independent of position and direction, assume also an independence of their length and direction from position. According to this conception changes or differences of position are complex magnitudes expressible in three independent units.

§2. In the course of our previous inquiries, we first distinguished between the relations of extension or partition and the relations of measure, and found that with the same extensive properties, different metric relations were conceivable; we then investigated the system of simple size-fixings by which the metric relations of space are completely determined, and of which all propositions about them are a necessary consequence; it remains to discuss the question how, in what degree, and to what extent these assumptions are borne out by experience. In this respect there is a real distinction between mere extensive relations, and metric relations; in so far as in the former, where the possible cases form a discrete manifold, the declarations of experience are indeed not quite certain, but still not inaccurate; while in the latter, where the possible cases form a continuous manifold, every determination from experience remains always inaccurate: be the probability ever so great that it is nearly exact. This consideration becomes important in the extensions of these empirical determinations beyond the limits of observation to the infinitely great and infinitely small; since the latter may clearly become more inaccurate beyond the limits of observation, but not the former.

In the extension of space-construction to the infinitely great, we must distinguish between *unboundedness* and *infinite extent*; the former belongs to the extension relations, the latter to the metric relations. That space is an unbounded threefold manifold, is an assumption which is developed by every conception of the outer world; according to which every instant the region of real perception is completed and the possible positions of a sought object are constructed, and which by these applications is forever confirming itself. The unboundedness of space possesses in this way a greater empirical certainty than any external experience. But its infinite extent by no means follows from this; on the other hand if we assume independence of bodies from position, and therefore ascribe to space constant curvature, it must necessarily be finite provided this curvature has ever so small a positive value. If we prolong all the geodesics starting in a given surface-element, we should obtain an unbounded surface of constant curvature, i.e., a surface which in a *flat* manifold of three dimensions would take the form of a sphere, and consequently be finite.

§3. The questions about the infinitely great are for the interpretation of nature useless questions. But this is not the case with the questions about the infinitely small. It is upon the exactness with which we follow phenomena into the infinitely small that our knowledge of their causal relations essentially depends. The progress of recent centuries in the knowledge of mechanics depends almost entirely on the exactness of the construction which has become possible through the invention of the infinitesimal calculus, and through the simple principles discovered by Archimedes, Galileo, and Newton, and used by modern physics. But in the natural sciences which are still in want of simple principles for such constructions, we seek to discover the causal relations by following the phenomena into great minuteness, so far as the microscope permits. Questions about the metric relations of space in the infinitely small are not therefore superfluous questions.

If we suppose that bodies exist independently of position, the curvature is everywhere constant, and it then results from astronomical measurements that it cannot be different from zero; or at any rate its reciprocal must be an area in comparison with which the range of our telescopes may be neglected. But if this independence of bodies from position does not exist, we cannot draw conclusions from metric relations of the great, to those of the infinitely small; in that case the curvature at each point may have an arbitrary value in three directions, provided that the total curvature of every measurable portion of space does not differ sensibly from zero. Still more complicated relations may exist if we no longer suppose the linear element expressible as the square root of a quadric differential. Now it seems that the empirical notions on which the metrical determinations of space are founded, the notion of a solid body and of a ray of light, cease to be valid for the infinitely small. We are therefore quite at liberty to suppose that the metric relations of space in the infinitely small do not conform to the hypotheses of geometry; and we ought in fact to suppose it, if we can thereby obtain a simpler explanation of phenomena.

The question of the validity of the hypotheses of geometry in the infinitely small is bound up with the question of the ground of the metric relations of space. In this last question, which we may still regard as belonging to the doctrine of space, is found the application of the remark made above; that in a discrete manifold, the ground of its metric relations is given in the notion of it, while in a continuous manifold, this ground must come from outside. Either therefore the reality which underlies space must form a discrete manifold, or we must seek the ground of its metric relations outside it, in binding forces which act upon it.

The answer to these questions can only be got by starting from the conception of phenomena which has hitherto been justified by experience, and which Newton assumed as a foundation, and by making in this conception the successive changes required by facts which it cannot explain. Researches starting from general notions, like the investigation we have just made, can only be useful in preventing this work from being hampered by too narrow views, and progress in knowledge of the interdependence of things from being checked by traditional prejudices.

This leads us into the domain of another science, of physics, into which the object of this work does not allow us to go to-day.

17
Hermann von Helmholtz (1821–1894)

Helmholtz was the most celebrated scientist of late nineteenth-century Germany. As the director of major research laboratories, as Rector of the University of Berlin, and as a brilliant popular lecturer on scientific subjects, he had an immense influence on German and European intellectual life; his style of thought influenced several generations of scientists. He laid the foundations for modern physiological optics and acoustics; he was closely studied by Pavlov; his work in theoretical physics influenced Maxwell, Kelvin, Planck, Boltzmann, and Heinrich Hertz; his writings on the nature of space and on non-Euclidean geometry prepared the way for Einstein.

Helmholtz was also a major figure in the revival of Kantianism. At the start of his career, German philosophy was under the spell of Schelling and Hegel and their '*Naturphilosophie*'; by the end, in large part as a result of his influence, an analytic and scientifically-minded neo-Kantianism was flourishing in the German universities, and a path had been prepared for the logical positivists, who looked back to Helmholtz as a major forerunner of their own ideas.

Moreover, Helmholtz was a powerful force in the structural and administrative transformation of German science. At the start of the nineteenth century, experimental science was pursued by talented individuals working on a small scale and often in isolation. By the end, German science was dominated by the great modern research laboratories and the large national universities—a transformation that also affected the world of mathematics, and led to the creation of influential research centres at Göttingen, Berlin, and elsewhere.

Intellectually, Helmholtz's career was dominated by a romantic passion to find the fundamental, unifying laws of nature—a passion that was in part inspired by his early reading of Kant. (His father, a close friend of Fichte's son, taught philosophy at the Potsdam *Gymnasium*, and exposed Helmholtz to the *Critique of pure reason* at an early age.) This exposure was a formative intellectual influence; like C.S. Peirce (who was similarly exposed) Helmholtz never drew a sharp distinction between his work in natural science and his work in philosophy. Even his earliest major scientific paper, the classic memoir 'On the conservation of energy' (*Helmholtz 1847*), which commences with a philosophical discussion, heavily influenced by Kant, of the law of causality and the nature of force, shows that he approached physics with the eye of an epistemologist, and epistemology with the eye of a physicist.

Helmholtz's scientific writings are emblematic of the best scientific writing of the age, and illustrate the fruitful way in which philosophy can enrich and give a common theme to such diverse pursuits as mathematics, physiology, and theoretical physics; indeed, for Helmholtz the sciences are part of a wider culture—a culture which embraces literature and the arts, as well as physics and mathematics. Helmholtz's seemingly divergent researches into human spatial perception, the foundations of geometry, and the basic laws of physics can all be viewed as attempts to answer, by means of science, the Kantian question, 'What can we know about the external world, and what limitations do our faculties of perception place upon us?'

Helmholtz's career can be divided into two halves. In the first half, from 1848 to 1870, Helmholtz (who had been trained as a physician) taught anatomy and physiology at the universities of Königsberg, Bonn, and Heidelberg. During this time he worked principally on the physiology of human sight and hearing. He invented the ophthalmoscope, investigated the speed of transmission of nerve impulses, made pioneering studies of colour vision, and wrote numerous articles on acoustics and optics. Many of his results are described in his *Handbook of physiological optics* (3 vols, *Helmholtz 1856–67*) and his *Theory of the sensation of tone as the physiological foundation for the theory of music* (*Helmholtz 1863*). (Lord Kelvin declared that each of these works was the *Principia* in its field.)

His work on human visual perception and his Kantian interest in epistemology led him, in the 1860s, to study the foundations of geometry and the human knowledge of space. He independently discovered many of the results in Riemann's paper 'On the hypotheses which lie at the foundation of geometry' (*Riemann 1868*), and he published four papers explaining and extending Riemann's ideas. His 'The facts in perception' (*1878b*), translated below, describes the connections between his Kantianism, his physiology, and his studies in non-Euclidean geometry.

Helmholtz's encounter with Riemann's work on geometry prompted him to abandon physiology and begin a second career in mathematical physics. In 1871, at the age of fifty, he accepted the chair of physics at Berlin. He inaugurated the second phase of his career by writing a series of influential papers on electrodynamics. These papers presented Maxwell's field theory to German physicists, and provided the starting-point for the work of Helmholtz's student, Heinrich Hertz. Helmholtz also worked on the theory of chemical thermodynamics, and attempted unsuccessfully to found all of physics on the principle of least action. An extensive discussion of his contributions to theoretical physics, as well as an illuminating background description of nineteenth-century German science, can be found in the biography of Helmholtz, *Koenigsberger 1902–3*.

A. THE ORIGIN AND MEANING OF GEOMETRICAL AXIOMS (*HELMHOLTZ 1876* AND *1878a*)

Helmholtz's first publication on geometry was his paper 'On the actual foundations of geometry' (*Helmholtz 1868a*). In the opening sentences he says

The investigations concerning the manner in which localization arises in the visual field have caused me to reflect as well on the origins of the general intuition of space ‖Raumanschauung‖. Now here there occurs a question whose answer falls within the domain of the exact sciences, namely, which propositions of geometry express truths of actual significance ‖von thatsächlicher Bedeutung‖, and which, on the other hand, are merely definitions or consequences of definitions and of the particular chosen terminology.

After undertaking these geometrical researches, he says, he learned of Riemann's 1854 inaugural dissertation, 'On the hypotheses which lie at the foundation of geometry' (*Riemann 1868*). Riemann had anticipated most of his results, but in one respect Helmholtz had gone farther. Riemann had assumed that the expression for length is the square root of a homogeneous second-degree function of the differentials of the coordinates:

$$ds^2 = \Sigma a_{ik} dx_i dx_k.$$

In his 'On the facts which lie at the foundation of geometry' (*Helmholtz 1868b*) Helmholtz showed that Riemann's assumption is a consequence of the deeper assumption that motions of rigid bodies are possible within the space. (This early paper is translated in *Helmholtz 1977*.)

The following two-part article contains Helmholtz's description of the epistemological consequences of his and Riemann's work on non-Euclidean geometry. The publishing history is complex, and the article exists in many different forms. The first part was originally delivered as a lecture to the *Dozenten-Verein* in Heidelberg in 1870, and then published in an authorized English translation by Edmund Atkinson in *Mind* in 1876, under the title, 'The origin and meaning of geometrical axioms'. The German version of the Heidelberg talk was apparently first published in 1876 in the third volume of Helmholtz's 'Popular scientific lectures' (*1865–76*); for this version, Helmholtz added six new opening paragraphs and a mathematical appendix. (Atkinson incorporated this new material when he reprinted his *Mind* translation in *Helmholtz 1881*; but he altered the title to 'On the origin and significance of geometrical axioms'.) The German version was later reprinted, with commentary and footnotes by Paul Hertz, in the influential selection of Helmholtz's *Schriften zur Erkenntnistheorie* edited by Hertz and Moritz Schlick (*Helmholtz 1921*); but this edition omits Helmholtz's appendix. *Helmholtz 1921*, together with the Hertz footnotes, has been translated as *Helmholtz 1977*.

In 1878 Helmholtz published a second article in *Mind*, 'The origin and meaning of geometrical axioms (II)'. This piece (*Helmholtz 1878a*) contains replies to critics of his first article. It consists of five introductory paragraphs and of sections numbered (I) to (III). Sections (II) and (III) were slightly reworked by Helmholtz, and reprinted as Appendix 3 to 'The facts in perception' (*Helmholtz 1878b*); section (II) became section 1 of that third Appendix, and section (III) became section 2. 'The facts in perception' is translated below, inclusive of the third appendix; so only the introductory paragraphs and section (I) of *Helmholtz 1878a* are reproduced here. The German versions of these works appear in two different collections of Helmholtz's papers. The German version of the second *Mind* article was published in the second volume of his 'Scientific essays' (*1882–95*); while 'The facts in perception' (with its appendices) was published in the collection entitled 'Lectures and speeches' (*1884*), which is in fact the second edition of the 'Popular scientific lectures', and which also contains the first *Mind* article.

The following translation of *Helmholtz 1876* is a lightly-edited version of the translation by Edmund Atkinson, and contains the additional material added by Helmholtz; the original article in *Mind* began at paragraph 7. The passage from *Helmholtz 1878a* is taken from the translation in *Mind*, which was probably made by Atkinson as well (although no translator is indicated). References should be to the paragraph numbers, which have been added in this edition.

⟦1⟧ The fact that a science like geometry can exist and can be developed as it has been has always attracted the closest attention among those who are interested in questions relating to the bases of the theory of cognition. Of all branches of human knowledge, there is none which, like it, has sprung as a completely armed Minerva from the head of Jupiter; none before whose death-dealing Aegis doubt and inconsistency have so little dared to raise their eyes. It escapes the tedious and troublesome task of collecting experimental facts, which is the province of the natural sciences in the strict sense of the word; the sole form of its scientific method is deduction. Conclusion is deduced from conclusion, and yet no one of common sense doubts that these geometrical principles must find their practical application in the real world about us. Land surveying as well as architecture, the construction of machinery no less than mathematical physics, are continually calculating relations of space of the most varied kind by geometrical principles; they expect that the success of their constructions and experiments shall agree with these calculations; and no case is known in which this expectation has been falsified, provided the calculations were made correctly and with sufficient data.

⟦2⟧ Indeed, the fact that geometry exists, and is capable of all this, has always been used as a prominent example in the discussion on that question, which forms, as it were, the centre of all antitheses of philosophical systems, that there can be a cognition of principles destitute of any bases drawn from experience. In the answer to Kant's celebrated question, 'How are synthetical principles *a priori* possible?' geometrical axioms are certainly those examples which appear to show most decisively that synthetical principles are *a priori* possible at all. The circumstance that such principles exist, and force themselves on our conviction, is regarded as a proof that space is an *a priori* mode of all external perception. He appears thereby to postulate, for this *a priori* form, not only the character of a purely formal scheme of itself quite unsubstantial, into which any given content of experience would fit; but also to include certain peculiarities of the scheme, which bring it about that only a certain content, and one which, as it were, is strictly defined, could occupy it and be apprehended by us.[1]

[3] It is precisely this relation of geometry to the theory of cognition which emboldens me to speak to you on geometrical subjects in an assembly of those who for the most part have limited their mathematical studies to the ordinary instruction in schools. Fortunately, the amount of geometry taught in our *gymnasia* will enable you to follow at any rate the tendency of the principles I am about to discuss.

[4] I intend to give you an account of a series of recent and closely connected mathematical researches which are concerned with the geometrical axioms, their relations to experience, with the question whether it is logically possible to replace them by others.

[5] Seeing that the researches in question are more immediately designed to furnish proofs for experts in a region which, more than almost any other, requires a higher power of abstraction, and that they are virtually inaccessible to the non-mathematician, I will endeavour to explain to such a layman the question at issue. I need scarcely remark that my explanation will give no proof of the correctness of the new views. He who seeks this proof must take the trouble to study the original researches.

[6] Anyone who has entered the gates of the first elementary axioms of geometry, that is, the mathematical doctrine of space, finds on his path that unbroken chain of conclusions of which I just spoke, by which ever more varied and more complicated figures are brought within the domain of law. But even in their first elements certain principles are laid down, with respect to which geometry confesses that she cannot prove them, and can only assume that anyone who understands the essence of these principles will at once admit their correctness. These are the so-called axioms.

[7] For example, the proposition that if the shortest line drawn between two points is called a *straight* line, there can be only one such straight line. Again, it is an axiom that through any three points in space, not lying in a straight line, a plane may be drawn, i.e. a surface which will wholly include every straight line joining any two of its points. Another axiom, about which there has been much discussion, affirms that through a point lying outside a straight line only one straight line can be drawn parallel to the first; two straight lines that lie in the same plane and never meet, however far they may be produced,

[1] In his book, *On the Limits of Philosophy*, Mr. W. Tobias maintains that axioms of a kind which I formerly enunciated are a misunderstanding of Kant's opinion. But Kant specially adduces the axioms, that the straight line is the shortest (*Kritik der reinen Vernunft*, Introduction, v; 2nd. ed. p. 16); that space has three dimensions (*Ibid.* part i. sect. i. §3, p. 41); that only one straight line is possible between two points (*Ibid.* part ii. sect. i. 'On the Axioms of Intuition'), as axioms which express *a priori* the conditions of intuition by the senses. It is not here the question, whether these axioms were originally given as intuition of space, or whether they are only the starting-points from which the understanding can develop such axioms *a priori* on which my critic insists.

being called parallel. There are also axioms that determine the number of dimensions of space and its surfaces, lines and points, showing how they are continuous; as in the propositions, that a solid is bounded by a surface, a surface by a line, and a line by a point; that the point is indivisible; that by the movement of a point a line is described, by that of a line a line or a surface, by that of a surface a surface or a solid; but by the movement of a solid a solid and nothing else is described.

‖8‖ Now what is the origin of such propositions, unquestionably true yet incapable of proof in a science where everything else is reasoned conclusion? Are they inherited from the divine source of our reason as the idealistic philosophers think, or is it only that the ingenuity of mathematicians has hitherto not been penetrating enough to find the proof? Every new votary, coming with fresh zeal to geometry, naturally strives to succeed where all before him have failed. And it is quite right that each should make the trial afresh; for, as the question has hitherto stood, it is only by the fruitlessness of one's own efforts that one can be convinced of the impossibility of finding a proof. Meanwhile solitary inquirers are always from time to time appearing who become so deeply entangled in complicated trains of reasoning that they can no longer discover their mistakes and believe they have solved the problem. The axiom of parallels especially has called forth a great number of seeming demonstrations.

‖9‖ The main difficulty in these inquiries is, and always has been, the readiness with which results of everyday experience become mixed up as apparent necessities of thought with the logical processes, so long as Euclid's method of constructive intuition is exclusively followed in geometry. It is in particular extremely difficult, on this method, to be quite sure that in the steps prescribed for the demonstration we have not involuntarily and unconsciously drawn in some most general results of experience, which the power of executing certain parts of the operation has already taught us practically. In drawing any subsidiary line for the sake of his demonstration, the well-trained geometer always asks if it is possible to draw such a line. It is well known that problems of construction play an essential part in the system of geometry. At first sight, these appear to be practical operations, introduced for the training of learners; but in reality they establish the existence of definite figures. They show that points, straight lines, or circles such as the problem requires to be constructed are possible under all conditions, or they determine any exceptions that there may be. The point on which the investigations turn, that we are about to consider, is essentially of this nature. The foundation of all proof by Euclid's method consists in establishing the congruence of lines, angles, plane figures, solids, &c. To make the congruence evident, the geometrical figures are supposed to be applied to one another, of course without changing their form and dimensions. That this is in fact possible we have all experienced from our earliest youth. But, if we proceed to build necessities of thought upon this assumption of the free translation of fixed figures, with unchanged form, to every part of space, we must see whether the assumption does not involve some presupposition of which no logical proof is given. We shall see later on that it does indeed contain one of the most serious import. But if so, every proof by congruence rests upon a fact which is obtained from experience only.

⟦10⟧ I offer these remarks, at first only to show what difficulties attend the complete analysis of the presuppositions we make in employing the common constructive method. We evade them when we apply, to the investigation of principles, the analytical method of modern algebraical geometry. The whole process of algebraical calculation is a purely logical operation; it can yield no relation between the quantities submitted to it that is not already contained in the equations which give occasion for its being applied. The recent investigations in question have accordingly been conducted almost exclusively by means of the purely abstract methods of analytical geometry.

⟦11⟧ However, after discovering by the abstract method what are the points in question, we shall best get a distinct view of them by taking a region of narrower limits than our own world of space. Let us, as we logically may, suppose reasoning beings of only two dimensions to live and move on the surface of some solid body. We will assume that they have not the power of perceiving anything outside this surface, but that upon it they have perceptions similar to ours. If such beings worked out a geometry, they would of course assign only two dimensions to their space. They would ascertain that a point in moving describes a line, and that a line in moving describes a surface. But they could as little represent to themselves what further spatial construction would be generated by a surface moving out of itself, as we can represent what would be generated by a solid moving out of the space we know. By the much-abused expression 'to represent' or 'to be able to think how something happens' I understand—and I do not see how anything else can be understood by it without loss of all meaning—the power of imagining the whole series of sensible impressions that would be had in such a case. Now as no sensible impression is known relating to such an unheard-of event, as the movement to a fourth dimension would be to us, or as a movement to our third dimension would be to the inhabitants of a surface, such a 'representation' is as impossible as the 'representation' of colours would be to one born blind, if a description of them in general terms could be given to him.

⟦12⟧ Our surface-beings would also be able to draw shortest lines in their superficial space. These would not necessarily be straight lines in our sense, but what are technically called *geodetic lines* of the surface on which they live; lines such as are described by a *taut* thread laid along the surface, and which can slide upon it freely. I will henceforth speak of such lines as the *straightest* lines of any particular surface or given space, so as to bring out their analogy with the straight line in a plane. I hope by this expression to make the conception more easy for the apprehension of my non-mathematical hearers without giving rise to misconception.

⟦13⟧ Now if beings of this kind lived on an infinite plane, their geometry would be exactly the same as our plane geometry. They would affirm that only one straight line is possible between two points; that through a third point lying outside this line only one line can be drawn parallel to it; that the ends of a straight line never meet though it is produced to infinity; and so on. Their space might be infinitely extended, but even if there were limits to their movement and perception, they would be able to represent to themselves a continuation beyond these limits; and thus their space would appear to them infinitely extended, just as ours does to us, although our bodies cannot leave the earth, and our sight only reaches as far as the visible fixed stars.

⟦14⟧ But intelligent beings of the kind supposed might also live on the surface of a sphere. Their shortest or straightest line between two points would then be an arc of the great circle passing through them. Every great circle, passing through two points, is by these divided into two parts; and if they are unequal, the shorter is certainly the shortest line on the sphere between the two points, but also the other or larger arc of the same great circle is a geodetic or straightest line, i.e. every smaller part of it is the shortest line between its ends. Thus the notion of the geodetic or straightest line is not quite identical with that of the shortest line. If the two given points are the ends of a diameter of the sphere, every plane passing through this diameter cuts semicircles on the surface of the sphere all of which are shortest lines between the ends; in which case there is an equal number of equal shortest lines between the given points. Accordingly, the axiom of there being only one shortest line between two points would not hold without a certain exception for the dwellers on a sphere.

⟦15⟧ Of parallel lines the sphere-dwellers would know nothing. They would maintain that any two straightest lines, sufficiently produced, must finally cut not in one only but in two points. The sum of the angles of a triangle would be always greater than two right angles, increasing as the surface of the triangle grew greater. They could thus have no conception of geometrical similarity between greater and smaller figures of the same kind, for with them a greater triangle must have different angles from a smaller one. Their space would be unlimited, but would be found to be finite or at least represented as such.

⟦16⟧ It is clear, then, that such beings must set up a very different system of geometrical axioms from that of the inhabitants of a plane, or from ours with our space of three dimensions, though the logical powers of all were the same; nor are more examples necessary to show that geometrical axioms must vary according to the kind of space inhabited by beings whose powers of reason are quite in conformity with ours. But let us proceed still farther.

⟦17⟧ Let us think of reasoning beings existing on the surface of an egg-shaped body. Shortest lines could be drawn between three points of such a surface and a triangle constructed. But if the attempt were made to construct congruent triangles at different parts of the surface, it would be found that two triangles, with three pairs of equal sides, would not have their angles equal. The sum of the angles of a triangle drawn at the sharper pole of the body would depart farther from two right angles than if the triangle were drawn at the blunter pole or at the equator. Hence it appears that not even such a simple figure as a triangle can be moved on such a surface without change of form. It would also be found that if circles of equal radii were constructed at different parts of such a surface (the length of the radii being always measured by shortest lines along the surface) the periphery would be greater at the blunter than at the sharper end.

[18] We see accordingly that, if a surface admits of the figures lying on it being freely moved without change of any of their lines and angles as measured along it, the property is a special one and does not belong to every kind of surface. The condition under which a surface possesses this important property was pointed out by Gauss in his celebrated treatise on the curvature of surfaces.[2] The 'measure of curvature', as he called it, i.e. the reciprocal of the product of the greatest and least radii of curvature, must be everwhere equal over the whole extent of the surface.

[19] Gauss showed at the same time that this measure of curvature is not changed if the surface is bent without distension or contraction of any part of it. Thus we can roll up a flat sheet of paper into the form of a cylinder, or of a cone, without any change in the dimensions of the figures taken along the surface of the sheet. Or the hemispherical fundus of a bladder may be rolled into a spindle-shape without altering the dimensions on the surface. Geometry on a plane will therefore be the same as on a cylindrical surface; only in the latter case we must imagine that any number of layers of this surface, like the layers of a rolled sheet of paper, lie one upon another, and that after each entire revolution round the cylinder a new layer is reached different from the previous ones.

[20] These observations are necessary to give the reader a notion of a kind of surface the geometry of which is on the whole similar to that of the plane, but in which the axiom of parallels does not hold good. This is a kind of curved surface which is, as it were, geometrically the counterpart of a sphere, and which has therefore been called the *pseudospherical surface* by the distinguished Italian mathematician E. Beltrami, who has investigated its properties.[3] It is a saddle-shaped surface of which only limited pieces or strips can be connectedly represented in our space, but which may yet be thought of as infinitely continued in all directions, since each piece lying at the limit of the part constructed can be conceived as drawn back to the middle of it and then continued. The piece displaced must in the process change its flexure but not its dimensions, just as happens with a sheet of paper moved about a cone formed out of a rolled up plane. Such a sheet fits the conical surface in every part, but must be more bent near the vertex and cannot be so moved over the vertex as to be at the same time adapted to the existing cone and to its imaginary continuation beyond.

[21] Like the plane and the sphere, pseudospherical surfaces have their measure of curvature constant, so that every piece of them can be exactly applied to every other piece, and therefore all figures constructed at one place on

---

[2] Gauss, *Werke*, Bd. IV. p. 215, first published in *Commentationes Soc. Reg. Scientt. Gottengensis recentiores*, vol. vi., 1828.

[3] *Saggio di Interpretazione della Geometria Non-Euclidea*, Napoli, 1868.—*Teoria fondamentale degli Spazii di Curvatura costante, Annali di Matematica*, Ser. II. Tom. II. pp. 232–55. Both have been translated into French by J. Hoüel, *Annales Scientifiques de l'Ecole Normale*, Tom V., 1869.

the surface can be transferred to any other place with perfect congruity of form, and perfect equality of all dimensions lying in the surface itself. The measure of curvature as laid down by Gauss, which is positive for the sphere and zero for the plane, would have a constant negative value for pseudospherical surfaces, because the two principal curvatures of a saddle-shaped surface have their concavity turned opposite ways.

⟦22⟧ A strip of a pseudospherical surface may, for example, be represented by the inner surface (turned towards the axis) of a solid anchor-ring. If the plane figure *aabb* (Fig. 1) is made to revolve on its axis of symmetry AB, the two arcs *ab* will describe a pseudospherical concave–convex surface like that of the ring. Above and below, towards *aa* and *bb*, the surface will turn outwards with ever-increasing flexure, till it becomes perpendicular to the axis, and ends at the edge with one curvature infinite. Or, again, half of a pseudospherical surface may be rolled up into the shape of a champagne-glass (Fig. 2), with tapering stem infinitely prolonged. But the surface is always necessarily bounded by a sharp edge beyond which it cannot be directly continued. Only by supposing each single piece of the edge cut loose and drawn along the surface of the ring or glass, can it be brought to places of different flexure, at which farther continuation of the piece is possible.

⟦23⟧ In this way too the straightest lines of the pseudospherical surface may be infinitely produced. They do not, like those on a sphere, return upon themselves, but, as on a plane, only one shortest line is possible between the two given points. The axiom of parallels does not, however, hold good. If a straightest line is given on the surface and a point outside it, a whole pencil of straightest lines may pass through the point, no one of which, though infinitely produced, cuts the first line; the pencil itself being limited by two straightest lines, one of which intersects one of the ends of the given line at an infinite distance, the other the other end.

⟦24⟧ Such a system of geometry, which excluded the axiom of parallels, was devised on Euclid's synthetic method, as far back as the year 1829, by N.J.

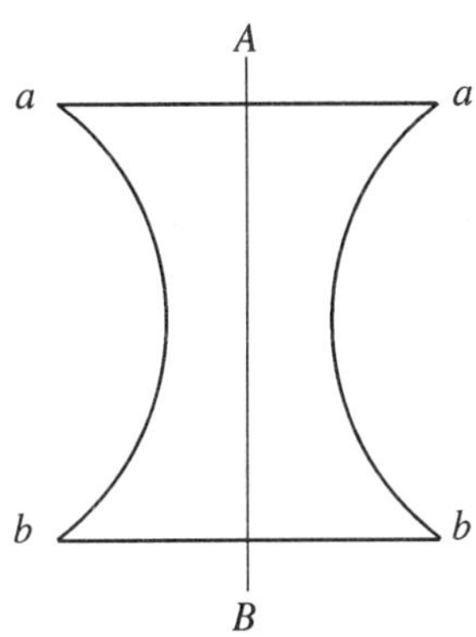

Fig. 1.

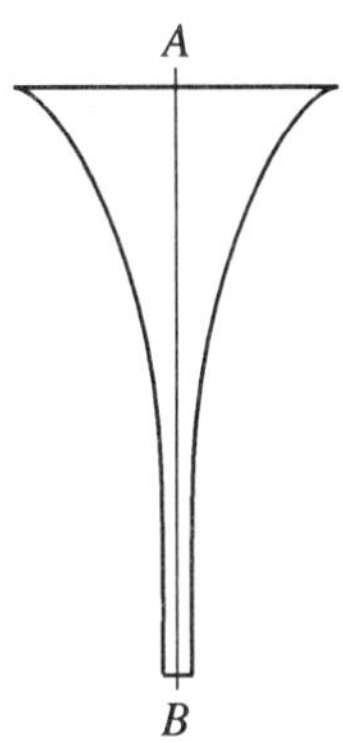

Fig. 2.

Lobatchevsky, professor of mathematics at Kasan,[4] and it was proved that this system could be carried out as consistently as Euclid's. It agrees exactly with the geometry of the pseudospherical surfaces worked out recently by Beltrami.

[25] Thus we see that in the geometry of two dimensions a surface is marked out as a plane, or a sphere, or a pseudospherical surface, by the assumption that any figure may be moved about in all directions without change of dimensions. The axiom, that there is only one shortest line between any two points, distinguishes the plane and the pseudospherical surface from the sphere, and the axiom of parallels marks off the plane from the pseudosphere. These three axioms are in fact necessary and sufficient, to define as a plane the surface to which Euclid's plane geometry has reference, as distinguished from all other modes of space in two dimensions.

[26] The difference between plane and spherical geometry has been long evident, but the meaning of the axiom of parallels could not be understood till Gauss had developed the notion of surfaces flexible without dilatation, and consequently that of the possibly infinite continuation of pseudospherical surfaces. Inhabiting, as we do, a space of three dimensions and endowed with organs of sense for their perception, we can represent to ourselves the various cases in which beings on a surface might have to develop their perception of space; for we have only to limit our own perceptions to a narrower field. It is easy to think away perceptions that we have; but it is very difficult to imagine perceptions to which there is nothing analogous in our experience. When, therefore, we pass to space of three dimensions, we are stopped in our power of representation, by the structure of our organs and the experiences got through them which correspond only to the space in which we live.

[27] There is however another way of treating geometry scientifically. All known space-relations are measureable, that is, they may be brought to determination of magnitudes (lines, angles, surfaces, volumes). Problems in geometry can therefore be solved, by finding methods of calculation for arriving at unknown magnitudes from known ones. This is done in *analytical geometry*, where all forms of space are treated only as quantities and determined by means of other quantities. Even the axioms themselves make reference to magnitudes. The straight line is defined as the *shortest* between two points, which is a determination of quantity. The axiom of parallels declares that if two straight lines in a plane do not intersect (are parallel), the alternate angles, or the corresponding angles, made by a third line intersecting them, are equal; or it may be laid down instead that the sum of the angles of any triangle is equal to two right angles. These, also, are determinations of quantity.

[28] Now we may start with this view of space, according to which the position of a point may be determined by measurements in relation to any given figure (system of co-ordinates), taken as fixed, and then inquire what are the

---

[4] *Principien der Geometrie*, Kasan, 1829–30.

special characteristics of our space as manifested in the measurements that have to be made, and how it differs from other extended quantities of like variety. This path was first entered by one too early lost to science, B. Riemann of Göttingen.[5] It has the peculiar advantage that all its operations consist in pure calculation of quantities, which quite obviates the danger of habitual perceptions being taken for necessities of thought.

⟦29⟧ The number of measurements necessary to give the position of a point, is equal to the number of dimensions of the space in question. In a line the distance from one fixed point is sufficient, that is to say, one quantity; in a surface the distances from two fixed points must be given; in space, the distances from three; or we require, as on the earth, longitude, latitude, and height above the sea, or, as is usual in analytical geometry, the distances from three co-ordinate planes. Riemann calls a system of differences in which one thing can be determined by *n* measurements an '*n*fold extended aggregate' or an 'aggregate of *n* dimensions.' Thus the space in which we live is a threefold, a surface is a twofold, and a line is a simple extended aggregate of points. Time also is an aggregate of one dimension. The system of colours is an aggregate of three dimensions, inasmuch as each colour, according to the investigations of Thomas Young and of Clerk Maxwell,[6] may be represented as a mixture of three primary colours, taken in definite quantities. The particular mixtures can be actually made with the colour-top.

⟦30⟧ In the same way we may consider the system of simple tones[7] as an aggregate of two dimensions, if we distinguish only pitch and intensity, and leave out of account differences of timbre. This generalization of the idea is well suited to bring out the distinction between space of three dimensions and other aggregates. We can, as we know from daily experience, compare the vertical distance of two points with the horizontal distance of two others, because we can apply a measure first to the one pair and then to the other. But we cannot compare the difference between two tones of equal pitch and different intensity, with that between two tones of equal intensity and different pitch. Riemann showed, by considerations of this kind, that the essential foundation of any system of geometry, is the expression that it gives for the distance between two points lying in any direction towards one another, beginning with the infinitesimal interval. He took from analytical geometry the most general form for this expression, that, namely, which leaves altogether open the kind of measurements by which the position of any point is given.[8] Then he showed that the kind of free mobility without change of form which belongs to bodies in our

---

[5] Ueber die Hypothesen welche der Geometrie zu Grunde liegen, Habilitationsschrift vom 10 Juni 1854. (*Abhandl. der königl. Gesellsch. zu Göttingen*, Bd. XIII.)

[6] Helmholtz's *Popular Lectures*, Series I. p. 213.

[7] Ibid. p. 86.

[8] For the square of the distance of two infinitely near points the expression is a homogeneous quadric function of the differentials of their co-ordinates.

space can only exist when certain quantities yielded by the calculation[9]—quantities that coincide with Gauss's measure of surface-curvature when they are expressed for surfaces—have everywhere an equal value. For this reason Riemann calls these quantities, when they have the same value in all directions for a particular spot, the measure of curvature of the space at this spot. To prevent misunderstanding,[10] I will once more observe that this so-called measure of space-curvature is a quantity obtained by purely analytical calculation, and that its introduction involves no suggestion of relations that would have a meaning only for sense-perception. The name is merely taken, as a short expression for a complex relation, from the one case in which the quantity designated admits of sensible representation.

⟦31⟧ Now whenever the value of this measure of curvature in any space is everywhere zero, that space everywhere conforms to the axioms of Euclid; and it may be called a *flat* (*homaloid*) space in contradistinction to other spaces, analytically constructible, that may be called *curved*, because their measure of curvature has a value other than zero. Analytical geometry may be as completely and consistently worked out for such spaces as ordinary geometry can for our actually existing homaloid space.

⟦32⟧ If the measure of curvature is positive we have *spherical* space, in which straightest lines return upon themselves and there are no parallels. Such a space would, like the surface of a sphere, be unlimited but not infinitely great. A constant negative measure of curvature on the other hand gives *pseudospherical* space, in which straightest lines run out to infinity, and a pencil of straightest lines may be drawn, in any flattest surface, through any point which does not intersect another given straightest line in that surface.

⟦33⟧ Beltrami[11] has rendered these last relations imaginable by showing that the points, lines, and surfaces of a pseudospherical space of three dimensions can be so portrayed in the interior of a sphere in Euclid's homaloid space, that every straightest line or flattest surface of the pseudospherical space is represented by a straight line or a plane, respectively, in the sphere. The surface itself of the sphere corresponds to the infinitely distant points of the pseudospherical space; and the different parts of this space, as represented in the sphere, become smaller, the nearer they lie to the spherical surface, diminishing more rapidly in the direction of the radii than in that perpendicular to them. Straight lines in the sphere, which only intersect beyond its surface, correspond to straightest lines of the pseudospherical space which never intersect.

⟦34⟧ Thus it appeared that space, considered as a region of measurable quantities, does not at all correspond with the most general conception of an aggregate of three dimensions, but involves also special conditions, depending

[9] They are algebraical expressions compounded from the coefficients of the various terms in the expression for the square of the distance of two contiguous points and from their differential quotients.

[10] As occurs, for instance, in the above-mentioned work of Tobias, pp. 70, etc.

[11] *Theoria fondamentale, &c., ut sup.*

on the perfectly free mobility of solid bodies without change of form to all parts of it and with all possible changes of direction; and, further, on the special value of the measure of curvature which for our actual space equals, or at least is not distinguishable from, zero. This latter definition is given in the axioms of straight lines and parallels.

[35] Whilst Riemann entered upon this new field from the side of the most general and fundamental questions of analytical geometry, I myself arrived at similar conclusions,[12] partly from seeking to represent in space the system of colours, involving the comparison of one threefold extended aggregate with another, and partly from inquiries on the origin of our ocular measure for distances in the field of vision. Riemann starts by assuming the above-mentioned algebraical expression which represents in the most general form the distance between two infinitely near points, and deduces therefrom the conditions of mobility of rigid figures. I, on the other hand, starting from the observed fact that the movement of rigid figures is possible in our space, with the degree of freedom that we know, deduce the necessity of the algebraic expression taken by Riemann as an axiom. The assumptions that I had to make as the basis of the calculation were the following.

[36] First, to make algebraical treatment at all possible, it must be assumed that the position of any point A can be determined, in relation to certain given figures taken as fixed bases, by measurement of some kind of magnitudes, as lines, angles between lines, angles between surfaces, and so forth. The measurements necessary for determining the position of A are known as its co-ordinates. In general, the number of co-ordinates necessary for the complete determination of the position of a point, marks the number of the dimensions of the space in question. It is further assumed that with the movement of the point A, the magnitudes used as co-ordinates vary continuously.

[37] Secondly, the definition of a solid body, or rigid system of points, must be made in such a way as to admit of magnitudes being compared by congruence. As we must not, at this stage, assume any special methods for the measurement of magnitudes, our definition can, in the first instance, run only as follows: Between the co-ordinates of any two points belonging to a solid body, there must be an equation which, however the body is moved, expresses a constant spatial relation (proving at last to be the distance) between the two points, and which is the same for congruent pairs of points, that is to say, such pairs as can be made successively to coincide in space with the same fixed pair of points.

[38] However indeterminate in appearance, this definition involves most important consequences, because with increase in the number of points, the number of equations increases much more quickly than the number of

[12] Ueber die Thatsachen die der Geometrie zum Grunde liegen (*Nachrichten von der königl. Ges. d. Wiss. zu Göttingen*, Juni 3, 1868).

co-ordinates which they determine. Five points, A, B, C, D, E, give ten different pairs of points

AB, AC, AD, AE,
BC, BD, BE,
CD, CE,
DE,

and therefore ten equations, involving in space of three dimensions fifteen variable co-ordinates. But of these fifteen, six must remain arbitrary, if the system of five points is to admit of free movement and rotation, and thus the ten equations can determine only nine co-ordinates as functions of the six variables. With six points we obtain fifteen equations for twelve quantities, with seven points twenty-one equations for fifteen, and so on. Now from *n* independent equations we can determine *n* contained quantities, and if we have more than *n* equations, the superfluous ones must be deducible from the first *n*. Hence it follows that the equations which subsist between the co-ordinates of each pair of points of a solid body must have a special character, seeing that, when in space of three dimensions they are satisfied for nine pairs of points as formed out of any five points, the equation for the tenth pair follows by logical consequence. Thus our assumption for the definition of solidity becomes quite sufficient to determine the kind of equations holding between the co-ordinates of two points rigidly connected.

⟦39⟧ Thirdly, the calculation must further be based on the fact of a peculiar circumstance in the movement of solid bodies, a fact so familiar to us that but for this inquiry it might never have been thought of as something that need not be. When in our space of three dimensions two points of a solid body are kept fixed, its movements are limited to rotations round the straight line connecting them. If we turn it completely round once, it again occupies exactly the position it had at first. This fact, that rotation in one direction always brings a solid body back into its original position, needs special mention. A system of geometry is possible without it. This is most easily seen in the geometry of a plane. Suppose that with every rotation of a plane figure its linear dimensions increased in proportion to the angle of rotation: the figure after one whole rotation through 360 degrees would no longer coincide with itself as it was originally. But any second figure that was congruent with the first in its original position might be made to coincide with it in its second position by being also turned through 360 degrees. A consistent system of geometry would be possible upon this supposition, which does not come under Riemann's formula.

⟦40⟧ On the other hand I have shown that the three assumptions taken together form a sufficient basis for the starting-point of Riemann's investigation, and thence for all his further results relating to the distinction of different spaces according to their measure of curvature.

⟦41⟧ It still remained to be seen whether the laws of motion, as dependent on moving forces, could also be consistently transferred to spherical or pseudospherical space. This investigation has been carried out by Professor Lipschitz of Bonn.[13] It is found that the comprehensive expression for all the laws of dynamics, Hamilton's principle, may be directly transferred to spaces of which the measure of curvature is other than zero. Accordingly, in this respect also, the disparate systems of geometry lead to no contradiction.

⟦42⟧ We have now to seek an explanation of the special characteristics of our own flat space, since it appears that they are not implied in the general notion of an extended quantity of three dimensions and of the free mobility of bounded figures therein. *Necessities of thought*, such as are involved in the conception of such a variety, and its measurability, or from the most general of all ideas of a solid figure contained in it, and of its free mobility, they undoubtedly are not. Let us then examine the opposite assumption as to their origin being empirical, and see if they can be inferred from facts of experience and so established, or if, when tested by experience, they are perhaps to be rejected. If they are of empirical origin, we must be able to represent to ourselves connected series of facts, indicating a different value for the measure of curvature from that of Euclid's flat space. But if we can imagine such spaces of other sorts, it cannot be maintained that the axioms of geometry are necessary consequences of an *à priori* transcendental form of intuition, as Kant thought.

⟦43⟧ The distinction between spherical, pseudospherical, and Euclid's geometry depends, as was above observed, on the value of a certain constant called, by Riemann, the measure of curvature of the space in question. The value must be zero for Euclid's axioms to hold good. If it were not zero, the sum of the angles of a large triangle would differ from that of the angles of a small one, being larger in spherical, smaller in pseudospherical, space. Again, the geometrical similarity of large and small solids or figures is possible only in Euclid's space. All systems of practical mensuration that have been used for the angles of large rectilinear triangles, and especially all systems of astronomical measurement which make the parallax of the immeasurably distant fixed stars equal to zero (in pseudospherical space the parallax even of infinitely distant points would be positive), confirm empirically the axiom of parallels, and show the measure of curvature of our space thus far to be indistinguishable from zero. It remains, however, a question, as Riemann observed, whether the result might not be different if we could use other than our limited base-lines, the greatest of which is the major axis of the earth's orbit.

⟦44⟧ Meanwhile, we must not forget that all geometrical measurements rest ultimately upon the principle of congruence. We measure the distance between points by applying to them the compass, rule, or chain. We measure angles by

[13] 'Untersuchungen über die ganzen homogenen Functionen von *n* Differentialen' (Borchardt's *Journal für Mathematik*, Bd. lxx. 3, 71; lxxiii. 3, 1); 'Untersuchung eines Problems der Variationsrechnung' (*Ibid.* Bd. lxxiv.).

bringing the divided circle or theodolite to the vertex of the angle. We also determine straight lines by the path of rays of light which in our experience is rectilinear; but that light travels in shortest lines as long as it continues in a medium of constant refraction would be equally true in space of a different measure of curvature. Thus all our geometrical measurements depend on our instruments being really, as we consider them, invariable in form, or at least on their undergoing no other than the small changes we know of, as arising from variation of temperature, or from gravity acting differently at different places.

⟦45⟧ In measuring, we only employ the best and surest means we know of to determine what we otherwise are in the habit of making out by sight and touch or by pacing. Here our own body with its organs is the instrument we carry about in space. Now it is the hand, now the leg, that serves for a compass, or the eye turning in all directions is our theodolite for measuring arcs and angles in the visual field.

⟦46⟧ Every comparative estimate of magnitudes or measurement of their spatial relations proceeds therefore upon a supposition as to the behaviour of certain physical things, either the human body or other instruments employed. The supposition may be in the highest degree probable and in closest harmony with all other physical relations known to us, but yet it passes beyond the scope of pure space-intuition.

⟦47⟧ It is in fact possible to imagine conditions for bodies apparently solid such that the measurements in Euclid's space become what they would be in spherical or pseudospherical space. Let me first remind the reader that if all the linear dimensions of other bodies, and our own, at the same time were diminished or increased in like proportion, as for instance to half or double their size, we should with our means of space-perception be utterly unaware of the change. This would also be the case if the distension or contraction were different in different directions, provided that our own body changed in the same manner, and further that a body in rotating assumed at every moment, without suffering or exerting mechanical resistance, the amount of dilatation in its different dimensions corresponding to its position at the time. Think of the image of the world in a convex mirror. The common silvered globes set up in gardens give the essential features, only distorted by some optical irregularities. A well-made convex mirror of moderate aperture represents the objects in front of it as apparently solid and in fixed positions behind its surface. But the images of the distant horizon and of the sun in the sky lie behind the mirror at a limited distance, equal to its focal length. Between these and the surface of the mirror are found the images of all the other objects before it, but the images are diminished and flattened in proportion to the distance of their objects from the mirror. The flattening, or decrease in the third dimension, is relatively greater than the decrease of the surface-dimensions. Yet every straight line or every plane in the outer world is represented by a straight line or a plane in the image. The image of a man measuring with a rule a straight line from the mirror would contract more and more the farther he went, but with his shrunken rule the man in the image would count out exactly the same number of centimetres as the real man. And, in general, all geometrical measurements of lines or angles made with regularly varying images of real instruments would yield exactly the same results as in the outer world, all congruent bodies would coincide on being applied to one another in the mirror as in the outer world, all lines of sight in the outer world would be represented by straight lines of sight in the mirror. In short I do not see how men in the mirror are to discover that their bodies are not rigid solids and their experiences good examples of the correctness of Euclid's axioms. But if they could look out upon our world as we can look into theirs, without overstepping the boundary, they must declare it to be a picture in a spherical mirror, and would speak of us just as we speak of them; and if two inhabitants of the different worlds could communicate with one another, neither, so far as I can see, would be able to convince the other that he had the true, the other the distorted, relations. Indeed I cannot see that such a question would have any meaning at all, so long as mechanical considerations are not mixed up with it.

⟦48⟧ Now Beltrami's representation of pseudospherical space in a sphere of Euclid's space is quite similar, except that the background is not a plane as in the convex mirror, but the surface of a sphere, and that the proportion in which the images as they approach the spherical surface contract, has a different mathematical expression.[14] If we imagine then, conversely, that in the sphere, for the interior of which Euclid's axioms hold good, moving bodies contract as they depart from the centre like the images in a convex mirror, and in such a way that their representatives in pseudospherical space retain their dimensions unchanged,—observers whose bodies were regularly subjected to the same change would obtain the same results from the geometrical measurements they could make as if they lived in pseudospherical space.

⟦49⟧ We can even go a step further, and infer how the objects in a pseudospherical world, were it possible to enter one, would appear to an observer, whose eye-measure and experiences of space had been gained like ours in Euclid's space. Such an observer would continue to look upon rays of light or the lines of vision as straight lines, such as are met with in flat space, and as they really are in the spherical representation of pseudospherical space. The visual image of the objects in pseudospherical space would thus make the same impression upon him as if he were at the centre of Beltrami's sphere. He would think he saw the most remote objects round about him at a finite distance,[15] let us suppose a hundred feet off. But as he approached these distant objects, they would dilate before him, though more in the third dimension than superficially, while behind him they would contract. He would know that his eye judged wrongly. If he saw two straight lines which in his estimate ran parallel for the hundred feet to his world's end, he would find on following them that the farther he advanced the more they diverged, because of the dilatation of

[14] Compare the Appendix at the end of this Lecture.

[15] The reciprocal of the square of this distance, expressed in negative quantity, would be the measure of curvature of the pseudo-spherical space.

all the objects to which he approached. On the other hand, behind him, their distance would seem to diminish, so that as he advanced they would appear always to diverge more and more. But two straight lines which from his first position seemed to converge to one and the same point of the background a hundred feet distant, would continue to do this however far he went, and he would never reach their point of intersection.

⟦50⟧ Now we can obtain exactly similar images of our real world, if we look through a large convex lens of corresponding negative focal length, or even through a pair of convex spectacles if ground somewhat prismatically to resemble pieces of one continuous larger lens. With these, like the convex mirror, we see remote objects as if near to us, the most remote appearing no farther distant than the focus of the lens. In going about with this lens before the eyes, we find that the objects we approach dilate exactly in the manner I have described for pseudospherical space. Now any one using a lens, were it even so strong as to have a focal length of only sixty inches, to say nothing of a hundred feet, would perhaps observe for the first moment that he saw objects brought nearer. But after going about a little the illusion would vanish, and in spite of the false images he would judge of the distances rightly. We have every reason to suppose that what happens in a few hours to any one beginning to wear spectacles would soon enough be experienced in pseudospherical space. In short, pseudospherical space would not seem to us very strange, comparatively speaking; we should only at first be subject to illusions in measuring by eye the size and distance of the more remote objects.

⟦51⟧ There would be illusions of an opposite description, if, with eyes accustomed to measure in Euclid's space, we entered a spherical space of three dimensions. We should suppose the more distant objects to be more remote and larger than they are, and should find on approaching them that we reached them more quickly than we expected from their appearance. But we should also see before us objects that we can fixate only with diverging lines of sight, namely, all those at a greater distance from us than the quadrant of a great circle. Such an aspect of things would hardly strike us as very extraordinary, for we can have it even as things are if we place before the eye a slightly prismatic glass with the thicker side towards the nose: the eyes must then become divergent to take in distant objects. This excites a certain feeling of unwonted strain in the eyes, but does not perceptibly change the appearance of the objects thus seen. The strangest sight, however, in the spherical world would be the back of our own head, in which all visual lines not stopped by other objects would meet again, and which must fill the extreme background of the whole perspective picture.

⟦52⟧ At the same time it must be noted that as a small elastic flat disk, say of india-rubber, can only be fitted to a slightly curved spherical surface with relative contraction of its border and distension of its centre, so our bodies, developed in Euclid's flat space, could not pass into curved space without undergoing similar distensions and contractions of their parts, their coherence being of course maintained only in as far as their elasticity permitted their bending without breaking. The kind of distension must be the same as in passing from a small body imagined at the centre of Beltrami's sphere to its pseudospherical or spherical representation. For such passage to appear possible, it will always have to be assumed that the body is sufficiently elastic and small in comparison with the real or imaginary radius of curvature of the curved space into which it is to pass.

⟦53⟧ These remarks will suffice to show the way in which we can infer from the known laws of our sensible perceptions the series of sensible impressions which a spherical or pseudospherical world would give us, if it existed. In doing so, we nowhere meet with inconsistency or impossibility any more than in the calculation of its metrical proportions. We can represent to ourselves the look of a pseudospherical world in all directions just as we can develop the conception of it. Therefore it cannot be allowed that the axioms of our geometry depend on the native form of our perceptive faculty, or are in any way connected with it.

⟦54⟧ It is different with the three dimensions of space. As all our means of sense-perception extend only to space of three dimensions, and a fourth is not merely a modification of what we have, but something perfectly new, we find ourselves by reason of our bodily organization quite unable to represent a fourth dimension.

⟦55⟧ In conclusion, I would again urge that the axioms of geometry are not propositions pertaining only to the pure doctrine of space. As I said before, they are concerned with quantity. We can speak of quantities only when we know of some way by which we can compare, divide, and measure them. All space-measurements, and therefore in general all ideas of quantities applied to space, assume the possibility of figures moving without change of form or size. It is true we are accustomed in geometry to call such figures purely geometrical solids, surfaces, angles, and lines, because we abstract from all the other distinctions, physical and chemical, of natural bodies; but yet one physical quality, rigidity, is retained. Now we have no other mark of rigidity of bodies or figures but congruence, whenever they are applied to one another at any time or place, and after any revolution. We cannot, however, decide by pure geometry, and without mechanical considerations, whether the coinciding bodies may not both have varied in the same sense.

⟦56⟧ If it were useful for any purpose, we might with perfect consistency look upon the space in which we live as the apparent space behind a convex mirror with its shortened and contracted background; or we might consider a bounded sphere of our space, beyond the limits of which we perceive nothing further, as infinite pseudospherical space. Only then we should have to ascribe to the bodies which appear to us to be solid, and to our own body at the same time, corresponding distensions and contractions, and we should have to change our system of mechanical principles entirely; for even the proposition that every point in motion, if acted upon by no force, continues to move with unchanged velocity in a straight line, is not adapted to the image of the world in the convex mirror. The path would indeed be straight, but the velocity would depend upon the place.

⟦57⟧ Thus the axioms of geometry are not concerned with space-relations only but also at the same time with the mechanical deportment of solidest bodies in motion. The notion of rigid geometrical figure might indeed be conceived as transcendental in Kant's sense, namely, as formed independently of actual experience, which need not exactly correspond therewith, any more than natural bodies do ever in fact correspond exactly to the abstract notion we have obtained of them by induction. Taking the notion of rigidity thus as a mere ideal, a strict Kantian might certainly look upon the geometrical axioms as propositions given, *à priori*, by transcendental intuition, which no experience could either confirm or refute, because it must first be decided by them whether any natural bodies can be considered as rigid. But then we should have to maintain that the axioms of geometry are not synthetic propositions, as Kant held them; they would merely define what qualities and deportment a body must have to be recognized as rigid.

⟦58⟧ But if to the geometrical axioms we add propositions relating to the mechanical properties of natural bodies, were it only the axiom of inertia, or the single proposition, that the mechanical and physical properties of bodies and their mutual reactions are, other circumstances remaining the same, independent of place, such a system of propositions has a real import which can be confirmed or refuted by experience, but just for the same reason can also be gained by experience. The mechanical axiom, just cited, is in fact of the utmost importance for the whole system of our mechanical and physical conceptions. That rigid solids, as we call them, which are really nothing else than elastic solids of great resistance, retain the same form in every part of space if no external force affects them, is a single case falling under the general principle.

⟦59⟧ In conclusion, I do not, of course, maintain that mankind first arrived at space-intuitions, in agreement with the axioms of Euclid, by any carefully executed systems of exact measurement. It was rather a succession of everyday experiences, especially the perception of the geometrical similarity of great and small bodies, only possible in flat space, that led to the rejection, as impossible, of every geometrical representation at variance with this fact. For this no knowledge of the necessary logical connection between the observed fact of geometrical similarity and the axioms was needed; but only an intuitive apprehension of the typical relations between lines, planes, angles, etc., obtained by numerous and attentive observations—an intuition of the kind the artist possesses of the objects he is to represent, and by means of which he decides with certainty and accuracy whether a new combination, which he tries, will correspond or not with their nature. It is true that we have no word but *intuition* to mark this; but it is knowledge empirically gained by the aggregation and reinforcement of similar recurrent impressions in memory, and not a transcendental form given before experience. That other such empirical intuitions of fixed typical relations, when not clearly comprehended, have frequently enough been taken by metaphysicians for *à priori* principles, is a point on which I need not insist.

⟦60⟧ To sum up, the final outcome of the whole inquiry may be thus expressed:—

(1.) The axioms of geometry, taken by themselves out of all connection with mechanical propositions, represent no relations of real things. When thus isolated, if we regard them with Kant as forms of intuition transcendentally given, they constitute a form into which any empirical content whatever will fit, and which therefore does not in any way limit or determine beforehand the nature of the content. This is true, however, not only of Euclid's axioms, but also of the axioms of spherical and pseudospherical geometry.

(2.) As soon as certain principles of mechanics are conjoined with the axioms of geometry, we obtain a system of propositions which has real import, and which can be verified or overturned by empirical observations, just as it can be inferred from experience. If such a system were to be taken as a transcendental form of intuition and thought, there must be assumed a pre-established harmony between form and reality.

Appendix ⟦From the German version of 1876.⟧

The elements of the geometry of spherical space are most easily obtained by putting for space of four dimensions the equation for the sphere

$$x^2 + y^2 + z^2 + t^2 = R^2 \;\ldots\ldots\ldots\; (1.)$$

and for the distance ds between the points (x, y, z, t) and $[(x + dx), (y + dy), (z + dz), (t + dt)]$ the value

$$ds^2 = dx^2 + dy^2 + dz^2 + dt^2 \;\ldots\ldots\; (2.)$$

It is easily found by means of the methods used for three dimensions that the shortest lines are given by equations of the form

$$\left.\begin{array}{l} ax + by + cz + ft = 0 \\ \alpha x + \beta y + \gamma z + \phi t = 0 \end{array}\right\} \;\ldots\ldots\ldots\; (3.)$$

in which a, b, c, f, as well as α, β, γ, ϕ, are constants.

The length of the shortest arc, s, between the points (x, y, z, t), and (ξ, η, ζ, τ) follows, as in the sphere, from the equation

$$\cos\frac{s}{R} = \frac{x\xi + y\eta + z\zeta + t\tau}{R^2} \;\ldots\ldots\ldots\; (4.)$$

One of the co-ordinates may be eliminated from the values given in 2 to 4, by means of equation 1, and the expressions then apply to space of three dimensions.

If we take the distances from the points

$$\xi = \eta = \zeta = 0$$

from which equation 1 gives $\tau = R$, then,

$$\sin\left(\frac{s_0}{R}\right) = \frac{\sigma}{R}$$

in which $$\sigma = \sqrt{x^2 + y^2 + z^2}$$

or, $$s_0 = R.\text{arc sin}\left(\frac{\sigma}{R}\right) = R.\text{arc tang}\left(\frac{\sigma}{t}\right) \quad \ldots (5.)$$

In this, s_0 is the distance of the point x, y, z, measured from the centre of the co-ordinates.

If now we suppose the point x, y, z, of spherical space, to be projected in a point of plane space whose co-ordinates are respectively

$$\mathfrak{x} = \frac{Rx}{t} \quad \mathfrak{y} = \frac{Ry}{t} \quad \mathfrak{z} = \frac{Rz}{t}$$

$$\mathfrak{x}^2 + \mathfrak{y}^2 + \mathfrak{z}^2 = \mathfrak{r}^2 = \frac{R^2\sigma^2}{t^2}$$

then in the plane space the equations 3, which belong to the straightest lines of spherical space, are equations of the straight line. Hence the shortest lines of spherical space are represented in the system of $\mathfrak{x}$, $\mathfrak{y}$, $\mathfrak{z}$, by straight lines. For very small values of x, y, z, $t = R$, and

$$\mathfrak{x} = x, \ \mathfrak{y} = y, \ \mathfrak{z} = z\,.$$

Immediately about the centre of the co-ordinates, the measurements of both spaces coincide. On the other hand, we have for the distances from the centre

$$s_0 = R.\text{arc tang}\left(\pm\frac{\mathfrak{r}}{R}\right) \quad \ldots (6.)$$

In this, $\mathfrak{r}$ may be infinite; but every point of plane space must be the projection of two points of the sphere, one for which $s_0 < \frac{1}{2}R\pi$, and one for which $s_0 > \frac{1}{2}R\pi$. The extension in the direction of $\mathfrak{r}$ is then

$$\frac{ds_0}{d\mathfrak{r}} = \frac{R^2}{R^2 + \mathfrak{r}^2}\,.$$

In order to obtain corresponding expressions for pseudospherical space, let R and t be imaginary; that is, $R = \Re\mathrm{i}$, and $t = \mathfrak{t}\mathrm{i}$. Equation 6 gives then

$$\text{tang}\frac{s_0}{\mathrm{i}\Re} = \pm\frac{\mathfrak{r}}{\mathrm{i}\Re}$$

from which, eliminating the imaginary form, we get

$$s_0 = \tfrac{1}{2}\Re\log.\text{ nat.}\frac{\Re + \mathfrak{r}}{\Re - \mathfrak{r}}\,.$$

Here s_0 has real values only as long as $\mathfrak{r} = R$; for $\mathfrak{r} = \mathfrak{R}$ the distance s_0 in pseudospherical space is infinite. The image in plane space is, on the contrary, contained in the sphere of radius R, and every point of this sphere forms only one point of the infinite pseudospherical space. The extension in the direction of $\mathfrak{r}$ is

$$\frac{ds_0}{d\mathfrak{r}} = \frac{\mathfrak{R}^2}{\mathfrak{R}^2 - \mathfrak{r}^2}.$$

For linear elements, on the contrary, whose direction is at right angles to $\mathfrak{r}$, and for which t is unchanged, we have in both cases

$$\frac{\sqrt{dx^2 + dy^2 + dz^2}}{\sqrt{d\mathfrak{x}^2 + d\mathfrak{y}^2 + d\mathfrak{z}^2}} = \frac{t}{R} = \frac{\mathfrak{t}}{\mathfrak{R}} = \frac{\sigma}{\mathfrak{r}}$$
$$= \frac{\sqrt{x^2 + y^2 + z^2}}{\sqrt{\mathfrak{x}^2 + \mathfrak{y}^2 + \mathfrak{z}^2}}.$$

The origin and meaning of geometrical axioms (II)

⟦From *1878a.*⟧

My article on 'The Origin and Meaning of Geometrical Axioms' in MIND No. III. was critically examined by Professor Land in No. V., and I will now try to answer his objections. We differ substantially on two points. I am of opinion that the recent mathematical investigations—or, as they have been called, 'metamathematical investigations'*—as to wider kinds of geometry, have established the following propositions:

(1) Kant's proof of the *a priori* origin of geometrical axioms, based on the assumption that no other space-relations can be mentally represented, is insufficient, the assumption being at variance with fact.

(2) If, in spite of the defective proof, it is still assumed hypothetically that the axioms are really given *a priori* as laws of our space-intuitions, two kinds of equivalence of space-magnitudes must be distinguished: *(a) Subjective equality* given by the hypothetical transcendental intuition; *(b) Objective equivalence* of the real substrata of space-relations, proved by the equality of physical states or actions, existing or going on in what appear to us as congruent parts of space. The coincidence of the second with the first could be proved only by experience; and as the second would alone concern us in our scientific or practical dealings with the objective world, the first, in case of discrepancy, must be discounted as a *false show*.

* The name has been given by opponents in irony, as suggesting 'metaphysic'; but as the founders of 'Non-Euclidian Geometry' have never maintained its objective truth, they can very well accept the name.

For the rest, it is a misunderstanding on Prof. Land's part if he thinks I wished to raise any objection to the notion of space as being for us an *a priori* and necessary, or (in Kant's sense) transcendental, form of intuition. I had no such intention. It is true, my view of the relation between this transcendental form and reality, as I shall set it forth in the third section of this paper, does not quite coincide with that of many followers of Kant and Schopenhauer. But space may very well be a form of intuition in the Kantian sense, and yet not necessarily involve the axioms. To cite a parallel instance, it undoubtedly lies in the organisation of our optical apparatus that everything we see can be seen only as a spatial distribution of colours. This is the innate form of our visual perceptions. But it is not in the least thereby predetermined how the colours we see shall co-exist in space and follow each other in time. And just so, in my view, the representation of all external objects in space-relations may be the only possible form in which we can represent the simultaneous existence of a number of discrete objects, though there is no necessity that a particular space-perception should co-exist with or follow upon certain others; e.g., that every rectilineal equilateral triangle should have angles of 60°, whatever the length of the sides. By Kant, indeed, the proof that space is an *a priori* form is based essentially on the position that the axioms are synthetic propositions *a priori*. But even if this assertion with the dependent inference is dropt, the space-representation might still be the necessary *a priori* form in which every co-extended manifold is perceived. This is not surrendering any essential feature of the Kantian system. On the contrary, the system becomes more consistent and intelligible, if the proof of the possibility of metaphysic derived from the evidence of geometrical axioms is seen to break down. Kant himself, as is well known, limited the scope of metaphysical science to the geometrical and physical axioms. But the physical axioms are either of doubtful validity, or they are mere consequences of the principle of causality, that is to say, of our intellectual impulse to view everything that happens as conforming to law and thus as conceivable. And as Kant's *Kritik* is otherwise hostile to all metaphysical reasoning, his system seems to be freed from inconsistency, and a clearer notion of the nature of intuition is obtained, if the *a priori* origin of the axioms is abandoned, and geometry is regarded as the first and most perfect of the natural sciences.

I pass accordingly to the proof of the two theses enunciated above.

I.

Kant's proof of the *a priori* origin of the geometrical axioms is based on the assertion that it is impossible to form a mental representation of space-relations at variance with Euclid's geometry. But the 'metamathematical' investigations passed under review in my former paper have shown that it is quite possible to devise and consistently work out systems of geometry that differ from Euclid's both in the number of space-dimensions and in their axioms, with their related systems of mechanics. I myself have tried to show what would be the sensible appearance of objects in spherical or in pseudospherical space. The mathematical correctness of those geometrical deductions (carried out for the most part analytically) is, as far as I can see, beyond question, and the like may be said as to the perfect validity of the corresponding systems of mechanics, which afford the same degree of free mobility for solid bodies, and the same independence of mechanical and physical processes on mere position, that are presupposed in the Euclidian geometry. Nor is there the least difficulty or uncertainty as to the nature of the space-perceptions that human beings would have in such other circumstances. In particular, Beltrami's discovery of the way of representing pseudospherical space in a sphere of Euclidian space shows directly what would be the appearance of optical images in pseudospherical or spherical space. Every optical image of objects at rest as seen by a spectator at rest would, in fact, be exactly the same as that of the corresponding representation in Beltrami's sphere as seen from the centre (supposing always that the distance of the two eyes may be neglected in comparison with the imaginary radius-of-curvature of the space). There would be a difference only in the order of succession of the images, according as the observer or the solid objects moved. Nothing would be changed but the rule for inferring what images would succeed others in case of movement. And, as I have maintained, such differences are not necessarily considerable, nor need they excite attention. Men lived for a long time on what they thought was the flat earth, before they discovered its spherical form, and they struggled long enough against this truth, just as our Kantians at the present day will not listen to the possibility of representing pseudospherical space. The discrepancies in pseudospherical space would be of a somewhat similar kind, and not necessarily more striking (if the measure-of-curvature tallied) than are those betrayed by the spherical surface of the earth to an observer whose movements are limited to a few miles.

In discussing the question whether space-relations can be imagined in meta-mathematical spaces, the first thing to settle is the rule by which we shall judge of the imaginability of an object that we have never actually seen.

I advanced a definition which was to the effect—that for this we need the power of fully representing the sense-impressions which the object would excite in us according to the known laws of our sense-organs under all conceivable conditions of observation, and by which it would be distinguished from other similar objects. I am of opinion that this definition contains stricter and more definite requirements for the possibility of imagination than any previous one, and, as far as I can see, Prof. Land does not contend that these requirements cannot be satisfied for objects in spherical or pseudospherical spaces. At the same time, the representation of objects that we have often perceived, or that resemble such in whole or in parts, will necessarily be superior in one respect to the representation of objects of which this cannot be said, namely, in the swiftness and ease with which we can imagine beforehand the various aspects of the objects under different conditions of observation, or run them over in memory. This ease and swiftness in the imagination of an object never actually seen, will be wanting just in proportion as the observer has more rarely perceived and less carefully apprehended anything like it. Now we have absolutely never had before us constructions of three dimensions in spherical or pseudospherical space. The geometer, however, who has trained himself in the power of representing surfaces that can be bent without stretching and without change of their measure-of-curvature, as also the figures that can be drawn upon them, finds relations in these that are closely analogous to the relations in those other spaces. The physiologist too who has studied the combinations of sense-impressions under every possible variety of conditions, such as never occur in daily experience, is more practised in representing unusual (but yet strictly determinate) series of sense-impressions than one who has never had the same training. I may perhaps be pardoned, then, if I do not see why the fact that I come 'fresh from the physiology of the senses' to epistemological inquires should be a positive bar to my dealing with such questions as the one before us.

Since, then, the metamathematical space-relations have never been actually perceived by us, we are not to expect to have that power of swift and easy representation of the varying aspects of objects in them that can come only from daily experience and practice. The utmost we can expect is to arrive by slow steps and careful reflection at a full and consistent representation of the corresponding series of sense-impressions. But in point of fact, we strike upon as great and similar difficulties of representation when we seek to figure to ourselves the course of a greatly knotted thread, or a many-sided crystal model, or a complex building that we have never seen, although the possibility of figuring all these is proved by the fact of actual perception.

Unfortunately, Professor Land does not say whether he has any objection to my definition of imaginative representation, nor does he himself offer any other, though he several times hints that he means something different by 'imaginability'. Thus, at p. 41, he says: 'We do not find that they [the non-Euclidians] succeed in this [making metamathematical spaces imaginable], unless the notion of imaginability be stretched far beyond what Kantians and others understand by the word.' At the same place, he asserts that only that which can be connectedly constructed in our space can be regarded as 'imagined'. He adds at p. 45: 'Non-Euclidians try to make imaginable that which is not so in the sense required for argumentation in this case.' If by 'argumentation' is here meant the discussion of the question whether our conviction of the actual validity of Euclid's axioms in our objective world justifies a conclusion as to their *a priori* origin, I am of opinion that my definition of imaginability is the only one that can decide the question. If we should define thus: 'Nothing is to be held as imaginable in space, of which we cannot actually construct a model with existing bodies,'—all discussion of the question in dispute is, no doubt, cut short; but then this imaginability, ascribed by the definition to Euclid's space alone, affords not the least ground for deciding whether its origin is to be sought in a law of the objective world, or in the constitution of our minds. Accordingly, I do not believe that Professor Land means to postulate this, though his words bear the interpretation. I can only suppose him to object to my definition of 'imaginability' that it does not include a reference to the apparently spontaneous readiness with which the various aspects of any common object are represented when we have sensible experience of some one of them. But we know that such an association of different impressions can be acquired and strengthened by frequent repetition; as, notably, between the sound of a word and its meaning. I therefore do not see that we have the right to consider this readiness of suggestion as essential to imaginability. The fact, moreover, that Lobatchevsky, in the way of pure synthesis, that is to say, by means of actual geometrical constructions, worked out a complete system of pseudospherical geometry, agreeing exactly with the results of analytical inquiry, shows that such a geometry can be grasped in all its details by the imagination.

As regards the use of analytical methods in metamathematical inquiries, this is justified by the circumstance that we have here to do with the representation of an object that has never been perceived—an object whose notion, or (so to speak) architectural plan, has first to be developed, to be shown inherently consistent, and to be elaborated so far in detail as that for every particular case it is made clear what the corresponding sensible suppression would be in the circumstances. Now, this ideal development of the ground-plan is best attained by the methods of analytical geometry, securing as these do most effectively universality and completeness of demonstration. No doubt a manipulation of notions by means of the calcalus does not suffice to prove the existence of the object so treated, but the process is sufficient to the extent of proving the possibility of a consistent series of sensible pictures; whence it follows that the space-relations actually perceived in a real world by organs analogous to our own might correspond with a geometry different from Euclid's.

Since then the relations obtaining in metamathematical spaces of three dimensions satisfy the conditions of imaginability required by my definition—and more cannot be demanded in the case of objects never actually perceived—Kant's proof of the transcendental character of the axioms and their *a priori* origin must be pronounced insufficient.

‖As explained above, the final two sections (II) and (III) of *Helmboltz 1878a* appear below as §1 and §2 of Appendix 3 to *Helmholtz 1878b*.‖

B. THE FACTS IN PERCEPTION (*HELMHOLTZ 1878b*)

The following essay, perhaps the subtlest nineteenth-century defence of psychologism in mathematics, was delivered as the Rector's address at the anniversary celebrations for the Friedrich Wilhelm University in Berlin on 3 August 1878. Helmholtz was at the apex of his fame and influence, and took the opportunity to give a comprehensive statement of his epistemological views and of their relation to his work in physiological optics and in geometry.

The essay is painted with broad strokes across a huge canvas; it shows Helmholtz at his most sweeping and suggestive, but also at his most elusive. Helmholtz's argument raises numerous difficulties of interpretation. His own position is difficult to pin down with confidence, as is his relationship to other thinkers, and in particular his relationship to Kant. The issues are too complex to be explored in the present Note; but there are important studies in the secondary literature, notably in the footnote commentary to this lecture by Moritz Schlick (*Helmholtz 1921*), and in the monograph *Hatfield 1990. Hatfield 1990* is particularly helpful on the background to Helmholtz's theory of spatial perception, and surveys the work of his predecessors both in philosophy and in psychology.

The translation below is by Malcolm F. Lowe, and is taken from *Helmholtz 1977*. In that edition some of Helmholtz's longer paragraphs have been broken up, and in one case (paragraph 46) two paragraphs have been elided. For ease of reading, the shorter paragraphs of *1977* have been retained here; however, references to *Helmholtz 1878b* should be to the original paragraphing by paragraph numbers, which have been added in this edition.

My distinguished audience!

⟦1⟧ Today on the birthday of the founder of our university, the sorely-tried King Friedrich Wilhelm III, we celebrate the anniversary of its foundation. The year of its foundation, 1810, fell in the period of the greatest external stress upon our country. A considerable part of its territory had been lost, the land was exhausted from the preceding war and the enemy occupation. The martial pride which had remained with it from the times of the great elector and the great king had been deeply humiliated. And yet this same period now seems to us, when we glance backwards, to have been so rich in possessions of a spiritual kind, in inspiration, energy, ideal hopes and creative thoughts, that we might, despite the relatively brilliant external situation of our country and nation today, look back upon it almost with envy. If in that distressing situation the king's first thought was of founding the university before other material claims, if he then staked throne and life so as to entrust himself to the resolute inspiration of the nation in the struggle against the conqueror, this all shows how deeply within him too, the simple man disinclined to lively expressions of feeling, acted a trust in the spiritual powers of his people.

⟦2⟧ At that time Germany could point to a magnificent series of praiseworthy names in both art and science, names whose bearers are in part to be counted amongst the greatest of all times and peoples in the history of human culture.

⟦3⟧ Goethe was alive and so was Beethoven; Schiller, Kant, Herder, and Haydn had survived the first years of the century. Wilhelm von Humboldt was outlining the new science of comparative linguistics; Niebuhr, Fr. Aug. Wolf, and Savigny were teaching how to permeate ancient history, poetry, and law with living understanding; Schleiermacher was seeking a profound understanding of the spiritual content of religion. Joh. Gottlieb Fichte, the second rector of our university, the powerful and fearless public speaker, was carrying his audience away with the stream of his moral inspiration and the bold intellectual flight of his idealism.

‖4‖ Even the aberrations of this mentality, which express themselves in the easily recognizable weaknesses of romanticism, have something attractive compared with dry, calculating egoism. One marvelled at oneself in the fine feelings in which one knew how to revel, one sought to develop the art of having such feelings. One thought oneself allowed to admire fantasy all the more as a creative force, the more it had freed itself from the rules of the understanding. Much vanity lay hidden in this, but all the same a vanity of enthusiasm for high ideals.

‖5‖ The older ones amongst us still knew the men of that period, who had once entered the army as the first volunteers, always ready to immerse themselves in the discussion of metaphysical problems, well-read in the works of Germany's great poets, men who still glowed with rage when talking of the first Napoleon, but with rapture and pride when of deeds in the war of liberation.

‖6‖ How things have changed! We may well exclaim thus with amazement in a period when a cynical contempt for every ideal possession of humankind is propagated, on the streets and in the press, and has reached its peak in two revolting crimes,[a] which were obviously only aimed at the head of our emperor because in him was united everything that humanity, up to now, has regarded as worthy of veneration and gratitude.

‖7‖ We must almost make an effort to recall that only eight years have passed since the great hour, when at the call of the same monarch every rank of our people, without hesitation and filled with self-sacrificing and inspired patriotism, went into a dangerous war against an opponent whose might and valour were not unknown to us. We must almost make an effort to take note of the wide extent to which the endeavours. political and humane, to give the poorer ranks too of our people an existence less troubled and more worthy of human beings, have captured the activity and thoughts of the educated classes. Or to think how much their lot in material and legal respects has actually been improved.

‖8‖ The nature of mankind seems simply to be such that next to much light one can always find much shadow. Political freedom initially gives the vulgar motives a greater licence to reveal themselves and to embolden each other, as long as they are not faced with a public opinion ready to offer energetic opposition. Even in the years before the war of liberation, when Fichte was preaching sermons calling upon his generation to repent, these elements were not lacking. He depicts conditions and sentiments as ruling which recall the worst of our times. 'The present age adopts in its basic principle a stance of haughtily looking down upon those who, from a dream of virtue, let themselves be torn away from pleasures, and rejoicing in the thought that one must get beyond such things, and not at all be imposed upon in this manner.'* The only pleasure,

---

[a] ‖In 1878, two attempts were made to assassinate the emperor.‖

* Fichte, *Werke*, vol. VII, p. 40.

going beyond the purely sensuous, which he concedes to be known to the representatives of that age, is what he calls 'delighting in one's own artfulness'. And yet, in this same period, there was being prepared a mighty upswing which belongs to the most glorious events in our history.

⟦9⟧ Although we therefore need not regard our period as beyond hope, we should surely not soothe ourselves too easily with the consolation that things were indeed not better in other times than now. It is nevertheless advisable, when such dubious processes are going on, that each person should make a review—in the sphere given him to work in and which he knows—of the situation of the work towards the eternal goals of mankind: whether they are being kept in view, whether one has got nearer to them. In the youthful days of our university science too was youthfully bold and strong in hopes, its view was directed pre-eminently towards the highest goals. Although these were not to be reached so easily as that generation hoped, although it also emerged that long drawn out particular labours had to prepare the path towards them, so that initially the nature itself of the tasks demanded another kind of work—less enthusiastic, less immediately directed towards the ideal goals—it would still doubtless be pernicious should our generation have lost sight of the eternal ideals of mankind, over and above subordinate and practically useful tasks.

⟦10⟧ In that period, the fundamental problem placed at the beginning of all science was the problem of epistemology: 'What is true in our intuition and thought? In what sense do our representations correspond to actuality?' Philosophy and natural science encounter this problem from two opposite sides; it is a task common to both.

The former, which considers the mental side, seeks to separate out from our knowledge and representation what originates from the influences of the corporeal world, in order to set forth unalloyed what belongs to the mind's own workings. Natural science, on the contrary, seeks to separate off that which is definition, symbolism, representational form or hypothesis, in order to have left over unalloyed what belongs to the world of actuality[b] whose laws it seeks. Both seek to execute the same separation, although each is interested in a different part of what is separated. In the theory of sense perceptions, and in investigations into the fundamental principles of geometry, mechanics, and physics, even the enquirer into nature cannot evade these questions. As my own studies have frequently entered both domains, I want to try to give you a survey of what has been done in this direction on the part of enquiry into nature.

Naturally, in the last analysis the laws of thought are no different in the man enquiring into nature from what they are in the man who philosophizes. In all cases where the facts of daily experience—whose profusion is after all already very great—sufficed to give a percipient thinker, with an unconstrained feeling for truth, in some measure enough material for a correct judgement,

[b] ⟦'Wirklichkeit'; 'wirklich' has been translated consistently as 'actual' rather than 'real', because below Helmholtz expressly distinguishes it from 'reell'.⟧

the enquirer into nature must satisfy himself with acknowledging that the methodically completed gathering of empirical facts simply confirms the result gained previously. But there also occur cases of the contrary kind. Such cases will justify the fact—if it needs to be justified at all—that in what follows the questions concerned are not everywhere given new answers, but to a great extent ones given long ago are repeated. Indeed, often enough even an old concept, measured against new facts, gets a more vivid illumination and a new look.

⟦11⟧ Shortly before the beginning of the present century, Kant had developed the doctrine of forms of intuiting and thinking given prior to all experience—or (as he therefore termed them) '*transcendental*' forms of intuiting and thinking—into which forms any content we may represent must necessarily be absorbed, if this content is to become a representation. Regarding the qualities of sensation, Locke had already established a claim for the share which our corporeal and mental makeup has in the manner in which things appear to us. In this direction, investigations into the physiology of the senses, which were in particular completed and critically sifted by Johannes Müller and then summarized by him in the law of *specific energies of sensory nerves*, have now brought the fullest confirmation, one can almost say to an unexpected degree. At the same time, they have thereby portrayed and made intuitive, in a very decisive and palpable manner, the essence and significance of such a subjective form, given in advance of sensation. This theme has already often been discussed, for which reason my presentation of it today can be brief.

⟦12⟧ There occur two distinct degrees of difference between the various kinds of sensation. The more deeply incisive difference is that between sensations belonging to different senses, such as between blue, sweet, warm, high-pitched: I have permitted myself to term this a difference in the *modality* of sensation. It is so incisive as to exclude any transition from the one to the other, any relationship of greater or lesser similarity. One cannot at all ask whether, e.g., sweet is more similar to blue or to red. On the other hand, the second kind of difference—the less incisive—is that between different sensations of the same sense: I restrict the term a difference of *quality* to this difference alone. Fichte groups together these qualities of a single sense as a quality range ⟦Qualitätenkreis⟧, and terms a *difference of quality ranges* what I just called a difference of modality. Within each such range, transition and comparison are possible. We can make the transition from blue through violet and crimson into scarlet, and, e.g., declare yellow to be more similar to orange than to blue.

What physiological investigations now show is that the deeply incisive difference does not depend, in any manner whatsoever, upon the kind of external impression whereby the sensation is excited, but is determined alone and exclusively by the sensory nerve upon which the impression impinges. Excitation of the optic nerve produces only light sensations, no matter whether objective light—i.e. aether vibrations—impinges upon it, or an electric current which we pass through the eye, or pressure on the eyeball, or straining of the nerve stem during rapid changes of the direction of vision. The sensation arising through the latter influences is so similar to that of objective light, that people for a long time believed in light actually developing in the eye. Johannes Müller showed that such a development does not on any account take place, that the sensation of light was indeed only there because the optic nerve was excited.

⟦13⟧ Just as on the one hand each sensory nerve, excited by however so many influences, always gives only sensations from the quality range proper to itself, so on the other hand are produced by the same external influences—when they impinge upon different sensory nerves—the most varied kinds of sensation, these always being taken from the quality range of the nerve concerned. The same aether vibrations as are felt by the eye as light, are felt by the skin as heat. The same air vibrations as are felt by the skin as a quivering motion, are felt by the ear as a note. Here the difference in kind of the impression is moreover so great, that physicists felt at ease with the idea that agents as apparently different as light and radiant heat are alike in kind, and in part identical, only after the complete alikeness in kind of their physical behaviour had been established, by laborious experimental investigations in every direction.

⟦14⟧ But even within the quality range of each individual sense, where the kind of object exerting an influence at least codetermines the quality of the produced sensation, there still occur the most unexpected incongruities. In this respect, the comparison of eye and ear is instructive. For the objects of both—light and sound—are oscillatory motions, each of which excites different sensations according to the rapidity of vibration: in the eye different colours, in the ear different pitches.

If we allow ourselves, for the sake of greater perspicuity, to refer to the frequency relationships of light in terms of the musical intervals formed by corresponding tone frequencies, then the result is as follows: the ear is sensitive to some ten octaves of different tones, the eye only to a sixth, although the frequencies lying beyond these limits occur for both sound and light, and can be demonstrated physically. The eye has in its short scale only three mutually distinct basic sensations, out of which all of its qualities are composed by addition, namely red, green, and bluish violet. These mix in sensation without interfering with one another. The ear, on the other hand, distinguishes between an enormous number of tones of different pitches. No two chords composed out of different tones ring alike, while yet with the eye precisely the analogue of this is the case. For a white which looks alike can be produced with red and greenish blue from the spectrum, with yellow and ultramarine, with greenish yellow and violet, ⟦with green, red, and violet,⟧[c] or with any two or three—or with all—of these mixtures together. Were the situation alike with the ear, the consonance of *C* and *F* would sound like that of *D* and *G*, *E* and *A*, or *C*, *D*, *E*, *F*, *G* and *A*, etc. And—what is notable as regards the objective significance of colour—apart from the effect on the eye, one has not been able to detect a single

[c] ⟦This phrase was erroneously omitted by Hertz and Schlick from *Helmholtz* 1921.⟧

physical connection in which light which looked alike was regularly alike in value.

The whole foundation, finally, of the musical effect of consonance and dissonance depends upon the peculiar phenomenon of beats. The basis of these is a rapid alternation in intensity of tone, which arises from the fact that two tones almost alike in pitch alternately interact with their phases alike and opposed, and correspondingly excite now strong and now weak vibrations in a resonating body. The physical phenomenon might equally well occur through the interaction of two light-wave trains as through the interaction of two sound-wave trains. But the nerve must firstly be capable of being affected by both wave trains, and it must secondly be able to follow quickly enough the alternation of strong and weak intensity. The auditory nerve is markedly superior in the latter respect to the optic nerve. At the same time, each fibre of the auditory nerve is sensitive only to tones from a narrow interval of the scale, so that only tones situated quite near to each other in it can interact at all. Ones far from each other cannot interact, or not directly. When they do, this originates from accompanying overtones or combination tones. There therefore occurs with the ear this difference between resounding and non-resounding intervals, i.e. between consonance and dissonance. Each fibre of the optic nerve, on the other hand, is sensitive throughout the whole spectrum, although with different strength in different parts. Could the optic nerve at all follow in sensation the enormously rapid beats of light oscillations, then every mixed colour would act as a dissonance.

⟦15⟧ You can see how all these differences in the manner of action of light and sound are conditioned by the way in which the nervous apparatus reacts to them.

⟦16⟧ Our sensations are indeed effects produced in our organs by external causes: and how such an effect expresses itself naturally depends quite essentially upon the kind of apparatus upon which the effect is produced. Inasmuch as the quality of our sensation gives us a report of what is peculiar to the external influence by which it is excited, it may count as a *symbol* of it, but not as an *image*. For from an image one requires some kind of alikeness with the object of which it is an image—from a statue alikeness of form, from a drawing alikeness of perspective projection in the visual field, from a painting alikeness of colours as well. But a sign need not have any kind of similarity at all with what it is the sign of. The relation between the two of them is restricted to the fact that like objects exerting an influence under like circumstances evoke like signs, and that therefore unlike signs always correspond to unlike influences.

⟦17⟧ To popular opinion, which accepts in good faith that the images which our senses give us of things are wholly true, this residue of similarity acknowledged by us may seem very trivial. In fact it is not trivial. For with it one can still achieve something of the very greatest importance, namely forming an image of lawfulness in the processes of the actual world. Every law of nature asserts that upon preconditions alike in a certain respect, there always follow consequences which are alike in a certain other respect. Since like things are indicated in our world of sensations by like signs, an equally regular sequence will also correspond in the domain of our sensations to the sequence of like effects by law of nature upon like causes.

⟦18⟧ If berries of a certain kind in ripening develop at the same time a red pigment and sugar, then a red colour and a sweet taste will always be found together in our sensation for berries of this type.

⟦19⟧ Thus although our sensations, as regards their quality, are only *signs* whose particular character depends wholly upon our own make-up, they are still not to be dismissed as a mere semblance, but they are precisely signs of *something*, be it something existing or happening, and—what is most important—they can form for us an image of the *law* of this thing which is happening.

⟦20⟧ So physiology too acknowledges the qualities of sensation to be a mere form of intuition. But Kant went further. He spoke not only of the qualities of sensations as given by the peculiarities of our intuitive faculty, but also of space and time, since we cannot perceive anything in the external world without its happening at a specific time and being situated at a specific place. Specification in time is even an attribute of every internal perception as well. He therefore termed time the given and necessary *transcendental form of inner intuition*, and space the corresponding form of *outer intuition*. Thus Kant considers spatial specifications too as belonging as little to the world of the actual—or to 'the thing in itself'—as the colours which we see as attributes of bodies in themselves, but ⟦which⟧ are introduced by our eye into them.

Even here the approach of natural science can take the same path, up to a certain limit. Suppose we namely ask whether there is a common characteristic, perceivable in immediate sensation, whereby every perception relating to objects in space is characterized for us. Then we in fact find such a characteristic in the circumstance that motion of our body places us in different spatial relations to the perceived objects, and thereby also alters the impression made by them upon us. But the impulse to motion, which we give through an innervation of our motor nerves, is something immediately perceivable. That we do something, when we give such an impulse, is felt by us. What we do, we do not know in an immediate manner. Only physiology teaches us that we put into an excited state—or *innervate*—the motor nerves, that their stimulation is passed on to the muscles, that these consequently contract and move the limbs. Yet all the same we know, even without scientific study, which perceivable effect follows each of the various innervations that we are able to initiate.

That we learn it by frequently repeated attempts and observations, may be demonstrated with assurance in a long series of cases. We can learn even as adults to find the innervations needed for pronouncing the letters of a foreign language, or for a particular kind of voice production in singing. We can learn innervations for moving our ears, for squinting with our eyes inwards or outwards, or even upwards and downwards, and so on. The difficulty in performing such things consists only in our having to seek, by making attempts, to find the as yet unknown innervations needed for such previously unexecuted movements. We ourselves, moreover, know of these impulses in no other form, and through no other definable feature, than precisely the fact that they produce the intended observable effect. Thus this effect also alone serves to distinguish the various impulses in our representation.

⟦21⟧ Now when we give impulses of this sort (turning our gaze, moving our hands, going back and forth), we find that the sensations belonging to certain quality ranges (namely those relating to spatial objects) can thereby be altered; other psychic states of which we are conscious—memories, intentions, wishes, moods—cannot be altered at all. A thoroughgoing difference between the former and the latter is thereby laid down in immediate perception.

Thus if we desire to call the relationship which we alter in an immediate manner by the impulses of our will—what kind of relationship this is might moreover be still quite unknown to us—a *spatial* one, then perceptions of *psychic* activities do not enter into such a relationship at all. But probably all sensations of the outer senses must proceed subject to some kind of innervation or another, i.e. have some spatial specification.[d] In this case space will also appear to us—imbued with the qualities of our sensations of movement—in a sensory manner, as that through which we move, through which we can gaze forth. Spatial intuition would therefore be in this sense a subjective *form of intuition*, like the sensory qualities red, sweet, and cold. Naturally, the sense of this would just as little be mere semblance for the former as for the latter; the place specified for a specific individual object is no mere *semblance*.

⟦22⟧ From this point of view, however, space would appear as the *necessary* form of outer intuition, because precisely what we perceive as having some spatial specification comprises for us the external world. We comprehend as the world of inner intuition, as the world of self-consciousness, that in which no spatial relation is to be perceived.

⟦23⟧ And space would be a *given form* of intuition, possessed *prior to all experience*, to the extent that its perception were connected with the possibility of motor impulses of the will the mental and corporeal capacity for which had to be given to us, by our make-up, before we could have spatial intuition.

⟦24⟧ It will hardly be a matter for doubt, that the characteristic which we have discussed, of altering during movement, is an attribute of all perceptions relating to spatial objects.* The question will need to be answered, on the other hand, as to whether every specific peculiarity of our spatial intuition is now to be derived from this source. To this end we must consider what can be attained with the aid of the features of perception which have so far been discussed.

⟦25⟧ Let us try to put ourselves back into the position of a man without any experience. We must assume, in order to begin without spatial intuition,

---

[d] ⟦'räumlich bestimmt sein', i.e. any such sensation has a feature which can be altered by our moving.⟧

* On the localization of sensations of internal organs, see Appendix I to this paper.

that such a man knows even the effects of his innervations only to the extent that he has learnt how, by remission of a first innervation or by execution of a second counterimpulse, he can put himself back into the state from which he has removed himself by the first impulse. As this mutual self-cancellation of different innervations is wholly independent of what is thereby perceived, the observer can find out, without yet having previously gained any understanding of the external world, how he has to do this.

⟦26⟧ Let the situation of the observer initially be that he is faced with an environment of objects at rest. This will make itself known to him in the first place by the fact that as long as he gives no motor impulse his sensations remain unaltered. If he gives such an impulse (e.g. if he moves his eyes or hands, or steps forward), the sensations alter; and if he then, by remission or the appropriate counterimpulse, returns to the earlier state, all his sensations will again be the earlier ones.

⟦27⟧ Let us call the whole group of sensation aggregates which can be brought about during the period of time under discussion, by a certain specific and limited group of impulses of the will, the *presentables* for that period; and call *present*, on the other hand, the sensation aggregate from this group which happens to be being perceived. Then our observer is tied at this time to a certain range of presentables, but any individual one of which he can make present at any moment he wishes by executing the relevant movement. Each individual presentable from this group thereby appears to him as *enduring at every moment* of this period of time. He has observed it at every individual moment that he wanted to. The assertion that he would have been able to observe it also at any other intervening moment that he might have wanted to, is to be regarded as an inductive inference, drawn from the case of every moment at which a successful attempt was made to that of every moment whatsoever in the relevant period of time. Therefore, the representation of an *enduring existence of different things at the same time one beside another* can in this manner be acquired.

'One beside another' is a spatial description. But it is justified, since we have defined as 'spatial' the relationship altered by impulses of the will. One does not yet need to think of substantial things as what are here supposed to exist one beside another. 'To the right it is bright, to the left it is dark, in front there lies resistance but not behind' could for example be said at this stage of knowledge, with right and left being only names for certain eye movements, in front and behind for certain movements of the hands.

⟦28⟧ Now at other times the range of presentables, for the same group of impulses of the will, is going to be a different one. This range, with the individual which it contains, will thereby confront us as something given, as an 'objectum'. Those alterations which we can produce and revoke by conscious impulses of the will, are distinct from ones which are not consequences of such impulses and cannot be eliminated by them. The latter specification is negative. Fichte's appropriate expression for this is that the 'I' is faced with a 'not-I' which exacts recognition.

⟦29⟧ In asking about the empirical conditions under which spatial intuition develops, we must in these considerations take account chiefly of touch, since the blind can develop spatial intuition completely without the help of sight. Although for them space will not turn out to be filled up with objects in such richness and detail as for sighted persons, it yet seems most highly improbable that the foundations of spatial intuition for the two classes of person should be wholly different. If we ourselves attempt to make observations by touching, in the dark or with our eyes closed, then we may very well touch with one finger—or even with a pencil held in the hand like the surgeon with his probe—and still ascertain, in detail and with assurance, the corporeal form of the object present.

When wanting to find our way in the dark, we usually feel over larger objects with five or ten finger-tips simultaneously. We then obtain five to ten times as many reports in the same time as with one finger, and also use the fingers, like the tips of an open pair of dividers, for measuring magnitudes in the objects. All the same, the circumstance that we have an extended sensitive skin surface, with many sensitive points, recedes wholly into the background when touching things. What we are capable of ascertaining from the skin feeling by gently applying our hand, say upon the face of a medal, is extraordinarily rough and scanty in comparison with what we discover by a groping motion, even if only with the point of a pencil. With sight this process becomes much more complicated, because of the fact that besides the most refinedly sensitive spot on the retina—its central fovea—which is as it were led all round the retinal image when we look at something, there also cooperate at the same time a great host of other sensitive points, in a much more fertile manner than is the case with touch.

⟦30⟧ By moving the touching finger along the objects, one comes to know the sequence in which their impressions offer themselves. This sequence shows itself to be independent of whether one touches with one finger or another. It is moreover not a uniquely determined sequence, whose elements one must always go through, forwards or backwards, in the same order in order to get from one to another; thus it is not a linear sequence, but a surfacelike 'one beside another', or in Riemann's terminology a second-order manifold. That all this is so is easily seen.

Of course, the touching finger can get from one point to another, in the touchable surface, also by other motor impulses than those which push it along the surface; and different touchable surfaces require different movements for sliding upon them. A higher manifold is thereby required for the space in which what touches moves, than for the touchable surface: the third dimension must be added. But this suffices for all available experiences. For a closed surface divides completely the space with which we are acquainted. Even gases and liquids, which after all are not tied to the form of the human faculty of representation, cannot escape through a surface closed all round. And just as only a surface, not a space—thus a spatial structure of two and not three dimensions—can be bounded by a closed line, so also can a surface close off precisely only a space of three dimensions, and not one with four.

⟦31⟧ Thus might one get to know the spatial order of what exist 'one beside another'. As a further step, magnitudes would be likened with one another, by observing congruence of the touching hand with parts or points of the surfaces of bodies, or congruence of the retina with parts and points of the retinal image.

⟦32⟧ Because this intuited spatial order of things stems originally from the sequence in which the qualities of sensation offered themselves to the moving sense organ, there finally persists a curious consequence even in the completed representation of an experienced observer. The objects extant in space namely appear to us clothed in the qualities of our sensations. To us they appear red or green, cold or warm, to have a smell or taste, etc., whereas after all these qualities of sensation belong only to our nervous system and do not reach out at all into external space. The semblance does not cease even when we know this, because in fact this semblance is the original truth: it is indeed sensations which first offer themselves to us in a spatial order.

⟦33⟧ You can see that the most essential features of spatial intuition can in this way be derived. However, to the consciousness of the general public an intuition appears as something simply given, which comes about without reflection and search, and which is by no means to be resolved further into other psychic processes. This popular belief has been adopted by some workers in physiological optics, and also by the strictly observant Kantians, at least as regards spatial intuition. As is well known, already Kant assumed not only that the general form of spatial intuition is transcendentally given, but that it also contains in advance, and prior to any possible experience, certain narrower specifications as expressed in the axioms of geometry. These can be reduced to the following propositions:

(1) Between two points only *one* shortest line is possible. We call such a line '*straight*'.

(2) Through any three points a *plane* can be placed. A plane is a surface which wholly includes any straight line if it coincides with two of its points.

(3) Through any point only one line parallel to a given straight line is possible. Two lines are *parallel* if they are straight lines lying in the same plane which do not intersect within any finite distance.

⟦34⟧ Indeed, Kant used the alleged fact that these geometrical propositions appeared to us as *necessarily* correct, and that we could never at all even represent to ourselves a deviating behaviour of space, directly as a proof that they had to be given prior to all experience, and that for this reason the spatial intuition contained in them was itself a transcendental form of intuition, independent of experience.

⟦35⟧ In view of the controversies which have been conducted, in recent years, about the question of whether the axioms of geometry are transcendental or empirical propositions, I should like here to emphasize firstly that this question is wholly to be separated from the one first discussed of whether space is in general a transcendental form of intuition or not.*

⟦36⟧ Everything our eye sees, it sees as an aggregate of coloured surfaces in the visual field—that is its form of intuition. The particular colours which appear on this or that occasion, their arrangement and sequence—this is the result of external influences and is not determined by any law of our make-up. Similarly from the fact that space is a form of intuiting, nothing whatever follows about the facts expressed by the axioms. If such propositions are taken to be not empirical ones, but to belong instead to the necessary form of intuition, then this is a further particular specification of the general form of space; and those grounds which allowed the conclusion that the form of intuition of space is transcendental, do not necessarily for that reason already suffice to prove, at the same time, that the axioms too are of transcendental origin.

⟦37⟧ When Kant asserted that spatial relationships contradicting the axioms of Euclid could never in any way be represented, he was influenced by the contemporary states of development of mathematics and the physiology of the senses, just as he was thus influenced in his whole conception of intuition in general as a simple psychic process, incapable of further resolution.

⟦38⟧ If one wishes to try to represent to oneself something which has never before been seen, one must know how to depict to oneself the series of sense impressions which, according to the known laws of the latter, would have to come about if one observed that object and its gradual alterations successively from every possible viewpoint and with all of one's senses. And at the same time, these impressions must be such that every other interpretation is thereby excluded. If this series of sense impressions can be formulated completely and unambiguously, then one must in my judgement declare that thing to be intuitably representable. Since by presupposition it is a thing which is considered never yet to have been observed, no previous experience can come to our help and guide our fantasy in seeking out the requisite series of impressions; instead, this can only happen by way of the *concept* of the object or relationship to be represented. Such a concept is thus first of all to be elaborated and to be made as specialized as the given purpose requires.

The concepts of spatial structures which are taken not to correspond to customary intuition can be reliably developed only by means of calculative analytic geometry. The analytic resources for our present problem were first given by Gauss in 1828 with his essay on the curvature of surfaces, and applied by Riemann in seeking out the logically possible self-consistent systems of geometry. These investigations have not unsuitably been termed *metamathematical.* One should also note that already in 1829 and 1840 Lobatschevsky worked out, in the customary synthetically intuitive manner, a geometry without the axiom of parallels, and one which concurs completely with the corresponding parts of the more recent analytic investigations. Finally, Beltrami

* See Appendix II below.

has formulated a method for forming images of metamathematical spaces in parts of Euclidean space, by means of which the specification of their manner of appearance in perspective vision is made fairly easy. Lipschitz has demonstrated that the general principles of mechanics can be carried over to such spaces, so that the series of sense impressions which would come about in them can be completely formulated. With this the intuitability of such spaces, in the sense of the definition of this concept given above, has been shown.*

⟦39⟧ But here is where disagreement occurs. I demand for the proof of intuitability only that one should be able to formulate, for every manner of observation, specifically and unambiguously the arising sense impressions, by using if necessary a scientific acquaintance with their laws, from which[e] it would ensue, at least for someone acquainted with these laws, that the thing concerned or relationship to be intuited was in fact present. The task of representing to oneself the spatial relationships in metamathematical spaces indeed demands some practice in understanding analytic methods, perspective constructions, and optical phenomena.

⟦40⟧ This is however in disagreement with the older concept of intuition, which only acknowledges something to be given through intuition if its representation enters consciousness at once with the sense impression, and without deliberation and effort. Our attempts to represent mathematical[f] spaces indeed do not have the ease, rapidity, and striking self-evidence with which we for example perceive the form of a room which we enter for the first time, together with the arrangement and forms of the objects contained in it, the materials of which these consist, and much else as well. Thus if this kind of self-evidence were an originally given and necessary peculiarity of all intuition, we could not up to now assert the intuitability of such spaces.

⟦41⟧ Yet we are now confronted with a host of cases, on further reflection, which show that assurance and rapidity for the occurrence of specific representations with specific impressions can also be acquired—even when no such connection is given by nature. One of the most striking examples of this kind is our understanding of our mother tongue. Its words are arbitrarily or accidentally chosen signs—every different language has different ones. Understanding of it is not inherited, since for a German child who was brought up amongst Frenchmen and has never heard German spoken, German is a foreign language. The child becomes acquainted with the meaning of the words and sentences only through examples of their use. In this process one cannot even make understandable to the child—until it understands the language—that the sounds it hears are supposed to be signs having a sense. Lastly, on growing up it understands these words and sentences without deliberation and effort, without knowing

---

* See my lecture on the axioms of geometry.

[e] ⟦Apparently meaning: 'from which sense impressions'.⟧

[f] ⟦Helmholtz presumably means 'metamathematical'.⟧

when, where, and through what examples it learnt them, and it grasps the finest variations of their sense—often ones where attempts at logical definition only limp clumsily behind.

‖42‖ It will not be necessary for me to multiply examples of such processes—they abound richly enough in daily life. This is precisely the basis of art, and most clearly that of poetry and the graphic arts. The highest manner of intuiting, as we find it in an artist's view, is this kind of apprehension of a new type of stationary or mobile appearance of man and nature. When the traces of like kind which are left behind in our memory by often repeated perceptions reinforce one another, it is precisely the law-like which repeats itself most regularly in like manner, while the incidental fluctuation is erased away. For the devoted and attentive observer, there grows up in this way an intuitive image of the typical behaviour of the objects which have interested him, and he knows as little afterwards how it arose as the child can give an account of the examples whereby it became acquainted with the meanings of words. That the artist has beheld something true emerges from the fact that it seizes us too with a conviction of its truth, when he presents it to us in an example purified from accidental perturbations. He is superior to us, however, in having known how to cull it from everything accidental, and from every confusion arising in the onward rush of the world.

‖43‖ Thus much just to recall how active this psychic process is in our mental life, from the latter's lowest to its highest stages of development. In previous studies I characterized as *unconscious inferences* the connections between representations which thereby occur—unconscious, inasmuch as their major premiss is formed from a series of experiences, each of which has long disappeared from our memory and also did not necessarily enter our consciousness formulated in words as a sentence, but only in the form of an observation of the senses. The new sense impression entering in present perception forms the minor premiss, to which there is applied the rule imprinted by the earlier observations. More recently I have avoided the name 'unconscious inferences', in order to escape confusion with the—as it seems to me—wholly unclear and unjustified conception thus named by Schopenhauer and his followers. Yet evidently we are dealing here with an elementary process lying at the foundation of everything properly termed thought, even though it still lacks critical sifting and completion of the individual steps, such as occurs in the scientific formation of concepts and inferences.

‖44‖ Thus as concerns firstly the issue of the origin of the axioms of geometry: the fact that the representation of metamathematical spatial relationships is not easy when experience is lacking, cannot be claimed as a ground against their intuitability. Moreover, the latter is completely demonstrable. Kant's proof for the transcendental nature of the axioms of geometry is thus inadequate. On the other hand, investigation of the empirical facts shows that the axioms of geometry, taken in the only sense in which one is allowed to apply them to the actual world, can be empirically tested and demonstrated, or even—if the case should arise—refuted.*

⟦45⟧ The memory vestiges of previous experiences also play a further and highly influential role in the observation of our visual field.

⟦46⟧ A no longer completely inexperienced observer receives even without movement of his eyes—whether by momentary illumination from an electric discharge or by deliberate rigid staring—a relatively rich image of the objects in front of him. Yet even an adult will still easily convince himself that this image becomes much richer, and especially much more precise, when he moves his glance around in the visual field, and thus employs that kind of spatial observation which I described earlier as the fundamental one. We are indeed also so used to letting our glance wander upon the objects we observe, that it requires a fair amount of practice before we succeed, for the purposes of experiments in physiological optics, in holding it fixed upon one point for a longish time without wavering.

In my works on physiological optics,** I have sought to explain how our acquaintance with the visual field can be acquired by observation of the images during the movements of our eyes, provided only that there exists, between otherwise qualitatively alike retinal sensations, some or other perceptible difference corresponding to the difference between distinct places on the retina. Such a difference should be called a *local sign*, according to Lotze's terminology; except that the fact that this sign is a local sign—i.e. that it corresponds to a difference of place and to which such difference—need not be known in advance.

Recent observations† have also reconfirmed that persons who were blind from youth onwards, and later regained their sight through an operation, could not at first distinguish by eye even between such simple forms as a circle and a square, until they had touched them.

Apart from this, physiological investigation teaches us that we can liken by visual estimation, in a relatively precise and assured manner, only such lines and angles in the visual field as can be brought, by normal eye motions, to form images in rapid succession at the same places on the retina. Indeed, we estimate much more assuredly the true magnitudes and distances of spatial objects situated not too far off, than the perspective ones, alternating according to viewpoint, in the visual field of the observer—although the former task concerns the three dimensions of space and is much more involved than the latter, which concerns only a surfacelike image. One of the greatest difficulties in drawing, as is well known, is to free oneself from the influence involuntarily exerted by our representation of the true magnitudes of the objects seen. Now it is precisely the situation described that we must expect, if our understanding of local signs was first acquired through experience. We can assuredly become

---

* See my *Wissenschaftliche Abhandlungen*, vol. II, p. 640; an excerpt is given as Appendix III below.

** *Handbuch der Physiologischen Optik* ⟦'Handbook of physiological optics'⟧; *Vorträge über das Sehen der Menschen* ⟦'Lectures on human sight'⟧, in *Vorträge und Reden*, vol. I, pp. 85 and 265.

† Dufour (Lausanne) in the *Bulletin de la Société médicale de la Suisse Romande*, 1876.

acquainted with the alternating sensory signs for what remains objectively constant, much more easily than with those for what alternates according to every single movement of our body, as the perspective images do.

⟦47⟧ There is, none the less, a large number of physiologists whose view we may term *nativist*, as opposed to the *empiricist* view which I myself have tried to defend, and for whom this conception of an acquired acquaintance with the visual field appears unacceptable. This is due to their not having got clear about what after all lies before us so plainly in the example of speech, namely how much can be achieved by accumulated memory impressions. For this reason, a host of various attempts have been made to reduce at least some part of visual perception to an innate mechanism, in the sense that certain impressions of sensation are supposed to release certain ready-made spatial representations.

I have demonstrated in detail* that all hypotheses of this kind proposed to date are inadequate, because in the end one can still always come up with cases where our visual perception is in more precise agreement with actuality than those assumptions would yield. One is then forced to add the further hypothesis that the experience gained during movements can in the end overcome the innate intuition, and thus achieve *in opposition* to the latter what it is supposed by the empiricist hypothesis to achieve *without* such an obstacle.

⟦48⟧ The nativist hypotheses about our acquaintance with the visual field thus *firstly* do not explain anything, but simply assume the existence of the fact to be explained while at the same time rejecting the possibility of reducing this fact to definitely ascertained psychic processes, although they themselves still have to appeal to the latter in other cases. *Secondly*, the assumption of every nativist theory—that ready-made representations of objects are elicited through our organic mechanism—appears much more audacious and doubtful than the assumption of the empiricist theory, which is that only the non-understood material of sensations originates from external influences, while all representations are formed from it in accordance with the laws of thought.

⟦49⟧ *Thirdly*, the nativist assumptions are unnecessary. The only objection that one has been able to bring against the empiricist explanation, is the assurance with which many animals move when newly born or just after crawling out of the egg. The less mentally endowed they are, the quicker they learn what they at all can learn. The narrower the paths are along which their thoughts must go, the more easily they find them. A newly born human child is extremely inept at seeing: it needs several days before learning to judge, from its visual images, in what direction it must turn its head in order to reach its mother's breast. Young animals are certainly much more independent of individual experience. But we still know practically nothing specific about what this instinct which guides them is: whether direct inheritance is possible of ranges of representations from the parents, or whether it is only a matter of desire or aversion—or of a motor impulse—which attach themselves to certain sensation

* See my *Handbuch der Physiologischen Optik* ⟦*op. cit.*⟧, 3. Abteilung, Leipzig, 1867.

aggregates. Vestiges of the last-mentioned phenomena still occur in a plainly recognizable manner with human beings. Properly and critically executed observations would be in the highest degree desirable in this domain.

⟦50⟧ Thus for the kind of set-up presupposed by the nativist hypothesis, one can at most claim a certain pedagogical merit which facilitates detection of the first lawlike relationships. The empiricist view too could be combined with presuppositions having this aim, for example that the local signs of neighbouring places on the retina are more similar to each other than are those of ones further apart, that those of corresponding places on the two retinas are more similar than those of disparate ones, and so on. For our present investigation it suffices to know that spatial intuition can come fully into being even with a blind person, and that with a sighted person—even should the nativist hypotheses prove partially correct—the final and most precise specification of spatial relationships is still conditioned by the observations made during movement.

⟦51⟧ I shall now return to our discussion of the initial, original facts of our perception. As has been seen, we do not merely have alternating sense impressions which come upon us without our doing anything about it. We rather observe during our own continuing activity, and thereby attain an acquaintance with the *enduring existence* of a lawlike relationship between our innervations and the becoming present of the various impressions from the current range of presentables. Each of our voluntary movements, whereby we modify the manner of appearance of the objects, is to be regarded as an experiment through which we test whether we have correctly apprehended the lawlike behaviour of the appearance before us, i.e. correctly apprehended the latter's[g] presupposed enduring existence in a specific spatial arrangement.

⟦52⟧ The chief reason, however, why the power of any experiment to convince is so much greater than that of observing a process going on without our assistance, is that with the experiment the chain of causes runs through our own self-consciousness. We are acquainted with one member of ⟦the chain of⟧ these causes—the impulse of our will—from inner intuition, and know through what motives it came about. From this, as from an initial member known to us and at a point in time known to us, there then begins to act that chain of physical causes which terminates in the outcome of the experiment. Yet the conviction to be attained has an essential presupposition, that the impulse of our will should neither itself already have been influenced by physical causes which at the same time also determined the physical process, nor should it for its own part have influenced the subsequent perceptions psychically.

⟦53⟧ The latter doubt can in particular be of relevance to our present topic. The impulse of the will for a specific movement is a psychic act, just as is the thereupon perceived alteration in the sensation. Could not then the first act bring about the second through purely psychic mediation?

It is not impossible. When we dream, something of the sort occurs. In our

---

[g] ⟦Or possibly 'their', referring to 'the objects' rather than 'the appearance'.⟧

dream we believe ourselves to execute a movement, and we then dream further that there occurs what should be its natural consequence. We dream of climbing into a boat, pushing it off from land, gliding out on the water, seeing the displacement of the surrounding objects, and so on. Here the dreamer's expectation that he will see the consequences of his conduct occur appears to bring about the dreamed perception in a purely psychic way. Who can say how long and finely spun out, how logically complete such a dream might be? Should everything in it occur in the most lawlike manner according to the order of nature, there would remain but one difference from the waking state—the possibility of being awakened, the rupture of this dreamed series of intuitions.

⟦54⟧ I do not see how we could refute a system of even the most extreme subjective idealism, if it regarded life as a dream. We might declare it to be as improbable and unsatisfying as possible—in this respect I would assent to the sharpest expressions of repudiation. But it could be implemented consistently, and it seems to me very important to keep this in view. It is well known how ingeniously Calderón implemented this theme in his 'Life a Dream'.

⟦55⟧ Fichte too assumes that the 'I' posits the 'not-I'—i.e. the world as it appears—for itself, because it needs it for developing its own thought-activity. Yet his idealism does indeed distinguish itself from that just referred to, in that he conceives of other human individuals not as dream images, but—starting from the assertion of the moral law—as essences alike with one's own 'I'. Since, however, the images whereby they each represent the 'non-I' must themselves all agree with one another, he conceives of all of the individual 'I's' as parts or emanations of the absolute 'I'. The world in which they found themselves was then that world of representations which the worldmind posited for itself, and could again receive the concept of reality, as happened with Hegel.

⟦56⟧ The *realist hypothesis*, on the other hand, trusts the testimony of ordinary self-observation, according to which the alterations of perception which follow some item of conduct have no psychic connection at all with the preceding impulse of the will. It regards as enduring, independent of the way in which we form representations, that which seems to prove to be thus in everyday perception—the material world outside us.

The realist hypothesis is the simplest we can form, it has been tested and confirmed in extraordinarily wide ranges of application, it is sharply defined for every individual specification, and it is therefore extraordinarily serviceable and fruitful as a basis for conduct. All this is without doubt. Even in the idealist manner of conceiving things, we would hardly know how else to express the lawlike in our sensations than by saying: 'Those acts of consciousness which occur with the character of perception proceed *as if* there actually existed the world of material things which is assumed by the realist hypothesis.' But we cannot get beyond this 'as if'. We cannot acknowledge the realist view to be more than an excellently serviceable and precise hypothesis. We are not allowed to ascribe to it necessary truth, since besides it idealist hypotheses not open to refutation are also possible.

⟦57⟧ It is good to keep this always before our eyes, so that we may not wish to infer more from the facts than there is to infer from them. The various gradations of the idealist and realist views are metaphysical hypotheses. As long as they are acknowledged to be such, they are ones which have their complete scientific justification, however harmful they may become when one wishes to present them as dogmas or alleged necessities of thought.

Science must discuss all admissible hypotheses, in order to retain a fully comprehensive view of the possible attempts at explanation. Hypotheses are even more necessary for conduct, because one cannot continually wait until an assured scientific decision has been reached, but must decide for oneself—whether according to probability or to aesthetic or moral feeling. In this sense one could have no objection even against metaphysical hypotheses. But it is unworthy of a thinker wishing to be scientific if he forgets the hypothetical origin of his propositions. When such concealed hypotheses are defended with pride and passionateness, the latter are the customary consequences of the unsatisfying feeling which their defender shelters, in the hidden depths of his conscience, about the justness of his cause.

⟦58⟧ Yet what we can find unambiguously, and as a fact without anything being insinuated hypothetically, is the lawlike in the phenomena. From the first step onwards, when we perceive before us the objects distributed in space, this perception is the acknowledgement of a lawlike connection between our movements and the therewith occurring sensations. Thus even the first elementary representations contain intrinsically some thinking, and proceed according to the laws of thought. Everything in intuition which is an addition to the raw material of sensations can be resolved into thinking, if we take the concept of thinking as broadly as has been done above.

⟦59⟧ For if 'comprehending' means forming *concepts*,[h] and in the concept of a class of objects we gather together and bind together whatever like characteristics they bear, it then results quite analogously, that the concept of a series of appearances alternating in time must seek to bind together that which remains alike in all of its stages. The wise man, as Schiller puts it:

> Seeks the familiar law
>   in chance's frightful miracles,
> Seeks the stationary pole
>   in the fleeting appearances.*

⟦60⟧ That which remains alike, without dependence upon anything else, through every alternation of time, we call *substance*. The relationship which

[h] ⟦Here there is a play on the words 'begreifen' and 'Begriff' which is untranslatable.⟧

* Sucht das vertraute Gesetz
  in des Zufalls grausenden Wundern,
Suchet den ruhenden Pol
  in der Erscheinungen Flucht.

⟦The contexts show that Helmholtz intends this and subsequent snatches of poetry in their most literal sense. So the relevant literal sense is given in this translation.⟧

remains alike between altering magnitudes, we call the *law* connecting them. What we perceive directly is only this law. The concept of substance can be gained only through exhaustive examination and always remains problematic, inasmuch as further examination is not ruled out. Formerly light and heat were counted as substances, until it later turned out that they are perishable forms of motion. And we must still always be prepared for new decompositions of the currently familiar chemical elements.

The first product of the thoughtful comprehension of the phenomena is *lawlikeness*. Should we have separated it out sufficiently purely, delimited its conditions with sufficient completeness and assurance, and also formulated them with sufficient generality that the outcome is unambiguously specified for all possibly occurring cases, and that we at the same time gain the conviction that it has proved true and will prove true at all times and in every case: then we acknowledge it as an existence enduring independent of the way in which we form representations, and call it the *cause*, i.e. that which primarily remains and endures behind what alternates. In my opinion, only the application of the word in this sense is justified, although it is applied in common speech in a very wishy-washy manner for whatever at all is the antecedent or occasion of something.

Inasmuch as we then acknowledge the law to be something compelling our perception and the course of natural processes, to be a power equivalent to our will, we call it '*force*'. This concept of a power confronting us is conditioned directly by the way in which our simplest perceptions come about. From the beginning, those changes which we make ourselves by acts of our will are distinct from ones which cannot be made by our will, and are not to be set aside by our will. Pain especially gives us the most penetrating lesson about the power of reality. Emphasis thereby falls upon the observational fact that the perceived range of presentables is not posited by a *conscious* act of our representation or will. Fichte's 'not-I' is here the exactly fitting negative expression. For the dreamer too, what he believes himself to see and feel appears not to be evoked by his will or by conscious concatenation of his representations, although the latter might in actuality often enough be unconsciously the case. For him too it is a 'not-I'. Likewise for the idealist, who regards it as the world-mind's world of representations.

⟦61⟧ In our language, we have a very fortunate way of characterizing that which lies behind the change of appearances and acts upon us, namely as 'the actual'. Here only action[i] is predicated. Absent is that secondary reference to what endures as substance which is included in the concept of the real, i.e. of the thinglike. As regards the concept of the objective, on the other hand, the concept of a ready-made image of an object usually finds its way into it, and

[i] ⟦Here the following equivalences are used: wirklich = actual, Wirken = action, (ein)wirken = to act, wirksam = active, reell = real, sachlich = thinglike, objektiv = objective, Gegenstand = object. Elsewhere 'Wirkung' is generally translated by 'effect' and 'Einwirkung' by 'influence'.⟧

one which does not suit the most primary perceptions.

Even with the logical dreamer, we must presumably characterize as active and actual those psychic states or motives which foist upon him, at the given time, the sensations corresponding in a lawlike manner to the present situation in his dream world. It is clear, on the other hand, that a division between what is thought and what is actual does not become possible until we know how to make the division between what the 'I' can and cannot alter. This does not become possible, however, until we discern what lawlike consequences the impulses of our will have at the given time. The lawlike is therefore the essential presupposition for the character of the actual.

⟦62⟧ I need not explain to you that it is a *contradictio in adjecto* to want to represent the real, or Kant's 'thing in itself', in positive terms but without absorbing it into the form of our manner of representation. This is often discussed. What we can attain, however, is an acquaintance with the lawlike order in the realm of the actual, admittedly only as portrayed in the sign system of our sense impressions:

Everything perishable
Is only a likeness.*

⟦63⟧ I take it as a favourable sign that we find Goethe, here and further on, together with us on the same path. Where it is a matter of broad panoramas, we may well trust his clear and unconstrained eye for truth. He demanded from science that it should be only an artistic arrangement of the facts and form no abstract concepts going beyond this, which to him seemed to be empty names and only to obscure the facts. In somewhat the same sense, Gustav Kirchhoff has recently characterized it as the task of the most abstract amongst the natural sciences, namely mechanics, to *describe completely and in the simplest manner* the motions occurring in nature.

As for 'obscuring', this indeed happens when we stay put in the realm of abstract concepts and do not explain to ourselves their factual sense, i.e. make clear to ourselves what observable new lawlike relationships between the appearances follow from them. Every correctly formed hypothesis sets forth, as regards its factual sense, a more general law of the appearances than we have until now directly observed—it is an attempt to ascend to something more and more generally and inclusively lawlike. Whatever factually new things it asserts must be tested and confirmed by observation and experiment. Hypotheses not having such a factual sense, or which in no way specify anything sure and unambiguous about the facts falling under them, are to be regarded only as worthless talk.

⟦64⟧ Every reduction of the appearances to the underlying substances and forces claims to have found something unchanging and final. An unconditional claim of this kind is something for which we never have a justification: this is

* Alles Vergängliche
Ist nur ein Gleichnis.

allowed neither by the fact that our knowledge is full of gaps, nor by the nature of the inductive inferences upon which all of our perception of the actual, from the first step onwards, is based.

⟦65⟧ Every inductive inference is based on trusting that an item of lawlike behaviour, which has been observed up to now, will also prove true in all cases which have not yet come under observation. This is a trust in the lawlikeness of everything that happens. However, lawlikeness is the condition of comprehensibility. Trust in lawlikeness is thus at the same time trust in the comprehensibility of the appearances of nature. While: should we presuppose that this comprehension will come to completion, that we shall be able to set forth something ultimate and finally unalterable as *the cause* of the observed alterations, then we call the regulative principle of our thought which impels us to this the *law of causality* ⟦*Kausalgesetz*⟧. We can say that it expresses a trust in the *complete comprehensibility* of the world.

Comprehension, in the sense in which I have described it, is the method whereby our thought masters the world, orders the facts and determines the future in advance. It is its right and its duty to extend the application of this method to everything that occurs, and it has already actually harvested great yields on this path. However, we have no further guarantee for the applicability of the law of causality than this law's success. We could live in a world in which every atom was different from every other one, and where there was nothing at rest. Then there would be no regularity of any kind to be found, and our thought-activity would have to be at a standstill.

⟦66⟧ The law of causality actually is an *a priori* given, a transcendental law. A proof of it from experience is not possible, since the first steps of experience, as we have seen, are not possible without employing inductive inferences, i.e. without the law of causality. But even suppose that complete experience could tell us—though we are still far from being entitled to affirm this—that everything so far observed had occurred in a lawlike manner. It would still follow from such experience only by an inductive inference, i.e. by presupposing the law of causality, that the law of causality would then also hold in the future. Here the only valid advice is: have trust and act!

> The inadequate
> It then takes place.*

⟦67⟧ That should perhaps be the answer given by us to the question: what is true in the way in which we form representations? Regarding what has always seemed to me to be the most essential advance in Kant's philosophy, we still stand on the ground of his system. In this regard I have also frequently emphasized in my previous studies the agreement between recent physiology of the senses and Kant's doctrines, although this admittedly does not mean that

---

* Das unzulängliche
Dann wird's Ereignis.

I had to swear by the master's words in all subordinate matters too. I believe the resolution of the concept of intuition into the elementary processes of thought to be the most essential advance in the recent period. This resolution is still absent in Kant, which is something that then also conditions his conception of the axioms of geometry as transcendental propositions. Here it was especially the physiological investigations on sense perceptions which led us to the ultimate elementary processes of cognition. These processes had to remain still unformulable in words, and unknown and inaccessible to philosophy, as long as the latter investigated only cognitions finding their expression in language.

‖68‖ Admittedly—for those philosophers who have retained the inclination for metaphysical speculations, the most essential thing in Kant's philosophy appears to be precisely what we have considered to be a defect hanging upon inadequate development in the specialized sciences of his time. Kant's proof, indeed, for the possibility of a metaphysics—and of course he himself did not know how to discover anything more about this alleged science—relies purely and simply on the belief that the axioms of geometry, and the related principles of mechanics, are transcendental propositions given *a priori*. Moreover his whole system properly speaking contradicts the existence of metaphysics, and the obscure points of his epistemology, about whose interpretation there has been so much controversy, derive from this root.

‖69‖ According to all this, science would seem to have its own secure territory standing firmly upon which it can seek for the laws of the actual—a marvellously rich and fruitful field of work. As long as it restricts itself to this activity, it will be unaffected by idealist doubts. Such work may appear modest in comparison with the soaring schemes of metaphysicians.

Yet with gods
Shall measure himself
No mortal.
If he raises himself up
And touches the stars
With the crown of his head,
Then nowhere cling
His uncertain soles,
And there play with him
Clouds and winds.

If he stands with firm
Pithy bones
On the well-founded
Lasting Earth:
In height he does not reach
Even with the oak
Or the vine
To liken himself.*

* Doch mit Göttern
Soll sich nicht messen
Irgendein Mensch.
Hebt er sich aufwärts

⟦70⟧ All the same, the example of the man who said this may teach us how a mortal, who had surely learnt how to stand, even when he touched the stars with the crown of his head, still retained a clear eye for truth and actuality. Something of the artist's vision, of the vision which led Goethe and also Leonardo da Vinci to great scientific thoughts, is what the genuine enquirer must always have. Both artist and enquirer strive, although with different approaches, towards the goal of discovering new lawlikeness. Only, one must not try to pass off idle enthusiasm and crazy fantasies as an artist's vision. The genuine artist and the genuine enquirer both know how to work genuinely, and to give their works a firm form and a convincing fidelity to truth.

⟦71⟧ Moreover, actuality has so far always revealed itself much more sublimely and richly to a science enquiring in a manner faithful to its laws, than the utmost efforts of mythical fantasy and metaphysical speculation had known how to depict it. What have all the monstrous offspring of Indian reverie, the piling up of gigantic dimensions and numbers, to say as against the actuality of the structure of the universe, as against the intervals of time in which sun and earth were formed, in which life evolved during geological history and adapted itself, in more and more perfect forms, to the more stable physical situations on our planet?

⟦72⟧ What metaphysics has prepared, in advance, concepts of effects such as magnets and moving electricity exert on each other? Physics at this moment is still striving to reduce them to well-specified elementary effects, without having reached a clear conclusion. But already light too seems to be nothing other than one more kind of motion of these two agencies, and the aether filling space is acquiring wholly new characteristic properties as a magnetizable and electrifiable medium.

⟦73⟧ And into what scheme of scholastic concepts shall we insert that supply of effective energy whose constancy the law of the conservation of force asserts? This supply, like a substance, cannot be destroyed or increased; it is at work as a driving force in every motion of both lifeless and living matter; it is a Proteus ever attiring itself in new forms, active throughout infinite space and yet not divisible by space without remainder, the effective factor in every effect, the motive factor in every motion—and yet not mind and not matter. Did the poet have a presentiment of it?

In life's currents, in a storm of deeds,
I float up and down,
Weave here and there!
Birth and grave,

Und berührt
Mit dem Scheitel die Sterne,
Nirgends haften dann
Die unsicheren Sohlen.
Und mit ihm spielen
Wolken und Winde.
Steht er mit festen
Markigen Knochen
Auf der wohlgegründeten
Deuernden Erde:
Reicht er nicht auf,
Nur mit der Eiche
Oder der Rebe
Sich zu vergleichen.

An eternal ocean,
An alternating weaving,
A glowing life,
Thus I create on the humming loom of time
And make the deity's living garment.*

⟦74⟧ We are particles of dust on the surface of our planet, which itself is barely to be called a grain of sand in the infinite space of the universe. We are the youngest generation of living things on earth, by the geological reckoning of time barely arisen from the cradle, still at the stage of learning, barely half-educated, and pronounced of age only out of mutual considerateness. Yet we have already—through the more powerful impulse of the law of causality—grown out above all of our fellow-creatures and are subduing them in the struggle for existence. We truly have sufficient ground to be proud that it has been given to us slowly to learn to understand, by faithful labour, 'the incomprehensibly high works'. And we need not feel in the least ashamed of not succeeding in this immediately at the first assault of a flight like that of Icarus.

APPENDICES

1. ON THE LOCALIZATION OF THE SENSATIONS OF INTERNAL ORGANS

The issue might arise here of whether the physiological and pathological sensations of internal organs of the body should not fall into the same category as psychic states, inasmuch as many of them are likewise not altered by movements, or at least not altered considerably.

Now there are indeed sensations of an ambiguous character, such as those of depression, melancholy, and anxiety, which may just as well arise from bodily causes as from psychic ones, and for which there is also lacking any representation of a particular localization. At most, in the case of anxiety, the region of the heart vaguely asserts a claim to be the seat of the sensation, as in general the older view making the heart the seat of many psychic feelings was obviously derived from the fact that the movement of this organ is often altered by such feelings, a movement which one feels partly directly and partly indirectly through superimposing one's hand. So there thus arises a kind of false bodily localization for what are actually psychic states. In states of illness this goes much further. I recall having seen, as a young doctor, a melancholic shoemaker who believed he could feel that his conscience had squeezed itself between his heart and his stomach.

---

* In Lebensfluten, in Tatensturm,
Wall' ich auf und ab,
Webe hin und her!
Geburt und Grab,
Ein ewiges Meer,
Ein wechselnd Weben,
Ein glühend Leben,
So schaff' ich am sausenden Webstuhl der Zeit,
Und wirke der Gottheit lebendiges Kleid.

On the other hand, there are of course a series of bodily sensations, such as hunger, thirst, satiety, neuralgic and inflammatory pains, which we localize as bodily ones—although uncertainly—and do not hold to be psychic, even though they are hardly altered by movements of the body. Admittedly, most inflammatory and rheumatic pains are considerably increased by pressure on the parts ⟦of the body⟧ in which they have their seat, or by movement of those parts. But even in the contrary case, as likewise neuralgic pains, they are probably to be regarded only as higher intensities of normally occurring feelings of pressure and strain on the parts concerned.

In this matter, the kind of localization often gives an indication of what occasioned our learning something about the place of the sensation. Thus almost all sensations of the abdominal viscera are transferred to particular places on the anterior abdominal wall, even for organs such as the duodenum, pancreas, spleen, and so on, which lie nearer to the posterior wall of the trunk. But pressure from outside can reach all of these organs almost only through the pliable anterior abdominal wall, not through the thick layers of muscle between the ribs, spine, and hipbone. It is further very noteworthy, that with toothache from an inflammation of the periosteum of a tooth the patients are usually uncertain at first which tooth is hurting—the upper or the lower—from a pair of teeth one above the other. One must first press forcefully upon the two teeth in order to find which is giving the pain. Would the origin of this not be that pressure on the periosteum of the root of the tooth in the normal state tends to occur only during chewing, and that both teeth of each pair then always suffer simultaneously a pressure of like strength?

The feeling of satiety is a sensation of fullness of the stomach, which is distinctly increased by pressure on the pit of the stomach, whereas the feeling of hunger is somewhat reduced by the same pressure. This may be what occasions their localization in the pit of the stomach. Besides which, if we assume that like local signs accrue to nerves ending at the same places in the body, then the clear-cut localization of one sensation of such an organ would suffice for its other sensations as well.

This probably holds for thirst too, to the extent that it is a sensation of dryness of the pharynx. The connected more general feeling of lack of water in the body, which is not eliminated by moistening the mouth and throat, is on the other hand not specifically localized.

The feeling of respiratory deficiency (or so-called shortness of breath), which has its own peculiar quality, is reduced by respiratory movements and accordingly localized. Yet the sensations for respiratory obstructions in the lungs are only imperfectly separated from those for circulatory obstructions, if the latter are not combined with palpable alterations in the heart beat. Perhaps this separation is so imperfect only because respiratory disturbances generally evoke an increased heart action as well, and a disturbed heart action makes it difficult to satisfy a respiratory deficiency.

It is moreover to be noted that we have no conception at all, without anatomical and physiological studies, of the form and movements of parts ⟦of the body‖ having such an extraordinary fine sensitivity, and thereby assured and adept movement, as our soft palate, epiglottis, and larynx, since we cannot see them without optical instruments and also cannot easily touch them. We do not indeed yet know, despite all the scientific investigations, how to describe all of their movements with assurance, e.g. not the movements of the larynx which occur in the production of a falsetto voice. Were we innately acquainted with localizations for those organs of ours which are equipped with tactile sensation, we would surely have to expect such an acquaintance just as much for our larynx as for our hands. But in fact our acquaintance with the form, magnitude, and movement of our own organs reaches only just as far as we can see and touch them.

The extraordinarily varied and finely executed movements of the larynx also teach us something else about the relation between the act of the will and its effect, namely that what we in the first place, and in an immediate manner, understand how to effect is not the innervation of a specific nerve or muscle, nor always a specific position of the mobile parts of our body, but instead the first observable external effect. To the extent that we can ascertain the position of the parts of the body with our eye or hand, this position is the first observable effect with which the conscious intention in the act of the will is concerned. When we cannot do that, as with the larynx and the rear parts of the mouth, the various modifications of the voice, of breathing, of swallowing, and so on are these nearest available effects.

Thus the movements of the larynx, although evoked by innervations which are completely alike in kind with those used for moving the limbs, are of no concern in the observation of spatial alterations. One might still ask, however, whether perhaps the very distinct and varied expression of movement which music produces is not reducible to the fact that alteration of pitch in singing is produced by muscle innervations—thus by the same kind of internal activity as is movement of the limbs.

A similar situation also exists with movements of the eyes. We all know very well how to direct our glance to a specific place in the visual field, i.e. how to make the image of that place fall upon the central fovea of the retina. But uneducated people do not know how they move their eyes in doing this, and do not always know how to obey the command of an eye doctor, say, to turn their eyes to the right, when it is expressed in this form. Indeed, even educated people—although they know how to look at an object held close before the nose and will then squint inwards—do not know how to obey the command to squint inwards without a corresponding object being there.

2. Space can be transcendental without the axioms' being so

Nearly all philosophical opponents of metamathematical investigations have treated the two statements as identical, which they by no means are. This has already been explained quite clearly by Benno Erdmann[1] in the manner of

[1] *Die Axiome der Geometrie*, Leipzig, 1877, ch. III.

expression current amongst philosophers. I have stressed it myself in an answer directed against the objections of Mr. Land of Leyden.[2]

Although Albrecht Krause, the author of the latest attack,[3] cites both of these discussions, with him too the first five of seven sections are still allotted again to defending the transcendental nature of the form of intuition of space, and only two deal with the axioms. This writer is to be sure not merely a Kantian, but an adherent of the most extreme nativist theories in physiological optics, and he considers the whole content of these theories to be included in Kant's system of epistemology, for which there is surely not the slightest justification, even if Kant's individual opinion—corresponding to the undeveloped state of physiological optics in his time—should have been something of the sort. The question of whether intuition was more or less resoluble into conceptual constructions had at that time not yet been raised.

Moreover, Mr. Krause ascribes to me conceptions of local signs, sensory memory, the influence of retinal magnitude, etc. which I have never had and never propounded, or which I have expressly taken pains to refute. As sensory memory ⟦Sinnengedächtnis⟧ I have always characterized only the memory for immediate sensory impressions which are not given a verbal formulation, but I would always have vigorously protested against the assertion that this sensory memory had its seat in the peripheral sense organs. I have performed and described experiments having the purpose of showing that even with falsified retinal images—e.g. when seeing through lenses or through converging, diverging, or laterally deflecting prisms—we quickly learn how to overcome the illusion and again see correctly, and then on p. 41 Mr. Krause insinuates to me the view that a child, because his eye is smaller, must see everything smaller than an adult does. Perhaps the present lecture will convince the author named that he has, up to now, completely misunderstood the sense of my empiricist theory of perception.

The objections made by Mr. Krause in his sections on the axioms have in part been disposed of in the present lecture, e.g. the grounds why intuitive representation of an object which has never yet been observed might be *difficult*. He then refers to my assumption, in the lecture on the axioms of geometry,[4] of surfacelike beings living on a plane or a sphere, which I made in order to render intuitive the relationship between the various geometries; and there follows an explanation that on the sphere two or many 'straightest'[5] lines could indeed exist between two points, but that Euclid's axiom speaks of the one 'straight' line. However, for the surface beings on the sphere the straight line connecting two points on the spherical surface has—according to the assumptions made—no real existence at all in their world. The 'straightest' line of their world would be for them precisely what the 'straight' one is for us.

---

[2] *Mind* **3** (1878), 212 ⟦*Helmholtz 1878a*, reprinted above⟧.
[3] *Kant und Helmholtz*, Lahr, 1878.
[4] See the beginning of this volume ⟦*Helmholtz 1876*, reprinted above⟧.
[5] I had thus named the shortest or geodetic lines.

Mr. Krause indeed makes an attempt to define a straight line as a line having only one direction. But how should one define direction?—surely only again by means of the straight line. Here we move in a vicious circle. Direction is indeed the less general concept, since every straight line contains two opposed directions.

There then follows an explanation that if the axioms were empirical propositions, we could not be absolutely convinced of their correctness, as surely we are. But this is precisely what the controversy hinges upon. Mr. Krause is convinced that we would not believe measurements testifying against the correctness of the axioms. Here he may well be right as regards a large number of people, who more willingly trust a proposition supported by ancient authority, and closely interwoven with all of their remaining information, than their own reflections. But it should surely be otherwise with a philosopher. People also behaved highly incredulously for long enough towards the spherical shape of the earth, its motion, and the existence of meteorites. There is incidentally something correct about his assertion, in that one is well advised to examine all the more rigorously the grounds of proof against propositions of old authority, the longer these propositions have so far proved to be factually correct in the experience of many generations. But the facts must surely decide in the end, and not preconceived opinions or Kant's authority.

It is furthermore correct that if the axioms are laws of nature, they naturally participate in the merely approximate provability of all laws of nature through induction. But the desire to be acquainted with exact laws is not yet a proof that such exist. Odd it is indeed that Mr. Krause rejects the results of scientific measurement because of their limited accuracy, whereas in order to prove, as regards transcendental intuition, that we have no need at all for measurements in convincing ourselves of the correctness of the axioms, he reassures himself with appraisals by visual estimation (p. 62). That is surely measuring friend and foe by different standards! As if any pair of dividers from the worst set of drawing instruments would not yield something more accurate than the best visual estimation, even disregarding the question—which my opponent does not ask himself at all—of whether visual estimation is innate and given *a priori* or whether it is not acquired too.

The expression 'measure of curvature', in its application to three-dimensional space, has given great offence to philosophical writers.[6] Now this name denotes a certain magnitude defined by Riemann, which when calculated for surfaces coincides with what Gauss called the measure of curvature of surfaces. The name has been retained by geometers as a brief term for the more general case of more than two dimensions. The controversy here concerns only the name, and nothing more than the name, for what is in any case the well-defined concept of a magnitude.

---

[6] e.g. in A. Krause, *op. cit.*, p. 84.

3. The Applicability of the Axioms to the Physical World

Here I want to develop the consequences which we would be forced to adopt if Kant's hypothesis of the transcendental origin of the axioms of geometry were correct, and to discuss what value this immediate acquaintance with the axioms would then have for how we judge the relationships in the objective world.[1]

§1.

I shall initially stay with the realist hypothesis, in this first section, and speak its language; thus I shall assume that the things we objectively perceive really exist and act upon our senses. I do this initially only in order to be able to speak the simple and understandable language of ordinary life and natural science, and thereby to express the sense of my opinion in a manner understandable to non-mathematicians too. I reserve the possibility of dropping the realist hypothesis in the following section, and repeating the corresponding account in abstract language and without any particular presupposition about the nature of the real.

First of all: the sort of alikeness or congruence of spatial magnitudes which might flow, according to the assumption made, from transcendental intuition, must be distinguished from the alikeness in value of such magnitudes which is to be ascertained by measurement with the aid of physical means.

I call spatial magnitudes *physically alike in value* if they are ones in which under like conditions, and in like periods of time, like physical processes can exist and run their course. The most frequent process, employed with suitable precautions, for determining spatial magnitudes physically alike in value is the transfer of rigid bodies, such as measuring rods or a pair of dividers, from one place to another. It is moreover a quite general outcome of all of our experiences, that if the alikeness in value of two spatial magnitudes has been shown by any adequate method of physical measurement, they show themselves to be alike in value also as regards all other known physical processes.

Physical alikeness in value is thus a completely determinate, unambiguous objective property of spatial magnitudes, and obviously nothing hinders us from ascertaining, by experiments and observations, how the physical alikeness in value of a specific pair of spatial magnitudes depends on the physical alikeness in value of other pairs of such magnitudes. This would give us a kind of geometry which, just for the purpose of our present investigation, I will call *physical geometry*, in order to distinguish it from the geometry which might be founded upon the hypothetically assumed transcendental intuition of space. Such a physical geometry, implemented in a pure and deliberate manner, would obviously be possible and have completely the character of a natural science.

Even its first steps would lead us to propositions corresponding to the axioms,

[1] So, in order to prevent fresh misunderstandings such as occur in A. Krause, *op. cit.*, p. 84: I am not the one 'who is acquainted with a transcendental space having laws proper to itself', but I am instead here seeking to draw the consequences of what I consider to be Kant's unproved and incorrect hypothesis, according to which the axioms are taken to be propositions given by transcendental intuition, and I do this in order to demonstrate that a geometry based upon such intuition would be wholly useless as regards objective knowledge.

if we only substitute for the transcendental alikeness of spatial magnitudes their physical alikeness in value.

Namely, as soon as we had found a suitable method for determining whether the separations of any two point pairs were alike (i.e. physically alike in value), we should also be able to distinguish the special case where three points a, b, c lie such that no other point, apart from b, can be found having the same separations from a and c as b does. We say in this case that the three points lie in a straight line.

We would then be in a position to seek three points A, B, C having all three like mutual separation, and which would thus be the corners of an equilateral triangle. Then we could seek two new points b and c, both having like separation from A, and such that b lies in a straight line with A and B, and c with A and C. Straight away the question would arise: is the new triangle Abc equilateral too as is ABC; thus do we have $bc = Ab = Ac$? *Euclidean* geometry answers: yes; *spherical* geometry asserts that $bc > Ab$ when $Ab < AB$, and *pseudospherical* geometry that $bc < Ab$ under the same condition. Already here the axioms would come up against a factual decision. I have chosen this simple example, because in it we deal only with measuring alikeness or non-alikeness of the separations between points, or as may be with the determinateness or non-determinateness of the situations of certain points, and because no spatial magnitudes of greater composition, straight lines or planes, need to be constructed at all. The example shows that this physical geometry would have its own propositions occupying the place of the axioms.

As far as I can see, even for an adherent of Kant's theory it cannot be a matter of doubt that it would be possible, in the manner described, to give the foundation of a purely empirical geometry, if we did not yet have such a geometry. In it we would deal only with observable empirical facts and their laws. And only inasmuch as the presupposition were correct, that physical alikeness in value always occurs simultaneously for all kinds of physical processes, would the science acquired in such a manner be a theory of space which was independent of the constitution of the physical bodies contained in space.

But Kant's adherents assert that there also exists, besides such a physical geometry, a *pure geometry* founded upon transcendental intuition alone, and that this indeed is the geometry which has been scientifically developed up to now. In this geometry we are said not to deal at all with physical bodies and their behaviour during motions. But we can instead—without knowing anything whatsoever about such ⟦bodies⟧ through experience—form for ourselves, by means of inner intuition, representations of absolutely unalterable and immobile spatial magnitudes, of bodies, surfaces, and lines, which even without ever being brought into coincidence through motion (which belongs only to physical bodies) could stand to one another in the relationship of alikeness and congruence.[2]

[2] Land in *Mind* **2** (1877), 41; A. Krause, *op cit.*, p. 62.

I will permit myself to emphasize that this inner intuition of the straightness of lines, and alikeness of separations or angles, must have absolute accuracy. Otherwise we would by no means have the right to decide whether two straight lines, when infinitely prolonged, would intersect only once or perhaps even twice (like great circles on a sphere) nor to assert that every straight line which intersects with one of two parallel lines lying in the same plane as itself, must also intersect with the other. One must not try to pass off visual estimation, which is so imperfect, as transcendental intuition, which latter demands absolute accuracy.

Suppose the case that we had such a transcendental intuition of spatial structures, and of their alikeness and congruence, and that we could convince ourselves on truly sufficient grounds that we had it. Then a system of geometry would of course be derivable from it, one which would be independent of all properties of physical bodies—a pure, transcendental geometry. This geometry too would have its axioms. But it is clear, even according to Kant's principles, that the propositions of this hypothetical pure geometry need not necessarily concur with those of physical geometry. For the one speaks of the alikeness of spatial magnitudes in inner intuition, the other of physical alikeness in value. This latter obviously depends on empirical properties of natural bodies and not merely on the make-up of our mind.

So one would then have to investigate whether the two mentioned kinds of alikeness necessarily always coincide. This is not to be decided by experience. Has it any sense to ask whether two pairs of dividers encompass, according to transcendental intuition, like or unlike lengths? I do not know how to connect any sense with this; and as far as I have understood recent adherents of Kant, I believe I may take it that they too would answer with a no. Visual estimation, as already said, is something we should not allow to be passed off on us in this respect.

Could one then perhaps infer from propositions of pure geometry that the separations of the tips of the two pairs of dividers are alike in magnitude? For that one would have to be acquainted with geometrical relations between these separations and other spatial magnitudes, and to know directly that these latter were alike in the sense of transcendental intuition. But since one can never know this directly, one also can never obtain it by geometrical inferences.

If one cannot acquire by experience the proposition that the two kinds of spatial alikeness are identical, it must be a metaphysical proposition and correspond to a necessity of thought. But this would then be a necessity of thought which determined not merely the form of items of empirical knowledge, but also their content, as for example with the above construction of two equilateral triangles—a consequence which would indeed contradict Kant's principles. Pure intuition and thought would then yield more than Kant is inclined to concede.

Suppose finally the case that physical geometry had found a series of universal empirical propositions which were identical in formulation with the axioms of pure geometry. It would at most follow from this that it was a permissible hypothesis, leading to no contradiction, that the physical alikeness in value of spatial magnitudes concurs with their alikeness in pure spatial intuition. But it would not be the only possible hypothesis. Physical space and the space of intuition could also be related to each other as is actual space to its image in a convex mirror.[3]

That physical geometry need not necessarily concur with transcendental geometry follows from our being in fact able to represent them to ourselves as not concurring.

The manner in which such an incongruence would make its appearance results already from what I have explained in an earlier paper.[4] Let us assume that the physical measurements corresponded to a pseudospherical space. With the observer and the observed objects at rest, the sensory impression of such a space would be the same as if we had before us Beltrami's spherical model in Euclidean space, and the observer were at its centre. But as soon as the observer changed his place, the centre of the sphere of projection would have to migrate with him, and the whole projection be displaced. Thus for an observer whose spatial intuitions and whose estimations of spatial magnitudes had been formed either from transcendental intuition, or as a result of his experience to date in respect of Euclidean geometry, there would arise the impression, that as soon as he himself moved, all objects seen by him were also displaced in a certain manner and variously expanded and contracted in various directions.

In a similar manner, and in accordance with only quantitatively divergent relationships, we see in our objective world too how the perspective-relative situation and the seeming magnitudes of objects at various distances change, as soon as the observer moves. Now we are indeed capable of discerning, from these changing visual images, that the objects surrounding us do not alter their relative mutual situation and magnitudes, as long as the perspective displacements correspond exactly to the law which—according to what has proved true in experience to date—governs them when the objects are at rest; given any deviation from this law, on the other hand, we infer a motion of the objects. In just the same way—as I believe myself permitted to presuppose as an adherent of the empiricist theory of perception—someone who went over from Euclidean to pseudospherical space would also at first indeed believe he saw seeming motions of the objects, but would very quickly learn how to adapt an estimation of the spatial relationships to the new conditions.

But the latter presupposition is one which has been formed only by analogy with what we otherwise know of sense perceptions, and which cannot be tested by making the experiment. So let us assume that the way in which such an observer judged spatial relationships could no longer be altered, because it was connected with innate forms of spatial intuition. Then he would nevertheless quickly ascertain that the motions which he believed he saw were only seeming motions, because of their always being reversed when he himself went back to

---

[3] See my lecture on the axioms of geometry ⟦*Helmholtz 1876*⟧.

[4] On the axioms of geometry ⟦*Helmholtz 1876*⟧.

his starting point; or a second observer could attest that everything remained at rest while the first one changed his place. Thus under scientific investigation—even if perhaps not for unconsidered intuition—it could quickly emerge which spatial relationships were the physically constant ones, somewhat as we ourselves know from scientific investigations that the sun stays fixed and the earth rotates, despite the persistence of the sensory semblance that the earth stays still and the sun goes round it once in 24 hours.

But then the whole of this presupposed transcendental intuition *a priori* would be demoted to the rank of a *sensory delusion*, of an *objectively false semblance*, which we should have to try to free ourselves from and forget, as is the case with the seeming motion of the sun. There would then be a contradiction between what appeared to be spatially alike in value according to innate intuition, and what proved to be such in objective phenomena. The whole of our scientific and practical interest would be attached to the latter. Spatial relationships which were physically alike in value would be portrayed by the transcendental form of intuition only in the way that a flat map portrays the surface of the earth, with very small portions and strips correct but larger ones necessarily false. It would then be not merely a matter of the *manner of appearance*, which of course necessarily occasions a modification of the content to be portrayed, but instead a matter of the relations between appearance and content making the two of them concur on a smaller scale, yet giving a *false semblance* when extended to a larger scale.

The conclusion which I draw from these considerations is the following: if there actually were innate in us an irradicable form of intuition of space which included the axioms, we should not be entitled to apply it in an objective and scientific manner to the empirical world until one had ascertained, by observation and experiment, that the parts of space made alike in value by the presupposed transcendental intuition were also physically alike in value. This condition coincides with Riemann's demand that the measure of curvature of the space in which we live should be determined empirically by measurement.

The measurements of this kind performed to date have not yielded any noticeable deviation from zero for the value of this measure of curvature. Thus we can of course regard Euclidean geometry as factually correct within the limits of accuracy of measurement which have been attained up to now.

§2.

The discussion in the first section remained wholly within the domain of the objective and of the realist viewpoint of the enquirer into nature, where the ultimate goal is to apprehend the laws of nature conceptually and where an intuitive acquaintance merely helps to ease matters, or as may be is a false semblance to be done away with.

Now Professor Land believes that in my account I had confused the concepts of the *objective* and of the *real*, that in my claiming that geometrical propositions could be tested and confirmed by experience it was presupposed without foundation (*Mind* **2**, 46) 'that empirical knowledge is acquired by simple importation or by counterfeit, and not by peculiar operations of the mind, solicited by varied impulses from an unknown reality'. Had Professor Land been acquainted with my works on sensations, he would have known that I myself, throughout my life, have combatted the kind of presupposition which he imputes to me. In my paper I did not speak of the distinction between the objective and the real, because it seemed to me that this distinction carried no weight at all in the investigation concerned. In order to give a foundation for this opinion of mine, we will now drop the hypothetical element in the realist view, and demonstrate that the propositions and proofs given so far will even then still have a perfectly correct sense, that one is even then still entitled to ask about the physical alikeness in value of spatial magnitudes and to decide this matter by experience.

The only presupposition we shall retain is that of the law of causality, namely that the representations having the character of perception which come about in us do so according to fixed laws, so that when different perceptions force themselves upon us, we are entitled to infer from this a difference in the real conditions under which they were formed. Concerning these conditions themselves—concerning what is properly real and underlies the appearances—we otherwise know nothing; all opinions which we may otherwise cherish about this are to be regarded as only more or less probable hypotheses. The presupposition we start with, on the other hand, is the basic law of our thought; if we wanted to give it up, we should then renounce altogether the possibility of comprehending these relationships in thought.

I emphasize, that no presuppositions at all about the nature of the conditions under which representations arise should be made here. The hypothesis of subjective idealism would be just as permissible as the realist view, whose language we have used up to now. We could assume that all of our perception was only a dream, although a most highly self-consistent dream in which representation developed out of representation according to fixed laws. In this case the ground for the occurrence of a new seeming perception would have to be sought only in the fact that it had been preceded in the dreamer's mind by the representations of certain other perceptions, and perhaps also by representations of a certain kind of impulses of his own will. That which in terms of the realist hypothesis we call laws of nature, would in terms of the idealist one be laws governing the sequence of successive representations having the character of perception.

Now we find it to be a fact of consciousness, that we believe ourselves to perceive objects which are at specific places in space. That an object should appear at a specific particular place, and not at another one, will have to depend on the kind of real conditions evoking the representation. We must infer that in order to effect the occurrence of the perception of another place for the same object, other real conditions would have had to be present. Thus in the real there must exist some or other relationships, or complexes of relationships, which specify at what place in space an object appears to us. I will call these, to use a brief term for them, *topogenous factors*. Of their nature we know nothing; we know only that the coming about of spatially different perceptions presupposes a difference in the topogenous factors.

Besides this there must, in the domain of the real, be other causes which effect our believing that at the same place we perceive at different times different material things having different properties. I will allow myself to give them the name of *hylogenous factors*. I chose these new names in order to exclude the involvement of any connotations which might be attached to customary words.

Now when we perceive and assert anything which states a mutual dependence of spatial magnitudes, then without doubt the factual sense of such a statement is merely that between certain topogenous factors, whose proper nature however remains unknown to us, there exists a certain lawlike connection, whose type is likewise unknown. For precisely this reason, Schopenhauer and many adherents of Kant have come to the incorrect conclusion that there is altogether no real content in our perceptions of spatial relationships—that space and its relationships are only a transcendental semblance without there corresponding anything actual to them. But we are at any rate entitled to apply the same considerations to our spatial perceptions as to other sensory signs, e.g. to colours. Blue is only a mode of sensation; but that we should see blue in a specific direction at a certain time is something which must have a real ground. Should we at some time see red there, this real ground will have to have altered.

When we observe that physical processes of various kinds can run their course in congruent spaces during like periods of time, this means that in the domain of the real, like aggregates and sequences of certain hylogenous factors can come about and run their course in combination with certain specific groups of different topogenous factors, such namely as give us the perception of parts of space physically alike in value. And when experience then instructs us, that any combination or sequence of hylogenous factors which can exist or run its course in combination with one group of topogenous factors, is also possible with any other physically equivalent group of topogenous factors—then this is at any rate a proposition having a real content, and thus topogenous factors undoubtedly influence the course of real processes.

In the example given above of the two equilateral triangles, it is only a matter of (1) the alikeness or non-alikeness, i.e. physical alikeness or non-alikeness in value, of distances between points; (2) the determinateness or non-determinateness of the topogenous factors of certain points. But these concepts of determinateness and of alikeness in value in respect of certain sequences can also be applied to objects whose nature is otherwise quite unknown. I infer from this that the science I have called physical geometry contains propositions having real content, and that its axioms are determined not by mere forms of representation, but by relationships in the real world.

This still does not entitle us to declare impossible the assumption of a geometry founded upon transcendental intuition. One could, e.g., assume that an intuition of the alikeness of two spatial magnitudes might be produced immediately, without physical measurement, by the influence of topogenous factors on our consciousness—thus that certain aggregates of topogenous factors might even be equivalent in respect of a psychic, immediately perceivable effect. The whole of Euclidean geometry is derivable from the formula giving the distance between two points as a function of their rectangular coordinates. Let us assume that the intensity of that psychic effect, whose alikeness appears in our representation as alikeness of distance between two points, depends in the same way on some set or other of three functions of the topogenous factors of any point, as does the distance in Euclidean space on the three coordinates of any point. Then the system of pure geometry of such a consciousness would have to satisfy the axioms of Euclid, however else the topogenous factors of the real world and their physical equivalence behaved.

It is clear, that in this case one could not decide from the form of intuition alone whether there was concurrence between psychic and physical alikeness in value for spatial magnitudes. And if it should turn out that there was concurrence, the latter would have to be regarded as a law of nature, or (as I have termed it in my popular lecture) as a pre-established harmony between the world of representations and the real world—just as much as it rests upon laws of nature that the straight line described by a light ray coincides with the one formed by a taut thread.

With this, I consider myself to have shown that the proof which I gave in Section 1.1 in the language of the realist hypothesis reveals itself to be valid even without the latter's presuppositions.

When we want to apply geometry to facts of experience, where it is always only a matter of physical alikeness in value, we can apply only the propositions of that science which I have termed physical geometry. For anyone who derives the axioms from experience, our geometry up to now has indeed been a physical geometry, only one which relies on a great host of experiences collected without any plan, instead of on a system of methodically pursued ones. One should moreover mention that this was already the view of Newton, who declares in the introduction to his *Principia*: 'Geometry itself has its foundation in mechanical practice, and is in fact nothing other than that part of the whole of mechanics which accurately states and founds the art of measurement.'[5]

On the other hand, the assumption that there is an acquaintance with the axioms which comes from transcendental intuition is:

(1) an *unproved* hypothesis;

(2) an *unnecessary* hypothesis, since it does not pretend to explain anything in what in fact is our world of representations which could not also be explained without its help;

(3) a *wholly unusable* hypothesis for explaining our acquaintance with the actual world, since the propositions laid down by it should only ever be applied to the relationships in the actual world after their objective validity has been experimentally tested and ascertained.

Kant's doctrine of the *a priori* given forms of intuition is a very fortunate

[5] Fundatur igitur Geometria in praxi Mechanica, et nihil aliud est quam Mechanicae universalis pars illa, quae artem mensurandi accurate proponit ac demonstrat.

and clear expression of the state of affairs; but these forms must be devoid of content and free to an extent sufficient for absorbing any content whatsoever that can enter the relevant form of perception. But the axioms of geometry limit the form of intuition of space in such a way that it can no longer absorb every thinkable content, if geometry is at all supposed to be applicable to the actual world. If we drop them, the doctrine of the transcendentality of the form of intuition of space is without any taint. Here Kant was not critical enough in his critique; but this is admittedly a matter of theses coming from mathematics, and this bit of critical work had to be dealt with by the mathematicians.

C. NUMBERING AND MEASURING FROM AN EPISTEMOLOGICAL VIEWPOINT (*HELMHOLTZ 1887*)

Helmholtz's most original and influential contributions to mathematics were written in the 1870s and treat the foundations of geometry. In the late 1870s and the 1880s, as Weierstrass and his students gave a rigorous arithmetical grounding to the calculus, it gradually became clear to mathematicians that the problems of real analysis were being pushed back on to the concept of natural number. These developments, and the discovery by Cantor (*Cantor 1874*) of non-denumerable sets, gave a new impetus to the study of logic and the foundations of arithmetic. The new tendency is evident in many of the selections that follow; see, for example, *Dedekind 1872*, *Peirce 1881*, *Kronecker 1887*, *Dedekind 1888*, and, of course, Frege's *Begriffschrift* (*Frege 1879*).

Helmholtz, too, in the following essay from 1887, wrote about the foundations of arithmetic. He was sixty-six, and his ideas were soon overwhelmed by the work of his younger contemporaries—in particular by Frege, who subjected his views to trenchant attack in *Frege 1891* and *1894*. But Helmholtz's essay was the most sophisticated effort in the nineteenth century to develop a broadly empiricist theory of arithmetic, far outstripping in mathematical and philosophical subtlety the writings of John Stuart Mill. In recent years there has been a revival of interest in empiricist philosophies of mathematics; Helmholtz's essay and Frege's criticisms are the natural starting-point for any historical discussion of the topic.

The translation is by Malcolm F. Lowe, and is taken from *Helmholtz 1977*. References to *Helmholtz 1887* should be to the sections (which were not numbered in the first German version).

Although numbering and measuring are the foundations of the most fruitful, sure and exact scientific methods known to us at all, relatively little work has been done on their epistemological foundations. On the philosophical side, strict disciples of Kant, who adhere to his system exactly as it developed historically amidst the views and knowledge of his time, had of course to regard the axioms of arithmetic as propositions given *a priori*, which narrow down the specification of the transcendental intuition of time in the same sense as the axioms of geometry do the intuition of space. Through this conception, the issue of a further foundation and derivation for these propositions was terminated in both cases.

In earlier writings I endeavoured to show that the axioms of geometry are not propositions given *a priori*, but that they are rather to be confirmed and refuted through experience. Here I emphasize once again that this does not eliminate Kant's view of space as a transcendental form of intuition; in my opinion this merely excludes just one unjustified particular specification of his view, although one which has become most fateful for the metaphysical endeavours of his successors.

It is then clear that if the empiricist theory—which I besides others advocate—regards the axioms of geometry no longer as propositions unprovable and without need of proof, it must also justify itself regarding the origin of the axioms of arithmetic, which are correspondingly related to the form of intuition of time.

Up to now, arithmeticians have placed the following propositions as axioms[a] at the head of their deductions:

AXIOM I. If two magnitudes are both alike with a third, they are alike amongst themselves.

AXIOM II. The *associative law of addition*, as it is termed by H. Grassmann

$$(a + b) + c = a + (b + c).$$

AXIOM III. The *commutative law of addition*:

$$a + b = b + a.$$

AXIOM IV. Like added to like gives like.

AXIOM V. Like added to unlike gives unlike.

Hermann and Robert Grassmann* have got further with this investigation than the other arithmeticians whose work is known to me, while at the same time

---

[a] [In English these axioms are normally stated in terms of 'equals'. But Helmholz will be seen to allude to the etymological connections between 'gleich' ('equal', 'alike'), 'vergleichen' ('compare', 'liken') and a host of other words. So we translate 'gleich' in most occurrences by 'alike' or 'like', to be understood in the sense 'exactly alike'.]

* Hermann Grassmann, *Die Ausdehnungslehre*, 1st edn, Leipzig, 1844 and 2nd edn, 1878. Robert Grassmann, *Die Formenlehre oder Mathematik*, Stettin, 1872.

pursuing philosophical viewpoints. In what follows, I shall have to go along their path throughout my presentation of the arithmetical deductions. Amongst other things, they reduce the two axioms II and III to a single one which we shall call Grassmann's Axiom, namely:

$$(a + b) + 1 = a + (b + 1),$$

from which they derive both of the above more general propositions by means of the so-called $(n + 1)$-proof. I hope to demonstrate, in what follows, that the correct basis for the theory of the addition of pure numbers has thereby indeed been obtained. Regarding the issue, however, of the objective application of arithmetic to physical magnitudes, there is added thereby[b] to the two concepts of *a magnitude* and of *alike in magnitude*, both of whose sense in the realm of facts remains unexplained, a third one as well—that of the *unit*. And at the same time, it seems to me an unnecessary restriction of the domain of validity of the propositions discovered, that one should from the outset treat physical magnitudes only as ones composed out of units.

Among more recent arithmeticians, E. Schröder* has also essentially attached himself to the Grassmann brothers, but in a few important discussions he has gone still deeper. As long as earlier arithmeticians habitually took the ultimate concept of number to be that of a cardinal number ⟦Anzahl⟧ of objects, they could not wholly free themselves from the laws of the behaviour of these objects, and they simply took it to be a fact that the cardinal number of a group of objects is ascertainable independent of the order in which they are numbered. To my knowledge, Mr. Schröder (*op. cit.*, p. 14) was the first to recognize that here a problem lies concealed: he also acknowledged—in my opinion justly—that there lies a task here for psychology, while on the other hand those empirical properties should be defined which the objects must have in order to be enumerable ⟦zählbar⟧.

Besides this there also are relevant discussions, in particular about the concept of a magnitude, to be found in Paul du Bois-Reymond's *Allgemeine Funktionenlehre*, Tübingen, 1882, part 1, ch. 1, and in A. Elsas' book *Über die Psychophysik*, Marburg, 1886, pp. 49ff. However, both of these books deal with more specific investigations, without discussing the complete foundations of arithmetic. Both believe that one may derive the concept of a magnitude from that of a line, the former in the empirical sense, the latter in the sense of strict Kantianism. I have already mentioned above and explained in previous writings what I have to object against the latter viewpoint. Paul du Bois-Reymond closes his investigation with a paradox according to which two opposed viewpoints, both of which lead to contradictions, are alike possible.

As the author just named is a most acute mathematician, who has sought

[b] ⟦i.e. in Grassmann's Axiom; in what follows 'magnitude' translates 'Grösse' and 'alike in magnitude' translates 'gleich gross'.⟧

**Lehrbuch der Arithmetik und Algebra*, Leipzig, 1873.

with particular interest after the deepest foundations of his discipline, the final result at which he arrived has encouraged me all the more to set out my own thoughts on this same problem.

In order to characterize, briefly at the outset, the viewpoint which leads to simple logical derivations and the solution of the mentioned contradictions, the following may serve. I consider arithmetic, or the theory of pure numbers, to be a method constructed upon purely psychological facts, which teaches the logical application of a symbolic system (i.e. of the numbers) having unlimited extent and an unlimited possibility of refinement. Arithmetic notably explores which different ways of combining these symbols (calculative operations) lead to the same final result. This teaches us, amongst other things, how to substitute simpler calculations even for extraordinarily complicated ones, indeed for ones which could not be completed in any finite time. Apart from thereby testing the internal logicality of our thought, such a procedure would admittedly be primarily a pure game of ingenuity with dreamt up objects—one which Paul du Bois-Reymond scornfully compares with the knight's move on the chess board—if it did not permit such extraordinarily useful applications. For by means of this symbolic system of the numbers we give descriptions of the relationships between real objects which, where applicable, can reach any required degree of exactitude, and by means of it in a large number of cases where natural bodies, governed by known laws of nature, meet or interact, the numerical values measuring the outcome are calculated in advance.

But then one must ask: what is the objective sense of our expressing relationships between real objects as magnitudes, by using denominate ⟦benannt⟧ numbers; and under what conditions can we do this? The question resolves itself, as we shall find, into two simpler ones, namely:

(1) What is the objective sense of our declaring two objects to be *alike* in a certain respect?

(2) What character must the physical connection between two objects have, in order that we may regard likenable attributes of these objects as *additively* combined, and consequently regard these attributes as *magnitudes* which can be expressed by using denominate numbers? We namely consider denominate numbers to be composed out of their parts, or as may be units, by addition.

1. The lawlike series of numbers

Numbering is a procedure based upon our finding ourselves capable of retaining, in our memory, the sequence in which acts of consciousness successively occurred in time. We may consider numbers initially to be a series of arbitrarily chosen symbols, for which we fix only a certain kind of succession as the lawlike or—as it is commonly put—the 'natural' one. Its being termed the 'natural' number series was probably connected merely with one specific application of numbering, namely the ascertaining of the *cardinal number* of given real things. As we throw these one after another onto the already numbered heap, the numbers follow one another by a natural process in their lawlike series. This has nothing to do with the sequence of number symbols; just as the symbols differ in different languages, their sequence too could be specified arbitrarily, so long as some or another specified sequence is immutably fixed as the normal or lawlike one. This sequence is in fact a norm or law given by human beings, our forefathers, who elaborated the language. I emphasize this distinction because the alleged 'naturalness' of the number series is connected with an incomplete analysis of the concept of number. The mathematicians term this lawlike number series that of the *positive whole numbers.*

The number series is impressed upon our memory extraordinarily much more firmly than any other series, which doubtless rests upon its much more frequent repetition. This is why we also prefer to use it in order to establish, through association with it, the recollection of other sequences in our memory; that is, we use the numbers as *ordinal numbers.*

2. *Unambiguity of the sequence*

In the number series, the processes of going forwards and going backwards are not equivalent but essentially different, as with the sequence of perceptions in time, whereas for lines—which exist in space permanently and without change in time—neither of the two possible directions of advance is distinguished as against the other.

In fact every present act of our consciousness, be it perception, feeling, or volition, works together in it with the memory images of past acts, but not of future ones, which are still not yet available in our consciousness at all; and we are conscious of the present act as specifically different from the memory images which exist beside it. The present representation is thereby contrasted, in an opposition pertaining to the form of intuition of time, as the succeeding one to the preceding ones, a relationship which is irreversible and to which every representation entering our consciousness is necessarily subject. In this sense, orderly insertion in the time sequence[c] is the inescapable form of our inner intuition.

3. *The sense of the symbolism*

According to the foregoing discussion, each number is determined only by its position in the lawlike series. We attach the symbol *one* to that member of the sequence with which we begin. *Two* is the number which immediately follows upon one in the lawlike series, i.e. without interposition of another number. *Three* is the number which likewise follows immediately upon two, and so on.

There is no reason for interrupting this sequence anywhere, or for returning in it to a symbol already used previously. The decimal system indeed makes it possible, through combining only ten different number symbols in a simple and

---

[c] ⌊die Einordnung in die Zeitfolge⌋

easily understandable way, to continue the series indefinitely without ever repeating a number symbol.*

We shall call the numbers which follow a given number in the lawlike series *higher* than that one, and those which precede it *lower*.** This gives a complete disjunction which is founded in the essence of the time sequence, and which we can express thus:

AXIOM VI: If two numbers are different, one of them must be higher than the other.

4. The addition of pure numbers

In order to formulate general propositions about numbers, I shall use the well-known symbolism of algebra. Each letter of the lower-case Latin alphabet shall be capable of symbolizing any arbitrary number, but always the same one within each individual theorem or each individual calculation.

Explanation of symbols. When I have symbolized any number by a letter, e.g. a, I shall symbolize the one immediately following it in the normal series by $(a + 1)$.

Thus here this symbol $(a + 1)$ shall have, to begin with, no other meaning than the one stated. Parentheses, however, will in general have their usual meaning: that the numbers enclosed within them should first be combined into a number before the remaining prescribed operations are carried out.

The alikeness symbol $a = b$ shall symbolize here, in the pure theory of numbers, only: 'a is the same number as b'. Therefore, from

$$a = b,$$

$$b = c$$

there follows immediately $a = c$, for the two above equations state that both a and c are the same number b. This establishes, in the pure theory of numbers, the validity of Axiom I for the series of the whole numbers.

5. Numbering the numbers

From now on we consider the normal number series to be established and given. We can now consider its members themselves to be a given series of representations in our consciousness, whose order starting from any arbitrarily chosen member we can again symbolize, using the normal number series beginning with *one*.

---

* 'Number theory' investigates number series in which a certain number is always followed again by one, which therefore repeat themselves periodically.

** At this stage I still avoid *greater* and *smaller*; this distinction more suitably attaches itself to the concept of cardinal number—of which later.

DEFINITION. I symbolize as $(a+b)$ that number in the principal series at which I arrive when numbering $(a+1)$ as one, $[(a+1)+1]$ as two, and so on until I have numbered up to b.

The description of this procedure can be summarized in the following equation (H. Grassmann's axiom of addition):

$$(a + b) + 1 = a + (b + 1). \tag{1}$$

Elucidation. This equation states that, having started from $(a + 1)$ as one and numbered up to b, and thus having found the number symbolized by $(a + b)$, if I continue numbering by one further step, I arrive in the former series at $(b + 1)$, and in the second series at the number following $(a + b)$, namely $[(a + b) + 1]$. So I thus symbolize $[(a + 1) + 1]$, mentioned in the definition, also by $[a + (1 + 1)]$ or $(a + 2)$, and further $[(a + 2) + 1]$ by $(a + 3)$, and so on without limit.

In the language of arithmetic we would call this procedure *addition* and the number $(a + b)$ the *sum* of a and b, and a and b themselves the *summands*. However, I draw attention to the fact that in the stated procedure the roles of the magnitudes a and b are not alike, so that proof must first be brought that they can be exchanged without altering the sum. This is to be done further on below. All the same, if we keep this reservation in mind, we can accept this terminology and say that the expression $(a + b)$ prescribes that b should be added to a, and that $(a + b)$ is the sum of a and b, whereby for the moment the order of b behind a must however be kept fixed. We may therefore call a the *first* and b the *second summand*. Correspondingly, each number $(a + 1)$ can be termed, by logically applying this terminology, the sum of the preceding a and the number one.

The stated procedure of addition must, in the lawlike number series, always yield a result, and indeed always the same one for the same numbers a and b. For each of the steps out of which we have composed the addition $(a + b)$ is an advance by one stage in the series of the positive whole numbers, from $(a + b)$ to $(a + b) + 1$, and from b to $(b + 1)$. Each single step can be carried out and, according to our presuppositions about the immutable preservation of the number series in our consciousness, each of them must always give the same result.

There will therefore certainly exist a number which corresponds to the number $(a + b)$, and only one. This proposition corresponds to the content of axiom IV, when the latter is applied to the pure numbers and to the kind of addition prescribed here.

On the other hand, it follows from the description given of the procedure that $(a + b)$ is necessarily different from a, and indeed higher than a, if b is one of the positive whole numbers.

If c is a higher number than a, I must necessarily reach c by numbering stage by stage upwards from a, and be able to number off what number c is as numbered from a. Let it be the bth such number, then

$$c = (a + b).$$

We wish for later quotation to term this proposition:

AXIOM VII. If a number c is higher than another one a, then I can portray c as the sum of a and a positive whole number b to be found.

THEOREM I. On the sequence of execution of several acts of addition, (*associative law* according to Grassmann).

When I have to add to a sum $(a + b)$ a number c, I obtain the same result as when I add the sum $(b + c)$ to the number a. Or written as an equation:

$$(a + b) + c = a + (b + c). \tag{2}$$

Proof. The proposition, as equation (1) states, holds for $c = 1$. It is to be shown that if it is correct for any particular value of c, it is also correct for the immediately following value $(c + 1)$.

Now according to equation (1):

$$[(a + b) + c] + 1 = (a + b) + (c + 1),$$
$$[a + (b + c)] + 1 = a + [(b + c) + 1] = a + [b + (c + 1)].$$

The latter according to proposition (1).

Thus if proposition (2) holds for the value of c occurring here, the expressions on the left hand side of the first two equations are the same numbers, and consequently also

$$(a + b) + (c + 1) = a + [b + (c + 1)],$$

which means that the proposition also holds for $(c + 1)$.

Then since, as stated before, it holds for $c = 1$, it holds also for $c = 2$. If it holds for $c = 2$, it holds also for $c = 3$, and so on without limit.

Corollary: Since the two expressions occurring in equation (2) have the same meaning, we may also omit the parentheses and introduce for both of them the symbolism:

$$a + b + c = (a + b) + c = a + (b + c). \tag{2a}$$

Only we must not alter for the moment the order of a, b, c in these expressions, until we have proved the admissibility of such a permutation.

6. *Generalization of the associative law*

Firstly we generalize the symbolism given in (2a).

$$R = a + b + c + d + \text{etc.} + k + l \tag{2b}$$

shall symbolize a sum in which the individual additions are executed in the order in which they are written, and to abbreviate the symbolism let

$$m + R = m + a + b + c + d + \text{etc.} + k + l,$$

whereas

$$m + (R) = m + (a + b + c + d + \text{etc.} + k + l).$$

On the other hand, according to the sense of this notation:

$$(R) + m = R + m.$$

Other capital Latin letters shall be used in the same sense as R.

Then:

$$R + b + c + S = [(R) + b + c] + S,$$

because they are expressions having the same meaning. On the other hand, by equation (2a):

$$(R) + b + c = (R) + (b + c).$$

Therefore

$$R + b + c + S = [R + (b + c)] + S = R + (b + c) + S,$$

i.e. instead of adding all the terms in their order, one can first combine two arbitrary intermediate ones into a sum. After this has happened, only one single number stands for the sum $(b + c)$ just formed, and one can go on in the same way to combine any other arbitrary pair of consecutive numbers, and so on.

Thus even with arbitrarily many terms one can alter the order in which the additions prescribed by the individual $+$ signs are carried out, without the total sum altering.

THEOREM II (*commutative law* according to H. Grassmann): If in a sum of two summands one of the summands is one, their order can be reversed. To this there corresponds the equation:

$$1 + a = a + 1. \tag{3}$$

Proof. The equation is correct for $a = 1$. It has to be shown besides, that if it is correct for any particular specified value of a it is also correct for $(a + 1)$. Now by equation (1):

$$(1 + a) + 1 = 1 + (a + 1).$$

By assumption, equation (3) is to hold for a, consequently

$$(1 + a) + 1 = (a + 1) + 1.$$

From these two equations it follows that

$$1 + (a + 1) = (a + 1) + 1, \tag{3a}$$

which was to be proved.

Since the proposition is correct for $a = 1$, it is also correct for $a = 2$, and since it is correct for $a = 2$ it is also correct for $a = 3$, and so on without limit.

THEOREM III. In any sum of two summands the order of the summands can be altered, without altering the number corresponding to the sum. Or written:

$$a + b = b + a. \tag{4}$$

The proposition is correct for $b = 1$ by Theorem II. If it is correct for a specific value of b, it is also correct for $(b + 1)$. For by Theorem I

$$(a + b) + 1 = a + (b + 1),$$

and by our presupposition

$$(a + b) + 1 = (b + a) + 1 = 1 + (b + a) = (1 + b) + a = (b + 1) + a.$$

Of the last three steps, the first and last are made by Theorem II, equation (3) and the intermediate one by Theorem I, equation (2). Consequently

$$a + (b + 1) = (b + 1) + a,$$

as was to be proved.

From the proposition

$$a + 1 = 1 + a$$

there therefore follows

$$a + 2 = 2 + a,$$

from this again

$$a + 3 = 3 + a,$$

and so on without end.

Proof of Axiom V. If a and f are different numbers, we can—as shown in Axiom VII—always specify a positive whole number b such that

$$(a + b) = f.$$

Then:

$$c + f = c + (a + b) = (c + a) + b.$$

According to this, $(c + a)$ is then necessarily different from $(c + f)$, that is: different numbers added to the same number give different sums.

Since, however, by Theorem III

$$c + f = f + c,$$

$$a + c = c + a,$$

this last conclusion can also be written

$$(f + c) = (a + c) + b,$$

that is: the same number added to different numbers gives different sums.

From this there follows the important proposition for the theory of subtraction and equations: that two numbers which yield the same sum when the same number is added to each of them, must be identical.

7. *Permuting the order of arbitrarily many elements*

In what we have described so far, as a method of numbering off for the purpose of addition, two series of numbers, which retained their normal sequence, were pairwise combined with each other, so that $(n + 1)$ was coordinated with 1, $(n + 2)$ with 2, etc. Only the starting-points of the two sequences in the number series were different.

We now wish to consider the more general case of coordinating the elements of two series, of which one is to preserve a certain sequence, and can therefore be portrayed by number symbols, while the other may have variable sequence. As symbols for the latter we wish to use here the letters of the Greek alphabet. These indeed also have a sequence impressed on our memory, as given in the ordinary arrangement of the alphabet. But we want to use this only as one sequence among many other possible ones, which is distinguished by accidental factors and whose habitual recollection facilitates an overall view, and want otherwise to permit ourselves to alter it arbitrarily. On the other hand, we demand that in the alterations to be made in the sequence of these elements, no element shall be omitted and none repeated. We can most easily check this in our memory, if we stipulate that the group should contain as elements all letters preceding a specific letter, e.g. κ, in the received order of the alphabet.

8. *Transposing two consecutive elements of a series*

If two elements, e.g. ε and ζ, are coordinated with two consecutive numbers n and $(n + 1)$, then n can be linked either with ε or with ζ; this gives the two following manners of coordination:

$$\begin{matrix} n & (n+1) \\ \varepsilon & \zeta \end{matrix}$$

or

$$\begin{matrix} n & (n+1) \\ \zeta & \varepsilon \end{matrix}$$

If we substitute for the first of these two orders the other one, and leave unaltered all the other coordinated pairs consisting each of one number and one letter, then no number goes without its coordinated letter and no letter without its coordinated number, and we do not repeat and also do not omit any letter. So if the series which contained the first two above-mentioned pairs was a group without gaps or repetitions before the transposition, it is also thus with the series yielded by the transposition.

By suitably repeating such transpositions of neighbouring elements, I can make *any arbitrarily chosen* element of the group the first in the series without producing repetitions or gaps in the series. For if the chosen element ξ is the nth, I can exchange it with the $(n - 1)$th, then with the $(n - 2)$th, etc., so that its location number becomes lower and lower, until I finally arrive at the lowest position number in the group, namely 1.

In the same way, I can make any element of the series whose position number is higher than m the m-th member of the group without producing gaps or repetitions. In this latter procedure those members of the series whose position numbers are lower than m will retain these numbers unaltered.

From this there ensues the following: by continued exchange of neighbouring members of a group, I can bring about *any possible sequence* of its members without omitting or repeating elements. For I can arbitrarily prescribe which shall become the first member of the series, and achieve this by the given procedure. Then I can prescribe which shall become the second and equally achieve this. The element which has just become first will not thereby be dislodged from its position. Then I can specify the third, etc., up to the last one.

THEOREM IV. Attributes of a series of elements which do not alter when arbitrary neighbouring elements are exchanged in order with each other, are not altered by any possible alteration of the order of the elements.

This leads us in the first place to the *generalization of the commutative law of addition*.

The capital letters may again, as with the generalization of the associative law, denote sums of arbitrarily many numbers. Then, by the generalized associative law, we have:

$$R + a + b + S = R + (a + b) + S.$$

According to the commutative law for two summands

$$a + b = b + a.$$

Therefore, since according to this $(a + b)$ is the same number as $(b + a)$:

$$R + a + b + S = R + (b + a) + S = R + b + a + S.$$

The latter again by the associative law.

Since, according to this, one can exchange with each other two arbitrary consecutive elements without altering the result of the sum, one can by Theorem IV permute all of them with one another, and bring them into any arbitrary order, without altering the sum.

With this one has proved the five basic axioms of addition for what we laid down as the basic concept of addition, and derived them from it. It now remains to show that this concept coincides with the one which issues from ascertaining the cardinal number of numerable objects.

This leads us first of all to the concept of the cardinal number of the elements of a group. If I need the complete number series from 1 to n in order to coordinate a number with each element of the group, then I call n the *cardinal number* of the members of the group. The discussion preceding the formulation of Theorem IV shows that the cardinal number of the members remains *unaltered by changes in their order*, when omissions and repetitions of them are avoided.

This theorem is now applicable to real objects for which α, β, γ, etc. may be regarded as temporarily given names. Only, these objects must—at least for as long as the result of an executed numbering is to be valid—in fact fulfil certain conditions in order to be numerable. They should not disappear, or merge with others, none of them should split into two, no new one should be added, so that each name given in the form of a Greek letter will also continue to correspond to one and only one circumscribed object, which remains recognizable as a single one.* Whether these conditions are obeyed for a specific class of objects can naturally only be determined by experience.

From the method of numbering described above it follows immediately, by the concept of addition already laid down, that *the total number of the members of two groups which have no member in common equals the sum of the cardinal numbers of the members of the two single groups*. In order to find the total number, one would be able to number firstly through the one group. If it has p members, the number $(p + 1)$ would fall upon the first member of the other group, $(p + 2)$ upon the second one, and so on. Thus the total number of members in the two groups will be found by exactly the same numbering procedure as will the sum—as defined above—of the two numbers which give each the cardinal number of elements in one group.

The concept of addition described above therefore indeed coincides with the concept of it which proceeds from determining the total cardinal number of several groups of numerable objects, but has the advantage of being obtainable without reference to external experience.

One has hereby proved, for the concepts of number and of a sum—taken only from inner intuition—from which we started out, the series of axioms of addition which are necessary for the foundation of arithmetic; and also proved, at the same time, that the outcome of this kind of addition coincides with the kind which can be derived from the numbering of external numerable objects.

Starting from here, the theory of subtraction and multiplication is developed without further difficulties, by defining the *difference* $(a - b)$ as that number which must be added to b in order to obtain a as sum, and multiplication as the addition of a cardinal number of like numbers. I need only refer here to Grassmann, who defines the multiplication of pure numbers by the two equations:

$$1 \cdot a = a,$$

$$(b + 1) \cdot a = b \cdot a + a.$$

With regard to subtraction, it should only be noted that one can also continue the numbers, as symbols of a sequence, without limit in a descending direction, by going backwards from 1 to 0, thence to (-1), (-2), etc. and treating these

* With this paper already in the press, I learn that Prof. L. Kronecker, in a lecture during the past winter semester, developed in a similar way the concepts of number and cardinal number.

new symbols just like the positive whole numbers which we previously used alone. Then the difference between two numbers always has one and indeed only one meaning; it is therefore determined unambiguously.

The *agreement* as well as the *difference* between the laws of addition and multiplication must also be discussed here, because of what follows.

The associative law and the commutative law hold for both operations. As we have seen:

$$(a + b) + c = a + (b + c)$$

$$a + b = b + a.$$

But also likewise:

$$(a \cdot b) \cdot c = a \cdot (b \cdot c)$$

$$a \cdot b = b \cdot a.$$

A difference between the basic properties of the two operations does not emerge until one combines, by each of them, a cardinal number n of like numbers a. Combined by addition, these give the product $n \cdot a$, which is itself again subject to the commutative law:

$$n \cdot a = a \cdot n.$$

By multiplication of n like factors, on the other hand, one gets the power a^n, in which—except in special cases—a and n cannot be exchanged without altering the value of the power.

An analogy equally emerges in one case when we combine each of these operations with the next higher one:

$$c \cdot (a + b) = (c \cdot a) + (c \cdot b)$$

$$c^{a+b} = c^a \cdot c^b.$$

But the analogy is lost under commutation, as the same relation no longer holds for $(a + b)^c$ on the one hand and $a^c \cdot b^c$ on the other.

9. Denominate ⟦benannte⟧ numbers

If we need to number unlike objects in the manner discussed above, it is generally only in order to check their full number. Of much greater importance and wider application is the numbering of like objects. Such objects, which in some or another specific respect are alike and get numbered, we call the *units* of the numeration, the cardinal number of them we term a *denominate number*, and the particular kind of units which it gathers together the *denomination* of the number.

A cardinal number, as we have seen above, is resolvable into parts which are *additively* gathered together in the whole. The sum of two denominated numbers of like denomination is the total number of all their units, thus necessarily again a denominate number of the same denomination. When we have to liken two different groups of different cardinal numbers, we term the one having the higher cardinal number the *greater*, and that having a lower cardinal number the *smaller*. If both have the same cardinal number, we term them *alike*.

Objects, or attributes of objects, which when likened with similar ones allow the distinction into greater, alike, or smaller, we call *magnitudes*. If we can express them by a denominate number, we call the latter the *value* of the magnitude, and the procedure by which we find the denominate number *measurement* of the magnitude. We incidentally, in many investigations carried out in practice, only succeed in reducing the measurement to units which are arbitrarily chosen or are given by the instrument used; the numbers which we find have then only the value of *proportional numbers*, until those units are reduced to universally known ones (*absolute units of physics*). However, these universally known units are not to be defined by their concept, but can only be displayed in particular natural bodies (weights, measuring rods) or particular natural processes (day, pendulum beat). The fact that they are more universally known by tradition among men does not alter the business and the concept of measuring, and appears in contrast with this as only an incidental feature.

In the following we shall have to investigate in which circumstances we can express magnitudes by denominate numbers, i.e. find their value, and what we attain thereby as regards factual knowledge. For that, however, we must first discuss the concepts of alikeness and of magnitude in respect of their objective meaning.

10. Physical alikeness

The special relationship which may exist between the attributes of two objects, and to which we give the name 'alikeness', is characterized by Axiom I as already adduced above: *if two magnitudes are both alike with a third, they are alike amongst themselves*. This implies at the same time that the relationship of alikeness is a mutual one. Because from

$$a = c,$$
$$b = c$$

it follows just as well that $a = b$ as that $b = a$.

Alikeness between the likenable attributes of two objects is an exceptional case, and can therefore be indicated by factual observation only in that the two like objects, when meeting or interacting in suitable conditions, allow the observation of a particular outcome which does not as a rule occur between other pairs of similar objects. We wish to term the procedure by which we place the two objects in suitable conditions for observing the said outcome, and ascertaining whether or not it takes place, the *method of likening* ⟦*Vergleichung*⟧.

If this procedure of likening is to give sure information on alikeness or difference for a specific attribute of the two objects, its outcome must exclusively and solely depend upon the condition that both objects possess the relevant attribute in the specific measure, always presupposing that the likening procedure is properly applied.

It follows, from the axiom adduced above, firstly that *the outcome of this likening must remain unaltered if the two objects are interchanged.* It moreover follows that if the two objects *a* and *b* prove to be alike, and it has been found by previous observation using the same method of likening that *a* is also alike with a third object *c*, then the corresponding likening of *b* and *c* must also show these to be alike.

These are requirements which we have to lay down for the relevant method of likening. *Only those kinds of procedure which fulfil the said requirements are capable of demonstrating alikeness.*

From these presuppositions, it follows that 'like magnitudes can be substituted for each other' in the first place for the outcome on whose observation we rely for ascertaining their alikeness. However, the alikeness of further effects or relationships of the relevant objects may also be connected, by natural laws, with alikeness in the case so far discussed, so that the relevant objects may be interchangeable also in these other respects. We are accustomed to expressing this linguistically as follows: we objectivize, as an attribute of the objects, the capacity they have for bringing about the outcome decisive for the first kind of likening; then we ascribe like magnitude of that attribute to the objects found to be alike, and characterize the further effects in which alikeness is preserved as effects of that attribute, or as empirically dependent upon that attribute alone. The sole meaning of such an assertion is always this: that objects which have proved to be alike for the kind of likening which decides concerning alikeness of this particular attribute are also mutually substitutable, without altering the outcome, in the further cases mentioned.

Magnitudes whose alikeness or non-alikeness is to be decided by the same method of likening are termed by us '*alike in kind*'. In our separating, by abstraction, the attribute whose alikeness or non-alikeness is hereby determined from everything whereby the objects otherwise differ, there remains for corresponding attributes of different objects only distinction by magnitude.*

Let me elucidate the sense of these abstract propositions by some well-known examples.

Weights. When I place two arbitrary bodies on the pans of a true balance, the balance will generally not be in equilibrium, but one pan will sink.

Exceptionally, I shall find certain pairs of bodies *a* and *b* which, when placed on the balance, will not disturb its equilibrium. If I then exchange *a* and *b*, the balance must remain in equilibrium. This is the well-known test of whether the

---

* H. Grassmann's definition of alikeness: 'Those things are alike about which the same can always be asserted or, more generally, which can be mutually substituted in every judgement' would demand that in every single case where alikeness is inferred this most general requirement, which is exposed to misinterpretation, should be applied.

balance is true, i.e. whether equilibrium of this balance indicates alikeness of the weights.

Finally it is confirmed that if the weight of a is alike not merely with the weight of b but also with that of c, then also $b = c$. The equilibrium of weights on a true balance, therefore, is indeed the basis of a method for determining a kind of alikeness.

The bodies whose weight we liken may incidentally consist of the most varied materials, and have the most varied forms and volumes. The weight which we make alike is only an attribute of them which has been separated by abstraction. If we call the bodies themselves weights, and call these weights magnitudes, then we should do this only when we can disregard all their other properties. This has its practical sense whenever we observe or bring about processes in the course of which only this one attribute of the bodies involved is under consideration, i.e. in which bodies which yield an equilibrium on the balance are mutually substitutable. This is for instance the case when we measure the inertia of the bodies concerned. That bodies of like weight are also of like inertia, and are substitutable for each other also in the latter connection, does not however follow from the concept of alikeness, but only from our knowledge of this particular law of nature for this particular connection.

11. Distances between two points[d]

The simplest geometrical structure for which a magnitude is specifiable is the distance between a pair of points. But to give a specific value to the distance for at least the time of the likening which measures it, the points must be fixedly linked as e.g. the tips of a pair of dividers. The well-known method of likening distance for two pairs of points consists in our investigating whether or not they can be brought into congruent coincidence. Experience confirms that this method is suitable for ascertaining alikeness, that the congruence always recurs in any situation and on exchanging the two point pairs in any arbitrary manner, that two point pairs which are congruent with a third are also congruent amongst themselves. In this way we can form the concept of like distances or separations.

Thence one can proceed to the concept of the straight line and its length. Think of two fixed points through which there is to pass a line. A *straight* line is one in which no point can alter its situation without altering at least one of its distances from the fixed points. A *curved* line, on the other hand, can be rotated about two of its points, whereby the situation of the remaining points is indeed altered, but not their distance from the two fixed points. We make the *length* of two bounded straight lines alike if the distance between the end points is alike, thus when the latter can be placed in congruence, whereby the lines too coincide congruently. To this extent the concept of length gives

[d] [In the original the remainder of section ten is erroneously made a separate section; in fact 'distances between two points' is merely one of the 'well-known examples' promised above by Helmholtz.]

something more than does the concept of distance. If we think of two point pairs *a*, *b* and *a*, *c*, of differing distance, which coincide at *a* and which are placed in a straight line, so that a portion of this line is common to both, then either *b* falls upon the line *ac* or *c* upon the line *ab*. This gives an opposition corresponding to that of greater and smaller, whereas the concept of distance gives immediately only that of like and unalike.

Time measurement presupposes the finding of physical processes which, repeating themselves in like manner exactly and under like conditions, if they have begun at the same moment also end simultaneously. For example: days, pendulum beats, the running out of sand and water clocks. Here the justification for assuming unaltered duration of repetition lies only in the circumstance that all the various methods of time measurement, if carefully executed, always yield concurring results. If two such processes *a* and *b* begin and end simultaneously, they take place not only in like time but in the same time. In respect of time there is no possible distinction between the two of them, and therefore also no possible exchange. And if a third process *c*, beginning simultaneously with *a*, ends in the same time, then it also does so with the simultaneously proceeding process *b*.

We liken the *brightness* of two visible areas by bringing them up to each other—so that any demarcation between them disappears apart from difference in brightness—and then checking whether there still remains a recognizable boundary between them.

We liken the *pitch* of tones, if it is a matter of small differences, by using the phenomenon of beats, which must be absent if the pitches are alike. We liken the *intensities of electric currents* with the differential galvanometer, which remains unaffected by them if they are alike. And so on.

Thus for the task of ascertaining alikeness in various respects, the most varied physical means have to be sought out. But all of them must fulfil the requirements laid down above, if they are to prove an alikeness.

The first axiom—'If two magnitudes are both alike with a third, they are alike amongst themselves'—is thus not a law having objective significance; it only determines which physical relations we are allowed to recognize as alikeness.

To quote some instances in which this axiom concerning the alikeness of the third object even underlies something executed mechanically, I may mention the grinding of plane glass surfaces. If two of these are ground down by rotating one of them continuously against the other, both may become spherical—the one concave and the other convex. If three are rotated against one another alternately, they must finally become plane. In the same manner, the edges of exact metal rulers are made straight by grinding them down against one another in threes.

12. On additive physical connection of magnitudes alike in kind*

Likening magnitudes, as discussed by us so far, answers the question of whether they are like or unalike; but should they be different, it does not yet give any measure for the magnitude of their difference. However, if the magnitudes concerned are to be completely specifiable by denominate numbers, the greater of the two numbers must be portrayable as the sum of the smaller and their difference. We have to investigate, therefore, under what conditions we may express a physical connection between magnitudes alike in kind as an addition.

The manner of connection involved here will in general depend upon the kind of magnitudes to be connected. We add e.g. weights simply by putting them in the same pan of a balance. We add periods of time by letting the second one begin at the exact moment when the first one ends; we add lengths by placing them one next to the other in a certain manner, namely in a straight line. And so on.

1. *Alikeness in kind of the sum and the summands.* As this connection is to involve magnitudes *of a specified kind*, its result should not alter if I exchange one or more of the magnitudes with magnitudes alike in magnitude and in kind. For they will then only be replaced with the same number having the same denomination as they already had themselves. But the result of connection, if it is to count as the sum of the connected magnitudes, must also be alike in kind with the parts, since the sum of two denominate numbers is again, as already mentioned above, a number of the same denomination. Therefore, *the issue of whether the result of connection stays alike when parts are exchanged, must be decided by the same method of likening with which we ascertained the alikeness of the parts to be exchanged.* This is the factual significance of the requirement that the sum of magnitudes alike in kind must be alike in kind with its summands.

Thus, e.g., in a sum of weights I can replace the individual pieces with ones of another material but like weight. Then the sum retains like weight, but its remaining physical attributes can alter.

2. *Commutative law.* The result of addition is independent of the order in which the summands are connected. The same must hold for physical connections which are to be regarded as additions.

3. *Associative law.* The conjoining of two magnitudes alike in kind can also take place physically by replacing them both with one undivided magnitude of the same kind and alike ⟦in magnitude⟧ with their sum. In this way, the two are then additively united before all the others. Thus e.g. in weighing, a five-gram piece can be substituted for five one-gram pieces. The result of connection should therefore not alter if I introduce, instead of some of the magnitudes to

* 'Connection' ⟦'Verknüpfung'⟧ is a Grassmannian term, though to be sure his use of it is predominantly subjective, while the use here is objective only.

be connected, others which are alike with the sum of these.

It can incidentally be shown that if the first two requirements are universally fulfilled, the third one is fulfilled too.

Think of the elements ordered succesively in the sequence in which, as a first case, they are to be connected with one another by affixing each one to the result of connecting those preceding it, as we prescribed above for the addition of $(a + b + c + \text{etc.})$. If now, as a second case, any of these magnitudes are required to be connected before the others, then by the commutative law—which is to hold by assumption—we can put them in first and second place, where they are then to be connected before the others, without our altering the result. Thereupon, according to our first condition above, we can also replace the result of this connection with another undivided object which, considered as a magnitude alike in kind, is alike in magnitude.

After that, we can take the next two magnitudes—or sums of magnitudes—to be connected, and bring them in their turn into the first two places, and so on, until all are connected in the prescribed sequence. By none of these operations do we alter the magnitude of the final result of the connections.

A physical method of connecting magnitudes alike in kind can thus be regarded as addition, if the result of the connection—when treated as a magnitude of the same kind—is not altered either by exchanging individual elements with each other, or by exchanging terms of the connection with like magnitudes of like kind.

With our having found the method of connecting the magnitudes concerned, in the way just described, it now also follows which are greater and which smaller. The product of additive connection, the whole, is greater than the parts of which it is composed. Regarding those simplest magnitudes which we have dealt with from earliest youth, such as times, lengths, and weights, we have never doubted about what was greater and what smaller, because we have indeed known additive methods of connecting them all along. It should however be borne in mind here that the method of likening, as described above, generally tells us only whether the magnitudes are like or unalike.

If two magnitudes x are alike, then all functions of them formed by calculation in like manner are also alike. Of these some will increase and others decrease as x increases. Which of these functions permit an additive physical connection is only to be decided by particular experience.

Those cases in which two kinds of additive connection are possible are then instructive. Thus, by exactly the same method of likening, we determine both whether two wires are of like electrical resistance w and whether they are of like conductance λ, because

$$w = \frac{1}{\lambda}.$$

We add resistance by joining the wires in succession so that the conducted electricity must flow through them all one after another. We add the conductance of the wires by placing them side by side and joining them all up together at the one end and at the other. Thus we objectivize here, as physical magnitudes, two different functions of the same variable. If a wire has greater resistance, then it has lower conductance, and conversely. Thus the question of what is greater and what smaller will be answered for the two of them conversely. In the same manner electric condensers (Leyden jars) can be joined side by side or in succession. For like charge, in the former case the capacity is added, in the latter one the voltage (potential). The former grows when the latter decreases.

All the same, we should not be surprised if the axioms of addition are verified in the course of nature, since we recognize as addition only those physical connections which satisfy the axioms of addition.

13. The divisibility of magnitudes and units

So far we have not yet needed to analyse magnitudes into units. The concept of magnitude, as of alikeness in magnitude and of addition of magnitudes, was obtainable without such an analysis. However, we in fact attain the greatest simplicity in portraying magnitudes only on resolving them into units and expressing them as denominate numbers.

Magnitudes which can be added are in general also divisible. If every occurring magnitude can be regarded as additively composed, by the addition procedure valid for magnitudes of this kind, out of a cardinal number of like parts, then by the associative law of addition each of these magnitudes can be replaced, wherever only its value is of account, with the sum of its parts. It is in this way then replaced with a denominate number, and other magnitudes of like kind with other cardinal numbers of the same parts. The description of the individual magnitudes of like kind can then be conveyed, to a listener acquainted with the like parts chosen as units, by simply enunciating the numbers.

If the occurring magnitudes are not expressible without remainder using the chosen units, then one divides the units again in the known manner, and can in this way give a specification of the value of any of the occurring magnitudes to any arbitrary precision. Complete precision is of course only attainable for rational proportions.

Irrational proportions may occur in real objects; yet one cannot ever portray them with complete accuracy in numbers, but only enclose their values between arbitrarily reducible limits. This narrowing down between limits suffices for computation of all functions such that if the changes in the magnitudes upon which they depend become smaller and smaller, then the values of the functions themselves undergo smaller and smaller changes, which in the end can fall below any statable finite amount. This holds in particular for the calculation of differentiable functions of irrational magnitudes. On the other hand, one can of course also form discontinuous functions for whose calculation it does not suffice to know the limits between which the irrational value lies, however narrowly these limits may be drawn. In respect of those functions, the portrayal of irrational magnitudes using our number system always remains insufficient.

But in geometry and physics we have not yet encountered such kinds of discontinuity.

14. Quantitative Determination of Properties (physical constants, coefficients)[e]

Apart from the magnitudes discussed so far, which are directly recognizable as such because they can be conjoined by addition, there still remains a series of other relationships, also expressible using denominate or non-denominate numbers, for which an additive conjunction with ones alike in kind is not yet known. They are found whenever there occurs a relation, obeying natural laws, between additive magnitudes in processes which are influenced by the peculiarities of some particular substance or particular body, or by the particular way in which the process concerned was initiated.

Thus e.g. the law of refraction of light declares that a particular relationship exists between the sines of the angles of incidence and refraction of a light ray of a particular wavelength, when this enters a given transparent substance from a vacuum. But the number expressing this relationship is different for different transparent substances, and thus denotes a property of them—their refractive index. Specific gravity, thermal conductivity, electrical conductivity, thermal capacity and so on are similar magnitudes. To these one may add those quantities (integration constants of dynamics) which remain unaltered during the undisturbed course of the motion, once it has been initiated, of a limited system of bodies.

Physics has gradually succeeded in reducing all of these quantities to units which can be composed—by multiplication, raising to a power, and the inverses of these operations—out of the three fundamental units of measurement for time, length, and mass.

The distinction between these quantities and additive magnitudes is not strictly adhered to in the language of physicists and mathematicians. The former too are often called magnitudes, as one expresses them using denominate numbers, though the term *coefficient* characterizes their physical nature relatively better. But this is not an essential distinction, for occasionally new discoveries can lead to ways of additively conjoining such coefficients, whereby they would move into the class of directly determinable magnitudes. To some extent this distinction indeed corresponds to the one which metaphysicians of an earlier period wished to state using the antithesis of *extensive* and *intensive* magnitudes. P. du Bois-Reymond calls the former *linear* and the latter *nonlinear* magnitudes.

On the other hand, however, it emerges from the given derivation that one must have first formed additive magnitudes before one can determine coefficients. For

[e] [In this section, 'quantity' translates 'Wert' (normally 'value'). In the original this section is misleadingly set as a subsection of the preceding one.]

the equation expressing a natural law cannot quantitatively determine a coefficient, until all other magnitudes occurring in it are already determined as magnitudes. The determination of additive magnitudes must therefore always come first, before one can find the values of non-additive ones.

15. The addition of magnitudes not alike in kind

A major role is played in physics by those objects which, when compared by different methods of likening, manifest simultaneously two, three, or even several magnitudes of different kinds, all of which are added when the objects are combined in the same physical way. Amongst these belong firstly the very large number of magnitudes oriented in space which occur in physics, i.e. magnitudes having a specific value and at the same time a specific orientation, yet ones which one can represent as a combination of several components (two in a plane, three in space) having fixed orientations. The overall view of the situation is made most simple, in general, by choosing as parallel to three rectangular coordinate axes those components which are to be connected, in the way prescribed by the law of the parallelogram of forces, to form a resultant. To this class there belong displacements of a point in space, its velocity and acceleration; corresponding to the latter the force propelling the point; moreover angular velocities and vanishingly small rotations, the current velocities of heavy fluids, electricity and heat, magnetic moments, and so on.

To conjoin additively one sums the components having like orientation; then these sums can be brought together to form a resultant. All physical connections of such magnitudes for which the outcome depends only upon the magnitude and orientation of the final resultant can be regarded as based upon additive connections of this kind. This was done for two dimensions by Gauss in his geometrical interpretation of imaginary magnitudes, for several dimensions by H. Grassmann as the addition of geometrical extents,[f] and by R. Hamilton in the theory of quaternions. The commutative law must be satisfied here too. Thus we can combine, into a resultant rotation, infinitely small rotations of a fixed body about two different axes, and combine angular velocities as well, but not finite rotations—because with these it is no longer a matter of indifference whether one rotates first about axis a and then about axis b, or conversely.

Yet even in mixing coloured light a similar relationship occurs. Any quantity of coloured light can be portrayed, in respect of the corresponding sense-impression, as a composition of three quantities of light of suitably chosen basic colours. Mixing several colours then has the same effect upon the eye as would a composition of three quantities of light, from the three basic colours, obtained for each particular basic colour by adding the corresponding quantities occurring in the composition of the various particular colours. This forms the basis

[f] [‘geometrische Strecken’; the addition of vectors is meant.]

for the possibility of portraying the laws of colour mixing geometrically with centre-of-gravity constructions, as first proposed by Newton.

16. The multiplication of denominate numbers

A denominate number $(a \cdot x)$, where x is to stand for the kind of unit and a for the cardinal number of such units, can be multiplied by a pure number n. This simply falls under the definition, adduced above, of the product as the sum of n like summands a. Since the sum of summands alike in kind is a magnitude alike in kind with them, so also is the product $(n \cdot a)$ a magnitude of the same denomination as a. The commutative law applies to this product, inasmuch as

$$n \cdot (a \cdot x) = a \cdot (n \cdot x)$$

i.e. one can also regard a as a pure number, and form new denominate summands $(n \cdot x)$.

In the same way, the law of multiplication of a sum is immediately given:

$$(m + n) \cdot (a \cdot x) = m \cdot (a \cdot x) + n \cdot (a \cdot x)$$

$$n \cdot (a \cdot x + b \cdot x) = n \cdot (a \cdot x) + n \cdot (b \cdot x).$$

Thus the multiplication of denominate numbers by pure numbers remains wholly within the framework of the definitions and propositions which were derived above for the multiplication of pure numbers by one another.

It is otherwise with the multiplication of two or more denominate numbers. This has a sense only in certain cases, when there can exist, between the units concerned, special physical connections subject to the three laws of multiplication:

$$a \cdot b = b \cdot a$$

$$a \cdot (b \cdot c) = (a \cdot b) \cdot c$$

$$a \cdot (b + c) = (a \cdot b) + (a \cdot c).$$

In geometry, the best-known examples of multiplicative combinations of this kind are the values of the area of a parallelogram and the volume of a parallelepiped, expressed by the product of two or three lengths, namely of a side and one or two altitudes. But physics forms a great number of such products of different units, and has correspondingly also examples of quotients, powers, and roots of them. If we denote a length by l, a time by t, and a mass by m, then some of these with their denominations are:

area	l^2
volume	l^3
velocity	l/t
motive force	$(m \cdot l)/t^2$
work	$(m \cdot l^2)/t^2$
pressure on a surface	$m/(l \cdot t^2)$

tension in a surface	m/t^2
density	m/l^3
quantity of magnetism	$(l/t)\cdot\sqrt{m\cdot l}$
magnetic force	$(l/t)\cdot\sqrt{m/l}$, etc.

Most of these combinations are based upon the determination of coefficients, but many of these magnitudes can indeed also give us additive physical connections, such as velocities, currents, forces, pressures, densities and so on. However, none of these multiplicatively defined units is alike in kind with those from which it is created, and it acquires a sense only through knowledge of particular geometrical or physical laws.

One should mention here the particular variety of multiplication which H. Grassmann laid down for oriented magnitudes in his theory of extension, and which is also made fundamental in the theory of quaternions. This lays down a different commutative law, namely

$$a\cdot b = -b\cdot a$$

and indeed gives such a great simplification in the symbolization, although not in the calculation of the values produced by the interaction of differently oriented magnitudes.

In this kind of calculation, the product of two extents[g] is the area of the parallelogram having the two of them as sides; however, one considers the surface bounded by the parallelogram to be positive on one side and negative on the other. When considering the one side of the surface, I must make a right-hand rotation in going from side a to side b through the angle between them; when considering the other side, I conversely make a left-hand rotation from b to a. This is the basis of the difference in the sequence $(b\cdot a)$ and $(a\cdot b)$.

It suffices to have mentioned here these forms of calculation and characterized their relation to the forms of calculation in pure number theory, since the present study has only been set the task of showing the significance and justification of calculation with pure numbers, and the possibility of applying them to physical magnitudes.

Our portraying some physical relationship as a magnitude, consequently, can only ever be based upon empirical knowledge of certain aspects of its physical behaviour in meeting and interacting with others. Thus I also conceive of the congruence of two spatial magnitudes which occur in bodies or are bounded by bodies, in the sense of my previous studies on the axioms of geometry, as a physical relationship to be ascertained empirically. We must know *firstly* the method of likening the magnitudes concerned, whereby they are characterized in kind, and *secondly* either the methods of additive connection or the natural laws in which they occur as coefficients, if we are to be able to express them using denominate numbers.

---

[g] [i.e. the vector product.]

The great simplification, and comprehensive clarity, of understanding which we achieve by reducing to quantitative relationships the variegated multiplicity of things and changes available to us, has a deep foundation in the nature of how we form concepts. When we form the concept of a class, we bring together in that concept everything that is alike in the objects belonging to the class. And also when we understand a physical relationship as a denominate number, we have expelled from the concept of its units everything of the differences which adhere to these in actuality. They are objects which we now only consider as instances of their class, and whose effectiveness in the respect under investigation also only depends upon their being such instances. In the magnitudes formed from them, there then remains only the most incidental of distinctions—that of cardinal number.

18
Julius Wilhelm Richard Dedekind (1831–1916)

Dedekind is known to philosophers for his work in the foundations of mathematics, and especially for his two essays on the real and the natural numbers—essays which were a major contribution to the arithmetization of analysis, to set theory, and to the axiomatic study of arithmetic. For this foundational work alone, Dedekind's accomplishments rank with those of Boole, Peirce, and Frege. But Dedekind was also a great mainstream mathematician, the founder of algebraic number theory; he introduced almost the entire apparatus of groups, fields, ideals, and modules and their morphisms that has dominated mathematics in the twentieth century.

The following selections will explore the connections between Dedekind's work in the foundations of mathematics and his work in algebra and number theory. It is important to stress that these two aspects of his thought are connected, for in contrast to Boole, De Morgan, and Peirce, who applied existing algebraic techniques to the analysis of logic, Dedekind's discoveries in algebra and his discoveries in foundations grew from the same soil: the germs of his later work can already be detected in his early *Habilitation* lecture. Many of the themes broached in the passages that follow overlap with philosophical topics treated by earlier writers: see Berkeley, Lambert, Boole, and De Morgan on the reference of algebraic symbols; Bolzano on the growth of mathematics and on the foundations of real analysis; Peirce on the logical foundations of the natural numbers; Gauss and Rowan Hamilton on extensions of the number-concept.

Dedekind entered the University of Göttingen in 1850; he studied mathematics and physics, attending Gauss's lectures on the method of least squares and on advanced geodesy. One of his friends was a fellow mathematics student, five years older than he, Bernhard Riemann. In 1852 Dedekind took his doctorate; the dissertation, written under the supervision of Gauss, was on the theory of Eulerian integrals. Both Riemann and Dedekind qualified as university lecturers in 1854. Dedekind's *Habilitation* thesis was 'On the formulae of transformations of systems of rectangular coordinates'. (The thesis is now in the Niedersächsische Staats- und Universitätsbibliothek, Göttingen, Cod. Ms. Dedekind VI, 7.) Riemann's celebrated inaugural lecture on the foundations of geometry (translated above) was delivered on 10 June, 1854; Dedekind's lecture was delivered on 30 June. Gauss (who had been present at both lectures) died on 23 February 1855, and Dedekind had the distinction of being one of the pallbearers at his funeral.

In 1855, P.G. Lejeune-Dirichlet left Berlin to succeed to Gauss's professorship in Göttingen; Dedekind, despite being a member of the faculty, attended Dirichlet's lectures in number theory, differential equations, and the theory of definite integrals. (Dedekind's famous edition of Dirichlet's lectures on the theory of numbers was assembled from the notes he took while attending these lectures.) Dedekind also attended Riemann's 1856 lectures on Abelian and elliptic functions. (Dedekind was later to be an editor of the works of Riemann, as well as to edit a volume of Gauss's works on number theory; Dedekind's own works were to be edited in turn by Emmy Noether.) At the same time, Dedekind was himself lecturing on Galois theory and on the theory of groups; as we shall see, he had already at this time formulated many of the concepts of abstract group theory. From 1858 to 1862 he taught at the Polytechnic in Zurich; it was during this time that he developed his ideas on the foundations of real analysis. In 1862 he was appointed to a professorship at the Polytechnic in his native city of Brunswick; he remained there until his death.

For contemporary assessments of Dedekind's life and work, see *Jourdain 1916*, *Landau 1917*, and the discussions in *Russell 1903*. *Wang 1957* analyses Dedekind's contributions to the foundations of arithmetic; *H.M. Edwards 1980* and *Stein 1988* discuss Dedekind's relationship to Kummer and Kronecker. (Edwards gives a detailed account of the origins of ideal theory.) For the relationship of Dedekind to Eudoxus, see *Stein 1990*; for a comparison of his theory of natural number with that of Frege, see *Boolos 1990*. *Dugac 1976* is a general survey of Dedekind's work in the foundations of mathematics, and reproduces many documents from the Göttingen *Nachlass*.

A. ON THE INTRODUCTION OF NEW FUNCTIONS IN MATHEMATICS (*DEDEKIND 1854*)

The following lecture was Dedekind's *Habilitationsrede*, delivered to qualify himself for admission to the rank of *Privat-Dozent* in Göttingen; it was first published in his *Werke* in 1932.[a] Mathematicians of the early nineteenth century had expanded the number concept by adding imaginary numbers, quaternions, and other unfamiliar objects to the world of mathematics; this topic has already been treated in the selections from Gauss and Rowan Hamilton. But

[a] The editors of Dedekind's *Werke* note that this lecture was 'held in the house of Prof. Hoeck on 30 June 1854, in the presence of Hoeck, Gauss, Weber, and Waitz'.

whereas their introduction of new numbers had been somewhat *ad hoc*, Dedekind is concerned in this lecture to develop a uniform account of the genesis of new classes of numbers in mathematics: to explain how the operation of adding positive numbers gives rise to the operation of subtraction, which in turn engenders the negative integers; how multiplication of integers then leads to the reciprocal operation of division, and division to the creation of the rational numbers. This early paper can be read as the announcement of a mathematical research programme that embraced his work both in foundations and in algebra. His foundational writings of *1872* and *1888* deepen the analysis he gives here: the first, by scrutinizing the generation of real numbers from the rationals; the second, by scrutinizing the set-theoretic origins of the positive integers. Moreover, the generating principle Dedekind discusses in this lecture (new classes of numbers are generated by taking the closure of an old class under some operation) and the point of view that regards *classes* of numbers as fundamental were to lead him to his discovery of the new structures in algebraic number theory—for example, to *fields*, which, for Dedekind, are classes of numbers closed under the operations of addition, subtraction, multiplication, and division.

The translation of *Dedekind 1854* is by William Ewald. References should be to the paragraph numbers, which have been added in this edition.

⟦1⟧ It is first necessary to say a few words about the meaning of this title.

⟦2⟧ This lecture is not about the introduction of a determinate class of new functions into mathematics (although one could perhaps interpret the title in that way); rather, it is about the general manner in which, in the progressive development of this science, new functions, or, as one can equally well say, new *operations*, are added to the chain of previous ones. It is easy to see that, on this interpretation, the chosen topic is concerned with a peculiarity of the systematic construction of mathematics—a peculiarity which will probably, in a more or less similar manner, recur in all the sciences. So allow me to make several general remarks in advance before I come back to mathematics and, in the end, to a quite specific part of mathematics.

⟦3⟧ If one finds the chief goal of each science to be the endeavour to fathom the *truth*—that is, the truth which is either wholly external to us, or which, if it is related to us, is not our arbitrary creation, but a necessity independent of our activity—then one declares the final results, the final goal (to which one can in any case usually only approach) to be invariable, to be unchangeable. In contrast, science itself, which represents the course of human knowledge up to these results, is infinitely manifold, is capable of infinitely different representations. This is because, as the work of man, science is subject to his arbitrariness and to all the imperfections of his mental powers. There would essentially be no more science for a man gifted with an unbounded understanding—a man for whom the final conclusions, which we attain through a long chain of inferences, would be immediately evident truths; and this would be so even if he stood in exactly the same relation to the objects of science as we do. This diversity in the conceptions of the object of a science finds its expression in the different forms, the different systems, in which one seeks to embed those conceptions. We can see this everywhere. For instance, in the descriptive natural sciences we see it in the grouping and classification of materials. Depending on the greater or lesser importance which the investigator of nature attaches to a criterion ⟦Merkmal⟧ as a concept suited to distinguishing and classification, he either elevates it to a chief touchstone of classification, or else he uses it to designate merely inessential differences. Thus in mineralogy we have two systems, one of which rests on the chemical composition of mineral bodies, and the other on their crystallographic, morphological nature. These two systems clash, and nobody has managed until now to bring them into complete harmony with each other. Each of these systems is perfectly justified, for science itself shows that similar bodies group themselves together most naturally in these ways. But it will occur to no mineralogist to advance, say, *differences of colour* as the characteristic features and to prefer a classification resting on them to all others. Of course, no reason can be given against this *a priori*; but *experience* teaches in the course of research that colour is not of such great significance for the true nature of bodies as are the criteria mentioned earlier; or, to put it another way, experience teaches that one cannot operate so certainly, so effectively with colour as with the other criteria. The introduction of such a concept as a *motif* for the arrangement of the system is, as it were, a hypothesis which one puts to the inner nature of the science; only in further development does the science answer; the greater or lesser *effectiveness* of such a concept determines its worth or worthlessness.

⟦4⟧ Something similar occurs in legal science. In his attempt both to systematize legal relationships and to systematize the laws themselves (that is, to represent some of them as logical consequences of others) the systematizer forms certain concepts—for instance, the concept of legal institutions. These concepts appear as definitions in the science, and with their help he is able to express the recognizable general truths that emerge from the infinite manifold of individual ones. But these truths themselves react upon the formation of definitions. So it may well happen that the concepts, whatever the motive for their introduction, were initially conceived either too narrowly or too broadly, and that they will therefore require modification in order to be able to extend their effectiveness, their implications, over a larger territory. The greatest art of the systematizer lies in this turning and manipulation of definitions for the sake of the discovered laws or truths in which they play a role.

⟦5⟧ It is probably superfluous to demonstrate analogous relationships in yet other sciences. It is not only an historical fact, but also a fact that rests on an inner necessity, that the further development of each science always reacts creatively on the system through which one tries to conceive the organism of the science.

⟦6⟧ Mathematics too, which is said to be the most certain of all the sciences, is no exception to this general law—even if it appears here in an entirely different form than in other sciences. In mathematics too, the definitions necessarily appear at the outset in a restricted form, and their generalization emerges only in the course of further development. But—and in this mathematics is distinguished from other sciences—these extensions of definitions no longer allow scope for arbitrariness; on the contrary, they follow with compelling necessity from the earlier restricted definitions, provided one applies the following principle: Laws which emerge from the initial definitions and which are characteristic for the concepts that they designate are to be considered as *of general validity*. Then these laws conversely become the source of the generalized definitions if one asks: How must the general definition be conceived in order that the discovered characteristic laws be always satisfied?—It is now my intention to illustrate this principle of induction with several examples.

⟦7⟧ Elementary arithmetic is based upon the formation of ordinal and cardinal numbers; the successive progress from one member of the sequence of positive integers to the next is the first and simplest operation of arithmetic; all other operations rest on it. If one collects into a single act the multiply-repeated performance of this elementary operation, one arrives at the concept of addition. From this concept that of multiplication is formed in a similar manner, and from multiplication that of exponentiation. But the definitions we thereby obtain for these fundamental operations no longer suffice for the further development of arithmetic, and that is because it assumes that the numbers with which it teaches us to operate are restricted to a very narrow domain. That is, arithmetic requires us, upon the introduction of each of these operations, to create the entire existing domain of numbers anew; or, more precisely, it demands that the indirect, inverse operations of subtraction, division, and the like be unconditionally applicable. And this requirement makes it necessary to create new classes of numbers, since with the original sequence of positive integers the requirement cannot be satisfied. Thus one obtains the negative, rational, irrational, and finally also the so-called imaginary numbers. Now, after the number domain has been extended in this manner it becomes necessary to define the operations anew; for until now their effectiveness was only determined for the sequence of positive integers, but we wish to be able to apply them to the newly created numbers as well. And these extensions of the definitions are not arbitrary, so long as one follows the general principle mentioned above—that is, one declares the laws which the operations obeyed in their restricted conception ⟦Auffassung⟧ to be valid in general, and then one turns around and derives the meaning of the operations for the new number domain.

⟦8⟧ We already have a definite example in multiplication. This operation arose from the requirement that a multiply-repeated performance of an operation of the next lower rank ⟦Ordnung⟧—namely the addition of a fixed positive or negative addend (the so-called multiplicand)—be collected together into a single act. The multiplier—that is, the number which states how often the addition of the multiplicand is to be thought of as repeated—is therefore at the outset necessarily a positive integer; a negative multiplier would, under this first definition of multiplication, make absolutely no sense. A special definition is therefore needed in order to admit negative multipliers as well, and thereby to liberate the operation from the initial constraint; but such a definition involves *a priori* complete arbitrariness, and it would only later be decided whether then this arbitrarily chosen definition would bring any real use to arithmetic; and even if the definition succeeded, one could only call it a lucky guess, a happy coincidence—the sort of thing a scientific method ought to avoid. So let us instead apply our general principle. We must investigate which laws govern the product if the multiplier undergoes in succession the same general alterations which led to the creation of the sequence of negative integers out of the sequence of positive integers. For this it suffices if we determine the alteration which the product undergoes if one makes the simplest numerical operation with the multiplier, namely, allowing it to go over into the next-following number. By successive repetition of this operation we obtain the familiar addition theorem for the multiplier: in order to multiply a number by a sum, one multiplies it by each summand and then adds these partial products together. From this theorem a subtraction theorem immediately follows for the case where the minuend is greater than the subtrahend. If one now declares this law to be valid in general (that is, to hold also when the difference which the multiplier represents is negative) then one obtains the definition of multiplication with negative multipliers; and it is then of course no accident that the general law which multiplication obeys is exactly the same for both cases.

⟦9⟧ I now wish to show how the definition of exponentiation completes itself in a similar manner. The successive repetition of the same multiplication—that is, the formation of a product out of a determinate number of identical factors (which are now rational numbers)—yields, when conceived as a single operation, the concept of exponentiation; but again it appears that, if the required operation is to make sense, the determining number (which shows how many such factors are to be taken, and which one calls the exponent of the power) can only be a positive integer. Once again, powers with exponents other than positive integers need a new definition; but instead of choosing the definition arbitrarily one must rather investigate how one has to frame it so that the laws following out of the original definition will turn out to be generally valid. One must therefore apply the same method as before—that is, one must ask about the alterations which the power undergoes when one subjects the exponent to the operations of addition, subtraction, multiplication, and division, while leaving the altered exponent a positive integer. Once the laws prevailing here are known, they yield in turn the generalized definition if one requires that these laws set the standard for the character of exponentiation in general. One achieves all this through the single theorem, which is clear immediately from the definition, namely, that multiplying by the same factor increases the exponent of the resulting power by one. In this way we have found the significance for the power of the simplest operations undertaken with the exponent; these operations are at the same time the source of all later operations. By the successive repetition of this alteration we obtain an addition theorem for the exponents, which, if it is assumed to be generally valid, eventually gives the definitions of powers with any arbitrarily chosen rational exponent. From the theorem that a power whose exponent is the sum of two positive integers equals the product of two powers of the same base whose exponents are these two integers, a subtraction theorem can be derived in which a power with an exponent that is expressed as a difference is reduced to the quotient of the two powers whose exponents are respectively the minuend and subtrahend of that difference, provided that the former is greater than the latter. If one holds this theorem to be of general validity, then one obtains at once the definition of powers with exponent zero and with negative exponents. In this way the first extension is found. In order now to obtain powers with fractional exponents one must first, as is easily done, derive from the addition theorem mentioned above the law that a multiplication of exponents corresponds to an equally high exponentiation of the power. From this it obviously follows that the division of an exponent requires us to perform the unique inverse of the original operation of exponentiation, namely, to split a given number into an (also given) number of equal factors. This way of proceeding repeatedly leads us to new number domains, because the previous domain no longer satisfies the demand for the general applicability of the arithmetical operations; one is thereby compelled to create the irrational numbers (with which the concept of limit appears) and finally also the imaginary numbers. These steps forward are so immense that it is difficult to decide which of the many different paths which open before us we should first pursue. But obviously, if one wants to apply to these new classes of numbers the operations of arithmetic as they have hitherto been developed, then repeated extensions of the earlier definitions are requisite, and here, at least with the appearance of imaginary numbers, the chief difficulties of systematic arithmetic begin. However, we may well hope that we will attain to a truly solid edifice of arithmetic if we here too persistently apply the fundamental principle: not to permit ourselves any arbitrariness, but always to allow ourselves to be led on by the discovered laws. As is well known, an unobjectionable theory of the imaginaries (not to mention the numbers newly invented by Hamilton) does not exist—or at any rate it has not yet been published.

⟦10⟧ One example closely related to our last example from the theory of exponentiation yields in addition, in a precisely analogous way, the geometrical theory of trigonometric functions. I mean here the development of the concepts of sine and cosine. Let them be defined as the ratios of the sides of a right-angled triangle to the hypotenuse, and let their arguments be the angles of the triangle; or, what comes to the same thing, define them as lines in a circle whose radius is unity. These definitions too are always restricted at the beginning, namely, to acute angles; they have sometimes been conceived more generally, but the style and manner strikes me as thoroughly arbitrary. In several textbooks the following procedure appears for generalizing these functions. A circle with radius = 1 is described and a radius is drawn from a fixed point of the periphery. Now imagine a second, movable radius. As it goes out from the fixed radius it describes in one direction positive angles, and in the other direction negative; we shall assume that the justification has been given for this introduction of negative angles. Then the perpendicular dropped from the end-point of the movable radius to the fixed radius is the *sine*, and the piece of the fixed radius lying between the base of this perpendicular and the centre of the circle is the *cosine* of the specified angle. It is then *inferred* from the different positions which these two lines assume that sine and cosine become *negative* for certain intervals. This manner and style of *inferring* must of course be rejected so long as it has not been generally proved that opposite directions always correspond to opposite signs. (This is in any case not possible in such generality, because it is not true.) Moreover, this entire derivation rests on the fact that that perpendicular is dropped from the end-point of the *movable* radius on to the *fixed* radius, and it immediately loses the last shreds of power as soon as one instead drops the perpendicular from the end-point of the fixed leg on to the movable one. No objection can be made to this, for the one procedure is as natural as the other.

⟦11⟧ Next, a fundamental theorem is proved by a geometrical construction, in which the sine and cosine of a sum of two angles are reduced to the sine and cosine of these angles themselves; but this construction is at most carried out for the case in which all three angles are acute, and one either considers it obvious and not worth the trouble to prove that this theorem is valid for all cases, no matter what value one may give to the angle, or, if this gap is felt, such a proof turns out to be extraordinarily complicated. To be sure, it always remains interesting to see that, even with the definitions that were formerly assumed the theorem coincidentally fits *all* cases. I say *coincidentally*, because one cannot call it anything else; but one can with little trouble discover a consistent and natural path to the same results. One really needs only the definitions of the sine and cosine of acute angles; and if the mentioned addition theorem is proved for these, then it yields, in the simplest way, when it is elevated to a general law, the extended definitions with compelling necessity. For if one allows one of the two angles to become equal to a right angle, while the other remains arbitrarily acute, then one obtains the definitions of the goniometric functions of obtuse angles; if one then in the same theorem gives one angle the value of two right angles and allows the other to run through the domain of concave angles, then one obtains the meaning of the sine and cosine of angles up to the size of four right angles, and if one continues so, one attains easily to the general definition for positive angles, and also in a similar manner, *via* subtraction, to that for negative angles.

⟦12⟧ These examples will suffice to show the peculiarity of the way in which, in mathematics, concepts develop from those which relate only to a restricted domain into more general ones. But in all these cases the original definition remains untouched for the restricted domain. However, in some parts of mathematics it happens that these original definitions must be utterly abandoned in order to make room for others. Such a case appears in higher mathematics. The operation of differentiation shows in a determinate manner how to form the so-called derived functions out of given functions; and integration is initially defined as the inverse of differentiation, as the operation through which the transition from the derived functions to the original functions is brought about. But one cannot here with the same justification from the outset demand the general applicability of integration, as we did earlier with the inverse of the operations of elementary arithmetic. For here it is not directly a question of arithmetical results, or of a proper calculation; rather, a *form* is to be found which, when subjected to the mechanical operation of differentiation, returns the given form. With this purely formal conception of differentiation and integration the general practicability of integration would be deeply problematic. But in the proper exploration of the relationship between differential and integral a connection appears that is fully independent of their formal expression. This connection is reciprocal in the following way: it always shows the transfer from differential to integral and from integral to differential to be possible. The concept of a limit that is necessary here ought not to provoke offence or give rise to the opinion that these operations ought to be rejected as impracticable; for this concept, as we have already seen, necessarily forces its way into even elementary arithmetic. But then, when one gives it this meaning (namely, to be the boundary-value of a sum of an ever-increasing number of ever-smaller parts) as a general definition, the operation of integration appears in complete self-sufficiency and independence of the differential calculus, without however losing the earlier connection with it. The history of modern mathematics teaches us the significance and fruitfulness of this conception of integral calculus as a self-sufficient operation. I need only remind you of the theory of Eulerian integrals and elliptic functions, both of which, in fact, found their first seeds in the integral calculus. But it is interesting to see that, as in the examples from elementary arithmetic which we mentioned earlier, so with these newly introduced functions a similar course of development has made itself visible ⟦bemerkbar⟧. The first, the so-called Γ-functions, were defined as definite integrals, and from this definition one developed the laws which they obey; but this definition is restricted to the domain of real positive numbers; or, at any rate, the effectiveness of this function ceases for every negative value of the argument, since the function then becomes infinitely great. But since one found that this function Γ is identical for all positive values of its argument with a function Π, which is defined as an infinite product and which also remains a variable function for all other values (even though at certain places it becomes infinite) one naturally abandons the earlier definition and transforms it into a theorem, and instead defines the function Π as the main function.

⟦13⟧ Something similar has occurred with the elliptic functions. While one earlier regarded the integral calculus as the true source of the theory of these functions—as in fact it was up until the most recent time—we have succeeded in giving them a more self-sufficient status by proceeding either from infinite series or from infinite products. It is however not possible here to enter further into the successive transformations of this theory, which continues to develop; even to acquire these new ideas demands significant study; I feel all the less up to the task of subjecting them to a critique from a secure standpoint, and I therefore break my lecture off here. I wish only to add that through a remarkable accident a recently published book by Prof. Apelt came into my hands a few hours ago, which, to judge by a hasty look, attempts to give not only a general theory of induction but also specific applications of it to mathematics.

B. *FROM THE* TENTH SUPPLEMENT TO DIRICHLET'S *LECTURES ON THE THEORY OF NUMBERS* (*DEDEKIND 1871*)

Dedekind's edition of Dirichlet's *Lectures on the theory of numbers* was based on classroom notes written up by Dedekind after he attended Dirichlet's lectures on number theory in Göttingen during the winter of 1856–7; the notes were later read and approved by Dirichlet. The *Lectures*, commonly referred to as 'Dirichlet–Dedekind', went through four editions: those of 1863, 1871, 1879, and 1894. The principal differences between the editions occur in the famous Tenth and Eleventh Supplements, which were entirely the work of Dedekind.

The Tenth Supplement, which first appeared in the second edition of 1871, created, at a single hammer-blow, modern algebraic number theory, and was a major step in the development of modern algebra. Dedekind in this Supplement introduced a series of new concepts, among them the concepts of field (§159), algebraic integer (§160), module (§161), ideal (§163), prime ideal (§163), and order (§165); this last concept of an order led directly to the modern concept of a ring. (The term 'ring' was later coined by Hilbert.[a]) In the Eleventh Supplement to the third edition of Dirichlet–Dedekind he introduced the notion of group characters; and in the fourth edition he introduced commutator subgroups. The concept of a group of arbitrary elements and the concept of a group homomorphism he had treated years earlier in his Göttingen lectures on Galois theory.[b] In introducing these concepts Dedekind generalized the discoveries of earlier number-theorists, and especially the discoveries of Kummer. In particular he showed that every ideal of an algebraic number field can be uniquely factored into prime ideals. But more important than such individual results is the fact that he gave number theory a new direction. Developing a line of

[a] Dedekind in effect uses rings in §170 of the fourth edition of Dirichlet–Dedekind. Kronecker used the terminology 'Integritätsbereich' to denote rings in §5 of his *Grundzüge* of 1881. Hilbert introduced the term 'Zahlring' in his *Zahlbericht* of 1894–5; and Fraenkel gave an abstract definition of a ring in Crelle's Journal for 1914.

[b] Dedekind discusses these lectures in the first footnote to §166 of the Eleventh Supplement to the fourth edition of Dirichlet–Dedekind. Some of Dedekind's group-theoretic studies from this period were posthumously published in *Dedekind 1930–2* (Vol. iii, pp. 439–46), with a comment by Emmy Noether on Dedekind's contributions to the theory of groups and their relationship to later research.

thought whose roots can ultimately be traced back to his Habilitation lecture, he aimed to present number theory as a unified system that is organized around the concepts of algebraic number and of field. For a full discussion of this aspect of his work, see *Haubrich 1992*.

Dedekind's Supplements not only opened the door to modern algebra and algebraic number theory: they were also an important stimulus to the study of set theory. Already in 1871 Dedekind was defining mathematical objects in terms of *sets* (or, in his terminology, 'systems'): a *field*, for example, was a subsystem of the real or complex numbers closed under the arithmetical operations. This conception was one of the influences prompting Cantor to his own study of subsets of the real line.[c] It is also noteworthy that for Dedekind a system could contain infinitely many elements, and that it need not be given by any decidable procedure; both of these features were novel in 1871.[d] Dedekind was to exploit this set-theoretic approach again in his 1872 definition of the irrational numbers, and he was to provide an account of the foundations of set theory in the opening sections of *Was sind und was sollen die Zahlen*?

It should be noticed that Dedekind's approach in the Tenth Supplement carries his ideas in the *Habilitationsvortrag* to a higher level of abstraction. In the inaugural lecture his concern had been with the creation of new number-classes by taking the *closure* of a set under a specified operation; and indeed, as the following passage shows, precisely this technique, applied to subsets of the real and complex numbers, led him to the creation of fields, ideals, and the other structures of algebraic number theory. But now he also defines operations *on the number-classes themselves*—for instance, taking the product of two ideals, or the quotient of one field by another. This technique of defining operations on sets would later be of service to him in his construction of the irrational numbers and in his analysis of the integers; and in the long footnote in the passage from *Dedekind 1877* translated below, Dedekind explicitly draws the connection between his work on ideal theory and on the irrational numbers.

Dedekind's *Tenth supplement* was one of the great turning points in the history of mathematics. Bourbaki calls it 'magisterial'; Landau says that it 'brought order to chaos, and light to the deepest darkness'; Noether comments that 'its style of thought now permeates the entirety of modern algebra'. It opened new vistas, and introduced a style of thought that was to dominate mathematics in the twentieth century. But Dedekind closed the preface to this revolutionary work with a plea for the study of the history of mathematics: 'Finally, I have taken pains to give references to the sources whenever I could,

---

[c] In the opening paragraphs of his early memoir *1878*, Cantor refers to Dedekind's Tenth Supplement of 1871, and observes that Dedekind's finite fields (i.e. fields with a finite basis) are an example of a denumerable set. For a discussion of the relationship between Cantor and Dedekind, see *Ferreiros 1993*.

[d] Both features drew harbs from Kronecker: see the passage from *Kronecker 1886* translated below. Kronecker's doubts (it should be noted) were expressed long before the discovery of the set-theoretic paradoxes.

to prompt the reader to study the original works, and to awaken in him a picture of the progress of science—whose truths, as deep as they are sublime, constitute a treasure which is the imperishable fruit of a truly noble struggle between the peoples of Europe.' This last phrase is explained by the published date of the Preface (1 March 1871), the precise day on which France formally acceded to the Treaty of Frankfurt and Prussian troops entered Paris.

The translation of *Dedekind 1871* is by William Ewald; references should be to the section number (§159).

---

§159.

The theory of binary quadratic forms, of their equivalence and composition, is only a special case of the theory of those homogeneous forms of the *n*th degree with *n* variables that can be decomposed into linear factors with algebraic coefficients. These forms were first considered by Lagrange;[1] later Dirichlet[2] repeatedly busied himself with this matter, but he only published those of his wide-ranging investigations that treated the transformation of such forms into themselves (see §§61, 62)—or, what comes to the same thing, that treated the theory of unities for the corresponding algebraic numbers. Finally Kummer[3] opened a new path by creating the ideal numbers, which leads not only to a very convenient terminology, but also to a deeper insight into the true nature of the algebraic numbers. In attempting to introduce the reader to these new ideas, we shall adopt a somewhat more elevated standpoint, and begin by introducing a concept which seems well suited to serve as a foundation for higher algebra and the portions of number theory that are connected to it.

By a *field* ⟦*Körper*⟧ we understand every system ⟦System⟧ of infinitely many real or complex numbers which is in itself so closed and complete that the addition, subtraction, multiplication, and division of any two of these numbers always yields a number of the same system. The simplest field is composed of all rational numbers; the largest, of all numbers. We call a field *A* a *divisor* of the field *M*, and the latter a *multiple* of the former, if all the numbers contained in *A* are also in *M*. One easily discovers that the field of rational numbers is a divisor of every other field. The totality ⟦Inbegriff⟧ of all numbers which are simultaneously contained in two fields *A*, *B* is also a field *D*, which can be called the *greatest* common divisor of the two fields *A*, *B*—for obviously every common divisor of *A* and *B* is necessarily a divisor of *D*. In the same

---

[1] *Sur la solution des problèmes indéterminés du second degré*. §VI. Mém de l'Ac. de Berlin. Vol. XXIII, 1769. (Œuvres de L⟦agrange⟧ Vol. II, 1868, p. 375.)—*Additions aux Eléments d'Algèbre* par L. Euler. §IX.

[2] See the note to §141.

[3] See the note to §16.

way there always exists a field M which shall be called the *smallest* common multiple of A and B because it is a divisor of every other common multiple of the two fields. Moreover, if to every number a of the field A there corresponds a number $b = \phi(a)$ in such a way that $\phi(a + a') = \phi(a) + \phi(a')$ and such that $\phi(aa') = \phi(a)\phi(a')$, then the numbers b (if they do not all vanish) also form a field $B = \phi(A)$, which is *conjugate* with A and which comes from A by the *substitution* ϕ; and then too, in reverse, $A = \psi B$ is conjugate with B. Two fields conjugate with a third are also conjugate with each other; and every field is conjugate with itself. Corresponding numbers in two conjugate fields A and B, such as $a = \phi(b)$, shall be called *conjugate* numbers.

The simplest fields are those that possess only a *finite* number of divisors. If one calls m determinate numbers $a_1, a_2, \ldots, a_m$ *dependent* or *independent of each other* according as the equation $x_1 a_1 + x_2 a_2 + \ldots + x_m a_m = 0$ is soluble or not in rational numbers $x_1, x_2, \ldots, x_m$ that do not all vanish, then one finds by very easy considerations that we shall not explore here that from a field Ω of the given sort[4] only a *finite* number n of independent numbers $\omega_1, \omega_2, \ldots, \omega_n$ can be selected, and therefore that every number ω of the field can always be uniquely represented in the form:

$$(1) \quad \omega = h_1\omega_1 + h_2\omega_2 + \ldots + h_n\omega_n = \Sigma h_i\omega_i$$

where $h_1, h_2, \ldots, h_n$ designate *rational* numbers. We shall call the number n the *degree*; the complex of n independent numbers ω_i a *basis for the field* Ω, and the n numbers h_i the *coordinates of the number* ω with respect to this basis; clearly every n numbers of the form (1) are also such a basis if the determinant formed from the corresponding n^2 coordinates is different from zero; to such a *transformation* of the basis by a linear substitution there corresponds a transformation of the coordinates by the so-called *transposed* substitution.

C. CONTINUITY AND IRRATIONAL NUMBERS (*DEDEKIND 1872*)

This article, whose central idea was worked out by Dedekind while he was teaching in Zurich in 1858, presents a rigorous arithmetical foundation for the theory of real numbers. Mathematicians other than Dedekind attempted at roughly the same time to develop a theory of the real numbers, employing a variety of approaches; these efforts filled a lacuna which had been left by Bolzano's great paper on the intermediate value theorem (*Bolzano 1817a*)—a

[4] If one replaces the rational numbers everywhere by numbers of a field R, then the following remarks hold as well for a field Ω which possesses only a finite number of such divisors that are at the same time multiples of R.

paper which was not yet generally known to the mathematicians of the time.

Sir William Rowan Hamilton, in his 'Theory of conjugate functions' (*Hamilton 1837*), had already gone a considerable distance towards Dedekind's conception: inspired by *Book V* of Euclid's *Elements*, he defined the irrational numbers as partitions of the rationals into two classes. But in contrast to Dedekind he did not then go on to investigate the properties of the partitions or to prove the basic theorems about the real numbers. (A similar criticism holds for the theory given by Joseph Bertrand in his *1849*.)

Weierstrass, too, had developed an arithmetical theory of the real numbers, which he presented, beginning in 1859, in his lectures in Berlin; but his results were not printed until years later (*Kossak 1872*). The theories of Eduard Heine (*1872*) and of Weierstrass's student Georg Cantor (*1872*) came to Dedekind's attention shortly before the publication of his own article. The merits and demerits of these various definitions of the irrational numbers are discussed at length below by Cantor in his *1883d*, §9. (Incidentally, the Cantor–Dedekind correspondence, translated below, began in earnest shortly after the appearance of these works.)

Dedekind remarks that his definition of an irrational number owes much to Book V of Euclid's *Elements*. The genesis of his definition and its relationship to the theories of Euclid and Eudoxus are discussed by Dedekind in his correspondence with Rudolf Lipschitz and Heinrich Weber. See in particular his letters to Lipschitz of 10 June and 27 July 1876, and his letter to Weber of 8 November 1878. (These letters are published in the third volume of Dedekind's *Werke* (pp. 464–90). The letters of Lipschitz and Weber to Dedekind are reproduced in the appendices to *Dugac 1976* (pp. 215–20, 272–6). Dugac also reproduces Dedekind's unpublished first draft for his *1872 (Dugac 1976*, pp. 203–9). The correspondence with Lipschitz can also be found in *Lipschitz 1986*. For a discussion, see *Stein 1990*.)

Dedekind's definition of the irrational numbers was adopted by many mathematicians—see, for example, *Dini 1878*, *Pasch 1882a*, *Jordan 1893*, and *Baire et alii 1905*. Despite Dedekind's assertion in the introductory paragraphs of *Continuity and irrational numbers* that he originally did not publish his theory because he did not regard it as being very fruitful, it laid the foundations for much of modern-day real analysis and point-set topology.

The translation is by W.W. Beman, and was originally published in 1901; it has been extensively revised for this edition by William Ewald. References to *Dedekind 1872* should be to the section numbers, which appeared in the original edition.

The considerations which form the subject of this pamphlet date from the autumn of 1858. I was then professor in the Polytechnic School in Zurich, and

I found myself for the first time obliged to lecture upon the elements of the differential calculus; I felt more keenly than ever before the lack of a truly scientific foundation for arithmetic. In discussing the concept of the approach of a variable magnitude to a fixed limiting value—in particular, in proving the theorem that every magnitude which grows continually, but not beyond all limits, must certainly approach a limiting value—I took refuge in geometrical evidence. Even now I regard such invocation of geometric intuition ⟦Anschauung⟧ in a first presentation of the differential calculus as exceedingly useful from a pedagogic standpoint, and indeed it is indispensable, if one does not wish to lose too much time. But no one will deny that this form of introduction into the differential calculus can make no claim to being scientific. For myself this feeling of dissatisfaction was so overpowering that I resolved to meditate on the question until I should find a purely arithmetical and perfectly rigorous foundation ⟦Begründung⟧ for the principles of infinitesimal analysis. The statement is frequently made that the differential calculus deals with continuous quantities, yet an explanation of this continuity is nowhere given; even the most rigorous expositions of the differential calculus do not base their proofs upon continuity but they either appeal more or less consciously to geometric representations or to representations suggested by geometry, or they depend upon theorems which are never established in a purely arithmetical manner. Among these, for example, belongs the above-mentioned theorem, and a more careful investigation convinced me that this theorem, or any one equivalent to it, can be regarded as a more or less sufficient foundation for infinitesimal analysis. It only remained to discover its true origin in the elements of arithemtic, and thereby to secure a real definition of the essence of continuity. I succeeded on November 24, 1858, and a few days afterward I communicated the results of my meditations to my dear friend Durège, which led to a long and lively discussion. Later I explained these views of a scientific basis of arithmetic to a few of my pupils, and here in Brunswick I read a paper upon the subject before the scientific club of professors, but I could not make up my mind to publish it, because, in the first place, the presentation is not altogether easy, and, second, the theory itself is not very fruitful. Nevertheless I had already half determined to select this theme as subject for this occasion, when a few days ago, on March 14, by the kindness of the author, the paper *Die Elemente der Funktionenlehre* by E. Heine (*Crelle's Journal*, Vol. 74) came into my hands and confirmed me in my decision. I fully agree with the substance of this memoir, and I could hardly do otherwise. But I will frankly say that my own presentation seems to me to be simpler in form and to bring out the vital point more clearly. While writing this preface (20 March 1872), I received the interesting paper *Über die Ausdehnung eines Stazes aus der Theorie der trigonometrischen Reihen*, by G. Cantor (*Math. Annalen*, Vol. 5), for which I owe the ingenious author my hearty thanks. After a hasty reading, it seems to me that the axiom given in Section II of that paper (except for the form of presentation) agrees with what I designate in Section III as the essence of continuity. But on my way of conceiving the real numbers they are complete in themselves; so I am unable to see the use of distinguishing real numbers of a yet higher type, even if this is only done conceptually.

§1.

Properties of rational numbers

The development of the arithmetic of rational numbers is here presupposed, but still I think it worth while to call attention to certain important matters without discussion, so as to show at the outset the standpoint assumed in what follows. I regard the whole of arithmetic as a necessary, or at least natural, consequence of the simplest arithmetical act, that of counting, and counting itself is nothing other than the successive creation of the infinite series of positive integers in which each individual is defined by the one immediately preceding; the simplest act is to pass from an already-created individual to its successor that is to be created. The chain of these numbers already forms in itself an exceedingly useful instrument for the human mind; it presents an inexhaustible wealth of remarkable laws which one obtains by introducing the four fundamental operations of arithmetic. Addition is the result of bringing together into a single act an arbitrary number of repetitions of the simplest act; multiplication arises from it in a similar way. While these two operations can always be carried out, the inverse operations, subtraction and division, are admissible only with restrictions. Whatever the immediate occasion may have been, whatever comparisons or analogies with experience, or intuition, may have led thereto, it is certainly true that just this limitation in performing the indirect operations has in each case been the real motive for a new creative act; thus negative and fractional numbers have been created by the human mind, and in the system of all rational numbers an instrument of infinitely greater perfection has been gained. This system, which I shall denote by R, possess first of all a completeness and self-containedness which I have designated in another place* as characteristic of a *field of numbers* ‖*Zahlkörper*‖ and which consists in this, that the four fundamental operations are always performable with any two individuals in R, i.e., the result is always an individual of R, the single case of division by the number zero being excepted.

For our immediate purpose, however, another property of the system R is still more important; it may be expressed by saying that the system R forms a well-ordered ‖wohlgeordnetes‖ domain of one dimension extending to infinity on two opposite sides. What is meant by this is sufficiently indicated by my use of expressions borrowed from geometric ideas; but just for this reason it will be necessary to bring out clearly the corresponding purely arithmetical properties so as to avoid even the appearance that arithmetic is in need of such foreign ideas.

---

* *Vorlesungen über Zahlentheorie*, by P.G. Lejeune Dirichlet. 2nd edn. §159.

To express that the symbols a and b represent one and the same rational number we put $a = b$ as well as $b = a$. Two rational numbers a, b are different just in case the difference $a - b$ has either a positive or negative value. In the former case a is said to be *greater* than b, and b *less* than a; this is also indicated by the symbols $a > b$, $b < a$.[†] Because $b - a$ has a positive value in the latter case, it follows that $b > a$, and $a < b$. The following laws hold for these two ways in which two numbers may differ:

I. If $a > b$, and $b > c$, then $a > c$. Whenever a, c are two different (or unequal) numbers, and b is greater than the one and less than the other, we shall not be inhibited by the echo of geometrical ideas, but shall briefly say: b lies between the two numbers a, c.

II. If a, c are two different numbers, there are infinitely many different numbers b lying between a, c.

III. If a is any definite number, then all numbers of the system R fall into two classes, A_1 and A_2, each of which contains infinitely many individuals; the first class A_1 comprises all numbers a_1 that are $<a$, the second class A_2 comprises all numbers a_2 that are $>a$; the number a itself may be assigned at pleasure to the first or second class, and it is then respectively the greatest number of the first class or the least of the second. In either case the decomposition of the system R into the two classes A_1, A_2 is such that every number of the first class A_1 is less than every number of the second class A_2.

§2.

Comparison of the rational numbers with the points of a straight line

These properties of rational numbers recall the corresponding relations of position between the points of a straight line L. If the two opposite directions existing upon it are distinguished by 'right' and 'left', and p, q are two different points, then either p lies to the right of q, and q to the left of p, or conversely q lies to the right of p and p to the left of q. A third case is impossible, provided that p, q are actually different points. The following laws hold for this difference in position:

I. If p lies to the right of q, and q to the right of r, then p lies to the right of r; we say that q lies between the points p and r.

II. If p, r are two different points, then there always exist infinitely many points q that lie between p and r.

III. If p is a definite point in L, then all points in L fall into two classes, P_1, P_2, each of which contains infinitely many individuals; the first class P_1 contains all the points p_1 that lie to the left of p, and the second class P_2 contains all the points p_2 that lie to the right of p; the point p itself may be

[†] Hence in what follows I always mean the so-called 'algebraic' greater and less, unless the word 'absolute' is added.

assigned at pleasure to the first or second class. In either case the division of the straight line L into two classes or portions P_1, P_2 is such that every point of the first class P_1 lies to the left of every point of the second class P_2.

As is well known, this analogy between rational numbers and the points of a straight line becomes a real correspondence when we select upon the straight line a definite origin or zero-point o and a definite unit of length for the measurement of segments. With the aid of the latter it is possible, for each rational number a, to construct a corresponding length; and if we lay this length upon the straight line to the right or left of o according as a is positive or negative, we obtain a definite end-point p, which may be regarded as the point corresponding to the number a; the point o corresponds to the rational number zero. Thus, to every rational number a, i.e., to every individual in R, there corresponds one and only one point p, i.e., an individual in L. If the two points p, q correspond to the two numbers a, b respectively, and if $a > b$, then p lies to the right of q. The laws I, II, III of the previous Section correspond exactly to the laws I, II, III of the present Section.

§3.

CONTINUITY OF THE STRAIGHT LINE

It is of the greatest importance that in the straight line L there are infinitely many points to which no rational number corresponds. If the point p corresponds to the rational number a, then, as is well known, the length op is commensurable with the invariable unit of measure used in the construction, i.e., there exists a third length, a so-called common measure, of which these two lengths are integral multiples. But the ancient Greeks already knew and had demonstrated that there are lengths incommensurable with a given unit of length, e.g., the diagonal of the square whose side is the unit of length. If we lay off such a length from the point o upon the line we obtain an end-point which corresponds to no rational number. Since it can furthermore be easily shown that there are infinitely many lengths which are incommensurable with the unit of length, we may affirm: The straight line L is infinitely richer in point-individuals than the domain R of rational numbers in number-individuals.

If now, as is our desire, we try to follow up arithmetically all phenomena in the straight line, the rational numbers are insufficient and it becomes unavoidable and necessary that the instrument R constructed by the creation of the rational numbers be essentially refined ‖verfeinern‖ by the creation of new numbers such that the domain of numbers shall have the same completeness, or as we may say at once, the same *continuity*, as the straight line.

The previous considerations are so familiar and well known that many will regard their repetition as quite superfluous. Still I regarded this recapitulation as necessary to prepare properly for the main question. For the way in which the irrational numbers are usually introduced is based directly upon the conception of extensive magnitudes—which itself is nowhere carefully defined—and number is explained as the result of measuring such a magnitude by another of the same kind.* Instead of this I demand that arithmetic shall develop out of itself.

It may in a general way be granted that such comparisons with non-arithmetical notions have furnished the immediate occasion for the extension of the number-concept (though this was certainly not the case in the introduction of complex numbers); but this is certainly no reason for introducing these foreign notions into arithmetic itself, the science of numbers. Just as negative and fractional rational numbers are formed by a free creation, and just as the laws of operating with these numbers must and can be reduced to the laws of operating with positive integers, so we must strive to give a complete definition of the irrational numbers using the rational numbers alone. The only question that remains is, how to do this.

When we compared the domain R of rational numbers with a straight line, we found in the former a gappiness, incompleteness, discontinuity; but to the straight line we ascribe absence of gaps, completeness, continuity. In what then does this continuity consist? Everything must depend on the answer to this question, and only through it shall we obtain a scientific basis for the investigation of *all* continuous domains. By vague remarks upon the unbroken connection in the smallest parts obviously nothing is gained; the problem is to indicate a precise characteristic of continuity that can serve as the basis for valid deductions. For a long time I pondered over this in vain, but finally I found what I was seeking. This discovery will, perhaps, be differently estimated by different people; but I believe the majority will find its content quite trivial. It consists of the following. In the preceding Section attention was called to the fact that every point p of the straight line produces a separation of the same into two portions ⟦Stücke⟧ such that every point of one portion lies to the left of every point of the other. I find the essence of continuity in the converse, i.e., in the following principle:

'If all points of the straight line fall into two classes such that every point of the first class ⟦Klasse⟧ lies to the left of every point of the second class, then there exists one and only one point which produces this division of all points into two classes, this severing of the straight line into two portions.'

As already said I think I shall not err in assuming that every one will at once grant the truth of this statement; the majority of my readers will be very much disappointed in learning that by this commonplace remark the secret of continuity is to be revealed. To this I may say that I am glad if every one finds the above principle so obvious and so in harmony with his own ideas of a line; for I am unable to adduce any proof of its correctness, nor has anyone the power. The assumption of this property of the line is nothing else than an axiom by which we attribute to the line its continuity, by which we think continuity

* The apparent advantage of the generality of this definition of number disappears as soon as we consider complex numbers. According to my view, on the other hand, the notion of the ratio between two numbers of the same kind can be clearly developed only after the introduction of irrational numbers.

into the line. If space has a real existence at all it is *not* necessary for it to be continuous; many of its properties would remain the same even if it were discontinuous. And if we knew for certain that space were discontinuous there would be nothing to prevent us, in case we so desired, from filling up its gaps in thought and thus making it continuous; this filling up would consist in a creation of new point-individuals and would have to be carried out in accordance with the above principle.

§4.

Creation of irrational numbers

From the last remarks it is clear how the discontinuous domain R of rational numbers must be completed so as to form a continuous domain. In §1 it was pointed out (III) that every rational number a separates the system R into two classes such that every number a_1 of the first class A_1 is less than every number a_2 of the second class A_2; the number a is either the greatest number of the class A_1 or the least number of the class A_2. If now any separation of the system R into two classes A_1, A_2 is given which possesses only *this* characteristic property that every number a_1 in A_1 is less than every number a_2 in A_2, then for brevity we shall call such a separation a *cut* ⟦*Schnitt*⟧ and we shall designate it by (A_1, A_2). We can then say that every rational number a produces one cut or, strictly speaking, two cuts, which, however, we shall not look upon as essentially different; *moreover* this cut possesses the property that either there exists a greatest number among the numbers of the first class, or there exists a least number among the numbers of the second class. And conversely, if a cut possesses this property, then it is produced by this greatest or least rational number.

But it is easy to show that there exist infinitely many cuts not produced by rational numbers. The following example is the most obvious.

Let D be a positive integer but not the square of an integer. Then there exists a positive integer λ such that

$$\lambda^2 < D < (\lambda + 1)^2.$$

If we assign to the second class A_2 every positive rational number a_2 whose square is $>D$, to the first class A_1 all other rational numbers a_1, this separation forms a cut (A_1, A_2), i.e., every number a_1 is less than every number a_2. For if $a_1 = 0$, or is negative, then a_1 is less than any number a_2, because, by definition, this last is positive; if a_1 is positive, then its square is $\leqslant D$, and hence a_1 is less than any positive number a_2 whose square is $>D$.

But this cut is produced by no rational number. To demonstrate this it must first be shown that there exists no rational number whose square $= D$. Although this is known from the first elements of the theory of numbers, the following indirect proof may nevertheless not be out of place. If there exists a rational number whose square $=D$, then there exist two positive integers t, u, that satisfy the equation

$$t^2 - Du^2 = 0,$$

and we may assume that u is the *least* positive integer possessing the property that its square, by multiplication by D, may be converted into the square of an integer t. Since evidently

$$\lambda u < t < (\lambda + 1)u,$$

the number $u' = t - \lambda u$ is a positive integer *less* than u. If further we put

$$t' = Du - \lambda t,$$

t' is likewise a positive integer, and we have

$$t'^2 - Du'^2 = (\lambda^2 - D)(t^2 - Du^2) = 0,$$

which is contrary to the assumption respecting u.

Hence the square of every rational number x is either $<D$ or $>D$. From this it easily follows that there is neither in the class A_1 a greatest, nor in the class A_2 a least number. For if we put

$$y = \frac{x(x^2 + 3D)}{3x^2 + D},$$

we have

$$y - x = \frac{2x(D - x^2)}{3x^2 + D}$$

and

$$y^2 - D = \frac{(x^2 - D)^3}{(3x^2 + D)^2}.$$

If in this we assume x to be a positive number from the class A_1, then $x^2 < D$, and hence $y > x$ and $y^2 < D$. Therefore y likewise belongs to the class A_1. But if we assume x to be a number from the class A_2, then $x^2 > D$, and hence $y < x$, $y > 0$, and $y^2 > D$. Therefore y likewise belongs to the class A_2. This cut is therefore produced by no rational number.

The incompleteness or discontinuity of the domain R of rational numbers consist in this property that not all cuts are produced by rational numbers. Thus, whenever we have a cut (A_1, A_2) produced by no rational number, we *create* a new number, an *irrational* number α, which we regard as completely defined by this cut (A_1, A_2); we shall say that the number α corresponds to this cut, or that it produces this cut. From now on, therefore, to every definite cut there corresponds a definite rational or irrational number, and we regard two numbers as *different* or *unequal* if and only if they correspond to essentially different cuts.

In order to obtain a basis for the orderly arrangement of all *real*, i.e., of all rational and irrational numbers we must investigate the relation between any two cuts (A_1, A_2) and (B_1, B_2) produced by any two numbers α and β. Obviously a cut (A_1, A_2) is given completely when one of the two classes, e.g., the first A_1 is known, because the second A_2 consists of all rational numbers not contained in A_1, and the characteristic property of such a first class lies in this, that if the number a_1 is contained in it, it also contains all numbers less than a_1. If now we compare two such first classes A_1, B_1 with each other, it may happen

1. That they are perfectly identical, i.e., that every number contained in A_1 is also contained in B_1, and that every number contained in B_1 is also contained in A_1. In this case A_2 is necessarily identical with B_2, and the two cuts are perfectly identical, which we denote in symbols by $\alpha = \beta$ or $\beta = \alpha$.

But if the two classes A_1, B_1 are not identical, then there exists in the one, e.g., in A_1, a number $a_1' = b_2'$ not contained in the other B_1 and consequently found in B_2; hence all numbers b_1 contained in B_1 are certainly less than this number $a_1' = b_2'$ and therefore all numbers b_1 are contained in A_1.

2. If now this number a_1' is the only one in A_1 that is not contained in B_1, then every other number a_1 contained in A_1 is also contained in B_1 and is consequently $<a_1'$, i.e., a_1' is the greatest among all the numbers a_1; hence the cut (A_1, A_2) is produced by the rational number $\alpha = a_1' = b_2'$. As for the other cut (B_1, B_2) we know already that all numbers b_1 in B_1 are also contained in A_1 and are less than the number $a_1' = b_2'$ which is contained in B_2; every other number b_2 contained in B_2 must, however, be greater than b_2', for otherwise it would be less than a_1', and so contained in A_1 and hence in B_1; thus b_2' is the least among all numbers contained in B_2, and consequently the cut (B_1, B_2) is produced by the same rational number $\beta = b_2' = a_1' = \alpha$. The two cuts are then only unessentially different.

3. If, however, there exist in A_1 at least two different numbers $a_1' = b_2'$ and $a_1'' = b_2''$, which are not contained in B_1, then there exist infinitely many of them; for all the infinitely many numbers lying between a_1' and a_1'' are obviously contained in A_1 (§1, II) but not in B_1. In this case we say that the numbers α and β corresponding to these two essentially different cuts (A_1, A_2) and (B_1, B_2) are *different*, and further that α is a *greater* than β, that β is *less* than α, which we express in symbols by $\alpha > \beta$ and by $\beta < \alpha$. I emphasize that this definition coincides completely with the one given earlier, when α, β are rational.

The remaining possible cases are these:

4. If there exists in B_1 one and only one number $b_1' = a_2'$ that is not contained in A_1 then the two cuts (A_1, A_2) and (B_1, B_2) are only unessentially different and they are produced by one and the same rational number $\alpha = a_2' = b_1' = \beta$.

5. But if there are in B_1 at least two numbers which are not contained in A_1, then $\beta > \alpha$, $\alpha < \beta$.

As this exhausts the possible cases, it follows that of two different numbers one is necessarily the greater, the other the less, so there are two possibilities.

A third is impossible. This was indeed already implied in the choice of the *comparative* (greater, less) to designate the relation between α, β; but this choice has only now been justified. In just such investigations one needs to exercise the greatest care so that even with the best intention to be honest one shall not, through a hasty choice of expressions borrowed from other notions that have already been developed, allow oneself to make inadmissible transfers from one domain to the other.

If we consider the case $\alpha > \beta$ again, it is obvious that the smaller number β, if rational, certainly belongs to the class A_1; for since there is in A_1 a number $a_1' = b_2'$ which belongs to the class B_2, it follows that the number β, whether it is the greatest number in B_1 or the least in B_2, is certainly $\leqq a_1'$ and hence contained in A_1. Likewise it follows from $\alpha > \beta$ that the greater number α, if rational, certainly belongs to the class B_2, because $\alpha \geqq a_1'$. Combining these two observations we get the following result: If a cut (A_1, A_2) is produced by the number α then any rational number belongs to the class A_1 or to the class A_2 according as it is less than or greater than α; if the number α is itself rational it may belong to either class.

From this we obtain finally the following: If $\alpha > \beta$, i.e., if there are infinitely many numbers in A_1 not contained in B_1, then there are infinitely many such numbers that are at the same time different from α and from β; every such rational number c is $<\alpha$, because it is contained in A_1; it is also $>\beta$ because it is contained in B_2.

§5.

Continuity of the Domain of Real Numbers

In consequence of the distinctions just laid down the system $\mathfrak{R}$ of all real numbers forms a well-ordered domain of one dimension; this means merely that the following laws prevail:

I. If $\alpha > \beta$, and $\beta > \gamma$, then $\alpha > \gamma$. We shall say that the number β lies between α and γ.

II. If α, γ are two different numbers, then there exist infinitely many different numbers β lying between α, γ.

III. If α is a definite number then all numbers of the system $\mathfrak{R}$ fall into two classes $\mathfrak{A}_1$ and $\mathfrak{A}_2$ each of which contains infinitely many individuals; the first class $\mathfrak{A}_1$ comprises all the numbers a_1 that are less than α, the second $\mathfrak{A}_2$ comprises all the numbers a_2 that are greater than α; the number α itself may be assigned at pleasure to the first class or to the second, and it is then respectively the greatest number of the first or the least of the second class. In each case the separation of the system $\mathfrak{R}$ into the two classes $\mathfrak{A}_1$, $\mathfrak{A}_2$ is such that every number of the first class $\mathfrak{A}_1$ is smaller than every number of the second class $\mathfrak{A}_2$, and we say that this separation is produced by the number α.

For brevity and in order not to weary the reader I suppress the proofs of these theorems which follow immediately from the definitions in the previous section.

Beside these properties, however, the domain $\mathfrak{R}$ possesses also *continuity*; i.e., the following theorem is true:

IV. If the system $\mathfrak{R}$ of all real numbers divides into two classes $\mathfrak{A}_1$, $\mathfrak{A}_2$ such that every number a_1 of the class $\mathfrak{A}_1$ is less than every number a_2 of the class $\mathfrak{A}_2$ then there exists one and only one number α by which this separation is produced.

Proof. By the separation or the cut of $\mathfrak{R}$ into $\mathfrak{A}_1$ and $\mathfrak{A}_2$ we obtain at the same time a cut (A_1, A_2) of the system R of all rational numbers which is defined thus: A_1 contains all rational numbers of the class $\mathfrak{A}_1$ and A_2 contains all other rational numbers, i.e., all rational numbers of the class $\mathfrak{A}_2$. Let α be the perfectly definite number which produces this cut (A_1, A_2). If β is any number different from α, there are always infinitely many rational numbers c lying between α and β. If $\beta < \alpha$, then $c < \alpha$; hence c belongs to the class A_1 and consequently also to the class $\mathfrak{A}_1$, and since we also have $\beta < c$ then β also belongs to the same class $\mathfrak{A}_1$, because every number in $\mathfrak{A}_2$ is greater than every number c in $\mathfrak{A}_1$. But if $\beta > \alpha$, then $c > \alpha$; hence c belongs to the class A_2 and consequently also to the class $\mathfrak{A}_2$, and since at the same time $\beta > c$, then β also belongs to the same class $\mathfrak{A}_2$, because every number in $\mathfrak{A}_1$ is less than every number c in $\mathfrak{A}_2$. Hence every number β different from α belongs to the class $\mathfrak{A}_1$ or to the class $\mathfrak{A}_2$ according as $\beta < \alpha$ or $\beta > \alpha$; consequently α itself is either the greatest number in $\mathfrak{A}_1$ or the least number in $\mathfrak{A}_2$, i.e., α is obviously the unique number by which the separation of R into the classes $\mathfrak{A}_1$, $\mathfrak{A}_2$ is produced. Which was to be proved.

§6.

Operations with Real Numbers

To reduce any operation with two real numbers α, β to operations with rational numbers, one need only proceed as follows: given the cuts (A_1, A_2), (B_1, B_2) produced by the numbers α and β in the system R, one defines the cut (C_1, C_2) which is to correspond to the result of the operation, γ. I confine myself here to the discussion of the simplest case, that of addition.

If c is any rational number, we put it into the class C_1, provided there are two numbers a_1 in A_1 and b_1 in B_1 such that their sum $a_1 + b_1 \geqq c$; all other rational numbers shall be put into the class C_2. This separation of all rational numbers into the two classes C_1, C_2 evidently forms a cut, since every number c_1 in C_1 is less than every number c_2 in C_2. If both α and β are rational, then every number c_1 contained in C_1 is $\leqq \alpha + \beta$, because $a_1 \leqq \alpha$, $b_1 \leqq \beta$, and therefore $a_1 + b_1 \leqq \alpha + \beta$; further, if there were contained in C_2 a number $c_2 < \alpha + \beta$, hence $\alpha + \beta = c_2 + p$, where p is a positive rational number, then we should have

$$c_2 = (\alpha - \tfrac{1}{2}p) + (\beta - \tfrac{1}{2}p),$$

which contradicts the definition of the number c_2, because $\alpha - \frac{1}{2}p$ is a number in A_1, and $\beta - \frac{1}{2}p$ a number in B_1; consequently every number c_2 contained in C_2 is $\geqq a + \beta$. Therefore in this case the cut (C_1, C_2) is produced by the sum $a + \beta$. Thus we shall not violate the definition which holds in the arithmetic of rational numbers if we always understand the sum $\alpha + \beta$ of any two real numbers α, β to be that number γ by which the cut (C_1, C_2) is produced. Further, if only one of the two numbers α, β is rational, e.g., α, it is easy to see that it makes no difference to the sum $\gamma = \alpha + \beta$ whether the number α is put into the class A_1 or into the class A_2.

The other operations of so-called elementary arithmetic (viz., the formation of differences, products, quotients, powers, roots, logarithms) can be defined just like addition, and in this way we arrive at real proofs of theorems (as, e.g., $\sqrt{2}\cdot\sqrt{3} = \sqrt{6}$), which to the best of my knowledge have never been established before. The excessive length that is to be feared in the definitions of the more complicated operations is partly inherent in the nature of the subject but can for the most part be avoided. The notion of an *interval* is very useful in this connection—i.e., a system A of rational numbers possessing the following characteristic property: if a and a' are numbers of the system A, then all rational numbers lying between a and a' are contained in A. The system R of all rational numbers, and also the two classes of any cut are intervals. If there exist a rational number a_1 which is less than, and a rational number a_2 which is greater than, every number of the interval A, then A is called a finite interval; there then exist infinitely many numbers having the same property as a_1 and infinitely many having the same property as a_2; the whole domain R breaks up into three parts A_1, A, A_2, and there enter two perfectly definite rational or irrational numbers α_1, α_2 which may be called respectively the lower and upper (or the lesser and greater) *limits* of the interval; the lower limit α_1 is determined by the cut for which the system A_1 forms the first class and the upper limit α_2 by the cut for which the system A_2 forms the second class. Of every rational or irrational number α lying between α_1 and α_2 it may be said that it lies *within* the interval A. If all numbers of an interval A are also numbers of an interval B, then A is called a portion of B.

Still lengthier considerations seem to loom up when we attempt to adapt the numerous theorems of the arithmetic of rational numbers (e.g., the theorem $(a + b)c = ac + bc$) to arbitrary real numbers. This, however, is not the case. It is easy to see that it all comes down to showing that the arithmetical operations possess a certain continuity. What I mean by this statement may be expressed in the form of general theorem:

'If the number λ is the result of an operation performed on the numbers α, β, γ, ... and λ lies within the interval L, then intervals A, B, C, ... can be found which contain the numbers α, β, γ, ... and such that the result of the same operation in which the numbers α, β, γ, ... are replaced by arbitrary numbers of the intervals A, B, C, ... is always a number lying within the interval L.' The forbidding clumsiness, however, which marks the statement of such a theorem convinces us that something must be brought in as an aid to expression; this is, in fact, attained in the most satisfactory way by introducing the ideas of *variable magnitudes*, *functions*, and *limiting values*, and it would be best to base the definitions of even the simplest arithmetical operations upon these ideas, a matter which, however, cannot be carried further here.

§7.

INFINITESIMAL ANALYSIS

Here at the close we ought to explain the connection between the preceding investigations and certain fundamental theorems of infinitesimal analysis.

We say that a variable magnitude x which passes through successive definite numerical values approaches a fixed *limiting value* α when in the course of the process x comes to lie between any two numbers between which α itself lies, or, what amounts to the same thing, when the difference $x - \alpha$, taken absolutely, becomes less than any given value different from zero.

One of the most important theorems may be stated as follows: 'If a magnitude x grows continually but not beyond all limits it approaches a limiting value.'

I prove it in the following way. By hypothesis there exists one number α_2 (and thus infinitely many) such that x remains continually $<\alpha_2$; I designate by $\mathfrak{A}_2$ the system of all these numbers α_2, by $\mathfrak{A}_1$ the system of all other numbers α_1; each of the latter possesses the property that in the course of the process x eventually becomes $\geqq\alpha_1$, hence every number α_1 is less than every number α_2, and consequently there exists a number α which is either the greatest in $\mathfrak{A}_1$ or the least in $\mathfrak{A}_2$(§5, IV). The former cannot be the case since x never ceases to grow, hence α is the least number in $\mathfrak{A}_2$. But for any α_1 we shall eventually have $\alpha_1 < x < \alpha$, i.e., x approaches the limiting value α.

This theorem is equivalent to the principle of continuity, i.e., it loses its validity as soon as we assume a single real number not to be contained in the domain $\mathfrak{R}$; or otherwise expressed: if this theorem is correct, then theorem IV in §5 is also correct.

Another theorem of infinitesimal analysis, also equivalent to this and more frequently employed, may be stated as follows: 'If in the variation of a magnitude x we can, for every given positive magnitude δ, assign a corresponding interval within which x changes by less than δ, then x approaches a limiting value.'

This converse of the easily demonstrated theorem that every variable magnitude which approaches a limiting value finally changes by less than any given positive magnitude can be derived both from the preceding theorem and directly from the principle of continuity. I take the latter course. Let δ be any positive magnitude (i.e., $\delta > 0$). Then by hypothesis a time will come after which x will change by less than δ, i.e., if at this time x has the value a, then afterwards we shall continually have $x > a - \delta$ and $x < a + \delta$. I now for a moment lay aside the original hypothesis and make use only of the theorem just proved that all later values of the variable x lie between two assignable finite values. Upon this I base a double separation of all real numbers. To the system

$\mathfrak{A}_2$ I assign a number α_2 (e.g., $a + \delta$) when in the course of the process x eventually becomes $\leqq \alpha_2$; to the system $\mathfrak{A}_1$ I assign every number not contained in $\mathfrak{A}_2$; if α_1 is such a number, then, however far the process may have advanced, it will still happen infinitely many times that $x > \alpha_2$. Since every number α_1 is less than every number α_2 there exists a perfectly definite number α which produces this cut $(\mathfrak{A}_1, \mathfrak{A}_2)$ of the system $\mathfrak{R}$ and which I will call the upper limit of the variable x which always remains finite. Likewise as a result of the behaviour of the variable x a second cut $(\mathfrak{B}_1, \mathfrak{B}_2)$ of the system $\mathfrak{R}$ is produced; a number β_1 (e.g., $a - \delta$) is assigned to $\mathfrak{B}_1$ when in the course of the process x eventually becomes $\geqq \beta_1$; every other number β_2, to be assigned to $\mathfrak{B}_2$, has the property that x never becomes $\geqq \beta_2$; therefore x is infinitely often $< \beta_2$; the number β by which this cut is produced I call the lower limiting value of the variable x. The two numbers α, β are obviously characterized by the following property: if ε is an arbitrarily small positive magnitude then we eventually have $x < \alpha + \varepsilon$ and $x > \beta - \varepsilon$, but never $x < \alpha - \varepsilon$ and never $x > \beta + \varepsilon$. Now two cases are possible. If α and β are different from each other, then necessarily $\alpha > \beta$, since we always have $\alpha_2 \geqq \beta_2$; the variable x oscillates, and, however far the process advances, always undergoes changes whose amount surpasses the value $(\alpha - \beta) - 2\varepsilon$ where ε is an arbitrarily small positive magnitude. But the original assumption contradicts this consequence; there remains only the second case $\alpha = \beta$, and since it has already been shown that, however small the positive magnitude ε may be, we always have eventually $x < \alpha + \varepsilon$ and $x > \beta - \varepsilon$, it follows that x approaches the limiting value α, which was to be proved.

These examples may suffice to bring out the connection between the principle of continuity and infinitesimal analysis.

D. *FROM* ON THE THEORY OF ALGEBRAIC INTEGERS (*DEDEKIND 1877*)

This selection is the introduction to Dedekind's '*Sur la théorie des nombres entiers algébriques*'. Dedekind was disappointed in the reception of his *Tenth supplement* of 1871. In his letter to Lipschitz of 29 April 1876 he declared that he had 'given up hope' that his presentation of ideal theory 'would interest anyone at all nowadays except myself'; and in the following year he told the translator of Dirichlet–Dedekind into Italian that it might be better to omit the *Tenth supplement* altogether because 'I am firmly convinced that not a single person reads this presentation of my theory of ideals' (*H.M. Edwards 1980*, p. 349). Instead, he prepared a new account of his ideas, publishing them first in a French version in 1876–7; the French version was then modified, translated into German, and worked into the *Eleventh supplement* to the third (1879) edition of Dirichlet–Dedekind. The passage that follows is the Introduction; it explains the origins of Dedekind's theory of ideals, and in a long footnote explicitly connects his work in algebra to his work in the foundations of real analysis.

The translation is by David Reed. References to *Dedekind 1877* should be to the paragraph numbers, which have been added in this edition.

INTRODUCTION

⟦1⟧ In response to the invitation which it has been my honour to receive, I propose, in this paper, to develop the *fundamental principles* of a general theory, admitting no exceptions, of algebraic integers [nombres entiers algebriques]—principles which I have previously published in the second edition of Dirichlet's *Vorlesungen über Zahlentheorie*. But, because of the extraordinary extent of this field of mathematical research, I shall limit myself here to the pursuit of a single specific objective which I shall attempt to set forth as clearly as possible in the following remarks.

⟦2⟧ The theory of divisibility of numbers, which serves as the foundation of the theory of numbers, was already established, in its essentials, by Euclid; at least, the principal theorem stating that any composite integer can be factored in one and only one way as a product of prime numbers is an immediate consequence of the theorem demonstrated by Euclid[1] that a product of two numbers cannot be divisible by a prime unless the prime also divides one of the numbers.

⟦3⟧ Two thousand years later Gauss gave, for the first time, an extension of the concept of integer; whereas, until his time, only the numbers $0, \pm 1, \pm 2, \ldots$ (which in what follows I shall calll *rational integers*) were considered to be integers, Gauss introduced[2] *complex* integers of the formm $a + b\sqrt{-1}$, a and b designating rational integers, and he showed that the general laws of divisibility for these numbers are identical with those governing the domain of rational integers.

⟦4⟧ The greatest generalization of the concept of an integer consists in the following. A number θ is said to be an algebraic number when it satisfies an equation

$$\theta^n + a_1\theta^{n-1} + a_2\theta^{n-2} + \ldots + a_{n-1}\theta + a_n = 0,$$

of finite degree n and with rational coefficients $a_1, a_2, \ldots, a_{n-1}, a_n$; θ is said to be an *algebraic integer* or simply an *integer* if it satisfies an equation of this type whose coefficients $a_1, a_2, \ldots, a_{n-1}, a_n$ are all rational integers. From this definition it follows at once that the sums, differences, and products of integers are also integers; consequently, an integer α will be said to be *divisible* by an integer β if we have $\alpha = \beta\gamma$, γ also being an integer. An integer ε will be called a *unit* if all integers are divisible by ε. By analogy, a *prime* number would be

[1] *Elements*, VII, 32.
[2] *Theoria residuorum biquadraticorum*, II; 1832.

an integer α not a unit and having only units ε and integers of the form $\varepsilon\alpha$ as divisors; but it is easy to see that in the domain of all integers which we are considering there are no such prime numbers, since any integer which is not a unit can be written as the product of two factors (or indeed any number of factors) which are themselves integers and not units.

⟦5⟧ Nevertheless the existence of prime numbers and the analogy with domains of rational integers or complex integers begins to reappear when, from the domain of all integers, an infinitely small part is selected as follows. If θ is a determinate algebraic number, then among the infinite number of equations with rational coefficients of which θ is a root there is one and only one

$$\theta^n + a_1\theta^{n-1} + a_2\theta^{n-2} + \ldots + a_{n-1}\theta + a_n = 0$$

which is of least degree, and this equation is therefore called *irreducible*. If x_0, $x_1, \ldots, x_{n-1}$ designate rational numbers taken arbitrarily, then all numbers of the form

$$\phi(\theta) = x_0 + x_1\theta + x_2\theta^2 + \ldots + x_{n-1}\theta^{n-1}$$

(the entire collection of which we shall represent by Ω) will also be algebraic numbers and will enjoy the basic property that sums, differences, products, and quotients of numbers will remain in Ω; I shall call such a collection a *finite field of degree n*. All numbers $\phi(\theta)$ belonging to the field Ω can be divided into two classes, integers (the collection of which is designated by $\mathfrak{o}$) and non-integers or fractional numbers. *The problem which we pose for ourselves is to establish the general laws of divisibility that govern such a system* $\mathfrak{o}$.

⟦6⟧ The system $\mathfrak{o}$ is clearly identical with the system of rational integers when $n=1$, and with the complex integers when $n=2$ and $\theta = \sqrt{-1}$. Certain phenomena which occur in these two specific domains $\mathfrak{o}$ occur in any domain $\mathfrak{o}$ of this type; it should be observed above all that the unlimited decomposition discussed above which occurs in the domain of all algebraic integers is not encountered in a domain $\mathfrak{o}$ of the type indicated, as one can easily see by looking at norms. If by the *norm* of any number $\mu = \phi(\theta)$ belonging to the field Ω one understands the product

$$N(\mu) = \mu\mu_1\mu_2 \ldots \mu_{n-1},$$

where the factors are the conjugate numbers

$$\mu = \phi(\theta),\ \mu_1 = \phi(\theta_1),\ \mu_2 = \phi(\theta_2),\ \ldots\ \mu_{n-1} = \phi(\theta_{n-1}),$$

$\theta, \theta_1, \theta_2, \ldots \theta_{n-1}$ designating all the roots of a single irreducible equation of degree n, then, as is well known, $N(\mu)$ will always be a rational number which is equal to 0 only if $\mu = 0$; at the same time, we always have

$$N(\alpha\beta) = N(\alpha)N(\beta),$$

α and β being any two numbers from the field Ω. If now μ is an integer and therefore a number in $\mathfrak{o}$, the other conjugate numbers $\mu_1, \mu_2, \ldots \mu_{n-1}$ will also be integers and therefore $N(\mu)$ will be a rational integer. This norm plays an extremely important role in the theory of the numbers of the domain $\mathfrak{o}$: any two numbers α, β from this domain are said to be *congruent* or *incongruent* with respect to a third number μ (taken as *modulus*) according as their difference $\pm(\alpha - \beta)$ is or is not divisible by μ. One can, exactly as in the theory of rational and complex numbers, separate the numbers of the system $\mathfrak{o}$ into *classes of numbers* such that each class consists of all those numbers which are congruent to a specified number which is the representative of the class; and a more detailed study shows that the number of such classes (with the sole exception of the case $\mu = 0$) is always finite and further equals the absolute value of $N(\mu)$. An immediate consequence is that $N(\mu) = \pm 1$ if and only if μ is a unit. If now a number of the system $\mathfrak{o}$ is said to be *decomposable* when it is the product of two numbers of the system neither of which is a unit, it evidently follows from the preceding discussion that any decomposable number can always be represented as the product of a finite number of *indecomposable* factors.

⟦7⟧ This result corresponds completely to the law which holds in the theory of rational and complex integers, namely, that any composite number can be represented as the product of a finite number of prime factors. But it is at this point that the analogy with the older theory, observed up to now, threatens to break down for good. In his investigations into the domain of numbers relating to the theory of the division of the circle, and which therefore correspond to equations of the form $\theta^m = 1$, Kummer noticed a phenomenon which so completely and essentially distinguishes the numbers of this domain in general from those previously considered that there appeared to be virtually no hope of maintaining the simple laws governing the older theory of numbers. In effect, while in the domains of rational and complex integers a composite number can only be represented *in a unique way* as a product of prime factors, it can be shown that, in the numerical domains considered by Kummer, a decomposable number can often be represented *in several entirely different ways* as a product of indecomposable numbers—or, what comes to the same thing, it can be shown that *indecomposable* numbers do not possess the full character of a *prime* number properly defined, viz. that a prime number cannot divide the product of two or more factors unless it divides at least one of those factors. But the more that further research into such numerical domains may have seemed hopeless,[3] the more do Kummer's persevering efforts deserve acknowledgement—efforts which were finally rewarded with a truly great and fertile discovery. This geometer finally succeeded[4] in bringing all the apparent irregularities under gen-

[3] In his memoir *De numeris complexis qui radicibus unitatis et numeris integris realibus constant* (*Vratislaviae*, 1844, §8), Kummer says: 'Maxime dolendum videtur, quod haec numerorum realium virtus, ut in factores primos dissolvi possint qui pro eodem numero semper iidem sint, non eadem est numerorum complexorum, quae si esset tota haec doctrina, quae magnis adhuc difficultatibus laborat, facile absolvi et ad finem perduci posset.' ⟦'It is greatly to be regretted that this property of the real numbers always to be uniquely decomposable into prime factors is not also shared by the complex numbers, since, if it were, this whole subject, which is currently labouring under enormous difficulties, could easily be resolved and brought to a conclusion.'⟧

[4] *Zur Theorie der complexen Zahlen* (Crelle's Journal, Vol. 35).

eral laws. By considering indecomposable numbers (which do not themselves have the attribute of true prime numbers) as the products of *ideal* prime factors which do not appear and do not manifest themselves except when combined together and not isolated, he obtained the surprising result that the laws of divisibility in the number-domains he studied now coincided completely with those that govern the domain of rational integers. Any number, not a unit, behaves, in all questions of divisibility, both actively and passively, either like a prime number or like a number formed as the product of completely determined prime factors, existing or ideal. Two ideal numbers, whether prime or composite, whose products with the same third ideal number are existing numbers are said to be *equivalent*, and all ideal numbers equivalent to a single existing number form a *class of ideal numbers*. The collection of all existing numbers, considered as a special case of ideal numbers, forms the *principal class*. To each class there corresponds an infinite system of homogeneous *forms* in n variables and of degree n which are decomposable into n linear factors with algebraic coefficients. The number of such classes is finite, and Kummer was able to extend to the determination of this number the principles Dirichlet had used to determine the number of classes of binary quadratic forms.

⟦8⟧ The great success of Kummer's work in the domain of the division of the circle led to the assumption that the same laws obtained in *all* numerical domains $\mathfrak{o}$ of the most general kind, as described above. In my investigations, whose objective was to bring the question to a definitive conclusion, I began by relying on the theory of higher-order congruences, as I had previously noticed that by the application of this theory Kummer's work could be considerably simplified. Even though this work did bring me close to the objective, I was unable by this route to bring certain exceptional cases under laws which applied to the other cases. I did not arrive at the general theory without exceptions (which I first published in the place indicated above) until after I had entirely abandoned the older formal approach and had replaced it with another method which begins from the simplest fundamental conception and proceeds directly towards the objective. In this method I have no need of a new creation such as Kummer's ideal numbers. Instead I use a *system of real, existing numbers* which I call an *ideal*; and this suffices completely. The power of this concept rests on its extreme simplicity, and, my goal being above all to inspire confidence in this notion, I shall try to develop the series of ideas which led me to this concept.

⟦9⟧ Kummer did not define ideal numbers themselves, but only divisibility by such numbers. If a number α possesses a certain property A, meaning in all cases that α satisfies one or more congruences, he says that α is divisible by a specific ideal number corresponding to the property A. Although this introduction of new numbers is entirely legitimate, one risks being drawn to precipitate conclusions and specious demonstrations, by the mode of expression chosen, in which one speaks of specific ideal numbers and their products, and also by the presumed analogy with the theory of natural numbers; and in fact these pitfalls have not been altogether avoided. On the other hand, an exact definition which would cover all of the ideal numbers that are encountered in a given numerical domain $\mathfrak{o}$ and, at the same time, a general definition of their multiplication, would seem to be all the more necessary in view of the fact that these ideal numbers do not exist within the given numerical domain $\mathfrak{o}$. To satisfy these requirements, it will be necessary and sufficient to establish once and for all the character common to all of these properties $A, B, C, \ldots$ which always (and only) serve to introduce specific ideal numbers, and then to indicate generally how, from two such properties A, B, to which there correspond two specific ideal numbers, one can deduce the property C which must correspond to the product of these ideal numbers.[5]

⟦10⟧ This problem is much simplified by the following reflections.

⟦11⟧ Since such a characteristic property A serves to define, not an ideal number itself, but only the divisibility of numbers in $\mathfrak{o}$ by an ideal number, we are led naturally to consider the collection $\mathfrak{a}$ of all the numbers α of the domain $\mathfrak{o}$ which are divisible by a specific ideal number. To shorten the terminology, from now on I shall call such a system $\mathfrak{a}$ an *ideal*, so that for each specific ideal

---

[5] The legitimacy or rather the necessity of such requirements, which should always be imposed when introducing new arithmetical elements, becomes even more evident when compared with the introduction of real irrational numbers—a matter which I have discussed in a special essay (*Stetigkeit und irrationale Zahlen*; Brunswick, 1872). In admitting that the arithmetic of *rational* numbers (the collection of which we shall designate by R) has been definitively established ⟦fondeé⟧, the next task is to learn in what manner one should introduce the irrational numbers and define the operations of addition, subtraction, multiplication, and division on such numbers. As a first requirement I recognize that arithmetic must be kept pure of any mixture of foreign elements, and for this reason I reject the definition according to which number would be the ratio of two magnitudes of the same kind; on the contrary, the definition or the creation of irrational numbers must be founded solely on phenomena which one can observe clearly *within the domain R*. In the second place, we must require that all real irrational numbers be generated at the same time from one common definition and not successively as roots of equations, logarithms, etc. The definition must, in the third place, be of such a nature as to permit as well a perfectly clear definition of the calculations (addition, etc.) that one will perform on the new numbers. This can be done as follows, although I shall only outline the process:

1. I call a *cut* ⟦*section*⟧ of the domain R any separation ⟦partage⟧ of all rational numbers into two categories such that each number of the first category is algebraically smaller than any number of the second category.

2. Each determinate rational number *a generates* ⟦*engendre*⟧ a determinate cut (or two cuts, not essentially different) as follows: a rational number will be classified as belonging to the first or the second category according as it is algebraically smaller or larger that a (a itself may be placed at will in either of the two categories).

3. There are an infinite number of cuts which *cannot* be generated by rational numbers in the manner indicated; for each cut of this type, we create and introduce into arithmetic a special *irrational* number corresponding to this cut (or generating it).

4. Let α, β be any two real numbers (rational or irrational); it is easy, by looking at the cuts they generate, to define whether one has $\alpha < \beta$ or $\alpha > \beta$; further, one can easily define, using the two cuts, the four cuts that correspond to sum, difference, product, and quotient of the two numbers α and β. In this way the four fundamental operations of arithmetic for two real numbers are defined without the slightest obscurity, and one can really *demonstrate* properties such as $\sqrt{2} \cdot \sqrt{3} = \sqrt{6}$—something which, so far as I am aware, has hitherto not been done rigorously.

5. The irrational numbers thus defined form, together with the rational numbers, a domain $\mathfrak{R}$ that is without gaps and *continuous*; each cut of the domain $\mathfrak{R}$ is produced by a specific number of the same domain; it is impossible to find new classes of numbers in this domain $\mathfrak{R}$.

number α there is a specific ideal $\mathfrak{a}$. Now since, conversely, the property A (that is, the divisibility of a number α by the ideal number) can be precisely expressed by saying that α belongs to the corresponding ideal $\mathfrak{a}$, we can, in place of the properties A, B, C, ... which were used to introduce ideal numbers, consider the ideals $\mathfrak{a}$, $\mathfrak{b}$, $\mathfrak{c}$, ... to establish their common and exclusive character. If we now recall that the introduction of ideal numbers had no other purpose than to bring the laws of divisibility in the numerical domain $\mathfrak{o}$ into complete conformity with the theory of rational numbers, it is evidently necessary that the numbers really existing in $\mathfrak{o}$, and which present themselves in the first instance as factors of composite numbers, be considered as only a special case of ideal numbers. If therefore μ is a specific number in $\mathfrak{o}$, the system $\mathfrak{a}$ of all the numbers $\alpha = \mu\omega$ in the domain $\mathfrak{o}$ divisible by μ will also have the essential character of an ideal and will be called a *principal ideal*; this system is evidently not altered if we replace μ by $\varepsilon\mu$, ε designating any unit in $\mathfrak{o}$. Now, from the notion of integer established above, there follow immediately two elementary theorems on divisibility:

1. If the two integers $\alpha = \mu\omega$, $\alpha' = \mu\omega'$ are divisible by the integer μ, their sum $\alpha + \alpha' = \mu(\omega + \omega')$ and their difference $\alpha - \alpha' = \mu(\omega - \omega')$ will also be divisible by μ; for the sum $\omega + \omega'$ and the difference $\omega - \omega'$ of two integers ω, ω' are themselves integers.

2. If $\alpha = \mu\omega$ is divisible by μ, any number of the form $\alpha\omega' = \mu(\omega\omega')$ divisible by α will also be divisible by μ; for any product $\omega\omega'$ of two integers ω, ω' is also an integer.

⟦12⟧ If we apply these theorems, valid for all integers, to the numbers ω of our numerical domain $\mathfrak{o}$, designating by μ one of these determinate numbers and by $\mathfrak{a}$ the principal ideal corresponding to it, we obtain the two fundamental properties of such a system $\mathfrak{a}$:

I. *Sums and differences of any two numbers of the system $\mathfrak{a}$ are also always numbers of the same system $\mathfrak{a}$.*

II. *Any product of a number of the system $\mathfrak{a}$ and a number of the system $\mathfrak{o}$ is a number of the system $\mathfrak{a}$.*

⟦13⟧ Now, as we are pursuing the objective of bringing the laws of divisibility holding in the domain $\mathfrak{o}$ into complete conformity with those which hold in the domain of rational integers through the introduction of ideal numbers and a corresponding style of language, it follows that the definitions of ideal numbers and divisibility by those numbers should be expressed in such a way that the two elementary theorems 1 and 2 noted above continue to hold even if μ is no longer an existing but an ideal number. Consequently the two properties (I) and (II) will apply not only to principal ideals, but to all ideals. We have thus found a characteristic common to all ideals; and to each number, whether existent or ideal, there corresponds a well-defined ideal $\mathfrak{a}$ enjoying the properties (I) and (II).

⟦14⟧ But a fact of the highest importance and whose truth I was unable to demonstrate rigorously until after many vain efforts and after surmounting the greatest difficulties is that, conversely, any system $\mathfrak{a}$ which has the properties (I) and (II) is also an ideal, that is, that $\mathfrak{a}$ is the collection of all numbers α of the domain $\mathfrak{o}$ which are divisible by an existing number or an ideal number. This fact is indispensable for completing the theory. The two properties (I) and (II) are therefore not merely necessary but also sufficient for a system $\mathfrak{a}$ to be an ideal. Any other condition to which one would subject a system $\mathfrak{a}$, if it is not a simple consequence of propositions (I) and (II), will render a complete explanation of all the divisibility phenomena within $\mathfrak{o}$ impossible.

⟦15⟧ This observation led my naturally to base the entire theory of numbers in the domain $\mathfrak{o}$ on this simple definition, which is clear and requires no reference to ideal numbers.[6]

Any system $\mathfrak{a}$ of integers of the field Ω which has properties (I) and (II) is called an ideal of this field.

⟦16⟧ The divisibility of a number α by a number μ consists in α being a number $\mu\omega$ of the principal ideal which corresponds to μ, and which can conveniently be designated by $\mathfrak{o}(\mu)$ or $\mathfrak{o}\mu$; and from property (II) or theorem 2 it follows at the same time that all the numbers of the principal ideal $\mathfrak{o}\alpha$ are also numbers of the principal ideal $\theta\mu$. Conversely, it is clear that α is certainly divisible by μ when all of the numbers of $\theta\alpha$ and therefore α itself are contained in the ideal $\mathfrak{o}\mu$. From this we are led to establish the following notion of divisibility, not only for principal ideals, but for all ideals:

An ideal $\mathfrak{a}$ is said to be divisible *by an ideal $\mathfrak{b}$ (or to be a multiple of $\mathfrak{b}$ and $\mathfrak{b}$ a divisor of $\mathfrak{a}$) provided all of the numbers of $\mathfrak{a}$ are at the same time contained in $\mathfrak{b}$. An ideal $\mathfrak{p}$ different from $\mathfrak{o}$ which has no divisors other than $\mathfrak{o}$ and $\mathfrak{p}$ is said to be a* prime ideal.[7]

⟦17⟧ From this divisibility of ideals, *which evidently includes that of numbers*, we must separate the following concept of the *multiplication* and the *products* of two ideals:

If α runs through all the numbers of an ideal $\mathfrak{a}$, and β through all the numbers of an ideal $\mathfrak{b}$, then all products of the form $\alpha\beta$ and all sums of these products will form an ideal which shall be called the product *of the ideals $\mathfrak{a}$, $\mathfrak{b}$ and which shall be designated by $\mathfrak{ab}$.*[8]

⟦18⟧ Then one sees immediately that the product $\mathfrak{ab}$ is divisible by both $\mathfrak{a}$ and $\mathfrak{b}$, but the establishment of a complete link between the two concepts of divisibility and multiplication of ideals is arrived at only after overcoming the characteristic difficulties deeply bound up with the nature of the subject. This link is expressed in its essentials in the following two theorems:

---

[6] It is of course permitted, although in no way necessary, to make correspond to each *ideal* such as $\mathfrak{a}$ an *ideal number* which generates it, if it is not a principal ideal.

[7] Similarly the ideal number corresponding to $\mathfrak{a}$ will be said to be *divisible* by the ideal number corresponding to the ideal $\mathfrak{b}$; to a prime ideal there corresponds a *prime* ideal number.

[8] The ideal number corresponding to $\mathfrak{ab}$ will be called the *product* of the ideal numbers corresponding to $\mathfrak{a}$ and $\mathfrak{b}$.

If the ideal $\mathfrak{c}$ *is divisible by the ideal* $\mathfrak{a}$, *there will always be one and only one ideal* $\mathfrak{b}$ *such that* $\mathfrak{ab} = \mathfrak{c}$.

All ideals other than $\mathfrak{o}$ *are either prime or can be represented in one and only one way as the product of prime ideals.*

⟦19⟧ In the essay which follows I limit myself to demonstrating these results with full rigour and in synthetic fashion. This provides the correct *foundation* for the complete theory of ideals and decomposable forms, which offers an inexhaustible field of research to mathematicians. Of all subsequent developments (for which I refer the reader to Dirichlet's *Vorlesungen über Zahlentheorie*, as well as to several articles which are to appear) I have included in the present essay only the partition of ideals into *classes*, and the demonstration that the number of these *classes of ideals* (or the classes of the corresponding forms) is finite. The first section contains only the propositions which are indispensable for the present task; they are extracted from an auxiliary theory, also important for other investigations, and of which I shall later publish a complete exposition. The second section, whose goal is to clarify by completely determinate numerical examples the general notions which are to be introduced later, could be entirely suppressed; but I have preserved it, because it is useful in facilitating the understanding of the following sections, where one finds the theory of integers of an arbitrary finite field developed to the point indicated above. For this purpose, it suffices to use only the first principles of the general theory of fields—a theory whose further development leads easily to the algebraic principles invented by Galois, which serve in their turn as the basis for deeper investigations into the theory of ideals.

E. WAS SIND UND WAS SOLLEN DIE ZAHLEN? (*DEDEKIND 1888*)

This celebrated essay is Dedekind's study of the foundations of the natural numbers. It explores the same territory as *Frege 1879*, *Frege 1884*, *Frege 1893*, *Frege 1903a*, and *Peirce 1881*, but to a greater mathematical depth; it exerted a strong influence on subsequent research in the foundations of mathematics, notably by influencing Peano in his axiomatization of arithmetic, and Zermelo in his axiomatization of set theory.

The first three sections of Dedekind's paper treat the general theory of sets (which Dedekind refers to as 'systems'). He begins by informally stating some of the basic principles of set theory.

A system S is fully determined by its elements, so that two systems are equivalent if and only if they have the same elements (*extensionality*). (Responding explicitly to Kronecker's criticisms in *Kronecker 1886*, Dedekind declares that the membership-relation need not be decidable.) *Subsets* are defined in terms of set-membership: $A \ni S$ if and only if every element of A is an element of S. (Dedekind's symbol $\ni$ would today be written either $\subseteq$ or ε. He did not here distinguish between the two concepts.) The *union* of systems $A, B, C, \ldots$ is designated by $\mathfrak{M}(A, B, C, \ldots)$; Dedekind says that this latter system is *compounded* out of the others. For *intersection*, Dedekind writes $\mathfrak{S}(A, B, C, \ldots)$; he calls the intersection (when it is non-empty) the *community* of the systems $A, B, C, \ldots$. (He did not recognize the existence of the empty set.)

Ernst Zermelo acknowledged the influence of this monograph on his formulation of the axioms for set theory; see *Zermelo 1908a*. But Zermelo made several important alterations to Dedekind. Unlike Dedekind, he distinguished between set-membership and the subset relation; he distinguished between a singleton set and its sole element; and he recognized the existence of the empty set.[a] Moreover, Zermelo explicitly added the axiom of *choice*,[b] the axiom of *infinity* (which Dedekind had tried to prove as a theorem in ¶ 66), and the axiom of *separation* (Aussonderung).

Dedekind's treatment of *mappings* (§2) and *one-to-one mappings* (§3)—Dedekind calls them 'similar mappings'—is straightforward and follows the lines laid down by Dirichlet.

The next three sections (§§4–6) use these set-theoretic ideas to define the natural numbers. Dedekind begins by defining the important notion of a *chain relative to a mapping* ϕ: a set K is a chain relative to ϕ if the image of K under ϕ is a subset of K. (The notion of *chain* is thus a variant of the idea, so often exploited by Dedekind, of taking the closure of a set with respect to a particular operation.) *The* chain of K relative to ϕ—$\phi_0(K)$ or K_0— is defined to be the intersection of all ϕ-chains containing K (¶ 45); Dedekind proves that this intersection is itself a ϕ-chain.[c] In §5, Dedekind gives his definition of infinity: a

---

[a] In an undated and unpublished three-page manuscript entitled, 'Gefahren der Systemlehre', written sometime after 1899, Dedekind noted that the system containing a single element must be distinguished from that element, and he modified *Was sind und was sollen die Zahlen*? so as to admit the empty set. The manuscript is in the Handschriftenabteilung of the Niedersächsische Staats- und Universitätsbibliothek, Göttingen, and has the catalogue number Cod. Ms. Dedekind III, 1.

[b] The axiom of choice is needed for a rigorous proof of several of the results in Dedekind's paper (see for example his proof of Theorem 159) and its use is more-or-less implicit in many of the set-theoretic works of the late nineteenth century. But its importance as a distinct mathematical principle was not recognized until Zermelo used it in his two proofs of the well-ordering theorem (*Zermelo 1904* and *1908a*). The 1908 proof is, as Zermelo stresses, based on ideas in *Dedekind 1888*, and generalizes to the transfinite case the notion of a chain and the induction theorem. (For an exposition of the proof and a discussion of its use of the axiom of choice see *Hallett 1984*, pp. 253–66.)

In a footnote to §27 of his *1908b*, Zermelo notes that his proof of the Schröder-Bernstein theorem 'rests solely on Dedekind's chain-theory'. As Emmy Noether points out, Dedekind had already proved this theorem, in precisely Zermelo's manner, in an unpublished note of 11 July 1887. Dedekind's note and Noether's comment appear in *Dedekind 1930–2* (iii, pp. 447–9). Zermelo, in a footnote to *Cantor 1883b*, says that Cantor did not appreciate the importance of the Schröder-Bernstein result (*Cantor 1932*, p. 209); but see Cantor's letter to Dedekind of October 1882, translated below.

[c] This notion is essentially the same as Frege's definition of the *ancestral*, which Frege first stated in §§26 and 29 of the *Begriffschrift* (*Frege 1879*). The term 'ancestral relation' was coined by

set is infinite if it is similar to a proper subset of itself.[d] Dedekind then attempts to prove the existence of an infinite set; the proof has been widely criticized, and subsequent writers have taken the existence of infinite sets as an axiom.

In §6 Dedekind defines the natural numbers as any simply infinite system. Simply infinite systems he defines in terms of chains: the system N is simply infinite if and only if N is the chain $\phi_0(1)$ of some 'base element'—designated by '1'—and 1 is not contained in $\phi(N)$. (This last condition guarantees that the system is infinite.) If one recasts Dedekind's definition, replacing ϕ by the operation $x \to x + 1$, one readily obtains the Peano axioms for the natural numbers. (Indeed, *Dedekind 1888* was Peano's source for his axioms, and was duly acknowledged in the preface to *Peano 1889*.)

Dedekind, like Peirce, does not insist that the number-sequence should be unique. His aim was to analyse the *structure* of the natural numbers, up to isomorphism, which is, mathematically speaking, all that he needs. He is not at all concerned to furnish a unique answer to Frege's question, 'What is the number one?' Perhaps as a consequence, Dedekind achieved the deep isomorphism theorem in §10: a result that eluded his contemporaries Peirce and Frege, who also investigated the logical foundations of the natural numbers.[e]

In §§7 and 8, Dedekind proves some basic theorems about < and about Z_n, the system of numbers $\{x: x \leqslant n\}$.

In §9, Dedekind establishes the legitimacy of definitions by primitive recursion, proving, in particular, his powerful Theorem 126: If θ is a mapping of

---

Whitehead and Russell (*1910*, part II, section E). *Quine 1940* calls it the 'ancestral'. Dedekind notes the equivalence of his definition to Frege's in his letter to Hans Keferstein of 27 February 1890 (*van Heijenoort 1967*, p. 101).)

[d] This definition had already been given by C.S. Peirce in his *1881* and, somewhat more clearly, in §IV of *Peirce 1885*.

[e] Peirce seems not to have grasped the full import of Dedekind's work, and even claimed (in a manuscript published after his death) that *Was sind und was sollen die Zahlen*? 'proves no difficult theorem that I had not proved or published years before' (*Peirce 1931–58*, iv, p. 268). Frege, in contrast, in the preface to his *Grundgesetze*, acknowledged that, 'In much less space it [*Dedekind 1888*] pursues the laws of arithmetic much further than is done here. To be sure, this brevity is attained only because a great deal is really not proved at all. Frequently Herr Dedekind merely says that the proof follows from such and such propositions; he makes use of dots, as in the expression "$\mathfrak{M}(A, B, C, \ldots)$"; an inventory of the logical or other laws taken by him as basic is nowhere to be found, and even if it were, there would be no way of telling whether no others were actually used; for that to be possible the proofs would have to be not merely indicated but carried out, without gaps. Herr Dedekind, like myself, is of the opinion that the theory of numbers is a part of logic; but his work hardly contributes to the confirmation of this, because the expressions "system", and "a thing belongs to a thing", which he uses, are not usual in logic and are not reduced to acknowledged logical notions. I do not say this as a reproach, for his procedure may have been the most appropriate for his purpose; I say it only to set my intention in a clearer light by way of contrast. The length of a proof ought not to be measured by the yard. It is easy to make a proof look short on paper by skipping over many intermediate links in the chain of inference and merely indicating large parts of it' (*Frege 1893*, p. viii).

The greater precision of Frege's system undoubtedly made his logical presuppositions explicit in a way that Dedekind's were not; and Frege, unlike Dedekind, provided an analysis, not just of arithmetic, but of quantificational logic as well. On the other hand, in Dedekind's defence it should be observed that Zermelo and Peano had little difficulty in extracting their axioms from his work, thereby supplying some of the rigour Frege demanded.

any set S into itself, and if $\omega \varepsilon S$, then there exists a unique map $\psi: N \rightarrow S$ such that $\psi(1) = \omega$ and $\psi(n+1) = \theta(\psi(n))$. The remaining five sections are devoted to applications of this theorem. In §10 he proves that all simply infinite systems are isomorphic—thus, as he notes, establishing the correctness of his earlier definition of the natural numbers.[f] In §§11–13 he gives the usual recursive definitions of addition, multiplication, and exponentiation, and proves the basic properties of these operations; in §14 he defines the cardinal number of a finite set, and proves that a non-empty set is infinite if and only if it is similar to no finite cardinal number. (By modern standards a fully rigorous proof of this result would require the axiom of choice.)

The title of Dedekind's paper is subtle: rigidly translated, it asks, 'What are, and what ought to be, the numbers?' But *sollen* here carries several senses—among them, 'What is the *best* way to regard the numbers?'; 'What is the *function* of numbers?'; and 'What are numbers *supposed* to be?' But perhaps Dedekind's title is famous enough to be left in the original.

The drafts for *Dedekind 1888* are in the possession of the Niedersächsische Staats- und Universitätsbibliothek, Göttingen, Cod. Ms. Dedekind III-1. An early draft of *Was sind und was sollen die Zahlen*? from 1872–8 is reproduced in *Dugac 1976* (p. 293–309).

For recent discussion see *Parsons 1990a* and *1990b*.

The translation is by W.W. Beman, and was originally published in 1901 under the title, 'The nature and meaning of numbers'; Beman's translation has been extensively revised for this edition by William Ewald.

References to *Dedekind 1888* should be to the section and paragraph numbers, which were printed in the original editions.

PREFACE TO THE FIRST EDITION

In science nothing capable of proof ought to be believed without proof. Though this demand seems reasonable, I cannot regard it as having been met even in the most recent methods of laying the foundations of the simplest science; viz., that part of logic which deals with the theory of numbers.* In speaking of arithmetic (algebra, analysis) as merely a part of logic I mean to imply that I consider the number-concept entirely independent of the notions or intuitions

[f] Dedekind's remarks in ¶ 134 should be read in conjunction with his important letter to Hans Keferstein (*van Heijenoort 1967*, pp. 99–103). For a discussion, see *Wang 1957*.

* Of the works which have come under my observation I mention the valuable *Lehrbuch der Arithmetik und Algebra* of E. Schröder (Leipzig, 1873), which contains a bibliography of the subject, and in addition the memoirs of Kronecker and von Helmholtz upon the Number-Concept and upon Counting and Measuring (in the collection of philosophical essays published in honour of E. Zeller, Leipzig, 1887). The appearance of these memoirs has induced me to publish my own views—in many respects similar but in foundation essentially different—which I formulated many years ago in absolute independence of the works of others.

of space and time—that I rather consider it an immediate product of the pure laws of thought. My answer to the problems propounded in the title of this paper is, then, briefly this: numbers are free creations of the human mind; they serve as a means of apprehending more easily and more sharply the difference of things. It is only through the purely logical process of building up the science of numbers and by thus acquiring the continuous number-domain that we are enabled accurately to investigate our notions of space and time by bringing them into relation with this number-domain created in our mind.* If we scrutinize closely what is done in counting a set or number of things, we are led to consider the ability of the mind to relate things to things, to let a thing correspond to a thing, or to represent a thing by a thing, an ability without which no thinking is possible. Upon this unique and therefore absolutely indispensable foundation, as I have already affirmed in an announcement of this paper,† the whole science of numbers must, in my opinion, be established. The design of such a presentation I had formed before the publication of my paper on continuity, but only after its appearance (and with many interruptions occasioned by increased official duties and other necessary labours) was I able in the years 1872 to 1878 to commit to paper a first rough draft which several mathematicians examined and partially discussed with me. It bears the same title and contains, though not arranged in the best order, all the essential fundamental ideas of my present paper, in which they are more carefully elaborated. As such main points I mention here the sharp distinction between finite and infinite (64), the notion of the number ⟦Anzahl⟧ of things (161), the proof that the form of argument known as complete induction (or the inference from n to $n + 1$) is really conclusive (59), (60), (80), and that therefore the definition by induction (or recursion) is determinate and consistent (126).

This memoir can be understood by any one possessing what is usually called good common sense; no technical philosophical, or mathematical, knowledge is in the least degree required. But I know that, in the shadowy forms which I bring before him, many a reader will scarcely recognize his numbers which all his life long have accompanied him as faithful and familiar friends; he will be frightened by the long series of simple inferences corresponding to our step-by-step understanding, by the matter-of-fact dissection of the chains of reasoning on which the laws of numbers depend, and he will become impatient at being compelled to follow out proofs of truths which to his supposed inner consciousness seem at once evident and certain. But in just this possibility of reducing such truths to others more simple, no matter how long and apparently artificial the series of inferences, I recognize a convincing proof that their possession (or belief in them) is never given by inner intuition but is always gained only by a (more or less complete) repetition of the individual inferences. I like to compare this action of thought (which is difficult to follow because of the rapidity

* See Section III of my memoir, *Continuity and Irrational Numbers*, Brunswick, 1872.

† Dirichlet's *Vorlesungen über Zahlentheorie*, third edition, 1879, § 163, note on page 470.

of its performance) with the action which an accomplished reader performs in reading; this reading always remains a more or less complete repetition of the individual steps which the beginner has to take in his wearisome spelling-out; but a very small number of these steps, and therefore a very small effort or exertion of the mind, is sufficient to enable the practised reader to recognize the correct, true word. (Only with very great probability, to be sure; for, as is well known, it occasionally happens that even the most practised proof-reader allows a typographical error to escape him, i.e., reads falsely—a thing which would be impossible if the chain of thoughts associated with spelling were fully repeated.) So from the time of birth we are continually and in increasing measure led to relate things to things and thus to exercise that faculty of the mind on which the creation of numbers depends; this exercise goes on continually, though without definite purpose, in our earliest years; the accompanying formation of judgements and chains of reasoning leads us to a store of real arithmetical truths to which our first teachers later refer as to something simple, self-evident, and given in the inner consciousness. Thus it happens that many very complicated notions (such as, for example, that of the number ⟦Anzahl⟧ of things) are erroneously regarded as simple. In this sense, which I express by words formed after a well-known saying *ἀεὶ ὁ ἄνθρωπος ἀριθμητίζει* ⟦'humanity always arithmetizes'⟧ I hope that the following pages, as an attempt to establish the science of numbers upon a uniform foundation, will find a generous welcome, and that other mathematicians will be led to reduce the long series of inferences to more moderate and attractive proportions.

In accordance with the purpose of this memoir I restrict myself to the consideration of the series of so-called natural numbers. In my earlier memoir on continuity (1872) I have already shown (at any rate for the example of irrational numbers) how the step-by-step extension of the number-concept is subsequently to be carried out—the creation of zero, of the negative, rational, irrational, and complex numbers—always by a reduction to earlier concepts, and indeed without any introduction of foreign conceptions (such as for example that of measurable magnitudes), which according to my view can attain perfect clearness only through the science of numbers; the other extensions may be carried out in the same way, as I have already shown in Section III of that memoir. I propose sometime to present this whole subject in systematic form. From just this point of view it appears as something self-evident and not new that every theorem of algebra and higher analysis, no matter how remote, can be expressed as a theorem about natural numbers—a declaration I have heard repeatedly from the lips of Dirichlet. But I see nothing meritorious—and this was just as far from Dirichlet's thought—in actually performing this wearisome circumlocution and insisting on the use and recognition of no other than natural numbers. On the contrary, the greatest and most fruitful advances in mathematics and other sciences have invariably been made by the creation and introduction of new concepts, rendered necessary by the frequent recurrence of complex phenomena which could be mastered by the old notions only with difficulty. I gave a lecture on this subject before the philosophical faculty in the summer of 1854 on the occasion of my habilitation as *Privat-Dozent* in Göttingen. The point of view of this lecture met with the approval of Gauss; but this is not the place to go into further detail.

Instead, I shall use the opportunity to make some remarks relating to my earlier work, mentioned above, on *Continuity and irrational numbers*. The theory of irrational numbers there presented, worked out in the fall of 1853, is based on the phenomenon (§4) occurring in the domain of rational numbers which I designate by the term cut ⟦Schnitt⟧ and which I was the first to investigate carefully; it culminates in the proof of the continuity of the new domain of real numbers (§5, iv). It appears to me to be somewhat simpler, I might say easier, than the two theories, different from it and from each other, which have been proposed by Weierstrass and G. Cantor, and which are also perfectly rigorous. It has since been adopted without essential modification by U. Dini in his *Fondamenti per la teorica delle funzioni di variabili reali* (Pisa, 1878). In the course of this exposition my name is mentioned, not in the description of the purely arithmetical phenomenon of the cut, but when the author discusses the existence of a measurable quantity corresponding to the cut. This might easily lead to the supposition that my theory rests upon the consideration of such quantities. But nothing could be further from the truth. Rather, in §3 of my paper I advanced several reasons why I wholly reject the introduction of measurable quantities. Indeed, at the end of the paper I pointed out that for a great part of the science of space the continuity of its configurations is not even a necessary condition—quite apart from the fact that in works on geometry continuity is only casually mentioned by name, but is never clearly defined, and therefore cannot be employed in demonstrations. To explain this matter more clearly I note the following example: If we select three non-collinear points A, B, C at pleasure, with the single limitation that the ratios of the distances AB, AC, BC are algebraic numbers,[†] and regard as existing in space only those points M, for which the ratios of AM, BM, CM to AB are likewise algebraic numbers, then it is easy to see that the space made up of the points M is everywhere discontinuous. But in spite of this discontinuity, and despite the existence of gaps in this space, all constructions that occur in Euclid's *Elements*, can, so far as I can see, be just as accurately effected here as in perfectly continuous space; the discontinuity of this space would thus not be noticed in Euclid's science, would not be felt at all. If anyone should say that we cannot conceive of space as anything else than continuous, I should venture to doubt it and to call attention to the fact that a far advanced, refined scientific training is demanded in order to perceive clearly the essence of continuity and to understand that besides rational quantitative relations, also irrational, and besides algebraic, also transcendental quantitative relations are conceivable. It appears to me all the more beautiful that, without any notion of measurable quantities and simply by a finite system of simple steps of

[†] Dirichlet's *Vorlesungen über Zahlentheorie*, § 159 of the second edition, § 160 of the third.

thought, man can advance to the creation of the pure continuous number-domain; and only by this means is it in my opinion possible for him to render the notion of continuous space clear and definite.

The same theory of irrational numbers founded upon the phenomenon of the cut is set forth in the *Introduction à la théorie des fonctions d'une variable* by J. Tannery (Paris, 1886). If I rightly understand a passage in the preface to this work, the author has thought out his theory independently, that is, at a time when not only my paper, but Dini's *Fondamenti* mentioned in the same preface, was unknown to him. This agreement seems to me a gratifying proof that my conception conforms to the nature of the case, a fact recognized by other mathematicians, e.g., by Pasch in his *Einleitung in die Differential- und Integralrechnung* (Leipzig, 1883). But I cannot quite agree with Tannery when he calls this theory the development of an idea due to J. Bertrand and contained in his *Traité d'arithmétique*—the idea, namely, that an irrational number is defined by the specification of all rational numbers that are less and all those that are greater than the number to be defined. As regards this statement (which is repeated by Stolz—apparently without careful investigation—in the preface to the second part of his *Vorlesungen über allgemeine Arithmetik* (Leipzig, 1886)) I venture to remark the following: The conviction that an irrational number is to be considered as fully defined by the specification just described, was the common property of all mathematicians who concerned themselves with the notion of the irrational long before the time of Bertrand. Just this manner of determining it is in the mind of every computer who calculates the irrational root of an equation by approximation. And if, as Bertrand does exclusively in his book, (the eighth edition, of the year 1885, lies before me) one regards the irrational number as the ratio of two measurable quantities, then this manner of determining it is already set forth in the clearest possible way in the celebrated definition which Euclid gives of the equality of two ratios (*Elements*, V, 5). This same ancient conviction was the source of my theory as well as that of Bertrand and many other more or less complete attempts to lay the foundations for the introduction of irrational numbers into arithmetic. But though one is so far in perfect agreement with Tannery, yet in an actual examination one cannot fail to observe that Bertrand's presentation, in which the phenomenon of the cut in its logical purity is not even mentioned, has no similarity whatever to mine, inasmuch as it resorts at once to the existence of a measurable quantity, a notion which for reasons mentioned above I wholly reject. Aside from this fact this method of presentation seems also in the succeeding definitions and proofs, which are based on the postulate of this existence, to contain such fundamental gaps that I still regard the statement made in my paper (§6), that the theorem $\sqrt{2} \cdot \sqrt{3} = \sqrt{6}$ has nowhere yet been strictly demonstrated, as justified with respect to this work as well—a work that is in many other respects excellent, and with which I was unacquainted at that time.

R. DEDEKIND

HARZBURG, 5 October 1887.

PREFACE TO THE SECOND EDITION

The present memoir soon after its appearance met with both favourable and unfavourable criticisms; indeed serious faults were charged against it. I have been unable to convince myself of the justice of these charges, and I now issue a new edition of the memoir, which for some time has been out of print. Because I do not have the time for a public defence, I reprint it without change, adding only the following notes to the first preface.

The property which I employed as the definition (64) of an infinite system had been pointed out before the appearance of my paper by G. Cantor (*Ein Beitrag zur Mannigfaltigkeitslehre, Crelle's Journal*, Vol. 84, 1878), and by Bolzano (*Paradoxien des Unendlichen*, §20, 1851). But neither of these authors made the attempt to use this property for the definition of the infinite and upon this foundation to establish with rigorous logic the science of numbers. But this is precisely the content of my wearisome labour, which in all its essentials I had completed several years before the appearance of Cantor's memoir and at a time when the work of Bolzano was unknown to me, even by name. For the benefit of those who are interested in and understand the difficulties of such an investigation, I add the following remark. We can lay down an entirely different definition of the finite and infinite, which appears still simpler since the notion of the similarity of a mapping ⟦Abbildung⟧ (26) is not even assumed, viz.:

'A system *S* is said to be finite when it may be so transformed into itself (36) that no proper part (6) of *S* is transformed into itself; otherwise *S* is called an infinite system.'

Now let us attempt to erect our edifice upon this new foundation! We shall soon meet with serious difficulties, and I believe myself warranted in saying that the proof of the perfect agreement of this definition with the former can be obtained only (and then easily) when we are permitted to assume the series of natural numbers as already developed and to make use of the final considerations in (131); and yet nothing is said of all these things in either the one definition or the other! From this we can see how very great is the number of steps in thought needed for such a remodelling of a definition.

About a year after the publication of my memoir I became acquainted with G. Frege's *Grundlagen der Arithmetik*, which had already appeared in the year 1884. However different the view of the essence of number adopted in that work is from my own, yet it contains, particularly from §79 on, points of very close contact with my paper, especially with my definition (44). The agreement, to be sure, is not easy to discover on account of the different form of expression; but the positiveness with which the author speaks of the logical inference from n to $n + 1$ (page 93, below) ⟦§9⟧ shows plainly that here he stands upon the same ground with me. In the meantime E. Schröder's *Vorlesungen über die Algebra der Logik* has almost been completed (1890–1891). Upon the importance of this extremely suggestive work, to which I pay my highest tribute, it is impossible here to enter further; I will simply confess that in spite of the remark made on p. 253 of Part I, I have retained my somewhat clumsy symbols (8) and (17);

they make no claim to be adopted generally but are intended simply to serve the purpose of this arithmetical paper, to which, in my view they are better adapted than sum and product symbols.

R. DEDEKIND

HARZBURG, 24 August, 1893.

PREFACE TO THE THIRD EDITION

When I was asked roughly eight years ago to replace the second edition of this work (which was already out of print) by a third, I had misgivings about doing so, because in the mean time doubts had arisen about the reliability ⟦Sicherheit⟧ of important foundations of my conception. Even today I do not underestimate the importance, and to some extent the correctness, of these doubts. But my trust in the inner harmony of our logic is not thereby shattered; I believe that a rigorous investigation of the power ⟦Schöpferkraft⟧ of the mind to create from determinate elements a new determinate, their system, that is necessarily different from each of these elements, will certainly lead to an unobjectionable formulation of the foundations of my work. But I am prevented by other tasks from completing such an investigation; so I beg for leniency if the paper now appears for the third time without changes—which can be justified by the fact that interest in it, as the persistent inquiries about it show, has not yet disappeared.

BRUNSWICK, 30 September 1911.

R. DEDEKIND

WAS SIND UND WAS SOLLEN DIE ZAHLEN?

’Αεὶ ὁ ἄνθρωπος ἀριθμητίζει.[a]

Dedicated to my sister
Julie
and my brother
Adolf,
Dr. jur., Provincial High Court of Brunswick
with heartfelt love.

§1.

SYSTEMS OF ELEMENTS

1. In what follows I understand by *thing* ⟦*Ding*⟧ every object of our thought. In order to be able conveniently to speak of things, we designate them by symbols,

[a] ⟦‘Humanity always arithmetizes.’⟧

e.g., by letters, and we venture to speak briefly of the thing a or of a simply, when we mean the thing denoted by a and not at all the letter a itself. A thing is completely determined by all that can be affirmed or thought concerning it. A thing a is the same as b (identical with b), and b the same as a, when all that can be thought concerning a can also be thought concerning b, and when all that is true of b can also be thought of a. That a and b are only symbols or names for one and the same thing is indicated by the notation $a = b$, and also by $b = a$. If further $b = c$, that is, if c as well as a is a symbol for the thing denoted by b, then $a = c$. If the above coincidence of the thing denoted by a with the thing denoted by b does not exist, then the things a, b are said to be different; a is another thing than b, b another thing than a; there is some property belonging to the one that does not belong to the other.

2. It very frequently happens that different things, a, b, c, ... for some reason can be considered from a common point of view, can be associated in the mind, and we say that they form a *system* S; we call the things a, b, c, ... *elements* of the system S, they are *contained* in S; conversely, S *consists* of these elements. Such a system S (an aggregate, a manifold, a totality) as an object of our thought is likewise a thing (1); S is completely determined when, for every thing, it is determined whether it is an element of S or not.* The system S is hence the same as the system T, in symbols $S = T$, when every element of S is also an element of T, and every element of T is also an element of S. For uniformity of expression it is advantageous to include also the special case where a system S consists of a *single* (one and only one) element a, i.e., the thing a is an element of S, but no thing different from a is an element of S. On the other hand, we intend here for certain reasons wholly to exclude the empty system which contains no element at all, although for other investigations it may be appropriate to imagine such a system.

3. Definition. A system A is said to be *part* of a system S when every element of A is also an element of S. Since this relation between a system A and a system S will occur continually in what follows, we shall express it briefly by the symbol $A \ni S$. For simplicity and clearness I shall avoid the inverse symbol $S \in A$, by which the same fact might be expressed. But for lack of a better word I shall sometimes say that S is the *whole* of A, by which I mean to express that among the elements of S are found all the elements of A. Since further every element s of a system S by (2) can be itself regarded as a system, we can hereafter employ the notation $s \ni S$.

4. Theorem. $A \ni A$, by reason of (3).

* How this determination is brought about, and whether we know a way of deciding upon it, is a matter of indifference for all that follows; the general laws to be developed in no way depend upon it; they hold under all circumstances. I mention this expressly because Kronecker not long ago (*Crelle's Journal*, Vol. 99, pp. 334–336) has endeavoured to impose certain limitations upon the free formation of concepts in mathematics which I do not believe to be justified; but there seems to be no call to enter upon this matter with more detail until the distinguished mathematician shall have published his reasons for the necessity or merely the expediency of these limitations.

5. Theorem. If $A \ni B$ and $B \ni A$, then $A = B$.

The proof follows from (3), (2).

6. Definition. A system A is said to be a *proper* ‖*echter*‖ part of S, when A is part of S, but different from S. According to (5) S is then not a part of A, i.e., there is in S an element which is not an element of A.

7. Theorem. If $A \ni B$ and $B \ni C$, which may be designated briefly by $A \ni B \ni C$, then $A \ni C$, and A is certainly a proper part of C if A is a proper part of B or if B is a proper part of C.

The proof follows from (3), (6).

8. Definition. By the system *compounded* out of any systems $A, B, C, \ldots$ (which shall be designated by $\mathfrak{M}(A, B, C, \ldots)$) we mean that system whose elements are determined by the following prescription: a thing is considered as an element of $\mathfrak{M}(A, B, C, \ldots)$ if and only if it is an element of some one of the systems $A, B, C, \ldots$, i.e., when it is an element of A, *or* B, *or* $C, \ldots$. We include also the case where only a single system A exists; then obviously $\mathfrak{M}(A) = A$. We observe further that the system $\mathfrak{M}(A, B, C, \ldots)$ compounded out of $A, B, C, \ldots$ is to be distinguished from the system whose elements are the systems $A, B, C, \ldots$ themselves.

9. Theorem. The systems $A, B, C, \ldots$ are parts of $\mathfrak{M}(A, B, C, \ldots)$.

The proof follows from (8), (3).

10. Theorem. If $A, B, C, \ldots$ are parts of a system S, then $\mathfrak{M}(A, B, C, \ldots) \ni S$.

The proof follows from (8), (3).

11. Theorem. If P is part of one of the systems $A, B, C, \ldots$ then $P \ni \mathfrak{M}(A, B, C, \ldots)$.

The proof follows from (9), (7).

12. Theorem. If each of the systems $P, Q, \ldots$ is part of one of the systems $A, B, C, \ldots$ then $\mathfrak{M}(P, Q, \ldots) \ni \mathfrak{M}(A, B, C, \ldots)$.

The proof follows from (11), (10).

13. Theorem. If A is compounded out of any of the systems $P, Q, \ldots$ then $A \ni \mathfrak{M}(P, Q, \ldots)$.

Proof. For every element of A is by (8) an element of one of the systems $P, Q, \ldots$, consequently by (8) also an element of $\mathfrak{M}(P, Q, \ldots)$, whence the theorem follows by (3).

14. Theorem. If each of the systems $A, B, C, \ldots$ is compounded out of any of the systems $P, Q, \ldots$ then

$$\mathfrak{M}(A, B, C, \ldots) \ni \mathfrak{M}(P, Q, \ldots)$$

The proof follows from (13), (10).

15. Theorem. If each of the systems $P, Q, \ldots$ is part of one of the systems $A, B, C, \ldots$, and if each of the latter is compounded out of any of the former, then

$$\mathfrak{M}(P, Q, \ldots) = \mathfrak{M}(A, B, C, \ldots).$$

The proof follows from (12), (14), (5).

16. Theorem. If

$$A = \mathfrak{M}(P, Q) \text{ and } B = \mathfrak{M}(Q, R)$$

then

$$\mathfrak{M}(A, R) = \mathfrak{M}(P, B).$$

Proof. For by the preceding theorem (15), $\mathfrak{M}(A, R)$ as well as $\mathfrak{M}(P, B) = \mathfrak{M}(P, Q, R)$.

17. Definition. A thing g is said to be a *common* element of the systems A, B, C, . . ., if it is contained in each of these systems (that is, in A *and* in B *and* in C, . . .). Likewise a system T is said to be a *common part* of A, B, C, . . . when T is part of each of these systems; and by the *intersection* ⟦*Gemeinheit*⟧ of the systems A, B, C, . . . we understand the perfectly determinate system $\mathfrak{G}(A, B, C, \ldots)$ which consists of all the common elements g of A, B, C, . . . and hence is likewise a common part of those systems. We again include the case where only a single system A occurs; then $\mathfrak{G}(A) = A$. But the case may also occur that the systems A, B, C, . . . possess no common element at all, therefore no common part, no intersection; they are then called systems *without* common part, and the symbol $\mathfrak{G}(A, B, C, \ldots)$ is meaningless (compare the end of (2)). We shall however almost always in theorems concerning intersections leave it to the reader to remember the condition of their existence and to discover the proper interpretation of these theorems for the case of non-existence.

18. Theorem. Every common part of A, B, C, . . . is part of $\mathfrak{G}(A, B, C, \ldots)$.
The proof follows from (17).

19. Theorem. Every part of $\mathfrak{G}(A, B, C, \ldots)$ is a common part of A, B, C, . . .
The proof follows from (17), (7).

20. Theorem. If each of the systems A, B, C, . . . is whole (3) of one of the systems P, Q, . . . then

$$\mathfrak{G}(P, Q, \ldots) \ni \mathfrak{G}(A, B, C, \ldots).$$

Proof. For every element of $\mathfrak{G}(P, Q, \ldots)$ is a common element of P, Q, . . ., therefore also a common element of A, B, C, . . ., which was to be proved.

§2.

Mapping of a system

21. Definition.* By a mapping ⟦*Abbildung*⟧ ϕ of a system S we understand a law according to which, to every determinate element s of S, there *belongs* a determinate thing which is called the *image* ⟦*Bild*⟧ of s and which is denoted by $\phi(s)$; we say also that $\phi(s)$ *corresponds* to the element s, that $\phi(s)$ *results* or is *produced* from s by the *mapping* ϕ, that s is *mapped* into $\phi(s)$ by the

* See Dirichlet's *Vorlesungen über Zahlentheorie*, third edition, 1879, §163.

mapping ϕ. If now T is any part of S, then the mapping ϕ of S likewise contains a determinate mapping of T, which for the sake of simplicity may be denoted by the same symbol ϕ; it consists in this: to every element t of the system T there corresponds the same image $\phi(t)$ which t possesses as an element of S; at the same time the system consisting of all images $\phi(t)$ shall be called the image of T and be denoted by $\phi(T)$; in this way $\phi(S)$ is also defined. As an example of a mapping of a system we may regard the mere assignment of determinate symbols or names to its elements. The simplest mapping of a system is that by which each of its elements is mapped into itself; it will be called the *identical* mapping of the system. For convenience, in the following theorems (22), (23), (24), which deal with an arbitrary mapping ϕ of an arbitrary system S, we shall denote the images of elements s and parts T respectively by s' and T'; in addition we agree that small and capital italics without accent shall always signify elements and parts of this system S.

22. Theorem.* If $A \ni B$, then $A' \ni B'$.

Proof. Every element of A' is the image of an element contained in A, and therefore also in B; it is therefore an element of B', which was to be proved.

23. Theorem. The image of $\mathfrak{M}$ $(A, B, C, \ldots)$ is $\mathfrak{M}$ $(A', B', C', \ldots)$.

Proof. If we denote the system $\mathfrak{M}$ $(A, B, C, \ldots)$ (which by (10) is likewise part of S) by M, then every element of its image M' is the image m' of an element m of M; since therefore by (8) m is also an element of one of the systems $A, B, C, \ldots$ and consequently m' is an element of one of the systems $A', B', C', \ldots$, and hence by (8) is also an element of $\mathfrak{M}$ $(A', B', C', \ldots)$, we have by (3)

$$M' \ni \mathfrak{M}(A', B', C', \ldots).$$

On the other hand, since $A, B, C, \ldots$ are by (9) parts of M, and hence $A', B', C', \ldots$ by (22) are parts of M', we have by (10)

$$\mathfrak{M}(A', B', C', \ldots) \ni M'.$$

By combination with the above we have by (5) the theorem to be proved

$$M' = \mathfrak{M}(A', B', C', \ldots).$$

24. Theorem.** The image of every common part of $A, B, C, \ldots$, and therefore that of the intersection $\mathfrak{G}$ $(A, B, C, \ldots)$ is part of $\mathfrak{G}$ $(A', B', C', \ldots)$.

Proof. By (22) A is a common part of $A', B', C', \ldots$, whence the theorem follows by (18).

25. Definition and theorem. If ϕ is a mapping of a system S, and ψ a mapping of the image $S' = \phi(S)$, there always results a mapping θ of S, *compounded*† out of ϕ and ψ, which consists in this: to every element s of S there corresponds the image

---

* See theorem 27.

** See theorem 29.

† A confusion of this compounding of mappings with that of systems of elements is hardly to be feared.

$$\theta(s) = \psi(s') = \psi(\phi(s)),$$

where again we have put $\phi(s) = s'$. This mapping θ can be denoted briefly by the symbol $\psi.\phi$ or $\psi\phi$, and the image $\theta(s)$ by $\psi\phi(s)$ where the order of the symbols ϕ, ψ is to be considered, since in general the symbol $\phi\psi$ has no interpretation and actually has a meaning only when $\psi(S') \ni S$. If now χ signifies a mapping of the system $\psi(s') = \psi\phi(s)$ and η the mapping $\chi\psi$ of the system S' compounded out of ψ and χ, then $\chi\theta(s) = \chi\psi(s') = \eta(s') = \eta\phi(s)$; therefore the compound mappings $\chi\theta$ and $\eta\phi$ coincide for every element s of S, i.e., $\chi\theta = \eta\phi$. In accordance with the meaning of θ and η this theorem can conveniently be expressed in the form

$$\chi.\psi\phi = \chi\psi.\phi,$$

and this mapping compounded out of ϕ, ψ, χ can be designated briefly by $\chi\psi\phi$.

§3.

SIMILARITY OF A MAPPING. SIMILAR SYSTEMS

26. Definition. A mapping ϕ of a system S is said to be *similar* ‖*ähnlich*‖ or *distinct*, when to different elements a, b of the system S there always correspond different images $a' = \phi(a)$, $b' = \phi(b)$. Since, in this case, if $s' = t'$ then $s = t$, every element of the system $S' = \phi(S)$ is the image s' of a single, perfectly determinate element s of the system S, and we can therefore set over against the mapping ϕ of S an *inverse* mapping of the system S', to be denoted by $\bar{\phi}$, which consists in this: to every element s' of S' there corresponds the image $\bar{\phi}(s') = s$; obviously this mapping is also similar. It is clear that $\bar{\phi}(S') = S$, that further ϕ is the inverse mapping belong to $\bar{\phi}$, and that the mapping $\bar{\phi}\phi$ compounded out of ϕ and $\bar{\phi}$ by (25) is the identical mapping of S (21). We obtain at once the following additions to §2, retaining the notation there given:

27. Theorem.* If $A' \ni B'$, then $A \ni B$.

Proof. If a is an element of A then a' is an element of A', therefore also of B', hence $= b'$, where b is an element of B; but since from $a' = b'$ we always have $a = b$, every element of A is also an element of B, which was to be proved.

28. Theorem. If $A' = B'$, then $A = B$.

The proof follows from (27), (4), (5).

29. Theorem.** If $G = \mathfrak{G}(A, B, C, \ldots)$, then

$$G' = \mathfrak{G}(A', B', C', \ldots).$$

---

* See theorem 22.

** See theorem 24.

Proof. Every element $\mathfrak{G}(A', B', C', \ldots)$ is certainly contained in S', and is therefore the image g' of an element g contained in S; but since g' is a common element of $A', B', C', \ldots$ then by (27) g must be a common element of $A, B, C, \ldots$ therefore also an element of G; hence every element of $\mathfrak{G}(A', B', C', \ldots)$ is an image of an element g of G, therefore an element of G', i.e. $\mathfrak{G}(A', B', C', \ldots) \ni G'$, and accordingly our theorem follows from (24), (5).

30. Theorem. The identical mapping of a system is always a similar mapping.

31. Theorem. If ϕ is a similar mapping of S and ψ a similar mapping of $\phi(S)$, then the mapping $\psi\phi$ of S, compounded of ϕ and ψ, is a similar mapping, and the associated inverse mapping $\overline{\psi\phi} = \bar{\phi}\bar{\psi}$.

Proof. For to different elements a, b of S correspond different images $a' = \phi(a)$, $b' = \phi(b)$, and to these again different images $\psi(a') = \psi\phi(a)$, $\psi(b') = \psi\phi(b)$. Therefore $\psi\phi$ is a similar mapping. Besides, every element $\psi\phi(s) = \psi(s')$ of the system $\psi\phi(S)$ is mapped by $\bar{\psi}$ into $s' = \phi(s)$ and this by $\bar{\phi}$ into s, therefore $\psi\phi(s)$ is mapped by $\bar{\phi}\bar{\psi}$ into s, which was to be proved.

32. Definition. The systems R, S are said to be *similar* when there exists a similar mapping ϕ of S such that $\phi(S) = R$, and therefore $\bar{\phi}(R) = S$. Obviously by (30) every system is similar to itself.

33. Theorem. If R, S are similar systems, then every system Q similar to R is also similar to S.

Proof. For if ϕ, ψ are similar mappings of S, R such that $\phi(S) = R$, $\psi(R) = Q$, then by (31) $\psi\phi$ is a similar mapping of S such that $\psi\phi(S) = Q$, which was to be proved.

34. Definition. We can therefore separate all systems into *classes* by putting into a determinate class all and only those systems $Q, R, S, \ldots$ that are similar to a determinate system R, called the *representative* of the class; according to (33) the class is not changed by taking as representative any other system belonging to it.

35. Theorem. If R, S are similar systems, then every part of S is also similar to a part of R, and every proper part of S is also similar to a proper part of R.

Proof. For if ϕ is a similar mapping of S, $\phi(S) = R$, and $T \ni S$, then by (22) the system similar to T, $\phi(T) \ni R$; if further T is a proper part of S, and s is an element of S not contained in T, then by (27) the element $\phi(s)$ contained in R cannot be contained in $\phi(T)$; hence $\phi(T)$ is a proper part of R, which was to be proved.

§4.

Mapping of a System into Itself

36. Definition. If ϕ is a similar or dissimilar mapping of a system S, and $\phi(S)$ is part of a system Z, then ϕ is said to be a mapping of S into Z, and we say S is mapped by ϕ into Z. Hence we call ϕ a mapping of the system S into *itself* when $\phi(S) \ni S$, and we propose in this paragraph to investigate the general laws of such a mapping ϕ. In doing this we shall use the same notations as in §2 and

again put $\phi(s) = s'$, $\phi(T) = T'$. These images s', T' are by (22), (7) themselves again elements or parts of S, like all things designated by italic letters.

37. Definition. K is called a *chain* ⟦*Kette*⟧ when $K' \ni K$. We remark expressly that this name does not in itself belong to the part K of the system S, but is given only with respect to the particular mapping ϕ; with reference to another mapping of the system S into itself K can very well not be a chain.

38. Theorem. S is a chain.

39. Theorem. The image K' of a chain K is a chain.

Proof. For from $K' \ni K$ it follows by (22) that $(K')' \ni K'$, which was to be proved.

40. Theorem. If A is part of a chain K, then $A' \ni K$.

Proof. For from $A \ni K$ it follows by (22) that $A' \ni K'$, and since by (37) $K' \ni K$, therefore by (7) $A' \ni K$, which was to be proved.

41. Theorem. If the image A' is part of a chain L, then there exists a chain K which satifies the conditions $A \ni K, K' \ni L$: and $\mathfrak{M}(A, L)$ is just such a chain K.

Proof. If we actually put $K = \mathfrak{M}(A, L)$, then by (9) the one condition $A \ni K$ is fulfilled. Since further by (23) $K' = \mathfrak{M}(A', L')$ and by hypothesis $A' \ni L$, $L' \ni L$, then by (10) the other condition $K' \ni L$ is also fulfilled, and hence it follows (because by (9) $L \ni K$) that $K' \ni K$, i.e., K is a chain, which was to be proved.

42. Theorem. A system M compounded solely out of chains A, B, C, ... is a chain.

Proof. Since by (23) $M' = \mathfrak{M}(A', B', C', \ldots)$, and by hypothesis $A' \ni B$, $B' \ni B$, $C' \ni C$, ..., it follows by (12) that $M' \ni M$, which was to be proved.

43. Theorem. The intersection G of chains A, B, C, ... is a chain.

Proof. Since by (17) G is a common part of A, B, C, ..., therefore by (22) G' is a common part of A', B', C', ..., and by hypothesis $A' \ni A$, $B' \ni B$, $C' \ni C$, ..., then by (7) G' is also a common part of A, B, C, ... and hence by (18) also a part of G, which was to be proved.

44. Definition. If A is any part of S, we will denote by A_o the intersection of all those chains (e.g., S) of which A is part; this intersection A_o exists (17) because A is itself a common part of all these chains. Since further by (43) A_o is a chain, we will call A_o the *chain of the system* A, or briefly the chain of A. This definition too is strictly related to the fundamental determinate mapping ϕ of the system S into itself, and if later, for the sake of clearness, it is necessary, we shall at pleasure use the symbol $\phi_o(A)$ instead of A_o, and likewise designate the chain of A corresponding to another mapping ω by $\omega_o(A)$. For this very important notion the following theorems hold true.

45. Theorem. $A \ni A_o$.

Proof. For A is a common part of all those chains whose intersection is A_o, whence the theorem follows by (18).

46. Theorem. $(A_o)' \ni A_o$.

Proof. For by (44) A_o is a chain (37).

47. Theorem. If A is part of a chain K, then $A_o \ni K$.

Proof. For A_o is the intersection and hence also a common part of all the chains K of which A is part.

48. Remark. One can easily convince oneself that the notion of the chain A_o defined in (44) is completely characterized by the preceding theorems, (45), (46), (47).

49. Theorem. $A' \ni (A_o)'$.

The proof follows from (45), (22).

50. Theorem. $A' \ni A_o$.

The proof follows from (49), (46), (7).

51. Theorem. If A is a chain, then $A_o = A$.

Proof. Since A is part of the chain A, then by (47) $A_o \ni A$, whence the theorem follows by (45), (5).

52. Theorem. If $B \ni A$, then $B \ni A_o$.

The proof follows from (45), (7).

53. Theorem. If $B \ni A_o$, then $B_o \ni A_o$, and conversely.

Proof. Because A_o is a chain, then by (47) from $B \ni A_o$ we also get $B_o \ni A_o$; conversely, if $B_o \ni A_o$, then by (7) we also get $B \ni A_o$, because by (45) $B \ni B_o$.

54. Theorem. If $B \ni A$, then $B_o \ni A_o$.

The proof follows from (52), (53).

55. Theorem. If $B \ni A_o$, then $B' \ni A_o$ as well.

Proof. For by (53) $B_o \ni A_o$, and since by (50) $B' \ni B_o$, the theorem to be proved follows by (7). The same result, as is easily seen, can be obtained from (22), (46), (7), or also from (40).

56. Theorem. If $B \ni A_o$, then $(B_o)' \ni (A_o)'$.

The proof follows from (53), (22).

57. Theorem and definition. $(A_o)' = (A')_o$, i.e., the image of the chain of A is at the same time the chain of the image of A. Hence we can designate this system in short by A_o' and at pleasure call it the *chain-image* or *image-chain* of A. With the clearer notation given in (44) the theorem might be expressed by $\phi(\phi_o(A)) = \phi_o(\phi(A))$.

Proof. If for brevity we put $(A')_o = L$, then L is a chain (44), and by (45) $A' \ni L$; hence by (41) there exists a chain K satisfying the conditions $A \ni K$, $K' \ni L$; hence from (47) we have $A_o \ni K$, therefore $(A_o)' \ni K'$, and hence by (7) also $(A_o)' \ni L$, i.e.,

$$(A_o)' \ni (A')_o.$$

Since further by (49) $A' \ni (A_o)'$, and by (44), (39) $(A_o)'$ is a chain, then by (47)

$$(A')_o \ni (A_o)',$$

whence the theorem follows by the preceding result (5).

58. Theorem. $A_o = \mathfrak{M}(A, A_o')$, i.e., the chain of A is compounded out of A and the image-chain of A.

Proof. If for brevity we again put

$$L = A_o' = (A_o)' = (A')_o \text{ and } K = \mathfrak{M}(A, L),$$

then by (45) $A' \ni L$, and since L is a chain, by (41) the same is true of K; since further $A \ni K$ (9), therefore by (47)

$$A_o \ni K.$$

On the other hand, since by (45) $A \ni A_o$, and by (46) also $L \ni A_o$, then by (10)

$$K \ni A_o,$$

whence the theorem to be proved $A_o = K$ follows by the preceding result (5).

59. Theorem of *complete induction*. To show that the chain A_o is part of any system Σ—be this latter part of S or not—it is sufficient to show,

ρ. that $A \ni \Sigma$, and

σ. that the image of every common element of A_o and Σ is likewise an element of Σ.

Proof. For if ρ is true, then by (45) the intersection $G = \mathfrak{G}(A_o, \Sigma)$ certainly exists, and by (18) $A \ni G$; since besides by (17)

$$G \ni A_o,$$

then G is also part of our system S, which is mapped into itself by ϕ; and by (55) we have $G' \ni A_o$. If σ is likewise true, i.e., if $G' \ni \Sigma$, then G' must, as a common part of the systems A_o, Σ, by (18), be part of their intersection G, i.e., G is a chain (37), and since, as above noted, $A \ni G$, then by (47)

$$A_o \ni G,$$

and therefore by combination with the preceding result $G = A_o$, hence by (17) $A_o \ni \Sigma$, which was to be proved.

60. The preceding theorem, as will be shown later, forms the scientific basis for the form of demonstration known by the name of complete induction (the inference from n to $n + 1$); it can also be stated in the following manner: In order to show that all elements of the chain A_o possess a certain property $\mathfrak{E}$ (or that a proposition $\mathfrak{S}$ about an indeterminate thing n actually holds good for all elements n of the chain A_o) it is sufficient to show,

ρ. that all elements a of the system A possess the property $\mathfrak{E}$ (or that $\mathfrak{S}$ holds for all a's), and

σ. that to the image n' of every element n of A_o which has the property $\mathfrak{E}$, there belongs the same property $\mathfrak{E}$ (or that the proposition $\mathfrak{S}$, as soon as it holds for an element n of A_o, must also hold for its image n').

Indeed, if we denote by Σ the system of all things possessing the property $\mathfrak{E}$ (or for which the proposition $\mathfrak{S}$ holds) the complete agreement of the present manner of stating the theorem with that employed in (59) is immediately obvious.

61. Theorem. The chain of $\mathfrak{M}(A, B, C, \ldots)$ is $\mathfrak{M}(A_o, B_o, C_o, \ldots)$.

Proof. If we designate by M the former, by K the latter system, then by (42)

K is a chain. Since then by (45) each of the systems $A, B, C, \ldots$ is part of one of the systems $A_o, B_o, C_o, \ldots$, and therefore by (12) $M \ni K$, then by (47) we also have

$$M_o \ni K.$$

On the other hand, since by (9) each of the systems $A, B, C, \ldots$ is part of M, and hence by (45), (7) is also part of the chain M_o, then by (47) each of the systems $A_o, B_o, C_o, \ldots$ must also be part of M_o, therefore by (10)

$$K \ni M_o$$

whence (by combination with the preceding result) follows the theorem to be proved, $M_o = K$ (5).

62. Theorem. The chain of $\mathfrak{G}(A, B, C, \ldots)$ is part of $\mathfrak{G}(A_o, B_o, C_o, \ldots)$.

Proof. If we designate by G the former, by K the latter system, then by (43) K is a chain. Since then each of the systems $A_o, B_o, C_o, \ldots$ by (45) is a whole of one of the systems $A, B, C, \ldots$, and hence by (20) $G \ni K$, therefore by (47) we obtain the theorem to be proved, $G_o \ni K$.

63. Theorem. If $K' \ni L \ni K$, and therefore K is a chain, L is also a chain. If it is a proper part of K, and U is the system of all those elements of K which are not contained in L, and if the chain U_o is a proper part of K, and if V is the system of all those elements of K which are not contained in U_o, then $K = \mathfrak{M}(U_o, V)$ and $L = \mathfrak{M}(U_o', V)$. If finally $L = K'$ then $V \ni V'$.

The proof of this theorem, of which (as of the two preceding) we shall make no use, may be left for the reader.

§5.

The Finite and the Infinite

64. Definition.* A system S is said to be *infinite* when it is similar to a proper part of itself (32); otherwise S is said to be a *finite* system.

65. Theorem. Every system consisting of a single element is finite.

Proof. For such a system possesses no proper part (2), (6).

66. Theorem. There exist infinite systems.

Proof.† My own realm of thoughts, i.e., the totality S of all things which can be objects of my thought, is infinite. For if s signifies an element of S, then

---

* If one does not care to employ the notion of similar sysems (32) one must say: S is said to be infinite, when there is a proper part of S (6) into which S can be distinctly (similarly) mapped (26), (36). In this form I submitted the definition of the infinite which forms the core of my whole investigation in September 1882 to G. Cantor and several years earlier to Schwarz and Weber. All other attempts that have come to my knowledge to distinguish the infinite from the finite seem to me to have met with so little success that I think I may be permitted to forgo any criticism of them.

† A similar consideration is found in §13 of the *Paradoxien des Unendlichen* by Bolzano (Leipzig, 1851).

the thought s', that s can be object of my thought, is itself an element of S. If we regard this as an image $\phi(s)$ of the element s, then the mapping ϕ of S, thus determined, has the property that the image S' is part of S; and S' is certainly a proper part of S, because there are elements in S (e.g., my own ego) which are different from such a thought s' and therefore are not contained in S'. Finally it is clear that if a, b are different elements of S, their images a', b' are also different, and that therefore the mapping ϕ is a distinct (similar) mapping (26). Hence S is infinite, which was to be proved.

67. Theorem. If R, S are similar systems, then R is finite or infinite according as S is finite or infinite.

Proof. If S is infinite, and therefore similar to a proper part S' of itself, then if R and S are similar, S' by (33) must be similar to R and by (35) likewise similar to a proper part of R, which therefore by (33) is itself similar to R; therefore R is infinite, which was to be proved.

68. Theorem. Every system S which possesses an infinite part is likewise infinite; or, in other words, every part of a finite system is finite.

Proof. If T is infinite (i.e. if there is a similar mapping ψ of T, such that $\psi(T)$ is a proper part of T) then, if T is part of S, we can extend this mapping ψ to a mapping ϕ of S in which, if s denotes any element of S, we put $\phi(s) = \psi(s)$ or $\phi(s) = s$ according as s is an element of T or not. This mapping ϕ is a similar one; for, if a, b denote different elements of S, then, if both are contained in T, the image $\phi(a) = \psi(a)$ is different from the image $\phi(b) = \psi(b)$, because ψ is a similar mapping; if further a is contained in T, but b not, then $\phi(a) = \psi(a)$ is different from $\phi(b) = b$, because $\psi(a)$ is contained in T; if, finally, neither a nor b is contained in T then $\phi(a) = a$ is different from $\phi(b) = b$, which was to be shown. Since further $\psi(T)$ is part of T (and therefore by (7) also part of S) it is clear that $\phi(S) \ni S$. Since finally $\psi(T)$ is a proper part of T there exists in T, and therefore also in S, an element t not contained in $\psi(T) = \phi(T)$; since then the image $\phi(s)$ of every element s not contained in T is equal to s (and hence is different from t) t cannot be contained in $\phi(S)$; hence $\phi(S)$ is a proper part of S and consequently S is infinite, which was to be proved.

69. Theorem. Every system which is similar to a part of a finite system is itself finite.

The proof follows from (67), (68).

70. Theorem. If a is an element of S, and if the aggregate T of all the elements of S different from a is finite, then S is finite.

Proof. We have by (64) to show that if ϕ denotes any similar mapping of S into itself, the image $\phi(S)$ or S' is never a proper part of S but always $=S$. Obviously $S = \mathfrak{M}(a, T)$ and hence by (23), if the images are again denoted by primes, $S' = \mathfrak{M}(a', T')$, and, on account of the similarity of the mapping ϕ, a' is not contained in T' (26). Since further by hypothesis $S' \ni S$, then a' and likewise every element of T' must either $=a$, or be an element of T. If then—the case which we will treat first—a is not contained in T', then $T' \ni T$ and hence $T' = T$, because ϕ is a similar mapping and because T is a finite system; and since a', as remarked, is not contained in T', i.e., not in T, then $a' = a$, and hence in this case we actually have $S' = S$ as was stated. In the opposite case when a is contained in T' and hence is the image b' of an element b contained in T, we will designate by U the aggregate of all those elements u of T which are different from b; then $T = \mathfrak{M}(b, U)$ and by (15) $S = \mathfrak{M}(a, b, U)$, hence $S' = \mathfrak{M}(a', a, U')$. We now determine a new mapping ψ of T by setting $\psi(b) = a'$, and generally $\psi(u) = u'$, whence by (23) $\psi(T) = \mathfrak{M}(a', U')$. Obviously ψ is a similar mapping, because ϕ was, and because a is not contained in U and therefore a' is not contained in U'. Since a and every element u is different from b, it follows (on account of the similarity of ϕ) that a' and every element u' must be different from a and consequently contained in T; hence $\psi(T) \ni T$, and, since T is finite, $\psi(T) = T$, and $\mathfrak{M}(a', U') = T$. From this by (15) we obtain

$$\mathfrak{M}(a', a, U') = \mathfrak{M}(a, T)$$

i.e., according to the above argument, $S' = S$. Therefore in this case too the proof demanded has been secured.

§6.

SIMPLY INFINITE SYSTEMS. SERIES OF NATURAL NUMBERS

71. Definition. A system N is said to be *simply infinite* when there exists a similar mapping ϕ of N into itself such that N appears as the chain (44) of an element not contained in $\phi(N)$. We call this element, which we shall denote in what follows by the symbol 1, the *base-element* of N, and say the simply infinite system N is *ordered* ⟦*geordnet*⟧ by this mapping ϕ. If we retain the earlier convenient symbols for images and chains (§4) then the essence of a simply infinite system N consists in the existence of a mapping ϕ of N and an element 1 which satisfy the following conditions α, β, γ, δ:

α. $N' \ni N$.

β. $N = 1_o$.

γ. The element 1 is not contained in N'.

δ. The mapping ϕ is similar.

Obviously it follows from α, γ, δ that every simply infinite system N is actually an infinite system (64), because it is similar to a proper part N' of itself.

72. Theorem. In every infinite system S a simply infinite system N is contained as a part.

Proof. By (64) there exists a similar mapping ϕ of S such that $\phi(S)$ or S' is a proper part of S; hence there exists an element 1 in S which is not contained in S'. The chain $N = 1_o$, which corresponds to this mapping ϕ of the system S into itself (44), is a simply infinite system ordered by ϕ; for the characteristic conditions α, β, γ, δ in (71) are obviously all fulfilled.

73. Definition. If in the consideration of a simply infinite system N ordered by a mapping ϕ we entirely neglect the special character of the elements, simply retaining their distinguishability and taking into account only the relations to one another in which they are placed by the ordering mapping ϕ, then these elements are called *natural numbers* or *ordinal numbers* or simply *numbers*, and the base-element 1 is called the *base-number* of the *number-series* N. With reference to this liberation of the elements from every other content (abstraction) we are justified in calling the numbers a free creation of the human mind. The relations or laws which are derived entirely from the conditions α, β, γ, δ in (71), and therefore are always the same in all ordered simply infinite systems, whatever names may happen to be given to the individual elements (compare 134), form the first object of the *science of numbers* or *arithmetic*. From the general concepts and theorems of §4 about the mapping of a system into itself we obtain immediately the following fundamental laws, where a, b, ... m, n, ... always denote elements of N (therefore numbers); A, B, C, ... parts of N; a', b', ... m', n', ... A', B', C' ... the corresponding images, which are produced by the ordering mapping ϕ and are always elements or parts of N; the image n' of a number n is also called the number *following* n.

74. Theorem. Every number n by (45) is contained in its chain n_0 and by (53) the condition $n \ni m_0$ is equivalent to $n_0 \ni m_0$.

75. Theorem. By (57) $n_0' = (n_0)' = (n')_0$.

76. Theorem. By (46) $n_0' \ni n_0$.

77. Theorem. By (58) $n_0 = \mathfrak{M}(n, n_0')$.

78. Theorem. $N = \mathfrak{M}(1, N')$, hence every number different from the base-number 1 is an element of N', i.e., an image of a number.

The proof follows from (77) and (71).

79. Theorem. N is the only number-chain containing the base-number 1.

Proof. For if 1 is an element of a number-chain K, then by (47) the associated chain $N \ni K$; hence $N = K$, because it is self-evident that $K \ni N$.

80. Theorem of complete induction (inference from n to n'). In order to show that a theorem holds for all numbers n of a chain m_0, it is sufficient to show,

ρ. that it holds for $n = m$, and

σ. that from the validity of the theorem for a number n of the chain m_0 its validity for the following number n' always follows.

This results immediately from the more general theorem (59) or (60). The most frequently occurring case is where $m = 1$ and therefore m_0 is the complete number-series N.

§7.

Greater and smaller numbers

81. Theorem. Every number n is different from the following number n'.

Proof by complete induction (80):

ρ. The theorem is true for the number $n = 1$, because it is not contained in N' (71), while the following number $1'$ as image of the number 1 contained in N is an element of N'.

σ. If the theorem is true for a number n and we put the following number $n' = p$, then n is different from p, whence by (26) on account of the similarity (71) of the ordering mapping ϕ it follows that n', and therefore p, is different from p'. Hence the theorem holds also for the number p following n, which was to be proved.

82. Theorem. In the image-chain n_o' of a number n by (74), (75) is contained its image n', but not the number n itself.

Proof by complete induction (80):

ρ. The theorem is true for $n = 1$, because $1_o' = N'$, and because by (71) the base-number 1 is not contained in N'.

σ. If the theorem is true for a number n, and we again put $n' = p$, then n is not contained in p_o, and therefore it is different from every number q contained in p_o, whence by reason of the similarity of ϕ it follows that n', and therefore p, is different from every number q' contained in p_o', and is hence not contained in p_o'. Therefore the theorem holds also for the number p following n, which was to be proved.

83. Theorem. The image-chain n_o' is a proper part of the chain n_o.

The proof follows from (76), (74), (82).

84. Theorem. From $m_o = n_o$ it follows that $m = n$.

Proof. Since by (74) m is contained in m_o, and

$$m_o = n_o = \mathfrak{M}(n, n_o')$$

by (77), then if the theorem were false and hence m different from n, m would be contained in the chain n_o', hence by (74) $m_o \ni n_o'$, i.e., $n_o \ni n_o'$; but this contradicts theorem (83). Hence our theorem is established.

85. Theorem. If the number n is not contained in the number-chain K, then $K \ni n_o'$.

Proof by complete induction (80):

ρ. By (78) the theorem is true for $n = 1$.

σ. If the theorem is true for a number n, then it is also true for the following number $p = n'$; for if p is not contained in the number-chain K, then by (40) n also cannot be contained in K and hence by our hypothesis $K \ni n_o'$; now by (77) $n_o' = p_o = \mathfrak{M}(p, p_o')$, hence $K \ni \mathfrak{M}(p, p_o')$ and p is not contained in K, thus $K \ni p_o'$, which was to be proved.

86. Theorem. If the number n is not contained in the number-chain K, but its image n' is, then $K = n_o'$.

Proof. Since n is not contained in K, then by (85) $K \ni n_o'$, and since $n' \ni K$, then by (47) $n_o' \ni K$, and hence $K = n_o'$, which was to be proved.

87. Theorem. In every number-chain K there exists one, and by (84) only one, number k whose chain $k_o = K$.

Proof. If the base-number 1 is contained in K, then by (79) $K = N = 1_o$. Otherwise let Z be the system of all numbers not contained in K; since the base-number 1 is contained in Z, but Z is only a proper part of the number-series N, then by (79) Z cannot be a chain, i.e., Z' cannot be part of Z; hence there exists in Z a number n whose image n' is not contained in Z, and is therefore certainly contained in K; since further n is contained in Z, and therefore not in K, then by (86) $K = n'_o$, and hence $k = n'$, which was to be proved.

88. Theorem. If m, n are different numbers then by (83), (84) one and only one of the chains m_o, n_o is a proper part of the other, and either $n_o \ni m'_o$ or $m_o \ni n'_o$.

Proof. If n is contained in m_o (and hence by (74) $n_o \ni m_o$) then m cannot be contained in the chain n_o (because otherwise by (74) we should have $m_o \ni n_o$, and therefore $m_o = n_o$, and hence by (84) also $m = n$) and thence it follows by (85) that $n_o \ni m'_o$. Otherwise, when n is not contained in the chain m_o, we must have by (85) $m_o \ni n'_o$, which was to be proved.

89. Definition. The number m is said to be *less* than the number n and n *greater* than m, in symbols

$$m < n,\ n > m,$$

when the condition

$$n_o \ni m'_o$$

is fulfilled, which by (74) may also be expressed

$$n \ni m'_o.$$

90. Theorem. If m, n are any numbers, then always one and only one of the following cases λ, μ, ν occurs:

$$\lambda.\ m = n,\ n = m, \text{ i.e., } m_o = n_o$$

$$\mu.\ m < n,\ n > m, \text{ i.e., } n_o \ni m'_o$$

$$\nu.\ m > n,\ n < m, \text{ i.e., } m_o \ni n'_o.$$

Proof. For if λ occurs (84) then neither μ nor ν can occur, because by (83) we never have $n_o \ni n'_o$. But if λ does not occur then by (88) one and only one of the cases μ, ν occurs, which was to be proved.

91. Theorem. $n < n'$.

Proof. For the condition for the case ν in (90) is fulfilled by $m = n'$.

92. Definition. To express that m is either $=n$ or $<n$, hence not $>n$ (90), we use the symbols

$$m \leqq n \text{ or also } n \geqq m$$

and we say m is *at most equal* to n, and n is *at least equal* to m.

93. Theorem. Each of the conditions

$$m \leqq n,\ m < n',\ n_o \ni m_o$$

is equivalent to each of the others.

Proof. For if $m \leqq n$, then from λ, μ in (90) we always have $n_o \ni m_o$, because by (76) $m_o' \ni m$. Conversely, if $n_o \ni m_o$, and therefore by (74) $n \ni m_o$, it follows from $m_o = \mathfrak{M}(m, m_o')$ that either $n = m$, or $n \ni m_o'$, i.e., $n > m$. Hence the condition $m \leqq n$ is equivalent to $n_o \ni m_o$. Besides it follows from (22), (27), (75) that this condition $n_o \ni m_o$ is again equivalent to $n_o' \ni m_o'$, i.e., by μ in (90) to $m < n'$, which was to be proved.

94. Theorem. Each of the conditions

$$m' \leqq n,\ m' < n',\ m < n$$

is equivalent to each of the others.

The proof follows immediately from (93), if we replace in it m by m', and from μ in (90).

95. Theorem. If $l < m$ and $m \leqq n$ or if $l \leqq m$, and $m < n$, then $l < n$. But if $l \leqq m$ and $m \leqq n$, then $l \leqq n$.

Proof. For from the corresponding conditions (89), (93) $m_o \ni l_o'$ and $n_o \ni m_o$, we have by (7) $n_o \ni l_o'$ and the same thing comes also from the conditions $m_o \ni l_o$ and $n_o \ni m_o'$, because in consequence of the former we have $m_o' \ni l_o'$. Finally, from $m_o \ni l_o$ and $n_o \ni m_o$ we have $n_o \ni l_o$, which was to be proved.

96. Theorem. In every part T of N there exists one and only one *least* number k, i.e., a number k which is less than every other number contained in T. If T consists of a single number, then is it also the least number in T.

Proof. Since T_o is a chain (44), by (87) there exists one number k whose chain $k_o = T_o$. Since from this it follows by (45), (77) that $T \ni \mathfrak{M}(k, k_o')$, then first k itself must be contained in T (because otherwise $T \ni k_o'$, hence by (47) also $T_o \ni k_o'$, i.e., $k \ni k_o'$, which by (83) is impossible), and, second, every number of the system T different from k must be contained in k_o', i.e., be $>k$ (89), whence it follows at once from (90) that there exists in T one and only one least number, which was to be proved.

97. Theorem. The least number of the chain n_o is n, and the base-number 1 is the least of all numbers.

Proof. For by (74), (93) the condition $m \ni n_o$ is equivalent to $m \geqq n$. Our theorem also follows immediately from the proof of the preceding theorem, because if we there assume $T = n_o$, evidently $k = n$ (51).

98. Definition. If n is any number, then we shall denote by Z_n the system of all numbers that are *not greater* than n, and hence *not* contained in n_o'. The condition

$$m \ni Z_n$$

by (92), (93) is obviously equivalent to each of the following conditions:

$$m \leqq n,\ m < n',\ n_o \ni m_o.$$

99. Theorem. $1 \ni Z_n$ and $n \ni Z_n$.

The proof follows from (98) or from (71) and (82).

100. Theorem. Each of the conditions equivalent by (98)

$$m \ni Z_n,\ m \leqq n,\ m < n',\ n_o \ni m_o$$

is also equivalent to the condition

$$Z_m \ni Z_n.$$

Proof. For if $m \ni Z_n$, and hence $m \leqq n$, and if $l \ni Z_m$, and hence $l \leqq m$, then by (95) also $l \leqq n$, i.e., $l \ni Z_n$; if therefore $m \ni Z_n$, then every element l of the system Z_m is also an element of Z_n, i.e., $Z_m \ni Z_n$. Conversely, if $Z_m \ni Z_n$, then by (7) $m \ni Z_n$, because by (99) $m \ni Z_m$, which was to be proved.

101. Theorem. The conditions for the cases λ, μ, ν in (90) may also be put in the following form:

$$\lambda.\ m = n,\ n = m,\ Z_m = Z_n$$

$$\mu.\ m < n,\ n > m,\ Z_{m'} \ni Z_n$$

$$\nu.\ m > n,\ n < m,\ Z_{n'} \ni Z_m.$$

The proof follows immediately from (90) if we observe that by (100) the conditions $n_o \ni m_o$ and $Z_m \ni Z_n$ are equivalent.

102. Theorem. $Z_1 = 1$.

Proof. For by (99) the base-number 1 is contained in Z_1, while by (78) every number different from 1 is contained in $1_o'$, hence by (98) not in Z_1, which was to be proved.

103. Theorem. By (98) $N = \mathfrak{M}(Z_n, n_o')$.

104. Theorem. $n = \mathfrak{G}(Z_n, n_o)$, i.e., n is the only common element of the systems Z_n and n_o.

Proof. From (99) and (74) it follows that n is contained in Z_n and n_o; but every element of the chain n_o different from n is by (77) contained in n_o', and hence by (98) is not in Z_n, which was to be proved.

105. Theorem. By (91), (98) the number n' is not contained in Z_n.

106. Theorem. If $m < n$, then Z_m is a proper part of Z_n, and conversely.

Proof. If $m < n$, then by (100) $Z_m \ni Z_n$, and since the number n, (by (99) contained in Z_n) can by (98) not be contained in Z_m because $n > m$, therefore Z_m is a proper part of Z_n. Conversely if Z_m is a proper part of Z_n then by (100) $m \leqq n$, and since m cannot be $=n$, because otherwise $Z_m = Z_n$, we must have $m < n$, which was to be proved.

107. Theorem. Z_n is a proper part of $Z_{n'}$.

The proof follows from (106), because by (91) $n < n'$.

108. Theorem. $Z_{n'} = \mathfrak{M}(Z_n, n')$.

Proof. For every number contained in $Z_{n'}$ by (98) is $\leqq n'$, hence either $=n'$ or $<n'$, and therefore by (98) is an element of Z_n. Therefore $Z_{n'} \ni \mathfrak{M}(Z_n, n')$. Since conversely by (107) $Z_n \ni Z_{n'}$ and by (99) $n' \ni Z_{n'}$, then by (10) we have

$$\mathfrak{M}(Z_n, n') \ni Z_{n'}$$

whence our theorem follows by (5).

109. Theorem. The image Z_n' of the system Z_n is a proper part of the system $Z_{n'}$.

Proof. For every number contained in Z_n' the image m' of a number m is contained in Z_n, and since $m \leqq n$, and hence by (94) $m' \leqq n'$, we have by (98) $Z_n' \ni Z_{n'}$. Since further the number 1 by (99) is contained in $Z_{n'}$, but by (71) is not contained in the image Z_n', Z_n' is a proper part of $Z_{n'}$, which was to be proved.

110. Theorem. $Z_{n'} = \mathfrak{M}(1, Z_n')$.

Proof. Every number of the system $Z_{n'}$ different from 1 by (78) is the image m' of a number m which must be $\leqq n$, and hence by (98) contained in Z_n (because otherwise $m > n$, hence by (94) also $m' > n'$ and consequently by (98) m' would not be contained in $Z_{n'}$); but from $m \ni Z_n$ we have $m' \ni Z_n'$, and hence

$$Z_{n'} \ni \mathfrak{M}(1, Z_n').$$

Since conversely by (99) $1 \ni Z_{n'}$, and by (109) $Z_n' \ni Z_{n'}$, by (10) we have $\mathfrak{M}(1, Z_n') \ni Z_{n'}$ and hence our theorem follows by (5).

111. Definition. If in a system E of numbers there exists an element g which is greater than every other number contained in E, then g is said to be the *greatest* number of the system E, and by (90) there can evidently be only one such greatest number in E. If a system consists of a single number, then this number is itself the greatest number of the system.

112. Theorem. By (98) n is the greatest number of the system Z_n.

113. Theorem. If there exists in E a greatest number g, then $E \ni Z_g$.

Proof. For every number contained in E is $\leqq g$, and hence by (98) is contained in Z_g, which was to be proved.

114. Theorem. If E is part of a system Z_n, or, what amounts to the same thing, if there exists a number n such that all numbers contained in E are $\leqq n$, then E possesses a greatest number g.

Proof. The system of all numbers p satisfying the condition $E \ni Z_p$—and by our hypothesis such numbers exist—is a chain (37), because by (107), (7) it follows also that $E \ni Z_{p'}$, and hence by (87) $= g_o$, where g signifies the least of these numbers (96), (97). Hence also $E \ni Z_g$, therefore by (98) every number contained in E is $\leqq g$, and we have only to show that the number g is itself contained in E. This is immediately obvious if $g = 1$, for then by (102) Z_g, and consequently also E, consists of the single number 1. But if g is different from 1 and consequently by (78) is the image f' of a number f, then by (108) $E \ni \mathfrak{M}(Z_f, g)$; if therefore g were not contained in E, then we would have $E \ni Z_f$, and there would consequently be among the numbers p a number f by (91) $< g$, which is a contradiction; hence g is contained in E, which was to be proved.

115. Definition. If $l < m$ and $m < n$ we say the number m *lies between* l and n (also between n and l).

116. Theorem. There exists no number lying between n and n'.

Proof. For as soon as $m < n'$, and hence by (93) $m \leqq n$, then by (90) we cannot have $n < m$, which was to be proved.

117. Theorem. If t is a number in T, but not the least (96), then there exists in T one and only one *next smaller* number s, i.e., a number s such that $s < t$,

and that there exists in T no number lying between s and t. Similarly, if t is not the greatest number in T (111) there always exists in T one and only one *next greater* number u, i.e., a number u such that $t < u$, and that there exists in T no number lying between t and u. At the same time t in T is next greater than s and next smaller than u.

Proof. If t is not the least number in T, then let E be the system of all those numbers of T that are $<t$; then by (98) $E \ni Z_{t}$, and hence by (114) there exists in E a greatest number s obviously possessing the properties stated in the theorem; it is the only such number. If further t is not the greatest number in T, then by (96) there certainly exists among all the numbers of T that are $>t$ a least number u, which, and which alone, possesses the properties stated in the theorem. In like manner the correctness of the last part of the theorem is obvious.

118. Theorem. In N the number n' is next greater than n, and n next less than n'.

The proof follows from (116), (117).

§8

Finite and infinite parts of the number sequence

119. Theorem. Every system Z_n in (98) is finite.

Proof by complete induction (80).

ρ. By (65), (102) the theorem is true for $n = 1$.

σ. If Z_n is finite, then from (108) and (70) it follows that $Z_{n'}$ is also finite, which was to be proved.

120. Theorem. If m, n are different numbers, then Z_m, Z_n are dissimilar systems.

Proof. By reason of symmetry we may by (90) assume that $m < n$; then by (106) Z_m is a proper part of Z_n, and since by (119) Z_n is finite, by (64) Z_m and Z_n cannot be similar, which was to be proved.

121. Theorem. Every part E of the number-sequence N which possesses a greatest number (111), is finite.

The proof follows from (113), (119), (68).

122. Theorem. Every part U of the number-sequence N which possesses no greatest number, is simply infinite (71).

Proof. If u is any number in U, there exists in U by (117) one and only one next greater number than u, which we will denote by $\psi(u)$ and regard as an image of u. The thus perfectly determined mapping ψ of the system U obviously has the property

$$\alpha. \qquad \psi(U) \ni U,$$

i.e., U is mapped into itself by ψ. If further u, v are different numbers in U, then by symmetry we may by (90) assume that $u < v$; thus by (117) it follows from the definition of ψ that $\psi(u) \leqq v$ and $v < \psi(v)$, and hence by (95)

$\psi(u) < \psi(v)$; therefore by (90) the images $\psi(u)$, $\psi(v)$ are different, i.e.,

δ. the mapping ψ is similar.

Further, if u_1 denotes the least number (96) of the system U, then every number u contained in U is $\geqq u_1$, and since generally $u < \psi(u)$, then by (95) $u_1 < \psi(u)$, and therefore by (90) u_1 is different from $\psi(u)$, i.e.,

γ. the element u_1 of U is not contained in $\psi(U)$.

Therefore $\psi(U)$ is a proper part of U and hence by (64) U is an infinite system. In agreement with (44) we denote by when $\psi_o(V)$, when V is any part of U, the chain of V corresponding to the mapping ψ. We wish to show finally that

β. $U = \psi_o(u_1)$.

In fact, since every such chain $\psi_o(V)$ (by reason of its definition (44)) is a part of the system U mapped into itself by ψ, evidently $\psi_o(u_1) \ni U$; conversely, it is obvious from (45) that the element u_1 contained in U is contained in $\psi_o(u_1)$. But if we assume that there exist elements of U that are not contained in $\psi_o(u_1)$, then there must be among them by (96) a least number w, and since (by what precedes) this is different from the least number u_1 of the system U, by (117) there must exist in U a number v which is next smaller than w, whence it follows at once that $w = \phi(v)$; since therefore $v < w$, v, by reason of the definition of w, must be contained in $\psi_o(u_1)$; but from this by (55) it follows that $\psi(v)$ (and hence w) must be contained in $\psi_o(u_1)$, and since this is contrary to the definition of w, our foregoing hypothesis is inadmissible; therefore $U \ni \psi_o(u_1)$ and hence also $U = \psi_o(u_1)$, as stated. From α, β, γ, δ it then follows by (71) that U is a simply infinite system ordered by ψ, which was to be proved.

123. Theorem. In consequence of (121), (122) any part T of the number-sequence N is finite or simply infinite, according as a greatest number exists or does not exist in T.

§9

Definition of a Mapping of the Number-Sequence by Induction

124. In what follows we denote numbers by small italics and retain throughout all symbols of the previous §§6 to 8, while Ω designates an arbitrary system whose elements are not necessarily contained in N.

125. Theorem. Given an arbitrary (similar or dissimilar) mapping θ of a system Ω into itself, and given a determinate element ω in Ω, then to every number n there corresponds one and only one mapping ψ_n of the associated number-system Z_n explained in (98), which satisfies the conditions:*

*For clearness here and in the following theorem (126) I have especially mentioned condition I, although properly it is a consequence of II and III.

I. $\psi_n(Z_n) \ni \Omega$

II. $\psi_n(1) = \omega$

III. $\psi_n(t') = \theta\psi_n(t)$, if $t < n$,

where the symbol $\theta\psi_n$ has the meaning given in (25).

Proof by complete induction (80).

ρ. The theorem is true for $n = 1$. For in this case by (102) the system Z_n consists of the single number 1, and the mapping ψ is therefore completely defined by II alone, so that I is fulfilled while III drops out entirely.

σ. If the theorem is true for a number n then we show that it is also true for the following number $p = n'$, and we begin by proving that there can be only a single corresponding mapping ψ_p of the system Z_p. In fact, if a mapping ψ_p satisfies the conditions

I′. $\psi_p(Z_p) \ni \Omega$

II′. $\psi_p(1) = \omega$

III′. $\psi_p(m') = \theta\psi_p(m)$, when $m < p$

then there is also contained in it by (21), because $Z_n \ni Z_p$ (107), a mapping of Z_n which obviously satisfies the same conditions I, II, III as ψ_n, and therefore coincides throughout with ψ_n. For all numbers contained in Z_n, and hence (98) for all numbers m which are $< p$, i.e., $\leqq n$, we therefore have:

$$\psi_p(m) = \psi_n(m) \qquad (m)$$

whence there follows, as a special case,

$$\psi_p(n) = \psi_n(n). \qquad (n)$$

Since further by (105), (108) p is the only number of the system Z_p not contained in Z_n, and since III′ and (n) we must also have

$$\psi_p(p) = \theta\psi_n(n) \qquad (p)$$

we have shown the correctness of our foregoing statement that there can be only one mapping ψ_p of the system Z_p satisfying the conditions I′, II′, III′, because by the conditions (m) and (p) just derived ψ_p is completely reduced to ψ_n. We have next to show conversely that this mapping ψ_p of the system Z_p completely determined by (m) and (p) actually satisfies the conditions I′, II′, III′. Obviously I′ follows from (m) and (p) with reference to I, and because $\theta(\Omega) \ni \Omega$. Similarly II′ follows from (m) and II, since by (99) the number 1 is contained in Z_n. The correctness of III′ follows first for those numbers m which are $< n$ from (m) and III, and for the single number $m = n$ yet remaining it results from (p) and (n). Thus it is completely established that the validity of our theorem for the number n implies its validity for the following number p, which was to be proved.

126. Theorem of *definition by induction*. Given an arbitrary (similar or

dissimilar) mapping θ of a system Ω into itself, and given a determinate element ω in Ω, then there exists one and only one mapping ψ of the number-series N which satisfies the conditions

$$\begin{aligned} &\text{I.} \quad \psi(N) \ni \Omega \\ &\text{II.} \quad \psi(1) = \omega \\ &\text{III.} \quad \psi(n') = \theta\psi(n), \text{ where } n \text{ represents every number.} \end{aligned}$$

Proof. Since, if there actually exists such a mapping ψ, it contains by (21) a mapping ψ_n of the system Z_n which satisfies the conditions I, II, III stated in (125), then, because there exists one and only one such mapping ψ_n,

$$\psi(n) = \psi_n(n). \qquad (n)$$

Since ψ is thus completely determined it follows also that there can exist only one such mapping ψ (see the closing remark in (130)). That conversely the mapping ψ determined by (n) also satisfies our conditions I, II, III, follows easily from (n) with reference to the properties I, II and (p) shown in (125), which was to be proved.

127. Theorem. Under the hypotheses made in the foregoing theorem,

$$\psi(T') = \theta\psi(T),$$

where T denotes any part of the number-sequence N.

Proof. If t denotes every number of system T, then $\psi(T')$ consists of all elements $\psi(t')$, and $\theta\psi(T)$ of all elements $\theta\psi(t)$; hence our theorem follows because (by III in (126)) $\psi(t') = \theta\psi(t)$.

128. Theorem. If we maintain the same hypotheses and denote by θ_o the chains (44) which correspond to the mapping θ of the system Ω into itself, then

$$\psi(N) = \theta_o(\omega).$$

Proof. We show first by complete induction (80) that

$$\psi(N) \ni \theta_o(\omega),$$

i.e., that every image $\psi(n)$ is also an element of $\theta_o(\omega)$. In fact,

ρ. this theorem is true for $n = 1$, because by (126, II) $\psi(1) = \omega$, and because by (45) $\omega \ni \theta_o(\omega)$.

σ. If the theorem is true for a number n, and hence $\psi(n) \ni \theta_o(\omega)$, then by (55) also $\theta(\psi(n)) \ni \theta_o(\omega)$, i.e., by (126, III) $\psi(n') \ni \theta_o(\omega)$. Hence the theorem is true for the following number n', which was to be proved.

In order further to show that every element ν of the chain $\theta_o(\omega)$ is contained in $\psi(N)$, and therefore that

$$\theta_o(\omega) \ni \psi(N)$$

we likewise apply complete induction, i.e., theorem (59) transferred to Ω and the mapping θ. In fact,

ρ. the element $\omega = \psi(1)$, and hence is contained in $\psi(N)$.

σ. If ν is a common element of the chain $\theta_o(\omega)$ and of the system $\psi(N)$, then $\nu = \psi(n)$, where n denotes a number, and by (126, III) we get $\theta(\nu) = \theta\psi(n) = \psi(n')$. Therefore $\theta(\nu)$ is contained in $\psi(N)$, which was to be proved.

From the theorems just established, $\psi(N) \ni \theta_o(\omega)$ and $\theta_o(\omega) \ni \psi(N)$, we get by (5) $\psi(N) = \theta_o(\omega)$, which was to be proved.

129. Theorem. Under the same hypotheses we have generally:

$$\psi(n_o) = \theta_o(\psi(n)).$$

Proof by complete induction (80). For

ρ. By (128) the theorem holds for $n = 1$, since $1_o = N$ and $\psi(1) = \omega$.

σ. If the theorem is true for a number n, then

$$\theta(\psi(n_o)) = \theta(\theta_o(\psi(n)));$$

since by (127), (75)

$$\theta(\psi(n_o)) = \psi(n_o'),$$

and by (57), (126, III)

$$\theta(\theta_o(\psi(n))) = \theta_o(\theta(\psi(n))) = \theta_o(\psi(n')),$$

we get

$$\psi(n_o') = \theta_o(\psi(n')),$$

i.e., the theorem is true for the number n' following n, which was to be proved.

130. Remark. Before we pass to the most important applications of the theorem of definition by induction proved in (126), (sections X–XIV), it is worth while to call attention to a circumstance by which it is essentially distinguished from the theorem of demonstration by induction proved in (80) or rather in (59), (60), however close the relation between the former and the latter may seem. For while the theorem (59) is true quite generally for every chain A_o where A is any part of a system S mapped into itself by any mapping ϕ (§4), the case is quite different with the theorem (126), which declares only the existence of a consistent (or one-to-one) mapping ψ of the simply infinite system 1_o. If in the latter theorem (still maintaining the hypotheses regarding Ω and θ) we replace the number-series 1_o by an arbitrary chain A_o out of such a system S, and define a mapping ψ of A_o into Ω in a manner analogous to that in (126, II, III) by assuming that

ρ. to every element a of A there is to correspond a determinate element $\psi(a)$ selected from Ω, and

σ. for every element n contained in A_o and its image $n' = \phi(n)$, the condition $\psi(n') = \theta\psi(n)$ is to hold, then it would very frequently happen that such a mapping ψ does not exist, since these conditions ρ, σ may prove incompatible, even though the freedom of choice contained in ρ be restricted at the outset to conform to the condition σ. An example will be sufficient to show this. If the system S consisting of the different elements a and b is so mapped into itself by ϕ that $a' = b$, $b' = a$, then obviously $a_o = b_o = S$; suppose further the system Ω consisting of the different elements α, β, and γ be so mapped into itself by θ that $\theta(\alpha) = \beta$, $\theta(\beta) = \gamma$, $\theta(\gamma) = \alpha$; if we now demand a mapping ψ of a_o in Ω such that $\psi(a) = \alpha$, and such that for every element n contained in a_o $\psi(n') = \theta\psi(n)$, we obtain a contradiction. For if $n = a$, we get $\psi(b) = \theta(\alpha) = \beta$, and hence for $n = b$, we must have $\psi(a) = \theta(\beta) = \gamma$, while we had assumed $\psi(a) = \alpha$.

But if there exists a mapping ψ of A_o into Ω which satisfies the foregoing conditions ρ, σ, without contradiction, then from (60) it follows easily that it is completely determined. For if the mapping χ satisfies the same conditions, then we have, generally, $\chi(n) = \psi(n)$, since by ρ this theorem is true for all elements $n = a$ contained in A, and since if it is true for an element n of A_o it must by σ be true also for its image n'.

131. In order to bring out clearly, the import of our theorem (126), we will here insert a consideration which is useful for other investigations also, e.g., for the so-called theory of groups.

We consider a system Ω, whose elements allow a certain combination such that from an element ν by the action of an element ω, there always results a determinate element of the same system Ω, which may be denoted by $\omega.\nu$ or $\omega\nu$, which in general is to be distinguished from $\nu\omega$. We can also consider this system in such a way that to every determinate element ω there corresponds a determinate mapping of the system Ω into itself (to be denoted by $\dot{\omega}$), in so far as to every element ν there corresponds the determinate image $\dot{\omega}(\nu) = \omega\nu$. If to this system Ω and its element ω we apply theorem (126), designating by $\dot{\omega}$ the mapping there denoted by θ, then there corresponds to every number n a determinate element $\psi(n)$ contained in Ω, which may now be denoted by the symbol ω^n and sometimes called the nth power of ω; this notion is completely defined by the conditions imposed upon it

$$\text{II.}\quad \omega^1 = \omega$$

$$\text{III.}\quad \omega^{n'} = \omega\omega^n,$$

and its existence is established by the proof of theorem (126).

If the foregoing combination of the elements is further so qualified that for arbitrary elements μ, ν, ω, we always have $\omega(\nu\mu) = \omega\nu(\mu)$, then we also have the theorems

$$\omega^{n'} = \omega^n\omega, \qquad \omega^m\omega^n = \omega^n\omega^m,$$

whose proofs can easily be effected by complete induction and may be left to the reader.

The foregoing general consideration may be immediately applied to the following example. If S is a system of arbitrary elements, and Ω the associated system whose elements are all the mappings ν of S into itself (36), then by (25) these elements can always be compounded; for $\nu(S) \ni S$, and the mapping $\omega\nu$ compounded out of such mappings ν and ω is itself again an element of Ω. Then all elements ω^n are mappings of S into itself, and we say they arise by repetition of the mapping ω. We will now call attention to a simple connection existing between this notion and the notion of the chain $\omega_o(A)$ defined in (44), where A again denotes any part of S. If for brevity we denote by A_n the image $\omega^n(A)$ produced by the mapping ω^n, then from III and (25) it follows that $\omega(A_n) = A_{n'}$. Hence it is easily shown by complete induction (80) that all these systems A_n are parts of the chain $\omega_o(A)$; for

ρ. by (50) this statement is true for $n = 1$, and

σ. if it is true for a number n, then from (55) and from $A_{n'} = \omega(A_n)$ it follows that it is also true for the following number n', which was to be proved. Since further by (45) $A \ni \omega_o(A)$, it follows from (10) that the system K compounded out of A and all images A_n is part of $\omega_o(A)$. Conversely, by (23) $\omega(K)$ is compounded out of $\omega(A) = A_1$ and all systems $\omega(A_n) = A_{n'}$ (therefore by (78) out of all systems A_n, which by (9) are parts of K) so by (10) $\omega(K) \ni K$, i.e., K is a chain (37). By (9) $A \ni K$, so by (47) it follows that $\omega_o(A) \ni K$. Therefore $\omega_o(A) = K$, i.e., the following theorem holds: If ω is a mapping of a system S into itself, and A is any part of S, then the chain of A corresponding to the mapping ω is compounded out of A and all the images $\omega^n(A)$ resulting from repetitions of ω. We advise the reader to return to the earlier theorems (57), (58) with this conception of a chain.

§10

The class of simply infinite systems

132. Theorem. All simply infinite systems are similar to the number-series N and consequently by (33) to one another.

Proof. Let the simply infinite system Ω be ordered (71) by the mapping θ, and let ω be the base-element of Ω. If we again denote by θ_o the chains corresponding to the mapping θ (44), then by (71) the following are true:

α. $\theta(\Omega) \ni \Omega$.

β. $\Omega = \theta_o(\omega)$.

γ. ω is not contained in $\theta(\Omega)$.

δ. The mapping θ is similar.

If ψ denotes the mapping of the number-series N defined in (126), then from β and (128) we get first

$$\psi(N) = \Omega,$$

and hence we have only to show that ψ is a similar mapping, i.e. (26) that to different numbers m, n there correspond different images $\psi(m)$, $\psi(n)$. On account of symmetry we may by (90) assume that $m > n$, hence $m \ni n_o'$, and the theorem to prove comes to this: that $\psi(n)$ is not contained in $\psi(n_o')$, and hence by (127) is not contained in $\theta\psi(n_o)$. This we establish for every number n by complete induction (80). In fact,

ρ. this theorem is true by γ for $n = 1$, since $\psi(1) = \omega$ and $\psi(1_o) = \psi(N) = \Omega$

σ. If the theorem is true for a number n, then it is also true for the following number n'; for if $\psi(n')$, i.e., $\theta\psi(n)$, were contained in $\theta\psi(n_o')$, then by δ and (27), $\psi(n)$ would also be contained in $\psi(n_o')$, while our hypothesis states just the opposite; which was to be proved.

133. Theorem. Every system which is similar to a simply infinite system and therefore by (132), (33) to the number-sequence N is simply infinite.

Proof. If Ω is a system similar to the number-sequence N, then by (32) there exists a similar mapping ψ of N such that

$$\text{I. } \psi(N) = \Omega;$$

then we put

$$\text{II. } \psi(1) = \omega.$$

If we denote, as in (26), the inverse by $\bar{\psi}$ (also a similar mapping of Ω) then to every element ν of Ω there corresponds a determinate number $\bar{\psi}(\nu) = n$, viz., that number whose image $\psi(n) = \nu$. Since to this number n there corresponds a determinate following number $\phi(n) = n'$, and to this again a determinate element $\psi(n')$ in Ω, there belongs to every element ν of the system Ω a determinate element $\psi(n')$ of that system which, as image of ν, we shall designate by $\theta(\nu)$. Thus a mapping θ of Ω into itself is completely determined,* and in order to prove our theorem we shall show that Ω is ordered by θ (71) as a simply infinite system, i.e., that the conditions α, β, γ, δ stated in the proof of (132) are all fulfilled. α is immediately obvious from the definition of θ. Further, to every number n there corresponds an element $\nu = \phi(n)$, for which $\theta(\nu) = \psi(n')$. So we have generally,

$$\text{III. } \psi(n') = \theta\psi(n),$$

and thence in connection with I, II, α we see that the mappings θ, ψ fulfil all the conditions of theorem (126). Therefore β follows from (128) and I. Further by (127) and I

$$\psi(N') = \theta\psi(N) = \theta(\Omega),$$

and thence in combination with II and the similarity of the mapping ψ we obtain γ; for otherwise $\psi(1)$ must be contained in $\psi(N')$, hence by (27) the number

*Evidently θ is the mapping $\psi\phi\bar{\psi}$ compounded by (25) out of $\bar{\psi}$, ϕ, ψ.

1 in N', which by (71, γ) is not the case. If finally μ, ν denote elements of Ω and m, n the corresponding numbers whose images are $\psi(m) = \mu$, $\psi(n) = \nu$, then from the hypothesis $\theta(\mu) = \theta(\nu)$ it follows by the foregoing that $\psi(m') = \psi(n')$. So (because of the similarity of ψ, ϕ) $m' = n'$, $m = n$. Therefore $\mu = \nu$; hence δ is true, which was to be proved.

134. Remark. By the two preceding theorems (132), (133) all simply infinite systems form a class in the sense of (34). At the same time, with reference to (71), (73) it is clear that every theorem regarding numbers, i.e., regarding the elements n of the simply infinite system N ordered by the mapping ϕ (and indeed every theorem in which we leave entirely out of consideration the special character of the elements n and discuss only such notions as arise from the arrangement ϕ) possesses perfectly general validity for every other simply infinite system Ω ordered by a mapping θ and its elements ν, and that the passage from N to Ω (e.g., the translation of an arithmetical theorem from one language into another) is effected by the mapping ψ considered in (132), (133), which changes every element n of N into an element ν of Ω, i.e., into $\psi(n)$. This element ν can be called the nth element of Ω and accordingly the number n is itself the nth number of the number-sequence N. The same significance which the mapping ϕ possesses for the laws in the domain N, in so far as every element n is followed by a determinate element $\phi(n) = n'$, is found, after the change effected by ψ, to belong to the mapping θ for the same laws in the domain Ω, in so far as the element $\nu = \psi(n)$ arising from the change of n is followed by the element $\theta(\nu) = \psi(n')$ arising from the change of n'. We are therefore justified in saying that ϕ is changed by ψ into θ, which is symbolically expressed by $\theta = \psi\phi\bar{\psi}$, $\phi = \bar{\psi}\theta\psi$. By these remarks, I believe, the definition of the notion of numbers given in (73) is fully justified. We now proceed to further applications of theorem (126).

§11
Addition of numbers

135. Definition. It is natural to apply the definition (set forth in theorem (126)) of a mapping ψ of the number-sequence N (or of the *function* $\psi(n)$ determined by it) to the case where the system (there denoted by Ω) in which the image $\psi(N)$ is to be contained is the number-series N itself. For this system Ω a mapping θ of Ω into itself already exists, viz., that mapping ϕ by which N is ordered as a simply infinite system (71), (73). Then $\Omega = N$, $\theta(n) = \phi(n) = n'$, hence

$$\text{I.} \quad \psi(N) \ni N,$$

and to determine ψ completely it remains only to select the element ω from Ω, i.e., from N, at pleasure. If we take $\omega = 1$, then evidently ψ becomes the identical mapping (21) of N, because the conditions

$$\psi(1) = 1, \; \psi(n') = (\psi(n))'$$

are generally satisfied by $\psi(n) = n$. If then we are to produce another mapping ψ of N, then for ω we must select a numbee m' different from 1 (and, by (78), contained in N) where m itself denotes any number. Since the mapping ψ is obviously dependent upon the choice of this number m, we denote the corresponding image $\psi(n)$ of an arbitrary number n by the symbol $m + n$, and call this number the *sum* which arises from the number m by the *addition* of the number n, or in short the sum of the numbers m, n. Therefore by (126) this sum is completely determined by the conditions*

$$\text{II.}\quad m + 1 = m',$$
$$\text{III.}\quad m + n' = (m + n)'.$$

136. Theorem. $m' + n = m + n'$.

Proof by complete induction (80). For

ρ. the theorem is true for $n = 1$, since by (135, II)

$$m' + 1 = (m')' = (m + 1)',$$

and by (135, III) $(m + 1)' = m + 1'$.

σ If the theorem is true for a number n, and we put the following number $n' = p$, then $m' + n = m + p$, hence also $(m' + n)' = (m + p)'$, whence by (135, III) $m' + p = m + p'$; therefore the theorem is true also for the following number p, which was to be proved.

137. Theorem. $m' + n = (m + n)'$.

The proof follows from (136) and (135, III).

138. Theorem. $1 + n = n'$.

Proof by complete induction (80). For

ρ. by (135, II) the theorem is true for $n = 1$.

σ. If the theorem is true for a number n and we put $n' = p$, then $1 + n = p$, therefore also $(1 + n)' = p'$, whence by (135, III) $1 + p = p'$, i.e., the theorem is true also for the following number p, which was to be proved.

139. Theorem. $1 + n = n + 1$.

The proof follows from (138) and (135, II).

140. Theorem $m + n = n + m$.

Proof by complete induction (80). For

ρ. by (139) the theorem is true for $n = 1$.

σ. If the theorem is true for a number n, then it follows that $(m + n)' = (n + m)'$, i.e., by (135, III) $m + n' = n + m'$, hence by (136) $m + n' = n' + m$; therefore the theorem is also true for the following number

---

*The above definition of addition based immediately upon theorem (126) seems to me to be the simplest. By the aid of the notion developed in (131) we can, however, define the sum $m + n$ by $\phi^n(m)$ or also by $\phi^m(n)$, where ϕ has again the foregoing meaning. In order to show the complete agreement of these definitions with the foregoing, we need by (126) only show that if $\phi^n(m)$ or $\phi^m(n)$ is denoted by $\psi(n)$, the condition $\psi(1) = m'$, $\psi(n') = \phi\psi(n)$are fulfilled. This is easily done with the aid of complete induction (80) and the help of (131).

n', which was to be proved.

141. Theorem $(l + m) + n = l + (m + n)$.

Proof by complete induction (80). For

ρ. the theorem is true for $n = 1$, because by (135, II, III, II) $(l + m) + 1 = (l + m)' = l + m' = l + (m + 1)$.

σ. If the theorem is true for a number n, then $((l + m) + n)' = (l + (m + n))'$, i.e., by (135, III)

$$(l + m) + n' = l + (m + n)' = l + (m + n'),$$

therefore the theorem is also true for the following number n', which was to be proved.

142. Theorem. $m + n > m$.

Proof by complete induction (80). For

ρ. by (135, II) and (91) the theorem is true for $n = 1$.

σ. If the theorem is true for a number n, then by (95) it is also true for the following number n'. For by (135, III) and (91)

$$m + n' = (m + n)' > m + n,$$

which was to be proved.

143. Theorem. The conditions $m > a$ and $m + n > a + n$ are equivalent.

Proof by complete induction (80). For

ρ. by (135, II) and (94) the theorem is true for $n = 1$.

σ. If the theorem is true for a number n, then it is also true for the following number n'. For by (94) the condition $m + n > a + n$ is equivalent to $(m + n)' > (a + n)'$, hence by (135, III) also equivalent to

$$m + n' > a + n',$$

which was to be proved.

144. Theorem. If $m > a$ and $n > b$, then

$$m + n > a + b.$$

Proof. For from our hypotheses we have by (143) $m + n > a + n$ and $n + a > b + a$ or, what by (140) is the same, $a + n > a + b$, whence the theorem follows by (95).

145. Theorem. If $m + n = a + n$, then $m = a$.

Proof. For if m does not $= a$ (i.e. by (90) either $m > a$ or $m < a$) then by (143) respectively $m + n > a + n$ or $m + n < a + n$. Therefore by (90) we cannot have $m + n = a + n$, which was to be proved.

146. Theorem. If $l > n$, then there exists one and (by (157)) only one number m which satisfies the condition $m + n = l$.

Proof by complete induction (80). For

ρ. the theorem is true for $n = 1$. In fact, if $l > 1$, i.e., (89) if l is contained in N', and hence is the transform m' of a number m, then by (135, II) it follows that $l = m + 1$, which was to be proved.

σ. If the theorem is true for a number n, then we show that it is also true for the following number n'. In fact, if $l > n'$, then by (91), (95) $l > n$, and hence there exists a number k which satisfies the condition $l = k + n$; since by (138) this is different from 1 (otherwise l would be $= n'$) then by (78) it is the image m' of a number m, and consequently $l = m' + n$, and therefore also by (136) $l = m + n'$, which was to be proved.

§12.

MULTIPLICATION OF NUMBERS

147. Definition. After having found in §11 an infinite system of new mappings of the number-sequence N into itself, we can by (126) use each of these in order to produce new mappings ψ of N. When we take $\Omega = N$, and $\theta(n) = m + n = n + m$, where m is a determinate number, we obtain

$$\text{I.} \quad \psi(N) \ni N,$$

and to determine ψ completely it remains only to select the element ω from N at pleasure. The simplest case occurs when we bring this choice into a certain agreement with the choice of θ, by putting $\omega = m$. Since the thus perfectly determinate ψ depends upon this number m, we designate the corresponding image $\psi(n)$ of any number n by the symbol $m \times n$ or $m.n$ or mn, and call this number the *product* arising from the number m by *multiplication* by the number n, or, for short, the product of the numbers m, n. This therefore by (126) is completely determined by the conditions

$$\text{II.} \quad m.1 = m$$

$$\text{III.} \quad mn' = mn + m,$$

148. Theorem. $m'n = mn + n$.
Proof by complete induction (80). For

ρ. by (147, II) and (135, II) the theorem is true for $n = 1$.

σ. If the theorem is true for a number n, we have

$$m'n + m' = (mn + n) + m'$$

and consequently by (147, III), (141), (140), (136), (141), (147, III)

$$m'n' = mn + (n + m') = mn + (m' + n) = mn + (m + n') =$$
$$(mn + m) + n' = mn' + n';$$

therefore the theorem is true for the following number n', which was to be proved.

149. Theorem. $1.n = n$.
Proof by complete induction (80). For

ρ. by (147, II) the theorem is true for $n = 1$.

σ. If the theorem is true for a number n, then we have $1.\ n + 1 = n + 1$, i.e., by (147, III), (135, II) $1.\ n' = n'$, therefore the theorem also holds for the following number n', which was to be proved.

150. Theorem. $mn = nm$.

Proof by complete induction (80). For

ρ. by (147, II), (149) the theorem is true for $n = 1$.

σ. If the theorem is true for a number n, then we have

$$mn + m = nm + m,$$

i.e., by (147, III), (148) $mn' = n'm$, therefore the theorem is also true for the following number n', which was to be proved.

151. Theorem. $l(m + n) = lm + ln$.

Proof by complete induction (80). For

ρ. by (135, II), (147, III), (147, II) the theorem is true for $n = 1$.

σ. If the theorem is true for a number n, we have

$$l(m + n) + l = (lm + ln) + l;$$

but by (147, III), (135, III) we have

$$l(m + n) + l = l(m + n)' = l(m + n'),$$

and by (141), (147, III)

$$(lm + ln) + l = lm + (ln + l) = lm + ln',$$

consequently $l(m + n') = lm + ln'$, i.e., the theorem is true also for the following number n', which was to be proved.

152. Theorem. $(m + n)l = ml + nl$.

The proof follows from (151), (150).

153. Theorem. $(lm)n = l(mn)$.

Proof by complete induction (80). For

ρ. by (147, II) the theorem is true for $n = 1$.

σ. If the theorem is true for a number n, then we have

$$(lm)n + lm = l(mn) + lm,$$

i.e., by (147, III), (151), (147, III)

$$(lm)n' = l(mn + m) = l(mn'),$$

hence the theorem is also true for the following number n', which was to be proved.

154. Remark. If in (147) we had assumed no relation between ω and θ, but had put $\omega = k$, $\theta(n) = m + n$, then by (126) we should have had a less simple mapping ψ of the number-series N; for the number 1 we should have had $\psi(1) = k$, and for every other number (which accordingly has the form n') we

should have had $\psi(n') = mn + k$. For in this way, as one easily sees, the condition $\psi(n') = \theta\psi(n)$, i.e., $\psi(n') = m + \psi(n)$ would be satisfied for all numbers n.

§13.

Exponentation of numbers

155. Definition. If in theorem (126) we again put $\Omega = N$, and further $\omega = a$, $\theta(n) = an = na$, we get a mapping ψ of N which still satisfies the condition

$$\text{I.}\quad \psi(N) \ni N;$$

the corresponding image $\psi(n)$ of any number n we denote by the symbol a^n, and call this number a *power of the base a*, while n is called the *exponent* of this power of a. Hence this notion is completely determined by the conditions

$$\text{II.}\quad a^1 = a$$

$$\text{III.}\quad a^{n'} = a.a^n = a^n.a.$$

156. Theorem. $a^{m+n} = a^m.a^n$.

Proof by complete induction (80). For

ρ. by (135, II), (155, III), (155, II) the theorem is true for $n = 1$.

σ. If theorem is true for a number n, we have

$$a^{m+n}.a = (a^m.a^n)a;$$

but by (155, III), (135, III) $a^{m+n}.a = a^{(m+n)'} = a^{m+n'}$, and by (153), (155, III) $(a^m.a^n)a = a^m(a^n.a) = a^m.a^{n'}$; hence $a^{m+n'} = a^m.a^{n'}$, i.e., the theorem is also true for the following number n', which was to be proved.

157. Theorem. $(a^m)^n = a^{mn}$.

Proof by complete induction (80). For

ρ. by (155, II), (147, II) the theorem is true for $n = 1$.

σ. If the theorem is true for a number n, we have

$$(a^m)^n.a^m = a^{mn}.a^m$$

but by (155, III) $(a^m)^n.a^m = (a^m)^{n'}$, and by (156), (147, III) $a^{mn}.a^m = a^{mn+m} = a^{mn'}$; hence $(a^m)^{n'} = a^{mn'}$, i.e., the theorem is also true for the following number n', which was to be proved.

158. Theorem. $(ab)^n = a^n.b^n$.

Proof by complete induction (80). For

ρ. by (155, II) the theorem is true for $n = 1$.

σ. If the theorem is true for a number n, then by (150), (153), (155, III) we have also $(ab)^n.a = a(a^n.b^n) = (a.a^n)b^n = a^{n'}.b^n$, and thus $((ab)^n.a)b = (a^{n'}.b^n)b$; but by (153), (155, III) $((ab)^n.a)b = (ab)^n.(ab) = (ab)^{n'}$, and likewise

$$(a^{n'}.b^n)b = a^{n'}.(b^n.b) = a^{n'}.b^{n'};$$

therefore $(ab)^{n'} = a^{n'}.b^{n'}$, i.e., the theorem is also true for the following number n', which was to be proved.

§14.

Number of the elements of a finite system

159. Theorem. If Σ is an infinite system, then every one of the number-systems Z_n defined in (98) is similarly mappable into Σ (i.e., similar to a part of Σ), and conversely.

Proof. If Σ is infinite, then by (72) there exists a part T of Σ which is simply infinite, and therefore by (132) similar to the number-sequence N. Consequently by (35) every system Z_n as part of N is similar to a part of T, and therefore also to a part of Σ, which was to be proved.

The proof of the converse—however obvious it may appear—is more complicated. If every system Z_n is similarly mappable into Σ, then to every number n there corresponds a similar mapping α_n of Z_n such that $\alpha_n(Z_n) \ni \Sigma$. From the existence of such a series of mappings α_n, regarded as given, but respecting which nothing further is assumed, we derive first by the aid of theorem (126) the existence of a new series of such mappings ψ_n possessing the special property that whenever $m < n$ (hence by (100) $Z_m \ni Z_n$) the mapping ψ_m of the part Z_m is contained in the mapping ψ_n of Z_n (21), i.e., the mappings ψ_m and ψ_n completely coincide with each other for all numbers contained in Z_m. Hence

$$\psi_m(m) = \psi_n(m).$$

In order to apply the theorem stated to gain this end we understand by Ω that system whose elements are all possible similar mappings of all systems Z_n into Σ, and by aid of the given elements α_n (likewise contained in Ω) we define in the following manner a mapping θ of Ω into itself. If β is any element of Ω (thus, e.g., a similar mapping of the determinate system Z_n into Σ) then the system $\alpha_{n'}(Z_{n'})$ cannot be part of $\beta(Z_n)$. For otherwise $Z_{n'}$ would be similar by (35) to a part of Z_n, and hence by (107) to a proper part of itself, and consequently infinite, which would contradict theorem (119). Therefore there exists in $Z_{n'}$ one number or several numbers p such that $\alpha_{n'}(p)$ is not contained in $\beta(Z_n)$. From these numbers p we select—simply to lay down something determinate—always the least k (96) and, since $Z_{n'}$ by (108) is compounded out of Z_n and n', we define a mapping γ of $Z_{n'}$ such that for all numbers m contained in Z_n the image $\gamma(m) = \beta(m)$ and $\gamma(n') = \alpha_{n'}(k)$. This obviously similar mapping γ of $Z_{n'}$ into Σ we consider then as an image $\theta(\beta)$ of the mapping β; thus a mapping θ of the system Ω into itself is completely defined. After the things named Ω and θ in (126) are determined we select finally for the element of Ω denoted by ω the given mapping α_1; thus by (126) a mapping ψ of the number-sequence N into Ω is determined, which, (if we denote the image belonging to an arbitrary number n, not by $\psi(n)$ but by ψ_n) satisfies the conditions

$$\text{II.} \quad \psi_1 = \alpha_1$$

$$\text{III.} \quad \psi_{n'} = \theta(\psi_n).$$

By complete induction (80) we find that ψ_n is a similar mapping of Z_n into Σ; for

ρ. by II this is true for $n = 1$.

σ. if this statement is true for a number n, it follows, from III and from the character of the above described transition θ from β to γ, that the statement is also true for the following number n', which was to be proved. Afterward we show likewise by complete induction (80) that if m is any number the above stated property

$$\psi_n(m) = \psi_m(m)$$

actually belongs to all numbers n, which are $\geqq m$, and therefore by (93), (74) belong to the chain m_o; in fact,

ρ. this is immediately evident for $n = m$, and

σ. if this property belongs to a number n it follows again from III and the nature of θ, that it also belongs to the number n', which was to be proved. After this special property of our new series of mappings ψ_n has been established, we can easily prove our theorem. We define a mapping χ of the number-sequence N, in which to every number n we let the image $\chi(n) = \psi_n(n)$ correspond; obviously by (21) all mappings ψ_n are contained in this one mapping χ. Since ψ_n was a mapping of Z_n into Σ, it follows first that the number-sequence N is likewise mapped by χ into Σ, hence $\chi(N) \ni \Sigma$. If further m, n are different numbers we may by symmetry according to (90) suppose $m < n$; then by the foregoing $\chi(m) = \psi_m(m) = \psi_n(m)$, and $\chi(n) = \psi_n(n)$. But since ψ_n was a similar mapping of Z_n into Σ, and m, n are different elements of Z_n, then $\psi_n(m)$ is different from $\psi_n(n)$, and hence also $\chi(m)$ is different from $\chi(n)$, i.e., χ is a similar mapping of N. Since further N is an infinite system (71), the same thing is true by (67) of the system $\chi(N)$ similar to it, and by (68) (because $\chi(N)$ is part of Σ) also of Σ, which was to be proved.

160. Theorem. A system Σ is finite or infinite, according as there does or does not exist a system Z_n similar to it.

Proof. If Σ is finite, then by (159) there exist systems Z_n which are not similarly mappable into Σ. Since by (102) the system Z_1 consists of the single number 1 (and hence is similarly mappable into every system) the least number k (96) to which a system Z_k not similarly mappable into Σ corresponds must be different from 1 and hence by (78) $= n'$. Since $n < n'$ (91) there exists a similar mapping ψ of Z_n into Σ; if then $\psi(Z_n)$ were only a proper part of Σ (i.e., if there existed an element α in Σ not contained in $\psi(Z_n)$) then since $Z_{n'} = \mathfrak{M}(Z_n, n')$ (108) we could extend this mapping ψ to a similar mapping ψ of $Z_{n'}$ into Σ by putting $\psi(n') = \alpha$. But by our hypothesis $Z_{n'}$ is not similarly mappable into Σ. Hence $\psi(Z_n) = \Sigma$, i.e., Z_n and Σ are similar systems. Conversely, if a system Σ is similar to a system Z_n, then by (119), (67) Σ is finite, which was to be proved.

161. Definition. If Σ is a finite system, then by (160) there exists one and (by (120), (33)) only one number n to which a system Z_n similar to the system Σ corresponds; this number n is called the *number* ⟦*Anzahl*⟧ of the elements contained in Σ (or also the *degree* of the system Σ) and we say Σ consists of or is a system of n elements, or that the number n shows *how many* elements are contained in Σ.* If numbers are used to express accurately this determinate property of finite systems they are called *cardinal numbers*. As soon as a determinate similar mapping ψ of the system Z_n is chosen by reason of which $\psi(Z_n) = \Sigma$, then to every number m contained in Z_n (i.e., every number m which is $\leqq n$) there corresponds a determinate element $\psi(m)$ of the system Σ, and conversely by (26) to every element of Σ by the inverse mapping $\bar{\psi}$ there corresponds a determinate number m in Z_n. Very often we denote all elements of Σ by a single letter, e.g., α, to which we append the distinguishing numbers m as indices, so that $\psi(m)$ is denoted by α_m. We say also that these elements are *counted and ordered* by ψ in a determinate manner, and we call α_m the mth element of Σ; if $m < n$ then $\alpha_{m'}$ is called the element *following* α_m, and α_n is called the *last* element. In this counting of the elements therefore the numbers m appear again as ordinal numbers (73).

162. Theorem. All systems similar to a finite system possess the same number of elements.

The proof follows immediately from (33), (161).

163. Theorem. The number of numbers contained in Z_n, i.e., of those numbers which are $\leqq n$, is n.

Proof. For by (32) Z_n is similar to itself.

164. Theorem. If a system consists of a single element, then the number of its elements $=1$, and conversely.

The proof follows immediately from (2), (26), (32), (102), (161).

165. Theorem. If T is a proper part of a finite system Σ, then the number of the elements of T is less than that of the elements of Σ.

Proof. By (68) T is a finite system, therefore similar to a system Z_m, where m denotes the number of the elements of T; if further n is the number of elements of Σ (therefore Σ is similar to Z_n) then by (35) T is similar to a proper part E of Z_n and by (33) Z_m and E are similar to each other. If then we were to have $n \leqq m$ (and hence $Z_n \ni Z_m$) by (7) E would also be a proper part of Z_m, and consequently Z_m would be an infinite system, which contradicts theorem (119). Hence by (90), $m < n$, which was to be proved.

166. Theorem. If $\Gamma = \mathfrak{M}(B, \gamma)$, where B denotes a system of n elements, and γ is an element of Γ not contained in B, then Γ consists of n' elements.

* For clearness and simplicity in what follows we restrict the notion of number throughout to finite systems; if then we speak of a number of certain things, it is always understood that the system whose elements these things are is a finite system.

Proof. For if $B = \psi(Z_n)$, where ψ denotes a similar mapping of Z_n, then by (105), (108) it may be extended to a similar mapping ψ of $Z_{n'}$, by putting $\psi(n') = \gamma$, and we get $\psi(Z_{n'}) = \Gamma$, which was to be proved.

167. Theorem. If γ is an element of a system Γ consisting of n' elements, then n is the number of all other elements of Γ.

Proof. For if B denotes the aggregate of all elements in Γ different from γ, then $\Gamma = \mathfrak{M}(B, \gamma)$; if b is the number of elements of the finite system B, by the foregoing theorem b' is the number of elements of Γ, and therefore $=n'$, whence by (26) we get $b = n$, which was to be proved.

168. Theorem. If A consists of m elements, and B of n elements, and if A and B have no common element, then $\mathfrak{M}(A, B)$ consists of $m + n$ elements.

Proof by complete induction (80). For

ρ. by (166), (164), (135, II) the theorem is true for $n = 1$.

σ. If the theorem is true for a number n, then it is also true for the following number n'. In fact, if Γ is a system of n' elements, then by (167) we can put $\Gamma = \mathfrak{M}(B, \gamma)$ where γ denotes an element and B the system of the n other elements of Γ. If then A is a system of m elements each of which is not contained in Γ (and therefore also not contained in B) and we put $\mathfrak{M}(A, B) = \Sigma$, by our hypothesis $m + n$ is the number of elements of Σ. And since γ is not contained in Σ, then by (166) the number of elements contained in $\mathfrak{M}(\Sigma, \gamma) = (m + n)'$, therefore by (135, III) $= m + n'$. But since by (15) obviously $\mathfrak{M}(\Sigma, \gamma) = \mathfrak{M}(A, B, \gamma) = \mathfrak{M}(A, \Gamma)$, then $m + n'$ is the number of the elements of $\mathfrak{M}(A, \Gamma)$, which was to be proved.

169. Theorem. If A, B are finite systems of m, n elements respectively, then $\mathfrak{M}(A, B)$ is a finite system and the number of its elements is $\leqq m + n$.

Proof. If $B \ni A$, then $\mathfrak{M}(A, B) = A$, and the number m of the elements of this system is by (142) $< m + n$, as was stated. But if B is not part of A, and T is the system of all those elements of B that are not contained in A, then by (165) their number is $p \leqq n$. And since obviously

$$\mathfrak{M}(A, B) = \mathfrak{M}(A, T),$$

then by (143) the number $m + p$ of the elements of this system is $\leqq m + n$, which was to be proved.

170. Theorem. Every system compounded out of a number n of finite systems is finite.

Proof by complete induction (80). For

ρ. by (8) the theorem is self-evident for $n = 1$.

σ. If the theorem is true for a number n, and if Σ is compounded out of n' finite systems, then let A be one of these systems and let B be the system compounded out of all the rest; since their number by (167) $= n$, then by our hypothesis B is a finite system. Since obviously $\Sigma = \mathfrak{M}(A, B)$, it follows from this and from (169) that Σ is also a finite system, which was to be proved.

171. Theorem. If ψ is a dissimilar mapping of a finite system Σ of n elements, then the number of elements of the image $\psi(\Sigma)$ is less than n.

Proof. If we select from all those elements of Σ that possess one and the same image always one and only one at pleasure, then the system T of all these selected elements is obviously a proper part of Σ; for ψ is a dissimilar mapping of Σ (26). At the same time it is clear by (21) that the transformation contained in ψ of this part T is a similar mapping, and that $\psi(T) = \psi(\Sigma)$; hence the system $\psi(\Sigma)$ is similar to the proper part T of Σ, and consequently our theorem follows by (162), (165).

172. Final remark. Although it has just been shown that the number m of the elements of $\psi(\Sigma)$ is less than the number n of the elements of Σ, yet in many cases we want to say that the number of elements of $\psi(\Sigma) = n$. The word number is then, of course, used in a different sense from that used hitherto (161); for if α is an element of Σ and a the number of all those elements of Σ that possess one and the same image $\psi(\alpha)$, then the latter as an element of $\psi(\Sigma)$ is still frequently regarded as representative of a elements, which at least from their derivation may be considered as different from one another, and is accordingly counted as an a-fold element of $\psi(\Sigma)$. In this way we reach the notion, very useful in many cases, of systems in which every element is endowed with a certain frequency-number which indicates how often it is to be reckoned as an element of the system. In the foregoing case, e.g., we would say that n is the number of the elements of $\psi(\Sigma)$ counted in this sense, while the number m of the actually different elements of this system coincides with the number of the elements of T. Similar deviations from the original meaning of a technical term which are simply extensions of the original notion occur very frequently in mathematics; but it does not lie in the line of this memoir to go further into their discussion.

F. *FROM THE* ELEVENTH SUPPLEMENT TO DIRICHLET'S *LECTURES ON THE THEORY OF NUMBERS* (*DEDEKIND 1894*)

The following extract is a comment on Dedekind's doctrine, expressed in the Preface to *Was sind und was sollen die Zahlen?*, that the numbers are a creation of the human mind; in the original text, the bulk of the extract occurs in a footnote. The translation is by William Ewald.

§161.

It often happens in mathematics and in other sciences that, given a system A of things or elements a, every particular element a is replaced, in accordance with a certain law, by a particular element a' corresponding to it (which can be contained in A or not); one calls such a law a *substitution*, and says that this substitution takes the element a into the element a', and similarly takes the system A into the system A' of elements a'. ‖Dedekind's footnote:‖ Already in the third edition of this work (1879, remark on p. 470) it is stated that no thought of any kind is possible without the capacity of the mind ‖Fähigkeit des Geistes‖ to compare a thing a with a thing a', or to relate a to a', or to allow an a' to correspond to a—and that the entire science of numbers also rests on this capacity. The elaboration of this thought has subsequently been published in my article, 'Was sind und was sollen die Zahlen?' (Brunswick, 1888); the notation employed there for mappings and their composition differs slightly in external features from the one used here.

G. LETTER TO HEINRICH WEBER (24 JANUARY 1888)

The following extract from Dedekind's correspondence with Weber was published in *Dedekind 1930–2*, Vol. iii, pp. 488–90. (Other letters to and from Weber are published in *Dedekind 1930–2* and in the appendices to *Dugac 1976*.) The translation is by William Ewald.

. . . I am delighted that you take such an interest in my article on numbers; not many do so. Cantor has called my attention to the fact that he already mentioned the difference between the finite and the infinite in 1877 (*Crelle*, Vol. 84, p. 242 ‖Cantor 1878‖), but says that he does not intend to claim priority. One could discuss this matter at length; he is right to a certain extent, but in 1882 he doubted the possibility of a simple definition, and was greatly surprised when I, as a result of his doubts, and at his request, communicated my definition to him; sometimes one possesses something without appreciating its value and significance. But I too have no wish to get involved in a conflict over priorities.—I have repeatedly read and thought through your remarks and suggestions; but whether they would lead to an essential simplification and shortening is difficult to say without seeing the new version in full detail. Besides, I must confess to you that I still regard the ordinal number and not the cardinal number (Anzahl)[a] as

[a] ‖'Cardinalzahl (Anzahl)'—in what follows, the German words in single parentheses are Dedekind's, not the translator's.‖

the original number-concept. It would perhaps have been better had I not mentioned these names (ordinal, cardinal) in my paper, since they are used in a different sense in ordinary grammar. My ordinal numbers, the abstract elements of the ordered simply-infinite system, have of course nothing to do with the adjectival form of the so-called (in grammar) ordinal numbers, from which form one could extract an argument for the conceptual priority of the cardinal numbers (Anzahlen). This adjectival form is also used where there is no question of an ordering (and consequently of my ordinal numbers), e.g. when one speaks of the fifth part of an interval. I hold the cardinal number (Anzahl) to be only an *application* of the ordinal number, and in our *ἀριθμητίζειν* ⟦'counting'⟧ too one reaches the concept five only via the concept four. But if one were to take your route—and I would strongly urge that it be explored once to the end—then I would advise that by number (Anzahl, cardinal number) one understand not the *class* itself (the system of all finite systems that are similar to each other) but something *new* (corresponding to this class) which the mind *creates*. We are a divine race and undoubtedly possess creative power, not merely in material things (railways, telegraphs) but especially in things of the mind. This is precisely the same question that you raise at the end of your letter in connection with my theory of irrationals, where you say that the irrational number is nothing other than the cut itself, while I prefer to create something *new* (different from the cut) that corresponds to the cut and of which I say that it brings forth, creates the cut. We have the right to ascribe such a creative power to ourselves; and moreover, because of the similarity ⟦Gleichartigkeit⟧ of all numbers, it is more expedient to proceed in this way. The rational numbers also produce cuts, but I would certainly not call the rational number identical with the cut it produces; and after the introduction of the irrational numbers one will often speak of cut-phenomena with such expressions, and ascribe to them such attributes, as would sound in the highest degree peculiar were they to be applied to the numbers themselves. Something quite similar holds for the definition of cardinal number (Anzahl) as a class; one will say many things about the class (e.g. that it is a system of infinitely many elements, namely, of all similar systems) that one would apply to the number only with the greatest reluctance; does anybody think, or won't he gladly forget, that the number four is a system of infinitely many elements? (But that the number four is the child of the number three and the mother of the number five is something that nobody will forget.) For the same reason, I always considered Kummer's *creation* of the ideal numbers to be thoroughly justified, if only it were rigorously carried out. Whether in addition the language of symbols ⟦Zeichensprache⟧ suffices to designate uniquely each individual that is to be created, is not important; it always suffices to designate the individuals that appear in any (limited) investigation. . . .

H. FELIX BERNSTEIN ON DEDEKIND AND CANTOR

At the request of Georg Cantor, Felix Bernstein visited Dedekind in Harzburg in 1897. Bernstein's account of the visit follows; it was published by Emmy Noether in *Dedekind 1930–2* (Vol. iii, p. 449). The translation is by William Ewald.

The visit in question was proposed by Cantor. He had shortly before found the paradox of the set of all ordinal numbers while attempting to prove that every set can be well-ordered—a proof which he tried to carry out with techniques roughly similar to the ones Zermelo, avoiding the inconsistent sets, later used in his first proof of well-ordering. Cantor was well aware that the discovered paradox could also be applied to the set of all things. Dedekind, in his paper, 'Was sind und was sollen die Zahlen?', had used this set to prove the existence of infinite sets—and in such a way that, by the construction of his paper, the definition of the numbers depends on the non-contradictory existence of these sets. Cantor had most likely already written and asked him for his reaction; and since no reaction materialized (probably because of Dedekind's severe illness in the winter 1896–97) he now commissioned me to obtain one by word of mouth.

Dedekind, however, was on that occasion unable to make a conclusive statement, and he said to me that in his ponderings he had almost begun to doubt whether human thought is fully rational.

The following episode should be of special interest. Dedekind said, with respect to the concept of set, that he imagined a set as a closed sack that contains completely determinate things—but things which one does not see, and of which one knows nothing except that they exist and are determinate. Somewhat later, Cantor gave his own conception of a set. He drew his colossal figure upright, made a magnificent gesture with his raised arm, and said, staring into the indeterminate [ins Unbestimmte], 'A set I imagine as an abyss.'

I. *FROM THE* NACHLASS

The following quotation is an undated fragment from the Dedekind *Nachlass* in the Niedersächsische Staats- und Universitätsbibliothek, Göttingen, where it has the classification Cod. Ms. Dedekind III, 2. The German text appears, by permission of the Göttingen Universitätsbibliothek, at the head of this book; it was also printed, with some errors of transcription, in *Dugac 1976* (p. 315). The translation is by William Ewald.

Of all the aids which the human mind has yet created to simplify its life—that is, to simplify the work in which thinking consists—none is so momentous and so inseparably bound up with the mind's most inward nature as the concept of *number*. Arithmetic, whose sole object is this concept, is already a science of immeasurable breadth, and there can be no doubt that there are absolutely no limits to its further development; and the domain of its application is equally immeasurable, for every thinking man, even if he does not clearly realize it, is a man of numbers, an arithmetician.

19

Georg Cantor (1845–1918)

Ever since Newton mathematicians have grappled with the problem of infinity as it arises in the foundations of the calculus and in the theory of the real numbers. This topic has arisen in many of the selections in Volume 1: recall, for instance, Berkeley's 'Of infinites' (*1707–8*); his critique of infinitesimals and fluxions in *The analyst* (*1734*); MacLaurin's defence of Newton in the *Treatise of fluxions* (*1742*); D'Alembert's comments on infinity, differentials, and limits (*1754*, *1765a*, and *1765b*); and Bolzano's paper on the intermediate value theorem (*1817a*), or his *Paradoxes of the infinite* (*1851*). In the course of the nineteenth century mathematicians like Gauss and Cauchy, Abel and Dirichlet, Fourier and Riemann, Weierstrass and Heine laboured to put real analysis on a rigorous foundation, and by the 1870s they had largely created the modern theory of the elementary calculus, giving precise ε and δ definitions of continuity, differentiability, limits, and the like.

Cantor's investigations grew out of this tradition, and mark a new era in the foundations of mathematics. Cantor, together with Dedekind, created the set-theoretic approach to mathematics that was to dominate the twentieth century; and for sheer unexpectedness, for sheer imaginative power, his theory of transfinite arithmetic has no equal in modern mathematics except possibly the theory of non-Euclidean geometry. The Continuum Hypothesis belongs to a different order of difficulty and depth than Boole's algebraic analysis of the syllogism, or De Morgan's studies of the logic of relations, or Frege and Peirce's discovery of the quantifiers. Those were important events, but in a sense inevitable; whereas the transfinite numbers burst on the world of mathematics entirely unexpectedly. Cantor's work raised profound new questions about the infinite, and was responsible for drawing mainstream mathematical intellects of the calibre of Hilbert, Poincaré, Hausdorff, Brouwer, Weyl, and von Neumann to the foundations of mathematics.

The following selections represent but a small part of Cantor's set-theoretical *œuvre*, and have been selected for the light they shed on the sources and development of his frequently murky thought. Those seeking a detailed examination of Cantor's researches and his struggles with the Continuum Hypothesis should consult the study by Michael Hallett, *Cantorian set theory and limitation of size* (*Hallett 1984*).

Cantor was born in St Petersburg, and schooled in Wiesbaden and Darmstadt. He entered the University of Berlin in 1863, where he received his mathematical education. His teachers at Berlin were Kummer, Kronecker, and above all Karl Weierstrass, whose lectures on the foundations of real analysis exerted a strong influence on Cantor's early papers. Cantor left Berlin in 1869 to assume a teaching position at the University of Halle; he remained in Halle for the rest of his life.

Cantor's early mathematical papers investigated Fourier series and the foundations of real analysis; it was this research that eventually led him to his research in the theory of sets. He was influenced in this early work by Riemann's 1854 *Habilitation* thesis, 'On the representation of a function by a trigonometric series', which had been posthumously published by Dedekind in 1868. (This paper, in addition to examining the convergence properties of Fourier series, also treated problems in infinitesimal analysis and stated the definition of the Riemann integral.) Cantor investigated the unique representability of functions by trigonometric series, proving in *Cantor 1870* that, if $f(x)$ is represented by a trigonometric series convergent for all x, then the representation is unique. In *Cantor 1871* he strengthened the result, showing that his theorem still holds even if the series diverges at a *finite* number of points in any given interval.

His next important paper, *Cantor 1872*, extended these results yet further; it made major contributions to classical analysis, and laid the groundwork for his study of infinite sets. Cantor begins in §1 of this paper by sketching his theory of the real numbers: a real number is an infinite series of rational numbers $a_1, a_2, \ldots, a_n, \ldots$ such that for any given ε there exists an n_1 such that, for $n \geq n_1$ and for any positive integer m, $|a_{n+m} - a_n| < \varepsilon$. (This theory is discussed by Cantor, and compared with the theories of Heine and Dedekind, in §9 of *Cantor 1883d*, translated below.) In §2, Cantor defines the notion of a 'boundary point' of a point-set P to be any point such that every neighbourhood of the point contains infinitely many points of P. The *first derivative* of P (designated by P') is the set of all boundary points of P; the second derivative P'' is the set of all boundary points of P', and so on. In §3 Cantor extends his earlier results on trigonometric series: uniqueness of representation holds even if the series diverges at an infinite number of points, provided the set of points is of finite order. (A point-set P is of finite order if, for some integer n, the nth derivative $P^{(n)}$ of P is a finite set.) The paper of 1872 laid the foundations of point-set topology (its techniques were adopted in particular by Haudsorff, Borel, and Fréchet); it also led Cantor into a deeper study of the cardinality of subsets of the real line and (in his *1880*) to the study of *infinite* derivatives of a point-set.

A. ON A PROPERTY OF THE SET OF REAL ALGEBRAIC NUMBERS (*CANTOR 1874*)

This article is Cantor's first published contribution to the theory of sets. The deep and epoch-making result of the paper is not the easy theorem alluded to in the title—the theorem that that the class of real algebraic numbers is countable—but rather the proof, in §2, that the class of real numbers is *not* countable. It was this result that first gave a clear sense to the idea that infinite sets could be of different sizes, and that marks the start of the theory of the transfinite. The background to Cantor's discovery can be found in the next selection below, in the opening letters of his correspondence with Dedekind.

It should be noted that Cantor's proof here is *not* the familiar 'diagonal argument', an argument he did not publish until 1891. (*Cantor 1891* is translated below.) Cantor's proof in this paper is rather based on the Bolzano–Weierstrass theorem, of which he had given a proof in *Cantor 1872*; unlike the diagonal argument, it does not extend to sets in general.

The translation is by William Ewald; references to *Cantor 1874* should be to the section numbers, which appeared in the original text.

By a real algebraic number one generally understands a real number ω which satisfies a non-constant equation of the form:

$$a_0\omega^n + a_1\omega^{n-1} + \ldots + a_n = 0, \tag{1}$$

where n, a_0, $a_1, \ldots a_n$ are integers: we can here imagine that the numbers n and a_0 are positive, the coefficients $a_0, a_1, \ldots a_n$ are without common parts, and the equation (1) is irreducible; with these stipulations it turns out that, by the well-known fundamental theorems of arithmetic and algebra, equation (1), which is satisfied by a real algebraic number, is fully determinate; conversely, if n is the degree of an equation of the form (1), then the equation is satisfied by at most n real algebraic numbers ω. The real algebraic numbers form in their totality a set ⟦Inbegriff⟧ of numbers, which shall be designated by (ω); as is readily seen, (ω) has the property that in every neighbourhood of any given number α there are infinitely many numbers from (ω). So it ought at first glance to be all the more striking that one can correlate the set (ω) one-to-one with the set (designated by the sign (ν)) of all positive integers ν—in such a way that to every algebraic number ω there corresponds a definite positive integer, and, conversely, to every positive integer ν there corresponds an entirely definite real algebraic number ω. Or, to say the same thing in different words, the set (ω) can be thought of in the form of an infinite lawlike sequence ⟦gesetzmäßigen Reihe⟧

$$\omega_1, \omega_2 \ldots \omega_\nu, \ldots \tag{2}$$

in which all the individuals of (ω) appear and each of which occurs in (2) at a definite position, which is given by the accompanying index. As soon as one has found a law by which such a correlation can be thought, it can be modified at will; so in §1 I shall describe the correlation which seems to me the least complicated.

In order to give an application of this property of the set of all real algebraic numbers I show in §2 that, given any arbitrarily chosen sequence of real numbers of the form (2), then, in any given interval ($\alpha \ldots \beta$), one can determine ⟦bestimmen⟧ numbers η which are *not* contained in (2); if one combines the results of these last two paragraphs, then one has a new proof of the theorem, first proved by Liouville, that in any given interval ($\alpha \ldots \beta$) there are infinitely many *transcendental* (that is, not algebraic) real numbers. Further, the theorem in §2 turns out to be the reason why sets of real numbers which form a so-called continuum (say, all real numbers which are ≥ 0 and ≤ 1) cannot be mapped one-to-one onto the set (ν); thus I have discovered the difference between a so-called continuum and any set like the totality of real algebraic numbers.

§1.

Let us return to equation (1), which is satisfied by an algebraic number ω and which, under the mentioned stipulations, is fully determinate. Then the sum of the absolute values of its coefficients, plus the number $n - 1$ (where n is the degree of ω) shall be called the *height* of the number ω. Let it be designated by N. That is, in the usual notation:

$$N = n - 1 + |a_0| + |a_1| + \ldots + |a_n|. \tag{3}$$

The height N is thus for every real algebraic number ω a definite positive integer; conversely, for every positive integral value of N there are only finitely many algebraic real numbers with height N; let this number be $\phi(N)$; then, for instance, $\phi(1) = 1$; $\phi(2) = 2$; $\phi(3) = 4$. The numbers of the set (ω), i.e. all the algebraic real numbers, can then be ordered in the following manner: one takes as the first number ω_1 the one number with the height $N = 1$; let the $\phi(2) = 2$ algebraic real numbers with the height $N = 2$ follow them according to size, and designate them by ω_2, ω_3; these are followed by the $\phi(3) = 4$ numbers with height $N{=}3$, according to size; in general, after all the numbers in (ω) up to a certain height $N = N_1$ have been enumerated ⟦abgezählt⟧ and assigned to a definite place, the real algebraic numbers with the height $N = N_1 + 1$ follow them according to size; thus one obtains the set (ω) of all real algebraic numbers in the form:

$$\omega_1, \omega_2, \ldots \omega_\nu, \ldots$$

One can, with respect to this ordering ⟦Anordnung⟧, speak of the νth real algebraic number; not a single member of the set has been omitted.

§2.

Suppose we have an infinite sequence of real numbers,

$$\omega_1, \omega_2, \ldots \omega_\nu, \ldots \tag{4}$$

where the sequence is given according to any law and where the numbers are distinct from each other. Then in any given interval ($\alpha \ldots \beta$) a number η (and consequently infinitely many such numbers) can be determined such that it does not occur in the series (4); this shall now be proved.

We go to the end of the interval $(\alpha \ldots \beta)$, which has been given to us arbitrarily and in which $\alpha < \beta$; the first two numbers of our sequence (4) which lie in the interior of this interval (with the exception of the boundaries), can be designated by α', β', letting $\alpha' < \beta'$; similarly let us designate the first two numbers of our sequence which lie in the interior of $(\alpha' \ldots \beta')$ by α'', β'', and let $\alpha'' < \beta''$; and in the same way one constructs the next interval $(\alpha''' \ldots \beta''')$, and so on. Here therefore α', α'' ... are by definition determinate numbers of our sequence (4), whose indices are continually increasing; the same goes for the sequence β', β'' ...; furthermore, the numbers α', α'', ... are always increasing in size, while the numbers β', β'', ... are always decreasing in size. Of the intervals $(\alpha \ldots \beta)$, $(\alpha' \ldots \beta')$, $(\alpha'' \ldots \beta'')$, ... each encloses all of those that follow.—Now here only two cases are conceivable.

Either the number of intervals so formed is finite; in which case, let the last of them be $(\alpha^{(\nu)} \ldots \beta^{(\nu)})$. Since in its interior there can be at most one number of the sequence (4), a number η can be chosen from this interval which is not contained in (4), thereby proving the theorem for this case.—

Or the number of constructed intervals is infinite. Then the numbers α, α', α'', ..., because they are always increasing in size without growing into the infinite, have a determinate boundary value α^∞; the same holds for the numbers β, β', β'', ... because they are always decreasing in size. Let their boundary value be β^∞. If $\alpha^\infty = \beta^\infty$ (a case that constantly occurs with the set (ω) of all real algebraic numbers), then one easily persuades oneself, if one only looks back to the definition of the intervals, that the number $\eta = \alpha^\infty = \beta^\infty$ *cannot* be contained in our sequence;[1] but if $\alpha^\infty < \beta^\infty$ then every number η in the interior of the interval $(\alpha^\infty \ldots \beta^\infty)$ or also on its boundaries satisfies the requirement that it not be contained in the sequence (4).—

The theorems proved in this article admit of extensions in various directions, of which I shall mention only one here:

'If $\omega_1, \omega_2, \ldots \omega_n, \ldots$ is a finite or infinite sequence of numbers which are linearly independent from one another (so that no equation of the form $a_1\omega_1 + a_2\omega_2 + \ldots + a_n\omega_n = 0$ is possible with integral coefficients which do not all vanish) and if one imagines the set (Ω) of all those numbers Ω which can be represented as rational functions with integral coefficients of the given numbers ω, then in every interval $(\alpha \ldots \beta)$ there are infinitely many numbers which are not contained in (Ω).'

In fact one persuades oneself through a method of proof similar to that in §1 that the set (Ω) can be conceived in the sequential form

$$\Omega_1, \Omega_2, \ldots \Omega_\nu, \ldots$$

from which, in view of §2, the correctness of the theorem follows.

[1] If the number η were contained in our sequence, then one would have $\eta = \omega_p$, where p is a definite index. But this is not possible, for ω_p does *not* lie in the interior of the interval $(\alpha^{(p)} \ldots \beta^{(p)})$, while by definition the number η does lie in the interior of the interval.

A quite special case of the theorem cited here (in which the sequence ω_1, $\omega_2, \ldots \omega_n, \ldots$ is finite and the degree of the rational functions, which yield the set (Ω), is stipulated in advance) has been proved, with recourse to Galoisian principles, by Herr B. Minnigerode (See *Math. Annalen*, Vol. 4, p. 497).

B. THE EARLY CORRESPONDENCE BETWEEN CANTOR AND DEDEKIND

Cantor and Dedekind exchanged letters over a period of many years. The mathematical portions of some of their letters were published by Emmy Noether and Jean Cavaillès in 1937; the mathematical portions of others were published (with many errors of transcription) by Ernst Zermelo in his edition of Cantor's writings (*Cantor 1932*). The Cantor *Nachlass*, containing the original letters from Dedekind, appears to have been destroyed in the Second World War. The letters from Cantor and drafts of some of the letters from Dedekind were taken to America by Emmy Noether, and are now kept at the library of the University of Evansville, Evansville, Indiana. (For an account of the history and the contents of the Evansville collection, see *Grattan-Guinness 1974*.) Portions of the non-mathematical correspondence, not published by Noether and Cavaillès or by Zermelo, are reproduced in *Grattan-Guinness 1974* and in the appendices to *Dugac 1976*.

The selection that follows is a translation of the correspondence published by Noether and Cavaillès, omitting only Cantor's letter to Dedekind of 22 December 1879. That letter gives a counterexample to a theorem of Appell in trigonometric series, and is of little importance today. (Cantor's late correspondence with Dedekind in 1899 is translated below as Selection E.) The correspondence spans the period from April 1872 (immediately after the publication of the theories of Dedekind and Cantor on the irrational numbers) to November 1882 (immediately before the publication of Cantor's *Foundations of a general theory of manifolds*). This was the period during which Cantor made his greatest discoveries in the theory of sets, and his correspondence with Dedekind provides rare documentation of the way in which a new mathematical theory was discovered.

The translation is by William Ewald; references should be to the dates of the correspondence.

Cantor to Dedekind

Halle a/S, 28 April 1872

I thank you most warmly for kindly sending me your treatise on continuity and irrational numbers. For many years, starting from arithmetical concerns, I have been training myself in a particular conception of this subject, and I have convinced myself that my conception substantially agrees with yours; the only difference is in the *conceptual introduction* ‖*begrifflichen Einführung*‖ of the numerical quantities. I agree wholeheartedly that the essence of continuity lies in what you emphasize.

Halle, 29 Nov. 73

Allow me to put a question to you. It has a certain theoretical interest for me, but I cannot answer it myself; perhaps you can, and would be so good as to write me about it. It is as follows.

Take the totality ‖Inbegriff‖ of all positive whole-numbered individuals n and designate it by (n). And imagine say the totality of all positive real numerical quantities x and designate it by (x). The question is simply, Can (n) be correlated to (x) in such a way that to each individual of the one totality there corresponds one and only one of the other? At first glance one says to oneself no, it is not possible, for (n) consists of discrete parts while (x) forms a continuum. But nothing is gained by this objection, and although I incline to the view that (n) and (x) permit no one-to-one correlation, I cannot find the explanation which I seek; perhaps it is very easy.

Would one not be inclined at first glance to maintain that (n) cannot be correlated one-to-one with the totality $\left(\frac{p}{q}\right)$ of all positive rational numbers $\frac{p}{q}$? And yet it is not difficult to show that (n) can be correlated one-to-one not only with this totality, but with the more general

$$(a_{n_1, n_2, \ldots n_\nu})$$

where n_1, n_2, ... n_ν are unrestricted positive integer indices in arbitrary number ν.

Halle, 2 December 73

I was exceptionally pleased to receive your answer to my last letter. I put my question to you because I had wondered about it already several years ago, and was never certain whether the difficulty I found was subjective or whether it was inherent in the subject. Since you write that you too are unable to answer it, I may assume the latter.—In addition, I should like to add that I have never seriously occupied myself with it, because it has no special practical interest for me. And I entirely agree with you when you say that for this reason it does not deserve much effort. But it would be good if it could be answered; e.g. if it could be answered with no, then one would have a new proof of Liouville's theorem that there are transcendental numbers.

Your proof that (n) can be correlated one-to-one with the field of all algebraic numbers is approximately the same as the way I prove my contention in the last letter. I take $n_1^2 + n_2^2 + \ldots + n_\nu^2 = \mathfrak{N}$ and order the elements accordingly.

Is it not excellent that, as you have characteristically stressed, one can speak of the nth algebraic number, so that every one appears once in the sequence?

As you quite rightly remark, our question admits the following formulation: 'Can (n) be one-to-one correlated with a totality:

$$(a_{n_1, n_2, \ldots})$$

where $n_1, n_2, \ldots$ are unrestricted, positive integral indices in infinite number.'

Halle, 7th December 73

In the last days I have had the time to pursue more thoroughly the conjecture I spoke to you about; only today do I believe myself to have finished with the thing; but if I should be deceiving myself, I should certainly find no more indulgent judge than you. So I take the liberty of presenting to your judgement what I have just written down in the imperfection of the first draft.

Assume that all positive ⟦real⟧ numbers $\omega < 1$ can be brought into the sequence:

(I) $$\omega_1, \omega_2, \omega_3, \ldots, \omega_n \ldots .$$

Let ω_a be the next largest term following ω_1 in the sequence, ω_β the next largest, and so on. One sets: $\omega_1 = \omega_1^1$, $\omega_a = \omega_1^2$, $\omega_\beta = \omega_1^3$ etc. One extracts from (I) the infinite sequence:

$$\omega_1^1, \omega_1^2, \omega_1^3, \ldots, \omega_1^n, \ldots$$

In the sequence that remains, one designates the first term by ω_2^1, the next greater following term by ω_2^2, etc. One extracts the sequence:

$$\omega_2^1, \omega_2^2, \omega_2^3, \ldots \omega_2^n, \ldots$$

If one continues in this way, one sees that the sequence (I) can be decomposed into the infinitely many:

(1) $$\omega_1^1, \omega_1^2, \omega_1^3, \ldots, \omega_1^n, \ldots$$

(2) $$\omega_2^1, \omega_2^2, \omega_2^3, \ldots, \omega_2^n, \ldots$$

(3) $$\omega_3^1, \omega_3^2, \omega_3^3, \ldots, \omega_3^n, \ldots$$

in each of which the terms increase continuously from left to right; we have:

$$\omega_k^\lambda < \omega_k^{\lambda+1}.$$

Now take an interval $(p \ldots q)$ so that no term of the sequence (1) lies in it; say within $(\omega_1^1, \ldots, \omega_1^2)$. Now, all terms of say the second sequence or the third could lie outside $(p \ldots q)$; however, there must be some sequence, which I shall call the kth, such that not all of its terms lie outside $(p \ldots q)$ (for otherwise the numbers within $(p \ldots q)$ would not be contained in (I), contrary to the hypothesis). Then one can fix an interval $(p' \ldots q')$ within $(p \ldots q)$ so that all terms of the kth sequence lie outside it. But then $(p' \ldots q')$ behaves in the same way with respect to the previous sequences; so eventually there must appear a k'^{th} sequence not all of whose members lie outside $(p' \ldots q')$, and then one takes inside $(p' \ldots q')$ a third interval $(p'' \ldots q'')$ so that all members of the k'^{th} sequence lie outside it.

So one sees that it is possible to form an infinite sequence of intervals:

$$(p \ldots q),\ (p' \ldots q'),\ (p'' \ldots q''),\ \ldots$$

such that each includes its successors, and such that they are related as follows to our sequences (1), (2), (3) . . . :

The members of the 1st, 2nd, ... $k - 1^{\text{st}}$ sequence lie outside $(p \ldots q)$

The members of the kth ... $(k' - 1)^{\text{st}}$ sequence lie outside $(p' \ldots q')$;

The members of the k'^{th} ... $(k'' - 1)^{\text{st}}$ sequence lie outside $(p'' \ldots q'')$.

Now, we can always find *at least* one number which lies inside each of these intervals; I shall call it η. One sees at once that this number η, which is clearly $\gtrless{}^0_1$, cannot be contained in any of our sequences (1), (2), . . ., (n). Thus from the assumption that all numbers $\gtrless{}^0_1$ are contained in (I), one arrives at the contrary conclusion that we can find a definite number $\eta \gtrless{}^0_1$ that is *not* in (I); consequently the assumption was incorrect.

So I finally believe myself to have found the reason why the totality designated by (x) in my earlier letters *cannot* be correlated one-to-one with the totality designated by (n).

Halle, 9 December 73

I have already found a simplified proof of the theorem just proved, so that the decomposition of the sequence (I) into (1), (2), (3), . . . is no longer necessary.

I show directly that if I start with a sequence

(I) $$\omega_1,\ \omega_2,\ \ldots,\ \omega_n,$$

then in *every* given interval $(\alpha \ldots \beta)$ I can determine a number η that is not contained in (I). From this it follows at once that the totality (x) cannot be correlated one-to-one with the totality (n); and I infer that there exist essential differences ⟦Wesensverschiedenheiten⟧ among the totalities and value-sets ⟦Inbegriffen und Werthmengen⟧ that I was until recently unable to fathom.

Now I must ask your forgiveness for having taken so much of your time with this question.

Halle, 10 December 73

Confirming the receipt of your friendly lines of the 8th of December, allow me to assure you that nothing can give me more pleasure than to have been lucky enough to arouse in you an interest for certain questions of analysis. And permit me to add that nothing can spur me to further efforts more than this; I should like to ask that you send me your comments in the future as well. Won't the general idea of $\omega_1, \omega_2, \ldots, \omega_\nu, \ldots$ yield good results for your theory of algebraic numbers[a]

Berlin, 25 December 73

Although I did not yet wish to publish the subject I recently for the first time discussed with you, I have nevertheless unexpectedly been caused to do so. I communicated my results to Herr Weierstrass on the 22nd; however, there was no time to go into details; already on the 23d I had the pleasure of a visit from him, at which I could communicate the proofs to him. He was of the opinion that I must publish the thing at least in so far as it concerns the algebraic numbers. So I wrote a short paper with the title: *On a property of the set of all real algebraic numbers*,[b] and sent it to Professor Borchardt to be considered for the *Journal für Math*.

As you will see, your comments (which I value highly) and your manner of putting some of the points were of great assistance to me. I wished to communicate this to you.

Berlin, 27 December 73

As a result of our recent correspondence, writing to you has become so natural that, without thinking about it, I jot down my replies and almost forget to apologize for their frequency. Perhaps it is caused by the similarity of our interests, and because the publicly beneficial development of science is dear to both our hearts.

The restriction which I have imposed on the published version of my investigations is caused in part by local circumstances (about which I shall perhaps later speak with you orally) and in part because I believe that it is important to apply my ideas at first to a single case (such as that of the real algebraic

---

[a] [Indeed, in his work *Über die Permutationen des Körpers aller algebraischen Zahlen* (*Ges. Werke*, Vol. ii, p. 278) Dedekind applies the theorem; but the existence of a finite basis makes well-ordering unnecessary for finite algebraic number fields.—Noether and Cavaillès.]

[b] [*Cantor 1874*, translated above.]

numbers); the extensions (whose possibility I can already see in plenty) ought then to cause no great trouble, and it will not much matter whether I make them or somebody else. So after a short introduction I have written two §§; in the first it is shown that the totality of real algebraic numbers can be correlated one-to-one with the totality of positive integers; in the second that when we have a sequence $\omega_1, \omega_2, \ldots, \omega_n, \ldots$ one can define numbers η in every interval that are not contained in it.

Dedekind's notes on the letters of 1873

1873.11.29

Herr G. Cantor (Halle) presented me with the question, whether the totality (n) of all positive whole-numbered individuals n (natural numbers) can be correlated with the totality (x) of all positive real numerical quantities x in such a way that to each individual of the one totality there corresponds one and only one of the other?—He ends with the words:

'Would one not be inclined at first glance to maintain that (n) cannot be correlated one-to-one with the totality $\left(\frac{b}{a}\right)$ of all positive rational numbers $\frac{b}{a}$? And yet it is not difficult to show that (n) can be correlated one-to-one not only with this totality, but with the more general

$$(a_{n_1, n_2, \ldots n_\nu})$$

where $n_1, n_2, \ldots n_\nu$ are unrestricted positive integer indices in arbitrary number ν.'

To this I replied by return of post that I could not answer the first question, but at the same time stated and fully proved the theorem that even the totality of all algebraic numbers can be correlated in the stated manner with the totality (n) of all natural numbers. (Shortly thereafter, this theorem and proof appeared almost word-for-word in Cantor's paper in *Crelle*, Vol. 77, even with the use of the artificial expression height ⟦*Höhe*⟧, but with the weakening, against my recommendation, that only the totality of all *real* algebraic numbers is considered.) But the opinion I expressed that the first question did not deserve too much effort because it has no particular practical interest was conclusively refuted by Cantor's proof of the existence of transcendental numbers (*Crelle*, Vol. 77).

1873.12.2

C. points out the importance of the first question: if it is answered negatively then the existence of transcendental numbers (Liouville) can be proved in a new way. He continues: 'Your proof that (n) can be one-to-one correlated with the field of all algebraic numbers is approximately the same as the way I prove my contention in the last letter. I take $n_1^2 + n_2^2 + \ldots + n_\nu^2 = \mathfrak{N}$ and order the elements accordingly. Is it not excellent that, as you have characteristically stressed, one can speak of the nth algebraic number, so that every one appears once in the sequence? As you quite rightly remark, our question admits the following formulation: "Can (n) be one-to-one correlated with a totality:

$$(a_{n_1, n_2, \ldots})$$

where n_1, n_2, ... are unrestricted, positive integral indices in infinite number." '

1873.12.7

C. communicates to me a rigorous proof, found on the same day, of the theorem that the totality of all positive numbers $\omega < 1$ *cannot* be one-to-one correlated with the totality (n).

I answer this letter, received on 8 December, on the same day with congratulations for the fine success; at the same time, I rephrase much more simply the core of the proof (which was still quite complicated). This presentation too went into Cantor's paper (*Crelle* Vol. 77) almost word-for-word; to be sure, the phrase I use, 'according to the principle of continuity', is avoided at the spot in question (p. 261, ll. 10–14)!

1873.12.9

C. writes to me in a hurry that he has found a simplified proof of the theorem. Since he does not mention my letter, it must have arrived later.

1873.12.10

C. confirms the receipt of my letter of 8 December, without mentioning the simplified presentation of the proof contained therein; thanks me for my interest in the subject.

1873.12.25

C. writes (from Berlin) that he has (on the prompting of Weierstrass) written a short article with the title: *On a property of the set of all real algebraic numbers*. 'As you will see, your remarks (which I value highly) and your manner of putting some of the points were of great assistance to me.'

I answer by return of post with the advice to drop the restriction to the field of all *real* algebraic numbers.

1873.12.27

C. writes (from Berlin): 'The restriction which I have imposed on the published version of my investigations is caused in part by local circumstances (about which I shall perhaps later speak with you orally) and in part because I believe that it is important to apply my ideas at first to a single case (such as that of the real algebraic numbers).'

I have never received an explanation of the 'Berlin circumstances'; we also later never dealt with the article (*Crelle*, Vol. 77).

Cantor to Dedekind

Halle, 5 January 74

As for the question with which I have recently occupied myself, it occurs to me that the same train of thought also leads to the following question:

Can a surface (say a square including its boundary) be one-to-one correlated to a line (say a straight line including its endpoints) so that to every point of the surface there corresponds a point of the line, and conversely to every point of the line there corresponds a point of the surface?

It still seems to me at the moment that the answer to this question is very difficult—although here too one is so impelled to say *no* that one would like to hold the proof to be almost superfluous.

Halle, 28 Jan. 74

In the *Comptes rendues* of last year I find an article by Hermite in which a simultaneous development of n exponential quantities e^{ax}, e^{bx}, . . . e^{nx} is treated with the most skilful handling of the analysis; and what seems most important, he bases on this a completely rigorous proof of the transcendence of the number e. Hermite confesses to have spent a great deal of time on the proof of the transcendence of π, but for this number he surrenders with the remark that he would be delighted if somebody else were to succeed.

Halle, 18 May 74

The wish to speak with you about scientific subjects and to have more personal contact with you makes me want to visit you occasionally in Brunswick this summer.

... When you get around to answering me, I should be grateful to hear whether you had the same difficulty as I in answering the question I sent to you in January about the correlation of a line and a surface, or whether I am deceiving myself. In Berlin a friend to whom I presented the same problem told me the subject was somewhat absurd, because it is self-evident that two independent variables cannot be reduced to one.

Dedekind to Cantor

... The objection of Herr I. which you communicated to me cannot, if I have correctly understood it, affect my presentation; the principle of continuity set forth in §3, p. 18 ⟦of *Continuity and irrational numbers*⟧ is of course only to be conceived as the necessary *completion* of the laws I, II, III stated in §2, p. 15; and II states exactly what you or Herr I. seem to miss. The objection would probably not have been made had I placed number IV before the principle on p. 18, as I do with the analogous laws in §5, p. 25. Or have I misunderstood the objection? Then I should be grateful for an explanation.

Brunswick, 11 May 1877

Cantor to Dedekind

Halle, 17 May 1877

Thank you for your answer. I must confess that on page 25 of your paper on irrational numbers you adduce, with I, II, III and IV,[c] properties that are thoroughly characteristic for the domain of all real numbers, so that *no* other value-system of real numbers simultaneously shares all these properties with that system.

However, allow me to remark that perhaps the stress which at various points in your paper you *expressly* lay on property IV as being the *essence* of continuity must lead to misunderstandings which, in my opinion, without that emphasis on IV (as the proper essence) could not accrue to your theory. In particular you say in the preface that my axiom fully agrees with what you state in §3

---

[c] ⟦I: Order; II: Density; III: Cut; IV: To every cut there corresponds a number. *Dedekind 1872*, §5.⟧

as the *essence* of continuity. But by that you understand the same property that is designated by IV on p. 25; but this property also holds of the system of all integers, which can however be regarded as a prototype of discontinuity.

In the interest of the subject, which has become important to me as well, I ask you, when you have the time, to go into my doubts more closely.

P.S. I tell myself that you lay special emphasis on IV because this property distinguishes the complete domain of numbers from the domain of all rational numbers; however it seems to me for the above reasons that one cannot give property IV the name 'essence of continuity' as you do.

Dedekind to Cantor

After your last letter it appears to me that we run the danger of arguing more about words than about things. Every attentive reader of my paper will certainly understand that my view about continuity is as follows: Domains with an opposition ⟦Gegensätzlichkeit⟧ and completeness of their elements—as completeness is expressed by I and II in §1 p. 14, §2 p. 15, §5 p. 25 (III is a consequence of I and only introduced to prepare the way for IV)—are not yet necessarily continuous domains; such domains obtain the property of continuity from the addition of property IV (on p. 18 (unnumbered) and on p. 25) and only from this property. And so this property is called the essence of continuity.

You tell me in your card of the 10th that my definition of continuity is not complete, and make a suggestion for removing this defect. I decline this suggestion, and call your attention to II, which contains what you are seeking. You concede in your last letter that my definition in fact overlooks nothing; e.g. when I say, 'Domains which possess properties I and II are called continuous if they also possess property IV' you have (if I understand your last letter correctly) nothing to object against the *completeness* of such an explanation. But it seems that you would prefer it if property II were moved from the relative clause into the conditional clause: 'Domains, whose elements possess an opposition ⟦Gegensätzlichkeit⟧ of the type defined by I are called continuous if they also possess properties II *and* IV'; and you worry that my exclusive *stressing* of IV as *the* property in which the *essence* of continuity is expressed could lead to misunderstandings. I do not share this concern; I am firmly convinced that every attentive reader of my paper understands my opinion as I have expressed it above; and so the example you adduce of the system of all rational integers fails to provide the motive for an objection. Moreover, as for the above rearrangement of the definition, I cannot say that it pleases me. And I infer from your postscript that you too, as soon as you were to attempt to *rework* my paper in this way, would certainly concede that this paper (which has as its chief object the advance of arithmetic from rational to irrational) would, with regard to precise exposition, lose its proper point—which consists solely in the emphasis on IV, since II is already available in the non-continuous rational domain. But if the revised definition pleases somebody better, I have nothing to say against its *legitimacy*—that is, not if it should be of advantage for certain investigations. But my original formulation pleases me much better, and I think it is more expedient in treating the essence of continuity to lay the emphasis solely on IV and to discuss property II earlier, before continuity or discontinuity is at issue. In any case, I utterly dispute the *necessity* of the reformulation of the definition; if one were really to demand this, one could just as well raise the question (with which I have already occupied myself) whether it is not expedient to move property I from the relative clause into the conditional clause as well. This is not at all an uninteresting question, but it would lead me too far astray if I were to go into it. I genuinely believe that our opinions diverge at most about expediencies, not about necessities; so a continuation of the debate will probably not yield very much.

Brunswick, 18 May 1877

Cantor to Dedekind

Halle, 20 June 1877

Thank you for your letter of 18 May. I completely agree with its contents; and I acknowledge that the difference in our points of view was merely external. And now I approach you again today with a request. (You see that our shared theoretical interests have a disadvantage for you: perhaps I bother you more than you would like.)

I should like to know whether you consider an inference-procedure that I use to be arithmetically rigorous.

The problem is to show that surfaces, bodies, indeed even continuous structures of ρ dimensions can be correlated one-to-one with continuous lines, i.e. with structures of only *one* dimension—so that surfaces, bodies, indeed even continuous structures of ρ dimensions have the same *power* [*Mächtigkeit*] as curves. This idea seems to conflict with the one that is especially prevalent among the representatives of modern geometry, who speak of simply infinite, doubly, triply, . . . ρ-fold infinite structures. (Sometimes you even find the idea that the infinity of points of a surface or a body is obtained by as it were squaring or cubing the infinity of points of a line.)

Since structures of the same number of dimensions can be related to one another *analytically*, it seems to me that those more general questions should be put in the following purely arithmetical form:

'Let $x_1, x_2, \ldots x_\rho$ be ρ independent real variables such that each can assume all values ≥ 0 and ≤ 1. Let y be a $\rho + 1^{\text{th}}$ real variable with the same range $\begin{pmatrix} y \geq 0 \\ \leq 1 \end{pmatrix}$.

Is it then possible to correlate the ρ variables $x_1, x_2, \ldots x_\rho$ with the one variable y so that to every determinate value-system $(x_1, x_2, \ldots x_\rho)$ there corresponds a determinate value y and conversely to every determinate value y there corresponds one and only one determinate value-system $(x_1, x_2, \ldots x_\rho)$?'

Although for years I held the opposite opinion to be true, it now seems to me that this question ought to be answered with *yes*, for the following reasons:

Every number $x \begin{smallmatrix}\geq 0\\ \leq 1\end{smallmatrix}$ can be expressed in one and only one way in the form of an *infinite* decimal fraction, so that:

$$x = \alpha_1 . \frac{1}{10} + \alpha_2 . \frac{1}{10^2} + \ldots + \alpha_\nu . \frac{1}{10^\nu} + \ldots$$

where α_ν are integers ≥ 0 and ≤ 9. So every number x determines an infinite sequence $\alpha_1, \alpha_2, \ldots$ and conversely.

So we can write:

$$x_1 = \alpha_{1,1} . \frac{1}{10} + \alpha_{1,2} . \frac{1}{10^2} + \ldots + \alpha_{1,\nu} . \frac{1}{10^\nu} + \ldots$$

$$x_2 = \alpha_{2,1} . \frac{1}{10} + \alpha_{2,2} . \frac{1}{10^2} + \ldots + \alpha_{2,\nu} . \frac{1}{10^\nu} + \ldots$$

$$x_\rho = \alpha_{\rho,1} . \frac{1}{10} + \alpha_{\rho,2} . \frac{1}{10^2} + \ldots + \alpha_{\rho,\nu} . \frac{1}{10^\nu} + \ldots$$

From these ρ numbers one can derive a $\rho + 1^{\text{st}}$ number y:

$$y = \beta_1 . \frac{1}{10} + \beta_2 . \frac{1}{10^2} + \ldots + \beta_\nu . \frac{1}{10^\nu} + \ldots$$

if one takes:

(I)
$$\beta_{(n-1)\rho+1} = \alpha_{1,n};\ \beta_{(n-1)\rho+2} = \alpha_{2,n};\ \ldots$$
$$\beta_{(n-1)\rho+\sigma} = \alpha_{\sigma,n};\ \ldots\ \beta_{(n-1)\rho+\rho} = \alpha_{\rho,n}$$

Since every positive integer ν can be expressed in one and only one way in the form:

$$\nu = (n-1)\rho + \sigma \quad \text{where} \quad \begin{matrix}\sigma > 0\\ \leq \rho\end{matrix},$$

one sees that the sequence $\beta_1, \beta_2, \ldots$ (and therefore also y) is completely determined by (I).

But also conversely, if one starts with the number y (and consequently with the sequence $\beta_1, \beta_2, \ldots$) then the ρ sequences:

$$\alpha_{1,1}, \alpha_{1,2}, \ldots$$

....

$$\alpha_{\sigma,1}, \alpha_{\sigma,2}, \ldots$$

....

$$\alpha_{\rho,1},\ \alpha_{\rho,2},\ \ldots$$

and consequently also the ρ numbers $x_1, x_2, \ldots, x_\rho$ are *uniquely* determined by the equations (I).

Dedekind to Cantor

The only objection that I can at the moment raise against your interesting theorem (which you will perhaps solve without difficulty) is the following. You say: 'Every number $x(\geq 0$ and $\leq 1)$ can be expressed in one and only one way in the form of an *infinite* decimal fraction, so that:

$$x = \alpha_1 . \frac{1}{10} + \alpha_2 . \frac{1}{10^2} + \ldots + \alpha_\nu . \frac{1}{10^\nu} + \ldots$$

where the α_ν are integers ≥ 0 and ≤ 9. So every number x determines an infinite sequence $\alpha_1, \alpha_2, \ldots$ and conversely.' The underlining of the word '*infinite*' makes me think that you exclude the case of an infinite fraction where an α_ν different from zero is followed only by the numerals $0 = \alpha_{\nu+1} = \alpha_{\nu+2} = \ldots$ and instead of writing

$$x = \frac{\alpha_1}{10} + \frac{\alpha_2}{10^2} + \ldots + \frac{\alpha_\nu}{10^\nu} + \frac{0}{10^{\nu+1}} + \frac{0}{10^{\nu+2}} + \ldots + \frac{0}{10^{\nu+\nu'}} + \ldots$$

always

$$x = \frac{\alpha_1}{10} + \frac{\alpha_2}{10^2} + \ldots + \frac{\alpha_\nu}{10^\nu} + \frac{9}{10^{\nu+1}} + \frac{9}{10^{\nu+2}} + \ldots + \frac{9}{10^{\nu+\nu'}} + \ldots$$

in order to exclude every possibility of a double representation of one and the same number. (The number $x = 0$ itself would have to be written in the form $0.0000 \ldots$; but $x = \frac{3}{10}$ as $0.29999 \ldots .$)

If this is your opinion (—one could of course equally well exclude the case that from a particular spot onwards only the numeral 9 appears; but then something similar would occur—) then my objection is the following. For the sake of simplicity, I restrict myself to the case $\rho = 2$, and set:

$$x = \frac{\alpha_1}{10} + \frac{\alpha_2}{10^2} + \ldots = 0.\alpha_1\alpha_2 \ldots \alpha_\nu \ldots$$

$$y = \frac{\beta_1}{10} + \frac{\beta_2}{10^2} + \ldots = 0.\beta_1\beta_2 \ldots \beta_\nu \ldots$$

and then like you form from the two numbers x, y the third number

$$z = 0.\gamma_1\gamma_2\gamma_3 \ldots$$

where

$$\gamma_1 = \alpha_1,\ \gamma_2 = \beta_1,\ \gamma_3 = \alpha_2,\ \gamma_4 = \beta_2 \ldots \gamma_{2\nu-1} = \alpha_\nu,\ \gamma_{2\nu} = \beta_\nu, \ldots .$$

Then z is a completely determinate function of the two continuous variables x, y and contained in the same interval $(0 \leq \tau \leq 1)$ But then there are infinitely many genuine fractions which are *never* equal to z, e.g.

$$0.478310507090\alpha_7 0\alpha_8 0\alpha_9 0 \ldots \alpha_\nu 0 \ldots$$

and by the same token every fraction $0.\gamma_1\gamma_2\gamma_3 \ldots$ in which from a definite point on either $\gamma_{2\nu-1}$ or $\gamma_{2\nu}$ is always equal to zero; for the converse derivation of x, y from such a z would lead to a non-existent (excluded) x or y.

I do not know if my objection goes to the essence of your idea, but I did not want to hold it back.

Brunswick, 22 June 1877

Cantor to Dedekind

⟦Postcard: Postmark 23.6.77⟧

Alas, you are entirely correct in your objection; but happily it concerns only the proof, not the content. For I proved *somewhat more* than I had realized, in that I bring a system $x_1, x_2, \ldots x_\rho$ of unrestricted real variables (that are ≥ 0 and ≤ 1) into one-to-one relationship with a variable y that does not assume all values of that interval, but rather all with the exception of certain y''. However, it assumes each of the corresponding values y' only *once*, and that seems to me to be the essential point. For now I can bring y' into a one-to-one relation with another quantity t that assumes all values ≥ 0 and ≤ 1.

I am delighted that you have found no other objections. I shall shortly write to you at greater length about this matter.

Halle, 25 June 1877

I sent you a postcard the day before yesterday, in which I acknowledged the gap you discovered in my proof, and at the same time remarked that I am able to fill it. But I cannot repress a certain regret that the subject demands more complicated treatment. However, this probably lies in the nature of the subject, and I must console myself; perhaps it will later turn out that the missing portion of that proof can be settled more simply than is at present in my power. But since I am at the moment concerned above all to persuade you of the correctness of my theorem, namely, the theorem:

(A) A continuous manifold extended in e dimensions can be correlated one-to-one with a continuous manifold of one dimension. Or (what is only another form of the same theorem): the points (elements) of a manifold extended in ρ dimensions can be determined by a real coordinate t in such a way that to every real value of t in the interval (0 . . . 1) there corresponds a point of the manifold, and conversely to every point of the M. there corresponds a definite value of t in the interval (0 . . . 1).

I allow myself to present another proof of it, which I found even earlier than the other.

I proceed from the theorem that every irrational number $e \gtrless {}^0_1$ can be expressed in a fully determinate manner in the form of an infinite continued fraction:

$$e = \cfrac{1}{\alpha_1 + \cfrac{1}{\alpha_2 + \cdots + \cfrac{1}{\alpha_\nu + \vdots}}} = (\alpha_1, \alpha_2, \ldots, \alpha_\nu, \ldots)$$

where α_ν is a positive integer. To every irrational number $e \gtrless {}^0_1$ there corresponds a definite infinite sequence α_ν and conversely to every definite infinite sequence α_ν there corresponds a definite irrational number $e \gtrless {}^0_1$.

Now, if $e_1, e_2, \ldots e_\rho$ are ρ quantities independent of one another, of which each can assume all irrational numerical values of the interval (0 . . . 1) and only these values, then set:

$$e_1 = (\alpha_{1,1}, \alpha_{1,2}, \ldots, \alpha_{1,\nu}, \ldots)$$
$$e_2 = (\alpha_{2,1}, \alpha_{2,2}, \ldots, \alpha_{2,\nu}, \ldots)$$
$$\cdots\cdots\cdots\cdots\cdots\cdots\cdots\cdots\cdots$$
$$e_\rho = (\alpha_{\rho,1}, \alpha_{\rho,2}, \ldots \alpha_{\rho,\nu}, \ldots);$$

and determine from them a $(\rho + 1)^{\text{st}}$ irrational number:

$$\partial = (\beta_1, \beta_2, \ldots, \beta_\nu, \ldots)$$

by the system of equations:

$$\text{(I)} \qquad \beta_{(n-1)\rho+1} = \alpha_{1,n}, \ldots, \beta_{(n-1)\rho+\sigma} = \alpha_{\sigma,n}, \ldots \beta_{n,\rho} = \alpha_{\rho,n}.$$

Then also conversely every irrational number $\partial \gtrless {}^0_1$ produces a definite system $e_1, e_2, \ldots e_\rho$ by means of (I).

It seems to me that here one does not run into the obstacle which you found in my earlier proof.

Now it is a matter of proving the following theorem:

(B) A variable number e which can assume all *irrational* numerical values of the interval (0 . . . 1) can be *one-to-one* correlated with a number x which takes on *all* values of this interval without exception.

For once this theorem (B) has been proven, one correlates one-to-one the variables previously designated by $e_1, e_2, \ldots e_\rho$ and ∂ to respective individual other variables

$$x_1, x_2, \ldots x_\rho, y$$

all of which have an unrestricted range in the interval $(0 \ldots 1)$. Then in this way too one establishes a one-to-one reciprocal relation of, on the one hand, the system

$$(x_1, x_2, \ldots x_\rho)$$

and, on the other, the single variable y; and this leads to the proof of theorem (A).

To demonstrate (B), one first brings all *rational* numbers of the interval $(0 \ldots 1)$ (including the end-points) into sequential form. They are:

$$r_1, r_2, \ldots, r_\nu, \ldots$$

The values that the variable e can assume are accordingly all values in $(0 \ldots 1)$ *with the exception of* the numbers r_ν.

In addition one takes some arbitrary infinite sequence ε_ν of *irrational* numbers in the interval $(0 \ldots 1)$, subject only to the conditions that $\varepsilon_\nu < \varepsilon_{\nu+1}$ and that $\lim(\varepsilon_\nu) = 1$ for $\nu = \infty$. Let f be a variable which can assume all real values $\substack{\geq 0 \\ \leq 1}$ except for the values ε_ν. Then the two limited variables e and f can be put into a reciprocal one-to-one relation to each other by the following definition:

If f is equal to no r_ν then the correspondence is:

$$e = f;$$

but if $f = r_\nu$, then the correspondence is $e = \varepsilon_\nu$; then one easily sees that also conversely: if e is equal to no ε_ν, then $f = e$ and if $e = \varepsilon_\nu$ then $f = r_\nu$.

The theorem (B) is now reduced to the following theorem:

(C) A number f that can assume all values of the interval $(0 \ldots 1)$ with the exception of certain ε_ν which satisfy the conditions: $\varepsilon_\nu < \varepsilon_{\nu+1}$ and $\lim \varepsilon_\nu = 1$ can be correlated one-to-one with a continuous variable x which takes on all values without exception of the interval $(0 \ldots 1)$.

Here it is significant that the points $\varepsilon_1, \varepsilon_2, \ldots$ form a *series* and that the interval $(0 \ldots 1)$ is therefore decomposed by it into infinitely many sub-intervals.

And so, as you will not fail to see, this theorem (C) can be proved by successive application of the following theorem:

(D) A number y that can assume all values of the interval $(0 \ldots 1)$ with the solitary exception of the value 0 can be correlated one-to-one with a number x that takes on all values of the interval $(0 \ldots 1)$ without exception.

This theorem (D) can be seen to be true by considering the following peculiar curve.

My quantities x, y are the coordinates of a moving point m, and y is a single-

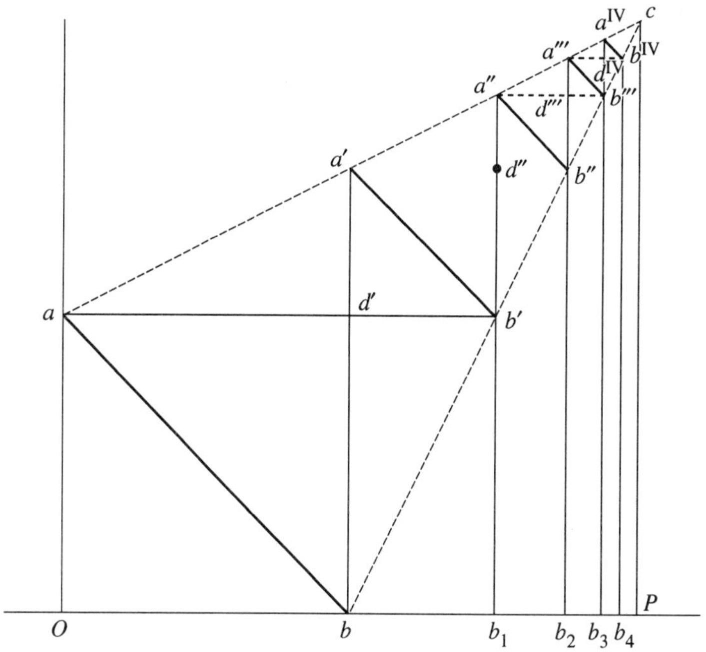

valued function of x; but while x assumes all values of the interval $(0 \ldots 1)$, the range of y is the same with *the solitary exception of the value 0.*

The curve consists of infinitely many parallel line segments $\overline{ab}$, $\overline{a'b'}$, $\overline{a''b''}$, and of the point c. *The end-points b, b′, b″, . . . are not regarded as belonging to the curve.* The lengths are:

$$\overline{op} = \overline{pc} = 1;\ \overline{ob} = \frac{1}{2};\ \overline{bb_1} = \frac{1}{4},\ \overline{b_1b_2} = \frac{1}{8},\ \overline{b_2b_3} = \frac{1}{16}\cdots$$

$$\overline{oa} = \frac{1}{2};\ \overline{a'd'} = \frac{1}{4};\ \overline{a''d''} = \frac{1}{8};\ \overline{a'''d'''} = \frac{1}{16},\ldots$$

For several years I have followed with interest the efforts that have been made, building on Gauss, Riemann, Helmholtz, and others, towards the clarification of all questions concerning the ultimate foundations of geometry. It struck me that all the important investigations in this field proceed from an unproven presupposition which does not appear to me self-evident, but rather to need a justification. I mean the presupposition that a ρ-fold extended continuous manifold needs ρ independent real coordinates for the determination of its elements, and that for a given manifold this number of coordinates can neither be increased nor diminished.

This presupposition had become my view as well, and I was almost convinced of its correctness. The only difference between my standpoint and all others was that I regarded that presupposition as a theorem which stood in great need of a proof; and I refined my standpoint into a question that I presented to several colleagues, in particular at the Gauss Jubilee in Göttingen. The question was the following:

'Can a continuous structure of ρ dimensions, where $\rho > 1$, be related one-to-one to a continuous structure of one dimension so that to each point of the former there corresponds one and only one point of the latter?'

Most of those to whom I presented this question were extremely puzzled that I should ask it, for it *is quite self-evident* that the determination of a point in an extension ⟦Ausgedehntheit⟧ of ρ dimensions always needs ρ independent coordinates. But whoever penetrated the sense of the question had to acknowledge that a proof was needed to show why the question should be answered with the 'self-evident' *no*. As I say, I myself was one of those who held it for the *most likely* that the question should be answered with a *no*—until quite recently I arrived by rather intricate trains of thought at the conviction that the answer to that question is an unqualified *yes*. Soon thereafter I found the proof which you see before you today.

So one sees what wonderful power lies in the ordinary real and irrational numbers, that one is able to use them to determine uniquely the elements of a ρ-fold extended continuous manifold *with a single coordinate*. I will only add at once that their power goes yet further, in that, as will not escape you, my proof can be extended without any great increase in difficulty to manifolds with an infinitely great dimension-number, provided that their infinitely-many dimensions have the form of a simple infinite sequence.

Now it seems to me that all philosophical or mathematical deductions that use that erroneous presupposition are inadmissible. Rather the difference that obtains between structures of *different* dimension-number must be sought in quite other terms than in the number of independent coordinates—the number that was hitherto held to be characteristic.

Halle, 29 June 1877

Please excuse my zeal for the subject if I make so many demands upon your kindness and patience; the communications which I lately sent you are even for me so unexpected, so new, that I can have no peace of mind until I obtain from you, honoured friend, a decision about their correctness. So long as you have not agreed with me, I can only say: *je le vois, mais je ne le crois pas*. And so I ask you to send me a postcard and let me know when you expect to have examined the matter, and whether I can count on an answer to my quite demanding request.

The proof of theorem (C) is considerably simplified by using the following symbolism:

If a, b are two variables that can be correlated one-to-one with each other, then one writes:

$$a \sim b.$$

And then, if $a \sim b$ and $b \sim c$, we also have:

$$a \sim c.$$

Moreover, if $a', a'', \ldots$ is a finite or infinite sequence of well-defined variables or constants which, taken pairwise, assume no common values, but, taken together, have a range ⟦Spielraum⟧ which is precisely that of the single variable a, then one sets:

$$a \equiv (a', a'', \ldots).$$

One then has the following theorem:

(E) If:
$$a \equiv (a', a'', \ldots)$$
$$b \equiv (b', b'', \ldots)$$

and moreover:

$$a' \sim b'$$
$$a'' \sim b''$$
$$a''' \sim b'''$$
$$\ldots\ldots\ldots\ldots$$

then:

$$a \sim b.$$

From (D) one obtains by the substitutions:

$$y = \frac{z - \alpha}{\beta - \alpha};\ x = \frac{u - \alpha}{\beta - \alpha}$$

the following generalization of (D):

(F) A number z which can assume all values of an interval $(\alpha \ldots \beta)$ except α can be correlated one-to-one with a number u that takes on all values of the interval $(\alpha \ldots \beta)$ without exception.

From this we obtain the following theorem:

(G) A number w that takes on all values of the interval $(\alpha \ldots \beta)$ except for the two end-values α, β can be correlated one-to-one with a variable number u that assumes all values of the interval $(\alpha \ldots \beta)$.

Proof. Let γ be a number $\substack{>\alpha \\ <\beta}$; w' a variable that assumes all values of the interval $(\alpha \ldots \gamma)$ with the exception of α and γ; w'' a variable that assumes all values of the interval $(\gamma \ldots \beta)$ with the exception of the one end-point β.

Then:

$$w \equiv (w', w'').$$

If now we designate by u'' a variable that assumes all values of the interval $(\gamma \ldots \beta)$ without exception, and by z a variable that assumes all values of the interval $(\alpha \ldots \beta)$ with the exception of α, then, by (F):

$$w'' \sim u'', \tag{2}$$

and so by (1) and (E):

$$w \sim (w', u'').$$

But $(w', u'') \equiv z$, so that:

$$w \sim z.$$

But by (F) one also has:

$$z \sim u, \text{ consequently also:}$$

$$w \sim u. \text{ QED.}$$

Now in order to prove (C) we decompose f into the variables $f', f'', \ldots$ and the isolated value 1, where f' assumes all values of the interval $(0 \ldots \varepsilon_1)$ with the exception of ε_1, $f^{(\nu)}$ all values of the interval $(\varepsilon_{\nu-1} \ldots \varepsilon_\nu)$ with the exception of the end-values $\varepsilon_{\nu-1}$ and ε_ν. Then one has:

$$f \equiv (f', f'', f''', \ldots f^{(\nu)}, \ldots, 1).$$

Now let x'' be a variable that assumes all values of $(\varepsilon_1 \ldots \varepsilon_2)$ without exception; let x^{IV} be a variable that assumes all values of $(\varepsilon_3 \ldots \varepsilon_4)$ without exception; let $x^{(2\nu)}$ be a var. that ass. all values of the int. $(\varepsilon_{2\nu-1} \ldots \varepsilon_{2\nu})$ without exception.
Then by (G):

$$f'' \sim x''$$

$$f^{\mathrm{IV}} \sim x^{\mathrm{IV}}$$

$$\cdots\cdots\cdots\cdots$$

$$f^{(2\nu)} \sim x^{(2\nu)}$$

$$\cdots\cdots\cdots\cdots$$

therefore:

$$f \sim (f', x'', f''', x^{\mathrm{IV}}, \ldots f^{(2\nu-1)}, x^{(2\nu)}, \ldots, 1).$$

But one has:

$$(f', x'', f''', x^{\mathrm{IV}}, \ldots f^{(2\nu-1)}, x^{(2\nu)}, \ldots, 1) \equiv x$$

and thus:

$$f \sim x.$$

Dedekind to Cantor

I have examined your proof once more, and I have discovered no gap in it; I am quite certain that your interesting theorem is correct, and I congratulate you on it. However, as I already indicated in the postcard, I should like to make a remark that counts *against* the conclusions concerning the concept of a manifold of ρ dimensions that you append in your letter of 25 June to the communication and the proof of the theorem. Your words make it appear—my interpretation may be incorrect—as though on the basis of your theorem you wish to cast doubt on the meaning ⟦Bedeutung⟧ or the importance of this concept; e.g. you say at the close of the letter, '*Now it seems to me* that all philosophical or mathematical deductions that use that erroneous presupposition [of the determinateness of the dimension-number] are inadmissible. Rather the difference that obtains between structures of *different* dimension-number must be sought in quite other terms than in the number of independent coordinates—the number that was hitherto held to be characteristic.'

Against this, I declare (despite your theorem, or rather in consequence of reflections which it stimulated) my conviction or my faith (I have not yet had time even to make an attempt at a proof) that the dimension-number of a continuous manifold remains its first and most important invariant, and I must defend all previous writers on this subject. To be sure, I gladly concede that the constancy of the dimension-number is thoroughly in need of proof, and so long as this proof has not been furnished one may doubt. But I do not doubt this constancy, although it appears to have been annihilated by your theorem. For all authors have clearly made the tacit, completely natural presupposition that in a new determination of the points of a continuous manifold by new coordinates, these coordinates should also (in general) be *continuous* functions of the old coordinates, so that whatever appears as continuously connected under the first set of coordinates remains continuously connected under the second. Now, for the time being I believe the following theorem: 'If it is possible to establish a reciprocal, one-to-one, and complete correspondence between the points of a continuous manifold A of a dimensions and the points of a continuous manifold B of b dimensions, then this *correspondence itself*, if a and b are *unequal*, is necessarily *utterly discontinuous*.' This theorem would also explain what happened with the first proof of your theorem, namely the incompleteness of the proof; the relation which you then wished to establish (by decimal fractions) between the points of a ρ-fold structure and the points of a unit interval would have been (if I do not deceive myself) *continuous*, if only it had also contained *all* points of the unit interval; similarly it seems to me that in your present proof the *initial* correspondence between the points of the ρ-interval (whose coordinates are all irrational) and the points of the unit interval (also with irrational coordinates) is, in a certain sense (smallness of the alteration) as continuous as possible; but to fill up the gaps, you are compelled to admit a frightful, dizzying discontinuity in the correspondence, which dissolves everything to atoms, so that every continuously connected part of the one domain appears in its image as thoroughly decomposed and discontinuous.

I hope I have expressed myself with sufficient clarity; the intent of my letter is only to ask you not to engage in public polemics against the article of faith that has hitherto been regarded as a fundamental truth of the theory of manifolds until you have thoroughly examined my objection.

Brunswick, 2 July 1877

Cantor to Dedekind

⟦Postcard: Postmark 2.7.77⟧

I am very happy that you have examined the proof and found it correct. I ask you to adhere to your plan, and to give me your views on the sense of the result more extensively and in greater depth; I should like to use them to form my judgement on the further pursuit of the subject.

Halle, 4 July 1877

I was very pleased by your letter of 2nd July, and thank you for your penetrating and exceptionally apt remarks.

In the conclusion of my letter of 25 June I unintentionally gave the appearance of wishing by my proof to oppose altogether the concept of a ρ-fold extended continuous manifold, whereas all my efforts have rather been intended to clarify it and to put it on the correct footing. When I said: 'Now it seems to me that all philosophical and mathem. deductions which use that erroneous presupposition—' I meant by this presupposition *not* 'the determinateness of the dimension-number' but rather the determinateness of the independent coordinates, whose number is assumed by certain authors to be in all circumstances equal to the number of dimensions. But if one takes the concept of coordinate *generally*, with no presuppositions about the nature of the intermediate functions, then the number of independent, one-to-one, complete coordinates, as I showed, can be set to any given number. I am also of your opinion that if we require that the correspondence be continuous, then only structures with the same number of dimensions can be related to each other one-to-one; and in this way we can find an invariant in the number of independent coordinates, which ought to lead to a definition of the dimension-number of a continuous structure.

However, I do not yet know how difficult this path (to the concept of dimension-number) will prove, because I do not know whether one is able to limit the concept of *continuous correspondence in general*. But everything in this direction seems to me to depend on the possibility of such a limiting ⟦Begrenzung⟧.

I believe I see a further difficulty in the fact that this path will probably fail if the structure ceases to be thoroughly continuous; but even in this case one wants to have something corresponding to the dimension-number—all the more so, given how difficult it appears to be to prove that the manifolds that occur in nature are thoroughly continuous.

By these lines, I wish only to indicate to you that, far from wishing to turn my result against the article of faith of the theory of manifolds, I rather wish to use it to secure its theorems, and, so far as possible, to make a contribution. For today I wish to take no more of your time, and I ask you, if you can find the time, to investigate the questions that press for an answer, not to despise them, and to let me know the results you obtain.

Halle, 23 Oct. 1877

. . . For a quarter year, Herr Borchardt has had a finished work of mine about the investigation I undertook last summer, entitled: *A contribution to the theory of manifolds* ⟦*Cantor 1878*⟧. I hope that it will appear shortly. Since I was able to use your friendly advice in writing it, perhaps you will be interested to know that I have found an even easier proof for one of my theorems. If two well-defined manifolds can be correlated to one another one-to-one and completely, element for element, then I say that they have the *same power* ⟦*Mächtigkeit*⟧ or that they are *equivalent*; similarly, I call two real variables *a* and *b equivalent* if they can be correlated to one another one-to-one and completely, and in this case I write, as you know,

$$a \sim b.$$

The theorem in question is as follows:

If *e* is a var. that assumes all irrational values $\begin{smallmatrix}>0\\<1\end{smallmatrix}$, and if *x* is a variable that takes on all rational and irrational values that are ≥ 0 and ≤ 1, then:

$$e \sim x.$$

Proof. Let ϕ_ν be the general term of a sequence that consists of all rational numbers ≥ 0 and ≤ 1; let η_ν be the gen. term of a sequence of any unequal irrational numbers $\begin{smallmatrix}>0\\<1\end{smallmatrix}$. E.g.:

$$n_\nu = \frac{\sqrt{2}}{2^\nu}$$

and let *h* denote the variable that takes on all values of the interval (0 . . . 1) with the exception of the ϕ_ν and the η_ν. Then:

$$x \equiv \{h, \eta_\nu, \phi_\nu\} \tag{1}$$

$$e \equiv \{h, \eta_\nu\}.$$

For the last formula we can also write:

$$e \equiv \{h, \eta_{2\nu-1}, \eta_{2\nu}\}. \tag{2}$$

If one compares the formulae (1) and (2) and notes that:

$$h \sim h; \ \eta_\nu \sim \eta_{2\nu-1}; \ \phi_\nu \sim \eta_{2\nu}$$

then it follows that:

$x \sim e$, as was to be shown.

Have you perhaps further investigated the question whether in the definition ⟦Begriffsbestimmung⟧ of an n-fold continuous manifold the condition of continuous correspondence suffices to make the concept safe against every contradiction and internally secure? . . .

P.S.—I have read some of Lipschitz's new *Textbook of analysis*; do you like it?

Dedekind to Cantor

1877.10.27

. . . However I still *believe* that the concept of dimension-number actually receives its invariant character from the condition of *continuous* correspondence.

The work of Lipschitz contains, so far as I have been able to see, much that is good and interesting; in accordance with my critical nature I have reservations about several points, but I find it very gratifying that he makes an earnest attempt to let mathematical rigour flourish even in a mathematical *textbook*. As for the justification of the theory of irrationals, in which he builds almost entirely upon your presentation (first published by Herr Heine), my greatest reservation is that (on p. 46) an entirely new *assumption* ('Then the boundary-value $\mathfrak{G}$ falls' etc.) is presented apparently as an obvious *consequence* of the earlier ones; moreover, I cannot grant the correctness of the final remark in §14 on the Greeks. . . .

Cantor to Dedekind

Halle a/Saale, 29th Dec. 1878

. . . You will probably also be in possession of the work, 'Fondamenti per la teorica delle funzioni di variabili reali di *Ulisse Dini*, Pisa 1878'. It seems to me to have been written with a deep knowledge of the subject and with great skill; he uses your method of introducing the numbers.—Although I agree with it completely, still I believe that it is equivalent to the method I indicated in a paper on trigonometric series[d] and that by the formal distinction of numerical quantities of different order—by which I *only* wished to express the different ways in which they could be given by simple infinite sequences (whose terms approach infinitely close *to each other* as the index increases)—the danger does not arise that one could believe that I wished to extend the domain of the real numbers. I *never even remotely* intended such a blunder; I expressly say in my paper that every number I designate by c can be set equal to a number b. Moreover, this blunder has actually been made by somebody else, outrageous though this may sound. I do not know whether you know of Thomae's *Outline of a theory of complex functions and theta-functions*; in the second edition, p. 9, one finds numbers which (*horribile dictu*) are smaller than every thinkable real number and yet are different from zero.

Thomae, Lüroth, Jürgens, and, a few days ago in Borchardt's *Journal*, von Netto have written about the question whether continuous manifolds with different dimension-number can be correlated to each other one-to-one and continuously—or rather, about the theorem that says that this is not possible; but the matter does not seem to me to be fully resolved.

Halle a/Saale, 17 Jan. 79

I think I have settled the question about the one-to-one and *continuous* mapping ⟦Abbildung⟧ of manifolds that was suggested by my investigations, and done so in the simplest and most rigorous manner, by reducing the question to the familiar fundamental theorem of analysis, according to which:

(I) a continuous function of *one* continuous variable t which has a negative value for $t = t_0$ and a positive value for $t = t_1$ assumes the value zero at least once in between.

The attempts of Thomae and Netto suffer, as you will perhaps have seen, from imperfections; thus, e.g., Thomae relies on a theorem* of Riemann's (*Gesammelte Werke* p. 450: A manifold ⟦Vielstreck⟧ of less than $n - 1$ dimensions, etc., etc.) which is, *for him*, unprovable; while, as I have now clearly seen, this theorem is as it were equivalent to the theorem to be proved for the case $\nu = n - 1$; but since $n - 1$ is as general as the number n, Thomae's proof revolves in a circle.

The general proof, which I shall give in a moment, has in fact been known to me for some time (over a year); but I did not formerly think it was rigorous, and refrained from speaking of it. The discovery which I made several days ago

---

[d] ⟦*Cantor 1872*.⟧

* He calls it an axiom.

is thus only that the proof is rigorous. My previous deception rests on the fact that the relations that appear in the proof are not one-to-one *on both sides*. The ambiguity that occurs is *not* however harmful, because it only occurs in the transition from a higher to a lower structure.

In the following I understand by a sphere of the ρth order the continuous structure of the ρth order which is extracted ⟦ausgesondert⟧ from a $\rho + 1$-fold manifold with coordinates $x_1, x_2, \ldots x_{\rho+1}$ by an equation:

$$(x_1 - a_1)^2 + (x_2 - a_2)^2 + \ldots (x_{\rho+1} - a_{\rho+1})^2 = r^2$$

so that e.g. the circle-line in the plane would be a sphere of the first order.

The theorem to be proved is, in the present extended version, this:

'A continuous M_μ and a continuous M_ν, where $\mu < \nu$, cannot be *continuously* correlated to each other so that to every element of M_μ there corresponds *a single* element of M_ν and to every element of M_ν there corresponds *one or more* elements of M_μ.'

This theorem is immediate for the case $\nu = 1$. To prove it in general, we suppose its correctness for $\nu = n - 1$ and show that it is then true for $\nu = n$.

To that end, we suppose a continuous relation between an M_μ and an M_n, where $\mu < n$, so that the transition from M_μ to M_n is one-to-one, and we show that this assumption contains an inner contradiction, or, better, contradicts the fundamental theorem (I) mentioned above.

Let a and b be two interior points of M_μ, A and B the corresponding points of M_n.

Taking A as the middle point, we construct in M_n a sphere K_{n-1} of the $n-1^{\text{st}}$ order which is so small that the point B lies outside the space which it encloses.

Taking a as the middle point, we likewise construct in M_μ a sphere $K_{\mu-1}$ of the $\mu-1^{\text{st}}$ order which is so small that

(1) the point b lies outside it; and,

(2) the continuous structure $G_{\mu-1}$ of the $\mu - 1^{\text{st}}$ order in M_n corresponding to this sphere lies entirely within the interior of the space bounded by the sphere K_{n-1}; which can be achieved by the continuity of the relation in the neighbourhood of the points a and A.

Let z be an arbitrary point of $K_{\mu-1}$, and ζ the corresponding point of $G_{\mu-1}$. We draw the straight line-segment $A\zeta$, whose continuation in this direction meets the sphere K_{n-1} in a single, determinate point Z.

Thus *by means of this construction* there corresponds to every point z of $K_{\mu-1}$ *one* point Z of K_{n-1} that varies continuously with z.

But the point Z cannot reach all points of K_{n-1}, *since this would violate our assumption for the case* $\nu = n - 1$.

We conclude with certainty that there are points P of the sphere K_{n-1} that are not reached by point Z; if one draws the ray AP from A out to such a point P, then it does *not* meet the structure $G_{\mu-1}$.

Now if we connect P by a continuous curve lying in the space M_n to the point B that lies outside K_{n-1} then we obtain:

(II) a linked continuous line APB, *which has no point in common with the structure* $G_{\mu-1}$.

To this line there correspond, by the presupposed continuous relation of M_μ and M_n, one or more curves in M_μ which run continuously from a to b, and thus:

(III) by the fundamental theorem (I) *must necessarily meet the sphere* $K_{\mu-1}$ *at least once.*

These two facts (II) and (III) contradict one another; so the assumption of a continuous relation of M_μ and M_ν is incorrect, and our theorem is proved for the case $\nu = n$.

Since the Göttingen Academy recently made me a corresponding member, and since Thomae published his proof in the *Göttinger Anzeigen*, I am considering sending them this proof as well. So I would be glad to hear your opinion on the matter first.

Dedekind to Cantor

I have studied your proof with attention, and have found only one small matter that might provoke doubt. After you have constructed a sphere K_{n-1} around the point A of M_n and a sphere $K_{\mu-1}$ around the point a of M_μ (whose image in M_n is designated by $G_{\mu-1}$) you give a mapping of $K_{\mu-1}$ to K_{n-1} that rests on this construction: to the point z of $K_{\mu-1}$ there corresponds a definite point ζ in $G_{\mu-1}$, and the intersection Z of the ray $A\zeta$ with K_{n-1} is regarded as a new image of z. But now it is conceivable that ζ coincides with A, because you admit that *several* points of M_μ can correspond to one and the same point of M_n; in this case the image Z would in general be *undetermined*. This difficulty is clearly easy to remove if the number of points a' in M_μ to which the same point A in M_n corresponds is *finite*, for one need only choose the radius of the sphere $K_{\mu-1}$ to be so small that the remaining points a' fall outside. But if the number of points is infinitely large, then for the present I see a real difficulty, and it is compounded by a problem in another portion of your proof. You say that one or more lines in M_μ leading from a to b correspond to the line APB; but this would in my opinion need more discussion and justification *even if B* is the image of a single or only a *finite* number of different points b' in M_μ; and if infinitely many points b' in M_μ possess the same image in M_n, then the existence of a continuous line from a to b whose image is the line APB is even more dubious. I *believe*, incidentally, that the theorem remains true even if it is conceded that every point in M_n can be the image of infinitely many distinct points in M_μ. Perhaps you will shortly eliminate the two mentioned difficulties. In a publication I would think it desirable if the names or technical expressions of the theory of manifolds were defined quite precisely (incidentally, I should like to see the shorter and equally Riemannian word 'domain' ⟦Gebiet⟧ given clear preference over the clumsy word 'manifold' ⟦Mannigfaltigkeit⟧); it would be very useful if this entire 'theory of domains' were presented *ab ovo*, without invoking geometric intuition, and in the process the concept of a continuous line leading from point *a* to point *b* within the domain *G* would have to be defined quite precisely and clearly. The definitions of Netto (whose treatise pleases me very much, and whose proof, I think, becomes quite accurate after a few modifications) contain a good kernel, but they seem to me capable of simplification and at the same time of being made more complete. I should not allow myself to make such a judgement were it not that many years ago, when I was editing Dirichlet's potential-lecture and wished to give a more rigorous justification of the so-called Dirichlet Principle, I had already occupied myself a great deal with such questions. I have several such definitions[e] which seem to me to give quite a good foundation, but later I let the whole matter drop and could for now give only an incomplete account, since I am fully occupied with the reworking of Dirichlet's number theory. But, as I hardly need to say, your communication interested me in the extreme, and, expressing my thanks, I remain. . . .

Brunswick, 19 January 1879

Cantor to Dedekind

⟦Postcard: Postmark 20.1.79⟧

In my proof which I sent you on the 17th of this month I find a point (albeit *not an important one*) that can be improved.

One proceeds better from two interior points *A* and *B* of M_n to which interior points of M_μ correspond.

Let *a* be *one* of the interior points that corresponds to *A*, while $b, b', b'', \ldots$ are *all* those that correspond to *B*. The sphere $K_{\mu-1}$ around *a* as the centre is chosen so small that $G_{\mu-1}$ lies in the interior of $K_{\mu-1}$ and that *all points* $b, b', b'' \ldots$ fall *outside* $K_{\mu-1}$. This is possible because of continuity, which *prevents* the points $b, b', b'', \ldots$ from approaching infinitely close to the point *a*.

Now there ought to be no more doubt that *one* of the curves in M_μ corresponding to the line *APB* leads from *a* to *one* of the points $b, b', b'', \ldots$ and must therefore meet the sphere $K_{\mu-1}$.

After I wrote this card I received a letter, for which many thanks. The answer comes later; I must be off to a meeting.

---

[e] ⟦See *Allgemeine Sätze über Räume* in Dedekind's *Nachlass* (*Gesammelte Werke*, Vol. ii, p. 352).—Noether and Cavaillès.⟧

[Postcard: Postmark 21.1.79]

Yesterday evening I was about to go out and was only able to note the receipt of your letter. One of your objections I anticipated in that card; as for the second, whereby the structure $G_{\mu-1}$ can contain the point A, it seems to me indeed to indicate a difficulty which I am not at the moment able to overcome, but which can perhaps be overcome if I make a suitable choice of the point A, for which there is a great deal of scope. In any case I had intended to allow the ambiguity in the transition from the higher to the lower M to be *so general* that to every point of M_n there also can correspond *infinitely many* points of M_μ; so it would *not suit me at all* if I had to *restrict* this assumption in order to save my proof. In any case, I am considering publishing *only in case* I should succeed in settling this point.

Berlin, 7 April 1882

... As a special treat I allow myself to call your attention to the following, which you have perhaps already noticed yourself.

If one has an n-fold extended continuously connected domain A and in it an everywhere-dense but *countable* point-set (M_ν), then for $n \geq 2$[f] the domain $\mathfrak{A}$ which is left over after removal of the set (M_ν) from A is also continuously connected in the sense that every two points N and N' of $\mathfrak{A}$ can be connected by uncountably many continuous, analytically definable lines which lie entirely within $\mathfrak{A}$—that is, which pass through not a single point of the set (M_ν). So *motion* is in a certain sense also possible in such a space.[g]

One recognizes the general correctness of this theorem most easily from my theorem that, given a sequence of real quantities:

$$\omega_1, \omega_2, \ldots, \omega_\nu, \ldots$$

there occur in every interval (α, β) quantities η which are equal to no ω_ν. Indeed, let us take $n=2$, i.e. A planar, then one can begin by connecting the points N, N' by a continuous line L without worrying about (M_ν), and choose on L a finite number of points $N_1, N_2, \ldots N_k$ in such a way that the line-segments $NN_1, N_1N_2, \ldots N_kN'$ fall entirely in the interior of A and then *replace* these line-segments with circular arcs that do not pass through a single point of (M_ν).

Take, e.g., the line-segment NN_1. A simply infinite set [Schaar] of circles goes through the two points N and N_1, and their centres lie on a straight line g and can be fixed on this line by a coordinate u. For u it is possible to give

f [Noether and Cavaillès have $n \leq 2$.]

g [Cantor gives a similar proof of this theorem in *Cantor 1882*.]

an interval (α, β) such that all the values of u in this interval correspond to circular arcs that lie entirely within A. The circles of the set which go not only through N and N_1 but also through the points

$$M_1, M_2, M_3, \ldots, M_\nu, \ldots$$

correspond to values of u that shall be designated by:

$$\omega_1, \omega_2, \ldots, \omega_\nu, \ldots$$

Now one takes a value η within the interval (α, β) that is not equal to any ω_ν, and one thereby obtains, by setting $u = \eta$, a circular arc lying wholly in A and connecting the points N and N_1—and on which *not a single point* M_ν *lies*. QED. . . .

W. Berlin, 15 April 1882

I should be grateful for your response to the mathem. content of my last letter—assuming you have received it, which I doubt.

My interest in the matter is the following: after the investigations and results about the concept of irrational number and its ideal nature (in Kummer's sense) which we both long ago obtained independently of each other, it is firmly established that in the formation of the concept of space there is no internal compulsion to imagine it as being everywhere continuous; you explicitly call attention to this fact in your paper on continuity.—So it was obvious that one ought to try to deduce the continuity of space from external reasons, namely from the fact of the continuous change of place, i.e. from motion, and in fact this was long my opinion.

But now this opinion is invalidated by the knowledge that, in a space so thoroughly discontinuous as the one I designated by $\mathfrak{A}$ in my letter, continuous change of place is possible from every point to every other point; and so one could perhaps imagine another, modified mechanics that would be valid for spaces $\mathfrak{A}$? . . .

Halle a/S, 15 Sept. 82

Enclosed is an attempt to formulate the question that has long interested me, namely, what we are to understand by a *continuum*. Please judge it leniently; if you do not find it wholly useless, we can discuss the matter orally.

An attempt to generalize your concept of cut and to use it for the general definition of the continuum did not succeed. In contrast, my starting-point of countable 'fundamental sequences' (which is what I now call sequences whose elements approach infinitely near to each other) seemed to accommodate itself to the investigation easily.

For linear point manifolds (i.e. those contained in a line) I can easily show with the help of my theorems that only a complete interval satisfies the conditions *A* and *B*. (+)

Similarly it ought to be possible to show, for point-manifolds that lie in a plane, that if *A* and *B* are satisfied, then they are either closed, single-piece lines or surfaces bounded by them. (+) . . .

(+) provided that for [*m, n*] one takes the distance between the two points.

P.S. A countable set ought not to be conceivable as a continuum under any ordering. In contrast, every non-countable set can probably be regarded as a continuum under certain orderings.

⟦The following remarks, on a separate sheet, are the enclosure referred to by Cantor:⟧

I. Let *M* be *any* well-defined manifold consisting of infinitely many elements *m, n, p*.

To every combination of two elements *m* and *n* of the manifold there is correlated a real, positive (*different from zero*) number in a determinate (*but wholly arbitrary*) manner; this number is designated by [*m, n*], and it can be as it were regarded as a function of the combination *m, n*. Conversely this function [*m, n*] induces certain relations among the elements of *M*, so that *m, n, p*, . . ., which originally had no relations to one another, now appear in a certain order *O*. We shall call this ordering *O* the ordering induced by the function [*m, n*]. Every other function [*m, n*]′ that differs from [*m, n*] in at least one term induces another ordering of *the same manif. M*.

II. *M* shall be called a *continuum* with respect to the ordering *O* induced by the function [*m, n*] if the following two conditions are satisfied:

A. If *m, n* are two arbitrary elements of *M* and ε is an arbitrary given quantity, then there can always be given a finite number of elements $m_1, m_2, \ldots, m_k$ of *M* such that:

$$[m, m_1],\ [m_1, m_2],\ \ldots,\ [m_{k-1}, m_k],\ [m_k, n]$$

are all smaller than ε.

B. If $m_1, m_2, \ldots, m_\nu$ is any countable infinite set of elements of *M* of such a nature that:

$$\lim\ [m_{\nu+\mu}, m_\nu] = 0 \text{ for } \nu = \infty$$

then there is always *one* and *only* one element *m* of *M* such that

$$\lim\ [m, m_\nu] = 0 \text{ for } \nu = \infty.$$

III. From this it clearly follows that one and the same set *M* can be a continuum with respect to *one* ordering *O*, and a discontinuum with respect to *another* ordering *O*′. Thus a square with the inclusion of its boundary is a continuum with respect to the ordering in which [*m, n*] is the straight-line distance of the points *m* and *n*; in contrast, the same square is a discontinuum with respect to an ordering that arises if one maps it one-to-one and completely onto an *interval*, so that the points m, n, p correspond to the points m', n', p' of the interval and one takes $[m, n]$ to be equal to the distance of the two corresponding points m', n'.

Halle, 15 Sept. 82

Halle a/S, 30 Sept. 1882

When the little sheet on the concept of the continuum falls into your hands, do not forget to strike the last passage, because it contains an error.

Clearly the square is also a continuum in the ordering I there conceived, albeit a continuum of only one dimension.

On the other hand, another ordering can easily be stated in which the square becomes a discontinuum.—The fundamental ideas are, that we can only talk of a continuum *with respect to* a particular ordering of the elements induced by a function $[m, n]$; and the two marks A and B of a continuum with respect to a given ordering. I hope to find time soon to explain these thoughts further, and perhaps to publish them.

⟦Postcard: Postmark 2.10.82⟧

I find that another condition C must be added to the conditions A and B for a continuum *with respect to a given ordering O*; it is not at all in general a consequence of the previous ones. Namely:

C. (Converse of B). If m is an element of M, m_ν a countable infinite set of elements such that:

$$\lim\ [m, m_\nu] = 0 \text{ for } \nu = \infty$$

then it must always be the case that: $\text{Lim}\ [m_{\nu+\mu}, m_\nu] = 0$ for $\nu = \infty$ and for arbitrary μ.

But with this everything that can be demanded of a continuum ought to be exhausted.

Halle a. S., 5th Nov. 1882

. . . The gap is for me all the more sensitive, because for many years I have been in the habit of submitting my inner mathematical experiences to your ripened judgement, and just after our latest visit in Harzburg and Eisenach God Almighty saw to it that I attained the most peculiar, unexpected results in the theory of manifolds and in the theory of numbers—or rather have found something which has been fermenting in me for years, and which I have long sought.—It is not a question of the general definition of a point-continuum, about which we spoke and in which I think I have made further progress, but rather about something much more general, and therefore more important.—

You remember I told you in Harzburg that I could not prove the following theorem:

If M' is part of a manif. M, M'' part of M', and if M and M'' can be reciprocally correlated one-to-one (i.e. M and M'' have the same power ⟦Mächtigkeit⟧) then M' has the same power as M and M''.

Now I have found the source of this theorem and can prove it rigorously and with the necessary generality; and this fills a large gap in the theory of manifolds.

I arrive at this result through a natural extension or continuation of the sequence of real integers, which leads me successively with the greatest certainty to the ascending powers, whose precise definition (apart from the first ones given by the number sequence $1, 2, 3, \ldots, \nu, \ldots$) had eluded me until now.—

I call the sequence $1, 2, 3, \ldots, \nu, \ldots$ the first whole real number-class ⟦ganze reale Zahlenclasse⟧ and I designate it by (ν).—

I arrive at the second class of real integers as follows:

Just as the number ν is the expression for a definite number ⟦Anzahl⟧ of settings-down of unity ⟦Setzungen der Einheit⟧ and for their union ⟦Vereinigung⟧, I begin by creating a new number ⟦Zahl⟧ ω which is to express the fact that the whole totality ⟦Inbegriff⟧ (ν) has been given; I can imagine ω as the limit of the numbers ν, if I thereby mean nothing more than that ω is the first integer created after *all* ν, i.e. the first that shall be called greater than all ν.

If I apply to ω the addition of a unit, as earlier was done to ν, then I obtain a new number $\omega + 1$, which expresses that first ω is set down, *then* the unit added and united with ω into a new number. I call the transition from a number ν or ω to the *next*-following the *first moment of generation* ⟦das *erste Erzeugungsmoment*⟧; and I call the transition from an increasing sequence of integers that has no greatest to the number that is next-largest to them all the *second* moment of generation.

So the formation of the number ω comes from the *second* moment of generation, that of the number $\omega + 1$ from the *first*.

If one now repeatedly applies *both* of these moments of generation, one arrives at a continuation of our sequence of numbers that progresses in a determinate succession:

$$1, 2, 3, \ldots \nu, \ldots \omega, \omega + 1, \ldots \omega + \nu \ldots, 2\omega, 2\omega + 1, \ldots, \mu\omega + \nu, \ldots$$
$$\lambda\omega^2 + \mu\omega + \nu, \ldots, \sigma\omega^k + \rho\omega^{k-1} + \ldots + \mu\omega + \nu, \ldots, \text{etc., etc.}$$

The first impression that this sequence ought to make on you is that one does not see how one can come to a sort of conclusion ⟦Abschluss⟧ in its continuation—a conclusion that would, however, be necessary if it is to furnish us a *new determinate* power, i.e. the power of the second class that *immediately* follows the power of the first class.—

In order to obtain this conclusion, yet a *third* moment is added to the two moments of generation defined above; I call it the *limiting moment* ⟦*Beschränkungsmoment*⟧. It consists in the requirement that one only undertake the creation of a new integer with the help of one of the two other moments if the totality of all preceding numbers is *countable* by ⟦*abzählbar* in⟧ a known number-class that already exists in its full extent.

In this way, observing all three moments, one can with complete certainty arrive at new number-classes and their powers, and the new numbers obtained in this way are then always of precisely the same reality and concrete determinateness as the older ones. I truly did not know what should hold us back from this activity of forming new integers, as soon as it appears that the introduction of a new one of these countless number-classes is desirable or even indispensable for the progress of science; and the latter appears to me to be in fact the case in the theory of manifolds, and perhaps in many other areas as well; at any rate *I* can make no further progress *without* this extension, and *with* it I am able to achieve many quite unexpected things.—

We must of course first stop to investigate the *first* number-class, which I call (α); it contains in its beginning the first (ν). By what was said above, (α) is completely characterized by the fact that the numbers $1, 2, 3, \ldots \omega, \omega + 1, \ldots, \omega^2, \ldots$ that *precede* a determinate number α *always* form a set ⟦Menge⟧ that (disregarding its natural ordering) is countable by the *first* class.

It can now be shown with complete rigour that the totality (α) is *not itself* countable by the first class.

So the number-totality (α) has a *different* power than (ν), indeed the *next* higher; for I can rigorously prove the following theorem:

'If (α') is any fragment of (α), then *either* (α') is finite, *or* it is countable by the first class, *or* (α') *can* be one-to-one correlated with the totality (α) itself, i.e. (α') is *countable* by the *second* class; *Quartum non datur*.'

Moreover, I believe I can rigorously prove that the totality of all our *real*, rational *and* irrational numbers can be one-to-one correlated with (α); and so, taking the preceding theorem into account, the *two-class theorem* has been proved for infinite linear ⟦manifolds⟧ (or for manifolds that can be reduced to these by a mapping).—

Perhaps you are surprised by my boldness in calling the things ⟦Dinge⟧ ω, $\omega + 1, \ldots, \alpha, \ldots$ *integers*, and even the *real integers* of the second class, while I gave them the more modest title of infinity-symbols when I used them previously (in the *Annalen*, vols. 17, 20, and 21).

But my freedom is explained by the remark that among the conceptual things ⟦Gedankendinge⟧ α that I call real integers of the *second* class there are relations that can be reduced to the basic operations.—

To be sure, the laws that obtain between them are essentially different and more complicated, more difficult to find by induction, than in our number theory which is based upon the older numbers. Already in addition it turns out that in general the *commutative* law does not obtain; $\alpha + \beta$ is in general *not* equal to $\beta + \alpha$.

One sees this very easily when one notices that $1 + \omega = \omega$ while $\omega + 1$ is quite different from ω.

If $\beta < \alpha$, one defines subtraction by the equation:

$$\alpha + (\beta - \alpha) = \beta$$

but *not* by the equation: $(\beta - \alpha) + \alpha = \beta$, which in general cannot be solved for $(\beta - \alpha)$. If α is the multiplicand and β the multiplier, then one has a determinate number in (α) that is the product

$$\alpha\beta;$$

but here as well one finds that in general $\beta\alpha$ is different from $\alpha\beta$.[i]

However the *associative* law is also valid in the second class, both for addition and for multiplication. One has:

$$\alpha + (\beta + \gamma) = (\alpha + \beta) + \gamma$$

$$\alpha(\beta\gamma) = (\alpha\beta)\gamma.$$

Only half of the distributive principle is valid, namely:

$$(\alpha + \beta)\gamma = \alpha\gamma + \beta\gamma.$$

Among the numbers (α) there is a natural distinction between the numbers α that arise with the help of the *first* moment of generation (and therefore have an *immediate* predecessor, which I call α_{-1}—which is in general different from $\alpha - 1$, because $\alpha - 1 = \alpha$ for $\alpha \geq \omega$) and those that have *no* immediate predecessor in the sequence and for which α_{-1} is therefore meaningless. Of the first sort those that are not factorable (which one can call prime numbers) are distinguished from the others.*

So far as I can see, our finite irrational numbers can be relatively easily deter-

---

[i] ‖Noether and Cavaillès: $\beta\alpha$.‖

* It seems to me (although I am not yet entirely certain) that every number α can be uniquely represented in the following form:

$$\alpha = c_0 \cdot \pi \cdot c_1 \cdot \pi' \cdot c_2 \cdot \pi'' \ \ldots \ c_{\nu - 1} \cdot \pi^{(\nu - 1)} \cdot c_\nu \cdot \rho$$

where $c_0, c_1, \ldots c_{\nu - 1}, c_\nu$ are positive integers from (ν), $\pi, \pi', \pi'', \ldots$ are prime numbers from (α), and ρ is a number from (α) of the second sort (for which ρ_{-1} ‖Noether and Cavaillès have e_{-1}‖ does not exist) such that it *is not divisible by any number of the first sort*. These numbers of the second sort have quite a special character, since, e.g., $\omega = \omega \cdot \nu$ where ν is a number from (ν). Therefore one cannot talk of a *determinate* factorization of ρ.—And therefore too, one cannot really talk of prime numbers of the *second* sort.

It should also be remarked that when I say that α is divisible by β, I maintain the possibility of the equation:

$$\alpha = \beta\gamma$$

where β is the multiplicand. This stipulation of the concept of divisibility seems to me required in the number-class (α).

mined with the help of the numbers α, which is a matter I intend to pursue further.—

Let P be the totality of numbers contained in the formula:

$$z = \alpha_1 \cdot \frac{1}{3} + \alpha_2 \cdot \frac{1}{3^2} + \ldots + \alpha_\nu \cdot \frac{1}{3^\nu} + \ldots$$

where α_ν can only assume the values 0 and 2.

P then has the two properties:

(1) $P \equiv P'$

(2) in no interval is P everywhere dense.

Task. Let M be an arbitrary manifold of elements; M_1 a proper part of M, M_2 a proper part of M_1; and suppose there is a reciprocal, one-to-one relation between M and M_2; to show, that M and M_1, and consequently also M_1, and M_2, have the *same power*.

⟦Postcard: Postmark 6.11.82⟧

I believe that I made an error in writing out my explanation yesterday, and it could lead to a misinterpretation.

In the product:

$$\alpha = \beta\gamma$$

β is for me the *multiplier* and γ the multiplicand; I say that α is divisible by β if α can be represented as a product in which β is the *multiplier*.

G.C.

The factorization which I suspect is generally valid is also to be understood in this sense:

$$\alpha = c_0 \pi c_1 \pi' \ \ldots \ c_{\nu-1} \pi^{(\nu-1)} c_\nu \rho.$$

C. FOUNDATIONS OF A GENERAL THEORY OF MANIFOLDS: A MATHEMATICO-PHILOSOPHICAL INVESTIGATION INTO THE THEORY OF THE INFINITE (*CANTOR 1883d*)

The following essay, Cantor's *Grundlagen*, is his most thorough explanation of the principles underlying his transfinite set theory.

After having discovered the existence of non-denumerable subsets of the real line (*Cantor 1874*), Cantor, in his *1878*, turned to a more general study of what he called *powers* (*Mächtigkeiten*) or *cardinal numbers* of point-sets of real numbers. He already knew of the existence of two powers: the power of the continuum, and the power of the integers. He seems to have been attempting to find point-sets of yet higher power when he proved the counter-intuitive theorem that the points of the unit interval can be correlated one-to-one with the points of any Euclidean interval of any dimension, even if the dimension is countably infinite. For reasons which are somewhat murky, Cantor proposed, at the end of *Cantor 1878*, his famous 'continuum hypothesis':

> [T]he question arises how the different parts of a continuous straight line, i.e. the different infinite manifolds of points that can be conceived in it, are related with respect to their powers. Let us divest this problem of its geometric guise, and understand (as has already been explained in §3) by a *linear* manifold of real numbers every conceivable totality ⟦Inbegriff⟧ of infinitely many, distinct real numbers. Then the question arises, into *how many* and which classes do the linear manifolds fall, if manifolds of the same power are placed into the same class, and manifolds of different power into different classes? By an inductive procedure, whose more exact presentation will not be given here, the theorem is suggested that the number of classes of linear manifolds that this principle of sorting gives rise to is finite, and indeed, equals *two*.

From 1878 on, Cantor's work was to be dominated by the attempt to prove this conjecture. He pursued two strategies.

Cantor's *direct* strategy was to use the derived sets $P^{(n)}$ (described above) of a point-set P to measure its cardinality. As Russell put it,

> Popularly speaking, the first derivative consists of all points in whose neighbourhood an infinite number of terms of the collection are heaped up; and subsequent derivatives give, as it were, different degrees of concentration in any neighbourbood. Thus, it is easy to see why derivatives are relevant to continuity; to be continuous, a collection must be as concentrated as possible in every neighbourhood containing any terms of the collection (*Russell 1903*, p. 324).

Because the process of taking derivatives does not necessarily terminate after a countably infinite number of iterations, Cantor, in his *1880*, continued the process into the transfinite, and introduced new 'symbols of infinity' for derivatives having an *infinite* order; these symbols of infinity were soon to evolve into the transfinite ordinal numbers (as Cantor notes in a footnote to §1 of the *Grundlagen*). In modern notation,

$$P^{(\infty)} = \bigcap_{n=1}^{\infty} P^{(n)}, \quad P^{(\infty+1)} = (P^{(\infty)})^{(1)}, \text{ etc.}$$

In *1884* Cantor used the direct strategy to obtain a partial result on the continuum problem, showing that the continuum hypothesis is true for sets such that $P^{(1)} \subseteq P$. Cantor briefly discusses his direct strategy in §10 of the *Grundlagen*.

Incidentally, the theorem he states, without proof, in §10 paragraphs 9–11 of the *Grundlagen* he withdrew as erroneous in §16 of *Cantor 1884a*. For a discussion see *Hallett 1984*, pp. 90–3. This was the starting-point for what has come to be known as descriptive set theory, the point of which is to investigate the behaviour of relatively simply described subsets of real numbers, and then in particular to study the continuum hypothesis for these 'simple' sets. Cantor's ideas were refined and extended by Young, Hausdorff, Alexandroff, Lusin, Souslin, Sierpinski, and others. That the results these mathematicians achieved were the best possible was shown by Gödel's method of proving the consistency of the GCH. Interest in descriptive set theory was revived with the large cardinals programme; for a brief survey of the basic results, see *Martin 1977* or *Hallett 1984*, pp. 103–14; Hallett also provides, on pp. 74–118, a detailed account of Cantor's attempts to prove the continuum hypothesis.

Cantor's second, indirect, strategy is the principal subject of the *Grundlagen*. The strategy was based on his ordinal theory of powers—that is, on the introduction of a class of transfinite ordinal numbers that can be used to 'count' the size of any infinite set (just as the finite ordinal numbers can be used to count the size of finite sets). On this strategy, the continuum hypothesis would be settled by determining where the power of the continuum lies on the scale of transfinite ordinal numbers: specifically, by showing that the continuum is counted by the *first* non-denumerable ordinal number. Cantor's *Grundlagen* introduced the transfinite ordinals and their arithmetical operations; used them to prove the existence of infinitely many powers; and introduced the concept of a well-ordered set. (Well-ordering is crucial to Cantor's strategy: to show that his ordinals suffice to count every set, and thus to count the continuum, he must establish that every set can be well-ordered (see his remarks at the beginning of §3). The first—controversial—proof of the well-ordering theorem was given in *Zermelo 1904* and *1908a*.)

Cantor's presentation of the mathematical details of his theory of transfinite numbers is informal and somewhat imprecise; he gave a more rigorous treatment later, in *Cantor 1897*. He also expanded on the philosophical underpinnings of his theory in *Cantor 1886a*, *1886b*, and especially *1887–8*.

There have been numerous studies of the origins of modern set theory; particularly useful are *Cavaillès 1962* and *Moore 1982*. The most extensive study of Cantor's philosophical and mathematical thought is Michael Hallett's *Cantorian set theory and limitation of size* (*Hallett 1984*); this work includes a valuable analysis of the *Grundlagen*, as well as extensive discussion of Cantor's influence on subsequent set theory.

The selection that follows is a translation of Cantor's *Grundlagen einer allgemeinen Mannigfaltigkeitslehre. Ein mathematisch-philosophischer Versuch in der Lehre des Unendlichen*, Teubner, Leipzig 1883. This is a *separate* printing of the fifth article in Cantor's six-part 'Über unendliche lineare Punktmannigfaltigkeiten', which appeared in the *Mathematische Annalen* between 1879 and 1884; the six articles are reprinted in the *Gesammelte Abhandlungen*, (*Cantor 1932*, pp. 139–246). The separate 1883 printing added a subtitle, a preface, and some footnotes; all of these additions were omitted from *Cantor 1932*. There are also some minor differences in punctuation between the two versions. The translation published here follows the fuller, separately published version. (The 1932 version, however, contains, at pp. 208–9, eight copyrighted footnotes by Zermelo commenting on aspects of Cantor's work.)

The translation is by William Ewald. References to *Cantor 1883d* should be to the section numbers (which occur in all editions) and to the paragraph numbers (which have been added here).

PREFACE

⟦1⟧ The present treatise will shortly appear in *Math. Annalen* as the fifth number of an article entitled 'Infinite Linear Point-Manifolds'; the first four numbers are contained in volumes XV, XVII, XX, and XXI of the same journal. All these works are connected to two articles that I have published in volumes LXXVII and LXXXIV of Borchardt's *Journal*,[a] in which the main ideas that have guided me in the theory of manifolds are already to be found. Since the present essay carries the subject much further, and since its main thesis is independent of the earlier articles, I decided to publish it separately under a title that corresponds more closely to its contents.

⟦2⟧ As I deliver these pages to the public, I must mention that I wrote them with two sorts of reader in view—for philosophers who have followed the developments in mathematics up to the most recent times, and for mathematicians who are familiar with the most important results, ancient and modern, of philosophy.

⟦3⟧ I know very well that the subject I discuss has at all times given rise to the most varied opinions and conceptions, and that neither mathematicians nor philosophers have achieved agreement on all points. So I do not believe that, in such a difficult, complicated, and all-embracing subject as the infinite, I shall have said the last word. But since after many years of research into this subject I have come to definite convictions, and since, in the course of my studies, these convictions have not wavered but have only become more firmly entrenched, I thought I had an obligation to put them in order and make them known.

⟦4⟧ May I thereby have found and expressed the objective truth, which I have been at pains to discover.

Halle, Christmas 1882

The Author

---

[a] ⟦The articles are *Cantor 1874* and *Cantor 1878*. The former is translated above.⟧

§1.

⟦1⟧ The foregoing account of my investigations in the theory of manifolds **[1]**[b] has reached a point where further progress depends on extending the concept of real integer beyond the previous boundaries; this extension lies in a direction which, so far as I know, nobody has yet attempted to explore.

⟦2⟧ I am so dependent on this extension of the number concept that without it I should be unable to take the smallest step forward in the theory of sets ⟦Mengen⟧; this circumstance is the justification (or, if need be, the apology) for the fact that I introduce seemingly exotic ideas into my work. For what is at stake is the extension or continuation of the sequence of integers into the infinite; and daring though this step may seem, I can nevertheless express, not only the hope, but the firm conviction that with time this extension will have to be regarded as thoroughly simple, proper, and natural. At the same time I do not conceal that with this undertaking I place myself in opposition to widespread intuitions ⟦Anschauungen⟧ about the mathematical infinite and to commonly-held opinions about the essence of number.

⟦3⟧ As for the mathematical infinite, to the extent that it has found a justified application in science and contributed to its usefulness, it seems to me that it has hitherto appeared principally in the role of a variable quantity, which either grows beyond all bounds or diminishes to any desired minuteness, but always remains *finite*. I call this infinite the *improper infinite* ⟦das *Uneigentlich-unendliche*⟧.

⟦4⟧ But in modern times, in geometry and in the theory of functions, another equally justified concept of the infinite has been developed. For example, in the investigation of an analytic function of a complex variable it has become necessary and common to imagine, in the plane that represents the complex variables, a single point lying at infinity (that is, an infinitely distant but definite point) and to examine the behaviour of the function in the neighbourhood of this point, just as in the neighbourhood of any point. It turns out that in the neighbourhood of the infinitely distant point the function displays exactly the same behaviour as at any other point lying in the finite region, so that in this case we are fully justified in thinking of the infinite as lying at a completely determined point.

⟦5⟧ When the infinite appears in such a definite form I call it the *proper infinite* ⟦*Eigentlich-Unendliches*⟧.

⟦6⟧ The mathematical infinite, in both forms, has led to great progress in geometry, in analysis, and in mathematical physics. But these two forms must be carefully distinguished if we are to understand what follows.

⟦7⟧ Infinity, in its first form (the improper-infinite) presents itself as a *variable finite* ⟦*veränderliches Endliches*⟧; in the other form (which I call the proper infinite) it appears as a thoroughly *determinate* ⟦*bestimmtes*⟧ infinite.

---

[b] ⟦Numerals in single brackets refer to Cantor's Endnotes at the end of this article.⟧

The infinite real integers, which I shall define later (and which I discovered many years ago, without becoming clearly aware that they are concrete numbers of real significance[1] ⟦Bedeutung⟧) have absolutely nothing in common with the first of these two forms, with the improper-infinite. Rather, they possess the same character of determinateness as we find in the infinitely-distant points in the theory of analytic functions; that is, they belong to the realm of the proper infinite.—But while the point at infinity of the complex-number plane stands isolated from all the points lying in the finite region, here we obtain not merely a single infinite integer but an infinite sequence of them; they are clearly differentiated from each other, and stand in lawlike number-theoretic relations both to each other and to the finite integers. But these relationships are not such as let themselves be reduced essentially to the relations of finite numbers to each other; the latter relations indeed appear frequently, but only in the different intensities ⟦Stärke⟧ and forms of the improper-infinite—for example, in functions of a variable x that grow infinitely large or infinitely small, in case they assume determinate finite order numbers in their becoming infinite. Such relations can in fact only be considered as veiled ratios of the finite (or at any rate as immediately reducible to the finite); in contrast, the laws of the (yet to be defined) proper-infinite integers are from the start different from the dependencies that reign in the finite, but it is not thereby ruled out that the finite real numbers themselves may receive certain new determinations with the help of the determinate-infinite ⟦bestimmt-unendlichen⟧ numbers.

⟦8⟧ The new determinate-infinite numbers will be defined with the help of two *principles of generation*; by the combined action of which it is possible to break through every obstacle ⟦Schranke⟧ in the concept-formation ⟦Begriffsbildung⟧ of real integers. But fortunately, as we shall see, a *third* principle stands against them. I call it the *constraining* or *limitation* ⟦*Hemmungs- oder Beschränkungs*⟧ principle; it imposes certain successive constraints on the thoroughly endless process of creation, so that we obtain natural segments ⟦Abschnitte⟧ in the absolutely infinite sequence of integers. I call these segments *number classes* ⟦*Zahlenklassen*⟧.

⟦9⟧ The *first* number class (I) is the set of finite integers 1, 2, 3, ..., ν, ...; it is followed by the *second* number-class (II), consisting of certain infinite integers following each other in determinate succession; as soon as the second number-class has been defined, one comes to the third, then to the fourth, and so on.

⟦10⟧ The introduction of these new integers seems to me of the greatest significance for the development and sharpening of the *concept of power* ⟦*Mächtigkeitsbegriff*⟧, which I introduced in my earlier articles (*Crelle's J.*, Vol. 77, p. 257; Vol. 84, p. 242) and frequently applied in the earlier numbers of this essay. According to this concept, every well-defined set has a determinate

[1] Hitherto I called them 'definite defined infinity symbols'; see *Math. Ann.*, Vol. xvii, p. 357; Vol. xx, p. 113; Vol. xxi, p. 54.

power; two sets have the same power if they can be, element for element, correlated with one another reciprocally and one-to-one.

⟦11⟧ For finite sets power coincides with the *Anzahl*[c] of elements, because, as everybody knows, such sets have the same Anzahl of elements in every ordering.

⟦12⟧ For infinite sets, on the other hand, until now nobody at all has talked of a precisely defined *Anzahl* of their elements—even though a determinate *power*, entirely independent of their ordering, can be ascribed to them.

⟦13⟧ It is easy to show that the *smallest* power of infinite sets must be ascribed to those sets which allow themselves to be reciprocally and one-to-one correlated with the *first* number class, and consequently have the same power as it. But an equally simple, natural definition of the *higher* powers has until now been lacking.

⟦14⟧ Our above-mentioned number classes of determinate-infinite integers now prove to be the natural representatives, occurring in a unified form, of the lawlike sequence of increasing powers of well-defined sets. I stress that the power of the second number-class (II) is not only different from the power of the first number-class, but that it is also in fact the *next highest* power; we can therefore call it the *second* power or the power of the *second class*. Similarly the third number class yields the definition of the third power or the power of the third class, and so on.

§2.

⟦1⟧ Another great gain ascribable to the new numbers consists, for me, in a *new* concept which has not yet been mentioned—namely, the concept of the *Anzahl* of the elements of a *well-ordered* infinite manifold. Because this concept is always expressed by a completely determinate number of our widened number-domain (provided only that the shortly-to-be-defined ordering of the elements of the set is determinate), and because on the other hand the Anzahl concept obtains an immediate concrete representation in our inner intuition ⟦Anschauung⟧, so, through this connection between Anzahl and number ⟦Zahl⟧, the reality (which I emphasize) of the latter is proved even in the cases where it is determinate-infinite.

⟦2⟧ A *well-ordered* set is a well-defined set in which the elements are bound to one another by a determinate given succession such that (i) there is a *first* element of the set; (ii) every single element (provided it is not the last in the succession) is followed by another determinate element; and (iii) for any desired finite or infinite set of elements there exists a determinate element which is *their immediate successor* in the succession (unless there is absolutely nothing in the succession following all of them). Two 'well-ordered' sets are now said to be

---

[c] ⟦No satisfactory uniform translation of this word exists which captures Cantor's use of the term. In what follows, 'Zahl' is always translated as 'number', and 'Anzahl' is left untranslated.⟧

of the same *Anzahl* (with respect to their given succesions) when a reciprocal one-to-one correlation of them is possible such that, if E and F are any two elements of the one set, and E_1 and F_1 are the corresponding elements of the other, then the position of E and F in the succession of the first set always agrees with the position of E_1 and F_1 in the succession of the second set (i.e. when E precedes F in the succession of the first set, then E_1 also precedes F_1 in the succession of the second set). This correlation, if it is possible at all, is, as one easily sees, always completely determinate; and since in the widened number sequence there is always one and only one number α such that the numbers preceding it (from 1 on) in the natural succession have the same Anzahl, one must set the 'Anzahl' of both of these 'well-ordered' sets directly to α, if α is an infinitely large number, and to the number $\alpha - 1$ which immediately precedes α, if α is a finite integer.

⟦3⟧ The essential difference between finite and infinite sets now turns out to be that a finite set presents the *same* Anzahl of elements for *every* succession which one can give its elements; in contrast, a set consisting of infinitely many elements will in general give rise to *different* Anzahlen, depending on the succession that one gives the elements. The *power* of a set is, as we saw, a property independent of the ordering; but the *Anzahl* of the set reveals itself to be, in general (as soon as one has anything to do with infinite sets) a property dependent on a given succession of elements. Nevertheless, there is even for infinite sets a certain connection between the *power* of a set and the *Anzahl* of its elements determined by a given succession.

⟦4⟧ If we take first a set which has the power of the first class and if we give the elements *any* determinate succession (so that it becomes a 'well-ordered' set), then its Anzahl is always a determinate number of the *second* number class and can never be determined by a number of any other number class than the second. On the other hand, every set of the first power can be ordered in such a succession that its Anzahl, with respect to this succession, becomes equal to any given number of the second number class. We can express these propositions as follows: every set of the power of the *first* class is *countable* ⟦*abzählbar*⟧ by numbers of the *second* number class and only by such numbers; and every set can always be given a succession of its elements such that it can be counted in this succession by an arbitrarily chosen number of the second number class, which number gives the *Anzahl* of the elements of the set with respect to that succession.

⟦5⟧ Analogous laws hold for sets of higher power. Thus every well-defined set of the power of the *second* class is countable *by* numbers of the *third* number class and only by such numbers, and the set can always be given such a succession of its elements that it is counted[2] in this succession by an *arbitrarily chosen*

[2] What in the earlier numbers of this article I called 'countable' is, according to the more precise and more general definition I have just introduced, precisely countability by numbers of the first number class (finite sets) or by numbers of the second class (sets of the first power).

number of the *third* number class, which number determines the Anzahl of the elements of the set with respect to that succession.

§3.

[1] The concept of *well-ordered set* turns out to be fundamental for the entire theory of manifolds. In a later article I shall discuss the law of thought that says that it is always possible to bring any *well-defined* set into the *form* of a *well-ordered* set—a law which seems to me fundamental and momentous and quite astonishing by reason of its general validity. Here I confine myself to the demonstration that from the concept of *well-ordered* set the fundamental operations for the integers, be they finite or determinate-infinite, arise in the simplest manner and that the laws governing them can be derived from immediate inner intuition with apodictic certainty. If two *well-ordered* sets M and M_1 are given whose Anzahlen correspond to the numbers α and β, then $M + M_1$ is also a *well-ordered* set, which arises if first the set M is given, and then (following it and united to it) the set M_1. There corresponds to the set $M + M_1$ (with respect to the given succession of its elements) a determinate Anzahl; let this number be called the *sum* of α and β and let it be designated by $\alpha + \beta$. We immediately see that if α and β are not both finite then $\alpha + \beta$ is in general different from $\beta + \alpha$. Already, therefore, the *commutative* law ceases to be valid in general for addition. It is now so simple to form the concept of the sum of *several* summands given in determinate sequence (where this sequence itself can be determinate-infinite) that I need not go into further detail here. I only remark that the *associative* law is in general valid. One has in particular $\alpha + (\beta + \gamma) = (\alpha + \beta) + \gamma$.

[2] If one takes a succession, determined by a number β, of various sets which are similar and which are similarly ordered and such that each has an Anzahl of its elements equal to α, then one obtains a new well-ordered set whose corresponding Anzahl supplies the definition for the product $\beta\alpha$, where β is the multiplier and α the multiplicand; here too it turns out that $\beta\alpha$ is in general different from $\alpha\beta$; thus also for the multiplication of numbers the commutative law is in general *invalid*. But the associative law for multiplication always holds, so that one has: $\alpha(\beta\gamma) = (\alpha\beta)\gamma$.

[3] Certain of the new numbers are distinguished in that they have the prime-number property, though this property must here be characterized in a somewhat more precise manner, so that by prime number one understands a number α for which the factoring $\alpha = \beta\gamma$, where β is the multiplier, is only possible where $\beta = 1$ or $\beta = \alpha$; but in general for prime numbers α the multiplicand has a certain range of indeterminacy, which in the nature of things cannot be altered. Nevertheless it will be shown in a later paper that the factoring of a number into its prime factors can always be achieved in an essentially *unique* manner, and even a *determinate* manner with respect to the sequence of factors (so far as these are not finite prime numbers appearing as neighbours in the product). Thus two kinds of determinate-infinite prime numbers emerge, of

which the first stands closer to the finite prime numbers while the prime numbers of the second kind have an entirely different character.

⟦4⟧ Further, with the help of this new knowledge I can now prove rigorously a proposition which is cited at the close of the article 'A contribution to the theory of manifolds' (*Crelle's J.*, Vol. 84, p. 257 ⟦*Cantor 1878*⟧) on the so-called linear infinite manifolds.

⟦5⟧ In the last number of this series I proved a theorem for point-sets P which are contained in an n-dimensional continuous region. This theorem can be expressed in the new terminology as follows: 'If P is a point-set whose derivative ⟦Ableitung⟧ $P^{(1)}$ vanishes identically, where α is any desired integer of the *first* or *second* number class, then the first derivative $P^{(\alpha)}$, and therefore also P itself, is a point-set of the power of the *first* class'. It seems to me most remarkable that this proposition has the following converse: 'If P is a point-set whose first derivative $P^{(1)}$ has the power of the *first* class, then there is an integer α belonging to the *first* or *second* number class, for which $P^{(\alpha)}$ identically vanishes, and of the numbers α for which this happens there is a smallest'.

⟦6⟧ In consequence of the kind request of my distinguished friend Prof. Mittag-Leffler in Stockholm, I shall very soon publish the proof of this proposition in the first volume of the new mathematical journal which he will edit. At the conclusion of this article Herr Mittag-Leffler will show how, on the basis of this theorem, his and Prof. Weierstrass's investigations into the existence of single-valued analytic functions with given singularities can be given an important generalization.

§4.

⟦1⟧ The extended sequence of integers can, if one wishes, be completed without further ado into a continuous set of numbers by adjoining to every integer α all real numbers x that are greater then zero and less than one.

⟦2⟧ Perhaps at this point the question will arise whether, since in this manner we have achieved a determinate extension of the domain of real numbers into the infinitely large, one cannot with equal success define determinate infinitely small numbers, or, what might come to the same thing, define finite numbers which do not coincide with the rational and irrational numbers (which latter appear as the limiting values of the sequence of rational numbers), but which might be inserted into supposed gaps amidst the real numbers, just as the irrational numbers are inserted into the chain of the rational numbers, or the transcendental numbers into the structure of the algebraic numbers.

⟦3⟧ The question of the establishment of such interpolations, on which some authors have expended much effort, can, in my opinion, only be clearly and distinctly answered with the help of our new numbers—in particular, with the general concept of the Anzahl of well-ordered sets. The previous attempts, it seems to me, partly rest on an erroneous confusion of the improper infinite with the proper infinite, and partly have been constructed on a thoroughly insecure and unstable foundation.

⟦4⟧ The improper infinite has often been called by recent philosophers a 'bad' infinite, in my opinion unjustly, since it has proved itself to be a very good and highly useful instrument in mathematics and the natural sciences. The infinitely small quantities have, so far as I know, until now in general been usefully developed *only* in the form of the improper-infinite, and are thus capable of all those differences, modifications, and relations which are found in infinitesimal analysis and in the theory of functions, and which are used to establish the rich profusion of analytic truths. But all attempts to force this infinitely small into a *proper* infinite must finally be given up as pointless. If proper-infinitely-small quantities exist at all, that is, are definable, then they certainly stand in no direct relationship to the familiar quantities which *become* infinitely small.

⟦5⟧ In contrast to these experiments with the infinitely small and the confusion of the two forms of the infinite, there is a frequently advocated view on the essence and the significance of numerical quantities according to which the only numbers that are conceived of as real are the *finite integers* of our number class (I).

⟦6⟧ At most a certain reality is conceded to the *rational* numbers which emerge directly from them. But as for the irrationals, they ought in pure mathematics to receive a purely *formal* meaning, in that they as it were only serve as instruments of calculation to fix properties of groups of integers and to describe these properties in a simple and uniform manner. According to this opinion, the true material of analysis is exclusively formed from the finite integers, and all the truths which have been found in arithmetic and analysis, or whose discovery is hoped for, ought to be conceived as relations of the finite integers to each other; infinitesimal analysis and with it the theory of functions are held to be legitimate only to the extent that their propositions can be interpreted as provable by laws governing finite integers. This conception of pure mathematics, although I cannot agree with it, has certain incontestable advantages, which I should here like to emphasize. Moreover, some of the most meritorious mathematicians of the present day are among its advocates, and this fact speaks in its favour.

⟦7⟧ If, as is here assumed, only the finite integers are actual, but all the others are nothing other that relation-forms ⟦Beziehungsformen⟧, then one can require that the proofs of analytic propositions be tested ⟦geprüft⟧ in accordance with their 'number-theoretic content' and that one fill in every gap in accordance with the principles of arithmetic; then the feasibility of such a filling-in is the true touchstone for the genuineness and complete rigour of proofs. It is not to be denied, that in this way we can perfect the justification for many theorems and also make other methodological improvements in various parts of analysis; moreover, by obeying the principles that flow out of this intuition, one obtains a safeguard against every manner of absurdity or error.

⟦8⟧ In this manner a definite (if also rather prosaic and obvious) principle is recommended to all as a guideline; it is supposed to indicate the true limits to the flight of mathematical speculation, to show the domain within which the passion for mathematical thought will run no danger of falling into the abyss of the 'transcendent'—where, it is said with fear and wholesome alarm, 'anything is possible'. But be that as it may, it is difficult to say (who knows?) whether it was not simply an attitude of expediency which caused the originators of this doctrine to recommend it to the rising powers, which so easily come into danger through high spirits and extravagance, as an effective counterbalance, a protection against all errors, although a *fruitful* principle cannot be found in it. I cannot believe that these mathematicians started from these principles and were led to the discovery of new truths; for although I grant that these maxims have many good aspects, I nevertheless hold them strictly speaking as *erroneous*. We are indebted to them for no true progress, and if they were actually to be followed, then science would be held back or banished into the narrowest confines. Fortunately in real life matters are not so dire, and those rules (which are in certain circumstances useful) have never been taken entirely literally, either in theory or in practice; it is striking that up until now, so far as I know, nobody has undertaken to formulate the rules more completely and better than I have here attempted to do.

⟦9⟧ If we look about in history, we find that similar opinions were often held; they are already to be found in Aristotle. As is well known, throughout the Middle Ages the proposition '*infinitum actu non datur*', taken from *Aristotle*, was held to be incontrovertible by all the *scholastics*. But if one considers the reasons which Aristotle **[2]** brought forward against the real existence of the infinite (see, e.g., his *Metaphysics*, Book XI, Chapter 10), it will be found that the main issue can be traced back to a presupposition which involves a *petitio principii*, namely, the presupposition that there are only finite numbers ⟦Zahlen⟧, from which he concluded that the only countings ⟦Zählungen⟧ are of finite sets. I believe however that I have proved above (and it will appear even more clearly in what follows) that determinate countings can be carried out just as well for infinite sets as for finite ones, provided that one gives the sets a determinate law that turns them into *well-ordered* sets. That without such a lawlike succession of the elements of a set it cannot be counted—this lies in the nature of the concept of *counting*. Finite sets too can be counted only if we have a determinate ordering ⟦Aufeinanderfolge⟧ of the counted elements; but here we encounter a particular property of finite sets, namely, that the result of the counting—the *Anzahl*—is *independent* of the particular ordering; while for infinite sets, as we have seen, such an independence does *not* in general hold. On the contrary, the Anzahl of an infinite set is an infinite integer which is *codetermined* by the law of the counting; it is precisely here, and here alone, that the essential difference, founded in nature itself and therefore never to be abolished, between the finite and the infinite is located. Never again will the existence of the infinite be denied because of this difference, but on the contrary the existence of the finite can now be upheld. If one allows the one to fall, then one must also do away with the other; and where would this path take us?

⟦10⟧ Another argument used by Aristotle against the reality of the infinite consists in the assertion that if the infinite existed then it would absorb

⟦aufheben⟧ the finite and destroy it. But, as will be clearly seen in what follows, the matter is in truth thus: to an infinite number, if it is thought of as determinate and completed ⟦vollendet⟧, a finite number can *very well* be adjoined and united to it *without* thereby effecting the cancellation of the latter; rather, the infinite number is modified by such an adjunction of a finite number to it. It is only the *reverse* procedure—the adjunction of an infinite number to a finite one, when the latter is placed first—that effects the cancellation of the latter without introducing any modifications of the former.—These facts concerning the finite and the infinite, entirely unrecognized by Aristotle, should give new stimulation not only to analysis but to other sciences, particularly the natural sciences.

⟦11⟧ In the course of scientific exertions and experiments which have lasted many years I have been logically compelled (almost against my will, because it is in opposition to traditions which had become precious to me) to the point of view which considers the infinitely great not merely in the form of something growing without limit ⟦unbegrenzt Wachseden⟧ (and in the closely related form of convergent infinite series first introduced in the seventeenth century) but also to fix it mathematically by numbers in the determinate form of the completed-infinite ⟦Vollendet-unendlichen⟧; and I believe that no arguments can be made against it which I would not know how to meet.

§5.

⟦1⟧ When I spoke just now of traditions, I understood them not merely in the narrower sense of that which has been experienced ⟦Sinne des Erlebten⟧, but trace them back to the founders of modern philosophy and natural science. Towards a judgement on the question which is raised here, I give some of the most important references. One should see:

Locke, *Essay on Human Understanding*, Bk. II, Chs. XVI and XVII.

Descartes, *Letters*, and the *Discussions* of his *Meditations*; also, *Principia* I, 26.

Spinoza, Letter XXIX, *Cogita. Metaph.*, parts I and II.

Leibniz, Erdmann edn., pp. 138, 244, 436, 744.[3]

⟦2⟧ Even today, one cannot find stronger arguments than these against the introduction of infinite integers; these arguments should be examined and compared with mine. I shall reserve for another occasion a detailed and thorough discussion of these works, and in particular of the highly important letter, rich in content, of Spinoza to L. Meyer; for the present I confine myself to the following.

⟦3⟧ However different the theories of these writers may be, in their judgement of the finite and infinite they essentially agree that finiteness is part of

[3] The following are also noteworthy: Hobbes, *De Corpore* ch. VII; Berkeley, *Treatise on the Principles of Human Knowledge*, §§128–131.

the concept of number and that the true infinite or Absolute, which is in God, permits no determination whatsoever. As to the latter point I fully agree, and cannot do otherwise; the proposition: '*omnis determinatio est negatio*' is for me entirely beyond question. But, as I have already said above in the discussion of the Aristotelian arguments against the '*infinitum actu*', I see in the first point a *petitio principii*, and this explains many contradictions to be found in all these writers, especially in Spinoza and Leibniz. I can find no justification for the assumption that besides the Absolute (which is not attainable by any determination) and the finite there should be no modifications ‖Modifikationen‖ which, although not finite, nevertheless are determinable by numbers and are therefore what I call proper-infinite.—Indeed, in my opinion this assumption stands in contradiction to certain propositions propounded by the last two philosophers. What I assert and believe to have proved in this work as well as in my earlier writings is this: that following the finite there is a *Transfinitum* (which one could also call *Suprafinitum*)—that is, there is an unbounded step-ladder of determinate modes which in their nature are not finite but infinite, but which, just like the finite,[d] can be determined by well-defined and distinguishable *numbers*. I am convinced that the domain of definable quantities is *not* exhausted by the finite quantities, and the bounds of our knowledge may accordingly be extended without violence to our nature. In place of the Aristotelian–scholastic proposition discussed in §4 I accordingly set another:

Omnia seu finita seu infinita *definita* sunt et excepto Deo ab intellectu determinari possunt.[e] **[3]**

‖4‖ The finiteness of human *understanding* is often adduced as the reason why only finite numbers can be thought; but again I see in this assertion the vicious circle already mentioned. Namely, it is tacitly meant by 'finiteness of understanding' that its capacity with respect to the formation of numbers ‖Zahlenbildung‖ is confined to finite numbers. But if it turns out that the understanding can also in a determinate sense define and distinguish infinite, that is, *superfinite* ‖*überendliche*‖ numbers, then either the words 'finite understanding' must be given an extended meaning (from which that conclusion can then no longer be drawn); or the predicate 'infinite' must in certain respects be conceded to the human understanding. The latter is in my opinion the only correct course. The words 'finite understanding' which one hears so frequently are in my opinion not at all apt; however limited in truth human nature may be, still *very much* of the infinite adheres to it, and I even assert that if it were not itself in many respects infinite, the solid confidence and certainty in the being ‖Sein‖ of the Absolute, about which we know we all agree, would be inexplicable. In particular I contend that the human understanding has an unbounded capacity for the stepwise formation of classes of integers which stand in a determinate relationship to the infinite modes ‖Modis‖ and whose *powers* are of increasing strength.

---

[d] ‖'Finite' is here a noun.‖

[e] ‖All things finite or infinite are *definite* and, God excepted, can be determined by the intellect.‖

⟦5⟧ The chief difficulties in the systems of the two last-named thinkers, which externally are indeed of different types but internally are closely related, can, I believe, be brought nearer to a solution by the methods I have introduced; indeed, some of the difficulties can already be satisfactorily solved and explained. These difficulties have understandably given rise to later criticism. But for all its merits this criticism has not, it seems to me, yielded an adequate substitute for the hampered development of the theories of Spinoza and Leibniz. For alongside of (or in place of) the mechanical explanation ⟦Erklärung⟧ of nature (which inside its proper domain has all the aids and advantages of mathematical analysis at its disposal, but whose one-sidedness and insufficiency have been strikingly exposed by Kant) there has until now not been even the start of an *organic* explanation of nature that would attempt to go further and that would be armed with the same mathematical rigour; this organic explanation can, I believe, be initiated only by taking up afresh and continuing the works and endeavours of those thinkers.

⟦6⟧ An especially difficult point in Spinoza's system is the relationship of the finite modes to the infinite one; it remains unexplained how and under what circumstances the finite can maintain its independence with respect to the infinite, or the infinite with respect to still higher infinities. The example already touched upon in §4 seems in its modest symbolism to indicate the way by which one can perhaps come nearer to the solution of this question. If ω is the first number of the second number class, then one has $1 + \omega = \omega$, but $\omega + 1 = (\omega + 1)$, where $(\omega + 1)$ is a number entirely distinct from ω. Therefore, as one here clearly sees, everything depends on the *placement* ⟦*Stellung*⟧ of the finite relative to the finite; if the former comes first, it merges into the infinite and vanishes therein; but if it *contents* itself to take its place *after* the infinite it is preserved and unites with it to form a new, because modified, infinite.

§6.

⟦1⟧ If difficulties should arise in conceiving of *infinitely large, self-contained* ⟦*abgeschlossene*⟧ integers, comparable to each other and to the finite numbers, and bound to each other and to the finite numbers by fixed laws, then these difficulties will be connected with the perception that, although the new numbers have in some respects the same character as the earlier ones, in many other respects they have a thoroughly idiosyncratic nature; indeed, it often happens that different characteristics ⟦Merkmale⟧ are united in one and the same infinite number although they never occur together in the finite numbers. In one of the passages cited in the previous section it is argued that an infinite integer, if it existed, would have to be *both* an even *and* an odd number, and since these two characteristics cannot occur together, it follows that *no* such number exists.

⟦2⟧ Obviously, one is here tacitly assuming that characteristics which are mutually exclusive for the traditional numbers must also be so for the new numbers; and one accordingly concludes that infinite numbers are impossible. Who fails to see the paralogism at a glance? For is not every generalization or extension of concepts bound up with an abandonment of special features ⟦Besonderheiten⟧—even unthinkable without such an abandonment? Hasn't one just recently had the idea of introducing the complex numbers—an idea of the greatest importance for the development of analysis and an idea that has led to the greatest progress—without seeing any obstacle in the fact that they can be called neither positive nor negative? And it is only a similar step that I venture here; indeed, it will probably be much easier for the general consciousness to follow me than was possible in going over from the real numbers to the complex; for the new integers, even though they are distinguished from the traditional numbers by a more intensive and substantial determinateness, have nevertheless as 'Anzahlen' exactly the same kind of reality. But difficulties long plagued the introduction of complex quantities until, after much trouble, one had found their geometrical representation by points or vectors in a plane.

⟦3⟧ To return briefly to the consideration of oddness and evenness, let us consider again the number ω in order to show how those characteristics which cannot be united in the finite numbers here occur together without any contradiction. In §3 general definitions of addition and multiplication were given, and I emphasized that in these operations the commutative law in general has *no* validity. In this fact I perceive an essential difference between the infinite and finite numbers. Recall that in a product $\beta\alpha$, β is the multiplier and α the multiplicand. The following two forms then at once arise for ω: $\omega = \omega.2$ and $\omega = 1 + \omega.2$. Accordingly ω can be conceived of both as an even and as an odd number. But from another point of view, namely, if 2 is taken as a multiplier, it can also be said that ω is neither an even nor an odd number, because, as one can easily prove, ω can be represented neither in the form $2.\alpha$ nor in the form $2.\alpha + 1$. Therefore, the number ω, in contrast to the traditional numbers, has in fact an entirely idiosyncratic nature, since all these characteristics and properties are united in it. And the remaining numbers of the second number class are even more peculiar, as I shall show.

§7.

⟦1⟧ In §5 I mentioned many passages in Leibniz's works where he speaks against infinite numbers. He says there, among other things, 'Il n'y a point de nombre infini ni de ligne ou autre quantité infinie, si on les prend pour des Touts veritables.' 'L'infini veritable n'est pas une modification, c'est l'absolu; au contraire, dès qu'on modifie on se borne ou forme un fini' (in the latter passage I agree with him in the first statement, but not in the second). But I am in the fortunate position of being able to quote passages of the same thinker in which he to a certain extent contradicts himself and declares himself in the most unambiguous way *for* the proper infinite (which is different from the Absolute). He says in the Erdmann edition, p. 118:

⟦2⟧ 'Je suis tellement pour l'infini actuel, qu'au lieu d'admettre que la nature l'abhorre, comme l'on dit vulgairement, je tiens qu'elle l'affecte partout, pour mieux marquer les perfections de son Autour. Ainsi je crois qu'il n'y a aucune partie de la matière qui ne soit, je ne dis pas divisible, mais actuellement divisée; et par conséquent la moindre particelle doit être considerée comme un monde plein d'une infinité de créatures différentes.'[f]

⟦3⟧ The proper-infinite, as we find it in, for example, well-defined point sets or in the construction of bodies from point-atoms ⟦punktuellen Atomen⟧ (I thus do not mean here the chemical–physical (Democritean) atoms, because I cannot hold them for existent, either in thought or in reality, although much that is useful has been achieved up to a certain limit by this fiction) has found its most determined defender in a very sharp-minded philosopher and mathematician of our century, Bernhard Bolzano, who has developed his views in the rich and beautiful work, 'Paradoxes of the infinite', Leipzig, 1851. His purpose is to demonstrate that the contradictions which the sceptics and peripatetics *of all times* have sought in the infinite do not exist at all, if one only takes the trouble (not, to be sure, always slight) to study in all seriousness the concepts of the infinite according to their true content. In this book one also finds a pertinent discussion of the mathematical improper-infinite as it appears in the form of differentials of the first and higher orders or in the infinite summation of series or other limiting processes. This infinite (called by some scholastics the 'syncategorematic infinite') is a mere helping-concept, a relation-concept of our thought; in his definition it includes variability, and so 'datur' can never be said of it in the proper sense.

⟦4⟧ It is remarkable that, with respect to *this* kind of infinite, absolutely no essential differences of opinion prevail even among present-day philosophers—if I may ignore the fact that certain modern schools of so-called positivists or realists **[4]** or materialists believe that in this *syncategorematic* infinite they see the *highest concept*, even though they themselves must concede that it has no *actual* being ⟦Sein⟧.

⟦5⟧ However, already in Leibniz we find in many places essentially the correct point of view; the following, for example, refers to the improper-infinite (Erdmann edition p. 436):

⟦6⟧ 'Ego philosophice loquendo non magis statuo magnitudines infinite parvas quam infinite magnas, seu non magis infinitesimas quam infinituplas. Utrasque enim per modum loquendi compendiosum pro mentis fictionibus habeo, ad calculum aptis, quales etiam sunt radices imaginariae in Algebra. Interim demonstravi, magnum has expressiones usum habere ad compendium cogitandi adeoque ad inventionem; et in errorem ducere non posse, cum pro infinite parvo substituere sufficiat tam parvum quam quis volet, ut error sit minor dato, unde consequitur errorem dari non posse.'[g]

---

[f] ⟦'I am so much in favour of the actual infinite that, rather than admit that nature abhors it (as one vulgarly says), I hold that it realizes it everywhere, in order the better to mark the perfections of its Author. Thus I believe that there is no part of matter that is not not only divisible, but actually divided; and in consequence the smallest particle ought to be considered as a world full of an infinity of different creatures.'⟧

[g] ⟦'Philosophically speaking, I no more set up infinitely small magnitudes than infinitely large, and no more infinitely few than infinitely many. For, in a concise manner of speaking, I consider

⟦7⟧ Bolzano is perhaps the only one for whom the proper-infinite numbers are legitimate (at any rate, he speaks about them a great deal); but I absolutely do *not* agree with the manner in which he handles them without being able to give a correct definition, and I regard, for example, §§29–33 of that book as unsupported and erroneous. The author lacks two things which are necessary for a genuine grasp of the concept of determinate-infinite numbers: both the general *concept of power* and the precise *concept of Anzahl.* To be sure, both appear in germ in isolated passages and as special cases. But he does *not* work his way through to full clarity and exactness, which explains many inconsistencies and even many errors in this valuable book.

⟦8⟧ *Without* these two concepts, I am convinced, one can *not* make further progress in the theory of manifolds. The same is true, I believe, for the fields that are a part of the theory of manifolds or that have the most intimate contact with it—for example, modern function theory on the one hand and logic and epistemology on the other. When I conceive the infinite, as I have done here and in my earlier investigations, there follows for me a genuine pleasure, to which I gratefully yield, in seeing how the concept of integer, which in the finite has only the background of *Anzahl,* as it were *splits* into *two* concepts when we ascend to the infinite—one of *power* (which is independent of the ordering which a set is given) and one of *Anzahl* (which is necessarily bound to a lawlike ordering of the set by virtue of which it becomes *well-ordered*). And when I descend again from the infinite to the finite I see clearly how the two concepts become one again and *flow together* to form the concept of finite integer.

§8.

⟦1⟧ We can speak of the actuality or existence of the integers, finite as well as infinite, in *two* senses; but strictly speaking they are the same two relationships in which in general the reality of any concepts and ideas can be considered. First, we may regard the integers as actual in so far as, on the basis of definitions, they occupy an entirely determinate place in our understanding, are well distinguished from all other parts of our thought, and stand to them in determinate relationships, and thus modify the substance of our mind in a determinate way; let us call this kind of reality of our numbers their *intrasubjective* or *immanent reality.* **[5]** But then, reality can also be ascribed to numbers to the extent that they must be taken as an expression or copy of the events and relationships in the external world which confronts the intellect, or to the extent that, for instance, the various number-classes (I), (II), (III), etc. are representatives of powers that actually

both to be fictions of the mind, suitable for calculation, like imaginary roots in algebra. However, I have shown that these expressions are very useful for conciseness of thought as well as for discovery; and it is not possible that they lead into error since, in order that the error be smaller than any given ⟦quantity⟧, it suffices to substitute for the infinitely small a quantity as small as you like; whence it follows that a ⟦fixed⟧ error cannot be given.'⟧

occur in physical and mental nature. I call this second kind of reality the *transsubjective* or the *transient reality* ⟦*transiente Realität*⟧ of the integers.

⟦2⟧ Because of the thoroughly realistic but, at the same time, no less idealistic foundation of my point of view, I have no doubt that these two sorts of reality always occur together in the sense that a concept designated in the first respect as existent always also possesses in certain, even infinitely many, ways a transient reality. **[6]** To be sure, the determination of this transient reality is often one of the most troublesome and difficult problems of metaphysics, and must frequently be left to the future, when the natural development of one of the other sciences will uncover the transient meaning of the concept in question.

⟦3⟧ This linking of both realities has its true foundation in the *unity* of the *all to which we ourselves belong*.—The mention of this linking has here only one purpose: that of enabling one to derive from it a result which seems to me of very great importance for mathematics, namely, that mathematics, in the development of its ideas has *only* to take account of the *immanent* reality of its concepts and has *absolutely no* obligation to examine their *transient* reality. Because of this remarkable feature—which distinguishes mathematics from all other sciences and provides an explanation for the relatively easy and unconstrained manner with which one may operate with it—it especially deserves the name of *free mathematics*, a designation which, if I had the choice, would be given precedence over the now usual 'pure' mathematics.

⟦4⟧ Mathematics is in its development entirely free and is only bound in the self-evident respect that its concepts must both be consistent with each other and also stand in exact relationships, ordered by definitions, to those concepts which have previously been introduced and are already at hand and established. **[7]** In particular, in the introduction of new numbers it is only obligated to give definitions of them which will bestow such a determinacy and, in certain circumstances, such a relationship to the older numbers that they can in any given instance be precisely distinguished. As soon as a number satisfies all these conditions it can and must be regarded in mathematics as existent and real. I think this is the reason, hinted at in §4, why one must regard the rational, irrational, and complex numbers as being every bit as existent as the finite positive integers.

⟦5⟧ It is not necessary, I believe, to fear, as many do, that these principles present any danger to science. For in the first place the designated conditions, under which alone the freedom to form numbers can be practised, are of such a kind as to allow only the narrowest scope for discretion. Moreover, every mathematical concept carries within itself the necessary corrective: if it is fruitless or unsuited to its purpose, then that appears very soon through its uselessness, and it will be abandoned for lack of success. But every superfluous constraint on the urge to mathematical investigation seems to me to bring with it a much greater danger, all the more serious because in fact absolutely no justification for such constraints can be advanced from the essence of the science—for the *essence* of *mathematics* lies precisely in its *freedom*.

⟦6⟧ If I had not discovered this property of mathematics by means of the reasoning I have described, then the entire development of the science itself, as we find it in our century, would have led me to exactly the same opinions.

⟦7⟧ Had Gauss, Cauchy, Abel, Jacobi, Dirichlet, Weierstrass, Hermite, and Riemann always been constrained to subject their new ideas to a metaphysical control, we should certainly not now enjoy the magnificent structure of the modern theory of functions which, although it was designed and erected in full freedom without ulterior purposes, nevertheless, in its applications to mechanics, astronomy, and mathematical physics, already discloses its transient meaning, as was to be expected; and we should not be seeing the great upswing in the theory of differential equations brought about by Fuchs, Poincaré, and many others, if these distinguished forces had been restrained and tied down by outside influences; and if Kummer had not exploited the freedom, rich in its consequences, of introducing so-called 'ideal' numbers into number theory, we should today not be able to admire the important and distinguished algebraic and arithmetical works of Kronecker and Dedekind.

⟦8⟧ Entitled though mathematics is to move in complete freedom from all metaphysical fetters, I am not, however, able to concede the same right to 'applied' mathematics, such as, for example, analytic mechanics and mathematical physics. These disciplines are in my opinion in their foundations as well as in their aims *metaphysical*; if they seek to make themselves free from metaphysics, as has been recently proposed by a celebrated physicist ⟦Gustav Kirchoff⟧, they degenerate into the form of a 'description of nature' in which the fresh breeze of free mathematical thought—as well as the power of *explaining* and *justifying* natural phenomena—must be absent.

§9.

⟦1⟧ Because of the great significance for the theory of manifolds of the so-called real, rational, and irrational numbers, I do not wish to omit to say the most important things about their definitions. I shall not discuss the introduction of the rational numbers, because rigorous arithmetic presentations of this have often been given; among those with which I am most familiar I emphasize those of H. Grassmann (*Lehrbuch der Arithmetik*, Berlin, 1861) and J.H.T. Müller (*Lehrbuch der allgemeinen Arithmetik*, Halle, 1855). But I should like to discuss briefly and in detail three principal ways (which are presumably in essence one) of giving a rigorous arithmetical introduction of the general real numbers. The *first* of these is the definition which Prof. Weierstrass has made use of for many years in his lectures on analytic functions; one can find some hints of it in the *Programmabhandlung* of E. Kossak (*Die Elemente der Arithmetik*, Berlin, 1872). *Second*, R. Dedekind has published in his article: *Stetigkeit und irrationale Zahlen* (Brunswick, 1872) ⟦translated above⟧ an idiosyncratic definition, and *third* in the year 1871 (*Math. Annalen*, Vol. v, p. 123) I gave a definition which superficially has a certain similarity to Weierstrass's, so much so that H. Weber (*Zeitschrift für Mathematik und Physik*, Vol. 27, historical-literary sect., p. 163) was able to confuse them; but in my opinion this third form of definition, later developed also by Lipschitz (*Grundlagen der Analysis*, Bonn, 1877) is the simplest and most natural of all, and it has the advantage that it can be immediately adapted to the analytic calculus.

⟦2⟧ These definitions agree that an irrational real number is given by a well-defined infinite set of rational numbers of the first power. But they differ over the way in which the set is linked with the number it defines, and in the conditions which the set has to fulfil in order to qualify as a foundation for the definition in question.

⟦3⟧ In the *first* definition a set of positive rational numbers a_ν is taken as a basis, is designated by (a_ν), and satisfies the condition that, whatever and however many of the a_ν are summed (so long as the Anzahl is finite) this sum always remains less than a specifiable limit. Now if one has two such aggregates (a_ν) and (a'_ν) it can be rigorously shown that only three cases can occur: either for every part of unity $1/n$ there exists a finite Anzahl m such that, for all $m' > m$, if one sums together the first m' elements of each aggregate, each sum contains the same number of copies of $1/n$; or, from a given n on, the first sum will always contain more copies of $1/n$ than the second; or, from a given n on, the second will always contain more than the first. Corresponding to these cases, if b and b' are the numbers to be defined by the aggregates (a_ν) and (a'_ν), in the first case we set $b = b'$, in the second $b > b'$, and in the third $b<b'$. If the two aggregates are combined to form a new one $(a_\nu + a'_\nu)$ this forms the basis for the definition $b+b'$; but if from each of the aggregates (a_ν) and (a'_ν) one forms the new aggregate $(a_\nu . a'_\mu)$ in which the elements are the products of all a_ν and all a'_μ, then this new aggregate forms the foundation for the definition of the product bb'.

⟦4⟧ One sees here that the creative element which binds the set with the number defined through it lies in the *formation of sums*; but it must be emphasized as *essential* that only the summation of an always *finite* Anzahl of rational elements is used and that the number b to be defined is *not* set at the beginning as equal to the sum Σa_ν of the infinite series (a_ν); this would be a *logical error*, because the definition of the sum Σa_ν is only reached by equating it with the *finished* number b which is necessarily defined earlier. I believe that this logical error, which was first avoided by Weierstrass, was in earlier times the universal practice, and was not noticed because it belongs to those rare cases in which actual errors can do no significant harm to calculations.—Nevertheless I am convinced that all the difficulties which have been found in the concept of the irrational are connected with the indicated error, whereas if we avoid this error the irrational numbers take root in our minds with the same precision, distinctness, and clarity as the rational numbers.

⟦5⟧ The definition of Dedekind takes as a basis the *totality of all* rational numbers, but divided into two groups in such a way that, if the numbers of the first group are designated by $\mathfrak{A}_\nu$, and those of the second group by $\mathfrak{B}_\mu$, then always $\mathfrak{A}_\nu < \mathfrak{B}_\mu$; such a partition of the set of rational numbers Dedekind calls a 'cut'. He designates it by $(\mathfrak{A}_\nu | \mathfrak{B}_\mu)$ and correlates with it a number b. If one compares two such cuts with one another one finds as with the *first* definition a total of *three* possibilities corresponding to which the two numbers b and b' represented by the two cuts are either set equal to each other, or $b > b'$, or $b < b'$. The first case occurs (apart from some easily handled exceptions which occur if the numbers to be defined are rational) only if the two cuts are completely identical. And here we see the undeniable advantage of this definition over the two others, namely, that every number b corresponds to only a *single* cut. But this definition has the great disadvantage that the numbers of analysis *never* occur as 'cuts', but must be brought into this form with a great deal of artificiality and effort.

⟦6⟧ Here too the definitions of the sum $b + b'$ and the product bb' are given on the basis of new cuts which are obtained from the old.

⟦7⟧ The disadvantage in the *first* and *third* definitions is that the same (i.e. equal) numbers occur infinitely often, and that accordingly an unambiguous overview of the totality of real numbers is not immediately obtainable. This disadvantage can be overcome with the greatest ease by a specialization of the underlying sets (a_ν) using one of the well-known unambiguous systems, such as, for instance, the decimal system or the simple development in continued fractions.

⟦8⟧ I come now to the *third* definition of real numbers. Here too an infinite set of rational numbers (a_ν) of the first power is taken as a basis, but it has other properties than in the Weierstrassian definition. I require that, after taking an arbitrarily small rational number ε, a finite Anzahl of elements of the set can be detached, so that the remaining ones have a pairwise difference which in absolute value is less than ε. Every such set (a_ν), which also can be characterized by the requirement:

$$\lim_{\nu=\infty} (a_{\nu+\mu} - a_\nu) = 0 \text{ (for arbitrary } \mu)$$

I call a *fundamental sequence* and correlate with it a number b, to be defined through it, for which one can expediently use the symbol (a_ν) itself (as Heine, after many conversations with me on the subject, has proposed) (see *Crelle's Journal*, Vol. 74, p. 172). Such a *fundamental sequence* presents three cases, as can be rigorously deduced from its definition: either the elements of the sequence a_ν for sufficiently large values of ν are smaller in absolute value than any arbitrarily given number; or, from a certain ν on they are greater than a fixed positive rational number ρ; or, from a certain ν on they are less than a fixed negative rational quantity $-\rho$. In the first case I say that b equals zero, in the second, that b is greater than zero or positive, in the third, that b is less than zero or negative.

⟦9⟧ Now come the elementary operations. If (a_ν) and (a'_ν) are two *fundamental sequences* that determine the numbers b and b', it can be shown that $(a_\nu \pm a'_\nu)$ and $(a_\nu . a'_\nu)$ are also *fundamental sequences* which therefore determine three new numbers which serve as definitions of the sum and difference $b \pm b'$ and the product bb'.

⟦10⟧ Moreover, if b is different from zero (the definition of which has just been given) one can prove that (a'_ν/a_ν) is also a *fundamental sequence* whose corresponding number furnishes the definition for the quotient b'/b.

⟦11⟧ The elementary operations between a number b given by a fundamental sequence (a_ν) and a directly given rational number a are included in the definitions just given by letting $a'_\nu = a$, $b' = a$.

⟦12⟧ Only now come the definitions of equality, greater than, and less than for two numbers b and b' (of which b' can also be $=a$), namely, one says that $b = b'$ or $b > b'$ or $b < b'$ according as $b - b'$ is equal to zero or greater than zero or less than zero.

⟦13⟧ After all these preparations we obtain as the first *rigorously provable* proposition that, if b is the number determined by a fundamental sequence (a_ν), then $b - a_\nu$ with increasing ν becomes less in absolute value than every conceivable rational number, or, what is the same thing, that:

$$\lim_{\nu=\infty} a_\nu = b.$$

Care must be taken on this cardinal point, whose significance can easily be overlooked: in the *third* definition the number b, say, is not defined as the 'limit' of the terms a_ν of a fundamental sequence (a_ν); for this would be a logical error similar to the one discussed for the *first* definition, i.e. we would be presuming the *existence* of the limit $\lim\limits_{\nu=\infty} a_\nu$. But the situation is rather the reverse. In our previous definitions the concept b has been thought of as having certain properties and certain relationships to the rational numbers, with the result that we can draw the conclusion with logical certainty that $\lim\limits_{\nu=\infty} a_\nu$ exists and is equal to b. Pardon me here the minuteness of detail, which I justify with the observation that most persons pass over this inconspicuous small detail and consequently tangle themselves in doubt and contradictions over the irrational; but by observing the facts emphasized here they would spare themselves these problems and would clearly discern that the irrational number, in virtue of the *property given to it by the definitions* has just as definite a reality in our minds as the rational numbers or even the integers, and that one does not even need to *gain* it through a limiting process, but that on the contrary by *possession* of it one becomes convinced of the practicability and evidence of limiting processes in general. **[8]** For now one extends with ease the proposition just mentioned as follows: If (b_ν) is any set of rational or irrational numbers with the property that $\lim\limits_{\nu=\infty} (b_{\nu+\mu} - b_\nu) = 0$ (whatever μ may be) then there exists a number b determined by a fundamental sequence (a_ν) such that

$$\lim_{\nu=\infty} b_\nu = b.$$

⟦14⟧ It also happens that the *same* numbers b, which are defined on the basis of fundamental sequences (a_ν) of this sort (I call them fundamental sequences of the *first* order) and which appear as limits of the a_ν are also representable in various ways as limits of sequences (b_ν) where every one of the b_ν is defined by a fundamental sequence of the first order $(a_\mu^{(\nu)})$ with fixed ν.

⟦15⟧ I accordingly call such a set (b_ν) a fundamental sequence of the *second* order provided it has the property that $\lim_{\nu=\infty} (b_{\nu+\mu} - b_\nu) = 0$ (for arbitrary μ).

⟦16⟧ In the same way we can construct not only fundamental sequences of the *third*, *fourth*, ... *n*th order, but also fundamental sequences of the αth order, where α is any number of the second number-class.

⟦17⟧ All these fundamental sequences perform for the determination of a real number b exactly the same thing as the fundamental sequences of the *first* order; the only difference lies in the more complicated, more spread-out form of what is given. None the less it seems to me in the highest degree appropriate, in so far as one takes the general standpoint of the third definition, to fix this distinction in the designated way, as I have already done in a similar manner elsewhere (*Math. Ann.*, Vol. 5, p. 123 ⟦*Cantor 1872*⟧). I therefore use the expression: the numerical quantity b is given by a fundamental sequence of the *n*th (respectively αth) order. If one agrees to this, one thereby acquires an extraordinarily flexible and at the same time intelligible idiom for describing in the simplest and most significant way the richness of the protean and often complicated webs of analysis. One also gains clarity and lucidity, which, in my opinion, is not to be undervalued. I herewith oppose the reservation which Dedekind, in the preface to his article 'Continuity and irrational numbers', expressed concerning these distinctions. It was not my intention to introduce *new* numbers by these fundamental sequences of the second, third order, etc., which could not already be represented by the fundamental sequences of the first order. Rather, I had in view only a conceptually different form of presentation; this appears clearly in various places in my work.

⟦18⟧ I should now like to call attention to a remarkable fact, namely, that in these orders of fundamental sequences (which I distinguish by numbers of the first and second number-classes) absolutely all thinkable forms in analysis with the usual sequential character, whether already found or yet unfound, are thoroughly exhausted—i.e. there exist no fundamental sequences whose order-number ⟦Ordnungszahl⟧ might be designated by a number of, say, the third number-class. I shall rigorously prove this fact on some other occasion.

⟦19⟧ I shall now attempt briefly to explain the utility of the *third* definition.

⟦20⟧ As a symbol that a number b is given by a fundamental sequence (e_ν) of any order n or α, I write

$$b \sim (e_\nu) \text{ or } (e_\nu) \sim b.$$

⟦21⟧ If, for instance, we have a convergent sequence with the general term c_ν, then the necessary and sufficient condition for convergence is known to be that $\lim_{\nu=\infty} (c_{\nu+1} + \ldots + c_{\nu+\mu}) = 0$ (for any μ). One therefore defines the sum of the sequence by the formula

$$\sum_{n=0}^{\infty} c_n \sim \left(\sum_{n=0}^{\nu} c_n\right).$$

If, for instance, all c_n are defined on the basis of fundamental sequences of the kth order, the same holds for $\sum_{n=0}^{\nu} c_n$ and the sum $\sum_{n=0}^{\nu} c_n$ appears here as defined by a fundamental sequence of the $(k+1)$st order.

⟦22⟧ For example, suppose the thought-content of the proposition $\sin(\pi/2) = 1$ is to be described. Then one can imagine, say $\pi/2$ and its powers as given by the formulae:

$$\pi/2 \sim (a_\nu), \; (\pi/2)^{2m+1} \sim (a_\nu{}^{2m+1}),$$

where we abbreviate

$$2\sum_{n=0}^{\nu} \frac{(-1)^n}{2n+1} = a_\nu.$$

Furthermore, it is the case that

$$\sin(\pi/2) \sim \left(\sum_{m=0}^{\mu} (-1)^m \frac{(\pi/2)^{2m+1}}{(2m+1)!}\right),$$

that is, $\sin \pi/2$ is defined on the basis of a fundamental sequence of the second order, and so that theorem expresses the equality of the rational number 1 with the number $\sin(\pi/2)$, which was given on the basis of a fundamental sequence of the second order.

⟦23⟧ Similarly, the thought-content of more complicated formulae (such as, for instance, those in the theory of theta-functions) can be precisely and comparatively simply described. But we usually end up with the greatest awkwardness when we attempt to reduce infinite sequences to those which converge unconditionally and are formed entirely of rational terms (especially when they have the same sign). Here we wholly avoid this awkwardness by using the *third* definition instead of the *first*. And it is clear that the awkwardness can be avoided, so long as it is not a question of numerical determination of sums of sequences by rational numbers but only one of finding absolutely sharp *definitions* for them. The *first* definition in any event seems to me not to be so easy to use when it is a question of the precise definition of the sums of sequences which do not converge unconditionally and in which the ordering of their terms, positive as well as negative, is precisely given. And even for unconditionally convergent sequences the production of the sum, even if it is independent of the ordering, can only actually be carried out for some definite ordering; one is therefore tempted in such cases as well to give the *third* definition preference over the first. Finally, the third definition is capable of being generalized to *super-finite* numbers, while such an extension of the *first* definition is utterly impossible. This difference lies simply in the fact that for super-finite numbers the commutative law even for addition is in general no longer valid; but the first definition is *inseparably bound* to this law, and stands or falls with it. But so long as we are dealing with types of numbers for which the commutative law of addition is valid the *first* definition, except for the points noted, is quite excellent.

§10.

[1] The concept of the 'continuum' has not only played an important role everywhere in the development of the sciences but has also always evoked the greatest differences of opinion and even vehement quarrels. This lies perhaps in the fact that, because the exact and complete definition of the concept has not been bequeathed to the dissentients, the underlying idea has taken on different meanings; but it may also be (and this seems to me the most probable) that the idea of the continuum had not been thought out by the Greeks (who may have been the first to conceive it) with the clarity and completeness which would have been required to exclude the possibility of different opinions among their posterity. Thus we see that Leucippus, Democritus, and Aristotle consider the continuum as a composite which consists *ex partibus sine fine divisibilibus*,[h] but Epicurus and Lucretius construct it out of their atoms considered as finite things. Out of this a great quarrel arose among the philosophers, of whom some followed Aristotle, others Epicurus; still others, in order to remain aloof from this quarrel, declared with Thomas Aquinas **[9]** that the continuum consisted neither of infinitely many nor of a finite Anzahl of parts, but of *absolutely no* parts. This last opinion seems to me to contain less an explanation of the facts than a tacit confession that one has not got to the bottom of the matter and prefers to get genteelly out of its way. Here we see the *medieval-scholastic origin* of a point of view which we still find represented today, in which the continuum is thought to be an unanalysable concept or, as others express themselves, a pure *a priori* intuition which is scarcely susceptible to a determination through concepts. Every arithmetical attempt at determination of this *mysterium* is looked on as a forbidden encroachment and repulsed with due vigour. Timid natures thereby get the impression that with the 'continuum' it is not a matter of a *mathematically logical concept* but rather of *religious dogma*.

[2] It is far from my intention to conjure up these controversial questions again; and besides, I lack the space for a detailed discussion of them. I feel obliged only to develop the concept of the continuum as soberly and logically and briefly as possible, and only with regard to the *mathematical* theory of sets. This treatment has not been easy for me because, among those mathematicians whose authority I should gladly invoke, not a single one has occupied himself with the continuum in the exact sense which I find necessary here.

[3] By taking as a basis one or several real or complex continuous quantities

---

[h] ['from parts divisible without end']

(or, more precisely, sets of continuous quantities) one indeed developed the concept of a single-valued or many-valued continuum dependent upon them, i.e. the concept of a continuous function; and in this way the theory of the so-called *analytic* functions arose, as well as of more general functions with remarkable properties (such as non-differentiability and the like); but the *independent* continuum itself has been assumed by mathematical writers only in its simplest guise, and has not been subjected to any more thorough inspection.

⟦4⟧ Next, I must explain that in my opinion to bring in *the concept of time* or the *intuition of time* in discussing the much more fundamental and more general concept of the continuum is *not* the correct way to proceed; *time* is in my opinion a representation ⟦Vorstellung⟧, and its clear explanation presupposes the concept of continuity upon which it depends and without whose assistance it cannot be conceived either objectively (as a substance) or subjectively (as the form of an *a priori* intuition), but is nothing other than a *helping and linking concept* ⟦ein *Hilfs- und Beziehungsbegriff*⟧, through which one ascertains the relation between various different motions that occur in nature and that are perceived by us. Such a thing as *objective* or *absolute time* never occurs in nature, and therefore *time* cannot be regarded as the measure of *motion*; far rather motion as the measure of *time*—were it not that *time*, even in the modest role of a *subjective necessary a priori* form of intuition ⟦Anschauungsform⟧, has not been able to produce any fruitful, incontestable success, although since Kant the time for this has not been lacking.

⟦5⟧ It is likewise my conviction that with the so-called *form of intuition* of *space* one cannot even begin to acquire knowledge about the *continuum*. For only with the help of a conceptually already *completed* continuum do *space* and the structure thought into it receive that content with which they can become the object, not merely of aesthetic contemplation or philosophical cleverness or imprecise comparisons, but of sober and exact mathematical investigations.

⟦6⟧ Consequently, there is nothing else remaining for me than, with the help of the concept of real number as defined in §9, to look for a pure arithmetical concept of point-continuum which will be as general as possible. As a foundation I use, as I must, n-dimensional plane *arithmetical* space G_n, that is, the aggregate of all value-systems

$$(x_1 | x_2 | \ldots | x_n)$$

in which each x, independently of the others, can take *all real* values from $-\infty$ to $+\infty$. Every particular value-system of this kind I call an *arithmetical* point of G_n. The distance between two such points is defined by the expression:

$$\left|\sqrt{(x_1' - x_1)^2 + (x_2' - x_2)^2 + \ldots + (x_n' - x_n)^2}\right|$$

By an *arithmetical* point-set P contained in G_n I understand every aggregate of points of the space G_n that is given in a lawlike way. The investigation thus leads towards the setting up of a sharp and completely general definition of *when P is to be called a continuum*.

⟦7⟧ I proved in *Crelle's Journal*, Vol. 84, p. 242 ⟦*Cantor 1878*⟧ that all spaces G_n, however great the dimension n may be, have the *same* power and that they are consequently of the *same power* as the linear continuum, and hence of the same power as, say, the aggregate of all real numbers in the interval (0 ... 1). Therefore the investigation and determination of the power of G_n reduces to the same question, specialized to the interval (0 ... 1), and I hope to be able soon, through a rigorous proof, to answer that the power sought is no other than that of our *second number-class* (II)[i]. It will follow that all infinite point-sets P have either the power of the first number class (I) or the power of the second number-class (II). We can also conclude that the aggregate of all functions of one or several variables which are representable by any given series *also only* has the power of the second number-class and is therefore *countable by* numbers of the third number-class (III). **[10]** This proposition will therefore apply, for example, to the aggregate of all 'analytic' functions, that is, functions of one or several variables produced by continuation of convergent power-series, or to the set of all functions of one or several real variables which are representable by trigonometric series.

⟦8⟧ In order now to inspect more closely the general concept of a continuum within G_n I use the concept of the derivative $P^{(1)}$ of any arbitrarily given point-set P. (I introduced this notion in *Math. Ann.*, Vol. 5 ⟦*Cantor 1872*⟧, then extended it ⟦in the earlier parts of this series⟧ to the concept of a derivative $P^{(\gamma)}$ where γ can be any integer of the number-classes (I), (II), (III), etc.)

⟦9⟧ The point-sets P may now be divided into two classes according to the power of their first derivative $P^{(1)}$. If $P^{(1)}$ has the power of (I), it turns out, as I have already said in §3 of this paper, that there is a first integer α of the *first* or *second* number-class (II), for which $P^{(\alpha)}$ vanishes. But if $P^{(1)}$ has the power of the second number-class (II) then $P^{(1)}$ can always be uniquely divided into two sets R and S so that

$$P^{(1)} \equiv R + S,$$

where R and S have an entirely different constitution:

⟦10⟧ R is so constituted that it can, by the repeated process of taking derivatives, eventually be annihilated; so there always exists a first integer γ of the number-class (I) or (II) for which

$$R^{(\gamma)} = 0.$$

I call such point-sets R *reducible*.

⟦11⟧ S, on the other hand, is so constituted that for this point-set the process of taking derivatives produces absolutely no change, in that

$$S \equiv S^{(1)}$$

and consequently

---

[i] ⟦This is Cantor's continuum hypothesis, first conjectured in *Cantor 1878*⟧.

$$S \equiv S^{(\gamma)};$$

such sets I call *perfect point-sets*. We can therefore say: if $P^{(1)}$ is of the power of the second number-class (II), then $P^{(1)}$ decomposes into a definite *reducible* and a definite *perfect* point-set.

⟦12⟧ Although these two predicates 'reducible' and 'perfect' cannot be applied to one and the same point-set, nevertheless with a bit of attentiveness one easily sees that irreducible is not identical to perfect and imperfect is not precisely the same as reducible.

⟦13⟧ In the terminology of my earlier works, the *perfect* point-sets S are not always 'everywhere dense' **[11]**; therefore they do not by themselves suffice for a complete definition of a point-continuum (although one must concede that a point-continuum must always be a *perfect* set).

⟦14⟧ To define the continuum, yet another concept is needed, namely, the concept of a *connected* ⟦*zusammenhängenden*⟧ point-set T.

⟦15⟧ We call T a *connected* point-set if, for any two of its points t and t' and for any arbitrarily small number ε there always exists a *finite* Anzahl of points $t_1, t_2, \ldots, t_\nu$ of T, such that the distances $\overline{tt_1}, \overline{t_1t_2}, \overline{t_2t_3}, \ldots, \overline{t_\nu t'}$ are all less than ε.

⟦16⟧ Now, all the geometric point-continua known to us fall under this concept of *connected* point-set, as it is easy to see; I believe that in these *two* predicates 'perfect' and 'connected' I have discovered the necessary and *sufficient* properties of a point-continuum. I therefore define a point-continuum inside G_n as a *perfect-connected set*. **[12]** Here 'perfect' and 'connected' are not merely words but completely general predicates of the *continuum*; they have been conceptually characterized in the sharpest way by the foregoing definitions.

⟦17⟧ Bolzano's definition of the continuum (*Paradoxien* §38) is certainly not right; it expresses one-sidedly just *one* property of the continuum, which is also satisfied by sets which result from G_n when one imagines an 'isolated' point-set at a distance from G_n (compare *Math. Ann.*, Vol. 21, p. 51 ⟦*Cantor 1883a*⟧); in the same way it is also satisfied by sets which are made up of several separated continua; obviously in such cases no continuum exists, although according to Bolzano this would be the case. We therefore see here a contravention of the proposition: 'Ad essentiam alicujus rei pertinet id, quo dato res necessario ponitur et quo sublato res necessario tollitur; vel id, sine quo res et vice versa quod sine re nec esse nec concipi potest.'[j]

⟦18⟧ Likewise, it seems to me that in the article of Dedekind (*Continuity and irrational numbers*) ⟦*Dedekind 1872*⟧ only *another* property of the continuum has been one-sidedly emphasized, namely, that property which it has in common with *all 'perfect'* sets.

---

[j] ⟦'To the essence of a thing pertains that which, if given, the thing is necessarily given, and, if lacking, the thing is necessarily lacking; or such that neither it without the thing nor the thing without it can either be or be conceived.'⟧

§11.

⟦1⟧ I shall now show how one is led to the definitions of the new numbers and how the natural cuts which I call the number-classes arise in the absolutely-infinite sequence of real integers. In this exposition I wish to add only the principal theorems on the *second* number-class and its relation to the first. The sequence (I) of positive integers 1, 2, 3, . . ., ν, . . . has its ground of origin ⟦Entstehungsgrund⟧ in the repeated positing ⟦Setzung⟧ and uniting ⟦Vereinigung⟧ of underlying unities ⟦Einheiten⟧, which are regarded as alike; the number ν is the expression for a definite finite Anzahl of such positings following one another in a sequence; it is also the expression for the unification of the posited unities into a whole. The formation of the finite real integers thus rests upon the principle of adding a unity to an already formed and existing number; I call this principle (which, as we shall soon see, also plays an essential role in the generation of the higher integers) the *first principle of generation*. The Anzahl of the numbers ν of class (I) formed in this way is infinite and there is no greatest among them. However contradictory it might be to speak of a greatest number of class (I), there is on the other hand nothing offensive in thinking of a *new* number which we shall call ω and which shall be the expression for the fact that the entire aggregate (I) is given in its natural succession according to a law.[4] (Similarly, ν is an expression for the fact that a certain finite Anzahl of unities is united into a whole.) We may even imagine the newly created number ω as a *limit* which the numbers ν approach, provided that we understand nothing more thereby than that ω is the *first* integer which follows all numbers ν: that is, it is to be called greater than each of the numbers ν. By permitting further positings to follow the positing of the number ω one obtains, with the help of the *first* principle of generation, the further numbers

$$\omega + 1,\ \omega + 2,\ \ldots,\ \omega + \nu,\ \ldots .$$

Since one comes to no greatest number in this way, one accordingly imagines a new one which we can call 2ω[k] and which is the first to follow all previous numbers ν and $\omega + \nu$; if one repeatedly applies the *first* principle of generation to the number 2ω one gets the continuation

$$2\omega + 1,\ 2\omega + 2,\ \ldots,\ 2\omega + \nu,\ \ldots$$

of the previous numbers.

⟦2⟧ The logical function which gave us the two numbers ω and 2ω is obviously different from the *first* principle of generation. I call it the *second principle of generation* of integers, and define it more exactly thus: if any definite succession of defined integers is put forward of which no greatest exists,

[4] From now on I replace the symbol ∞ (which I used in no. 2 of this essay) by ω, because the symbol ∞ is already variously used to designate indefinite infinites.

[k] ⟦In modern notation one would write $\omega \cdot 2$.⟧

a new number is created by means of this second principle of generation, which is thought of as the *limit* of those numbers; that is, it is defined as the next number greater than all of them.

⟦3⟧ By combined application of the two principles of generation one successively obtains the following continuation of the numbers which we have hitherto acquired

$$3\omega, 3\omega + 1, \ldots, 3\omega + \nu, \ldots$$

$$\ldots\ldots\ldots\ldots\ldots$$

$$\mu\omega, \mu\omega + 1, \ldots, \mu\omega + \nu, \ldots$$

$$\ldots\ldots\ldots\ldots\ldots$$

⟦4⟧ However here too we reach no end, for none of the numbers $\mu\omega + \nu$ is the greatest.

⟦5⟧ The second principle of generation therefore causes us to introduce a number immediately following all the numbers $\mu\omega + \nu$, which can be called ω^2. It is followed in definite succession by the numbers

$$\lambda\omega^2 + \mu\omega + \nu,$$

and one then obviously comes, by following the two principles of generation, to numbers of the following form

$$\nu_0\omega^\mu + \nu_1\omega^{\mu-1} + \ldots + \nu_{\mu-1}\omega + \nu_\mu;$$

however, the second principle of generation then drives us to set up a new number which is the next number greater than all these numbers; it is fittingly designated by

$$\omega^\omega.$$

⟦6⟧ The formation of new numbers has, as one sees, no end; by following *both* principles of generation one obtains ever again new numbers and sequences of numbers which have a fully determinate succession.

⟦7⟧ Thus at this point the *appearance* is given that by this way of forming new determinate-infinite integers we would have to lose ourselves in the *limitless* ⟦*Grenzenlose*⟧ and that we are unable to give this endless process a *certain temporary* termination, in order thereby to achieve a limitation similar to the one that, in a certain sense, was actually available for the older number-class (I); there we used only the *first* principle of generation, and so it was impossible to leave the sequence (I). But the *second* principle of generation must not only lead us beyond the former domain of the numbers; it also proves to be a means which, in combination with the *first* principle of generation, gives us the capacity *to break through every obstacle* in the concept-formation ⟦Begriffsbildung⟧ of integers.

⟦8⟧ But if we now notice that all the previously obtained numbers and their immediate successors fulfil a certain condition, then this condition, *if it is imposed as a requirement on all numbers yet to be formed*, proves to be a new *third* principle which I shall call a *restricting or limiting principle*; as I shall show, it means that the second number-class (II) defined with its assistance not only has a higher power ⟦*Mächtigkeit*⟧ than (I), but precisely the *next higher*, that is, the *second power*.

⟦9⟧ The mentioned condition, which each of the previously defined infinite numbers α fulfils, is, as one immediately persuades oneself, that the set of numbers which precede this number in the sequence is of the *power of the first number-class* (I). If, for instance, we take the number ω^ω, then its predecessors are obtained from the formula:

$$\nu_0\omega^\mu + \nu_1\omega^{\mu-1} + \ldots + \nu_{\mu-1}\omega + \nu_\mu,$$

where μ, ν_0, ν_1, . . ., ν_μ assume all finite, positive integral values including zero and excluding the case where $\nu_0 = \nu_1 = \ldots = \nu_\mu = 0$.

⟦10⟧ As is well known, this set can be brought into the form of a simply infinite sequence; it therefore has the power of (I).

⟦11⟧ Moreover, since every sequence of sets such that both the sets and the sequence itself are of the *first* power always yields a set which has the power of (I), it is clear that by continuing our sequence of numbers we *only obtain numbers* that *in fact* fulfil the condition.

⟦12⟧ We accordingly define the second number-class (II) as *the aggregate of all numbers* α *formable with the help of the two principles of generation and proceeding in a definite succession*

$$\omega,\ \omega + 1,\ \ldots,\ \nu_0\omega^\mu + \nu_1\omega^{\mu-1} + \ldots + \nu_{\mu-1}\omega + \nu_\mu,\ \ldots,\ \omega^\omega,\ \ldots,$$
$$\ldots\ \alpha,\ \ldots,$$

which are subject to the requirement that all the numbers preceding α, *from* 1 *on, form a set of the power of number-class (I).*

§12.

⟦1⟧ The first thing which we now have to prove, is the theorem *that the new number-class (II) has a power which is different from that of the first number-class (I).*

⟦2⟧ This proposition follows from the following proposition:

⟦3⟧ 'Let α_1, α_2, . . ., α_ν, . . . be any set which consists of different numbers of the *second* number-class and which has the first power (so that we are able to assume that they are in the form of a simple sequence (α_ν)). Then there are two possibilities. Either one of the numbers is the largest, in which case let it be γ. Or else there exists a definite number β of the second number-class (II) such that: (i) β does not occur among the numbers α; (ii) β is larger than all the α_ν; (iii) every integer $\beta' < \beta$ is surpassed in size by certain numbers of the sequence (α_ν). The numbers γ and β respectively can legitimately be called the "upper bound" of the set (α_ν).'

⟦4⟧ The proof of this proposition is simply as follows: let α_{χ_2} be the first number occurring in the series (α_ν) which is larger than α_1, α_{χ_3} the first which is larger than a a_{χ_2}, and so on.

⟦5⟧ One then has

$$1 < \chi_2 < \chi_3 < \chi_4 < \ldots$$

$$\alpha_1 < \alpha_{\chi_2} < \alpha_{\chi_3} < \alpha_{\chi_4} < \ldots$$

and

$$\alpha_\nu < \alpha_{\chi_\lambda}$$

whenever

$$\nu < \chi_\lambda.$$

⟦6⟧ Now here it can happen that, from a certain number α_{χ_ρ} on, all the numbers which follow it in the sequence (α_ν) are smaller than it. Then clearly it is the largest of all, and we have $\gamma = \alpha_{\chi_\rho}$. Otherwise, if one imagines the set of all integers from 1 on which are smaller than α_1, and if one appends to this set the set of all integers which are $\geq \alpha_1$ and $< \alpha_{\chi_2}$, then the set of all numbers which are $\geq \alpha_{\chi_2}$ and $< \alpha_{\chi_3}$, and so on, then one obtains a definite subset of successive numbers of our first two number-classes. This set of numbers is obviously of the *first* power, and there consequently exists (by the definition of (II)) a definite number β of the aggregate (II) which is the immediate successor of those numbers. Therefore $\beta > \alpha_{\chi_\lambda}$, and therefore too $\beta > \alpha_\nu$, because χ_λ can always be assumed to be so large that it becomes larger than any given ν and because then $\alpha_\nu < \alpha_{\chi_\lambda}$.

⟦7⟧ On the other hand, one easily sees that every number $\beta' < \beta$ is surpassed in size by certain numbers α_{χ_ν}. With this, all parts of the proposition have now been proved.

⟦8⟧ From this, the theorem now follows that the totality of all numbers of the second number-class (II) does *not* have the power of (I). For otherwise we could imagine the entire aggregate (II) in the form of a simple sequence

$$\alpha_1, \alpha_2, \ldots, \alpha_\nu, \ldots$$

And by the proposition we just proved, either this sequence would have a greatest term γ, or else all the numbers α_ν would be surpassed in size by a certain number β from (II). In the first case the number $\gamma + 1$ (which belongs to (II)), and in the second case the number β, would on the one hand belong to class (II) and on the other hand would not occur in the series (α_ν); which, by the presupposed identity of the sets (II) and (α_ν), is a contradiction. Therefore the number-class (II) has a *different* power than the number-class (I).

⟦9⟧ That of the two powers of the number-classes (I) and (II) the second is actually the *immediate successor* of the first, that is, that no other power exists between these two, is implied by a theorem which I shall shortly state and prove.

⟦10⟧ However, if we first cast a glance backwards and recall the means which have led not only to an extension of the concept of real integer but also to the concept of a new power, differing from the first, of well-defined sets, we see that there were three conspicuous logical steps ⟦Momente⟧ which brought about the extension and which ought to be distinguished from one another. They were the *two* previously defined *principles of generation* and besides these a *restricting* or *limiting principle* which consisted in the demand that a new integer could be made with the help of one of the two other principles of creation *only* if the totality of all previous numbers had the power of a defined number-class which was already *in existence* over its entire extent. In this way, by observing these three principles, one can attain with the greatest certainty and obviousness ever newer number-classes, and with them all the different, successive, ascending powers occurring in physical or mental nature, and the new numbers obtained in this way are then always of the same determinacy and objective reality as the earlier ones. I therefore do not see anything that should hold us back from this activity of forming such new numbers, as soon as it seems that the introduction of a new number-class from these innumerable number-classes is desirable or even indispensable for the progress of the sciences.

§13.

⟦1⟧ I come now to the promised proof that the powers of (I) and of (II) follow each other immediately, so that no other powers lie between.

⟦2⟧ If one chooses a set (α') of different numbers α' from the aggregate (II) according to any law—that is, if one imagines any set (α') contained in (II)—then such a set always has certain peculiarities which can be expressed in the following theorems:

⟦3⟧ 'Among the numbers of the set (α') there is always a *smallest*.'

⟦4⟧ 'If, in particular, one has a sequence of numbers of the aggregate (II): $\alpha_1, \alpha_2, \ldots, \alpha_\beta, \ldots$ which are constantly decreasing in size (so that $\alpha_{\beta'} > \alpha_\beta$ if $\beta > \beta'$) then this sequence necessarily terminates with a finite number of the sequence and ends with the smallest of the numbers; the sequence cannot be infinite.'

⟦5⟧ It is remarkable that this proposition (which is immediately clear if the numbers α_β are finite integers) can also be proven for infinite numbers α_β. In fact, by the preceding proposition (which easily follows from the definition of the number-class (II)) there is, among the numbers α_ν, if one considers only those for which the index ν is finite, a smallest. If this, say, is $= \alpha_\rho$, then it is clear that, since $\alpha_\nu > \alpha_{\nu+1}$, the sequence α_ν and thus the entire sequence α_β must consist of precisely ρ members and is consequently finite.

⟦6⟧ One now obtains the following fundamental theorem:

⟦7⟧ 'If (α') is any set of numbers contained in the aggregate (II), then only the following three cases can arise: either (α') is a finite aggregate (that is, it consists of a finite Anzahl of numbers) , or (α') has the power of the first class, or (α') has the power of (II); Quartum non datur.'

⟦8⟧ The proof, which is simple, goes as follows: let Ω be the first number of the *third* number-class (III). Then all the numbers α' of the set (α') are smaller than Ω, because the latter set (α') is contained in (II).

⟦9⟧ We now imagine the numbers α' ordered according to their size. Let α_ω be the smallest among them, $\alpha_{\omega+1}$ the next larger, and so on. We thus obtain the set (α') in the form of a 'well-ordered' set α_β, where β runs through numbers of our naturally extended number-sequence from ω on. Clearly, β always remains less than or equal to α_β,, and because $\alpha_\beta < \Omega$, we also have $\beta < \Omega$. The number β can therefore not go beyond the number-class (II) but remains within its domain. Therefore, only three cases can arise: either β remains smaller than a specifiable number of the sequence $\omega + \nu$ (and then (α') is a finite set); or β takes on all values of the sequence $\omega + \nu$ but remains smaller than a specifiable number of the sequence (II) (and then (α') is clearly a set of the *first* power); or, third, β takes on arbitrarily large values in (II) (and then β runs through *all* numbers of (II)). In this case the aggregate (α_β), that is, the set (α'), obviously has the power of (II), as was to be proved.

⟦10⟧ The following are immediate consequences of this theorem:

⟦11⟧ 'If one has any well-defined set M of the power of the number-class (II) and if one takes any infinite subset M' of M, then either the aggregate M' can be thought of in the form of a simple infinite sequence, or it is possible to map the two sets M and M' on to each other in a one-to-one fashion.'

⟦12⟧ 'If one has any well-defined set M of the second power, a subset M' of M and a subset M'' of M', and if one knows that M'' can be put into a one-to-one reciprocal correspondence with M, then M' is related in the same way to M and therefore also to M''.'

⟦13⟧ I mention this last theorem here because of its connection to the preceding theorem, under the presupposition that M has the power of (II). It is clearly also correct if M has the power of (I). But that this theorem has *general* validity, regardless of the power of the set M, seems to me highly remarkable. I shall go further into this matter in a later article and then indicate the peculiar interest which attaches to this general theorem.

§14.

⟦1⟧ I shall now in conclusion consider the numbers of the second number-class (II) and the operations that are possible with them; but I shall on this occasion confine myself to only the most obvious matters. I reserve the publication of more thorough investigations for later.[1]

⟦2⟧ I have defined the operations of addition and multiplication in general in §1, and have shown that for the infinite integers the commutative law does *not* in general hold, but that the associative law does; this consequently holds

[1]⟦Cantor developed his theory of the transfinite numbers in more detail and with greater rigour in *Cantor 1897*.⟧

in particular also for the numbers of the second number-class. With respect to the distributive law, it is in general valid only in the following form:

$$(\alpha + \beta)\gamma = \alpha\gamma + \beta\gamma$$

(where $\alpha + \beta$, α, β appear as multipliers)

as one immediately recognizes from inner intuition ⟦aus der inneren Anschauung unmittelbar erkennt⟧.

⟦3⟧ Subtraction can be considered from two points of view. If α and β are any two integers, $\alpha < \beta$, one easily persuades oneself that the equation

$$\alpha + \xi = \beta$$

admits one and only one solution for ξ, where, if α and β are numbers from (II), ξ will be a number from (I) or (II). This number ξ is to be set equal to $\beta - \alpha$.

⟦4⟧ If on the other hand one considers the following equation:

$$\xi + \alpha = \beta$$

it turns out that this can often not be solved for ξ at all; this case, for example, occurs in the following equation:

$$\xi + \omega = \omega + 1.$$

⟦5⟧ But also in those cases where the equation $\xi + \alpha = \beta$ is solvable for ξ it often happens that it is satisfied by infinitely many numerical values of ξ; but of these different solutions one will always be least.

⟦6⟧ We choose the sign $\beta_{-\alpha}$ to stand for this least root of the equation

$$\xi + \alpha = \beta$$

in case it is solvable at all; $\beta_{-\alpha}$ is therefore in general different from $\beta - \alpha$, which latter number always exists provided $\alpha < \beta$.

⟦7⟧ If, further, the equation

$$\beta = \gamma\alpha$$

holds between these integers β, α, γ (where γ is the multiplier), one easily persuades oneself that the equation

$$\beta = \xi\alpha$$

has no other solution with respect to ξ than $\xi = \gamma$, and in this case one designates γ by β/α.

⟦8⟧ On the other hand, one finds that the equation

$$\beta = \alpha\xi$$

(where ξ is the multiplicand), if it is at all solvable with respect to ξ, often has several (or even infinitely many) roots. But one of these is always the smallest; this smallest solution of the equation $\beta = \alpha\xi$, in case the latter is solvable at all, is to be designated by

$$\frac{\beta}{\alpha}.$$

[9] The numbers α of the second number-class are of two sorts: (1) numbers α which have an immediate predecessor (that is, α_{-1}) in the sequence—I call these numbers of the *first* sort; and (2) numbers α which have no immediate predecessors in the sequence (and for which α_{-1} therefore does not exist)—I call these numbers of the *second* sort. The numbers ω, 2ω, $\omega^{\nu} + \omega$, ω^{ω} are, for example, of the second sort, but $\omega + 1$, $\omega^2 + \omega + 2$, $\omega^{\omega} + 3$ are in contrast of the first sort.

[10] Correspondingly, *prime numbers* of the second number-class, which I defined in §1, divide into primes of the second and of the first sort.

[11] Prime numbers of the second sort are (in the order of their appearance in the number-class (II)) the following:

$$\omega, \omega^{\omega}, \omega^{\omega^2}, \omega^{\omega^3}, \ldots,$$

so that among all numbers of the form

$$\phi = \nu_0\omega^{\mu} + \nu_1\omega^{\mu-1} + \ldots + \nu_{\mu-1}\omega + \nu_{\mu}$$

only the *one* prime number ω of the *second* sort exists. But one must not infer from this relatively sparse distribution of primes of the second sort that the aggregate of them all has a smaller power than the number-class (II) itself; it happens that this aggregate has the same power as (II).

[12] The prime numbers of the first sort are, to begin with

$$\omega + 1, \omega^2 + 1, \omega^{\mu} + 1, \ldots.$$

These are the sole prime numbers of the first sort which occur among the numbers just designated by ϕ; the totality of all prime numbers of the first sort in (II) has the power of (II) as well.

[13] The prime numbers of the second sort have a property which gives them a completely distinctive character. If η is a prime number of the second sort, then we always have $\eta\alpha = \eta$, if η is any number smaller than α. From this it follows that, if α and β are any two numbers which are both smaller than η, then the product $\alpha\beta$ is always smaller than η.

[14] If we confine ourselves for the present to the numbers of the second number-class which have the form ϕ then we have the following rules for addition and multiplication. Let

$$\phi = \nu_0\omega^{\mu} + \nu_1\omega^{\mu-1} + \ldots + \nu_{\mu},$$

$$\psi = \rho_0\omega^{\lambda} + \rho_1\omega^{\lambda-1} + \ldots + \rho_{\lambda},$$

where we presuppose that ν_0 and ρ_0 are different from zero.

Addition

(1) If $\mu < \lambda$, one has

$$\phi + \psi = \psi.$$

(2) If $\mu > \lambda$, one has

$$\phi + \psi = \nu_0\omega^{\mu} + \ldots + \nu_{\mu-\lambda-1}\omega^{\lambda+1} + (\nu_{\mu-\lambda} + \rho_0)\omega^{\lambda} + \rho_1\omega^{\lambda-1} + \rho_2\omega^{\lambda-2} + \ldots + \rho_{\lambda}.$$

(3) For $\mu = \lambda$,

$$\phi + \psi = (\nu_0 + \rho_0)\omega^{\lambda} + \rho_1\omega^{\lambda-1} + \ldots + \rho_{\lambda}.$$

Multiplication

(1) If ν_μ is different from zero, one has

$$\phi\psi = \nu_0\omega^{\mu+\lambda} + \nu_1\omega^{\mu+\lambda-1} + \ldots + \nu_{\mu-1}\omega^{\lambda+1} + \nu_\mu\rho_0\omega^{\lambda} + \rho_1\omega^{\lambda-1} + \ldots + \rho_{\lambda}.$$

In case $\lambda = 0$, the last term on the right-hand side is $\nu_\mu\rho_0$.

(2) If $\nu_\mu = 0$, one has

$$\phi\psi = \nu_0\omega^{\mu+\lambda} + \nu_1\omega^{\mu+\lambda-1} + \ldots + \nu_{\mu-1}\omega^{\lambda+1} = \phi\omega^{\lambda}.$$

⟦15⟧ To break a number ϕ down into its prime factors we proceed as follows. If one has

$$\phi = c_0\omega^{\mu} + c_1\omega^{\mu_1} + c_2\omega^{\mu_2} + \ldots + c_\sigma\omega^{\mu_\sigma},$$

where

$$\mu > \mu_1 > \mu_2 > \ldots > \mu_\sigma$$

and

$$c_0,\ c_1,\ \ldots,\ c_\sigma$$

are positive finite numbers different from zero, then

$$\phi = c_0(\omega^{\mu-\mu_1} + 1)c_1(\omega^{\mu_1-\mu_2} + 1)c_2 \ldots c_{\sigma-1}(\omega^{\mu_{\sigma-1}-\mu_\sigma} + 1)c_\sigma\omega^{\mu_\sigma};$$

if one imagines the $c_0, c_1, \ldots, c_{\sigma-1}, c_\sigma$ as broken down into their prime factors (following the rules of the first number-class), then one has broken ϕ down into its prime factors; for the factors $\omega^{\kappa} + 1$ and ω are, as noted above, themselves prime numbers. This factoring of numbers of the form ϕ is uniquely determined, also with respect to the sequence of the factors if one abstracts from the commutability of the prime factors within the individual c and if it is determined that the last factor shall be a power of ω or equal to one, and that ω may only be a factor in the last position. I shall at a later opportunity write about the generalization of this factorization of an arbitrary number α of the second number-class into prime factors.[m]

---

[m] ⟦See *Cantor 1897*, §19.⟧

Cantor's endnotes

[1] *Theory of manifolds.* I use this word to designate a very broad theoretical concept which I have hitherto used only in the special form of a theory of geometric or arithmetical sets. In general, by a 'manifold' or 'set' I understand every multiplicity ‖jedes Viele‖ which can be thought of as one, i.e. every aggregate ‖Inbegriff‖ of determinate elements which can be united ‖verbunden‖ into a whole by some law. I believe that I am defining something akin to the Platonic *εἶδος* or *ἰδέα* as well as to that which Plato called *μικτόν* in his dialogue 'Philebus or the Supreme Good'. He contrasts this to the *ἄπειρον* (i.e. the unbounded, undetermined, which I call the improper infinite) as well as to the *πέρας*, i.e the boundary; and he explains it as an ordered 'jumble' of both. Plato himself indicates that these concepts are of Pythagorean origin; see A. Boeckh, *Philolaos des Pythagoreers Lehren*, Berlin 1819.

[2] *Aristotle.* See Zeller's account in his great work: *Die Philosophie der Griechen*, 3rd edn, pt. II, sec. 2, pp. 393–403. Plato's conception of the infinite is completely different from Aristotle's; see Zeller, pt. II, sec. 1, pp. 628–46. I also find points of contact between my conception and that in the philosophy of Nicholas of Cusa. (See R. Zimmermann, *Der Cardinal Nicolaus von Cusa als Vorgänger Leibnizens* (*Sitzungsberichte der Wiener Akademie der Wissenschaften*, year 1852.) Similarly for Giordano Bruno, the successor of Nicholas. See Brunnhofer, *Giordano Brunos Weltanschauung und Verhängnis*, Leipzig 1882.

But there is an essential difference in the fact that I use the number-classes (I), (II), (III), etc. to fix the different levels of the proper infinite once and for all, and only then do I regard it as my task to investigate the relationships of the supra-finite numbers not only mathematically, but also to pursue them and establish them wherever they occur in nature. I have no doubt that, as we pursue this path ever further, we shall never reach a boundary that cannot be crossed; but that we shall also never achieve even an approximate conception of the absolute. The absolute can only be acknowledged ‖anerkannt‖ but never known ‖erkannt‖—and not even approximately known. For just as in number-class (I) every finite number, however great, always has the same power of finite numbers greater than it, so every supra-finite number, however great, of any of the higher number-classes (II) or (III), etc. is followed by an aggregate of numbers and number-classes whose power is not in the slightest reduced compared to the entire absolutely infinite aggregate of numbers, starting with 1. As Albrecht von Haller says of eternity: 'I attain to the enormous number, but you, O eternity, lie always ahead of me.' The absolutely infinite sequence of numbers thus seems to me to be an appropriate symbol of the absolute; in contrast, the infinity of the first number-class (I), which has hitherto sufficed, because I consider it to be a graspable idea (not a representation ‖Vorstellung‖), seems to me to dwindle into nothingness by comparison. It also seems to me remarkable that to each of the number classes (and hence to each of the powers) there corresponds a completely determinate member of the absolutely infinite aggregate of numbers—and in such a way that for every supra-finite number γ there is a power that is the γth; so the different powers also form an absolutely infinite sequence. This is all the more peculiar in that the number γ which gives the order of a power (in case the number γ has an immediate predecessor) stands to the number of the number-class that has that power in a relationship of size whose smallness mocks all description; and all the more so, the greater we take γ to be.

[3] *determinari possunt.* I cannot ascribe any being ⟦Sein⟧ to the improper infinite, the indeterminate, or the variable—no matter what form they take. For they are nothing but either relational concepts or purely subjective representations of intuitions (*imaginationes*); they are never adequate ideas. Thus if only the improper infinite were meant in the proposition '*infinitum actu non datur*' I could accept it; but it would then be a pure tautology. But the meaning of this proposition in the places where it occurs seems to me to be that there can be no conceptual positing of a determinate infinite; I hold this proposition, so interpreted, for false.

[4] *Realists.* One finds the positivistic and realistic point of view concerning the infinite in, for example, Dühring *Natürliche Dialektik*, Berlin 1865, pp. 109–35, and in von Kirchmann, *Katechismus der Philosophie*, pp. 124–30. See also Ueberweg's remarks on Berkeley's *Treatise on the principles of human knowledge* in von Kirchmann's *Philosophische Bibliothek*. I can only repeat that I basically agree with all these authors in the evaluation of the improper infinite; the point of difference is, that they regard this syncategorematic infinite as the *only* infinite that can be grasped by phrases ⟦Wendungen⟧ or concepts or even by mere relational concepts. Dühring's proofs against the proper infinite could be expressed in fewer words. They seem to me to amount either to the contention that a finite number, however large we conceive it to be, can never be an infinite number (which follows at once from the concept), or to the contention that a variable finite number of indeterminately large size cannot be imagined to have the predicate of determinateness and therefore cannot be imagined to have the predicate of being; but again, this follows at once from the essence of variability. It seem to me obvious that these arguments do not tell in the slightest against the conceivability of determinate supra-finite numbers; nevertheless, these proofs are widely held to be proofs against the reality of supra-finite numbers. This argumentation seems to me to be similar to the reasoning: because there are infinitely many shades of green, there can be no red. It is peculiar that Dühring himself admits on page 126 of his paper that for the explanation of the 'possibility of unlimited synthesis' there must be a reason, which he designates as 'for understandable reasons, utterly unknown'. This seems to me a contradiction.

But we also find that thinkers who stand close to idealism, or who praise it in the highest terms, deny the legitimacy of the determinate-infinite numbers.

Chr. Sigwart, in his distinguished *Logik*, Vol. II, *Die Methodenlehre* (Tübingen 1878) argues just like Dühring and says on p. 47: 'an infinite number is a *contradictio in adjecto*'.

One finds similar remarks in Kant and in J.F. Fries: see the latter's *System der Metaphysik* (Heidelberg 1824) at §§51 and 52. And the philosophers of Hegel's school also do not admit the proper-infinite numbers; I mention only the useful work of K. Fischer, his *System der Logik und Metaphysik oder Wissenschaftslehre*, 2nd edn (Heidelberg 1865), p. 275.

[5] What I here call the 'immanent' or 'intrasubjective' reality of concepts or ideas ought to agree with the adjective 'adequate' in the sense in which Spinoza uses this word when he says (*Ethica*, part II, def. IV): 'Per ideam adequatam intelligo ideam, quae, quatenus in se sine relatione ad objectum consideratur, omnes verae ideae propietates sive denominationes intrinsecas habet.'[n]

[6] This conviction agrees essentially both with the principles of the Platonic system and with an essential tendency of the Spinozistic system; as for the former, see Zeller, *Philos. der Griechen*, 3rd edn, pt. 2, sec. 1, pp. 541–602. At the beginning of this passage, it is said: 'Only conceptual knowledge ‖Wissen‖ yields (according to Plato) true understanding ‖eine wahre Erkenntnis‖. But (and this presupposition Plato shares with others (Parmenides)) to the extent that our representations are true, their object must be real, and conversely. What can be known, is; what cannot be known, is not, and to the same extent that something is, it is also knowable.'

As for Spinoza, I need only mention his statement in *Ethica*, part II, prop. VII: 'ordo et connexio idearum idem est ac ordo et connexio rerum.'[o]

The same epistemological principle can also be found in the philosophy of Leibniz. Only since the growth of modern empiricism, sensualism, and scepticism, as well as of the Kantian criticism that grows out of them, have people believed that the source of knowledge and certainty is to be found in the senses or in the so-called pure form of intuition of the world of appearances, and that they must confine themselves to them. But in my opinion these elements do not furnish us with any secure knowledge. For this can be obtained only from concepts and ideas that are stimulated by external experience, and are essentially formed by inner induction and deduction as something that, as it were, was already in us and is merely awakened and brought to consciousness.

[7], [8] The procedure in the correct formation of concepts is in my opinion everywhere the same. One posits ‖setzt‖ a thing with properties that at the outset is nothing other than a name or a sign A, and then in an orderly fashion gives it different, or even infinitely many, intelligible predicates whose meaning is known on the basis of ideas that are already at hand, and which may not contradict one another. In this way one determines the connection of A to the concepts that are already at hand, in particular to related concepts. If one has reached the end of this process, then one has met all the preconditions for awakening the concept A which slumbered inside us, and it comes into being accompanied by the intrasubjective reality which is all that can be demanded

---

[n] ‖'By "adequate idea" I understand an idea which, to the extent that it is considered *in se*, without relation to an object, has all the properties or intrinsic denominations of a true idea.'‖

[o] ‖'The order and connection of ideas is the same as the order and connection of things.'‖

of a concept; to determine its transient meaning is then a matter for metaphysics.

[9] Thomas Aquinas, *Opuscula*, XLII de natura generis, cap. 19 et 20; LII de natura loci; XXXII de natura materiae et de dimensionibus interminatis. See also C. Jourdin, *La Philosophie de Saint Thomas d'Aquin*, pp. 303–29; K. Werner, *Der Heilige Thomas von Aquino* (Regensburg 1859), Vol. 2, pp. 177–201.

[10] Even the totality of all continuous functions (and also the totality of all integrable functions of one or several variables) should have the power only of the second number-class (II). If one drops all constraints and considers the totality of all continuous and non-continuous functions of one or several variables, then this set has the power of the third number-class (III).

[11] It can be proved that perfect sets can never have the power of (I).

As an example of a perfect point-set which is not everywhere dense in any interval, however small, consider the totality of all real numbers given by the formula

$$z = \frac{c_1}{3} + \frac{c_2}{3^2} + \ldots + \frac{c_\nu}{3^\nu} + \ldots$$

where the coefficients c_ν may arbitrarily assume the values 0 or 2 and where the series may consist of either a finite or an infinite number of terms.

[12] Observe that this definition of a continuum is free from every reference to that which is called the *dimension* of a continuous structure; the definition includes also continua that are composed of connected pieces of different dimensions, such as lines, surfaces, solids, etc. On a later occasion I shall show how it is possible to proceed in an orderly fashion from this general continuum to the special continua with definite dimension. I know very well that the word 'continuum' has previously *not* had a precise meaning in mathematics; so my definition will be judged by some as too *narrow*, by others as too *broad*. I trust that I have succeeded in finding a proper *mean* between the two.

In my opinion, a *continuum* can *only* be a *perfect* and *connected* structure. So, for example, a straight line segment lacking one or both of its end-points, or a disc whose boundary is excluded, are not *complete* continua; I call such point-sets *semi-continua*.

In general I understand by a semi-continuum an *imperfect, connected* point-set belonging to the *second class* which is so constituted that every two points of it can be connected by a complete continuum which is a part of this point-set. For instance, the space which I designate by $\mathfrak{A}$ in *Mathematische Annalen*, Vol. 20, p. 119 [*Cantor 1932*, p. 156] (i.e. the space which arises from G_n by removing any point-set of the first power) is a *semi-continuum*.

The derivative of a connected point-set is *always* a *continuum*, in which case it is of no importance whether the connected set has the first or second power.

If a connected point-set is of the *first* power it can be called neither a continuum nor a semi-continuum.

Having placed the foregoing concepts at the heart of the theory of manifolds,

I now have an obligation to investigate all the possible ways in which they can be applied to the structures of algebraic as well as transcendental geometry; it seems unlikely that the universality and sharpness of the results can be surpassed by other methods.

D. ON AN ELEMENTARY QUESTION IN THE THEORY OF MANIFOLDS (*CANTOR 1891*)

This article announces Cantor's 'diagonal argument' for proving the existence of non-denumerable sets, and, more generally, for proving that, for any set X, the cardinality of the power-set $\mathfrak{P}(X)$ is greater than the cardinality of X. The translation is by William Ewald; references to *Cantor 1891* should be to the paragraph numbers, which have been added in this edition.

⟦1⟧ In the article entitled: 'On a property of the set of all real algebraic numbers' (*Journ. Math.* Vol. 77, p. 258)[a] a proof is given, probably for the first time, of the theorem that there are infinite manifolds which cannot be correlated in a reciprocal one-to-one way with the totality ⟦Gesamtheit⟧ of all finite integers 1, 2, 3, ..., ν, ...; or, as I am accustomed to saying, which do not have the power of the number-sequence 1, 2, 3, ..., ν, ... That is, from the propositions proved in §2 it follows immediately that, for example, the totality of all real numbers of an arbitrary interval $(\alpha \ldots \beta)$ *cannot* be presented in the sequential form

$$\omega_1, \omega_2, \ldots, \omega_\nu, \ldots$$

⟦2⟧ But it is possible to give a much simpler proof of that theorem which does not depend on considering the irrational numbers.

⟦3⟧ Namely, if m and w are any two mutually exclusive characters ⟦Charaktare⟧ we consider a set ⟦Inbegriff⟧ M of elements

$$E = (x_1, x_2, \ldots, x_\nu, \ldots)$$

which depend on infinitely many coordinates $x_1, x_2, \ldots, x_\nu, \ldots$ where each of these coordinates is either m or w. Let M be the totality of all elements E.

⟦4⟧ The following elements, for example, belong to M:

[a] ⟦*Cantor 1874*; translated above.⟧

$$E^{\mathrm{I}} = (m, m, m, m, \ldots),$$
$$E^{\mathrm{II}} = (w, w, w, w, \ldots),$$
$$E^{\mathrm{III}} = (m, w, m, w, \ldots).$$

I now maintain that such a manifold M does not have the power of the sequence $1, 2, \ldots, \nu, \ldots$.

⟦5⟧ This follows from the following proposition:

'If $E_1, E_2, \ldots, E_\nu, \ldots$ is any simply infinite ⟦einfach unendliche⟧ sequence of elements of the manifold M, then there is always an element E_0 of M which corresponds to no E_ν.'

For proof, let

$$E_1 = (a_{1,1}, a_{1,2}, a_{1,\nu}, \ldots),$$
$$E_2 = (a_{2,1}, a_{2,2}, a_{2,\nu}, \ldots),$$
$$\ldots\ldots\ldots\ldots\ldots$$
$$E_\mu = (a_{\mu,1}, a_{\mu,2}, a_{\mu,\nu}, \ldots).$$
$$\ldots\ldots\ldots\ldots\ldots$$

Here the $a_{\mu,\nu}$ are determinately m or w. We now define a sequence $b_1, b_2, b_\nu, \ldots$, such that b_ν is equal to m or to w but is different from $a_{\nu,\nu}$.

⟦6⟧ That is, if $a_{\nu,\nu} = m$, then $b_\nu = w$, and if $a_{\nu,\nu} = w$, then $b_\nu = m$.

⟦7⟧ If we then consider the element

$$E_0 = (b_1, b_2, b_3, \ldots)$$

of M one sees at once that the equation

$$E_0 = E_\mu$$

can be fulfilled by no integral value of μ, since otherwise for the μ in question and for all integral values of ν,

$$b_\nu = a_{\mu,\nu},$$

and so in particular we would have $b_\mu = a_{\mu,\mu}$, which is excluded by the definition of b_ν. From this proposition it follows immediately that the totality of elements of M cannot be brought into the sequential form:

$$E_1, E_2, \ldots, E_\nu, \ldots;$$

otherwise, we would have the contradiction that a thing ⟦Ding⟧ E_0 would be an element of M as well as not an element of M.

⟦8⟧ This proof is remarkable not only because of its great simplicity, but more importantly because the principle followed therein can be extended immediately to the general theorem that the powers of well-defined manifolds have no maximum, or, what is the same thing, that for any given manifold L

we can produce a manifold M whose power is greater than that of L.

[9] Let, for instance, L be a linear continuum, say the totality of all real numbers which are ≥ 0 and ≤ 1.

[10] Let M be the totality of all single-valued functions $f(x)$ which take only the values 0 or 1, while x runs through all real values which are ≥ 0 and ≤ 1.

[11] That M does *not* have a *smaller* power than L follows from the fact that subsets of M can be given which have the same power as L—for instance, the subset which consists of all functions of x which have the value 1 for a single value x_0 of x, and for all other values of x have the value 0.

[12] But M does *not* have *the same* power as L: for otherwise the manifold M could be brought into a reciprocal one-to-one correspondence with the variable z, and M could be thought of in the form of a single-valued function of the two variables x and z

$$\phi(x, z)$$

such that to every value of z there corresponds an element $f(x) = \phi(x, z)$ of M, and, conversely, to every element $f(x)$ of M there corresponds a single determinate value of z such that $f(x) = \phi(x, z)$. But this leads to a contradiction. For if one understands by $g(x)$ the single-valued function of x which takes on only the values 0 and 1 and is different from $\phi(x, x)$ for every value of x, then on the one hand $g(x)$ is an element of M, and on the other hand $g(x)$ cannot arise from any value $z = z_0$ of $\phi(x, z)$, because $\phi(z_0, z_0)$ is different from $g(z_0)$.

[13] But if the power of M is neither smaller than nor equal to that of L, it follows that it is greater than the power of L (see *Crelle's Journal*, Vol. 84, p. 242).[b]

[14] In the 'Foundations of a general theory of manifolds'[c] I have already shown, in an entirely different manner, that the powers have no maximum; there it was even proved that the totality [Inbegriff] of all powers, if we think of these as ordered according to their size, forms a 'well-ordered set', so that in Nature there is for every power a next greater, and moreover every infinite ascending set of powers is followed by a next-greater.

[15] The 'powers' represent the unique and necessary generalization of the finite 'cardinal numbers'. They are none other than the actual-infinite cardinal numbers, and they have the same reality and determinateness as the others, except that the lawlike relations among them—their 'number theory'—is in part of a different sort than in the domain of the finite.

[16] The further development of this field is a task for the future.

---

[b] [*Cantor 1878*; reprinted in *Cantor 1932*, pp. 119–33.]

[c] [*Cantor 1883d*; translated above.]

E. CANTOR'S LATE CORRESPONDENCE WITH DEDEKIND AND HILBERT

The letters that follow were written by Cantor in the late 1890s; they deal with his distinction between 'consistent' and 'inconsistent' multiplicities, and therefore with what later became known as the 'paradoxes of set theory'.

The central events leading to the discovery of the set-theoretic paradoxes are as follows.

In an article dated February 1897 Cesare Burali-Forti published a purported proof of the following theorem: 'There exist *transfinite ordinal numbers* (or *order types*) *a* and *b* such that *a* is not equal to *b*, not smaller than *b*, and not larger than *b*.' He did *not* claim to have found a paradox in Cantor's theory, but merely to have proved his theorem by a routine *reductio ad absurdum* argument. (Burali-Forti's paper (*1897a*) in fact contained a misreading of Cantor's definition of well-ordering, as Burali-Forti realized when he read Cantor's proof of the trichotomy theorem (*Cantor 1897*); he acknowledged the misreading in *Burali-Forti 1897b*. Both Burali-Forti papers are translated in *van Heijenoort 1967*.).

Cantor's letter to Hilbert of 26 September 1897 (translated below) contains a *reductio* argument that every power is an aleph; the argument shows that he was aware of the necessity for distinguishing between *transfinite* and *absolutely infinite* sets (which he was soon to call *consistent* and *inconsistent* multiplicities).[a] But it does not treat this problem as a paradox endangering his theory of sets.

By 16 April 1902 (at the latest) Ernst Zermelo had discovered a version of Russell's paradox; he stated it to Edmund Husserl in the form of a proof that a set which contains each of its subsets as elements is inconsistent. (The dated fragment, in Husserl's hand, containing Zermelo's proof was published by *Rang and Thomas 1981*.) In a footnote to his *1908a*, Zermelo says that he had discovered Russell's antinomy independently of Russell 'and had communicated it prior to 1903 to Professor Hilbert among others'.

In 1903, Russell published his paradox in *The principles of mathematics*; for the first time, the Burali-Forti problem and the Russell paradox were treated as paradoxes rather than as *reductio* arguments. (For a detailed account of the way in which Russell came to regard his paradox *as* a paradox, see *Moore and Garciadiego 1981*.)

---

[a] This letter seems to be the earliest mention of the paradoxes in Cantor's correspondence with Hilbert. Felix Bernstein says that Cantor discovered his 'contradiction' in 1895 and communicated it by letter to Hilbert in 1896 and to Dedekind in 1899. (*Bernstein 1905a*, p. 187). Others have repeated this assertion. The original published source (cited by Bernstein) appears to be *Jourdain 1904*. Jourdain in turn was relying on information supplied by Cantor, who wrote to Jourdain on 4 November 1903 that he had conveyed the proof of the aleph theorem to Hilbert 'about 7 years ago'; but the letter in fact stems from September 1897. (The Cantor–Jourdain correspondence is published in *Grattan-Guinness 1971*.) See also Bernstein's remarks translated above, p. 836.

Russell communicated his paradox to Frege in a letter of 16 June 1902, receiving a reply on 22 June; both letters are reprinted in *van Heijenoort 1967*. Frege thereupon added an appendix to the second volume of the *Grundgesetze der Arithmetik* (*Frege 1903a*), mentioning Russell's paradox and proposing the solution which is discussed in *Quine 1955*. On being sent a copy of the *Grundgesetze*, Hilbert wrote to Frege on 7 November 1903 that:

Your example at the end of the book (p. 253) is known to us here; I found even more persuasive contradictions 4–5 years ago (and I believe that Dr Zermelo found it 3–4 years ago when I communicated my examples to him). They persuaded me that traditional logic is insufficient; that the doctrine of concept-formation ⟦Begriffsbildung⟧ needs to be sharpened and refined. I see the essential gap in the traditional construction of logic to be the assumption that a concept is already there ⟦ein Begriff bereits da sei⟧ if one can say, for any object, whether it falls under it or not—which all logicians and mathematicians have assumed until now. This seems to me to be insufficient. Rather, the decisive point is the knowledge of the consistency of the axioms.

Russell's publication of the paradox in 1903 (and Zermelo's controversial proof (*1904*) that every set can be well-ordered) gave rise, in the first decade of the twentieth century, to a spirited and well-known discussion of 'the paradoxes of set theory' by Russell, Jourdain, König, Richard, Hilbert, Poincaré, Hadamard, Borel, Zermelo, Brouwer, and others; and the view became widespread that the paradoxes had shaken the foundations of Cantorian set-theory. But Cantor himself did not see the paradoxes as a great surprise, nor did he view them as a threat to his theory. Many commentators have asserted that Cantor, in the years before the discovery of the paradoxes, operated with a 'naïve conception of set', and specifically accuse him of having accepted the unrestricted principle of comprehension (which holds that to any property P there corresponds the set of objects with P).[b] But historically this is inaccurate. Cantor's late letters to Hilbert and Dedekind—written before the discovery of the paradoxes—reject the idea that every property determines a (consistent) set. Indeed, in his *1885* review of Frege's *Grundlagen der Arithmetik* (*Frege 1884*), Cantor had already taken Frege to task for operating too loosely with the principle of comprehension:

[H]e fails utterly to see that quantitatively the "extension of a concept" is something wholly indeterminate; only in certain cases is the "extension of a concept" quantitatively determined; and then, to be sure, if it is finite it has a determinate number; if it is infinite, a determinate power. But for such a quantitative determination of the "extension of a concept" the concepts "number" and "power" must previously be given from the other side; it is a *twisting of the correct procedure* if one attempts to ground the latter concepts on the concept of the "extension of a concept".

Indeed, the roots of Cantor's distinction between consistent and inconsistent collections go back even further, to his *1883d* at least. In a letter to Jourdain

[b] See, e.g., *Fraenkel et alii. 1973*, pp. 15 and 30–1.

of 4 November 1903 he says (speaking of his letters to Hilbert on the proof of the theorem that every power is an aleph), 'I have intuitively known the undoubtedly correct proposition that there are no cardinal numbers other than alephs for about 20 years (since the discovery of the alephs themselves)' (*Grattan-Guinness 1971*, p. 116). And more explicitly, in a letter to G.C. Young of 9 March 1907, Cantor says:

> Do not let yourself be led astray by those who believe they should doubt the reality and consistency of the aleph-numbers; these numbers have *the same fixed thinghood* ⟦*feste Dinglichkeit*⟧ as the traditional, well-known, finite cardinal numbers. What Herr Schönflies calls *W* is not a 'set' in my sense of the word, but an 'inconsistent multiplicity'. When I wrote the *Grundlagen* ⟦*Cantor 1883d*⟧ I already realized this clearly, as can be seen from remarks (1) and (2) at the conclusion ⟦i.e. Cantor's end-notes⟧ where I call *W* the 'absolutely infinite sequence of numbers'.
>
> In (1) I expressly say that I only call multiplicities 'sets' if they can be conceived without contradiction as *unities*, that is, as *things*. . . . What Burali-Forti has put forward is utterly foolish. If you will look at his paper in the *Circolo Mathematico* you will see that *he has not even correctly understood the concept of a 'well-ordered set'* (*Moore and Garciadiego 1981*, pp. 345–6).

A detailed discussion of Cantor's conception of absolute infinity—a conception that, from the beginning, was central to his thought about the infinite—can be found in *Hallett 1984*, especially pp. 32–48.

The bulk of Cantor's surviving correspondence with Dedekind is now in the possession of the Clifford Memorial Library, University of Evansville, Evansville, Indiana. Excerpts from the letters translated here were published by Ernst Zermelo in his edition of the writings of Cantor (*Cantor 1932*, pp. 443–51); a partial translation of Zermelo's excerpts appeared in *van Heijenoort 1967* (pp. 113–17). *Grattan-Guinness 1974* describes the rediscovery of the Cantor–Dedekind correspondence, and lists many errors of transcription in Zermelo's edition. (In particular, Zermelo amalgamated two of Cantor's letters—those of 28 July and 3 August 1899—into a single letter, dated 28 July; he also introduced many alterations to Cantor's notation, omitted several sentences, and added one minor technical error.) The translations that follow were made from photocopies of the original correspondence; the photocopies were supplied by the Clifford Memorial Library. Accurate German versions of the correspondence with Dedekind have been published by *Grattan-Guinness 1974* and by *Dugac 1976*.

The correspondence with Hilbert is in the possession of the Handschriften-abteilung of the Niedersächsische Staats- und Universitätsbibliothek, Göttingen, by whose permission it is reproduced here. (The correspondence has the catalogue number Cod. Ms. Hilbert 54; the letters published here are a selection from Hilbert's extensive correspondence with Cantor. The translations were made from the originals in Göttingen; Cantor's scientific correspondence has subsequently been published as *Cantor 1991*, which includes the German text of the letters translated here.)

The translations are by William Ewald. The translation of Cantor's letter to Dedekind of 3 August 1899 follows the translation of Stefan Bauer-Mengelberg in *van Heijenoort 1967* (with some changes to correct the errors in Zermelo's transcription and to provide a uniform rendering of Cantor's terminology). References should be to the dates of the letters.

Cantor to Hilbert

26 Sept. 1897

Dear Colleague,

We left Brunswick early yesterday and went to Wolfenbüttel, where I spent three hours in the magnificent library while my wife inspected the town. There, at a stroke, I was transported from the mathematical present into the more beautiful poetical past of the sixteenth and seventeenth centuries. In the afternoon we travelled to Harzburg, settled into the Hotel Bellevue, and then immediately went on a walk to the Burgberg, where we arrived at 5:15 and met Privy Councillor Dedekind and our colleague Schubert over a cup of coffee. The former returned to Brunswick in the evening; the latter also spent the night in the Bellevue, and is now sitting opposite me by the Radau Waterfall and travels home today after dining. We are not planning to leave Harzburg for home until next Tuesday evening.

Since you have told me that you are now an editor of the *Math. Annalen*, I should like to ask your opinion about the publication of the third article of my ongoing treatise on 'Contributions to the foundation of transfinite set theory'. I hope to complete the manuscript during this vacation; but I wonder if it would not be expedient to publish the most important results beforehand in the *Göttingen Nachrichten*?

When is the first meeting of your Scientific Society (to which I have belonged as a corresponding member for twenty years)?

Unfortunately, because the afternoon was so far advanced, I had to break off our conversation about set theory in the Brunswick Polytechnic the day before yesterday—and precisely when you expressed a doubt whether all transfinite cardinal numbers or powers are contained in the alephs; in other words, whether every determinate $\mathfrak{a}$ or $\mathfrak{b}$ is also always an aleph.

It can be *rigorously proved* that the answer to this question is *yes*.

For the totality of all alephs is one that cannot be conceived as a determinate, well-defined, *finished* set. If this were the case, then this totality would be *followed* in size by a *determinate aleph*, which would therefore both *belong* to this totality (as an element) and *not belong*, which would be a contradiction.

Having made this observation, I can rigorously prove: 'If a determinate, well-defined, *finished* set had a cardinal number different from any aleph, then it would have to contain subsets whose cardinal number is *any* aleph—in other words, the set would have to contain the totality of all alephs'.

From this it is easy to show that, given the assumption (of a *determinate set* whose cardinal number is not an aleph), the totality of all alephs could then be grasped[a] as a determinate, well-defined, finished set. But I have just proved that this is not so. Therefore, every $\mathfrak{a}$ is always a determinate aleph.

In particular, the power $\mathfrak{o}$ of the linear continuum is always equal to a determinate aleph (I hope to show that $\mathfrak{o} = \aleph_1$).

But from this we can already see that the linear continuum, taken out of its context, is *countable in a higher sense*, i.e. can be represented as a *well-ordered set*.

Totalities that cannot be grasped by us as 'sets' (of which an example is the totality of alephs, as I just showed) I already called 'absolutely infinite' many years ago, and distinguished them sharply from the *transfinite sets*.

Do you possess my collection of letters on the theory of the transfinite that appeared some seven years ago in the *Zeitschrift für Philosophie*? If not, I shall gladly send you a copy.

Best wishes from my wife and me.

---

2 October 1897

Dear Colleague,

In returning to your letter of 27 September, I see that you there say, *entirely correctly*, 'The totality of alephs can be conceived as a determinate, well-defined set, since if any thing is given, it must always be possible to decide whether this thing is an aleph or not; and nothing more belongs to a well-defined set.'

All right ⟦in English⟧.

But you overlook the fact that in my letter from Harzburg I also used the characteristic 'finished', and said:

Theorem: 'The totality of all alephs cannot be conceived as a determinate, a well-defined, and *also a finished* set'. This is the *punctum saliens*, and I venture to say that this *completely certain theorem, provable rigorously from the definition of the totality of all alephs*, is the most important and noblest theorem of set theory. One must only understand the expression 'finished' correctly. I say of a set that it can be thought of as *finished* (and call such a set, if it contains infinitely many elements, 'transfinite' or 'super-finite') if it is possible without contradiction (as can be done with finite sets) to think of *all its elements as existing together*, and so to think of the set itself as *a compounded thing for itself*; or (in other words) if it is *possible* to imagine the set as *actually existing* with the totality of its elements.

So the 'transfinite' coincides with what has since antiquity been called the 'actual infinite', and is to be considered as an *ἀφωρισμένον* ⟦'something determined'⟧.

---

[a] ⟦Cantor has deleted the words, 'by a mortal mind.'⟧

So too I have translated the word 'set' ⟦Menge⟧ (when it is finite or transfinite) into French with 'ensemble' and into Italian with 'insieme'. And so too, in the first article of the work, 'Contributions to the founding of transfinite set theory', I define a 'set' (meaning thereby only the finite or transfinite) at the very beginning as an 'assembling together' ⟦Zusammenfassung⟧. But an 'assembling together' is only possible if an '*existing together*' ⟦*Zusammensein*⟧ is *possible*.

In contrast, infinite sets such that the *totality* of their elements cannot be thought of as 'existing together' or as a 'thing for itself' or an *ἀφωρισμένον*, and that therefore also *in this totality* are absolutely not an object of further *mathematical* contemplation, I call '*absolutely infinite* sets', and to them belongs the 'set of all alephs'.

So much for today.

---

11 October 1897

Dear Colleague,

In your letter of Sept. 27 you kindly offer to communicate results of my work to the Royal Academy of Sciences for publication in the 'Nachrichten'. I should like to make use of this offer by presenting excerpts from the third article of my forthcoming *Annalen* piece to one of the next meetings of your society.

Will the first meeting take place on the 23rd or the 30th of October? And when will the second be?

Have you received my letter of October 2nd, and the pamphlet 'On the theory of the transfinite'?

I see from your letter of September 27th that I must still answer a question it contains.

The theorem that the totality of all *continuous* (but also of all integrable) functions of one or more real variables has the power $\mathfrak{o}$ of the linear continuum is already proved in my *Grundlagen*,[b] albeit expressed rather hesitantly (with 'should').

However, for many years I have possessed a rigorous proof of this theorem that rests on the one hand on the theory of Fourier series, and on the other on the theorem contained in the formula

$$\mathfrak{o}^{\aleph_0} = \mathfrak{o}$$

(*Math. Annalen*, Vol. 46, p. 488; *Crelle's Journal*, Vol. 84, p. 257. ⟦*Cantor 1932*, p. 289.⟧)

In order to give you an idea of the proof, I shall for the sake of brevity confine myself to proving a simpler, similar theorem, namely:

---

[b] ⟦*Cantor 1883b*, Note 10 to §10; translated above.⟧

'The totality of all *analytic* functions $f(u)$ of an independent complex variable u has the power $\mathfrak{o}$'.

Proof. Every *individual* analytic function can, in Weierstrass's definition, be expressed in terms of infinitely many 'function elements' of the form:

$$S = (a_0 + b_0 i) + (a_1 + b_1 i)(u - (c + di)) + \dots$$
$$+ (a_n + b_n i)(u - (c + di))^n + \dots$$

where $i = \sqrt{-1}$ and where c, d, a_n, b_n are real quantities satisfying the condition that the series S converges in a definite but arbitrary neighbourhood of $u = e + di$.

Therefore the totality

$$\{f(u)\}$$

of all analytic functions has in any event *no greater* power than the totality

$$\{\mathcal{E}\}$$

of all collections

$$\mathcal{E} = (c, d, a_0, b_0, a_1, b_1, \dots, a_n, b_n, \dots),$$

where c, d, a_n, b_n can assume any real value independently of each other and *without any restriction*.

But this totality $\{\mathcal{E}\}$ has the power

$$\{\overline{\overline{\mathcal{E}}}\} = \mathfrak{o}^{\aleph_0} = \mathfrak{o}.$$

But since the power of $\{f(u)\}$ obviously cannot be *smaller* than $\mathfrak{o}$, it follows that:

$$\{\overline{\overline{f(u)}}\} = \mathfrak{o} \text{ Q.E.D.}$$

With best wishes,

---

20 February 1900

Dear Friend,

You will be puzzled that I have not yet finished the promised article.

But that is the way I am. It's not the thing itself—that's been done for a long time. It's only the presentation: I keep finding new things to add, and so keep putting off the conclusion. Basically it is only trivialities that are holding me up.

At the moment I am searching in earlier authors to see if they have not already hit upon the axioms of *finite* number theory.

It appears not. Write to me what you know on this point.

Grassmann is the only one I have spoken to about this, and the strange thing is, that on this matter I immediately converted him from his father and to my side. He showed me the passages of the *Ausdehnungslehre* in which the opinion is virtually expressed that arithmetic is a *formal* science, not a *real* science. This is probably the same as what Dedekind means when he declares number theory to be a *part of logic*.

The way I now see the matter, the following *two axioms* are the *necessary* and *sufficient* foundation of our finite number theory:

I. *There exist things* (i.e. objects of our thought). (Without this axiom the concept of 1 would not be possible. One can imagine a scepticism so *outré* that it rejects even this axiom.)

II. If $\mathfrak{V}$ is a *consistent* multiplicity of things and ∂ is a thing not included as part of $\mathfrak{V}$, then the multiplicity $\mathfrak{V} + \partial$ is *also consistent*.

These *two axioms* furnish me with the unbounded number-sequence

$$1, 2, 3, 4, \ldots$$

of finite cardinal integers, and all their laws can be *proved without the assistance of further axioms*.

Please write me your opinion on these points; that will spur me to make my article ready for publication!

I should very much like to visit you, but I have travelled so much this winter, and have spent so much money *on myself*, that I owe it to my family to restrain myself for a while. But in August I hope to meet with you in Paris! Could you not then come over here for a few days beforehand?! The marriage of my second daughter Gertrude with Dr Ernst Vahlen is scheduled for the end of March.

It is good for my wife that she is occupied with the preparations; this distracts her somewhat from sorrow over the death of our Rudolf.

Many greetings to you, your wife, and your Franz.

Cantor to Dedekind

Halle, 28 July 1899

Esteemed friend,

Your kind letter with the congratulations on our two bridal couples touched us both very deeply; and for me in particular it was so beneficial to receive a sign of life from you and your sister that I cannot refrain from expressing my feelings to you. I was also delighted to hear that you are breathing the strong, pure air of Harzburg this summer, and are refreshing yourself on the old, dear, and trusted forest paths where I repeatedly had the pleasure of accompanying you. I was also there at Pentecost of this year with my youngest daughter, Margarethe; we stayed in the Hotel Bellevue for five days; you had not yet arrived, but I asked your house-people about you, and they said they expected you within a week.

I should like us to remain in regular correspondence from now on, during the time granted to us; for with the 54 years that I now have behind me, one already thinks about the end. How many others have finished far earlier! In order to fill this gap (which has been bitter to me for many years) I should like to begin, with your consent, by reporting on the progress of my thoughts in set theory and to ask for your opinion on the chief points.

As you know, many years ago I had already arrived at a well-ordered sequence ⟦Folge⟧ of powers or transfinite cardinal numbers, which I call 'alephs':

$$\aleph_0, \ \aleph_1, \ \ldots \ \aleph_\nu, \ \ldots, \ \aleph_\omega, \ \aleph_{\omega+1}, \ \ldots .$$

$\aleph_0$ denotes ⟦bedeutet⟧ the power of what are usually called 'denumerable' sets; $\aleph_1$ is the next largest cardinal number, $\aleph_2$ the next largest, and so forth; $\aleph_\omega$ is the cardinal number immediately following all $\aleph_\nu$ (i.e. the next largest) and equals

$$\aleph_0 + \aleph_1 + \ldots + \aleph_\nu + \ldots$$

etc.

The big question was whether, besides the alephs, there are also other powers of sets; for two years now I have been in possession of a proof that there are no others; so that, for example, the arithmetic linear continuum (the totality ⟦Gesamtheit⟧ of all real numbers) has a determinate aleph as its cardinal number.

The Bacon–Shakespeare question, on the other hand, is for me completely finished; it cost me a great deal of time and money; to pursue it further I should have to make much greater sacrifices, travel to England, study the archives there, etc.

With warm greetings to you and your sister,

Your most devoted,

Georg Cantor

Halle, 3 August 1899

Esteemed friend,

As I wrote to you last week, it is important to me to know your opinion on certain fundamental points in set theory; please forgive me for the trouble this causes you.

If we start from the notion of a definite multiplicity ⟦Vielheit⟧ (a system, a totality) of things, it is necessary, as I discovered, to distinguish two kinds of multiplicities (by this I always mean *definite* multiplicities).

For a multiplicity can be such that the assumption that *all* of its elements 'are together' leads to a contradiction, so that it is impossible to conceive of the multiplicity as a unity, as 'one finished thing'. Such multiplicities I call *absolutely infinite* or *inconsistent multiplicities.*

As we can readily see, the 'totality of everything thinkable', for example, is such a multiplicity; later still other examples will turn up.

If on the other hand the totality of the elements of a multiplicity can be thought of without contradiction as 'being together', so that they can be gathered together into '*one* thing', I call it a *consistent multiplicity* or a 'set'. (In French and in Italian this notion is aptly expressed by the words 'ensemble' and 'insieme'.)

Two equivalent multiplicities either are both 'sets' or are both inconsistent.

Every submultiplicity ⟦Teilvielheit⟧ of a set is a set.

Whenever we have a set of sets, the elements of these sets again form a set.

If a set M is given, I call the general notion that applies to it and to all sets equivalent to it, and to these alone, its *cardinal number* or also its power, and I denote it by $\bar{\bar{m}}$. I then arrive at the system of all powers—which will later turn out to be an *inconsistent* multiplicity—in the following way.

A multiplicity is said to be 'simply ordered' if between its elements there exists a rank order such that, for any two of its elements, one is the earlier and the other the later, and that, for any three of its elements, one is the earliest, another is the middle one, and the remaining one is the last by rank among them.

If a simply ordered multiplicity is a *set*, then by its *type* $\bar{m}$ I understand the general notion that applies to it and to all ordered sets *similar* to it, and to these alone. (I use the notion of *similarity* in a more restricted sense than you do;[a] I say that two simply ordered multiplicities are *similar* if they can be brought into a one-to-one relation such that the rank order of corresponding elements is the same in both.)

A multiplicity is said to be *well-ordered* if it satisfies the condition that every *sub-multiplicity* of it has a *first* element; I call such a multiplicity a 'sequence' for short.

Every part ⟦Teil⟧ of a sequence is a sequence.

If now a sequence F is a set, I call the type $\bar{F}$ its *ordinal number* or, more briefly, its *number*; thus, when in what follows I speak simply of numbers, I shall have in mind only ordinal numbers, that is, types of well-ordered sets.

I now consider the system of *all numbers* and denote it by Ω.

I proved ⟦*1897*, p. 216⟧ that, of two distinct numbers α and β, one is always the smaller, the other the greater, and that, if for three numbers we have $\alpha < \beta$ and $\beta < \gamma$, we also have $\alpha < \gamma$.

Ω is therefore a simply ordered system.

But it also follows easily from the theorems on well-ordered sets proved in §13 that every multiplicity of numbers, that is, every part of Ω, contains a *least* number.

Hence the system Ω, when naturally ordered according to magnitude, forms a sequence.

---

[a] ⟦Dedekind uses 'ähnlich' in the sense of 'equivalent'.⟧

If we then add 0 to this sequence as an element—putting it first, of course—we obtain a sequence Ω',

$$0, 1, 2, 3, \ldots, \omega_0, \omega_0 + 1, \ldots, \gamma, \ldots,$$

in which, as we can readily see, *every* number is the *type* of the *sequence of all elements preceding it* (including 0). (The sequence Ω has this property only from $\gamma \geq \omega_0$.)

Ω' *cannot* be a *consistent* multiplicity (and therefore neither can Ω); if Ω' were consistent, then, since it is a well-ordered set, there would correspond to it a number δ greater than all numbers of the system Ω; but the number δ also occurs in the system Ω, because this system contains *all* numbers; δ would thus be greater than δ, which is a contradiction. Therefore

A. *The system Ω of all numbers is an inconsistent, absolutely infinite multiplicity.*

Since the *similarity* of well-ordered sets establishes at the same time their *equivalence*, to every number γ there corresponds a definite cardinal number $\mathfrak{c} = \bar{\gamma}$, namely, the cardinal number of any well-ordered set whose type is γ.

The cardinal numbers that correspond in this sense to the *transfinite* numbers of the system Ω I call 'alephs', and the *system of all alephs* is denoted by ת (*tav*, the last letter of the Hebrew alphabet).

I call the system of all numbers γ corresponding to one and the same cardinal number $\mathfrak{c}$ a 'number class', and, more specifically, the number class $Z(\mathfrak{c})$. We readily see that in every number class there occurs a least number γ_0 and that there is a number γ_1 falling outside $Z(\mathfrak{c})$ such that the condition

$$\gamma_0 \leqq \gamma < \gamma_1$$

is equivalent to the fact that the number γ belongs to the number class $Z(\mathfrak{c})$. Every number class is therefore a definite 'segment' of the sequence Ω.*

Certain numbers of the system Ω form, each one by itself, a number class; they are the *finite* numbers, $1, 2, 3, \ldots, \nu, \ldots$, to which correspond the various 'finite' cardinal numbers, $\bar{1}, \bar{2}, \bar{3}, \ldots, \bar{\nu}, \ldots$.

Let ω_0 be the least transfinite number; I call the aleph corresponding to it $\aleph_0$, so that

$$\aleph_0 = \bar{\omega}_0;$$

$\aleph_0$ is the *least* aleph and determines the number class

$$Z(\aleph_0) = \Omega_0.$$

The numbers α of Ω_0 satisfy the condition

$$\omega_0 \leqq \alpha < \omega_1$$

* Here we constantly use the theorem mentioned a few paragraphs above according to which *every* totality of numbers, and hence *every* submultiplicity of Ω, has a *minimum*, a *least number*.

and are characterized by it; here ω_1 is the least transfinite number whose cardinal number is not equal to $\aleph_0$. If we put

$$\bar{\omega}_1 = \aleph_1,$$

then $\aleph_1$ is not only distinct from $\aleph_0$, but it is also the next greater aleph, for we can prove that there is no cardinal number at all that would lie between $\aleph_0$ and $\aleph_1$. We thus obtain the number class $\Omega_1 = Z(\aleph_1)$, which immediately follows Ω_0. It contains all numbers β that satisfy the condition

$$\omega_1 \leqq \beta < \omega_2;$$

here ω_2 is the least transfinite number whose cardinal number differs from $\aleph_0$ and $\aleph_1$.

$\aleph_2 = \omega_2$ is the next greater aleph after $\aleph_1$; it determines the number class $\Omega_2 = Z(\aleph_2)$ that immediately follows Ω_1 and consists of all numbers γ that are $\geqq \omega_2$ and $< \omega_3$, where ω_3 is the least transfinite number whose cardinal number differs from $\aleph_0$, $\aleph_1$, and $\aleph_2$; and so on.

I would still like to stress the following:

$$\bar{\bar{\Omega}}_0 = \aleph_1,\ \bar{\bar{\Omega}}_1 = \aleph_2,\ \ldots,\ \bar{\bar{\Omega}}_\nu = \aleph_{\nu+1},$$

$$\sum_{\nu' = 0,1,2,\ldots,\nu} \aleph_{\nu'} = \aleph_\nu;$$

all this is easy to prove.

Among the transfinite numbers of the system Ω to which no $\aleph_\nu$ ‖with finite ν‖ corresponds as a cardinal number, there is again a least, which we call ω_{ω_0}, and with it we obtain a new aleph,

$$\aleph_{\omega_0} = \bar{\omega}_{\omega_0},$$

which is also definable by means of the equation

$$\aleph_{\omega_0} = \sum_{\nu = 1,2,3,\ldots} \aleph_\nu$$

and which we recognize as the next greater cardinal number after all the $\aleph_\nu$.

We see that this process of formation of the alephs and of the number classes of the system Ω that correspond to them is *absolutely* limitless.

B. *The system* ⊓ *of all alephs, when ordered according to magnitude,*

$$\aleph_0,\ \aleph_1,\ \ldots,\ \aleph_{\omega_0},\ \aleph_{\omega_0+1},\ \ldots,\ \aleph_{\omega_1},\ \ldots,$$

forms a sequence that is similar to the system Ω *and therefore likewise inconsistent, or absolutely infinite.*

The question now arises whether *all transfinite cardinal numbers* are contained in the system ⊓. In other words, is there a *set* whose power is not an aleph?

This question is to be answered *negatively*, and the reason for this lies in the *inconsistency* that we discerned in the systems Ω and ⊓.

Proof. If we take a definite multiplicity V and assume that *no aleph* corresponds to it *as its cardinal number*, we conclude that V must be *inconsistent*.

For we readily see that, on the assumption made, the whole system Ω is projectible into the multiplicity V, that is, there must exist a submultiplicity V' of V that is equivalent to the system Ω.[b]

V' is *inconsistent* because Ω is, and the same must therefore be asserted of V.

Accordingly, every transfinite *consistent multiplicity*, that is, every transfinite set, must have a *definite aleph* as its cardinal number. Hence

C. *The system* ℶ *of all alephs is nothing but the system of all transfinite cardinal numbers.*

All sets, and in particular all '*continua*', are therefore '*denumerable*' in an *extended sense*.

Furthermore C makes it clear that I was right when I stated ⟦*1895*, p. 484⟧ the theorem:

'If $\mathfrak{a}$ and $\mathfrak{b}$ are arbitrary cardinal numbers, then $\mathfrak{a} = \mathfrak{b}$ or $\mathfrak{a} < \mathfrak{b}$ or $\mathfrak{a} > \mathfrak{b}$.'

For, as we have seen, these relations of magnitude obtain between the alephs.

So much *in brief* for today.

Monday we travel to Hahnenklee a. Harz, where we shall celebrate our silver wedding anniversary on 9 August with our children and the two sons-in-law. We shall stay there three or four weeks.

You are probably already dwelling with your sister on the Burgberg; may the stay bring you both much joy.

With warm greetings,

Your most devoted,

Georg Cantor

---

[b] ⟦It is precisely at this point that the weakness of the proof sketched here lies. It has *not* been proved that the entire number sequence Ω would necessarily be 'projectible' into every multiplicity V that has no aleph as its cardinal number. Cantor apparently thinks that successive and arbitrary elements of V are assigned to the numbers of Ω in such a way that every element of V is used only *once*. *Either* this procedure would of necessity come to an end once all elements of V had been exhausted, and then V would be mapped on to a *segment* of the number sequence and its power would be an aleph, contrary to the assumption, *or* V would remain inexhaustible, and hence contain a constituent part that is equivalent to all of Ω and therefore inconsistent. Thus the intuition of time is applied here to a process that goes beyond all intuition, and a fictitious entity is posited of which it is assumed that it could make *successive* arbitrary choices and thereby define a subset V' of V that, by the conditions imposed, is precisely *not* definable. Only through the use of the 'axiom of choice', which postulates the possibility of a *simultaneous* choice and which Cantor uses unconsciously and instinctively everywhere but does not formulate explicitly anywhere, could V' be defined as a subset of V. But even then there would still remain a doubt: perhaps the proof involves 'inconsistent' multiplicities, indeed possibly contradictory notions, and is logically inadmissible already because of that. It is precisely doubts of this kind that impelled the editor a few years later to base his own proof of the well-ordering theorem ⟦*1904*⟧ purely upon the axiom of choice without using inconsistent multiplicities.—footnote of Ernst Zermelo.⟧

Berlin, 16 August 1899

Esteemed friend,

The celebrations continue for us without end. Yesterday we both—my wife and I—left your wonderfully beautiful Hahnenklee in order to be present here (or rather in Charlottenburg) at the wedding of my niece Margarethe Noberling ⟦?⟧, the daughter of my sister (whom you also know). The day after tomorrow, Friday, we return to Hahnenklee to our children.

We send our most heartfelt thanks for your congratulations on our silver wedding anniversary. The days which we spent with you in Interlaken on our honeymoon twenty-five years ago remain in our memory always.

I was delighted to learn from your letter that you are publishing a long article in *Crelle's Journal* on cubic fields.

Have you found the time to work through my recent letter on the system of all transfinite powers? I an anxious to see what you make of this matter.

I wish to add that when I spoke there of multiplicities, I tacitly had in view multiplicities of *unconnected things* ⟦Vielheiten *unverbundener Dinge*⟧, that is, multiplicities such that the removal of any one or several elements has no influence on the remaining-in-existence ⟦Bestehenbleiben⟧ of the remaining elements.

Above all, what do you think of the distinction between 'consistent' and 'inconsistent' multiplicities of *unconnected* things?

My wife asks me to send best wishes to you and your sister, in which I join.

Your most devoted,

Georg Cantor

P.S. My youngest son-in-law is the only son of the poet *Wilhelm Jensen*, which will probably interest your sister as a writer.

Hahnenklee, 28 August 1899

Esteemed friend,

In the hope that you have found the time to immerse yourself in my communications on the system of all cardinal numbers or powers, and to weigh and examine their contents in your mind, I should like to touch on another essential point which I intentionally did not broach in the first presentation, but which you will certainly have noticed (in part because of its absence).

One must raise the question How I know that the well-ordered multiplicities or sequences to which I ascribe the cardinal numbers

$$\aleph_0, \aleph_1, \ldots \aleph_{\omega_0}, \ldots \aleph_{\omega_1}, \ldots$$

are also actual 'sets' in the stated sense of the word, i.e. 'consistent multiplicities'. Is it not conceivable that even *these* multiplicities are 'inconsistent', and that the contradiction in supposing a 'coexistence ⟦Zusammensein⟧ of all their elements' has merely *not yet been noticed*? My answer is that this question must *also be extended to finite multiplicities*, and that a precise examination of the issue leads to the following result: even for finite multiplicities a 'proof' of their 'consistency' cannot be given. In other words: the fact of the 'consistency' of finite multiplicities is a simple, unprovable truth; it is '*The axiom* of arithmetic (in the old sense of the word)'. And in the same way the 'consistency' of the multiplicities to which I assign the alephs as cardinal numbers is 'the axiom of extended, of transfinite arithmetic'.

I should very much like to discuss these matters with you orally and in more detail; it would be easy for me to seize the opportunity. But I do not know if your work would be disturbed at the moment if I were to visit you from here for a few days?

With warm greetings,
Yours,
Georg Cantor

I hope the letter I sent you about 11 days ago from Berlin has reached your hands.

Dedekind to Cantor

Brunswick, 29 August 1899
Kaiser Wilhelm Strasse 87

Esteemed friend!

A visit from you will always be welcome to me and my sister, but I am not at all ripe for a discussion of your communication: it would for the time being be quite fruitless! You will certainly sympathize with me if I frankly confess that, although I have read through your letter of 3 August many times, I am utterly unclear about your distinction of totalities ⟦Inbegriffe⟧ into consistent and inconsistent; I do not know what you mean by the 'co-existence of all elements of a multiplicity', and what you mean by its opposite. I do not doubt that with a more thorough study of your letter a light will go on for me; for I have great trust in your deep and perceptive research. But until now, because of the uninterrupted flood of corrections that must be attended to, I have not had the time or the necessary mental energy to immerse myself in these things. But now only revisions lie ahead, and I promise to use the greater peace for this 'immersion'.

When the young Bernstein visited me in Harzburg at Pentecost of 1897 he spoke about Theorem B on page 7 of the translation of Marotte,[c] and was

[c] ⟦'If two sets M and N are such that M is equivalent to a part N_1 of N and N is equivalent to a part M_1 *of* M, then M and N are equivalent as well.' Dedekind is referring to the French translation of *Cantor 1897*; the theorems A, B, and C are stated by Cantor at the end of §2 of *Cantor 1897*.⟧

somewhat startled when I expressed my conviction that this theorem was *easy* to prove with my methods (*Was sind und was sollen die Zahlen*?); but we did not talk further about his or my proof. After his departure I sat down and composed the enclosed proof of the theorem C,[d] which is clearly the same as B. But whether the theorem A[e] can also be proved with the same ease by my methods is a question I have not investigated. In general, for years I have occupied myself extremely seldom with these interesting things, and since my step-by-step mind ‖Treppen-Verstand‖ has always been slow, it will not now be easy for me to work my way into your research.

With best greetings to you and your wife,

Your devoted,

R. Dedekind

‖Dedekind's enclosure was reprinted, with some errors of transcription, in *Cantor 1932* (p. 449). Dedekind had written down a proof of this theorem years before, on 11 July 1887, and apparently forgot that he had done so. The 1887 chain-theoretic proof is reproduced in *Dedekind 1930–2* (Vol. 3, pp. 447–9). This proof is substantially the same as the proof Zermelo gave, independently of Dedekind, in his *1908a*. Dedekind's 1899 proof follows:‖

PROPOSITION OF THE THEORY OF SYSTEMS

If the system U is a part of the system T, and T a part of the system S, and S is similar to U, then S is also similar to T.

Proof: The theorem is evidently trivial if T is identical with S or with U. In the contrary case, if T is a proper part of S, let A be the system of all those elements of S which are not contained in T—that is (in the notation of Dedekind, Cantor, Schröder)

$$S = \mathfrak{M}(A, T) = (A, T) = A + T.$$

By assumption, S is similar to the (proper) part U of T, so there is a similar mapping ϕ of S into itself, by which S is mapped into $S' = \phi(S) = U$; let A_0 be the 'chain of A' (§4 of my paper, *Was sind und was sollen die Zahlen*?), so (§8):

$$A_0 = A + A_0';$$

since A_0 is a part of S, $A_0' = \phi(A_0)$ is a part of $S' = \phi(S) = U$, and therefore A_0' is also a proper part of T; so A and A_0' have no common element, and A_0 is also a proper part of S. Let B be the system of all those elements of S which are not contained in A_0, i.e.

$$S = A + T = A_0 + B,\ T = A_0' + B,$$

[d] ‖'If M_1 is a part of a set M, M_2 a part of the set M_1, and if the sets M and M_2 are equivalent, then M_1 is also equivalent to the sets M and M_2.'‖

[e] ‖'If $\mathfrak{a}$ and $\mathfrak{b}$ are any two cardinal numbers, then always either $\mathfrak{a} = \mathfrak{b}$ or $\mathfrak{a} < \mathfrak{b}$ or $\mathfrak{a} > \mathfrak{b}$.'‖

where A_0' as a part of A_0 has no element in common with B. Now we define a mapping ψ of S by setting

$$\psi(s) = \phi(s) \text{ or } \psi(s) = s$$

according as the element s of S is contained in A_0 or in B. This mapping ψ of S is similar, for if s_1 and s_2 designate different elements of S, then they are

either contained in A_0; for $\psi(s_1) = \phi(s_1)$ is different from $\psi(s_2) = \phi(s_2)$, because ϕ is a similar mapping of S (which is used here for the first time, and only here);

or they are contained in B; for $\psi(s_1) = s_1$ is different from $\psi(s_2) = s_2$;

or one element s_1 is contained in A_0 and the other element s_2 is contained in B; then $\psi(s_1) = \phi(s_1)$ is different from $\psi(s_2) = s_2$, for $\psi(s_1)$ is contained in A_0' while s_2 is contained in B.

This similar mapping ψ maps $S = A_0 + B$ into

$$\psi(S) = \psi(A_0) + \psi(B) = T$$

because $\psi(A_0) = \phi(A_0) = A_0'$ and $\psi(B) = B$. *Q.E.D.*

Cantor to Dedekind

Hahnenklee, 31 August 1899

Esteemed friend,

You will have received the letter I sent you yesterday. If you find me zealous ⟦eifrig⟧ to persuade you of the necessity of dividing 'systems' into two sorts, I hope thereby to show my gratitude for the repeated inspiration and instruction I have received from your classic writings.

You will understand the essential, profound, and significant difference between 'consistent' and 'inconsistent' systems most quickly if you will follow me in the following simple reflection, which is quite independent of the apparatus I described on 3 August.

We assign equivalent 'sets' to one and the same power-*class* and non-equivalent sets to different classes. We consider the system

S of all thinkable classes $\mathfrak{a}$.

By $\mathfrak{a}$ I understand both the cardinal number and the power of the sets of the relevant class, which is one and the same for all these sets.

Let $M_\mathfrak{a}$ be any determinate set of class $\mathfrak{a}$.

I maintain that the fully-determinate well-defined ⟦völlig bestimmte wohldefinierte⟧ system S is *not a 'set'*.

Proof. If S were a set, then

$$T = \sum M_{\mathfrak{a}}$$

would also be a *set*, where the sum is taken over all classes $\mathfrak{a}$; so T would belong to a determinate class—say, $\mathfrak{a}_0$.

But we also have the following theorem:

'If M is any set with the cardinal number $\mathfrak{a}$, then one can always derive from it another set M' whose cardinal number $\mathfrak{a}'$ is greater than $\mathfrak{a}$.'

I proved this theorem by a *uniform procedure* for the case that is closest to us, where $\mathfrak{a}$ equals $\aleph_0$ (denumerability in the usual sense of the word) and where it equals $\mathfrak{o}$, where $\mathfrak{o}$ is the power of the arithmetic continuum. (The proof is in the *first* volume of the Reports of the 'Deutschen Mathematiker-vereinigung.' ⟦*Cantor 1891*, translated above.⟧) This procedure can be extended *with no difficulty* to an arbitrary $\mathfrak{a}$. The essence of this method can be simply expressed by the formula

$$2^{\mathfrak{a}} > \mathfrak{a}.$$

The definition of $2^{\mathfrak{a}}$ is given in §4 of my treatise ⟦*Cantor 1895*⟧.

Now let $\mathfrak{a}_0'$ be any cardinal number greater than $\mathfrak{a}_0$. Then T (which has the power $\mathfrak{a}_0$) contains as a part the set $M_{\mathfrak{a}_0'}$ (which has the power $\mathfrak{a}_0'$); this is a contradiction.

So the system T, and consequently also the system S, are *not sets*. *Thus there exist* determinate multiplicities that are *not also unities*—i.e. multiplicities such that a real 'coexistence of all their elements' is *im*possible. These are the ones I call 'inconsistent systems'; the others I call 'sets'.

Our presence here unfortunately ends next Monday.

In accordance with the kind permission of you and your sister, I plan to visit you on that day at noon, to spend a few hours with you, and then in the evening to travel towards Halle, where my family will already have journeyed.

Until we meet again,

Your most devoted

Georg Cantor

If I should be unable to come, I shall write so to you on a postcard.

20
Leopold Kronecker (1823–1891)

Kronecker appears here somewhat out of chronological sequence: although he belonged to an earlier generation than Dedekind and Cantor, the selections reproduced here were written in reaction against their work.

Earlier selections, in particular those from Berkeley, d'Alembert, Bolzano, Gauss, and Cantor, have confronted the problem of the infinite in mathematics, especially as it occurs in the foundations of the calculus. The late nineteenth century witnessed a spate of new mathematical results designed to put the calculus on a rigorous foundation: various purely arithmetical theories of the real numbers were propounded by Weierstrass, Heine, Cantor, and Dedekind; this work led to point-set topology, new theories of measure and integration, and the transfinite set theory of Cantor. But although these discoveries eliminated many of the old difficulties surrounding infinitesimals, limits, derivatives, and infinite series, they brought new problems in their wake—problems that were in evidence even before the discovery of the set-theoretic paradoxes, and that are the descendants of the problems raised by Berkeley in his criticisms of Newton. The following group of selections from Kronecker, Klein, and Poincaré show mathematicians reacting, from widely differing points of view, against various aspects of recent work on the foundations of real analysis.

Kronecker, the earliest, most radical, and most influential critic of the new infinitary methods of the Weierstrass school, studied both philosophy and mathematics in Berlin; he attended lectures by Friedrich Schelling in philosophy and by Dirichlet in mathematics. He made a thorough study of Descartes, Spinoza, Leibniz, Kant, and Hegel; in his doctoral examination he was examined by Friedrich Trendelenburg on the history of legal philosophy. While a student in Berlin he became a friend of Jacobi and Eisenstein; he took his doctorate in 1845 with a dissertation, supervised by Dirichlet, on number theory.

In 1855, having spent several years managing a family estate in Liegnitz, the independently-wealthy Kronecker returned to Berlin as a private scholar and amateur mathematician. That year was the year of Gauss's death, and a year of transition for German mathematics. Dirichlet went to Göttingen as Gauss's successor. The number theorist Ernst Eduard Kummer (who, incidentally, had been Kronecker's mathematics teacher years earlier at the Liegnitz *Gymnasium*) was appointed to Dirichlet's chair in Berlin. A few months later, in 1856, Karl Weierstrass was also appointed to a chair in mathematics in Berlin; he, Kronecker, and Kummer were to dominate mathematics there for the next three decades. In 1861 Kronecker was elected to the Berlin Academy, and thereby became entitled to teach in the University; he succeeded to Kummer's chair in 1883.

Kronecker held strong opinions about mathematical ontology: in his view, only the natural numbers genuinely exist; all other legitimate mathematical objects must be constructed from them in a finite number of steps. He had no tolerance for completed infinite collections or for non-constructive definitions. This philosophical standpoint led him into intellectual and even personal conflict with some of the most eminent of his contemporaries. Kronecker clashed with Weierstrass and with Cantor over the foundations of real analysis and the theory of transfinite numbers; lurid details of the dispute can be found in *Schoenflies 1922* and *1927*. And when in 1880 Kronecker nominated Dedekind for membership in the Berlin Academy, he submitted a report which described Dedekind's work in detail—but did not mention Dedekind's 'Continuity and the irrational numbers' (*Dedekind 1872*), which made essential use of completed infinite sets of rational numbers (see *Biermann 1960* and *1966*).

Despite his deeply held philosophical convictions, Kronecker's writings on the philosophy of mathematics are scanty and contain little more than a sketch of his position; even his most famous aphorism—'God created the integers; everything else is the work of man'—comes to us at second hand, as part of the folklore of mathematics.[a]

A. HILBERT AND KRONECKER (FROM *WEYL 1944b*)

Hermann Weyl, in his obituary of Hilbert, provides the following description of the relations between Hilbert and Kronecker. To put this passage in perspective, it should be noted that Hilbert, in his *1931a*, explicitly acknowledged the affinities between his proof theory and the ideas of Kronecker.

The following selection is from p. 613 of *Weyl 1944b* (= *Weyl 1968*, Vol. 4, p. 131).

[a] The first occurrence I have found in print of this saying is in Heinrich Weber's article, 'Leopold Kronecker' (*Weber 1893*, p. 15). Weber says that the quotation occurred in a lecture to the Berlin Naturforscher Versammlung in 1886; but this lecture seems never to have found its way into print. Weber gives the quotation in the form *Die ganzen Zahlen hat der liebe Gott gemacht, alles andere ist Menschenwerk*. The quotation occurs again in the preface to Hilbert's *Zahlbericht* of 1897, whence it would certainly have passed into general currency. (Hilbert's version, in contrast to Weber's, has 'integer' rather than 'integers': *Die ganze Zahl schuf der liebe Gott, alles übrige ist Menschenwerk*.)

When one inquires into the dominant influences acting upon Hilbert in his formative years one is puzzled by the peculiarly ambivalent character of his relationship to Kronecker: dependent upon him, he rebels against him. Kronecker's work is undoubtedly of paramount importance for Hilbert in his algebraic period. But the old gentleman in Berlin, so it seemed to Hilbert, used his power and authority to stretch mathematics upon the Procrustean bed of arbitrary philosophical principles and to suppress such developments as did not conform: Kronecker insisted that existence theorems should be proved by explicit construction, in terms of integers, while Hilbert was an early champion of Georg Cantor's general set-theoretic ideas. Personal relations added to the bitter feeling.[1] A late echo of this old feud is the polemic against Brouwer's intuitionism with which the sexagenarian Hilbert opens his first article on 'Neubegründung der Mathematik' ⟦*Hilbert 1922a*⟧: Hilbert's slashing blows are aimed at Kronecker's ghost whom he sees rising from his grave. But inescapable ambivalence even here—while he fights him he follows him: reasoning along strictly intuitionistic lines is found necessary by him to safeguard non-intuitionistic mathematics.

B. EXTRACT FROM HILBERT'S GÖTTINGEN LECTURES

The following account of Kronecker's philosophical views occurs in the unpublished transcription of Hilbert's lectures *Probleme der mathematischen Logik*. These lectures were held in Göttingen in the Summer Semester, 1920; the transcription was made by Moses Schönfinkel and Paul Bernays, and contains annotations in Hilbert's hand. The original is in the possession of the Mathematics Faculty of the University of Göttingen; the translation, by William Ewald, is printed here by their permission.

⟦Hilbert has just finished discussing the paradoxes of set theory. He continues:⟧

From the last paradox in particular we can now see that arbitrary definitions and inferences, made in the manner that was hitherto usual, are not allowed. And this leads to the expedient of prohibitions, of dictatorship.

The first, most far-reaching, and most radical dictator in this area was

[1] How Georg Cantor himself in his excitability suffered from Kronecker's opposition is shown by his violent outbursts in letters to Mittag-Leffler; see *Schoenflies 1927*. ⟦Weyl's footnote.⟧

Kronecker. His point of view is characterized by the maxim 'God created the integers; the rest is the work of man.' In particular, he rejected set theory as a mere game of fantasy containing nothing but illegitimate combinations that are no longer mathematical concepts. In number theory all truths are indubitable, the proofs incontestable and immediately comprehensible to common sense. This rests on their enduring checkability. In fact, the deepest laws of number theory can be clearly demonstrated by examples to any man one happens to meet on the street—for instance, the law of reciprocity or the theorem that every prime number of the form $4n + 1$ can be represented as $a^2 + b^2$. To understand such theorems and to check them in individual cases requires nothing more than the ordinary one-times-one, the sort of mathematics used by a cook.

On the basis of his way of looking at things, Kronecker forbids already the simplest irrational number $\sqrt{2}$; he introduces the concept of the modulus $x^2 - 2$ in place of this 'inadmissible' concept. He polemicizes against all the standard-setting mathematicians, in particular against Weierstrass—although it was precisely by the rigour of his concepts that Weierstrass liberated the infinitesimal calculus from all dross.

Of the intuitive method of Riemann, which at the time had the most splendid success, or of the newly-arisen Cantorian set theory, he wishes to know nothing; he remains closed to their accomplishments. Here he pursues an ostrich-politics.

His efforts to save at least those things that he needs in algebra become ever more cramped, and his theory of modules ever abstruser.

Kronecker fights against every concept to the extent that it makes statements possible whose correctness is not decidable in a finite number of operations. For example he allows the concept of the irreducibility of an entire rational function (with integral coefficients) only under the condition that a finite process is given for deciding the irreducibility. Kronecker told me personally that the statement that there are infinitely many prime numbers makes no sense until one has shown that after every prime number there is another prime number within a determinable numerical interval—as Euclid did, when he proved that there must be at least one prime number between a prime p and the number $p! + 2$. Propositions like the theorem of Dirichlet, that in every arithmetic progression (where the difference is not divisible by the initial element) there are infinitely many prime numbers, he declares to be inadmissible and in need of supplementation.

And Kronecker restricts logic as well. Just as he forbids arbitrary operation with the concepts 'reducible', 'irreducible', etc., so he stands towards the purely logical propositions like the *tertium non datur*, whose applicability he admits only under the condition that there is the possibility of deciding the existential question by a finite procedure.

Kronecker's ideas terminate in such an extreme position; and this consistency is a service he has rendered, which one may acknowledge without following him. To proceed in this way is to throw the baby out with the bathwater. The irrational number; the theory of functions; the whole of analysis with its finely developed results; physics; and finally the theory of sets—these are the most magnificent fruits of mathematical knowledge; they are indispensable. We rather say that the prohibitions must be formulated in such a way that the contradictions are eliminated but everything valuable remains—and not only must all the valuable results remain standing, but the freedom of concept-formation and of the methods of inference ought not to be limited beyond what is necessary.

The successors of Kronecker recognized this. Nevertheless, the Kroneckerian tendency towards a far-reaching restriction of mathematical concept-formation and inference is still frequently advocated by figures of authority, and his method of erecting prohibitions is still very popular.

In particular, Poincaré—perhaps the most illustrious representative of mathematics in the last generation—must be mentioned as an advocate of this approach. His strength lay in productivity. Fantasy, the power of creativity, belonged to him more than to any other mathematician. But on these matters he conducted himself in a manner that was merely carping, negative, and wholly unproductive. He produced no new ideas, and the new, fruitful scientific approach of Cantor he branded as 'Cantorism'. Like Kronecker, he dictated prohibitions. He declared that every statement about the infinite is admissible only to the extent that it can be transformed into a statement about the finite; every theorem is empty that cannot be verified. From this point of view he utterly rejected the method of set theory. Zermelo's theorem that every set can be well-ordered has no sense for him; and the same goes for almost all proofs that rest on set-theoretic principles, such as the proof than the functional equation $f(x + y) = f(x) + f(y)$ possesses other solutions than the trivial solution, $f(x) = \text{const. } x$.

To be sure, Poincaré does not go so far in his negativism as Kronecker, who also sought to exclude everything intuitive, and who allowed geometry at most to be valid as a heuristic aid.

One could reproach these dictators with many inconsistencies, and show that they do not go far enough with their prohibitions. The doctrine that they advocate cannot possibly be consistently carried through. Otherwise one would have to let go of even the simplest mathematical theorems. Even the commutative law of addition cannot be checked by finitely many operations. And the same goes for all number-theoretic propositions (for example Fermat's proposition about the unsolvability of the equation $x^p + y^p = z^p$ by positive integers x, y, z for an odd prime number p). Furthermore, many modes of inference that one needs for many purposes in algebra and number theory could not be applied if one were consistently to adhere to that standpoint, for example the principle that in every sequence of positive integers $n_1, n_2, n_3, \ldots$ there is always a first number that is less than all of its successors.

The influence of the Kroneckerian inclination was for a time very strong, and when I was a student it was almost disreputable to occupy oneself with set theory. [The aversion to it was made even stronger by the mystical embellishment that Cantor initially gave to his theory.][a]

[The first in the younger generation who seriously took Cantor's side were Minkowski and I; and] the person who, in the most recent times, has re-founded the theory [in my opinion, in the most precise manner] and at the same time in keeping with its spirit, is Zermelo.

C. TWO FOOTNOTES (FROM *KRONECKER 1881* AND *1886*)

Kronecker sometimes stated his philosophical opinion in brief asides in his mathematical work. The following comment occurs in §4 of *Kronecker 1881*.

The definition of irreducibility formulated in §1 lacks a secure foundation so long as no method has been given by means of which it can be decided, for any determinate given function, whether that function is reducible or not according to the definition. ⟦**Footnote**:⟧ The analogous requirement (which, to be sure, is often ignored) arises in many other places, both in definitions and in proofs; I shall treat this matter more generally and thoroughly elsewhere.

The following passage occurs in *Kronecker 1886*; it is reprinted in *Kronecker 1895–1930* (Vol. 3, pp. 155–6). Dedekind replied to Kronecker in the first footnote to §1 of *Was sind und was sollen die Zahlen*? (translated above).

Without the preconditions discussed here—that is, without the possibility from the outset of being able to replace module-systems with infinitely many elements by ones with a finite number ⟦Anzahl⟧ of elements—the concept-formation of

[a] ‖The bracket-signs are a pencilled addition to the typescript, and appear to have been inserted by Hilbert.‖

a 'module system with infinitely many elements' is not applicable ⟦nicht anwendbar⟧. But if one nevertheless wishes to allow it as a purely logical concept-formation, this may be allowed only on condition that, in every single case where we have a particular arithmetical application of the arithmetically insufficiently specified concept, a proof is given that the presupposition has been satisfied—i.e. that, in the individual case, the introduction of module systems with infinitely many elements is shown not to be necessary.

If one adds differentiation to the usual 'rational' operations (namely, addition, subtraction, multiplication, and division) then of course we can no longer make do with module systems having a fixed number of elements determined solely by the number of the variables: one then needs module systems having an 'indefinite' number of elements—that is, a number that can be increased at will, and that consequently can be called 'infinite'. But precisely because the introduction of such module systems seems to become necessary only when we step beyond the proper limits of algebra, they must be avoided in arithmetical–algebraic theories. ⟦**Footnote**:⟧ The foregoing considerations, it seems to me, are opposed to the introduction of such Dedekindian concept-formations as 'module', 'ideal', etc.; and similarly opposed to the introduction of the various concept-formations by whose aid people (Heine being probably the first) have repeatedly attempted of late to conceive and to ground the 'irrational' in general. Even the *general* concept of an infinite sequence—for example one that proceeds according to determinate powers of variables—is in my opinion, as above, only admissible on condition that in every particular case, on the basis of the arithmetical law for the formation of the terms (or of the coefficients), certain presuppositions are shown to be satisfied, which permit the series to be applied like finite expressions, and which accordingly make it unnecessary to go beyond the concept of a *finite* series.

D. ON THE CONCEPT OF NUMBER (*KRONECKER 1887*)

The following selection is Kronecker's most extensive presentation of his views on the philosophy of mathematics.[a] It originally appeared in the *Festschrift* for Eduard Zeller; a lengthened version, containing additional footnotes and with section headings, was published in *Kronecker 1895–1930* (Vol. 3, pp. 251–74). The translation below is of the text of the original article, but incorporates the added footnotes and section headings. (The remainder of the longer version, which eleborates on the mathematical details, has not been translated.) The

[a] Kronecker also discusses the nature of the integers in the introductory chapter to his *Lectures on the theory of numbers* (*Kronecker 1901*).

translation is by William Ewald; references to *Kronecker 1887* should be to the section numbers.

In the open fields of preliminary philosophical labour, from which one reaches the fenced-in domains of the various sciences, the concepts of number, space, and time, which are used in mathematics, are also to be developed. And it seems expedient to carry the development so far that, when the treatment in the special sciences begins, the concepts will already have been equipped with their basic properties.

This shall be done here for the concept of number—the simplest of those three concepts, whose dominant position Jacobi beautifully stressed in one of his letters to Alexander v. Humboldt.[1]

'An old man'—thus begins one of these letters—'compares mathematicians to the *lotophagi*. Whoever once, he says, tastes the sweetness of mathematical ideas can never desist. So ascribe my earlier letter[2] to the frenzy which those lotos-eaters sink into when they suspect that their cult has been neglected or is valued only for its practical applications. And does not Schiller say something similar in the *Xenien* in his little poem

Archimedes und der Jüngling

Zu Archimedes kam ein wissbegieriger Jüngling,
Weihe mich, sprach er zu ihm, ein in die göttliche Kunst,
Die so herrliche Dienste der Sternenkunde geleistet,
Hinter dem Uranos noch einen Planet entdeckt.
Göttlich nennst Du die Kunst, sie ist's, versetzte der Weise,
Aber sie war es, bevor noch sie den Kosmos erforscht,
Ehe sie herrliche Dienste der Sternenkunde geleistet,
Hinter dem Uranos noch einen Planeten entdeckt.
Was Du im Kosmos erblickst, ist nur der Göttlichen Abglanz,
In der Olympier Schaar thronet die ewige Zahl.'[a]

---

[1] The letters are found in G. Lejeune Dirichlet's *Nachlass*.

[2] This 'earlier letter' bears the date, 'Berlin, 26 Dec. 1846', and completely fills three octavo pages with Jacobi's small and dense handwriting. On the first page Jacobi writes 'So, you want to know what ideas had to be developed before Leverrier could in 1846 calculate the planet that lies beyond Neptune?' And on the third page: 'In these circumstances it is really quite extraordinary that Leverrier, with his ability to calculate, had the mathematical breadth of vision that is necessary in order to tackle an utterly new problem in a clever way. But one cannot measure the work of the human spirit in the necessary homeopathic doses.'

[a] ['An inquisitive youth came to Archimedes/Initiate me, he said to him, into the divine art/Which has rendered such magnificent services to astronomy,/And discovered another planet beyond Uranus./You call the art divine, and it is, replied the wise man,/But it was so even before it explored the cosmos,/Before it rendered magnificent services to astronomy,/And discovered another planet beyond Uranus./What you see in the cosmos is only a reflection of the divine/In the Olympian host the eternal number sits enthroned.']

In this witty parody of Schiller's poem 'Archimedes and the Student', Jacobi characterizes the position of the number concept in the whole of mathematics, and he does it with genuine poetry as well as with exact truth—precisely like Gauss in the words: 'Mathematics is the queen of the sciences, and arithmetic the queen of mathematics. She frequently condescends to render a service to astronomy and other natural sciences, but in all circumstances the first rank is due to her.'[3]

Indeed, arithmetic stands to the other two mathematical disciplines, geometry and mechanics, as the whole of mathematics does to astronomy and the other natural sciences. Arithmetic too renders manifold services to geometry and mechanics, and in return receives from her sister disciplines an abundance of stimulation. The word 'arithmetic' is here not to be understood in the usual restricted sense, but rather as including all mathematical disciplines with the exception of geometry and mechanics—especially, therefore, algebra and analysis. And I also believe that we shall one day succeed in 'arithmetizing' the entire content of all these mathematical disciplines—that is, in grounding them solely on the number-concept taken in its narrowest sense, and thus in casting off the modifications and extensions of this concept,[4] which were mostly occasioned by the application to geometry and mechanics. The difference in principle between geometry and mechanics on the one hand and the remaining mathematical disciplines (here gathered together under the term 'arithmetic') on the other is, according to Gauss, that the object of the latter, number, is *merely* our mind's product, while space as well as time also have *outside* of our mind a *reality*, whose laws we cannot completely prescribe *a priori*.[5]

§1.

Definition of the concept of number

The *ordinal numbers* ⟦*Ordnungszahlen*⟧ are the natural point of departure for the development of the concept of number. In them we possess a stock of signs ⟦Bezeichnungen⟧, ordered in a fixed succession, which we can adjoin to a collection ⟦Schaar⟧ of distinct objects that we are able to tell apart.[6] We combine *the totality of the signs thus applied* into the concept of the 'Anzahl of

[3] See 'Gauss zum Gedächtniss', by W. Sartorius v. Waltershausen, Leipzig 1856, p. 79. In this article on p. 97 Ὁ *θεὸς ἀριθμητίζει* is mentioned as a motto of Gauss, which is confirmed by a letter, found in Dirichlet's *Nachlass*, from Gauss's doctor, Baum, to Humboldt.

[4] I mean here especially the addition of irrational as well as continuous quantities.

[5] Gauss's words (in a letter to Bessel of 9 April 1830) are: 'It is my deepest conviction that the theory of space has a completely different position in our *a priori* knowledge than does the pure theory of quantity. Our knowledge of the former utterly lacks the complete conviction of necessity (and also of absolute truth) that belongs to the latter; we must in humility grant that, although number is *merely* the product of our mind, space also possesses a reality outside of our mind, and that we cannot entirely prescribe its laws *a priori*.' See also Herr Ernst Schering's address, held in the public session of the Royal Society of Science in Göttingen on 30 April 1877, p. 9.

[6] The objects can in a certain sense be similar to one another, and only spatially, temporally, or mentally distinguishable—for example, two equal lengths, or two equal temporal intervals.

objects' of which the collection is composed; and we attach the expression for this concept unambiguously to the *last* of the applied signs, since their succession is rigidly determined. Thus, for example, in the collection of letters (a, b, c, d, e) the sign 'first' can be adjoined to the letter a, the sign 'second' to the letter b, etc., and finally the sign 'fifth' to the letter e. The totality of the ordinal numbers thus applied, or the 'Anzahl' of the letters a, b, c, d, e, can accordingly, in keeping with the last of the applied ordinal numbers, be designated by the number 'five'.

The stock of signs that we possess in the ordinal numbers is always adequate because it is not so much an actual stock as an ideal one. In the laws of *formation* of our designations for the numbers we have the capacity to satisfy every demand. To be sure, only because, in the expression of a number, certain signs are repeated arbitrarily often. But if repetitions are permitted, then a single sign already suffices to express every number—namely, by repeating the sign as many times as the number. But such a primitive manner of representation would be unworkable, and the other primitive manner of representation (one sign for each number) would be just as impracticable. So in designating numbers by words one wanted to express as many numbers as possible with the smallest possible number of root words; and one did this by using a double-entry table as the schema of designation. If one thinks of a table of the following sort:

	V	IV	III	II	I
1					
2		•			
3	•				
4			•		
5				•	
6					•
7					
8					
9					

then by placing dots in the 45 squares one can represent all numbers up to 99,999—and in the same way that Greek word-formation accomplished the task. Column I contains the ones, Column II the tens, Column III the hundreds, Column IV the thousands, and Column V the ten thousands. So the five dots in the table represent the number 32,456. Its verbal Greek designation—*τρισμύριοι δισχίλιοι τετρακόσιοι πεντήκοντα ἕξ*—can be derived at once from the table if, for each of the five number-words, one takes the beginning from the row and the end from the Column. So the first dot, which is in row 3 (*τρεῖς*) and Column V (*μύριοι*) yields the number-word *τρισμύριοι*; the second dot, which is in row 2 (*δύο*) and in Column IV (*χίλιοι*), yields the number-word *δισχίλιοι*, etc., and for the fifth dot, which is in row 6 (*ἕξ*) and in Column I, yields the number-word *ἕξ* itself, without an ending. So Greek number-word formation enables one, with the help of 13 different signs (9 initial and 4 end) to distinguish all numbers up to 99,999.

One can form a collection of objects out of the ordinal numbers themselves. In accordance with the definition given above, for the collection which consists of a given ordinal number (the nth) and of all preceding ordinal numbers, the 'Anzahl' is expressed by the 'cardinal number' n corresponding to the nth ordinal number, and it is these cardinal numbers which are designated simply as 'numbers' ⟦Zahlen⟧. A number m is called 'smaller' than another number n when the ordinal number belonging to m precedes that belonging to n. The so-called natural succession of numbers is nothing other than the succession of the corresponding ordinal numbers.

§2.

The independence of number from the ordering followed by counting

If one 'counts' ⟦zählt⟧ a collection of objects—that is, if one adjoins the ordinal numbers in succession as signs to the individual objects—then one thereby gives the objects themselves a fixed ordering. Now if we retain this ordering of the objects but take a new succession of the ordinal numbers that have been applied as signs (by taking any permutation of them) and if we then adjoin as signs to the first object the ordinal number that comes first in the new succession, to the second object the second ordinal number, and so on, each succeeding object receiving the succeeding ordinal number, then the objects thereby obtain again a definite ordering, different from the earlier one, through the ordinal numbers that have been assigned to them, and they are therefore 'counted' in another ordering.[7] But in so doing the 'totality' ⟦Gesamtheit⟧ of the ordinal numbers used as signs (which according to the above definition gives the concept of the 'Anzahl of objects') remains unaltered; and consequently this Anzahl, that is, the *result* of the counting, is independent of the order followed or given by the counting. The 'Anzahl' of the objects of a collection is therefore a property of the collection as such, that is, of the totality of objects, thought of independently of any particular ordering.

If one mentally gathers together into a system any elements, which may be designated by the letters a, b, c, d, . . ., in such a way that the succession of the elements is fixed, then, for example, the two systems (a, b, c) and (c, a, b) are different from each other. And in fact if instead of a, b, c one takes any three distinct numbers and then takes a point in space, designated by the system (a, b, c), whose three orthogonal coordinates are determined by the values $x = a$, $y = b$, $z = c$, then the two points (a, b, c) and (c, a, b) are different from each other. But now let us call any two systems $(a, b, c, d, \ldots)$,

[7] In order to explain the possibility of counting objects in various orderings, I intentionally use a permutation, not of the objects themselves, but of the number-signs. In this way, I need no other presupposition about the objects than that, as in §1, they are 'distinguishable'.

$(a', b', c', d', \ldots)$ *equivalent* when it is possible to transform the one into the other by replacing in sequence every element of the first system by exactly one of the second. Then the necessary and sufficient condition for the equivalence of two systems is the equality of the Anzahl of their elements, and the Anzahl of the elements of a system $(a, b, c, d, \ldots)$ is accordingly characterized as the only '*invariant*' of all systems equivalent to one another.[8]

§3.

THE ADDITION OF NUMBERS

One can take the numbers themselves ⟦Zahlen⟧ as objects of counting ⟦Zählen⟧. One can thus, for example, starting from the number $n_1 + 1$, count a further n_2 times—that is, collect together into a collection exactly so many of the numbers immediately following the number n_1 that its Anzahl comes to n_2. This 'counting further' is called '*adding the number n_2 to the number n_1*', and the number s which one reaches by this counting further is called the 'result of the addition' or the 'sum of n_1 and n_2' and is represented by $n_1 + n_2$. But one reaches precisely the same result s when one adds the number n_1 to the number n_2, that is, when one counts a further n_1 times starting from the number $n_2 + 1$, and consequently $n_1 + n_2 = n_2 + n_1$. Likewise, in general: $n_1 + n_2 + n_3 + \ldots + n_r = n_\alpha + n_\beta + n_\gamma + \ldots + n_\rho$ if $\alpha, \beta, \gamma, \ldots, \rho$ denote the numbers $1, 2, 3, \ldots, r$ in any order. For, if one forms the whole collection of systems of two numbers (h, k), which arises if one successively equates:

$$h = 1 \text{ and } k = 1, 2, \ldots, n_1,$$

$$h = 2 \text{ and } k = 1, 2, \ldots, n_2,$$

$$h = 3 \text{ and } k = 1, 2, \ldots, n_3,$$

$$\ldots\ldots\ldots\ldots\ldots\ldots$$

$$\ldots\ldots\ldots\ldots\ldots\ldots$$

$$h = r \text{ and } k = 1, 2, \ldots, n_r$$

then we obtain the number $n_1 + n_2 + n_3 + \ldots + n_r$ as the Anzahl of the systems of collections if we count them in the sequence in which they have been formed here. But if one orders them so that they follow one another thus:

$$h = \alpha \text{ and } k = 1, 2, \ldots, n_\alpha,$$

$$h = \beta \text{ and } k = 1, 2, \ldots, n_\beta,$$

[8] I believe this makes precise the content of the sentence with which Lipschitz begins his textbook on analysis. This sentence reads: 'If in the contemplation of separate things one disregards the marks that distinguish the things, one is left with the concept of the *Anzahl* of the contemplated things.'

$$h = \gamma \text{ and } k = 1, 2, \ldots, n_\gamma,$$

. . . .

. . . .

$$h = \rho \text{ and } k = 1, 2, \ldots, n_\rho$$

then we obtain the number $n_\alpha + n_\beta + n_\gamma + \ldots + n_\rho$ as the Anzahl of the systems of collections, and the same Anzahl is therefore represented on the one hand by the sum: $n_1 + n_1 + n_3 + \ldots + n_r$ and on the other hand by the sum: $n_\alpha + n_\beta + n_\gamma + \ldots + n_\rho$.

§4.

The multiplication of numbers

If the individual summands $n_1, n_2, n_3, \ldots n_r$ are all equal to precisely the same number n, then one calls the addition the 'multiplication of the number n by the multiplier r' and sets:

$$n_1 + n_2 + n_3 + \ldots + n_r = rn.$$

One calls the result of the thus-defined multiplication the product of the numbers r and n. One gets precisely the same result when one multiplies the number r by the multiplier n, and the product of arbitrarily many numbers n_1 n_2 $n_3 \ldots n_r$ is entirely independent of the order in which the multiplications are carried out. For if one imagines all the systems $(h_1, h_2, h_3, \ldots, h_r)$ formed of r numbers which arise when one sets

for h_1 all values $1, 2, 3, \ldots, n_1$,

for h_2 all values $1, 2, 3, \ldots, n_2$,

for h_3 all values $1, 2, 3, \ldots, n_3$,

. . . .

. . . .

for h_r all values $1, 2, 3, \ldots, n_r$,

then these systems can be ordered according to the sizes of the values of

$$h_r + h_{r-1}g + h_{r-2}g^2 + \ldots + h_1 g^{r-1}$$

where g denotes a number that is larger than all of the numbers $n_1, n_2, n_3, \ldots, n_r$. The systems then follow one another in the way that they would follow one another with respect to size if $h_1 h_2 h_3 \ldots h_r$ represented a number with the *numerals* $h_1, h_2, h_3, \ldots, h_r$ in the number system with base g. The principle of such an ordering is, by the way, none other than the lexicographic ordering in the case where the letters of an alphabet appear in sequence rather than the numbers $1, 2, 3, \ldots$.

The different partitions of the systems $(h_1, h_2, h_3, \ldots, h_r)$ (which are characterized by the different values of h_1 and whose Anzahl is n_1) follow one another in the given ordering according to the size of the value of h_1; within each partition the n_2 different sub-partitions characterized by the value of h_2 follow one another according to the size of *this* value, and so on. If one designates by s_1 the Anzahl of those systems in which $h_1 = 1$, then s_1 is also the Anzahl of the systems in *each* of the n_1 partitions, which are characterized by the values $h_1 = 1, 2, 3, \ldots, n_1$. The total-Anzahl of all systems is accordingly expressed by the product $n_1 s_1$.

If one now designates by s_2 the Anzahl of those systems in which $h_1 = 1$ and $h_2 = 1$, then s_2 is also the Anzahl of the systems in each of the n_2 sub-partitions which (holding the value $h_1 = 1$ fixed) are characterized by the n_2 values $h_2 = 1, 2, 3, \ldots, n_2$. The Anzahl, designated by s_1, of all systems of the partition in which $h_1 = 1$ is therefore expressed by the product $n_2 s_2$, and the Anzahl of *all* systems equals $n_1 n_2 s_2$. If one continues in this manner one acquires the product $n_1 n_2 n_3 \ldots n_r$ as the expression for the Anzahl of all the systems $(h_1, h_2, h_3, \ldots, h_r)$.

If now, as above, $\alpha, \beta, \gamma, \ldots, \rho$ denote the numbers $1, 2, 3, \ldots, r$ in any ordering, and if one orders all the systems $(h_1, h_2, h_3, \ldots, h_r)$ in the way that they would follow one another if $h_\alpha h_\beta h_\gamma \ldots h_\rho$ represented a number with the numerals $h_\alpha, h_\beta, h_\gamma, \ldots, h_\rho$ in the number system with base g, then one obtains by the procedure I have explained the product $n_\alpha n_\beta n_\gamma \ldots n_\rho$ as the expression for the Anzahl of all the systems $(h_1, h_2, h_3, \ldots, h_r)$, and therefore in fact it must be the case that:

$$n_1 n_2 n_3 \ldots n_r = n_\alpha n_\beta n_\gamma \ldots n_\rho.$$

The product of arbitrarily many numbers is accordingly independent of the succession of the factors, that is, from the succession in which the multiplications are carried out.

§5.

Calculations with variables

The laws of addition and multiplication of numbers have thus been completely developed out of the definitions. The same laws had to be assumed as authoritative as soon as one began to apply letters to designate numbers whose determination can or should remain in reserve. But with the introduction *in principle* of 'indeterminates', which is due to Gauss, the special theory of whole numbers widened into the general arithmetical theory of total whole-numbered functions of indeterminates. This general theory allows us to discard all the concepts that, properly speaking, are foreign to arithmetic—for instance, that of irrational algebraic numbers. Even the concept of *negative* numbers can be avoided if we replace the factor -1 in formulae with an indeterminate x and if we replace the equality-sign with the Gaussian congruence-sign *modulo* $(x + 1)$. Thus the equation:

$$7 - 9 = 3 - 5$$

is transformed into the congruence:

$$7 + 9x \equiv 3 + 5x \pmod{x + 1};$$

it thereby gains in content as well, since the congruence has a meaning for every positive whole number x, namely, that $7 + 9x$ yields the same remainder when divided by $x + 1$ as $3 + 5x$. Moreover, this congruence immediately yields the equation as soon as one conceives x no longer as an indeterminate but as a 'quantity' determined by the equation $x + 1 = 0$, and thus introduces the 'negative unit'. (It is, by the way, clearly pointed out in Dr Hermann Schubert's textbook that the meaning of the formula $7 - 9 = 3 - 5$ is in need of a more exact explanation, and that 'properly speaking, a new use of equality-signs' is being made.[9])

In the results of 'general arithmetic' or of the 'arithmetical theory of entire whole-numbered functions of indeterminates' one can see only a summary of those results that arise when one attributes integral values to the indeterminates. So the results of *general* arithmetic also belong properly to the special, ordinary theory of numbers, and all the results of the profoundest mathematical research must in the end be expressible in the simple forms of the properties of integers. But to let these forms appear simply, one needs above all a suitable, surveyable manner of expression and representation for the numbers themselves. The human spirit has been working on this project persistently and laboriously since the greyest prehistory, more or less successfully, and in different ways in the different national groupings. The fruit of this work, our words and numerals for the numbers, was just as much the precondition for the discovery of the treasure of knowledge that is the possession of modern arithmetic, as for the erection of those 'laws into which we compress our knowledge of the motion of the heavenly bodies'; but it was also the precondition for the entire shape of modern practical life, for the immense extension of trade and communication which so essentially distinguishes the modern world from the ancient.

---

[9] *System der Arithmetik und Algebra, als Leitfaden für den Unterricht in höheren Schulen.* By Dr Hermann Schubert, Potsdam 1885, p. 26. I have in the preceding analysis used much of what is contained in §5 of this work on the development of the 'Concept of number'.

21

Christian Felix Klein (1849–1925)

Like Kronecker, Felix Klein was a major voice raising misgivings about the tendency in nineteenth-century mathematics known as the *arithmetizing of the continuum*. Gauss and Cauchy, Bolzano and Dirichlet, Weierstrass, Heine, Cantor, Dedekind, Peano, and others had laboured throughout the century to put the theory of real numbers on a strictly arithmetical (and ultimately set-theoretic) foundation, and to exclude geometric intuition from the foundations of the calculus. Now, at the end of the century, both Kronecker and Klein raise objections to this arithmetizing project. But the difference in outlook between them is considerable, Kronecker's objections being those of a number-theorist, while Klein's are those of a geometer: so that on some points they are further from each other than either is from Weierstrass. Broadly speaking, they are reacting against the two different halves of the arithmetical continuum: Kronecker against the continuum, and Klein against what he considers an over-emphasis upon arithmetic.

Kronecker, the algebraist and number-theorist *par excellence*, has no objection to arithmetizing: on the contrary. His objection is precisely that the continuum, conceived of as a completed infinite set of irrational numbers, each of which is in turn a completed infinite set of rational numbers, goes beyond the bounds of true arithmetic, which, for him, is the arithmetic that can be grounded in operations on the finite natural numbers. *Menschenwerk* is of course allowed in mathematics, but it must not outstrip the capacities of *Menschen*; and the infinitary principles employed by Cantor and Dedekind seem to Kronecker to do just that. This point of view led Kronecker to reject the work of the modern analysts virtually in its entirety.

Klein, on the other hand, is more tolerant, and he raises no objections to completed infinite sets *per se*. Indeed, in contrast to Kronecker, Klein even regards the arithmetical continuum with admiration—for him, it represents a partial truth, rather than an outright falsehood. His concern is to defend the role of intuition in mathematics, to urge the merits of a less algebraic, a more visual and *anschaulich* style. He argues that geometric intuition, far from deserving to be banished from mathematics, is essential to its development; he looks back above all to the accomplishments of Riemann, whose works and whose mode of reasoning had been eclipsed by the new trend towards Weierstrassian rigour. Klein admires the work of the modern analysts, but he protests against their tendency to reduce mathematics exclusively to arithmetic. And this attitude pits him against Kronecker as well as against Weierstrass; for nowhere is the tendency towards *exclusive* arithmetizing more evident than in Kronecker.

Klein (who incidentally was married to a granddaughter of Hegel) was Professor of Mathematics in Göttingen from 1886 until his death; earlier, he taught at the Universities of Erlangen, Munich, and Leipzig. He received his mathematical education in Bonn under the supervision of Plücker.

Klein's mathematical research was in the theory of automorphic functions, mathematical physics, number theory, topology, and, above all, the foundations of geometry. In 1871–3, building on work of Cayley, Klein published a series of papers classifying non-Euclidean geometries into various kinds of elliptic and hyperbolic geometries; he then described projective-geometrical models for each of them. This work was a major step in the understanding and the acceptance of non-Euclidean geometry. Klein also, in 1872, put forward his Erlanger Programme, in which he classified geometries in terms of their invariants under particular groups of transformations. This conception of the subject dominated research in geometry for decades, and gave a boost to group theory as well; for a recent account, see *Birkhoff and Bennett 1988*. Klein also did fundamental research on the theory of orientable surfaces; one such surface, described by him in 1882, is familiar to undergraduates as the 'Klein bottle'.

Klein was an influential figure in German mathematical education, a side of his interests that is evident in the selections that follow. He restored Göttingen to its position as a world centre of mathematics, inviting first Hilbert and then Minkowski to join the faculty; he also occupied himself with the teaching of mathematics in the secondary schools. Several of his books are concerned with this subject, and with the popular presentation of mathematical ideas; see, for instance, his *1924–8*.

A. KLEIN ON THE SCHOOLS OF MATHEMATICS (FROM *KLEIN 1911*)

The following passage is taken from Klein's 1893 Evanston Colloquium lectures on mathematics ('Held', according to the title page, 'before the Congress of Mathematics in connection with the World's Fair in Chicago'). The present passage is the fourth paragraph of the first lecture (which was on Alfred Clebsch); it was delivered on 28 August 1893. The passage seems to be the first in which a distinction is drawn between logicians, formalists, and intuitionists. But Klein's classification differs from the one in modern use, first, in distinguishing between individual styles of thought rather than between rival schools; and, secondly, in applying to mathematics generally, rather than merely to the foundations of mathematics. The present title has been supplied by the editor; the title of the original lecture was simply 'Clebsch'.

Among mathematicians in general, three main categories may be distinguished; and perhaps the names *logicians, formalists*, and *intuitionists* may serve to characterize them. (1) The word *logician* is here used, of course, without reference to the mathematical logic of Boole, Peirce, etc.; it is only intended to indicate that the main strength of the men belonging to this class lies in their logical and critical power, in their ability to give strict definitions, and to derive rigid deductions therefrom. The great and wholesome influence exerted in Germany by *Weierstrass* in this direction is well known. (2) The *formalists* among the mathematicians excel mainly in the skilful formal treatment of a given question, in devising for it an 'algorithm'. *Gordan*, or let us say *Cayley* and *Sylvester*, must be ranged in this group. (3) To the *intuitionists*, finally, belong those who lay particular stress on geometrical intuition (*Anschauung*), not in pure geometry only, but in all branches of mathematics. What Benjamin Peirce has called 'geometrizing a mathematical question' seems to express the same idea. Lord *Kelvin* and *von Staudt* may be mentioned as types of this category.

Clebsch must be said to belong both to the second and third of these categories, while I should class myself with the third, and also the first. For this reason my account of Clebsch's work will be incomplete; but this will hardly prove a serious drawback, considering that the part of his work characterized by the second of the above categories is already so fully appreciated here in America.

B. ON THE MATHEMATICAL CHARACTER OF SPACE-INTUITION AND THE RELATION OF PURE MATHEMATICS TO THE APPLIED SCIENCES (FROM *KLEIN 1911)*

The following selection was Klein's sixth Evanston Colloquium lecture, and was delivered on 2 September 1893. References should be to the paragraph numbers, which have been added in this edition.

⟦1⟧ In the preceding lectures I have laid so much stress on geometrical methods that the inquiry naturally presents itself as to the real nature and limitations of geometrical intuition.

⟦2⟧ In my address before the Congress of Mathematics at Chicago I referred to the distinction between what I called the *naïve* and the *refined* intuition. It is the latter that we find in Euclid; he carefully develops his system on the basis of well-formulated axioms, is fully conscious of the necessity of exact proofs, clearly distinguishes between the commensurable and incommensurable, and so forth.

⟦3⟧ The naïve intuition, on the other hand, was especially active during the period of the genesis of the differential and integral calculus. Thus we see that Newton assumes without hesitation the existence, in every case, of a velocity in a moving point, without troubling himself with the inquiry whether there might not be continuous functions having no derivative.

⟦4⟧ At the present time we are wont to build up the infinitesimal calculus on a purely analytical basis, and this shows that we are living in a *critical* period similar to that of Euclid. It is my private conviction, although I may perhaps not be able to fully substantiate it with complete proofs, that Euclid's period also must have been preceded by a 'naïve' stage of development. Several facts that have become known only quite recently point in this direction. Thus it is now known that the books that have come down to us from the time of Euclid constitute only a very small part of what was then in existence; moreover, much of the teaching was done by oral tradition. Not many of the books had that artistic finish that we admire in Euclid's 'Elements'; the majority were in the form of improvised lectures, written out for the use of the students. The investigations of Zeuthen* and Allman† have done much to clear up these historical conditions.

⟦5⟧ If we now ask how we can account for this distinction between the naïve and refined intuition, I must say that, in my opinion, the root of the matter lies in the fact that *the naïve intuition is not exact, while the refined intuition is not properly intuition at all, but arises through the logical development from axioms considered as perfectly exact.*

⟦6⟧ To explain the meaning of the first half of this statement it is my opinion that, in our naïve intuition, when thinking of a point we do not picture to our mind an abstract mathematical point, but substitute something concrete for it. In imagining a line, we do not picture to ourselves 'length without breadth', but a *strip* of a certain width. Now such a strip has of course *always* a tangent (Fig. 9); i.e. we can always imagine a straight strip having a small portion (element) in common with the curved strip; similarly with respect to the osculating circle. The definitions in this case are regarded as holding only approximately, or as far as may be necessary.

⟦7⟧ The 'exact' mathematicians will of course say that such definitions are not definitions at all. But I maintain that in ordinary life we actually operate with such inexact definitions. Thus we speak without hesitancy of the direction

* *Die Lehre von den Kegelschnitten im Altertum*, übersetzt von R. v. Fischer-Benzon, Kopenhagen, Höst, 1886.

† *Greek geometry from Thales to Euclid*, Dublin, Hodges, 1889.

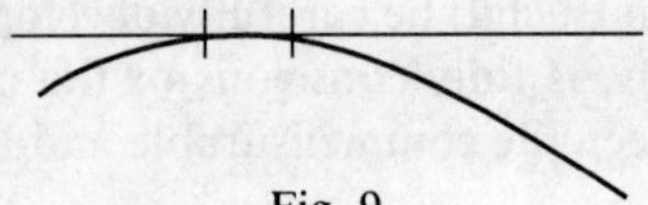

Fig. 9.

and curvature of a river or a road, although the 'line' in this case has certainly considerable width.

⟦8⟧ As regards the second half of my proposition, there actually are many cases where the conclusions derived by purely logical reasoning from exact definitions can no more be verified by intuition. To show this, I select examples from the theory of automorphic functions, because in more common geometrical illustrations our judgement is warped by the familiarity of the ideas.

⟦9⟧ Let any number of non-intersecting circles 1, 2, 3, 4, . . ., be given (Fig. 10), and let every circle be reflected (i.e. transformed by inversion, or reciprocal radii vectores) upon every other circle; then repeat this operation again and again, *ad infinitum*. The question is, what will be the configuration formed by the totality of all the circles, and in particular what will be the position of the limiting points. There is no difficulty in answering these questions by purely logical reasoning; but the imagination seems to fail utterly when we try to form a mental image of the result.

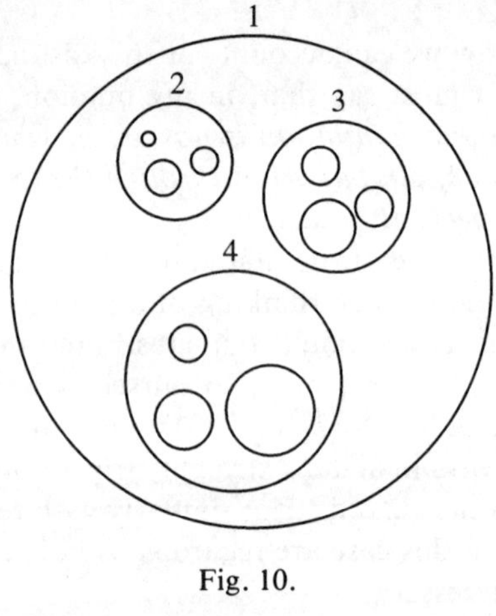

Fig. 10.

⟦10⟧ Again, let a series of circles be given, each circle touching the following, while the last touches the first (Fig. 11). Every circle is now reflected upon every other just as in the preceding example, and the process is repeated indefinitely. The special case when the original points of contact happen to lie on a circle being excluded, it can be shown analytically that the continuous curve which is the locus of all the points of contact *is not an analytic curve*. The points of contact form a manifoldness that is everywhere dense on the curve (in the sense of G. Cantor), although there are intermediate points between

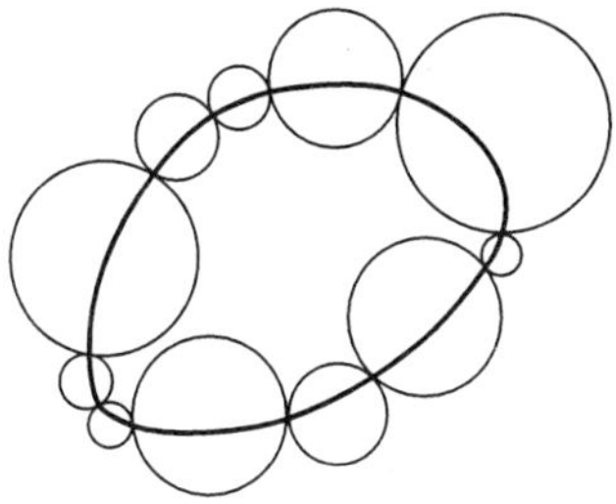

Fig. 11.

them. At each of the former points there is a determinate tangent, while there is none at the intermediate points. Second derivatives do not exist at all. It is easy enough to imagine a *strip* covering all these points; but when the width of the strip is reduced beyond a certain limit, we find undulations, and it seems impossible to clearly picture to the mind the final outcome. It is to be noticed that we have here an example of a curve with indeterminate derivatives arising out of purely geometrical considerations, while it might be supposed from the usual treatment of such curves that they can only be defined by artificial analytical series.

⟦11⟧ Unfortunately, I am not in a position to give a full account of the opinions of philosophers on this subject. As regards the more recent mathematical literature, I have presented my views as developed above in a paper published in 1873, and since reprinted in the *Math. Annalen.** Ideas agreeing in general with mine have been expressed by Pasch, of Giessen, in two works, one on the foundations of geometry,† the other on the principles of the infinitesimal calculus.‡ Another author, Köpcke, of Hamburg, has advanced the idea that our space-intuition is exact as far as it goes, but so limited as to make it impossible for us to picture to ourselves curves without tangents.§

⟦12⟧ On one point Pasch does not agree with me, and that is as to the exact value of the axioms. He believes—and this is the traditional view—that it is possible finally to discard intuition entirely, basing the whole science on the axioms alone. I am of the opinion that, certainly, for the purposes of research it is always necessary to combine the intuition with the axioms. I do not believe, for instance, that it would have been possible to derive the results discussed in my former lectures, the splendid researches of Lie, the continuity of the shape of algebraic curves and surfaces, or the most general forms of triangles, without the constant use of geometrical intuition.

---

* *Über den allgemeinen Functionsbegriff und dessen Darstellung durch eine willkürliche Curve*, Math. Annalen, Vol. 22 (1883), pp. 249–259.

† *Vorlesungen über neuere Geometrie*, Leipzig, Teubner, 1882.

‡ *Einleitung in die Differential- und Integralrechnung*, Leipzig, Teubner, 1882.

§ *Ueber Differentiirbarkeit und Anschaulichkeit der stetigen Functionen*, Math. Annalen, Vol. 29 (1887), pp. 123–140.

⟦13⟧ Pasch's idea of building up the science purely on the basis of the axioms has since been carried still farther by Peano, in his logical calculus.

⟦14⟧ Finally, it must be said that the degree of exactness of the intuition of space may be different in different individuals, perhaps even in different races. It would seem as if a strong naïve space-intuition were an attribute pre-eminently of the Teutonic race, while the critical, purely logical sense is more fully developed in the Latin and Hebrew races. A full investigation of this subject, somewhat on the lines suggested by Francis Galton in his researches on heredity, might be interesting.[a]

---

[a] ⟦This passage has elicited considerable negative comment. Jacques Hadamard quotes the passage, and observes:

> That such an assertion is not in agreement with facts will appear clearly when we come to examples. It is hardly doubtful that, in stating it, Klein implicitly considers intuition, with its mysterious character, as being superior to the prosaic way of logic (we have already met with such a tendency in section III) and is evidently happy to claim that superiority for his countrymen. We have heard recently of that special kind of ethnography with Nazism: we see that there was already something of this kind in 1893 (*Hadamard 1945*, p. 107).

Hadamard (who wrote this passage in 1944) was attempting to make a point about the baleful effects of nationalism on science; and he is as critical of Duhem's 1915 anti-German propaganda as he is of Klein. Hadamard's point is laudable; but it is not clear that in making it he has fairly interpreted Klein.

Klein, in fact, is careful to insist that mathematics needs *both* logical *and* intuitive thought: he does not favour one at the expense of the other. (See, for example, the concluding three paragraphs of *Klein 1895* below; but the point is clear in the present article.) Nor do Klein's *Lectures on the development of mathematics* (*Klein 1926–7*), originally delivered during the years 1914–19, display any discernible nationalistic or racial bias: Klein admired the great French analysts, the geometrical profundity of Poincaré, and above all the accomplishments of Einstein. (Klein was one of the earliest and most vigorous champions of the theory of relativity.)

Hadamard seems to have based his interpretation of Klein solely on this one paragraph; other writers, relying entirely on Hadamard, have baldly asserted that Klein was a proto-Nazi. But Klein appears to have mentioned racial issues in print only twice: in the present passage, and in some remarks in the *Lectures*. The remarks come in the introductory, biographical portion of the lecture on Sylvester, and may safely be said not to be typical of the Nazis' manner of expressing their views. After first describing Sylvester's career in glowing terms, Klein continues:

> Sylvester was an extraordinarily lively, many-faceted spirit, who penetrated with the greatest intensity into everything he encountered, and who brought everything into interconnection; but he was less inclined to polish the things he had picked up, to turn them systematically and deliberately into rounded works in the grand style. In science, his proper domain was the very abstract, combinatorial manner of looking at things. And in this spirit he worked not only on the problem of invariants but on problems in the various domains of mathematics—e.g. on questions of mechanics—and did so in a splendid manner. A remark he once made to me about how chemical formulae are to be conceived is typical of his style of thought. These formulae at the time had attracted the interest of mathematicians, and he later, in an article in the *American Journal*, paralleled them in an interesting way with the symbolic processes of binary invariant theory. He wanted to see in them solely the logical relation of two concepts, and he dismissed with a smile the idea of two concrete atoms bound to one another. But—this is in any case not the mental foundation on which natural science makes its progress.
>
> As a personality, Sylvester was outstandingly stimulating, witty, and effervescent. He was a brilliant speaker, and often distinguished himself, to the pleasure of all, by his striking, skilful poetic art. In the brilliance and flexibility of his mind, he is a true representative of his race; he came from a purely Jewish family that, hitherto nameless, only took the name Sylvester in his generation (*Klein 1926–7*, Vol. i, 163).

The passage attacked by Hadamard should be compared with a similar passage about styles of mathematical thought in ¶11 of *Klein 1895* (below); there, Klein's conjecture is couched in terms of distinct *psychological* types rather than in terms of *national* types, and may be found more plausible. But the motivation in both passages seems to have been the same: to put forward an offhand conjecture in the light of currently-fashionable Darwinian or psychological theory, and not to express malice.⟧

⟦15⟧ What has been said above with regard to geometry ranges this science among the applied sciences. A few general remarks on these sciences and their relation to pure mathematics will here not be out of place. From the point of view of pure mathematical science I should lay particular stress on the *heuristic value* of the applied sciences as an aid to discovering new truths in mathematics. Thus I have shown (in my little book on Riemann's theories) that the Abelian integrals can best be understood and illustrated by considering electric currents on closed surfaces. In an analogous way, theorems concerning differential equations can be derived from the consideration of sound-vibrations; and so on.

⟦16⟧ But just at present I desire to speak of more practical matters, corresponding as it were to what I have said before about the inexactness of geometrical intuition. I believe that the more or less close relation of any applied science to mathematics might be characterized by the degree of exactness attained, or attainable, in its numerical results. Indeed, a rough classification of these sciences could be based simply on the number of significant figures averaged in each. Astronomy (and some branches of physics) would here take the first rank; the number of significant figures attained may here be placed as high as seven, and functions higher than the elementary transcendental functions can be used to advantage. Chemistry would probably be found at the other end of the scale, since in this science rarely more than two or three significant figures can be relied upon. Geometrical drawing, with perhaps 3 to 4 figures, would rank between these extremes; and so we might go on.

⟦17⟧ The ordinary mathematical treatment of any applied science substitutes exact axioms for the approximate results of experience, and deduces from these axioms the rigid mathematical conclusions. In applying this method it must not be forgotten that mathematical developments transcending the limit of exactness of the science are of no practical value. It follows that a large portion of abstract mathematics remains without finding any practical application, the amount of mathematics that can be usefully employed in any science being in proportion to the degree of accuracy attained in the science. Thus, while the astronomer can put to good use a wide range of mathematical theory, the chemist is only just beginning to apply the first derivative, i.e. the rate of change at which certain processes are going on; for second derivatives he does not seem to have found any use as yet.

⟦18⟧ As examples of extensive mathematical theories that do not exist for applied science, I may mention the distinction between the commensurable and incommensurable, the investigations on the convergency of Fourier's series, the theory of non-analytical functions, etc. It seems to me, therefore, that Kirchhoff makes a mistake when he says in his *Spectral-Analyse* that absorption takes place only when there is *exact* coincidence between the wave-lengths. I side with Stokes, who says that absorption takes place *in the vicinity* of such coincidence. Similarly, when the astronomer says that the periods of two planets must be exactly commensurable to admit the possibility of a collision, this holds only abstractly, for their mathematical centres; and it must be remembered that such things as the period, the mass, etc., of a planet cannot be exactly defined, and are changing all the time. Indeed, we have no way of ascertaining whether two astronomical magnitudes are incommensurable or not; we can only inquire whether their ratio can be expressed approximately by two *small* integers. The statement sometimes made that there exist only analytic functions in nature is in my opinion absurd. All we can say is that we restrict ourselves to analytic, and even only to simple analytic, functions because they afford a sufficient degree of approximation. Indeed, we have the theorem (of Weierstrass) that any continuous function can be approximated to, with any required degree of accuracy, by an analytic function. Thus if $\phi(x)$ be our continuous function, and δ a small quantity representing the given limit of exactness (the width of the strip that we substitute for the curve), it is always possible to determine an *analytic* function $f(x)$ such that

$$\phi(x) = f(x) + \varepsilon, \text{ where } |\varepsilon| < |\delta|,$$

within the given limits.

‖19‖ All this suggests the question whether it would not be possible to create a, let us say, *abridged* system of mathematics adapted to the needs of the applied sciences, without passing through the whole realm of abstract mathematics. Such a system would have to include, for example, the researches of Gauss on the accuracy of astronomical calculations, or the more recent and highly interesting investigations of Tchebycheff on interpolation. The problem, while perhaps not impossible, seems difficult of solution, mainly on account of the somewhat vague and indefinite character of the questions arising.

‖20‖ I hope that what I have here said concerning the use of mathematics in the applied sciences will not be interpreted as in any way prejudicial to the cultivation of abstract mathematics as a pure science. Apart from the fact that pure mathematics cannot be supplanted by anything else as a means for developing the purely logical powers of the mind, there must be considered here as elsewhere the necessity of the presence of a few individuals in each country developed in a far higher degree than the rest, for the purpose of keeping up and gradually raising the *general* standard. Even a slight raising of the general level can be accomplished only when some few minds have progressed far ahead of the average.

‖21‖ Moreover, the 'abridged' system of mathematics referred to above is not yet in existence, and we must for the present deal with the material at hand and try to make the best of it.

‖22‖ Now, just here a practical difficulty presents itself in the teaching of mathematics, let us say of the elements of the differential and integral calculus. The teacher is confronted with the problem of harmonizing two opposite and almost contradictory requirements. On the one hand, he has to consider the limited and as yet undeveloped intellectual grasp of his students and the fact that most of them study mathematics mainly with a view to the practical applications; on the other, his conscientiousness as a teacher and man of science would seem to compel him to detract in no wise from perfect mathematical rigour and therefore to introduce from the beginning all the refinements and niceties of modern abstract mathematics. In recent years the university instruction, at least in Europe, has been tending more and more in the latter direction; and the same tendencies will necessarily manifest themselves in this country in the course of time. The second edition of the *Cours d'analyse* of Camille Jordan may be regarded as an example of this extreme refinement in laying the foundations of the infinitesimal calculus. To place a work of this character in the hands of a beginner must necessarily have the effect that at the beginning a large part of the subject will remain unintelligible, and that, at a later stage, the student will not have gained the power of making use of the principles in the simple cases occurring in the applied sciences.

⟦23⟧ It is my opinion that in teaching it is not only admissible, but absolutely necessary, to be less abstract at the start, to have constant regard to the applications, and to refer to the refinements only gradually as the student becomes able to understand them. This is, of course, nothing but a universal pedagogical principle to be observed in all mathematical instruction.

⟦24⟧ Among recent German works I may recommend for the use of beginners, for instance, Kiepert's new and revised edition of Stegemann's text-book;* this work seems to combine simplicity and clearness with sufficient mathematical rigour. On the other hand, it is a matter of course that for more advanced students, especially for professional mathematicians, the study of works like that of Jordan is quite indispensable.

⟦25⟧ I am led to these remarks by the consciousness of a growing danger in the higher educational system of Germany—the danger of a separation between abstract mathematical science and its scientific and technical applications. Such separation could only be deplored; for it would necssarily be followed by shallowness on the side of the applied sciences, and by isolation on the part of pure mathematics.

C. THE ARITHMETIZING OF MATHEMATICS (*KLEIN 1895*)

The following lecture was delivered to the Royal Academy of Sciences of Göttingen, 2 November 1895. The translation, which appeared in the May, 1896 *Bulletin of the American Mathematical Society*, is by Isabel Maddison. References should be to the paragraph numbers, which have been added in this edition.

* *Grundriss der Differential- und Integral-Rechnung*, 6te Auflage, herausgegeben von Kiepert, Hannover, Helwing, 1892.

⟦1⟧ Though the details of mathematical science, by their very nature, elude the comprehension of the layman, and therefore fail to arouse his interest, yet the mathematician may profitably indicate certain general points of view from which he surveys his science, especially if these points of view determine his attitude to kindred subjects. I propose therefore on the present occasion to explain my position in regard to an important mathematical tendency which has as its chief exponent Weierstrass, whose eightieth birthday we have lately celebrated. I refer to the *arithmetizing* of mathematics. Some account of this tendency and its origin may be given by way of preface.

⟦2⟧ The popular conception of mathematics is that of a strictly logically coordinated system, complete in itself, such as we meet with, for instance, in Euclid's geometry; but as a matter of fact, modern mathematics in its origin was of a totally different character. With the contemplation of nature as starting-point, and its interpretation as object, a philosophical principle, the principle of continuity, was made fundamental; and the use of this principle characterizes the work of the great pioneers, Newton and Leibnitz, and the mathematicians of the whole of the eighteenth century—a century of discoveries in the evolution of mathematics. Gradually, however, a more critical spirit asserted itself and demanded a logical justification for the innovations made with such assurance, the establishment, as it were, of law and order after the long and victorious campaign. This was the time of Gauss and Abel, of Cauchy and Dirichlet. But this was not the end of the matter. Gauss, taking for granted the continuity of space, unhesitatingly used space intuition as a basis for his proofs; but closer investigation showed not only that many special points still needed proof, but also that space intuition had led to the too hasty assumption of the generality of certain theorems which are by no means general. Hence arose the demand for exclusively arithmetical methods of proof; nothing shall be accepted as a part of the science unless its rigorous truth can be clearly demonstrated by the ordinary operations of analysis. A glance at the more modern textbooks of the differential and integral calculus suffices to show the great change in method; where formerly a diagram served as proof, we now find continual discussions of quantities which become smaller than, or which can be taken smaller than, any given small quantity. The continuity of a variable, and what it implies, or does not imply, are discussed, and a question is brought forward whether we can, properly speaking, differentiate or integrate a function at all. This is the Weierstrassian method in mathematics, the 'Weierstrass'sche Strenge', as it is called.

⟦3⟧ Of course even this assigns no absolute standard of exactness; we can introduce further refinements if still stricter limitations are placed on the association of the quantities. This is exemplified in Kronecker's refusal to employ irrational numbers, and consequent reduction of mathematics to relations between whole numbers only; and in another way in the efforts made to introduce symbols for the different logical processes, in order to get rid of the association of ideas, and the lack of accuracy which creeps in unnoticed, and therefore not allowed for, when ordinary language is used. In this connection special mention must be made of an Italian mathematician, Peano, of Turin, to whom we are indebted for various interesting notes on other points.

⟦4⟧ Summing up all these developments in the phrase, *the arithmetizing of mathematics*, I pass on to consider the influence of the tendency here described on parts of the science outside the range of analysis proper. Thus, as you see, while voluntarily acknowledging the exceptional importance of the tendency, I do not grant that the arithmetized science is the essence of mathematics; and my remarks have therefore the twofold character of positive approbation, and negative disapproval. For since I consider that the essential point is not the mere putting of the argument into the arithmetical form, but the more rigid logic obtained by means of this form, it seems to me desirable—and this is the positive side of my thesis—to subject the remaining divisions of mathematics to a fresh investigation based on the arithmetical foundation of analysis. On the other hand I have to point out most emphatically—and this is the negative part of my task—that it is not possible to treat mathematics exhaustively by the method of logical deduction alone, but that, even at the present time, intuition has its special province. For the sake of completeness I ought also to deal with the algorithmic side of mathematics, discussing the importance of symbolic methods; but as this subject does not appeal to me personally, I shall not enter upon it. It must be understood that I have not much that is new to say on special points; my object is rather to collect and arrange material already familiar, justifying its existence where necessary.

⟦5⟧ In the short time at my disposal I must content myself with presenting the most important points; I begin therefore by tracing the relation of the positive part of my thesis to the domain of geometry. The arithmetizing of mathematics began originally, as I pointed out, by ousting space intuition; the first problem that confronts us as we turn to geometry is therefore that of reconciling the results obtained by arithmetical methods with our conception of space. By this I mean that we accept the ordinary principles of analytical geometry, and try to find from these the geometrical interpretation of the more modern analytical developments. This problem, while presenting no special difficulty, has yet many ramifications, as I have had the opportunity of showing during the past year in a seminar devoted to this subject. The net result is, on the one hand, a refinement of the process of space intuition; and on the other, a clearer view of the analytical results considered, with the consequent elimination of the paradoxical character that is otherwise apt to attach itself to them. What is the most general idea of a curve, of a surface? What is meant by saying of a curve, etc., that it is 'analytic' or 'non-analytic'? These and similar questions must be thoroughly sifted and clearly explained. The next point is that we must subject the fundamental principles of geometry to a fresh investigation. As far as the theory of the matter is concerned, this might very well be done, as it was originally, on purely geometrical lines; but in practice on account of the overwhelming complications that present themselves, recourse must be had to the processes of analysis, that is to the methods of analytical geometry. The investigation of the formulæ by means of which we represent the different forms in space (that is, the so-called non-Euclidian geometry, and all that is connected with it) disposes of only one side, and that the more obvious one, of the inquiry; there still remains the more important question: What justification have we for regarding the totality of points in space as a number-manifold in which we interpolate the irrational numbers in the usual manner between the rational numbers arranged in three directions? We ultimately perceive that space intuition is an inexact conception, and that in order that we may subject it to mathematical treatment, we idealize it by means of the so-called axioms, which actually serve as postulates. Kerry, who died at an early age, dealt with the philosophical side of these questions, and I agree with his results in the main and especially as regards his criticism of DuBois Reymond. Conversely this fresh determination of our conception of space has in its turn given rise to new refinements of our analytical ideas. We picture before us in space an infinite number of points and forms composed of them; from this idea have sprung the fundamental investigations on masses of points and transfinite numbers with which G. Cantor has opened up new spheres of thought to arithmetical science. Finally it is much to be desired that full use should be made of the new point of view in the further exposition of geometry, especially infinitesimal geometry; this result will be most easily attained by treating the subiect analytically. Of course, I do not mean by this a blind calculation with x, y, and z, but merely a subsidiary use of these quantities when the question concerns the precise determination of boundary conditions.

⟦6⟧ From this outline of the new geometrical programme you see that it differs greatly from any that was accepted during the first half of this century, when the prevailing tendencies led to the development of projective geometry, which has long been established as a permanent constituent of our subject. Projective geometry has opened up for us with the greatest ease many new tracts of country in our science, and has been rightly called a royal road to its own particular branch of learning; our new road is on the contrary arduous and thorny, and unremitting care is needed to clear a way through the obstacles. It leads us back to what is more nearly the geometry of the ancients, and in the light of our modern ideas we learn to understand precisely the true nature of the latter, as Zeuthen has lately shown in the most brilliant manner.

⟦7⟧ Moreover we must introduce the same process of reasoning into mechanics and mathematical physics. To avoid going too much into detail I will merely explain this by two examples. Throughout applied mathematics, as in the case of space intuition, we must idealize natural objects before we can use them for purposes of mathematical argument; but we find continually that in one and the same subject we may idealize objects in different ways, according to the purpose that we have in view. To mention only a single instance, we treat matter either as continuous throughout space, or as made up of separate molecules, which we may consider to be either at rest or in rapid motion. How and to what degree are these different hypotheses equivalent in regard to the mathematical consequences that can be deduced from them? The earlier expositions of Poisson and others, as also the developments of the Kinetic Theory of Gases, are not sufficiently thorough in this respect for the modern mathematician; the problem requires to be investigated afresh *ab initio*. I expect that a publication by Boltzmann, which will shortly appear, will contain some interesting conclusions on this subject.

⟦8⟧ Another question is this: Practical physics provides us plentifully with experimental results, which we unconsciously generalize and adopt as theorems about the idealized objects. The existence of the so-called Green's function on any closed surface with an arbitrarily chosen pole, corresponding to the fact in electricity that a conductor under the influence of a charged point is in a state of electrical equilibrium, belongs to this category; as also the theorem that every finite elastic body is capable of an infinite series of harmonic oscillations, and my deduction of the fundamental propositions of Riemann's Theory of Abelian Functions from our knowledge of the electric currents started on any conductor when the poles of a galvanic battery are applied to it. Are these indeed, taken in the abstract, exact mathematical theorems, or how must they be limited and defined in order that they may become so? Mathematicians have successfully sought to investigate this; first, C. Neumann and Schwarz with their theory of Potential, and later the French school, following on the lines of the German, with the result that the theorems taken from physics have been shown to hold good to a very considerable extent. You see here what is the precise object of these renewed investigations; not any new physical insight, but abstract mathematical argument in itself, on account of the clearness and precision which will thereby be added to our view of the experimental facts. If I may use an expression of Jacobi's in a somewhat modified sense, it is merely a question of intellectual integrity, 'die Ehre des menschlichen Geistes' ⟦'the honour of the human spirit'⟧.

⟦9⟧ After expressing myself thus it is not easy, without running counter to the foregoing conclusions, to secure to intuition her due share in our science; and yet it is exactly on this antithesis that the point of my present statements depends. I am now thinking not so much of the cultivated intuition just discussed, which has been developed under the influence of logical deduction and might almost be called a form of memory; but rather of the naïve intuition, largely a natural gift, which is unconsciously increased by minute study of one branch or other of the science. The word intuition ⟦Anschauung⟧ is perhaps not well chosen; I mean it to include that instinctive feeling for the proportion of the moving parts with which the engineer criticizes the distribution of power in any piece of mechanism he has constructed; and even that indefinite conviction the practiced calculator possesses as to the convergence of any infinite process that lies before him. I maintain that mathematical intuition—so understood—is always far in advance of logical reasoning and covers a wider field.

⟦10⟧ I might now introduce a historical excursus, showing that in the development of most of the branches of our science, intuition was the starting-point, while logical treatment followed. This holds in fact, not only of the origin of the infinitesimal calculus as a whole, but also of many subjects that have come into existence only in the present century. For example, I may remind you of Riemann's Theory of the Functions of a Complex Variable; and I am glad to add also that the Theory of Numbers, a subject which for a long time seemed to be most unsuited for intuitive methods of treatment, appears to have received a fresh impetus from the application of intuition in the hands of Minkowski and others. After this it would be a matter of great interest to trace from the present standpoint the development, not of particular mathematical subjects, but of the individual mathematician; but in regard to this it must suffice to mention that the two most active mathematical investigators of the present day, Lie in Leipzig, and Poincaré in Paris, both originally made use of intuitive methods. But all this, if I pursued it further, would lead us too much into detail, and finally bring us only to particular cases. I prefer to sketch the everyday results of this somewhat refined intuition, as regards the quantitative, rather than the merely arithmetical or constructive, treatment of physical or technical problems. Let me refer again to the two examples from the theory of electricity already adduced; any physicist would be able to trace, without further difficulty, and with tolerable accuracy, the form of the surface of Green's function, or, in the second experiment, the shape of the lines of force in a given case. Again, consider any given differential equation, I will say, to take the most simple instance, a differential equation of the first order in two variables. Most probably the analytical method of solution fails; nevertheless we can at once find graphically the general form of the integral curves, as has recently been done for the renowned differential equation of the Problem of Three Bodies by Lord Kelvin, a master in the art of mathematical intuition. The question in all such cases, to use the language of analysis, is one of interpolation, in which less stress is laid on exactness in particular details than on a consideration of the general conditions. I will once more emphasize the fact that in stating all our laws of nature, or in trying to formulate mathematically any actual occurrence, the art lies in making a similar use of interpolation; for we have to consider the simple laws connecting the essential quantities, apart from the multitude of fortuitous disturbances. This is ultimately what I have termed above the process of idealization. Logical investigation is not in place until intuition has completed the task of idealization.

⟦11⟧ I beg that you will consider these remarks as a description, not as an explanation, of what actually occurs. The mathematician can do no more than state the character of each particular psychical operation from observations of his own mental process. Perhaps some day physiology and experimental psychology will enable us to draw more accurate conclusions as to the relation between the processes of intuition and those of logical thought. The great differences shown by observations of different individuals confirm the supposition that it is indeed a question of distinct, that is, not necessarily connected, mental activities. Modern psychologists distinguish between visual, motor, and auditory endowments; mathematical intuition, as above defined, appears to belong more closely to the first two classes, and the logical method to the third class. In common with many of my fellow mathematicians I gladly welcome these investigations which psychologists have only just undertaken, for it is to be hoped that with the increase of accurate information about the psychological conditions of mathematical thought and their particular varieties, many of the differences of opinion which necessarily remain unsettled at present will disappear.

⟦12⟧ I must add a few words on mathematics from the point of view of pedagogy. We observe in Germany at the present day a very remarkable condition of affairs in this respect; two opposing currents run side by side without affecting one another appreciably. Among the teachers in our *Gymnasia* the need of mathematical instruction based on intuitive methods has now been so strongly and universally emphasized that one is compelled to enter a protest, and vigorously insist on the necessity for strict logical treatment. This is the central thought of a small pamphlet on elementary geometrical problems which I published last summer. Among the university professors of our subject exactly the reverse is the case; intuition is frequently not only undervalued, but as much as possible ignored. This is doubtless a consequence of the intrinsic importance of the arithmetizing tendency in modern mathematics. But the result reaches far beyond the mark. It is high time to assert openly once for all that this implies, not only a false pedagogy, but also a distorted view of the science. I gladly yield the utmost freedom to the preferences of individual academic teachers, and have always discouraged the laying-down of general rules for higher mathematical teaching, but this shall not prevent me from saying that two classes at least of mathematical lectures must be based on intuition; the elementary lectures which actually introduce the beginner to higher mathematics—for the scholar must naturally follow the same course of development on a smaller scale, that the science itself has taken on a larger—and the lectures which are intended for those whose work is largely done by intuitive methods, namely, natural scientists and engineers. Through this one-sided adherence to logical form we have lost among these classes of men much of the prestige properly belonging to mathematics, and it is a pressing and urgent duty to regain this prestige by judicious treatment.

⟦13⟧ To return to theoretical considerations, the general views which I uphold in regard to the present problems of mathematical science need scarcely be specially formulated. While I desire in every case the fullest logical working out of the material, yet I demand at the same time an intuitive grasp and investigation of the subject from all sides. Mathematical developments originating in intuition must not be considered actual constituents of the science till they have been brought into a strictly logical form. Conversely, the mere abstract statement of logical relations cannot satisfy us until the extent of their application to every branch of intuition is vividly set forth, and we recognize the manifold connections of the logical scheme, depending on the branch which we have chosen, to the other divisions of our knowledge. The science of mathematics may be compared to a tree thrusting its roots deeper and deeper into the earth and freely spreading out its shady branches to the air. Are we to consider the roots or the branches as its essential part? Botanists tell us that the question is badly framed, and that the life of the organism depends on the mutual action of its different parts.

22
Jules Henri Poincaré (1854–1912)

Henri Poincaré, Professor of Mathematics at the University of Paris from 1881 until his death, was the most prolific and influential mathematician of his generation. Although he was only 58 when he died, his technical articles fill eleven large volumes; he also published four volumes of philosophical essays on the foundations of mathematics and physics. His technical writings cover large tracts of pure and applied mathematics: algebraic topology, non-Euclidean geometry, Lie groups, number theory, differential equations, the theory of functions, algebraic geometry, mathematical physics, celestial mechanics. Indeed, Poincaré virtually created several new mathematical disciplines from the whole cloth. He discovered and brought to maturity the theory of automorphic functions of a single complex variable; his studies of homotopy and homology virtually founded the subject of algebraic topology; his work in differential equations was the most profound in over a century. Since his time only Hilbert has exhibited comparable breadth and originality and has had as much influence: between them, they laid the groundwork for much of twentieth-century mathematics.

Despite his importance, the secondary literature on Poincaré is sparse. *Mooij 1966* and *Folina 1992* appear to be the only monographs in existence on his philosophy of mathematics. *Hadamard 1922* and *1933* give a general account of his mathematical work; there is also important supplementary material on Poincaré in volumes 2 and 11 of his *Œuvres*, and in *Dieudonné 1975*. For an illuminating account of Poincaré's controversy with Cantor, Russell, and Hilbert, see *Goldfarb 1988*.

A. ON THE NATURE OF MATHEMATICAL REASONING (*POINCARÉ 1894*)

The following article is Poincaré's first major exposition of his philosophical views. Foreshadowing his later controversies with the logicists and with Hilbert, Poincaré argues that mathematics cannot be reduced to syllogistic reasoning, but that it relies on intuition: specifically, on the sort of intuition exemplified by the principle of mathematical induction. (Poincaré does not claim that *all* of mathematics can be reduced to applications of this one principle; induction is offered here merely as the most important example of an intuitive principle.) Poincaré's term 'intuition' is not very precise, and, despite the deliberate Kantian overtones, is only distantly related to Kant's *Anschauung*; certainly the heavy philosophical machinery of the Kantian system is absent. But the two thinkers agree that intuition is distinct from sensibility, and that mathematics belongs to the realm of the synthetic *a priori*—a doctrine which Poincaré advocates by arguing that the principle of induction can be derived neither from logic nor from experience.

The translation is by George Bruce Halsted, and follows the revised text published in *Poincaré 1902*. References should be to the section numbers, which appeared in the original version.

I

The very possibility of the science of mathematics seems an insoluble contradiction. If this science is deductive only in appearance, whence does it derive that perfect rigour no one dreams of doubting? If, on the contrary, all the propositions it enunciates can be deduced one from another by the rules of formal logic, why is not mathematics reduced to an immense tautology? The syllogism can teach us nothing essentially new, and, if everything is to spring from the principle of identity, everything should be capable of being reduced to it. Shall we then admit that the enunciations of all those theorems which fill so many volumes are nothing but devious ways of saying A is A?

Without doubt, we can go back to the axioms, which are at the source of all these reasonings. If we decide that these cannot be reduced to the principle of contradiction, if still less we see in them experimental facts which could not partake of mathematical necessity, we have yet the resource of classing them among synthetic *a priori* judgements. This is not to solve the difficulty, but only to baptize it; and even if the nature of synthetic judgements were for us no mystery, the contradiction would not have disappeared, it would only have moved back; syllogistic reasoning remains incapable of adding anything to the data given it; these data reduce themselves to a few axioms, and we should find nothing else in the conclusions.

No theorem could be new if no new axiom intervened in its demonstration; reasoning could give us only the immediately evident verities borrowed from direct intuition; it would be only an intermediary parasite, and therefore should we not have good reason to ask whether the whole syllogistic apparatus did not serve solely to disguise our borrowing?

The contradiction will strike us the more if we open any book on mathematics; on every page the author will announce his intention of generalizing some proposition already known. Does the mathematical method proceed from the particular to the general, and, if so, how then can it be called deductive?

If finally the science of number were purely analytic, or could be analytically derived from a small number of synthetic judgements, it seems that a mind sufficiently powerful could at a glance perceive all its truths; nay more, we might even hope that some day one would invent to express them a language sufficiently simple to have them appear self-evident to an ordinary intelligence.

If we refuse to admit these consequences, it must be conceded that mathematical reasoning has of itself a sort of creative virtue, and consequently differs from the syllogism.

The difference must even be profound. We shall not, for example, find the key to the mystery in the frequent use of that rule according to which one and the same uniform operation applied to two equal numbers will give identical results.

All these modes of reasoning, whether or not they be reducible to the syllogism properly so called, retain the analytic character, and just because of that are powerless.

II

The discussion is old; Leibniz tried to prove 2 and 2 make 4; let us look a moment at his demonstration.

I will suppose the number 1 defined and also the operation $x + 1$ which consists in adding unity to a given number x.

These definitions, whatever they be, do not enter into the course of the reasoning.

I define then the numbers 2, 3, and 4 by the equalities

$$(1)\ 1 + 1 = 2; \quad (2)\ 2 + 1 = 3; \quad (3)\ 3 + 1 = 4.$$

In the same way, I define the operation $x + 2$ by the relation:

$$(4)\ x + 2 = (x + 1) + 1.$$

That presupposed, we have

$$\begin{aligned} 2 + 1 + 1 &= 3 + 1 && \text{(Definition 2)}, \\ 3 + 1 &= 4 && \text{(Definition 3)}, \\ 2 + 2 &= (2 + 1) + 1 && \text{(Definition 4)}, \end{aligned}$$

whence

$$2 + 2 = 4 \quad \text{Q.E.D.}$$

It cannot be denied that this reasoning is purely analytic. But ask any mathematician: 'That is not a demonstration properly so called', he will say to you: 'that is a verification'. We have confined ourselves to comparing two purely conventional definitions and have ascertained their identity; we have learned nothing new. *Verification* differs from true demonstration precisely because it is purely analytic and because it is sterile. It is sterile because the conclusion is nothing but the premisses translated into another language. On the contrary, true demonstration is fruitful because the conclusion here is in a sense more general than the premisses.

The equality $2+2=4$ is thus susceptible of a verification only because it is particular. Every particular statement in mathematics can always be verified in this same way. But if mathematics could be reduced to a series of such verifications, it would not be a science. So a chess-player, for example, does not create a science in winning a game. There is no science apart from the general.

It may even be said the very object of the exact sciences is to spare us these direct verifications.

III

Let us, therefore, see the geometer at work and seek to catch his process.

The task is not without difficulty; it does not suffice to open a work at random and analyse any demonstration in it.

We must first exclude geometry, where the question is complicated by arduous problems relative to the role of the postulates, to the nature and the origin of the notion of space. For analogous reasons we cannot turn to the infinitesimal analysis. We must seek mathematical thought where it has remained pure, that is, in arithmetic.

A choice still is necessary; in the higher parts of the theory of numbers, the primitive mathematical notions have already undergone an elaboration so profound that it becomes difficult to analyse them.

It is, therefore, at the beginning of arithmetic that we must expect to find the explanation we seek, but it happens that precisely in the demonstration of the most elementary theorems the authors of the classic treatises have shown the least precision and rigour. We must not impute this to them as a crime; they have yielded to a necessity; beginners are not prepared for real mathematical rigour; they would see in it only useless and irksome subtleties; it would be a waste of time to try prematurely to make them more exacting; they must pass over rapidly, but without skipping stations, the road traversed slowly by the founders of the science.

Why is so long a preparation necessary to become habituated to this perfect rigour, which, it would seem, should naturally impress itself upon all good minds? This is a logical and psychological problem well worthy of study.

But we shall not take it up; it is foreign to our purpose; all I wish to insist on is that, not to fail of our purpose, we must recast the demonstrations of the most elementary theorems and give them, not the crude form in which they are left, so as not to harass beginners, but the form that will satisfy a skilled geometer.

DEFINITION OF ADDITION.—I suppose already defined the operation $x + 1$, which consists in adding the number 1 to a given number x.

This definition, whatever it be, does not enter into our subsequent reasoning.

We now have to define the operation $x + a$, which consists in adding the number a to a given number x.

Supposing we have defined the operation

$$x + (a - 1),$$

the operation $x + a$ will be defined by the equality

$$x + a = [x + (a - 1)] + 1. \tag{1}$$

We shall know then what $x + a$ is when we know what $x + (a - 1)$ is, and as I have supposed that to start with we knew what $x + 1$ is, we can define successively and 'by recurrence' the operations $x + 2$, $x + 3$, etc.

This definition deserves a moment's attention; it is of a particular nature which already distinguishes it from the purely logical definition; the equality (1) contains an infinity of distinct definitions, each having a meaning only when one knows the preceding.

PROPERTIES OF ADDITION.—*Associativity*.—I say that

$$a + (b + c) = (a + b) + c.$$

In fact the theorem is true for $c = 1$; it is then written

$$a + (b + 1) = (a + b) + 1,$$

which, apart from the difference of notation, is nothing but the equality (1), by which I have just defined addition.

Supposing the theorem true for $c = \gamma$, I say it will be true for $c = \gamma + 1$. In fact, supposing

$$(a + b) + \gamma = a + (b + \gamma),$$

it follows that

$$[(a + b) + \gamma] + 1 = [a + (b + \gamma)] + 1$$

or by definition (1)

$$(a + b) + (\gamma + 1) = a + (b + \gamma + 1) = a + [b + (\gamma + 1)],$$

which shows, by a series of purely analytic deductions, that the theorem is true for $\gamma + 1$.

Being true for $c = 1$, we thus see successively that so it is for $c = 2$, for $c = 3$, etc.

Commutativity.—1° I say that

$$a + 1 = 1 + a.$$

The theorem is evidently true for $a = 1$; we can *verify* by purely analytic reasoning that if it is true for $a = \gamma$ it will be true for $a = \gamma + 1$; for then

$$(\gamma + 1) + 1 = (1 + \gamma) + 1 = 1 + (\gamma + 1);$$

now it is true for $a = 1$, therefore it will be true for $a = 2$, for $a = 3$, etc., which is expressed by saying that the enunciated proposition is demonstrated by recurrence.

2° I say that

$$a + b = b + a.$$

The theorem has just been demonstrated for $b = 1$; it can be *verified* analytically that if it is true for $b = \beta$, it will be true for $b = \beta + 1$.

The proposition is therefore established by recurrence.

DEFINITION OF MULTIPLICATION.—We shall define multiplication by the equalities:

$$a \times 1 = a. \tag{1}$$

$$a \times b = [a \times (b - 1)] + a. \tag{2}$$

Like equality (1), equality (2) contains an infinity of definitions; having defined $a \times 1$, it enables us to define successively: $a \times 2$, $a \times 3$, etc.

PROPERTIES OF MULTIPLICATION.—*Distributivity*.—I say that

$$(a + b) \times c = (a \times c) + (b \times c).$$

We verify analytically that the equality is true for $c = 1$; then that if the theorem is true for $c = \gamma$, it will be true for $c = \gamma + 1$.

The proposition is, therefore, demonstrated by recurrence.

Commutatitvity.—1° I say that

$$a \times 1 = 1 \times a.$$

The theorem is evident for $a = 1$.

We verify analytically that if it is true for $a = \alpha$, it will be true for $a = \alpha + 1$.

2° I say that

$$a \times b = b \times a.$$

The theorem has just been proved for $b = 1$. We could verify analytically that if it is true for $b = \beta$, it will be true for $b = \beta + 1$.

IV

Here I stop this monotonous series of reasonings. But this very monotony has the better brought out the procedure, which is uniform and is met again at each step.

This procedure is the demonstration by recurrence. We first establish a theorem for $n = 1$; then we show that if it is true of $n = 1$, it is true of n, and thence conclude that it is true for all the whole numbers.

We have just seen how it may be used to demonstrate the rules of addition and multiplication, that is to say, the rules of the algebraic calculus; this calculus is an instrument of transformation, which lends itself to many more differing combinations that does the simple syllogism; but it is still an instrument purely analytic, and incapable of teaching us anything new. If mathematics had no other instrument, it would therefore be forthwith arrested in its development; but it has recourse anew to the same procedure, that is, to reasoning by recurrence, and it is able to continue its forward march.

If we look closely, at every step we meet again this mode of reasoning, either in the simple form we have just given it, or under a form more or less modified.

Here then we have the mathematical reasoning *par excellence*, and we must examine it more closely.

V

The essential characteristic of reasoning by recurrence is that it contains, condensed, so to speak, in a single formula, an infinity of syllogisms.

That this may the better be seen, I will state one after another these syllogisms which are, if you will allow me the expression, arranged in 'cascade'.

These are of course hypothetical syllogisms.

The theorem is true of the number 1.

Now, if it is true of 1, it is true of 2.

Therefore it is true of 2.

Now, if it is true of 2, it is true of 3.

Therefore it is true of 3, and so on.

We see that the conclusion of each syllogism serves as minor to the following.

Furthermore the majors of all our syllogisms can be reduced to a single formula.

If the theorem is true of $n - 1$, so it is of n.

We see, then, that in reasoning by recurrence we confine ourselves to stating the minor of the first syllogism, and the general formula which contains as particular cases all the majors.

This never-ending series of syllogisms is thus reduced to a phrase of a few lines.

It is now easy to comprehend why every particular consequence of a theorem can, as I have explained above, be verified by purely analytic procedures.

If instead of showing that our theorem is true of all numbers, we only wish to show it true of the number 6, for example, it will suffice for us to establish the first 5 syllogisms of our cascade; 9 would be necessary if we wished to prove the theorem for the number 10; more would be needed for a larger number; but, however great this number might be, we should always end by reaching it, and the analytic verification would be possible.

And yet, however far we thus might go, we could never rise to the general theorem, applicable to all numbers, which alone can be the object of science. To reach this, an infinity of syllogisms would be necessary; it would be necessary to overleap an abyss that the patience of the analyst, restricted to the resources of formal logic alone, never could fill up.

I asked at the outset why one could not conceive of a mind sufficiently powerful to perceive at a glance the whole body of mathematical truths.

The answer is now easy; a chess-player is able to combine four moves, five moves, in advance, but, however extraordinary he may be, he will never prepare more than a finite number of them; if he applies his faculties to arithmetic, he will not be able to perceive its general truths by a single direct intuition; to arrive at the smallest theorem he cannot dispense with the aid of reasoning by recurrence, for this is an instrument which enables us to pass from the finite to the infinite.

This instrument is always useful, for, allowing us to overleap at a bound as many stages as we wish, it spares us verifications, long, irksome, and monotonous, which would quickly become impracticable. But it becomes indispensable as soon as we aim at the general theorem, to which analytic verification would bring us continually nearer without ever enabling us to reach it.

In this domain of arithmetic, we may think ourselves very far from the infinitesimal analysis, and yet, as we have just seen, the idea of the mathematical infinite already plays a preponderant role, and without it there would be no science, because there would be nothing general.

VI

The judgement on which reasoning by recurrence rests can be put under other forms; we may say, for example, that in an infinite collection of different whole numbers there is always one which is less than all the others.

We can easily pass from one statement to the other and thus get the illusion of having demonstrated the legitimacy of reasoning by recurrence. But we shall always be arrested, we shall always arrive at an undemonstrable axiom which will be in reality only the proposition to be proved translated into another language.

We cannot therefore escape the conclusion that the rule of reasoning by recurrence is irreducible to the principle of contradiction.

Neither can this rule come to us from experience; experience could teach us that the rule is true for the first ten or hundred numbers; for example, it cannot attain to the indefinite series of numbers, but only to a portion of this series, more or less long but always limited.

Now if it were only a question of that, the principle of contradiction would suffice; it would always allow of our developing as many syllogisms as we wished; it is only when it is a question of including an infinity of them in a single formula, it is only before the infinite that this principle fails, and there too, experience becomes powerless. This rule, inaccessible to analytic demonstration and to experience, is the veritable type of the synthetic *a priori* judgement. On the other hand, we cannot think of seeing in it a convention, as in some of the postulates of geometry.

Why then does this judgement force itself upon us with an irresistible evidence? It is because it is only the affirmation of the power of the mind which knows itself capable of conceiving the indefinite repetition of the same act when once this act is possible. The mind has a direct intuition of this power, and experience can only give occasion for using it and thereby becoming conscious of it.

But, one will say, if raw experience cannot legitimate reasoning by recurrence, is it so of experiment aided by induction? We see successively that a theorem is true of the number 1, of the number 2, of the number 3 and so on; the law is evident, we say, and it has the same warranty as every physical law based on observations, whose number is very great but limited.

Here is, it must be admitted, a striking analogy with the usual procedures of induction. But there is an essential difference. Induction applied to the physical sciences is always uncertain, because it rests on the belief in a general order of the universe, an order outside us. Mathematical induction, that is, demonstration by recurrence, on the contrary, imposes itself necessarily because it is only the affirmation of a property of the mind itself.

VII

Mathematicians, as I have said before, always endeavour to *generalize* the propositions they have obtained, and, to seek no other example, we have just proved the equality:

$$a + 1 = 1 + a$$

and afterwards used it to establish the equality

$$a + b = b + a$$

which is manifestly more general.

Mathematics can, therefore, like the other sciences, proceed from the particular to the general.

This is a fact which would have appeared incomprehensible to us at the outset of this study, but which is no longer mysterious to us, since we have ascertained the analogies between demonstration by recurrence and ordinary induction.

Without doubt recurrent reasoning in mathematics and inductive reasoning in physics rest on different foundations, but their march is parallel, they advance in the same sense, that is to say, from the particular to the general.

Let us examine the case a little more closely.

To demonstrate the equality

$$a + 2 = 2 + a$$

it suffices to twice apply the rule

$$a + 1 = 1 + a \tag{1}$$

and write

$$a + 2 = a + 1 + 1 = 1 + a + 1 = 1 + 1 + a = 2 + a. \tag{2}$$

The equality (2) thus deduced in a purely analytic way from the equality (1) is, however, not simply a particular case of it; it is something quite different.

We cannot therefore even say that in the really analytic and deductive part of mathematical reasoning we proceed from the general to the particular in the ordinary sense of the word.

The two members of the equality (2) are simply combinations more complicated than the two members of the equality (1), and analysis only serves to separate the elements which enter into these combinations and to study their relations.

Mathematicians proceed therefore 'by construction', they 'construct' combinations more and more complicated. Coming back then by the analysis of these combinations, of these aggregates, so to speak, to their primitive elements, they perceive the relations of these elements and from them deduce the relations of the aggregates themselves.

This is a purely analytical proceeding, but it is not, however, a proceeding from the general to the particular, because evidently the aggregates cannot be regarded as more particular than their elements.

Great importance, and justly, has been attached to this procedure of 'construction', and some have tried to see in it the necessary and sufficient condition for the progress of the exact sciences.

Necessary, without doubt; but sufficient, no.

For a construction to be useful and not a vain toil for the mind, that it may serve as stepping-stone to one wishing to mount, it must first of all possess a sort of unity enabling us to see in it something besides the juxtaposition of its elements.

Or, more exactly, there must be some advantage in considering the construction rather than its elements themselves.

What can this advantage be?

Why reason on a polygon, for instance, which is always decomposable into triangles, and not on the elementary triangles?

It is because there are properties appertaining to polygons of any number of sides and that may be immediately applied to any particular polygon.

Usually, on the contrary, it is only at the cost of the most prolonged exertions that these could be found by studying directly the relations of the elementary triangles. The knowledge of the general theorem spares us these efforts.

A construction, therefore, becomes interesting only when it can be ranged beside other analogous constructions, forming species of the same genus.

If the quadrilateral is something besides the juxtaposition of two triangles, this is because it belongs to the genus polygon.

Moreover, one must be able to demonstrate the properties of the genus without being forced to establish them successively for each of the species.

To attain that, we must necessarily mount from the particular to the general, ascending one or more steps.

The analytic procedure 'by construction' does not oblige us to descend, but it leaves us at the same level.

We can ascend only by mathematical induction, which alone can teach us something new. Without the aid of this induction, different in certain respects from physical induction, but quite as fertile, construction would be powerless to create science.

Observe finally that this induction is possible only if the same operation can be repeated indefinitely. That is why the theory of chess can never become a science, for the different moves of the same game do not resemble one another.

B. ON THE FOUNDATIONS OF GEOMETRY (*POINCARÉ 1898*)

Poincaré wrote extensively on the foundations of geometry. His most influential philosophical articles deal with non-Euclidean geometry and the role of geometric hypotheses in the physical sciences; these famous essays, gathered together in *Poincaré 1902*, *1905a*, *1908*, and *1913a*, exerted a considerable force on early twentieth-century philosophy of science, and in particular on discussions, by Carnap, Reichenbach, and others, of the theory of relativity. The following article is less well known. It deals with the foundations of geometry from the point of view of groups of transformations, and follows in the footsteps of Klein and Lie rather than of Riemann and Lobatchevsky. The article was first published in *The Monist*, in a translation from Poincaré's manuscript by T.J. McCormack; that translation is reprinted here. References should be to the titles of the sections, which appeared in the original version. (Roman numerals for the sections have also been added in this edition.)

Although I have already had occasion to set forth my views on the foundations of geometry,[1] it will not, perhaps, be unprofitable to revert to the question with new and ampler developments, and seek to clear up certain points which the reader may have found obscure. It is with reference to the definition of the point and the determination of the number of dimensions that new light appears to me most needed; but I deem it opportune, nevertheless, to take up the question from the beginning.

[I] SENSIBLE SPACE

Our sensations cannot give us the notion of space. That notion is built up by the mind from elements which pre-exist in it, and external experience is simply

[1] Both in the *Revue Générale des Sciences* and in the *Revue de Métaphysique et de Morale*.

the occasion for its exercising this power, or at most a means of determining the best mode of exercising it.

Sensations by themselves have no spatial character.

This is evident in the case of isolated sensations—for example, visual sensations. What could a man see who possessed but a single immovable eye? Different images would be cast upon different points of his retina, but would he be led to classify these images as we do our present retinal sensations?

Suppose images formed at four points *A*, *B*, *C*, *D* of this immovable retina. What ground would the possessor of this retina have for saying that, for example, the distance *AB* was equal to the distance *CD*? We, constituted as we are, have a reason for saying so, because we know that a *slight* movement of the eye is sufficient to bring the image which was at *A* to *C*, and the image which was at *B* to *D*. But these slight movements of the eye are impossible for our hypothetical man, and if we should ask him whether the distance *AB* was equal to the distance *CD*, we should seem to him as ridiculous as would a person appear to us who should ask us whether there was more difference between an olfactory sensation and a visual sensation than between an auditive sensation and a tactual sensation.

But this is not all. Suppose that two points *A* and *B* are very near to each other, and that the distance *AC* is very great. Would our hypothetical man be cognizant of the difference? We perceive it, we who can move our eyes, because a very slight movement is sufficient to cause an image to pass from *A* to *B*. But for him the question whether the distance *AB* was very small as compared with the distance *AC* would not only be insoluble, but would be devoid of meaning.

The notion of the contiguity of two points, accordingly, would not exist for our hypothetical man. The rubric, or category, under which he would arrange his sensations, if he arranged them at all, would consequently not be the space of the geometer and would probably not even be continuous, since he could not distinguish small distances from large. And even if it were continuous, it could not, as I have abundantly shown elsewhere, be either homogeneous, isotropic, or tridimensional.

It is needless to repeat for the other senses what I have said for sight. Our sensations differ from one another qualitatively, and they can therefore have no common measure, no more than can the gramme and the metre. Even if we compare only the sensations furnished by the same nerve-fibre, considerable effort of the mind is required to recognize that the sensation of to-day is of the same kind as the sensation of yesterday, but greater or smaller; in other words, to classify sensations according to their character, and then to arrange those of the same kind in a sort of scale, according to their intensity. Such a classification cannot be accomplished without the active intervention of the mind, and it is the object of this intervention to refer our sensations to a sort of rubric or category pre-existing in us.

Is this category to be regarded as a 'form of our sensibility'? No, not in the sense that our sensations, individually considered, could not exist without it. It becomes necessary to us only for comparing our sensations, for reasoning upon our sensations. It is therefore rather a form of our understanding.

This, then, is the first category to which our sensations are referred. It can be represented as composed of a large number of scales absolutely independent of one another. Further, it simply enables us to compare sensations of the same kind and not to measure them, to perceive that one sensation is greater than another sensation, but not that it is twice as great or three times as great.

How much such a category differs from the space of the geometer! Shall we say that the geometer admits a category of quite the same kind, where he employs three scales such as the three axes of co-ordinates? But in our category we have not three scales only, but as many as there are nerve-fibres. Further, our scales appear to us as so many separate worlds fundamentally distinct, while the three axes of geometry all fulfil the same office and may be interchanged one for another. In fine, the co-ordinates are susceptible of being measured and not simply of being compared. Let us see, therefore, how we can rise from this rough category which we may call sensible space to geometric space.

⟦II⟧ The feeling of direction

It is frequently said that certain of our sensations are always accompanied by a peculiar feeling of direction, which gives to them a geometric character. Such are visual and muscular sensations. Others on the contrary, like the sensations of smell and taste, are not accompanied by this feeling, and consequently are void of any geometric character whatever. On this theory the notion of direction would be pre-existent to all visual and muscular sensations and would be the underlying condition of the same.

I am not of this opinion; and let us first ask if the feeling of direction really forms a constituent part of the sensation. I cannot very well see how there can be anything else *in* the sensation than the sensation itself. And be it further observed that the same sensation may, according to circumstances, excite the feeling of different directions. Whatever be the position of the body, the contraction of the *same* muscle, the biceps of the right arm, for example, will always provoke the *same* muscular sensation; and yet, through being apprised by other concomitant sensations that the position of the body has changed, we also know perfectly well that the direction of the motion has changed.

The feeling of direction, accordingly, is not an integrant part of the sensation, since it can vary without the sensation being varied. All that we can say is that the feeling of direction is associated with certain sensations. But what does this signify? Do we mean by it that the sensation is associated with a certain indescribable something which we can represent to ourselves but which is still not a sensation? No, we mean simply that the various sensations which correspond to the same direction are associated *with one another*, and that one of them calls forth the others in obedience to the ordinary laws of association of ideas. Every association of ideas is a product of habit merely, and it would be necessary for us to discover how the habit was formed.

But we are still far from geometric space. Our sensations have been classified in a new manner: those which correspond to the same direction are grouped together; those which are isolated and have reference to no direction are not considered. Of the innumerable scales of sensations of which our sensible space was formed some have disappeared, others have been merged into one another. Their number has been diminished.

But the new classification is still not space; it involves no idea of measurement; and, furthermore, the restricted category so reached would not be an isotropic space, that is to say, different directions would not appear to us as fulfilling the same office and as interchangeable with one another. And so this 'feeling of direction' far from explaining space would itself stand in need of explanation.

But will it help us even towards the explanation we seek? No, because the laws of that association of ideas which we call the feeling of direction are extraordinarily complex. As I explained above, the same muscular sensation may correspond to a host of different directions according to the position of the body, which is made known to us by other concomitant sensations. Associations so complex can only be the result of an extremely long process. This, therefore, is not the path which will lead us most quickly to our goal. Therefore we will not regard the feeling of direction as something attained but will revert to the 'sensible space' with which we started.

⟦III⟧ Representation of space

Sensible space has nothing in common with geometric space. I believe that few persons will be disposed to contest this assertion. It would be possible, perhaps, to refine the category which I set up at the beginning of this article, and to construct something which would more resemble geometric space. But whatever concession we might make, the space so constructed would be neither infinite, nor homogeneous, nor isotropic: it could be such only by ceasing to be accessible to our senses.

Seeing that our representations are simply the reproductions of our sensations, therefore, we cannot image geometric space. We cannot represent to ourselves objects in geometric space, but can merely reason upon them as if they existed in that space.

A painter will struggle in vain to construct an object of three dimensions upon canvas. The image which he traces, like his canvas, will never have more than two. When we endeavour, for example, to represent the sun and the planets in space, the best we can do is to represent the visual sensations which we experience when five or six tiny spheres are set revolving in close proximity.

Geometric space, therefore, cannot serve as a category for our representations. It is not a form of our sensibility. It can serve us only in our reasonings. It is a form of our understanding.

⟦IV⟧ Displacement and alteration

We at once perceive that our sensations vary, that our impressions are subject to change. The laws of these variations were the cause of our creating geometry and the notion of geometric space. If our sensations were not variable, there would be no geometry.

But that is not all. Geometry could not have arisen unless we had been led to distribute into two classes the changes which can arise in our impressions. We say, in one case, that our impressions have changed because the objects causing them have undergone some alterations of character, and again that these impressions have changed because the objects have suffered displacement. What is the foundation of this distinction?

A sphere of which one hemisphere is blue and the other red, is rotating before our eyes and shows first a blue hemisphere and then a red hemisphere. Again, a blue liquid contained in a vase suffers a chemical reaction which causes it to turn red. In both cases the impression of blue has given way to the impression of red. Now why is the first of these changes classed among displacements, and the second among alterations? Evidently because in the first case it is sufficient for me merely to go around the globe to bring myself face to face again with the other hemisphere, and so to receive a second time the impression of blue.

An object is displaced before my eye, and its image which was first formed on the centre of the retina is now brought to the edge of the retina. The sensation which was carried to me by a nerve-fibre proceeding from the centre of the retina is succeeded by another which is carried to me by a fibre proceeding from the edge. These sensations are conducted to me by two different nerves. They ought to appear to me different in character, and if they did not, how could I distinguish them?

Why, then, do I come to conclude that the *same* image has been displaced? Is it because one of these sensations frequently succeeds the other? But similar successions are frequent. These it is that produce all our associations of ideas, and we do not ordinarily conclude that they are due to displacement of an object which is invariable in character.

But what happens in this case is that we can *follow the object with the eye*, and by a displacement of our eye which is generally voluntary and accompanied by muscular sensations, we can bring the image back to the centre of the retina and so *re-establish the primitive sensation*. The following, therefore, is my conclusion.

Among the changes which our impressions undergo, we distinguish two classes:

(1) The first are independent of our will and not accompanied by muscular sensations. These are *external changes* so called.

(2) The others are voluntary and accompanied by muscular sensations. We may call these *internal changes*.

We observe next that in certain cases when an external change has modified our impressions, we can, by voluntarily provoking an internal change, re-establish our primitive impressions. The external change, accordingly, can be *corrected* by an internal change. External changes may consequently be subdivided into the two following classes:

1. Changes which are susceptible of being corrected by an internal change. These are *displacements*.

2. Changes which are not so susceptible. These are *alterations*.

An immovable being would be incapable of making this distinction. *Such a being, therefore, could never create geometry*,—even if his sensations were variable, and even if the objects surrounding him were movable.

⟦V⟧ Classification of displacements

A sphere of which one hemisphere is blue and the other red, is rotating before me and presents to me first its blue side and then its red side. I regard this external change as a displacement because I can correct it by an internal change, namely, by going around the sphere. Let us repeat the experiment with another sphere, of which one hemisphere is green and the other yellow. The impression of the yellow hemisphere will succeed that of the green, as before that of the red succeeded that of the blue. For the same reason I shall regard this new external change as a displacement.

But this is not all. I also say that these two external changes are due to the *same* displacement, that is to say, to a rotation. Yet there is no connection between the impression of the yellow hemisphere and that of the red, any more than there is between that of the blue and that of the green, and I have no reason for saying that the same relation exists between the yellow and the green as exists between the red and the blue. No, I say that these two external changes are due to the same displacement because I have 'corrected' them by the same internal change. But how am I to know that the two internal changes by which I corrected first the external change from the blue to the red, then that from the green to the yellow, are to be considered identical? Simply because they have provoked the *same* muscular sensations; and for this it is not necessary for me to know geometry in advance and to represent to myself the movements of my body in geometric space.

Thus several external changes which in themselves have no common relation may be corrected by the same internal change. I collect these into the same class and consider them as the same displacement.

An analogous classification may be made with respect to internal changes. Not all internal changes are capable of correcting an external change. Only those which are may be called displacements. On the other hand the same external change may be corrected by several different internal changes. A person knowing geometry might express this idea by saying that my body can go from the position A to the position B by several different paths. Each of these paths corresponds to a series of muscular sensations; and at present I am cognizant of nothing but these muscular sensations. No two of these series have a common resemblance, and if I consider them nevertheless as representing the *same* displacement, it is because they are capable of correcting the same external change.

The foregoing classification suggests two reflections:

1. The classification is not a crude datum of experience, because the aforementioned compensation of the two changes, the one internal and the other external, is never exactly realized. It is, therefore, an active operation of the mind, which endeavours to insert the crude results of experience into a pre-existing form, into a category. This operation consists in identifying two changes because they possess a common character, and in spite of their not possessing it exactly. Nevertheless, the very fact of the mind's having occasion to perform this operation is due to experience, for experience alone can teach it that the compensation has approximately been effected.

2. The classification further brings us to recognize that two displacements are identical, and it hence results that a displacement can be *repeated* twice or several times. It is this circumstance that introduces number, and that permits measurement where formerly pure quality alone held sway.

⟦VI⟧ Introduction of the Notion of Group

That we are able to go farther is due to the following fact, the importance of which is cardinal.

It is obvious that if we consider a change A, and cause it to be followed by another change B, we are at liberty to regard the *ensemble* of the two changes A followed by B as a single change which may be written $A + B$ and may be called the resultant change. (It goes without saying that $A + B$ is not necessarily identical with $B + A$.) The conclusion is then stated that if the two changes A and B are displacements, the change $A + B$ also is a displacement. Mathematicians express this by saying that *the ensemble, or aggregate, of displacements is a group*. If such were not the case there would be no geometry.

But how do we know that the *ensemble* of displacements is a group? Is it by reasoning *a priori*? Is it by experience? One is tempted to reason *a priori* and to say: if the external change A is corrected by the internal change A', and the external change B by the internal change B', the resulting external change $A + B$ will be corrected by the resulting internal change $B' + A'$. Hence this resulting change is by definition a displacement, which is to say that the *ensemble* of displacements forms a group.

But this reasoning is open to several objections. It is obvious that the changes A and A' compensate each other; that is to say, that if these two changes are made in succession, I shall find again my original impressions,—a result which I might write as follows:

$$A + A' = 0.$$

I also see that $B + B' = 0$. These are hypotheses which I made at the outset and which served me in defining the changes A, A', B, and B'. But is it certain that we shall still have $B + B' = 0$—*after* the two changes A and A'? Is it certain that these two changes compensate in such a manner that not only shall I recover my original impressions, but that the changes B and B' shall recover all their original properties, and in particular that of mutual compensation? If we admit this, we may conclude from it that I shall recover my primitive impressions when the four changes follow in the order

$$A, A', B, B';$$

but not that the same will still be the case when they succeed in the order

$$A, B, B', A'.$$

Nor is this all. If two external changes α and α' are regarded as identical on the basis of the convention adopted above, or in other words, are susceptible of being corrected by the same internal change A; if, on the other hand, two other external changes β and β' can be corrected by the same internal change B, and consequently may also be regarded as identical, have we the right to conclude that the two changes $\alpha + \beta$ and $\alpha' + \beta'$ are susceptible of being corrected by the same internal change, and are consequently identical? Such a proposition is in no wise evident, and if it be true it cannot be the result of *a priori* reasoning.

Accordingly, this set of propositions, which I recapitulate by saying that displacements form a group, is not given us by *a priori* reasoning. Are they then a result of experience? One is inclined to admit that they are; and yet one has a feeling of real misgiving in so doing. May not more precise experience prove some day that the law above enunciated is only approximate? What, then, will become of geometry?

But we may rest assured on this point. Geometry is safe from all revision; no experience, however precise, can overthrow it. If it could have done it, it would have done so long ago. We have long known that all the so-called experimental laws are approximations, and rough approximations at that.

What, then, is to be done? When experience teaches us that a certain phenomenon does not correspond *at all* to these laws, we strike it from the list of displacements. When it teaches us that a certain change obeys them *only approximately*, we consider the change, *by an artificial convention*, as the resultant of two other component changes. The first component is regarded as a displacement *rigorously* satisfying the laws of which I have just spoken, while the second component, which is small, is regarded as a qualitative alteration. Thus we say that natural solids undergo not only great changes of position but also small flexions and small thermal dilatations.

By an external change α we pass, for example, from the *ensemble* of impressions A to the *ensemble* B. We correct this change by a voluntary internal change β and are carried back to the *ensemble* A. A new external change α' causes us to pass again from the *ensemble* A to the *ensemble* B. We ought to expect then that this change α' could in its turn be corrected by another voluntary internal change β' which would provoke the same muscular sensations as β and which would call forth again the *ensemble* of impressions A. If experience does not confirm this prediction, we shall not be embarrassed. We say that the change α', although like α it has been the cause of my passing from the *ensemble A* to the *ensemble B*, is nevertheless not identical with the change α. If our prediction is confirmed only approximately we say that the change α' is a displacement identical with the displacement α but accompanied by a slight qualitative alteration.

In fine, these laws are not imposed by nature upon us but are imposed by us upon nature. But if we impose them upon nature, it is because she suffers us to do so. If she offered too much resistance, we should seek in our arsenal for another form which would be more acceptable to her.

⟦VII⟧ Consequences of the existence of the group

This first fact, that displacements form a group, contains in germ a host of important consequences. Space must be homogeneous; that is, all its points are capable of playing the same part. Space must be isotropic; that is, all directions which issue from the same point must play the same part.

If a displacement D transports me from one point to another, or changes my orientation, I must after such displacement D be still capable of the same movements as before the displacement D, and these movements must have preserved their fundamental properties, which permitted me to classify them among displacements. If it were not so, the displacement D followed by another displacement would not be equivalent to a third displacement; in other words, displacements would not form a group.

Thus the new point to which I have been transported plays the same part as that at which I was originally; my new orientation also plays the same part as the old; space is homogeneous and isotropic.

Being homogeneous, it will be unlimited; for a category that is limited cannot be homogeneous, seeing that the boundaries cannot play the same part as the centre. But this does not say that it is infinite; for the sphere is an unbounded surface, and yet it is finite. All these consequences, accordingly, are germinally contained in the fact which we have just discovered. But we are as yet unable to perceive them, because we do not yet know what a direction is or even what a point is.

⟦VIII⟧ Properties of the group

We have now to study the properties of the group. These properties are purely formal. They are independent of any quality whatever, and in particular of the qualitative character of the phenomena which constitute the change to which we have given the name displacement. We remarked above that we could regard two changes as representing the same displacement, although the phenomena were quite different in qualitative nature. The properties of this displacement remain the same in the two cases; or rather the only ones which concern us, the only ones which are susceptible of being studied mathematically, are those in which quality is in no wise concerned. A brief digression is necessary here to render my thought comprehensible. What mathematicians call a group is the *ensemble* of a certain number of operations and of all the combinations which can be made of them. In the group which is occupying us our operations are displacements. It sometimes happens that two groups contain operations which are entirely different as to character, and that these operations nevertheless combine according to the same laws. We then say that the two groups are *isomorphic.*

The different permutations of six objects form a group and the properties of this group are independent of the character of the objects. If in place of the six material objects we take six letters, or even the six faces of a cube, we obtain groups which differ as to their component materials, but which are all isomorphic with one another.

The formal properties are those which are common to all isomorphic groups. If I say, for example, that such and such an operation repeated three times is equivalent to such and such an other repeated four times, I have announced a formal property entirely independent of quality. These formal properties are susceptible of being studied mathematically. They should be enunciated, therefore, in *precise* propositions. On the other hand, the experiences which serve to verify them can never be more than approximate; that is to say, the experiences in question can never be the true foundation of these propositions. We have within us, in a potential form, a certain number of models of groups, and experience merely assists us in discovering which of these models departs least from reality.

⟦IX⟧ Continuity

It is observed first that the group is *continuous.* Let us see what this means, and how the fact can be established.

The same displacement can be repeated twice, three times, etc. We obtain thus different displacements which may be regarded as *multiples* of the first. The multiples of the same displacement D form a group; for the succession of two of these multiples is still a multiple of D. Further, all these multiples are interchangeable (a truth which is expressed by saying that the group which they form is a *sheaf*); that is, it is indifferent whether we repeat D first three times and then four times, or first four times and then three times. This is an analytical judgement *a priori*; an out-and-out tautology. This group of the multiples of D is only a part of the total group. It is what is called a *sub-group.*

Now we soon discover that any displacement whatever can always be divided into two, three, or any number of parts whatever; I mean that we can always find another displacement which, repeated two, three times will reproduce the given displacement. This divisibility to infinity conducts us naturally to the notion of mathematical continuity; yet things are not so simple as they appear at first sight.

We cannot prove this divisibility to infinity, directly. When a displacement is very small, it is inappreciable for us. When two displacements differ very little, we cannot distinguish them. If a displacement D is extremely small, its consecutive multiples will be indistinguishable. It may happen then that we cannot distinguish $9D$ from $10D$, nor $10D$ from $11D$, but that we can nevertheless distinguish $9D$ from $11D$. If we wanted to translate these crude facts of experience into a formula, we should write

$$9D = 10D,\ 10D = 11D,\ 9D < 11D.$$

Such would be the formula of physical continuity. But such a formula is repugnant to reason. It corresponds to none of the models which we carry about in us. We escape the dilemma by an artifice; and for this physical continuity—or, if you prefer, for this sensible continuity, which is presented in a form unacceptable to our minds—we substitute mathematical continuity. Severing our sensations from that something which we call their cause, we assume that the something in question conforms to the model which we carry about in us, and that our sensations deviate from it only in consequence of their crudeness.

The same process recurs every time we apply measurement to the data of the senses; it is notably applicable to the study of displacements. From the point which we have now reached, we can render an account of our sensations in several different ways.

(1) We may suppose that each displacement forms part of a sheaf formed of all the multiples of a certain small displacement far too small to be appreciated by us. We should then have a discontinuous sheaf which would give us the illusion of physical continuity because our gross senses would be unable to distinguish any two consecutive elements of the sheaf.

(2) We may suppose that each displacement forms part of a more complex and richer sheaf. All the displacements of which this sheaf is composed would be interchangeable. Any two of them would be multiples of another smaller displacement which likewise formed part of the sheaf and which might be regarded as their greatest common divisor. Finally, any displacement of the sheaf could be divided into two, three, or any number of parts, in the sense which I have given to this word above, and the divisor would still be part of the sheaf. The different displacements of the sheaf would be, so to speak, commensurable with one another. To every one of them would correspond a commensurable number, and *vice versa*. This therefore would be already a sort of mathematical continuity, but this continuity would still be imperfect, for there would be nothing corresponding to incommensurable numbers.

(3) We may suppose, finally, that our sheaf is perfectly continuous. All its displacements are interchangeable. To every commensurable or incommensurable number corresponds a displacement and *vice versa*. The displacement corresponding to the number na is nothing else than the displacement corresponding to the number a repeated n times.

Why has the last of these three solutions been adopted? The reasons for the choice are complicated.

(1) It has been established by experience that displacements which are sufficiently large can be divided by any number whatever; and as the means of measurement increased in precision, this divisibility was demonstrated for displacements much smaller, with respect to which it first seemed doubtful. We have thus been led by induction to suppose that this divisibility is a property of all displacements, however small, and consequently to reject the first solution and to decide in favour of divisibility to infinity.

(2) The first solution, like the second, is incompatible with the other properties of the group which we know from other experience. I shall explain this further on. The third solution, accordingly, is imposed upon us by this fact alone. The contrary might have happened. It might have been that the properties of the group were incompatible with continuity. Then we should undoubtedly have adopted the first solution.

⟦X⟧ Sub-groups

The most important of the formal properties of a group is the existence of sub-groups. It must not be supposed that there can be as many sub-groups formed as we like, and that it is sufficient to cut up a group in an arbitrary manner, as one would inert clay, in order to obtain a sub-group. If two displacements be taken at random in a group, it will be necessary, in order to form a sub-group from them, to conjoin with them all their combinations; and in the majority of cases it happens that in combining these two displacements in all possible manners we arrive ultimately at the primitive group again in its original intact form. It may happen thus that a group contains no sub-group.

But groups are distinguished from one another, in a formal point of view, by the number of sub-groups which they contain and by the mutual relations of the sub-groups. A superficial examination of the group of displacements renders it patent that it contains some sub-groups. A more minute examination will disclose them all. We shall see that among these sub-groups there are some that are: (1) continuous, i.e., have all their displacements divisible to infinity; (2) discontinuous, i.e., have no displacements that are divisible to infinity; (3) mixed, i.e., have displacements divisible to infinity and in addition others that are not so divisible.

From another point of view we distinguish among our sub-groups sheaves whose displacements are all interchangeable and those which do not possess this property.

The following is another manner of classing displacements and sub-groups.

Let us consider two displacements D and D'. Let D'' be a third displacement, defined to be the resultant of the displacement D' followed by the displacement D followed itself by the inverse displacement of D'. This displacement D'' is called the *transformation* of D by D'.

From the formal point of view all the transformations of the same displacement are equivalent, so to speak; they play the same part; the Germans say that they are *gleichberechtigt*. Thus (if I may be permitted for an instant to use in advance the ordinary language of geometry which we are supposed not yet to know) two rotations of 60° are *gleichberechtigt*, two helicoidal displacements of the same step and same fraction of spiral are *gleichberechtigt*.

The transformations of all displacements of a sub-group g by the same displacement D' form a new sub-group which is called the transformation of the sub-group g by the displacement D'. The different transformations of the same sub-group, playing the same part in a formal point of view, are *gleichberechtigt*.

It happens generally that many of the transformations of the same sub-group are identical; it will sometimes even happen that all the transformations of a sub-group are identical with one another and with the primitive sub-group. It is then said this sub-group is *invariant* (which happens, for example, in the case of the sub-group formed of all translations). The existence of an invariant sub-group is a formal property of the highest importance.

⟦XI⟧ Rotative sub-groups

The number of sub-groups is infinite; but they may be divided into a rather limited number of classes of which I do not wish to give here a complete enumeration. But these sub-groups are not all perceived with the same facility. Some among them have been only recently discovered. Their existence is not an intuitive truth. Unquestionably it can be deduced from the fundamental properties of the group, from properties which are known to everybody, and which are, so to speak, the common patrimony of all minds. Unquestionably it is contained there in germ; yet those who have demonstrated their existence have justly felt that they had made a discovery and have frequently been obliged to write long memoirs to reach their results.

Other sub-groups, on the contrary, are known to us in much more immediate manner. Without much reflection everyone believes he has a direct intuition of them, and the affirmation of their existence constitutes the axioms of Euclid. Why is it that some sub-groups have directly attracted attention, whilst others have eluded all research for a much longer time? We shall explain it by a few examples.

A solid body having a fixed point is turning before our eyes. Its image is depicted on our retina and each of the fibres of the optic nerve conveys to us an impression; but owing to the motion of the solid body this impression is variable. One of these fibres, however, conveys to us a constant impression. It is that at the extremity of which the image of the fixed point has been formed. We have, thus, a change which causes certain sensations to vary, but leaves others invariable. This is a property of the displacement, but at first blush it does not appear that it is a formal property. It seems to belong to the qualitative character of the sensations experienced. We shall see, however, that we can disengage a formal property from it, and to render my thought clear I shall compare what takes place in this case with what happens in another instance which is apparently analogous.

I suppose that a certain body is moving before my eyes in any manner, but that a certain region of this body is painted in a colour sufficiently uniform to leave no shades discernible. Let us say it is red. If the movements are not of too great compass and if the red region is sufficiently large in extent, certain parts of the retina will remain constantly in the image of that region, certain nerve-fibres will convey to us constantly the impression of the red, the displacement will have left certain sensations invariable.

But there is an essential difference between the two cases. Let us go back to the first one. We witnessed there an external change in which certain sensations A did not change, whilst other sensations B did change. We are able to correct this external change by an internal change, and in this correction the sensations A still remain invariable.

But now here is a new solid body which is turning before our eyes and is experiencing the same rotations as the first. This is a new external change which may be different altogether from the first from a qualitative point of view, because the new body which is turning may be painted in new colours, or because we are apprised of its rotation by touch and not by sight. We discover, however, that it is the *same* displacement, because it can be corrected by the same internal change. And we also discover that certain sensations A' in this new external change (totally different perhaps from A) have remained invariable, whilst other sensations B' varied. Thus, this property of conserving certain sensations ultimately appears to us as a formal property independent of the qualitative character of these sensations.

We pass to the second example. We have, first, an external change in which a certain sensation C, a sensation of red, has remained constant. Let us suppose that another solid body, differently painted, undergoes the same displacement. Here is a new external change, and we know that it represents the same displacement because we can correct it by the same internal change. We discover generally that in this new external change certain sensations have not remained constant. Thus the conservation of the sensation C will appear to us as an accidental property only, connected with the qualitative nature of the sensation.

We are thus led to distinguish among displacements those which conserve certain sensations. The *ensemble* of the displacements which thus conserve a given system of sensations evidently forms a sub-group which we may call a *rotative sub-group*.

Such is the conclusion which we draw from experience. It is needless to point out how crude is the experience and how precise on the other hand is the conclusion. Therefore experience cannot impose the conclusion upon us, but it suffices to suggest it to us. It suffices to show that of all the groups of which the models pre-exist in us, the only ones which we can accept with a view of referring to them our sensations, are those which contain such a sub-group.

By the side of the rotative sub-group, we should consider its transformations, which also may be called rotative sub-groups. (Sub-groups of rotations about a fixed point.) By new experiences, always very crude, it is then shown:

(1) That any two rotative sub-groups have common displacements.

(2) That these common displacements, all interchangeable among one another, form a sheaf, which may be called a rotative sheaf. (Rotations about a fixed axis.)

(3) That any rotative sheaf forms part not only of two rotative sub-groups, but of an infinity of them.

Here is the origin of the notion of the straight line, as the rotative sub-group was the origin of the notion of the point.

Let us now look at all the displacements of a rotative sheaf. If we look at any displacement whatever, it will not in general be interchangeable with all the displacements of the sheaf, but we shall discover very soon that there exist displacements which are interchangeable with all those of the rotative sheaf, and that they form a more extensive sub-group which may be called the helicoidal sub-group (combinations of rotations about an axis and of translations parallel to that axis). This will be evident when it is observed that a straight line can slide along itself.

Finally, we derive from the same crude observations such propositions as the following:

Any displacement sufficiently small and forming part of a given rotative sub-group, can always be decomposed into three others belonging respectively to three given rotative sheaves. Every displacement interchangeable with a rotative sub-group forms part of this sub-group.

Any displacement sufficiently small can always be decomposed into two others belonging respectively to two given rotative sub-groups, or to *six given rotative sheaves*.

Later on I shall revert in detail to the origin of these various propositions.

⟦XII⟧ Translative sub-groups

With these propositions we have sufficient material, not to construct the geometry of Euclid, but to limit the choice between that of Euclid and the geometries of Lobatchevski and Riemann. In order to go farther, we are in need of a new proposition to take the place of the postulate of parallels. The proposition substituted will be the existence of an *invariant* sub-group, of which all the displacements are interchangeable and which is formed of all translations.

It is this that determines our choice in favour of the geometry of Euclid, because the group that corresponds to the geometry of Lobatchevski does not contain such an invariant sub-group.

⟦XIII⟧ Number of dimensions

In the ordinary theory of groups, we distinguish order and degree. Let us suppose the simplest case first, that of a group formed by different permutations between certain objects. The number of the objects is called the degree; the number of the permutations is called the order of the group. Two such groups may be isomorphic and their permutations may combine according to the same laws without their degree being the same. Thus let us consider the different ways in which a cube can be superposed upon itself. The vertices may be interchanged one with another, as may also be the faces and the edges; whence result three groups of permutations which are evidently isomorphic among themselves; but their degree may be either eight, six, or twelve, since there are eight vertices, six faces, and twelve edges.

On the other hand, two mutually isomorphic groups have always the same order. The degree is, so to speak, a material element, and the order a formal element, the importance of which is far greater. The theory of two groups of different degree may be the same so far as its formal properties are concerned; just as the mathematical theory of the addition of three cows and four cows is identical with that of three horses and four horses.

When we pass to continuous groups, the definitions of order and degree must be modified, though without sacrificing their spirit. Mathematicians suppose ordinarily that the object of the operations of the group is an *ensemble* of a certain number n of quantities susceptible of being varied in a continuous manner, which quantities are called *co-ordinates*. On the other hand, every operation of the group may be regarded as forming part of a sheaf analogous to the rotative sheaf and as a multiple of a very high order of an infinitesimal operation belonging to the same sheaf. Then, every infinitesimal operation of the group can be decomposed into k other operations belonging to k given sheaves. The number n of the co-ordinates (or of the dimensions) is then the *degree*, and the number k of the components of an infinitesimal operation is the *order*. Here again two isomorphic groups may have different degrees, but must be of the same order. Here again the degree is an element relatively material and secondary, and the order a formal element. According to the laws established above, our group of displacements is here of the sixth order, but its degree is yet unknown. Is the degree given us immediately?

Displacements, we have seen, correspond to changes in our sensations, and if we distinguish in the present group between form and material, the material can be nothing else than that which the displacements cause to change, viz., our sensations. Even if we suppose that what we have above called sensible space has already been elaborated, the material would then be represented by as many continuous variables as there are nerve-fibres; the 'degree' of our group would then be extremely large. Space would not have three dimensions but as many as there are nerve-fibres. Such is the consequence to which we come if we accept as the material of our group what is immediately given us. How shall we escape the difficulty? Evidently by replacing the group which is given us, together with its form and its material, by another *isomorphic* group, the material of which is simpler.

But how is this to be done? It is precisely owing to this circumstance, that the displacements which conserve certain elements are the same as those which conserve certain other elements. Then all those elements which are conserved by the same displacements we agree to replace by a single element which has a purely schematic value only. Whence results a considerable reduction of degree.

For example, I see a solid body rotating about a fixed point. The parts near the fixed point are painted red. Here is a displacement, and within this displacement I perceive that something remains invariable—namely, the sensation of red conveyed to me by a certain optical nerve-fibre. Some time afterward I see another solid body turning about a fixed point. But the parts near the fixed point are painted green. The sensations experienced are in themselves quite different, but I perceive that it is the same displacement because it can be corrected by the same internal change. Here again something remains invariable; but this something is totally different from the material point of view; it is the sensation of green conveyed by a certain nerve-fibre.

These two things, which materially are so different, I replace schematically by a single thing which I call a point, and I express my thought by saying that in the one case as in the other, a point of the body has remained fixed. Thus every one of our new elements will be what is conserved by all the displacements of a sub-group; to every sub-group there will then correspond an element and *vice versa*.

Let us consider the different transformations of the same sub-group. They are infinite in number and may form a simple, double, triple, continuous infinity. To each one of these transformations an element can be made to correspond; I have then a simple, double, triple, etc., infinity of them, and the degree of our continuous group is 1, 2, 3, ...

Suppose that we choose the different transformations of a rotative sub-group. We have here a triple infinity. The material of our group is accordingly composed of a triple infinity of elements. The degree of the group is three. We have then chosen the point as the element of space and given to space three dimensions.

Suppose we choose the different transformations of a helicoidal sub-group. Here we have a quadruple infinity. The material of our group is composed of a quadruple infinity of elements. Its degree is four. We then have chosen the straight line as the element of space—which would give to space four dimensions.

Suppose, finally, that we choose the different transformations of a rotative sheaf. The degree would then be five. We have chosen as the element of space the figure formed by a straight line and a point on that straight line. Space would have five dimensions.

Here are three solutions, which are each logically possible. We prefer the first because it is the simplest, and it is the simplest because it is that which gives to space the smallest number of dimensions. But there is another reason which recommends this choice. The rotative sub-group first attracts our attention because it conserves certain sensations. The helicoidal sub-group is known to us only later and more indirectly. The rotative sheaf on the other hand is itself merely a sub-group of the rotative sub-group.

‖XIV‖ The notion of point

I feel that I am here touching on the most delicate spot of this discussion, and I am compelled to stop for a moment to justify more completely my previous assertions, which some persons may be disposed to doubt. Many persons, indeed, would consider the notion of a point of space as so immediate and so clear that any definition of it is superfluous. But I believe it will be granted me that so subtle a notion as that of the mathematical point, without length, breadth, or thickness, is not immediate, and that it needs to be explained.

But is it the same with the vaguer and less precisely defined, yet more empirical notion, of *place*? Is there any one who does not fancy he knows perfectly well what he is talking about when he says: this object occupies the place which was just occupied by that object. To determine the range of such an assertion, and the conclusions which can be drawn from it, let us seek to analyse its signification. If I have moved neither my body, my head, nor my eye, and if the image of the object B affects the same retinal fibres that the image of the object A previously affected; if again, although I have moved neither my arm nor my hand, the same sensory fibres which extend to the end of the finger, and which formerly conveyed to me the impression which I attributed to the object A now convey to me the impression which I attribute to the object B; if both these conditions are fulfilled—then ordinarily we agree to say that the object B occupies the place which previously the object A occupied.

Before analysing so complicated a convention as that just stated I shall first make a remark. I have just enunciated two conditions: one relating to sight, and one relating to touch. The first is necessary but not sufficient, for we say in ordinary language that the point on the retina where an image is formed gives us knowledge only of the direction of the visual ray, but that the distance from the eye remains unknown. The second condition is at once necessary and sufficient, because we assume that the action of touch is not exercised at a distance, and that the object A like the object B cannot act upon the finger except by immediate contact. All this agrees with what experience has taught us; namely, that the first condition can be fulfilled without the second being realized, but that the second cannot be fulfilled without the first. Let it be remarked that we have here something which we could not know *a priori*, that experience alone is able to demonstrate it to us.

Nor is this all. To determine the place of an object I made use only of an eye and a finger. I could have made use of several other means—for example, of all my other fingers. Having been made aware that the object A has produced upon my first finger a tactual impression, suppose that by a series of movements S my second finger comes into contact with the same object A. My first tactual impression ceases and is replaced by another tactual impression which is conveyed to me by the nerve of the second finger, and which I still attribute to the action of the object A. Some time afterwards, and without my having moved my hand, the same nerve of the second finger conveys to me another tactual impression, which I attribute to the action of another object B. I then say that the object B has taken the place of the object A.

At this moment I make a series of movements S' the inverse of the series S. How do I know that these two series are inverse to one another? Because experience has taught me that when the internal change S that corresponds to certain muscular sensations is followed by an internal change S' which corresponds to other muscular sensations, a compensation is effected and my primitive impressions, originally modified by the change S, are reestablished by the change S'.

I execute the series of movements S'. The effect ought to be to take back my first finger to its initial position and so to put it into contact with the object B, which has taken the place of the object A. I ought, therefore, to expect that the nerve of my first finger should convey to me a tactual sensation attributable to the object B. In fact this is what happens.

But would it therefore be absurd to suppose the contrary? And why would it be absurd? Shall I say that the object B having taken the place of the object A, and my first finger having resumed its original place, it ought to touch the object B just as before it touched the object A? This would be an outright begging of the question. And to show this let us attempt to apply the same reasoning to another example, or rather let us return to the example of sight and touch which I cited at the outset.

The image of the object A has made an impression on one of my retinal fibres. At the same time the nerve of one of my fingers conveys to me a tactual impression which I attribute to the same object. I move neither my eye nor my hand. And a moment after the image of the object B has impressed the same retinal fibre. By a course of reasoning perfectly similar to that which precedes, I should be tempted to conclude that the object B had taken the place of the object A, and I should expect that the nerve of my finger would convey to me a tactual impression attributable to B. And yet I should be deceived. For the image of B may chance to be formed upon the same point of the retina as the image of A, although the distance to the eye may not be the same in the two cases.

Experience has refuted my reasoning. I extricate myself by saying that it is not sufficient for two bodies to cast their image upon the same retinal fibre in order to justify me in saying that the two bodies are in the same place; and I should extricate myself in a similar manner in the case of the two fingers, if the indications of the second finger had not been in accord with those of the first, and if experience had been at variance with my reasoning. I should still say that two objects A and B can make an impression upon the same finger by means of touch and yet not be in the same place; in other words, I should conclude that touch could be effected at a distance. Or, again, I should agree to consider *A* and *B* as being in the same place only on the condition of there being concordance not only between their effects upon the first finger, but also between their effects upon the second finger. One might almost say, in a certain point of view, that one more dimension would be attributed to space in this manner.

To sum up, there are certain laws of *concordance*, which can be revealed to us only by experience, and which are at the basis of the vague notion of place.

But even taking these laws of concordance for granted, can we deduce from them the much more precise notion of point and the notion of number of dimensions? This remains to be examined.

First an observation. We have spoken of two objects *A* and *B*, which have cast one after another their image on the same point of the retina. But these two images are not identical; otherwise how could I distinguish them? They differ, for example, in colour. The one is red, the other is green. We have, accordingly, two sensations which differ in quality and which are doubtless conveyed to me by two different though contiguous nerve-fibres. What have they in common with one another, and why am I led to associate them together? I believe that if the eye were immovable, we should never have thought of this association. It is the movements of the eye that have taught us that there is the same relation on the one hand between the sensation of green at the point *A* of the retina and the sensation of green at the point *B* of the retina, and on the other hand between the sensation of red at the point *A* of the retina and the sensation of red at the point *B* of the retina. We have found, in fact, that the same movements, corresponding to the same muscular sensations, cause us to pass from the first to the second, or from the third to the fourth. Were this not so, these four sensations would appear qualitatively distinct, and we should no more think of establishing a sort of proportion between them than we should between an olfactory, a gustatory, an auditive, and a tactual sensation.

Yet whatever be the origin of this association, it is implied in the notion of place, which could not have grown up without it. Let us analyse, therefore, its laws. We can only conceive them under two different forms equally remote from mathematical continuity; namely, under the form of discontinuity or under the form of physical continuity.

Under the first form, our sensations will be divided into a very large number of 'families'; all the sensations of one family being associated with one another and not being associated with those of other families. Since to every family there would correspond a place, we should have a finite but very large number of places, and the places would form a discrete aggregate. There would be no reason for classifying them in a table of three dimensions rather than in one of two or four; and we could not deduce from them either the mathematical point or space.

Under the second form, which is more satisfactory, the different families interpenetrate one another. *A*, for example, will be associated with *B*, and *B* with *C*. But *A* will not appear to us as associated with *C*. We shall find that

A and *C* do not belong to the same family, although on the one hand *A* and *B*, and on the other hand *B* and *C*, will appear to us as belonging to the same family. Thus we cannot distinguish between a weight of nine grammes and one of ten grammes, or between the latter weight and a weight of eleven grammes. But we can readily tell the difference between the first weight and the third. This is always the formula of physical continuity.

Let us picture to ourselves a series of wafers partially covering one another in such a way that the plane is totally covered; or better, let us picture to ourselves something analogous in a space of three dimensions. If these wafers were to form by their superposition only a sort of one-dimensional ribbon, we should recognize it, because the associations of which I have just been speaking obey a law that may be stated as follows: if *A* is associated at once with *B*, *C*, and *D*, *D* is associated with *B* or with *C*. This law would not be true if our wafers covered by their superposition a plane or a space of more than two dimensions. When I say, therefore, that all possible places constitute an aggregate of one dimension or of more than one dimension, I mean to say simply that this law is true or that it is false. When I say that they constitute an aggregate of two or three dimensions, I simply affirm that certain analogous laws are true.

Such are the foundations on which we may attempt to construct a *static* theory of the number of dimensions. It will be seen how complicated is this manner of defining the number of dimensions, how imperfect it is; and it is useless to remark upon the distance which still separates the physical continuity of three dimensions as thus understood from the real mathematical continuity of three dimensions.

⟦XV⟧ Discussion of the preceding theory

Without dwelling upon the multitude of difficult details, let us see in what those associations consist upon which the notion of place rests. We shall see that we are finally led back, after a long detour, to the notion of group, which appeared to us at the outset the best fitted for elucidating the question of the number of dimensions.

By what means are different 'places' distinguished from one another? How, for example, are two places occupied successively by the extremity of one of my fingers to be distinguished? Evidently by the movement which my body has made in the interval, movements which are made known to me by a certain series of muscular sensations. These two places correspond to two distinct attitudes and positions of the body which are known solely by the movements which I have had to make in changing a certain initial attitude and a certain initial position; and these movements themselves are known to me only by the muscular sensations which they have provoked.

Two attitudes of the body, or two corresponding places of the finger, appear to me identical if the two movements which I must make to reach them differ so little from each other that I cannot distinguish the corresponding muscular sensations. They will appear to me non-identical, without some new convention, if they correspond to two series of distinguishable muscular sensations.

But in this manner we have engendered not a physical continuity of three dimensions but a physical continuity of a much larger number of dimensions; for I can cause the muscular sensations corresponding to a very large number of muscles to vary, and I do not on the other hand consider a single muscular sensation only, nor even an aggregate of simultaneous sensations, but a series of successive sensations, and I can make the laws by which these sensations succeed one another vary in an arbitrary manner.

Why is the number of dimensions reduced, or, what is the same thing, why do we consider two places as identical when the two corresponding attitudes of the body are different? Why do we say in certain cases that the place occupied by the extremity of a finger has not changed, although the attitude of the body has changed?

It is because we discover that very *frequently*, in the movement which causes the passage from the one to the other of these two attitudes, the tactual sensation attributable to the contact of this finger with an object A persists and remains constant. We *agree* then, to say that these two attitudes shall be placed in the same class and that this class shall embrace all attitudes corresponding to the same place occupied by the same finger. We agree that these two attitudes shall still be placed in the same class even when they are accompanied by no tactual sensation, or by variable tactual sensations.

This convention has been evoked by experience, because experience alone informs us that certain tactual sensations are frequently persistent. But in order that conventions of this kind shall be permissible, they must satisfy certain conditions which it now remains for us to analyse.

If I place the attitudes A and B in the same class, and also the attitudes B and C in the same class, it follows necessarily that the attitudes A and C must be regarded as belonging to the same class. If, then, we agree to say that the movements which cause the passage from the attitude A to the attitude B do not change the place of the finger, and if the same holds true of the movements which cause the passage from the attitude B to the attitude C, it follows necessarily that the same must again be true of those which cause the passage from the attitude A to the attitude C. In other words, the aggregate of the movements causing a passage from one attitude to another attitude of the same class constitutes a group. It is only when such a group exists that the convention above laid down is acceptable. To every class of attitudes, and consequently to every place, there will therefore correspond a group, and we are here led back again to the notion of group, without which there would be no geometry.

Nevertheless, there is a difference between the principle here under discussion and the theory which I developed above. Here each place appears to me associated with a certain group which is introduced as the sub-group S of the group G formed by the movements which can give to the body all possible positions and all possible attitudes, the relative situations of the different parts of the body being allowed to vary in any manner whatsoever. In our other theory, on the contrary, every point was associated with a sub-group S' of the group

G' formed by the displacements of the body viewed as an invariable solid, that is to say, by displacements such that the relative situations of the different parts of the body do not vary.

Which of the two theories is to be preferred? It is evident that G' is a sub-group of G and S' a sub-group of S. Further, G' is much simpler than G, and for this reason the theory which I first propounded and which is based upon the consideration of the group G' appears to me simpler and more natural, and consequently I shall hold to it.

But be this as it may, the introduction of a group, more or less complicated, appears to be absolutely necessary. Every purely statical theory of the number of dimensions will give rise to many difficulties, and it will always be necessary to fall back upon a dynamical theory. I am happy to be in accord on this point with the ideas set forth by Professor Newcomb in his *Philosophy of hyperspace*.

‖XVI‖ The reasoning of Euclid

But in order to show that the idea of displacement, and consequently the idea of group, has played a preponderant part in the genesis of geometry, it remains to be shown that this idea dominates all the reasoning of Euclid and of the authors who after him have written upon elementary geometry.

Euclid begins by enunciating a certain number of axioms; but it must not be imagined that the axioms which he enunciates explicitly are the only ones to which he appeals. If we carefully analyse his demonstrations we shall find in them, in a more or less masked form, a certain number of hypotheses which are in reality axioms disguised; and we may say almost as much of some of his definitions.

His geometry begins with declaring that two figures are equal if they are superposable. This assumes that they can be displaced and also that among all the changes which they may undergo, we can distinguish those which may be regarded as displacements without deformation. Again, this definition implies that two figures which are equal to a third are equal to each other. And that is tantamount to saying that if there be a displacement which puts the figure A upon the figure B, and a second displacement which superposes the figure B upon the figure C, there will also be a third, the resultant of the first two, which will superpose the figure A upon the figure C. In other words, it is presupposed that the displacements form a group. The notion of a group, accordingly, is introduced from the outset, and inevitably introduced.

When I pronounce the word 'length', a word which we frequently do not think necessary to define, I implicitly assume that the figure formed by two points is not always superposable upon that which is formed by two other points; for otherwise any two lengths whatever would be equal to each other. Now this is an important property of our group.

I implicitly enunciate a similar hypothesis when I pronounce the word 'angle'.

And how do we proceed in our reasonings? By displacing our figures and causing them to execute certain movements. I wish to show that at a given point in a straight line a perpendicular can always be erected, and to accomplish this I conceive a movable straight line turning about the point in question. But I presuppose here that the movement of this straight line is possible, that it is continuous, and that in so turning it can pass from the position in which it is lying on the given straight line, to the opposite position in which it is lying on its prolongation. Here again is a hypothesis touching the properties of the group.

To demonstrate the cases of the equality of triangles, the figures are displaced so as to be superposed one upon the other.

Finally, what is the method employed in demonstrating that from a given point one and only one perpendicular can always be drawn to a given straight line? The figure is turned 180° around the given straight line, and in this manner the point symmetrical to the given point with respect to the given straight line is obtained. We have here a feature most characteristic, and here appears the part which the straight line most frequently plays in geometric demonstrations, namely, that of an axis of rotation.

There is implied here the existence of the sub-group which I have called the rotative sheaf. When—which also frequently happens—a straight line is made to slide along itself (for we shall, of course, continue to suppose that it can serve as an axis of rotation), we implicitly take the existence of the helicoidal sub-group for granted. In fine, the principal foundation of Euclid's demonstrations is really the existence of the group and its properties.

Unquestionably he appeals to other axioms which it is more difficult to refer to the notion of group. An axiom of this kind is that which some geometers employ when they define a straight line as the shortest distance between two points. *But it is precisely such axioms that Euclid enunciates.* The others, which are more directly associated with the idea of displacement and with the idea of groups, are the very ones which he implicitly admits, and which he does not deem it even necessary to enunciate. This is tantamount to saying that the former are the fruit of a later experience, that the others were first assimilated by us, and that consequently the notion of group existed prior to all the others.

⟦XVII⟧ The geometry of Staudt

It is known that Staudt attempted to base geometry upon different principles. Staudt admits the following axioms only:

1. Through two points a straight line can always be drawn.

2. Through three points a plane can always be drawn.

3. Every straight line which has two of its points in a plane lies entirely in that plane.

4. If three planes have one point in common, and one only, any straight line will cut at least one of these three planes.

These axioms are sufficient to establish all the *descriptive* properties relating to the intersections of straight lines and planes. To obtain the metrical properties we begin with *defining* a harmonic pencil of four straight lines, taking as definition the well-known descriptive property. Then the anharmonic ratio of four points is *defined*, and finally, supposing that one of these four points has been relegated to infinity, the ratio of two lengths is *defined*.

This last is the weak point of the foregoing theory, attractive though it be. To arrive at the notion of length by regarding it merely as a particular case of the anharmonic ratio is an artificial and repugnant detour. This evidently is not the manner in which our geometric notions were formed.

Let us see now whether we can conceive, without the introduction of the notion of group and of movement, how the notions which serve as the foundations of this ingenious geometry have taken their rise. Let us see what experiences might have led us to formulate the axioms enunciated above.

If the straight line is not given as an axis of rotation, it can be given only in one way, namely, as the trajectory of a ray of light. I mean, that the experiences, always more or less crude, which serve us as our point of departure, should all be applicable to the ray of light, and that we must define the straight line as a line for which the simple laws which the ray of light approximately obeys will be rigorously true. The following is the experience which must be made in order to verify the most important of our axioms, namely, the third.

Let two threads be stretched. Let the eye be placed at the extremity of one of these threads. We see that the thread is entirely hidden by its extremity, which teaches us that the thread is rectilinear, that is to say, is the direction of the trajectory of a ray of light. Let the same be done for the second thread. The following is then observed: either there will be no position of the eye for which one of the threads is entirely hidden by the other, or there will be an infinity of such positions.

How is the question of the number of dimensions presented in this order of ideas? Let us consider all the positions of the eye for which one of the strings is hidden by the other. Let us suppose that in one of these positions the point A of the first string is hidden by the point A' of the second, the point B by the point B', the point C by the point C'. We then discover that if the body is so displaced that the point A is always hidden by the point A' and the point B by the point B', that the point C always remains hidden by the point C', and that in general any point whatsoever of the first thread remains hidden by the same point of the second thread by which it was hidden before the body was displaced. We express this fact by saying that although the body is displaced, the position of the eye has not changed.

We see thus that the position of the eye is defined by two conditions, viz., that A is hidden by A' and B by B'. We express this fact by saying that the *locus* of the points such that the two threads mutually hide each other has two dimensions.

Similarly, let us suppose that in a certain position of the body four threads A, B, C, D, hide four points A', B', C', D'; let us suppose that the body is displaced, but in such a manner that A, B, and C continue to hide A', B', and C'. We shall then discover that D continues to hide D', and we shall again express this fact by saying that the position of the eye has not changed. This position will therefore be defined by three conditions, and this is why we say that space has three dimensions.

It will be remarked that the law as thus experimentally discovered, is only approximately true. But this is not all. It is not even always true, because D or D' may have moved at the same time that my body was being displaced. We then simply declare that this law is often approximately true.

But we are desirous of arriving at geometrical axioms which are rigorously and always true, and we always escape the dilemma by the same artifice, namely, by saying that we agree to consider the change observed as the resultant of two others, viz., of one which rigorously obeys the law and which we attribute to the displacement of the eye, and of a second one which is generally very small and which we attribute either to qualitative alterations or to the movements of external bodies.

We have not been able to avoid the consideration of movements of the eye and of the body, yet we may say, that from a certain point of view the geometry of Staudt is predominantly a visual geometry, while that of Euclid is predominantly muscular.

Undoubtedly unconscious experiences analogous to those of which I have just spoken may have played a part in the genesis of geometry; but they are not sufficient. If we had proceeded, as the geometry of Staudt supposes us to have done, some Apollonius would have discovered the properties of polars. But it would have been only long after that the progress of science would have made clear what a length or an angle is. We should have had to wait for some Newton to discover the various cases of the equality of triangles. And this is evidently not the way that things have come to pass.

⟦XVIII⟧ The axiom of Lie

It is Sophus Lie who has contributed most towards making prominent the importance of the notion of group and laying the foundations of the theory that I have just expounded. It is he, in fact, who gave the present form to the mathematical theory of continuous groups. But to render possible its application to geometry, he regards a new axiom as necessary, which he enunciates by declaring that space is a *Zahlenmannigfaltigkeit*; that is, that to every point of a straight line there corresponds a number and *vice versa*.

Is this axiom absolutely necessary? And could not the other principles which Lie has laid down dispense with it? We have seen above in connection with continuity, that the best known groups may be distributed from a certain point of view into three classes; all the operations of the group can be divided into sheaves; for 'discontinuous' groups the different operations of the same sheaf are only a single operation repeated once, twice, three times, etc.; for 'continuous' groups properly so called the different operations of the same sheaf correspond to different whole numbers, commensurable or incommensurable; finally, for groups which may be called 'semi-continuous', these operations correspond to different commensurable numbers.

Now it may be demonstrated that no discontinuous or semi-continuous group exists possessing other properties than those which experience has led us to adopt for the fundamental group of geometry, and which I here briefly recall: The group contains an infinity of sub-groups, all *gleichberechtigt*, which I call rotative sub-groups. Two rotative sub-groups have a sheaf in common which I call rotative and which is common not only to two but also to an infinity of rotative sub-groups. Finally, every very small displacement of the group may be regarded as the resultant of six displacements belonging to six given rotative sheaves. A group satisfying these conditions can be neither discontinuous nor semi-continuous.

Unquestionably this is an exceedingly recondite property, and not easy to demonstrate. Geometers who were ignorant of it have not the less hit upon its consequences, as for example, when they learned that the ratio of a diagonal to the side of a square is incomensurable. It was for this reason that the introduction of incommensurables into geometry became necessary.

The group, therefore, must be continuous, and it seems as if the axiom of Lie were useless.

Nevertheless, we are obliged to remark that the classification of groups above sketched is not complete; groups may be conceived which are not included in it. We might, therefore, suppose that the group is neither discontinuous, semi-continuous, nor continuous. But this would be a complex hypothesis. We reject it, or rather we never think of it, for the reason that it is not the simplest compatible with the axioms adopted.

The foundation of the axiom of Lie remains to be supplied.

⟦XIX⟧ Geometry and contradiction

In following up all the consequences of the different geometric axioms, are we never led to contradictions? The axioms are not analytical judgments *a priori*; they are conventions. Is it certain that all these conventions are compatible?

These conventions, it is true, have all been suggested to us by experiences, but by crude experiences. We discover that certain laws are approximately verified, and we decompose the observed phenomenon conventionally into two others: a purely geometric phenomenon which exactly obeys these laws; and a very minute disturbing phenomenon.

Is it certain that this decomposition is always permissible? It is certain that these laws are *approximately* compatible, for experience shows that they are all approximately realized at one and the same time in nature. But is it certain that they would be compatible if they were absolutely rigorous?

For us the question is no longer doubtful. Analytical geometry has been securely established, and *all* the axioms have been introduced into the equations which serve as its point of departure; we could not have written these equations if the axioms had been contradictory. Now that the equations are written, they can be combined in all possible manners; analysis is the guarantee that contradictions shall not be introduced.

But Euclid did not know analytical geometry, and yet he never doubted for a moment that his axioms were compatible. Whence came his confidence? Was he the dupe of an illusion? And did he attribute to our unconscious experiences more value than they really possess? Or perhaps, since the idea of the group was potentially pre-existent in him, did he have some obscure instinct for it, without reaching a distinct notion of it? I shall leave the question undecided, although I incline towards the second solution.

⟦XX⟧ The use of figures

It may be asked why geometry cannot be studied without figures. This is easy to account for. When we commence studying geometry, we have already had in innumerable instances the fundamental experiences which have enabled our notion of space to originate. But they were made without method, without scientific attention and unconsciously, so to speak. We have acquired the ability *to represent to ourselves* familiar geometric experiences without being obliged to resort to material reproductions of them; but we have not yet deduced from them logical conclusions. How is this to be done? Before enunciating the law, the experience in question is perceptually represented by stripping it as completely as possible of all accessory or disturbing circumstances—just as a physicist eliminates the sources of systematic error in his experiments. It is here that figures are necessary, but they are an instrument only slightly less crude than the chalk which is employed in drawing them; and, like material objects, it is beyond our power to represent them in the geometric space which forms the object of our studies; we can only represent them in sensible space. We accordingly do not study material figures, but simply make use of them in studying something which is higher and more subtle.

⟦XXI⟧ Form and matter

We owe the theory which I have just sketched to Helmholtz and Lie. I differ from them in one point only, but probably the difference is in the mode of expression only and at bottom we are completely in accord.

As I explained above, we must distinguish in a group the form and the matter. For Helmholtz and Lie the matter of the group existed previously to the form, and in geometry the matter is a *Zahlenmannigfaltigkeit* of three dimensions. The number of dimensions is therefore posited prior to the group. For me, on the contrary, the form exists before the matter. The different ways in which a cube can be superposed upon itself, and the different ways in which the roots of a certain equation may be interchanged, constitute two isomorphic groups. They differ in matter only. The mathematician should regard this difference as superficial, and he should no more distinguish between these two groups than he should between a cube of glass and a cube of metal. In this view the group exists prior to the number of dimensions.

We escape in this way also an objection which has often been made to Helmholtz and Lie. 'But your group', say these critics, 'presupposes space; to construct it you are obliged to assume a continuum of three dimensions. You proceed as if you already knew analytical geometry.' Perhaps the objection was not altogether just; the continuum of three dimensions which Helmholtz and Lie posited was a sort of non-measurable magnitude analogous to magnitudes concerning which we may say that they have grown larger or smaller, but not that they have become twice or three times as large.

It is only by the introduction of the group, that they made of it a measurable magnitude, that is to say a veritable space. Again, the origin of this non-measurable continuum of three dimensions remains imperfectly explained.

But, it will be said, in order to study a group even in its formal properties, it is necessary to construct it, and it cannot be constructed without matter. One might as well say that one cannot study the geometric properties of a cube without supposing this cube to be of wood or of iron. The complexus of our sensations has without doubt furnished us with a sort of matter, but there is a striking contrast between the grossness of this matter and the subtle precision of the form of our group. It is impossible that this can be, properly speaking, the matter of such a group. The group of displacements such as it is given us directly by experience, is something more gross in character; it is, we may say, to continuous groups proper what the physical continuum is to the mathematical continuum. We first study its form agreeably to the formula of the physical continuum, and since there is something repugnant to our reason in this formula we reject it and substitute for it that of the continuous group which, potentially, pre-exists in us, but which we originally know only by its form. The gross matter which is furnished us by our sensations was but a crutch for our infirmity, and served only to force us to fix our attention upon the pure idea which we bore about in ourselves previously.

[XXII] Conclusions

Geometry is not an experimental science; experience forms merely the occasion for our reflecting upon the geometric ideas which pre-exist in us. But the occasion is necessary; if it did not exist we should not reflect; and if our experiences were different, doubtless our reflections would also be different. Space is not a form of our sensibility; it is an instrument which serves us not to represent things to ourselves, but to reason upon things.

What we call geometry is nothing but the study of formal properties of a certain continuous group; so that we may say, space is a group. The notion of this continuous group exists in our mind prior to all experience; but the assertion is no less true of the notion of many other continuous groups; for example, that which corresponds to the geometry of Lobatchevski. There are, accordingly, several geometries possible, and it remains to be seen how a choice is made between them. Among the continuous mathematical groups which our mind can construct, we choose that which deviates least from that rough group, analogous to the physical continuum, which experience has brought to our knowledge as the group of displacements.

Our choice is therefore not imposed by experience. It is simply guided by experience. But it remains free; we choose this geometry rather than that geometry, not because it is more *true*, but because it is the more *convenient*.

To ask whether the geometry of Euclid is true and that of Lobatchevski is false, is as absurd as to ask whether the metric system is true and that of the yard, foot, and inch, is false. Transported to another world we might undoubtedly have a different geometry, not because our geometry would have ceased to be true, but because it would have become less convenient than another. Have we the right to say that the choice between geometries is imposed by reason, and, for example, that the Euclidean geometry is alone true because the principle of the relativity of magnitudes is inevitably imposed upon our mind? It is absurd, they say, to suppose a length can be equal to an abstract number. But why? Why is it absurd for a length and not absurd for an angle? There is but one answer possible. It appears to us absurd, because it is contrary to our habitual way of thinking. Unquestionably reason has its preferences, but these preferences have not this imperative character. It has its preferences for the simplest because, all other things being equal, the simplest is the most convenient. Thus our experiences would be equally compatible with the geometry of Euclid and with a geometry of Lobatchevski which supposed the curvature of space to be very small. We choose the geometry of Euclid because it is the simplest. If our experiences should be considerably different, the geometry of Euclid would no longer suffice to represent them conveniently, and we should choose a different geometry.

Let it not be said that the reason why we deem the group of Euclid the simplest is because it conforms best to some pre-existing ideal which has already a geometric character; it is simpler because certain of its displacements are interchangeable with one another, which is not true of the corresponding displacements of the group of Lobatchevski. Translated into analytical language, this means that there are fewer terms in the equations, and it is clear that an algebraist who did not know what space or a straight line was would nevertheless look upon this as a condition of simplicity.

In fine, it is our mind that furnishes a category for nature. But this category is not a bed of Procrustes into which we violently force nature, mutilating her as our needs require. We offer to nature a choice of beds among which we choose the couch best suited to her stature.

C. INTUITION AND LOGIC IN MATHEMATICS (*POINCARÉ 1900*)

The following essay treats many of the same issues as do the selections from Felix Klein: styles of mathematical reasoning; the arithmetizing of mathematics; the psychology of mathematical invention; the varieties of intuition, and their role in mathematics. The translation is by George Bruce Halsted, and follows the revised and shortened text published in *Poincaré 1905a*. References should be to the section numbers, which appeared in the original edition.

I

It is impossible to study the works of the great mathematicians, or even those of the lesser, without noticing and distinguishing two opposite tendencies, or rather two entirely different kinds of minds. The one sort are above all preoccupied with logic; to read their works, one is tempted to believe they have advanced only step by step, after the manner of a Vauban who pushes on his trenches against the place besieged, leaving nothing to chance. The other sort are guided by intuition and at the first stroke make quick but sometimes precarious conquests, like bold cavalrymen of the advance guard.

The method is not imposed by the matter treated. Though one often says of the first that they are *analysts* and calls the others *geometers*, that does not prevent the one sort from remaining analysts even when they work at geometry, while the others are still geometers even when they occupy themselves with pure analysis. It is the very nature of their mind which makes them logicians of intuitionalists, and they cannot lay it aside when they approach a new subject.

Nor is it education which has developed in them one of the two tendencies and stifled the other. The mathematician is born, not made, and it seems he is born a geometer or an analyst. I should like to cite examples and there are surely plenty; but to accentuate the contrast I shall begin with an extreme example, taking the liberty of seeking it in two living mathematicians.

M. Méray wants to prove that a binomial equation always has a root, or, in ordinary words, that an angle may always be subdivided. If there is any truth that we think we know by direct intuition, it is this. Who could doubt that an angle may always be divided into any number of equal parts? M. Méray does not look at it that way; in his eyes this proposition is not at all evident, and to prove it he needs several pages.

On the other hand, look at Professor Klein: he is studying one of the most abstract questions of the theory of functions: to determine whether on a given Riemann surface there always exists a function admitting of given singularities. What does the celebrated German geometer do? He replaces his Riemann surface by a metallic surface whose electric conductivity varies according to certain laws. He connects two of its points with the two poles of a battery. The current, says he, must pass, and the distribution of this current on the surface will define a function whose singularities will be precisely those called for by the statement.

Doubtless Professor Klein well knows he has given here only a sketch; nevertheless he has not hesitated to publish it; and he would probably believe he finds in it, if not a rigorous demonstration, at least a kind of moral certainty. A logician would have rejected with horror such a conception, or rather he would not have had to reject it, because in his mind it would never have originated.

Again, permit me to compare two men, the honour of French science, who have recently been taken from us, but who both entered long ago into immortality. I speak of M. Bertrand and M. Hermite. They were scholars of the same school at the same time; they had the same education, were under the same influences; and yet what a difference! Not only does it blaze forth in their writings; it is in their teaching, in their way of speaking, in their very look. In the memory of all their pupils these two faces are stamped in deathless lines; for all who have had the pleasure of following their teaching, this remembrance is still fresh; it is easy for us to evoke it.

While speaking, M. Bertrand is always in motion; now he seems in combat with some outside enemy, now he outlines with a gesture of the hand the figures he studies. Plainly he sees and he is eager to paint, this is why he calls gesture to his aid. With M. Hermite, it is just the opposite; his eyes seem to shun contact with the world; it is not without, it is within he seeks the vision of truth.

Among the German geometers of this century, two names above all are illustrious, those of the two scientists who founded the general theory of functions, Weierstrass and Riemann. Weierstrass leads everything back to the consideration of series and their analytic transformations; to express it better, he reduces analysis to a sort of prolongation of arithmetic; you may turn through all his books without finding a figure. Riemann, on the contrary, at once calls geometry to his aid; each of his conceptions is an image that no one can forget, once he has caught its meaning.

More recently, Lie was an intuitionalist; this might have been doubted in reading his books, but no one could doubt it after talking with him; you saw at once that he thought in pictures. Madame Kovalevski was a logician.

Among our students we notice the same differences; some prefer to treat their problems 'by analysis', others 'by geometry'. The first are incapable of 'seeing in space', the others are quickly tired of long calculations and become perplexed.

The two sorts of minds are equally necessary for the progress of science; both the logicians and the intuitionalists have achieved great things that others could not have done. Who would venture to say whether he preferred that Weierstrass had never written or that there had never been a Riemann? Analysis and synthesis have then both their legitimate roles. But it is interesting to study more closely in the history of science the part which belongs to each.

II

Strange! If we read over the works of the ancients we are tempted to class them all among the intuitionalists. And yet nature is always the same; it is hardly probable that it has begun in this century to create minds devoted to logic. If we could put ourselves into the flow of ideas which reigned in their time, we should recognize that many of the old geometers were in tendency analysts. Euclid, for example, erected a scientific structure wherein his contemporaries could find no fault. In this vast construction, of which each piece however is due to intuition, we may still to-day, without much effort, recognize the work of a logician.

It is not minds that have changed, it is ideas; the intuitional minds have remained the same; but their readers have required of them greater concessions.

What is the cause of this evolution? It is not hard to find. Intuition cannot give us rigour, nor even certainty; this has been recognized more and more. Let us cite some examples. We know there exist continuous functions lacking derivatives. Nothing is more shocking to intuition than this proposition, which is imposed upon us by logic. Our fathers would not have failed to say: 'It is evident that every continuous function has a derivative, since every curve has a tangent.'

How can intuition deceive us on this point? It is because when we seek to imagine a curve we can not represent it to ourselves without width; just so, when we represent to ourselves a straight line, we see it under the form of a rectilinear band of a certain breadth. We well know these lines have no width; we try to imagine them narrower and narrower and thus to approach the limit; so we do in a certain measure, but we shall never attain this limit. And then it is clear we can always picture these two narrow bands, one straight, one curved, in a position such that they encroach slightly one upon the other without crossing. We shall thus be led, unless warned by a rigorous analysis, to conclude that a curve always has a tangent.

I shall take as a second example Dirichlet's principle on which rest so many theorems of mathematical physics; to-day we establish it by reasoning very rigorous but very long; heretofore, on the contrary, we were content with a very summary proof. A certain integral depending on an arbitrary function can never vanish. Hence it is concluded that it must have a minimum. The flaw in this reasoning strikes us immediately, since we use the abstract term *function* and are familiar with all the singularities functions can present when the word is understood in the most general sense.

But it would not be the same had we used concrete images, had we, for example, considered this function as an electric potential; it would have been thought legitimate to affirm that electrostatic equilibrium can be attained. Yet perhaps a physical comparison would have awakened some vague distrust. But if care had been taken to translate the reasoning into the language of geometry, intermediate between that of analysis and that of physics, doubtless this distrust would not have been produced, and perhaps one might thus, even to-day, still deceive many readers not forewarned.

Intuition, therefore, does not give us certainty. This is why the evolution had to happen; let us now see how it happened.

It was not slow in being noticed that rigour could not be introduced in the reasoning unless first made to enter into the definitions. For the most part the objects treated of by mathematicians were long ill defined; they were supposed to be known because represented by means of the senses or the imagination; but one had only a crude image of them and not a precise idea on which reasoning could take hold. It was there first that the logicians had to direct their efforts.

So in the case of incommensurable numbers. The vague idea of continuity, which we owe to intuition, resolved itself into a complicated system of inequalities referring to whole numbers.

By that means the difficulties arising from passing to the limit, or from the consideration of infinitesimals, are finally removed. Today in analysis only whole numbers are left or systems, finite or infinite, of whole numbers bound together by a net of equality or inequality relations. Mathematics, as they say, is arithmetized.

III

A first question presents itself. Is this evolution ended? Have we finally attained absolute rigour? At each stage of the evolution our fathers also thought they had reached it. If they deceived themselves, do we not likewise cheat ourselves?

We believe that in our reasonings we no longer appeal to intuition; the philosophers will tell us this is an illusion. Pure logic could never lead us to anything but tautologies; it could create nothing new; not from it alone can any science issue. In one sense these philosophers are right; to make arithmetic, as to make geometry, or to make any science, something else than pure logic is necessary. To designate this something else we have no word other than *intuition*. But how many different ideas are hidden under this same word?

Compare these four axioms: (1) Two quantities equal to a third are equal to one another; (2) if a theorem is true of the number 1 and if we prove that it is true of $n + 1$ if true for n, then it will be true of all whole numbers; (3) if on a straight line the point C is between A and B and the point D between A and C, then the point D will be between A and B; (4) through a given point there is not more than one parallel to a given straight line.

All four are attributed to intuition, and yet the first is the enunciation of one of the rules of formal logic; the second is a real synthetic *a priori* judgement, and is the foundation of rigorous mathematical induction; the third is an appeal to the imagination; the fourth is a disguised definition.

Intuition is not necessarily founded on the evidence of the senses; the senses would soon become powerless; for example, we can not represent to ourselves a chiliagon, and yet we reason by intuition on polygons in general, which include the chiliagon as a particular case.

You know what Poncelet understood by the *principle of continuity*. What is true of a real quantity, said Poncelet, should be true of an imaginary quantity; what is true of the hyperbola whose asymptotes are real, should then be true of the ellipse whose asymptotes are imaginary. Poncelet was one of the most intuitive minds of this century; he was passionately, almost ostentatiously, so; he regarded the principle of continuity as one of his boldest conceptions, and yet this principle did not rest on the evidence of the senses. To assimilate the hyperbola to the ellipse was rather to contradict this evidence. It was only a sort of precocious and instinctive generalization, which, moreover, I have no desire to defend.

We have then many kinds of intuition; first, the appeal to the senses and the imagination; next generalization by induction, copied, so to speak, from the procedures of the experimental sciences; finally, we have the intuition of pure number, whence arose the second of the axioms just enunciated, which is able to create the real mathematical reasoning. I have shown above by examples that the first two cannot give us certainty; but who will seriously doubt the third, who will doubt arithmetic?

Now in the analysis of today, when one cares to take the trouble to be rigorous, there can be nothing but syllogisms or appeals to this intuition of pure number, the only intuition which cannot deceive us. It may be said that today absolute rigour is attained.

IV

The philosophers make still another objection: 'What you gain in rigour,' they say, 'you lose in objectivity. You can rise toward your logical ideal only by cutting the bonds which attach you to reality. Your science is infallible, but it can only remain so by imprisoning itself in an ivory tower and renouncing all relation with the external world. From this seclusion it must go out when it would attempt the slightest application.'

For example, I seek to show that some property pertains to some object whose concept seems to me at first indefinable, because it is intuitive. At first I fail or must content myself with approximate proofs; finally I decide to give to my object a precise definition, and this enables me to establish this property in an irreproachable manner.

'And then,' say the philosophers, 'it still remains to show that the object which corresponds to this definition is indeed the same made known to you by intuition; or else that some real and concrete object whose conformity with your intuitive idea you believe you immediately recognize corresponds to your new definition. Only then could you affirm that it has the property in question. You have only displaced the difficulty.'

That is not exactly so; the difficulty has not been displaced, it has been divided. The proposition to be established was in reality composed of two different truths, at first not distinguished. The first was a mathematical truth, and it is now rigorously established. The second was an experimental verity.

Experience alone can teach us that some real and concrete object corresponds or does not correspond to some abstract definition. This second verity is not mathematically demonstrated, but neither can it be, no more than can the empirical laws of the physical and natural sciences. It would be unreasonable to ask more.

Well, is it not a great advance to have distinguished what long was wrongly confused? Does this mean that nothing is left of this objection of the philosophers? That I do not intend to say; in becoming rigorous, mathematical science takes a character so artificial as to strike every one; it forgets its historical origins; we see how the questions can be answered, we no longer see how and why they are put.

This shows us that logic is not enough; that the science of demonstration is not all science and that intuition must retain its role as complement, I was about to say as counterpoise or as antidote to logic.

I have already had occasion to insist on the place intuition should hold in the teaching of the mathematical sciences. Without it young minds could not make a beginning in the understanding of mathematics; they could not learn to love it and would see in it only a vain logomachy; above all, without intuition they would never become capable of applying mathematics. But now I wish before all to speak of the role of intuition in science itself. If it is useful to the student it is still more so to the creative scientist.

V

We seek reality, but what is reality? The physiologists tell us that organisms are formed of cells; the chemists add that cells themselves are formed of atoms. Does this mean that these atoms or these cells constitute reality, or rather the sole reality? The way in which these cells are arranged and from which results the unity of the individual, is not it also a reality much more interesting than that of the isolated elements, and should a naturalist who had never studied the elephant except by means of the microscope think himself sufficiently acquainted with that animal?

Well, there is something analogous to this in mathematics. The logician cuts up, so to speak, each demonstration into a very great number of elementary operations; when we have examined these operations one after the other and ascertained that each is correct, are we to think we have grasped the real meaning of the demonstration? Shall we have understood it even when, by an effort of memory, we have become able to repeat this proof by reproducing all these elementary operations in just the order in which the inventor had arranged them? Evidently not; we shall not yet possess the entire reality; that I know not what, which makes the unity of the demonstration, will completely elude us.

Pure analysis puts at our disposal a multitude of procedures whose infallibility it guarantees; it opens to us a thousand different ways on which we can embark in all confidence; we are assured of meeting there no obstacles; but of all these ways, which will lead us most promptly to our goal? Who shall tell us which to choose? We need a faculty which makes us see the end from afar, and intuition is this faculty. It is necessary to the explorer for choosing his route; it is not less so to the one following his trail who wants to know why he chose it.

If you are present at a game of chess, it will not suffice, for the understanding of the game, to know the rules for moving the pieces. That will only enable you to recognize that each move has been made conformably to these rules, and this knowledge will truly have very little value. Yet this is what the reader of a book on mathematics would do if he were a logician only. To understand the game is wholly another matter; it is to know why the player moves this piece rather than that other which he could have moved without breaking the rules of the game. It is to perceive the inward reason which makes of this series of successive moves a sort of organized whole. This faculty is still more necessary for the player himself, that is, for the inventor.

Let us drop this comparison and return to mathematics. For example, see what has happened to the idea of continuous function. At the outset this was only a sensible image, for example, that of a continuous mark traced by the chalk on a blackboard. Then it became little by little more refined; before long it was used to construct a complicated system of inequalities, which reproduced, so to speak, all the lines of the original image; this construction finished, the centring of the arch, so to say, was removed, that crude representation which had temporarily served as support and which was afterward useless was rejected; there remained only the construction itself, irreproachable in the eyes of the logician. And yet if the primitive image had totally disappeared from our recollection, how could we divine by what caprice all these inequalities were erected in this fashion one upon another?

Perhaps you think I use too many comparisons; yet pardon still another. You have doubtless seen those delicate assemblages of siliceous needles which form the skeleton of certain sponges. When the organic matter has disappeared, there remains only a frail and elegant lace-work. True, nothing is there except silica, but what is interesting is the form this silica has taken, and we could not understand it if we did not know the living sponge which has given it precisely this form. Thus it is that the old intuitive notions of our fathers, even when we have abandoned them, still imprint their form upon the logical constructions we have put in their place.

This view of the aggregate is necessary for the inventor; it is equally necessary for whoever wishes really to comprehend the inventor. Can logic give it to us? No; the name mathematicians give it would suffice to prove this. In mathematics logic is called *analysis*, and analysis means *division, dissection*. It can have, therefore, no tool other than the scalpel and the microscope.

Thus logic and intuition have each their necessary role. Each is indispensable. Logic, which alone can give certainty, is the instrument of demonstration; intuition is the instrument of invention.

VI

But at the moment of formulating this conclusion I am seized with scruples. At the outset I distinguished two kinds of mathematical minds, the one sort logicians and analysts, the others intuitionalists and geometers. Well, the analysts also have been inventors. The names I have just cited make my insistence on this unnecessary.

Here is a contradiction, at least apparently, which needs explanation. And first, do you think these logicians have always proceeded from the general to the particular, as the rules of formal logic would seem to require of them? Not thus could they have extended the boundaries of science; scientific conquest is to be made only by generalization.

In one of the chapters of *Science and hypothesis* I have had occasion to study the nature of mathematical reasoning, and I have shown how this reasoning, without ceasing to be absolutely rigorous, could lift us from the particular to the general by a procedure I have called *mathematical induction*. It is by this procedure that the analysts have made science progress, and if we examine the detail itself of their demonstrations, we shall find it there at each instant beside the classic syllogism of Aristotle. We, therefore, see already that the analysts are not simply makers of syllogisms after the fashion of the scholastics.

Besides, do you think they have always marched step by step with no vision of the goal they wished to attain? They must have divined the way leading thither, and for that they needed a guide. This guide is, first, analogy. For example, one of the methods of demonstration dear to analysts is that founded on the employment of dominant functions. We know it has already served to solve a multitude of problems; in what consists then the role of the inventor who wishes to apply it to a new problem? At the outset he must recognize the analogy of this question with those which have already been solved by this method; then he must perceive in what way this new question differs from the others, and thence deduce the modifications necessary to apply to the method.

But how does one perceive these analogies and these differences? In the example just cited they are almost always evident, but I could have found others where they would have been much more deeply hidden; often a very uncommon penetration is necessary for their discovery. The analysts, not to let these hidden analogies escape them, that is, in order to be inventors, must, without the aid of the senses and imagination, have a direct sense of what constitutes the unity of a piece of reasoning, of what makes, so to speak, its soul and inmost life.

When one talked with M. Hermite, he never evoked a sensuous image, and yet you soon perceived that the most abstract entities were for him like living beings. He did not see them, but he perceived that they are not an artificial assemblage, and that they have some principle of internal unity.

But, one will say, that still is intuition. Shall we conclude that the distinction made at the outset was only apparent, that there is only one sort of mind, and that all the mathematicians are intuitionalists, at least those who are capable of inventing?

No, our distinction corresponds to something real. I have said above that there are many kinds of intuition. I have said how much the intuition of pure number, whence comes rigorous mathematical induction, differs from sensible intuition to which the imagination, properly so called, is the principal contributor.

Is the abyss which separates them less profound than it at first appeared? Could we recognize with a little attention that this pure intuition itself could not do without the aid of the senses? This is the affair of the psychologist and the metaphysician, and I shall not discuss the question. But the thing's being doubtful is enough to justify me in recognizing and affirming an essential difference between the two kinds of intuition; they have not the same object and seem to call into play two different faculties of our soul; one would think of two search-lights directed upon two worlds strangers to one another.

It is the intuition of pure number, that of pure logical forms, which illumines and directs those we have called *analysts*. This it is which enables them not only to demonstrate, but also to invent. By it they perceive at a glance the general plan of a logical edifice, and that too without the senses appearing to intervene. In rejecting the aid of the imagination, which, as we have seen, is not always infallible, they can advance without fear of deceiving themselves. Happy, therefore, are those who can do without this aid! We must admire them; but how rare they are!

Among the analysts there will then be inventors, but they will be few. The majority of us, if we wished to see afar by pure intuition alone, would soon feel ourselves seized with vertigo. Our weakness has need of a more solid staff, and, despite the exceptions of which we have just spoken, it is none the less true that sensible intuition is in mathematics the most usual instrument of invention.

Apropos of these reflections, a question comes up that I have not the time either to solve or even to enunciate with the developments it would admit of. Is there room for a new distinction, for distinguishing among the analysts those who above all use pure intuition and those who are first of all preoccupied with formal logic?

M. Hermite, for example, whom I have just cited, cannot be classed among the geometers who make use of sensible intuition; but neither is he a logician, properly so called. He does not conceal his aversion to purely deductive procedures which start from the general and end in the particular.

D. MATHEMATICS AND LOGIC: I (*POINCARÉ 1905b*)

The next three selections appeared as a series of articles in the *Revue de métaphysique et de morale* under the title '*Les mathématiques et la logique*'. The three articles contain Poincaré's reaction against the recent foundational work of a number of mathematicians, among them Cantor, Zermelo, Russell, Peano, Couturat, and Hilbert. The articles are noteworthy, not only for spectacular polemics, but also for their introduction of two influential arguments into the foundations of mathematics. The first is that the various logical analyses of fundamental mathematical concepts (such as number or the principle of mathematical induction) commit a *petitio principii* by presupposing the very concepts they are supposed to define. This argument was particularly powerful at the time, since Hilbert had not yet formulated a crisp distinction between formal languages and metalanguages. Once the distinction was in place, it became possible to distinguish between the (strong) induction principles expressed in the formal language and the (weaker) induction principles accepted in the metalanguage; but in 1905 it looked as though Hilbert might have to presuppose fully-fledged mathematical induction to prove the consistency of arithmetic (including the principle of induction).

Poincaré's second argument, introduced in the third essay in the series, is his most influential contribution to foundations of mathematics. This is the discussion of the 'vicious circle principle', or of *impredicative definitions*. In the interval between the first two articles and the third, Poincaré had read *Russell 1906a*, which explores various possible solutions to the set-theoretical paradoxes; moreover, he had received a letter from Zermelo which attempted to answer his criticisms in the first two articles of the logicist definition of number. Poincaré's response is an intricate simultaneous argument against all his opponents. Russell had used the term 'predicative' to designate properties that define a class—that is, for Russell Φ is predicative if $\{x:\Phi(x)\}$ exists. Poincaré, in discussing Richard's paradox of the class of decimal numbers definable in a finite number of words, adopts Richard's solution of the paradox: the definition of the would-be class is illegitimate because it rests upon a *vicious circle*. Poincaré then proposes a change in Russell's terminology: 'The definitions that ought to be regarded as impredicative are those that contain a vicious circle.' In effect, Poincaré here proposes to deploy the vicious circle principle not only against the so-called 'semantical paradoxes' (like Richard's), but also against the set-theoretical paradoxes (like Burali-Forti's and Zermelo's). And in an argumentative *tour de force* he also tries to use the new conception of impredicativity against Zermelo and the logicists, arguing that their definitions of natural number are illegitimately impredicative. This last issue led to several exchanges with Zermelo: see *Zermelo 1908a* and *1909b*, and *Poincaré 1909a* and *1909b*. For recent discussions of the delicate issues raised by predicativity see, for example, *Feferman 1964*, *Parsons 1965*, *Chihara 1973*, *Goldfarb 1988*, and *Goldfarb 1989*.

A textual note on the selection that follows. In *Science et méthode* (*Poincaré 1908*) the first two articles in the series were heavily abridged, and divided into two chapters: Chapter III ('Mathematics and logic') and Chapter IV ('The new logics'). The division took place between sections IX and X of the first article; so the following selection, after abridgement, became Chapter III of *Science et méthode* and sections I to V of Chapter IV. In the following translation the section-numbers are those of the original, longer article; but the section-numbers of the version of 1908 are also given in double square brackets. Passages that appeared in the original version but that were *deleted* by Poincaré in the 1908 version are enclosed in single square brackets. Passages that he *added* in the 1908 version are given in the footnotes. The translation of the passages that appeared in 1908 is by George Bruce Halsted, with revisions by William Ewald; the translation of the remainder is by William Ewald. References to *Poincaré 1905b* should be to the section numbers, which appeared in the original edition.

I[a]

In recent years numerous works have been published on pure mathematics and the philosophy of mathematics, trying to separate and isolate the logical

[a] ⟦In the revised version of 1908 Poincaré added the following introductory paragraphs, which preceded I⟧:
Can mathematics be reduced to logic without having to appeal to principles peculiar to mathematics? There is a whole school, abounding in ardour and full of faith, striving to prove it. They have their own special language, which is without words, using only signs. This language is understood only by the initiates, so that commoners are disposed to bow to the trenchant affirmations of the adepts. It is perhaps not unprofitable to examine these affirmations somewhat closely, to see if they justify the peremptory tone with which they are presented.

But to make clear the nature of the question it is necessary to enter upon certain historical details and in particular to recall the character of the works of Cantor.

Since long ago the notion of infinity had been introduced into mathematics; but this infinite was what philosophers call a *becoming*. The mathematical infinite was only a quantity capable of increasing beyond all limit: it was a variable quantity of which it could not be said that it *had passed* all limits, but only that it *could pass* them.

Cantor has undertaken to introduce into mathematics an *actual infinite*, that is to say a quantity which not only is capable of passing all limits, but which is regarded as having already passed them. He has set himself questions like these: Are there more points in space than whole numbers? Are there more points in space than points in a plane? etc.

And then the number of whole numbers, that of the points of space, etc., constitutes what he calls a *transfinite cardinal number*, that is to say a cardinal number greater than all the ordinary cardinal numbers. And he has occupied himself in comparing these transfinite cardinal numbers. In arranging in a well-ordering the elements of an aggregate containing an infinity of them, he has also imagined what he calls transfinite ordinal numbers, upon which I shall not dwell.

Many mathematicians have followed his lead and set a series of questions of the sort. They so familiarized themselves with transfinite numbers that they have come to make the theory of finite elements of mathematical reasoning. These works have been analysed and expounded very clearly by M. Couturat in a series of articles entitled: The principles of mathematics ‖*Couturat 1905a*‖.

[I mention in the first instance the writings of Hilbert and his disciples, those of Whitehead, of B. Russell, those of Peano and his school. One should not be surprised that I do not mention Veronese: although he has struck on the same points as Hilbert, he takes a different point of view and in contrast is constantly preoccupied with preserving a legitimate place for intuition.

[I have lately had the opportunity to praise Hilbert's book and to emphasize its importance, and in general all his works seem to me to possess very great interest. Henceforth they will play an important role in all investigations of this genre, and one may wonder if they do not once again place in question several conclusions that certain philosophers believed settled.]

For M. Couturat, the new works, and in particular those of Russell and Peano, have finally settled the controversy, so long pending between Leibnitz and Kant. They have shown that there are no synthetic judgements *a priori* (Kant's phrase to designate judgements which can neither be demonstrated analytically, nor reduced to identities, nor established experimentally); they have shown that mathematics is entirely reducible to logic and that intuition here plays no role.

This is what M. Couturat has set forth in the work just cited; this he says still more explicitly in his Kant jubilee discourse, so that I heard my neighbour whisper: 'I well see this is the centenary of Kant's *death*.'

Can we subscribe to this conclusive condemnation? I think not, and I shall try to show why.

---

numbers depend upon that of Cantor's cardinal numbers. In their eyes, to teach arithmetic in a truly logical fashion one should begin by establishing the general properties of transfinite cardinal numbers, and then distinguish among them a very small class, that of the ordinary whole numbers. Thanks to this detour, one might succeed in proving all the propositions relative to this little class (that is to say all our arithmetic and our algebra) without using any principle foreign to logic. This method is evidently contrary to all sane psychology; it is certainly not in this way that the human mind proceeded in constructing mathematics; so its authors do not dream, I think, of introducing it into secondary teaching. But is it at least logic, or, better, is it correct? This may be doubted.

The geometers who have employed it are however very numerous. They have accumulated formulae and they have thought to free themselves from what was not pure logic by writing memoirs where the formulae no longer alternate with explanatory discourse as in the books of ordinary mathematics, but where this discourse has completely disappeared.

Unfortunately they have reached contradictory results, what are called the *Cantorian antinomies*, to which we shall have occasion to return. These contradictions have not discouraged them and they have tried to modify their rules so as to make those disappear which had already shown themselves, without being sure, for all that, that new ones would not manifest themselves.

It is time to administer justice on these exaggerations. I do not hope to convince them; for they have lived too long in this atmosphere. Besides, when one of their demonstrations has been refuted, we are sure to see it resurrected with insignificant alterations, and some of them have already risen several times from their ashes. Such long ago was the Lernaean hydra with its famous heads which always grew again. Hercules got through, since his hydra had only nine heads, or eleven; but here there are too many, some in England, some in Germany, in Italy, in France, and he would have to give up the struggle. So I appeal only to unprejudiced men of good judgement.‖

II

What strikes us first in the new mathematics is its purely formal character: 'We think,' says Hilbert, 'three sorts of *things*, which we shall call points, lines, and planes. We stipulate that a line shall be determined by two points, and that in place of saying this line is determined by these two points, we may say it passes through these two points, or that these two points are situated on this line.' What these *things* are, not only we do not know, but we should not seek to know. We have no need to, and one who never had seen either point or line or plane could geometrize as well as we. That the phrase *to pass through*, or the phrase *to be situated upon* may arouse in us no image, the first is simply a synonym of *to be determined* and the second of *to determine*.

Thus, be it understood, to demonstrate a theorem, it is neither necessary nor even advantageous to know what it means. The geometer might be replaced by the *logic piano* imagined by Stanley Jevons; or, if you choose, a machine might be imagined where the assumptions were put in at one end, while the theorems came out at the other, like the legendary Chicago machine where the pigs go in alive and come out transformed into hams and sausages. No more than these machines need the mathematician know what he does.

I do not make this formal character of his geometry a reproach to Hilbert. This is the way he should go, given the problem he set himself. He wished to reduce to a minimum the number of the fundamental assumptions of geometry and completely enumerate them; now, in reasonings where our mind remains active, in those where intuition still plays a part, in living reasonings, so to speak, it is difficult not to introduce an assumption or a postulate which passes unperceived. It is therefore only after having carried back all the geometric reasonings to a purely mechanical form that he could be sure of having accomplished his design and finished his work.

What Hilbert did for geometry, others have tried to do for arithmetic and analysis. Even if they had entirely succeeded, would the Kantians be finally condemned to silence? Perhaps not, for in reducing mathematical thought to an empty form, it is certainly mutilated. Even admitting it were established that all the theorems could be deduced by purely analytic procedures, by simple logical combinations of a finite number of assumptions, and that these assumptions are only conventions; the philosopher would still have the right to investigate the origins of these conventions, to see why they have been judged preferable to the contrary conventions.

And then, the logical correctness of the reasonings leading from the assumptions to the theorems is not the only thing which should occupy us. The rules of perfect logic, are they the whole of mathematics? As well say the whole art of playing chess reduces to the rules of the moves of the pieces. Among all the constructs which can be built up of the materials furnished by logic, choice must be made; the true geometer makes this choice judiciously because he is guided by a sure instinct, or by some vague consciousness of I know not what more profound and more hidden geometry, which alone gives value to the edifice constructed.

To seek the origin of this instinct, to study the laws of this deep geometry, felt, not stated, would also be a fine employment for the philosophers who do not want logic to be all. But this is not the point of view I wish to adopt; it is not thus I wish to consider the question. The instinct mentioned is necessary for the inventor, but it would seem at first we might do without it in studying the science once created. Well, what I wish to investigate is if it be true that, the principles of logic once admitted, one can, I do not say discover, but demonstrate, all the mathematical verities without making a new appeal to intuition.

III

I once said no to this question. Should my reply be modified by the recent works? My saying no was because 'the principle of complete induction' seemed to me at once necessary to the mathematician and irreducible to logic. The statement of this principle is: 'If a property be true of the number 1, and if we establish that it is true of $n + 1$ provided it be of n, it will be true of all the whole numbers.' Therein I see mathematical reasoning *par excellence*. I did not mean to say, as has been supposed, that all mathematical reasonings can be reduced to an application of this principle. Examining these reasonings closely, we there should see applied many other analogous principles, presenting the same essential characteristics. In this category of principles, that of complete induction is only the simplest of all and this is why I have chosen it as type.[b]

IV

DEFINITIONS AND AXIOMS

The existence of such principles is a difficulty for the uncompromising logicians; how do they pretend to get out of it? The principle of complete induction, they say, is not an assumption properly so called or a synthetic judgement *a priori*; it is just simply the definition of whole number. It is therefore a simple convention. To discuss this way of looking at it, we must examine a little closely the relations between definitions and axioms.

Let us go back first to an article by M. Couturat on mathematical definitions which appeared in *l'Enseignement mathématique*, a magazine published by Gauthier-Villars and by Georg at Geneva. We shall see there a distinction between the *direct definition* and the *definition by postulates*.

'The definition by postulates', says M. Couturat, 'applies, not to a single

[b] ⌊In the version of 1908, Poincaré added the following two-sentence paragraph at this point: The current name, principle of complete induction, is not justified. This mode of reasoning is none the less a true mathematical induction which differs from ordinary induction only by its certitude.⌋

notion, but to a system of notions; it consists in enumerating the fundamental relations which unite them and which enable us to demonstrate all their other properties; these relations are postulates.'

If all these notions but one have previously been defined, then this last will be by definition the thing which verifies these postulates.

Thus certain indemonstrable axioms of mathematics would be only disguised definitions. This point of view is often legitimate; and I have myself admitted it in regard for instance to Euclid's postulate.

The other axioms of geometry do not suffice to define distance completely; the distance then will be, by definition, among all the magnitudes which satisfy these other axioms, that which is such as to make Euclid's postulate true.

Well, the logicians suppose true for the principle of complete induction what I admit for Euclid's postulate; they want to see in it only a disguised definition.

But to give them this right, two conditions must be fulfilled. Stuart Mill says every definition implies an axiom, that by which the existence of the defined object is affirmed. According to that, it would no longer be the axiom which might be a disguised definition, it would on the contrary be the definition which would be a disguised axiom. Stuart Mill meant the word existence in a material and empirical sense; he meant to say that in defining the circle we affirm there are round things in nature.

Under this form, his opinion is inadmissible. Mathematics is independent of the existence of material objects; in mathematics the word 'exist' can have only one meaning; it means free from contradiction. Thus rectified, Stuart Mill's thought becomes exact; in defining a thing, we affirm that the definition implies no contradiction.

If therefore we have a system of postulates, and if we can demonstrate that these postulates imply no contradiction, we shall have the right to consider them as representing the definition of one of the notions entering therein. If we cannot demonstrate that, it must be admitted without proof, and that then will be an axiom; so that, seeking the definition under the postulate, we should find the axiom under the definition.

Usually, to show that a definition implies no contradiction, we proceed *by example*; we try to make an *example* of a thing satisfying the definition. Take the case of a definition by postulates; we wish to define a notion A, and we say that, by definition, an A is anything for which certain postulates are true. If we can prove directly that all these postulates are true of a certain object B, the definition will be justified; the object B will be an *example* of an A. We shall be certain that the postulates are not contradictory, since there are cases where they are all true at the same time.

But such a direct demonstration by example is not always possible.

To establish that the postulates imply no contradiction, it is then necessary to consider all the propositions deducible from these postulates considered as premisses, and to show that, among these propositions, no two are contradictory. If these propositions are finite in number, a direct verification is possible. This case is infrequent and uninteresting.

If these propositions are infinite in number, this direct verification can no longer be made; recourse must be had to procedures where in general it is necessary to invoke just this principle of complete induction which is precisely the thing to be proved.

This is an explanation of one of the conditions the logicians should satisfy, *and further on we shall see they have not done it.*

V

There is a second. When we give a definition, it is to use it.

We therefore shall find in the sequel of the exposition the word defined; have we the right to affirm, of the thing represented by this word, the postulate which has served for definition? Yes, evidently, if the word has retained its meaning, if we do not attribute to it implicitly a different meaning. Now this is what sometimes happens, and it is usually difficult to perceive it; it is needful to see how this word comes into our discourse, and if the gate by which it has entered does not imply in reality a definition other than that stated.

This difficulty presents itself in all the applications of mathematics. The mathematical notion has been given a very refined and very rigorous definition; and for the pure mathematician all doubt has disappeared; but if one wishes to apply it to the physical sciences for instance, it is no longer a question of this pure notion, but of a concrete object which is often only a rough image of it. To say that this object satisfies, at least approximately, the definition, is to state a new truth, which experience alone can put beyond doubt, and which no longer has the character of a conventional postulate.

But without going beyond pure mathematics, we also meet the same difficulty.

You give a subtle definition of numbers; then, once this definition has been given, you think no more of it; because, in reality, it is not it which has taught you what number is; you long ago knew that, and when the word number further on is found under your pen, you give it whatever sense first comes to hand. To know what is this meaning and whether it is the same in this phrase or that, it is needful to see how you have been led to speak of number and to introduce this word into these two phrases. I shall not for the moment dilate upon this point, because we shall have occasion to return to it.

Thus consider a word of which we have given explicitly a definition A; afterwards in the discourse we make a use of it which implicitly supposes another definition B. It is possible that these two definitions designate the same thing. But that this is so is a new truth which must either be demonstrated or admitted as an independent axiom.

We shall see farther on that the logicians have not fulfilled the second condition any better than the first.

VI

The definitions of number are numerous and very different; I forgo the enumeration even of the names of their authors. We should not be astonished that there are so many. If one among them was satisfactory, no new one would be given. If each new philosopher occupying himself with this question has thought he must invent another one, this was because he was not satisfied with those of his predecessors, and he was not satisfied with them because he thought he saw a *petitio principii.*

I have always felt, in reading the writings devoted to this problem, a profound feeling of discomfort; I was always expecting to run against a *petitio principii*, and when I did not immediately perceive it, I feared I had overlooked it.

This is because it is impossible to give a definition without using a sentence, and difficult to make a sentence without using a number word, or at least the word several, or at least a word in the plural. And then the declivity is slippery and at each instant there is risk of a fall into *petitio principii.*

I shall devote my attention in what follows only to those of these definitions where the *petitio principii* is most ably concealed.

VII

Pasigraphy

The symbolic language created by Peano plays a very grand role in these new researches. It is capable of rendering some service, but I think M. Couturat attaches to it an exaggerated importance which must astonish Peano himself.

The essential element of this language is certain algebraic signs which represent the different conjunctions: if, and, or, therefore. That these signs may be convenient is possible; but that they are destined to revolutionize all philosophy is a different matter. It is difficult to admit that the word *if* acquires, when written $\supset$, a virtue it had not when written *if*.

This invention of Peano was first called *pasigraphy*, that is to say the art of writing a treatise on mathematics without using a single word of ordinary language. This name defined its range very exactly. Later, it was raised to a more eminent dignity by conferring on it the title of logistic. This word is, it appears, employed at the Military Academy, to designate the art of the quartermaster of cavalry, the art of marching and cantoning troops; but here no confusion need be feared, and it is at once seen that this new name implies the design of revolutionizing logic.

We may see the new method at work in a mathematical memoir by Burali-Forti, entitled: *Una Questione sui numeri transfiniti*, inserted in Volume XI of the *Rendiconti del circolo matematico di Palermo* ⟦*Burali-Forti 1897*⟧.

I begin by saying this memoir is very interesting, and my taking it here as example is precisely because it is the most important of all those memoirs written in the new language. Besides, the uninitiated may read it, thanks to an Italian interlinear translation.

What constitutes the importance of this memoir is that it has given the first example of those antinomies met in the study of transfinite numbers, and making since some years the despair of mathematicians. The aim, says Burali-Forti, of this note is to show there may be two transfinite numbers (ordinals), a and b, such that a is neither equal to, greater than, nor less than b.

To reassure the reader: to comprehend the considerations which follow, he has no need of knowing what a transfinite ordinal number is.

Now, Cantor had precisely proved that between two transfinite numbers as between two finite, there can be no other relation than equality, or inequality in one sense or the other. But it is not of the substance of this memoir that I wish to speak here; that would carry me much too far from my subject; I only wish to consider the form, and just to ask if this form makes it gain much in rigour and whether it thus compensates for the efforts it imposes upon the writer and the reader.

First we see Burali-Forti define the number 1 as follows:

$$1 = \iota T' \{ Ko \frown (u, h) \varepsilon (u \varepsilon \mathrm{Un}) \},$$

a definition eminently fitted to give an idea of the number 1 to persons who had never heard speak of it.

I understand Peanian too ill to dare risk a critique, but still I fear this definition contains a *petitio principii*, considering that I see the figure 1 in the first member and Un in letters in the second.

However that may be, Burali-Forti starts from this definition and, after a short calculation, reaches the equation:

$$(27) \qquad 1 \varepsilon \mathrm{No},$$

which tells us that One is a number.

And since we are on these definitions of the first numbers, we recall that M. Couturat has also defined 0 and 1.

What is zero? It is the number of elements of the null class. And what is the null class? It is that containing no element.

To define zero by null, and null by no, is really to abuse the wealth of language; so M. Couturat has introduced an improvement in his definition by writing:

$$0 = \iota \Lambda : \phi x = \Lambda . \supset . \Lambda = (x \varepsilon \phi x),$$

which means: zero is the number of things satisfying a condition never satisfied.

But as never means *in no case* I do not see that the progress is great.

I hasten to add that the definition M. Couturat gives of the number 1 is more satisfactory.

One, he says in substance, is the number of elements in a class in which any two elements are identical.

It is more satisfactory, I have said, in this sense that to define 1, he does not use the word *one*; in compensation, he uses the word *two*. But I fear, if asked what is *two*, M. Couturat would have to use the word *one*.

VIII

But to return to the memoir of Burali-Forti; I have said his conclusions are in direct opposition to those of Cantor. Now, one day M. Hadamard came to see me and the talk fell upon this antinomy.

'Burali-Forti's reasoning,' I said, 'does it not seem to you irreproachable?' 'No, and on the contrary I find nothing to object to in that of Cantor. Besides, Burali-Forti had no right to speak of the aggregate of *all* the ordinal numbers.'

'Pardon, he had the right, since he could always put

$$\Omega = T'(\text{No}, \bar{\varepsilon} >).$$

I should like to know who was to prevent him, and can it be said a thing does not exist, when we have called it Ω?'

It was in vain, I could not convince him (which besides would have been sad, since he was right). Was it merely because I do not speak the Peanian with enough eloquence? Perhaps; but between ourselves I do not think so.

Thus, despite all this pasigraphic apparatus, the question was not solved. What does that prove? In so far as it is a question only of proving one a number, pasigraphy suffices, but if a difficulty presents itself, if there is an antinomy to solve, pasigraphy becomes impotent.

IX

[Now, what use to us is pasigraphy? If the thesis of the logicians is true, all mathematical reasonings are nothing but mechanical combinations of the rules of logic. I do not wish to say that mathematics could be created by an utterly unintelligent being. But the role of intelligence is limited to choosing from a limited arsenal of rules given in advance, without any right to invent new ones. In this case, all its reasonings can be pasigraphed. Consequently, pasigraphy can furnish us with a criterion for deciding the question that occupies us. If every treatise of mathematics can be translated into Peanian, then the logicians are correct. If that translation is impossible, or if it cannot be given without introducing premisses irreducible to logic, then the Kantians triumph.

[It will be helpful to examine translation more closely. To make us yield, it does not suffice to present us with a page containing nothing but formulae and not a single word of ordinary language. Burali-Forti's adventure illustrates the necessity for circumspection. Burali-Forti and Cantor arrived at contradictory conclusions; but one or the other was wrong. The former employed pasigraphy; the latter could have done so just as easily, and moreover he was the one who was right. So pasigraphy does not preserve us from error. Why? Is it because the rules of logic are deceptive? Evidently not: it is because one has made an appeal to intuition, and one did so incorrectly. This appeal took place, otherwise one would not have been deceived; and it was hidden, otherwise one would not have been able to use the Peanian language. So it is possible, even if one speaks this language fluently, to appeal to intuition in it without being aware that you are doing so. And it will be necessary when one is in the presence of a pasigraphic reasoning, even if the reasoning is correct, to examine it to see whether a similar appeal is hidden in some corner.

X ⟦I⟧

Russell's logic

[Russell begins by developing the fundamental principles of logic, and Couturat begins his exposition in the same way.[c]] It might seem that there is nothing new to be said about logic, and that Aristotle had seen to the bottom of it. But the domain that Russell attributes to logic is infinitely more extended than that of classical logic, and he has put forward views on this subject that are both original and at times correct.

First, Russell subordinates the logic of classes to that of propositions, while the logic of Aristotle was above all the logic of classes, and took as its point of departure the relation of subject to predicate. The classic syllogism, 'Socrates is a man,' etc., gives place to the hypothetical syllogism: 'If *A* is true, *B* is true; now if *B* is true, *C* is true,' etc. And this is, I think, a most happy idea, because the classic syllogism is easy to reduce ⟦ramener⟧ to the hypothetical syllogism, while the inverse transformation is not without difficulty.

And then this is not all. Russell's logic of propositions is the study of the laws of combination of the conjunctions *if*, *and*, and *or*, and the negation *not*. This is a considerable extension of ancient logic. [The properties of the classical syllogism extend without difficulty to the hypothetical syllogism, and in the forms of the latter one easily recognizes the scholastic forms; one finds again the essentials of classical logic. But the theory of the syllogism is nothing but the syntax of the conjunction *if* and perhaps of negation.]

In adding here two other conjunctions *and* and *or*, Russell opens to logic a new field. The symbols *and*, *or* follow the same laws as the two signs $\times$ and $+$, that is to say the commutative, associative, and distributive laws. Thus *and* represents logical multiplication, while *or* represents logical addition. This also is very interesting.

Russell reaches the conclusion that any false proposition implies all other propositions, true or false. M. Couturat says this conclusion will at first seem paradoxical. It is sufficient however to have corrected a bad thesis in mathematics to recognize how right Russell is. The candidate often is at great pains to get the first false equation; but that once obtained, it is only sport then for him to accumulate the most surprising results, some of which may even be true.

[Another felicitous invention is that of *propositional function*: one calls by this label every proposition that depends on something variable, and one

---

[c] ⟦In *Poincaré 1908* a new chapter (Chapter IV, 'The new logics') begins here. The first sentence of this section in 1908 was altered to read: 'To justify its pretensions, logic had to change. We have seen new logics arise, of which the most interesting is that of Russell.'⟧

designates it by $\phi(x)$, x being the variable. The proposition $\phi(x)$ can be true or false. It can happen that it is true for certain choices of x and false for certain others, and this is the origin of the notion of *class* and of the logic of classes; for $\phi(x)$ defines the class of x for which the proposition $\phi(x)$ is true. Moreover, every class can be defined in this fashion; the class man, for example, is the class of x for which the proposition 'x is a man' is true, and this proposition is a propositional function of x.

[It can also happen that the propositional function $\phi(x)$ is true for every choice of x, or at least for every x which belongs to a given class, e.g. to the class defined by the propositional function $\psi(x)$. Then one has

$$\text{If } \psi(\mathrm{x}),\ \phi(\mathrm{x}).$$

This is the sort of proposition that the scholastics designate by A.

[Or again, it can happen that the proposition $\phi(x)$ is true for *at least* one x of the class $\psi(x)$. Then one has a proposition of the sort the scholastics designated by I.

[But what I wish to note is the relation between propositions of the form A and the form I and logical addition and multiplication.

[For if, say, the class $\psi(x)$ contains four elements x_1, x_2, x_3, x_4, then the proposition

$$\text{If } \psi(x),\ \phi(x) \text{ for any } x \text{ (form } A)$$

means:

$$\phi(x_1) \text{ and } \phi(x_2) \text{ and } \phi(x_3) \text{ and } \phi(x_4) \text{ (logical multiplication)}$$

and the proposition

$$\text{If } \psi(x),\ \phi(x) \text{ for at least one } x \text{ (form } I)$$

means

$$\phi(x_1) \text{ or } \phi(x_2) \text{ or } \phi(x_3) \text{ or } \phi(x_4) \text{ (logical addition).}$$

[And this furnishes us with a means of applying logical addition or multiplication to an infinite number of propositions.

[One can also envisage propositional functions of two variables x and y; Russell does so, and thus he constructs the logic of *relations*, but he writes xRy instead of $\phi(x, y)$. Here too we can imagine that $\phi(x, y)$ is true for every choice of x and y, or for at least one pair of x and y, or that for any x one can find a y such that the proposition is true.

[There is no reason not to envisage functions of three variables, and this is in essence what Russell does when he speaks of *classes of relations*, for then R varies, and xRy is a propositional function which depends on three variables, x, y, and R.]

XI ‖II‖

We see how much richer the new logic is than the classic logic; the symbols are multiplied and allow of varied combinations *which are no longer limited in number*. Has one the right to give this extension to the meaning of the word *logic*? It would be useless to examine this question and to seek with Russell a mere quarrel about words. Grant him what he demands; but be not astonished if certain verities declared irreducible to logic in the old sense of the word find themselves now reducible to logic in the new sense—something very different.

A great number of new notions have been introduced, and these are not simply combinations of the old. Russell knows this, and not only at the beginning of the first chapter, 'The logic of propositions', but at the beginning of the second and third, 'The logic of classes' and 'The logic of relations', he introduces new words that he declares indefinable.

And this is not all; he likewise introduces principles he declares indemonstrable. But these indemonstrable principles are appeals to intuition, synthetic judgements *a priori*. We regard them as intuitive when we meet them more or less explicitly enunciated in mathematical treatises; have they changed character because the meaning of the word logic has been enlarged and we now find them in a book entitled *Treatise on logic? They have not changed nature; they have only changed place.*

‖III‖

Could these principles be considered as disguised definitions? It would then be necessary to have some way of proving that they imply no contradiction. It would be necessary to establish that, however far one followed the series of deductions, he would never be exposed to contradicting himself.

We might attempt to reason as follows: We can verify that the operations of the new logic applied to premisses exempt from contradiction can only give consequences equally exempt from contradiction. If therefore after n operations we have not met contradiction, we shall not encounter it after $n + 1$. Thus it is impossible that there should be a moment when contradiction *begins*, which shows we shall never meet it. Have we the right to reason in this way? No, for this would be to make use of complete induction; and *remember, we do not yet know the principle of complete induction.*

We therefore have not the right to regard these axioms as disguised definitions, and only one resource remains for us, to admit a new act of intuition for each of them. Moreover I believe this is indeed the thought of Russell and M. Couturat.

['One defines in an analogous manner', says Couturat, 'the logical sum and product not just of two relations, but of all the relations of a class: these new definitions are necessary, because the preceding ones cannot be extended (by complete induction) except to a finite class of relations, while the new ones hold for any class, infinite as well as finite.

['One is obliged to postulate by special axioms the existence of the logical sum and product thus defined for an entire class of relations.']

Thus each of the nine indefinable notions and of the twenty indemonstrable propositions (I believe if it were I that did the counting, I should have found some more) which are the foundation of the new logic, logic in the broad sense, presupposes a new and independent act of our intuition and (why not say it?) a veritable synthetic judgement *a priori*. On this point all seem agreed, but what Russell claims, and *what seems to me doubtful, is that after these appeals to intuition, that will be the end of it; we need make no others and can build all mathematics without the intervention of any new element.*

XII ‖IV‖

Couturat often repeats that this new logic is altogether independent of the idea of number. I shall not amuse myself by counting how many numeral adjectives his exposition contains, both cardinal and ordinal, or indefinite adjectives such as several. We may, however, cite some examples:

'The logical product of *two* or *more* propositions is . . .';

'All propositions are capable only of *two* values, true and false';

'The relative product of *two* relations is a relation';

'A relation exists between *two* terms,' etc., etc.

Sometimes this inconvenience would not be unavoidable, but sometimes also it is essential. A relation is incomprehensible without two terms; it is impossible to have the intuition of the relation, without having at the same time that of its two terms, and without noticing that they are two, because, if the relation is to be conceivable, it is necessary that there be two and only two.

XIII

[CARDINAL NUMBER

[Now we enter into the domain of arithmetic; we meet first what Couturat calls the theory of cardinal number. It rests on the idea of *correspondence*. Two classes have the same cardinal number if one can establish a bi-uniform correspondence between their elements. I shall not examine whether the idea of correspondence constitutes a new notion; P. Boutroux has studied the question at the Geneva Congress (cf. also *Revue de métaphysique*, July 1905), and the discussion which his communication provoked proves at least that matters are not as clear as the logicians believe.

[Next come the definitions of addition and multiplication. If two classes have no common member, the sum of their cardinal numbers is the cardinal number of their logical sum. Now let $\phi(x)$ and $\psi(y)$ be two propositional functions defining two classes; then the logical product of these two classes [$\phi(x)$ *and*

$\psi(y)$] can be regarded as a propositional function where the variable is represented by the pair x, y; so this propositional function defines a class; and *if the two variables are independent*, the cardinal number of that class is the product of the cardinal numbers of the two classes $\phi(x)$ and $\psi(y)$.

[I shall not examine here the question whether the notion of the independence of two variables is capable of definition. We can concede the point.

[Arithmetical addition and multiplication are thus deduced from logical addition and multiplication, and if the operations symbolized by the signs + and × satisfy the commutative, associative, and distributive laws, this is simply because logical addition and multiplication, characterized by the signs *or* and *and*, do so as well.

XIV

[These definitions and proofs offer an important advantage: they apply to infinite cardinal numbers as well as to finite cardinal numbers. So do the proofs of Cantor; so, too, if one examines them closely, do the proofs one finds in elementary treatises of arithmetic. And yet, when I studied that question in my article *On the nature of mathematical reasoning*, I believed it necessary to reject them, or at least to set them aside.

[Why so? Because they seemed to me to require a too-direct and too-evident appeal to intuition. Today they reappear and are envisaged as the exemplar of a purely logical proof; but what in them has changed?

[Why has the synthetic judgement that seemed necessary to me vanished? Quite simply, *because it has already been performed once*, in the chapter entitled *Logic*, and it is inexpedient to start from the beginning.

[And similarly: what distinguishes the intuition of logical addition from that of arithmetical addition? In the latter, the elements to be added are considered simply as individuals, stripped by abstraction of all their qualitative differences. In logical addition, one dispenses with this abstraction; so the act of intuition is more complex, but apart from that it is the same.

[It will be objected that Russell, in contrast to the usual procedure, takes at first the point of view of comprehension, and later solely the point of view of extension. And this is assuredly a very important innovation. But to pass from one point of view to the other, an act of intuition is again necessary.

XV

[Up to this point, the logisticians have succeeded in avoiding, not every appeal to intuition (they have made them repeatedly), but at least all recourse to the principle of complete induction. The question is whether they can go any further. They believe they can; I believe they cannot; and this is the point that divides us. *So it is only here that the true debate begins.*

[It is certain that if they cannot go further, mathematics will be significantly reduced. Several algebraic identities and, beyond that, not a single general theorem—that will be everything. With great difficulty one will be able to show by example that the numbers are not all equal to each other. But no number-theory, no analysis, no geometry. Mathematical treatises will be far shorter, and one will be able to reduce considerably the programmes of secondary education.]

XVI ‖V‖

ARITHMETIC

I reach what M. Couturat calls the *ordinal theory*, which is the foundation of arithmetic properly so called. M. Couturat begins by stating Peano's five axioms, which are independent, as has been proved by Peano and Padoa.

1. Zero is an integer.

2. Zero is not the successor of any integer.

3. The successor of an integer is an integer. (To this it would be proper to add, Every integer has a successor.)

4. Two integers are equal if their successors are.

5. If s is a class that (i) contains 0; and, (ii) contains the successor of x if it contains the integer x, then s contains all integers.

The fifth axiom is the principle of complete induction.

M. Couturat considers these axioms as disguised definitions; they constitute the definition by postulates of zero, of successor, and of integer.

But we have seen that for a definition by postulates to be acceptable we must be able to prove that it implies no contradiction.

Is this the case here? Not at all.

The demonstration cannot be made *by example*. We cannot take a part of the integers, for instance the first three, and prove they satisfy the definition.

If I take the series 0, 1, 2, I see it fulfils the axioms 1, 2, 4, and 5; but to satisfy axiom 3 it still is necessary that 3 be an integer, and consequently that the series 0, 1, 2, 3, fulfil the axioms; we might prove that it satisfies axioms 1, 2, 4, 5, but axiom 3 requires besides that 4 be an integer and that the series 0, 1, 2, 3, 4 fulfil the axioms, and so on.

It is therefore impossible to demonstrate the axioms for certain integers without proving them for all; we must give up proof by example.

It is necessary then to take all the consequences of our axioms and see if they contain no contradictions. If these consequences were finite in number, this would be easy; but they are infinite in number; they are the whole of mathematics, or at least all of arithmetic.

[So what is to be done? Perhaps one could find a rigorous way of showing that a new reasoning could not introduce a contradiction, provided one supposes that, in the preceding series of reasonings, we have not yet encountered a contradiction.

[If this could be done, we should be certain that we never had to fear contradiction.

[But *this is to use complete induction*, and it is precisely the principle of complete induction that is to be justified.

[And one must not say: it is a question of verifying that the principle of complete induction does not carry with it contradictory consequences; so I ought to study the consequences of this principle, and consequently I have the right to let it play a role in my reasonings.

[This would be a paralogism for two reasons:

[1. If I rely on the principle itself to show that it does not imply a contradiction, I prove solely that if it is true it is not contradictory; and this tells us nothing. It does not suffice to compare *certain* consequences of the principle; it is necessary to compare them *all*.

[2. The proposition does not have the same sense in the statement and in the use we make of it. In the statement, it means: there are numbers which satisfy the principle, and these numbers, by definition, I call integers. And what do I do in the application? I say that, whatever the *number* of my successive reasonings, I shall never be led to contradictory conclusions, because this *number*, being an integer, satisfies the principle. But how do I know that the number of my reasonings is an integer? If I give the vulgar sense to this word, that will not be difficult; but if I define it as I just did, how do I know that the number of my reasonings is one of those that satisfy the principle?

XVII

[I shall later examine Hilbert's attempts to escape these difficulties, but first I wish to refute the proof of Russell and Couturat. What hinders me is that that proof does not exist.

['This definition', says Couturat simply, 'assures neither the existence nor the uniqueness of the defined object. Above all uniqueness is not evident.' They do not make a detailed investigation of uniqueness and existence.

[And this raises a psychological question: how could two such astute logicians not have noticed this gap?

[The first answer is that Couturat followed Russell step by step; but it remains to explain how Russell was led astray. Perhaps we shall find the explanation in another of his writings. 'The principle of mathematical induction', he says (*Mind*, July 1905), 'says not merely that the addition of 1 will always give a number, but that every natural number can be obtained by such additions starting from 0.' But this is not right at all; the principle of induction does not say that every integer can be obtained by successive additions; it says that, for every number that can be obtained by successive additions, one can prove an arbitrary property by recurrence.

[A number can be defined by recurrence; one can reason about this number by recurrence; these are two distinct propositions. The principle of induction does not tell us that the former is true; it tells us that the former implies the latter.

[This is Russell's confusion, and this explains why he was able inadvertently to advance a definition that he was incapable of justifying by showing that it is free of contradiction.

[(*To be continued*)]

E. MATHEMATICS AND LOGIC: II (*POINCARÉ 1906a*)

The following selection is the second of Poincaré's three articles entitled *Les mathématiques et la logique* that appeared in the *Revue de métaphysique et de moral* in 1905 and 1906. The second article was intended as a continuation of the first, and its sections were accordingly numbered XVIII to XXXI. In *Science et méthode* (*Poincaré 1908*) these two articles were heavily abridged and divided into two chapters: Chapter III ('Mathematics and logic') and Chapter IV ('The new logics'). The division took place between sections IX and X of the first article. Thus, Chapter IV of *Science et méthode* was taken from sections X to XVII of the first article (which sections were renumbered I to V), and sections XVIII to XXXI of the second article; those sections were renumbered VI to XIII. In the following translation the section-numbers are those of the original, longer article; but the section-numbers of the version of 1908 are given in double square brackets. Passages that appeared in the original version but that were *deleted* by Poincaré in the 1908 version are enclosed in single square brackets. Passages that he *added* in the 1908 version are given in the footnotes. The translation is by William Ewald, and incorporates, with revisions, the translation by George Bruce Halsted of *Poincaré 1908*, Chapter IV. References to *Poincaré 1906a* should be to the section numbers, which appeared in the original edition.

XVIII ⟦VI⟧

HILBERT'S LOGIC

I come now to the important work of Hilbert which he communicated to the Congress of Mathematicians at Heidelberg, and of which a French translation by M. Pierre Boutroux appeared in *L'Enseignement mathématique*, while an English translation due to Halsted appeared in *The Monist*. In this work, which contains profound thoughts, the author's aim is analogous to that of Russell, but on many points he diverges from his predecessor.

'But', he says, 'on attentive consideration we become aware that in the usual exposition of the laws of logic certain fundamental concepts of arithmetic are already employed; for example, the concept of aggregate, in part also the concept of number. We thus fall into a circle, and therefore to avoid paradoxes a partly simultaneous development of the laws of logic and arithmetic is requisite.'

We have seen above that what Hilbert says of the principles of logic *in the usual exposition* applies likewise to the logic of Russell. So for Russell logic is prior to arithmetic; for Hilbert they are 'simultaneous'. [It is helpful to observe that this is in part because Hilbert considers the notion of set to be arithmetical, while Russell calls it the notion of class and regards it as logical.] We shall find further on other still greater differences, but we shall point them out as we come to them. I prefer to follow the development of Hilbert's thought step-by-step, quoting the most important passages.

'Let us take as the basis of our consideration first of all a thought-thing 1 (one).' Notice that in so doing we in no wise imply the notion of number, because it is understood that 1 is here only a symbol and that we do not at all seek to know its meaning. 'The taking of this thing together with itself respectively two, three, or more times ....' Ah! this time it is no longer the same; if we introduce the words 'two', 'three', and above all 'more', 'several', we introduce the notion of number; and then the definition of finite whole number which we shall presently find, will come too late. [Our author was far too astute not to notice this begging of the question. And, at the end of his work, he seeks to patch things up; we shall have to examine how well he has done so.]

Hilbert then introduces two simple objects 1 and =, and considers all the combinations of these two objects, all the combinations of their combinations, etc. It goes without saying that we must forget the ordinary meaning of these two signs and not attribute any to them. Afterwards he separates these combinations into two classes, the class of the existent and the class of the non-existent; and for the time being this separation is entirely arbitrary. Every affirmative statement tells us that a certain combination belongs to the class of the existent; every negative statement tells us that a certain combination belongs to the class of the non-existent.

[We then see him introduce, like Russell, the conjunctions *if*, *and*, *or*—i.e. logical addition and multiplication. We shall also meet with propositional functions; but here we should note an important difference. For Russell, the variable x is absolutely indeterminate; for Hilbert it is one of the combinations formed with 1 and =.]

XIX ⟦VII⟧

Note now a difference of the highest importance. For Russell any object whatsoever, which he designates by x, is an object absolutely undetermined and about which he supposes nothing; for Hilbert it is one of the combinations formed with the symbols 1 and =; he could not conceive of the introduction of anything other than combinations of objects already defined. Moreover Hilbert formulates his thought in the most lucid way, and I think I must reproduce his statement *in extenso*:

II. In the axioms the arbitrary objects—taking the place of the notion "every" or "all" in ordinary logic—represent only those thought-objects and their mutual combinations that at this stage are taken as primitive or are to be newly defined. In the derivation of consequences from the axioms the arbitrary objects that occur in the axioms may therefore be replaced only by such thought-objects and their combinations. We must also duly note that, when a new thought-object is added and taken as primitive, the axioms previously assumed apply to a larger class of objects or must be suitably modified.

The contrast with Russell's viewpoint is complete. For this philosopher we may substitute for *x* not only objects already known, but anything. Russell is faithful to his point of view, which is that of comprehension. He starts from the general idea of being, and enriches it more and more while restricting it, by adding new qualities. Hilbert on the contrary recognizes as possible beings only combinations of objects already known; so that (looking at only one side of his thought) we might say he takes the viewpoint of extension.

[Why, now, is Hilbert led to such a disagreement with Russell? To understand this, it is necessary to recall what he said in the beginning, and to emphasize it:

[Frege finds himself disarmed before the paradoxes of the theory of sets, such as the paradox of the set of all sets; these paradoxes show, it seems to me, that the conceptions and means of investigation prevalent in logic, taken in the traditional sense, do not measure up to the rigorous demands that set theory imposes. Rather, from the very beginning a major goal of the investigation into the notion of number should be to avoid such contradictions and to clarify these paradoxes.

[And at the end, 'Principles II (which we have stated above) and III permit us to escape the paradoxes mentioned at the beginning of the article . . . .'

[Thus, in Hilbert's eyes, to take, in an intransigent fashion, the point of view of comprehension (as Russell does) is to be lacking in precision and rigour, and to expose oneself to contradiction.

[Who is right? I do not wish to examine this question here; a deeper discussion of this question, interesting though it might be, would delay us far too long. However, the example of Burali-Forti (which we discussed above with regard to pasigraphy) inclines me to say that Hilbert is right.

[Burali-Forti reasoned precisely without conforming to Hilbert's principle II, and it seems that he was deceived; a felicitous error, and very instructive.

[In any case, logic is unable to decide between Hilbert and Russell.

XX[a]

[Let us continue with the exposition of Hilbert's ideas. He introduces the two following axioms:

(1) $$x = x;$$

(2) If $x = y$ and $\omega(x)$, $\omega(y)$.

He considers them as representing the definition by postulates of the symbol =, which until now was devoid of any meaning. But to justify this definition it is necessary to show that these two axioms do not lead to a contradiction.

[And so Hilbert says that all the propositions that one can deduce from the axioms are of the form $\alpha = \alpha$ (which in vulgar language one calls identities); so these propositions cannot be contradictory.

[But how does he know that all these propositions are identities? We consider a series of consequences deduced from our axioms, and we stop at a certain stage in this series; if at this stage we have so far obtained nothing but identities, we can verify that, by applying to these identities any of the operations permitted by logic, we can obtain only new identities.

[One concludes that one can never obtain anything but identities; but *to reason thus is to employ complete induction.*

XXI

[Hilbert next introduces three new symbols, u, f, and f', which he defines by three axioms:

(3) $$f(ux) = u(f'x)$$

(4)[b] $$f(ux) = f(uy) \,|\, ux = uy$$

(5) $$\overline{f(ux) = u1}$$

[These axioms are nothing other than Peano's axioms 3, 4, and 2 (see above, §XVI). The author would have done well to mention this fact, which may escape

---

[a] ⟦In 1908 Section XX was replaced by the following short section, which was numbered Section VIII:

Let us continue with the exposition of Hilbert's ideas. He introduces two assumptions which he states in his symbolic language but which signify, in the language of the uninitiated, that every quantity is equal to itself, and that every operation performed upon two identical quantities gives identical results.

So stated, they are evident, but thus to present them would be to misrepresent Hilbert's thought. For him mathematics has to combine only pure symbols, and a true mathematician should reason upon them without preconceptions as to their meaning. So his assumptions are not for him what they are for the common people.

He considers them as representing the definition by postulates of the symbol = heretofore devoid of all signification. But to justify this definition we must show that these two assumptions lead to no contradiction. For this Hilbert used the reasoning of our number III, without appearing to perceive that he is using complete induction.⟧

[b] ⟦Hilbert's '|' in this equation is the symbol he employed in *Hilbert 1904* for material implication.⟧

some readers. Axiom 5 is left to one side; axiom 1 is also missing, but it should be taken as understood, or as implied by the last of our equations.

[Be that as it may, it is necessary to justify this definition by showing that these equations can never lead to a contradiction. And to do so, Hilbert undertakes to prove that the first two equations can never lead to anything but *homogeneous propositions*, i.e. to equalities whose two sides contain the same number of letters; indeed, he says, the first equation, when one replaces x with any object, never yields anything but a homogeneous equation; likewise for the second, *on condition that the premiss is itself a homogeneous equality*.

[Again, this involves complete induction, as can be seen from the phrase I have just emphasized. So here too *Hilbert is obliged to have recourse to the principle of complete induction*.

XXII[c]

[Now comes a quite enigmatic passage:

['If we translate the well-known axioms for mathematical induction into the language I have chosen, we see that these axioms can be adjoined to the others without contradiction.' How do we see this? The matter remains mysterious; there is indeed a footnote to an address at the Paris Congress, but if one reads the address one does not learn that the problem *has been* solved, but only that it would be immensely desirable if it *were* solved.

[Moreover, even if Hilbert had succeeded in justifying the principle of complete induction, *that justification would come too late, since he has already applied this principle twice*.

[The following lines only increase our perplexity:

['There is no difficulty in grounding the notion of finite ordinal number . . . .' Then comes the statement of an axiom analogous to Peano's Axiom 5, and it is shown by an 'example' that it does not imply a contradiction.

[One might believe that this is the start of the announced proof. The author, one thinks, having defined finite ordinal number and having shown that his definition is free of contradiction, will prove that every finite ordinal number has a successor that is also a finite ordinal number, and he will ascend to the notion of the ordinal type of the set of integers, i.e. to the *smallest infinite*.

[But not at all; Hilbert adds, 'After which we can prove, *using the fact that*

---

[c] [In 1908 Section XXII was replaced by the following short section, which was numbered Section IX:

The end of Hilbert's memoir is altogether enigmatic and I shall not lay stress upon it. Contradictions accumulate; we feel that the author is dimly conscious of the *petitio principii* he has committed, and that he seeks vainly to patch up the holes in his argument.

What does this mean? *At the moment of proving that the definition of integer by means of complete induction implies no contradiction, Hilbert withdraws, as Russell and Couturat withdrew, because the difficulty is too great.*]

the smallest infinite exists, that, for any given finite ordinal number, a still greater one can be found.'

[Thus the notion of the smallest infinite is not deduced from that of finite ordinal number, but is rather prior to it; and we must consider this sentence. 'We see that these axioms (those of complete induction) can be adjoined to the others without contradiction, which establishes the existence of the smallest infinite.' We must, I say, consider this sentence, and especially its two words, 'we see', as constituting the entire proof.

[Or rather no, it is not even that, for at an earlier stage in his reasoning Hilbert says: 'To give a complete proof, it would be necessary to appeal to the notion of finite ordinal number.' So is it the case that this last concept is prior to the other? One does not know what to conclude.

[What does this mean? *At the moment of proving that the definition of integer by means of complete induction implies no contradiction, Hilbert withdraws, as Russell and Couturat withdrew, because the difficulty is too great.*

XXIII

[But even if we admit that the principle has been justified, would it follow that one has the right to make use of it as he does? As Hilbert explains it, the process is always the same; to introduce a new proposition, one seeks to show that that introduction will not lead to a contradiction. After one has given this proof, the new axiom is regarded as legitimate.

[But how to give this proof? According to Hilbert, it is necessary either to show that, if there is a contradiction at a given moment, then this contradiction would have to have appeared at an earlier stage of the theory (this is the direct application of the principle of induction), or one proceeds by *reductio ad absurdum* (which in general involves an indirect application of the same principle).

[Thus one envisages a series of reasonings succeeding one another, and one applies to this succession, regarded as an ordinal type, a principle that is true for certain ordinal types, called finite ordinal numbers, and which is true for these types precisely because these types are by definition those for which it is true.

[But what proves that the ordinal type that corresponds to the succession of our reasonings is precisely one of the 'finite ordinal numbers' thus defined? Have we proved that this type corresponds to the definition? No; and if we had done so, the principle of induction would no longer be a postulate serving as a definition. It would be a theorem like the others, capable of proof; and this entire detour would be unnecessary.

[Is it the case that this succession has no existence except by an arbitrary convention, in which case we are free to choose whatever definition we please? In other words, to conceive of that succession, do we need the definition of 'finite ordinal number' or of 'the smallest infinite'?

[Not at all, and the proof is that Hilbert already applied the principle of induction twice, long before having spoken of 'finite ordinal number' or of 'the smallest infinite'.

[So he had from that moment the direct intuition of that succession of reasonings and of the corresponding ordinal type; while what he defined afterwards is nothing but a combination of empty symbols, of which we know only that they satisfy certain conditions. By what right do we apply to *that* what has been proved for *this*?

[However one manages to justify the principle of induction, the application remains illegitimate because the principle applied is different from the one justified. The same words have a different sense.

[*Hilbert, like his precursors, has not satisfied the second condition (that of §IV) any better than the first (that of §V).*

XXIV

[In following Hilbert, I now arrive at a point that is a little outside my subject, but about which I shall say a few words because of its importance. It concerns the way in which Hilbert conceives the relation of set to element; contrary to established usage, he says, he regards the notion of element as subsequent to the notion of set.

[So it seems that Hilbert considers the genus as prior and not as subsequent to the species, and that consequently he, like Russell, takes the point of view of comprehension and not that of extension. But this seems to me only half right.

[To define a set, Hilbert introduces (following his usual procedure) a new symbol m, which at first is devoid of meaning. Then, given an arbitrary object x, he forms the combination mx which is destined to characterize the relation of the object x to the set m. So he lays down as an axiom constituting a definition by postulates

$$mx = x$$

whenever the vulgar would say that the object x belongs to the set m, and

$$mx = a$$

in the contrary case, where a is an object chosen once and for all in an arbitrary fashion from the objects that belong to the set m.

[Hence the set m is neither subsequent nor prior to the objects x which can be its elements; they are *simultaneous*, since at first that element and those objects are nothing but symbols devoid of meaning and independent of each other; their mutual dependence dates only from the moment when one lays down the axioms, which comes later.

[Thus the set is not prior to its elements; it is prior solely to mx, i.e. to its relation with its elements.

[So this is not at all Russell's point of view, and the contrast is all the more striking since, for the English philosopher, objects are capable of being classified precisely because they are endowed with qualities, while for the German scholar, they are nothing but combinations of symbols that one regiments arbitrarily.

[We shall nevertheless find in Hilbert a reminder of Russell's logic; indeed, he introduces a propositional function $a(x)$ which enters in the definition of the set m such that m is the set of objects x for which the proposition $a(x)$ is true. But we must remember that Hilbert's propositions are nothing but combinations of symbols.

XXV

[I shall end the discussion of this remarkable memoir, so full of original and interesting ideas, by saying several words on what I earlier called the *attempt at patching-up*. The difficulties mentioned above could not escape a man like Hilbert; indeed, at the end of his article he is full of scruples, and he attempts to extricate himself from the affair by several lines which I must cite *in extenso*:

V. Whenever in the preceding we spoke of *several* thought-objects, of *several* combinations, of *various* kinds of combinations, or of *several* arbitrary objects, a bounded number of such objects was to be understood. Now that we have established the definition of finite number we are in a position to comprehend the general meaning of this way of speaking. The meaning of the "arbitrary" consequence and of the "differing" of one proposition from all propositions of a certain kind is also now, on the basis of the definition of finite number (corresponding to the idea of mathematical induction) susceptible of an exact description by means of a recursive procedure. It is in this way that we can carry out completely the proof, sketched above, that the proposition $f(ux^0) = u1$ differs from every proposition obtained as a consequence of Axioms 1–4 by a finite number of steps; we need only consider the proof itself to be a mathematical object, namely, a finite set whose elements are connected by propositions stating that the proof leads from 1–4 to 6, and we must then show that such a proof contains a contradiction and therefore does not exist consistently in the sense defined by us.

[This is not very satisfying. The word several, at the beginning of the work, did not have its usual sense; it did not mean a 'finite number' *as large as one wishes*, but a 'limited number', e.g. 4 or 5. But then what is the significance of the proofs? They can show that after four or five syllogisms the axioms do not lead to a contradiction. But that is not what was in question.

[It was necessary to show that we do not meet a contradiction *however far* we pursue the chain of reasonings; only then can we say that the axioms are not contradictory.

[And this is not all. The fundamental proof needs to be 'completed', and for the completion it is necessary 'to rely on the notion of finite number'. Now, that definition itself rests on that of the smallest infinite, and the latter in turn on the proof under dispute. But this is a vicious circle.

[How to reconcile all this? By regarding 'proof itself as a mathematical object', i.e. in Hilbert's language as a symbol which is undefined except for a certain number of relations with other symbols. The word proof loses its meaning and is only defined by the postulates. But one does not escape from the dilemma:

[Either you know in advance what a proof is and how a proof can lead to contradictions, and then you have no need for this definition by postulates. Moreover, nothing guarantees that this proof which you knew in advance is the same thing as the empty symbol which you agree to call a proof, but which, by definition, is only something that satisfies a certain formula;

[Or you do not know it in advance, and then the question which you raised at the beginning: 'Can a proof based on these axioms lead me to contradictions?' is absolutely destitute of sense. And then why do you ask it? You will find it difficult to explain.

[When you say quite rightly: 'For a conventional definition to be acceptable, it is necessary that it not imply a contradiction', the sense of that rule itself is not at all conventional.

XXVI

Infinite number

[The principle of induction, says Couturat, characterizes the finite numbers in the sense that all reasonings based on this principle are valid only for finite numbers. Whence he concludes that this principle, valid at most for the arithmeticians who do not ascend to the idea of infinity, cannot be of any use in the theory of infinite number. 'The theory of infinite cardinal numbers can therefore be entirely constituted in a direct and independent manner on a purely logical base, without making appeal to the idea of order, and without even invoking the distinction between finite and infinite numbers, nor consequently the principle of induction.' This is what we shall see.

[What is the fundamental theorem of the theory of infinite cardinal numbers? It is Bernstein's theorem; let us recall its statement.

[We consider two sets A and B; if one can make the elements of these two sets correspond in such a manner that, to each element of one set there corresponds one and only one element of the other, then one says that the two sets have the same *cardinal number* and one writes:

$$A \equiv B.$$

This is the definition of cardinal number. On the other hand, one says that a set A' is *a part* of a set A, if A contains all the elements of A' and A' does not contain all those of A.

[Bernstein's theorem now says that if one has:

$$A_0 \equiv B_1 \text{ and } A_1 \equiv B_0$$

where A_1 is a part of A_0 and B_1 is a part of B_0, then one has as well:

$$A_0 \equiv B_0.$$

[Let us examine the proof. The relation $A_0 \equiv B_1$ tells us that to every element of A_0 there corresponds an element of B_1, and since A_1 is a part of B_0, to diverse elements of A_1 there correspond elements of B_1, of which the set B_2 will be a part of B_1. And one will have $A_1 \equiv B_2$ and $A_0 - A_1 \equiv B_1 - B_2$.

[One defines in the same way a set A_2 which will be a part of A_1 and which will be such that $B_1 \equiv A_2$ and $B_0 - B_1 \equiv A_1 - A_2$.

[Now since one has $A_1 \equiv B_2$ and since A_2 is a part of A_1, one finds in the same way a set B_3 which will be a part of B_2 and satisfy the conditions

$$A_2 \equiv B_3 \text{ and } A_1 - A_2 \equiv B_2 - B_3.$$

[One now defines A_3 *and so on* such that one has a series of sets $A_0, A_1, \ldots, A_n, \ldots, B_0, B_1, \ldots, B_n, \ldots$ such that A_{n+1} is a part of A_n and B_{n+1} is a part of B_n and such that

$$A_n \equiv B_{n+1},\ A_{n-1} - A_n \equiv B_n - B_{n+1}$$

$$B_n \equiv A_{n+1},\ B_{n-1} - B_n \equiv A_n - A_{n+1}.$$

[Now let C be the set of all elements common to the various sets $A_0, A_1, \ldots, A_n, \ldots$; and let D be the set of all elements common to the various sets $B_0, B_1, \ldots, B_n, \ldots$. The one has:

$$A_0 = \Sigma(A_n - A_{n-1}) + C$$

$$B_0 = \Sigma(B_n - B_{n+1}) + D.$$

[Since, *when, in an indefinite series of sets, each is a part of its predecessor, the first is formed of all the elements that belong to all the sets and of all those which belong to one of them without belonging to the successor.*

[This principle which I have just emphasized is evident; but it seems to presuppose a special appeal to intuition. However, I do not insist on the point. We now show that

$$C \equiv D.$$

[For to an element of C, a part of A_0, there corresponds an element of B, in virtue of the correspondence defined by the relation:

$$A_0 \equiv B.$$

[Since this element is a part of A_n and since the correspondence defined by $A_0 \equiv B_1$ is the same as that defined by $A_n \equiv B_{n+1}$, the corresponding element is a part of B_{n+1}; so it is a part of *all* the B and consequently of D. Conversely, to every element α of D there corresponds an element β of A_0 in virtue of the same correspondence; and since that element β is part of B_{n+1}, the element α is part of A_n and of *all* the A_n and consequently of C. So one has:

$$C \equiv D$$

and on putting together all our equations:

$$A_0 \equiv B_0.$$

Q.E.D.

[I emphasized above the words *and so on* to make evident the application of the principle of induction. Our sets A_n and B_n are defined *by recurrence* and we reason about them *by recurrence*.

[If Couturat knows another proof of Bernstein's theorem, may he hasten to publish it, for this would be an important mathematical discovery. But if he does not, he should cease to claim that the theory of infinite numbers can be set up without the principle of induction. Nor should he write that Russell and Whitehead have proved purely formally, relying on purely logical principles, all the propositions of that theory, and have purged it of every postulate and every appeal to intuition. If they had been able to purge it at the same time of all contradiction, they would have rendered us a signal service; alas!, mathematicians still discuss that theory without being close to agreement.

XXVII

[The preceding should give us food for thought. We have to prove a theorem in whose proof we introduce a postulate which is the definition of an object A. Then one of two things must hold:

[Either the name of the object A occurs in the statement of the theorem. In this case it is clear that the definition of that object must occur among our premisses; otherwise, not only would it be impossible to prove the theorem, but the theorem would have no meaning;

[Or else, on the contrary, the name A does not occur in the statement. Then one can prove the theorem without introducing the postulate which defines the object; it suffices, whenever one meets the name of A in the proof, to replace it with its definition. Accordingly this name plays no part in the proof, which becomes independent of the definition. That definition will no longer be one of our premisses.

[Now, what happens in the case of Bernstein's theorem? One relies on the principle of induction which, according to the logisticians, is the definition of finite number. Moreover, in the statement of the theorem, it is not a question of finite numbers, but solely of infinite numbers. So we must be able to prove the theorem without relying on the principle.

[But that is impossible; for this principle is not the definition of finite integer that figured in the proof, i.e. of the index n of the set A_n; and indeed, if we examine the way in which one was led to speak of that index, we should see that this principle was not involved.]

XXVII ‖X‖

Geometry

Geometry, says M. Couturat, is a vast body of doctrine wherein the principle of complete induction does not enter. That is true in a certain measure; we cannot say it is entirely absent, but it enters very slightly. If we refer to the *Rational geometry* of Dr Halsted (New York, John Wiley and Sons, 1904) built up in accordance with the principles of Hilbert, we see the principle of induction enter for the first time on page 114 (unless I have made on oversight, which is quite possible).[d]

So geometry, which only a few years ago seemed the domain where the reign of intuition was uncontested, is today the realm where the logicians seem to triumph. Nothing could better measure the importance of the geometric works of Hilbert and the profound impress they have left on our conceptions.

But be not deceived. *What is after all the fundamental theorem of geometry? It is that the assumptions of geometry imply no contradiction, and this we can not prove without the principle of induction.*

How does Hilbert demonstrate this essential point? By leaning upon analysis and through it upon arithmetic and through it upon the principle of induction.

And if ever one invents another demonstration, it will still be necessary to lean upon this principle, since the possible consequences of the assumptions, of which it is necessary to show that they are not contradictory, are infinite in number.

XXIX ‖XI‖

Conclusion

Our conclusion straightway is that *the principle of induction cannot be regarded as the disguised definition of the whole number.*

Here are three truths: (1) the principle of complete induction; (2) Euclid's postulate; (3) the physical law according to which phosphorus melts at 44° (cited by M. Le Roy).

These are said to be three disguised definitions: the first, that of the whole number; the second, that of the straight line; the third, that of phosphorus.

I grant it for the second; I do not admit it for the other two. I must explain the reason for this apparent inconsistency.

First, we have seen that a definition is acceptable only on condition that it implies no contradiction. We have shown likewise that for the first definition this demonstration is impossible; on the other hand, we have just recalled that for the second Hilbert has given a complete proof.

As to the third, evidently it implies no contradiction. Does this mean that the definition guarantees, as it should, the existence of the object defined? We are here no longer in the mathematical sciences, but in the physical, and the word existence has no longer the same meaning. It no longer signifies absence of contradiction; it means objective existence.

You already see a first reason for the distinction I made between the three cases; there is a second. In the applications we have to make of these three concepts, do they present themselves to us as defined by these three postulates?

[d] ‖Second edn, 1907, p. 86; French edn, 1911, p. 97.‖

The possible applications of the principle of induction are innumerable; take, for example, one of those we have expounded above, and where it is sought to prove that a set of axioms can lead to no contradiction. For this we consider one of the series of syllogisms we may employ in starting from these axioms as premisses.

When we have finished the *n*th syllogism, we see we can make still another, and this is the $n + 1th$. Thus the number n serves to count a series of successive operations; it is a number obtainable by successive additions. This therefore is a number from which we may go back to unity by *successive subtractions*. Evidently we could not do this if we had $n = n - 1$, since then by subtraction we should always obtain again the same number. So the way we have been led to consider this number n implies a definition of the finite whole number, and this definition is the following: *A finite whole number is that which can be obtained by successive additions; it is such that n is not equal to n* − 1.

That granted, what do we do? We show that if there has been no contradiction up to the *n*th syllogism, no more will there be up to the $n + 1th$, and we conclude there never will be. You say: I have the right to draw this conclusion, since the whole numbers are by definition those for which a like reasoning is legitimate. But that implies another definition of the whole number, which is as follows: *A whole number is that on which we may reason by recurrence*. In the particular case it is that of which we may say that, if the absence of contradiction up to the time of a syllogism of which the number is an integer carries with it the absence of contradiction up to the time of the syllogism whose number is the following integer, we need fear no contradiction for any of the syllogisms whose number is an integer.

The two definitions are not identical; they are doubtless equivalent, but only in virtue of a synthetic judgement *a priori*; we cannot pass from one to the other by a purely logical procedure. Consequently we have no right to adopt the second, after having introduced the whole number by a way that presupposes the first.

On the other hand, what happens with regard to the straight line? I have already explained this so often that I hesitate to repeat it again, and shall confine myself to a brief recapitulation of my thought.

We have not, as in the preceding case, two equivalent definitions logically irreducible one to the other. We have only one, expressible in words. Will it be said there is another which we feel without being able to word it, since we have the intuition of the straight line or since we represent to ourselves the straight line? First of all, we cannot represent it to ourselves in geometric space, but only in representative space, and then we can represent to ourselves just as well the objects which possess the other properties of the straight line, save that of satisfying Euclid's postulate. These object are 'the non-Euclidean lines', which from a certain point of view are not entities void of sense, but circles (true circles of true space) orthogonal to a certain sphere. If, among these objects equally capable of representation, it is the first (the Euclidean lines) which we call lines, and not the latter (the non-Euclidean lines), this is properly by definition.

And arriving finally at the third example, the definition of phosphorus, we see the true definition would be: Phosphorus is the bit of matter I see in that flask.

XXX ⟦XII⟧

And since I am on this subject, still another word. Of the phosphorus example I said: 'This proposition is a real verifiable physical law, because it means that all bodies having all the other properties of phosphorus, save its point of fusion, melt like it at 44°.' And it was answered: 'No, this law is not verifiable, because if it were shown that two bodies resembling phosphorus melt one at 44° and the other at 50°, it might always be said that doubtless, besides the point of fusion, there is some other unknown property by which they differ.'

That was not quite what I meant to say. I should have written, 'All bodies possessing such and such properties finite in number (to wit, the properties of phosphorus stated in the books on chemistry, the fusion-point excepted) melt at 44°.'

And the better to make evident the difference between the case of the line and that of phosphorus, one more remark. The line has in nature many images, more or less imperfect, of which the chief are the light rays and the rotation axis of the solid. Suppose we find the ray of light does not satisfy Euclid's postulate (for example by showing that a star has a negative parallax), what shall we do? Shall we conclude that the line, being by definition the trajectory of light, does not satisfy the postulate; or, on the other hand, that the line by definition satisfying the postulate, the ray of light is not straight?

Assuredly we are free to adopt the one or the other definition and consequently the one or the other conclusion; but to adopt the first would be stupid, because the ray of light probably satisfies only imperfectly not merely Euclid's postulate, but the other properties of the straight line, so that if it deviates from the Euclidean line, it deviates no less from the rotation axis of solids, which is another imperfect image of the straight line; while finally it is doubtless subject to change, so that such a line which yesterday was straight will cease to be straight to-morrow if some physical circumstance has changed.

Suppose now we find that phosphorus does not melt at 44°, but at 43.9°. Shall we conclude that phosphorus being by definition that which melts at 44°, this body that we did call phosphorus is not true phosphorus, or, on the other hand, that phosphorus melts at 43.9°? Here again we are free to adopt the one or the other definition and consequently the one or the other conclusion; but to adopt the first would be stupid because we cannot be changing the name of a substance every time we determine a new decimal of its fusion-point.

XXXI ⟦XIII⟧

To sum up, Russell and Hilbert have each made a vigorous effort; they have each written a work full of original views, profound and often well warranted.

These two works give us much to think about and we have much to learn from them. Among their results, some, many even, are solid and destined to live.

But to say that they have finally settled the debate between Kant and Leibniz and ruined the Kantian theory of mathematics is evidently incorrect. I do not know whether they really believed they had done it, but if they believed so, they deceived themselves.

F. MATHEMATICS AND LOGIC: III (*POINCARÉ 1906b*)

The following selection is the last of Poincaré's three articles entitled *Les mathématiques et la logique* that appeared in the *Revue de métaphysique et de moral*. In *Science et méthode* (*Poincaré 1908*) this article was abridged to form Chapter V, and was given the new title, 'The latest efforts of the logisticians'. In the following translation the section-numbers are those of the original, longer article; the section-numbers of the version of 1908 are given in double square brackets. Passages that appeared in the original version but that were *deleted* by Poincaré in the 1908 version are enclosed in single square brackets. Passages that he *added* in the 1908 version are given in the footnotes. The translation of the passages that appeared in 1908 is by George Bruce Halsted, with revisions by William Ewald; the translation of the remainder is by William Ewald. References to *Poincaré 1906b* should be to the section numbers, which appeared in the original edition.

I[a]

[The definition of number

[Since my last essay on the relations between mathematics and logic several articles have been published on that question. To start with, we have the articles

[a] ⟦In the 1908 version Poincaré replaced §I with the following paragraph:

The logicians have attempted to answer the preceding considerations. For that, a transformation of logistic was necessary, and Russell in particular has modified his original views on certain points. Without entering into the details of the debate, I should like to return to the two questions that are, to my mind, most important: Have the rules of logistic demonstrated their fruitfulness and infallibility? Is it true they afford means of proving the principle of complete induction without any appeal to intuition?⟧

of Pieri and Couturat in the March issue of the *Révue de métaphysique*; these have the aim of replying to my critiques.

[Then there is an important article by Russell, 'On some difficulties in the theory of transfinite numbers and order types', in the *Proceedings of the London Mathematical Society* for 7 March 1906. Finally, I have received a letter from Zermelo.

[I should like to reply to the objections of Couturat and Pieri, and to examine whether the new studies of Russell lead us to change the status of the question. I shall begin with the article of Couturat, but will be excused if I do not linger over the first part of that article, and in particular over the portions concerning the definitions of integers. For I have too many things to say on the essential point of the debate to linger over questions that seem to me less important.

[I persist in thinking that Couturat defines the clear by the obscure, and that one cannot postulate ⟦poser⟧ x and y without thinking *two*; but perhaps there are readers who wish to follow the discussion with interest, and I do not wish to inflict on them the tedious spectacle of an interminable guerrilla war. So I concede to Couturat without discussion:

[1. That *always false* is not the same thing as *never true*.

[2. That before the works of Burali-Forti it was permissible to doubt that one was a number (at any rate, an ordinal).

[3. That the idea of unity does not imply the number one.

[And now I proceed to the questions which are, in my opinion, the most important. Have the rules of logic proved their fruitfulness and infallibility? Is it true that they permit a demonstration of the principle of complete induction without any appeal to intuition?]

II

The infallibility of logistic

On the question of fertility, it seems M. Couturat has naïve illusions. Logistic, according to him, lends invention 'stilts and wing', and on the next page: '*Ten years ago*, Peano published the first edition of his *Formulaire*.'

How is that, ten years of wings and not to have flown!

I have the highest esteem for Peano, who has done very pretty things (for instance his 'space-filling curve', a phrase now discarded); but after all he has not gone further nor higher nor quicker than the majority of wingless mathematicians, and would have done just as well with his legs.

On the contrary I see in logistic only shackles for the inventor. It is no aid to conciseness—far from it, and if twenty-seven equations were necessary to establish that 1 is a number, how many would be needed to prove a real theorem? If we distinguish, with Whitehead, the individual x; the class of which the only member is x, which shall be called ιx; the class of which the only member is the class of which the only member is x, which shall be called *ux*—do you think these distinctions, useful as they may be, go far to quicken our pace?

Logistic forces us to say all that is ordinarily left to be understood; it makes us advance step by step; this is perhaps surer but not quicker.

It is not wings you logisticians give us, but leading-strings. And then we have the right to require that these leading-strings prevent our falling. This will be their only excuse. When a bond does not bear much interest, it should at least be an investment for a father of a family.

Should your rules be followed blindly? Yes, else only intuition could enable us to distinguish among them; but then they must be infallible; for only in an infallible authority can one have a blind confidence. This, therefore, is for you a necessity. Infallible you shall be, or not at all.

You have no right to say to us: 'It is true we make mistakes, but so do you.' For us to blunder is a misfortune, a very great misfortune; for you it is death.

Nor may you ask: Does the infallibility of arithmetic prevent errors in addition? The rules of calculation are infallible, and yet we see those blunder *who do not apply these rules*; but in checking their calculation it is at once seen where they went wrong. Here it is not at all the case; the logicians *have applied* their rules, and they have fallen into contradiction; and so true is this, that they are preparing to change those rules and to 'sacrifice the notion of class'. Why change them if they were infallible?

'We are not obliged', you say, 'to solve *hic et nunc* all possible problems.' Oh, we do not ask so much of you. If, in face of a problem, you would give *no* solution, we should have nothing to say; but on the contrary you give us *two* of them, and those contradictory, and consequently at least one false; this it is which is failure.

Russell seeks to reconcile these contradictions, which can only be done, according to him, 'by restricting or even sacrificing the notion of class'. And M. Couturat, banking on the success of his attempt, adds: 'If the logicians succeed where others have failed, M. Poincaré will do well to remember this phrase, and give the honour of the solution to logistic.'

But no! Logistic exists; it has its code which has already had four editions; or rather this code is logistic itself. Is Mr Russell preparing to show that one at least of the two contradictory reasonings has transgressed the code? Not at all; he is preparing to change these laws and to abrogate a certain number of them. If he succeeds, I shall give the honour of it to Russell's intuition and not to the Peanian logistic, which he will have destroyed.

III

THE LIBERTY OF CONTRADICTION

[I raised, in the article mentioned above, two chief objections to the definition of integer adopted by the logisticians. The first of these objections is discussed by Couturat on pages 231–241, and by Pieri on pages 194–203. I shall examine and compare the points of view of Couturat and Pieri.[b]]

What does the word *exist* mean in mathematics? It means, I said, to be free from contradiction. This M. Couturat contests. 'Logical existence', he says, 'is quite another thing from the absence of contradiction. It consists in the fact that a class is not empty.' And doubtless to affirm that the class a is not null is, by definition, to affirm that a's exist. But one of the two affirmations is as denuded of meaning as the other, if they do not both signify, either that one may see or touch a's, which is the meaning physicists or naturalists give them, or that one may conceive an a without being drawn into contradictions, which is the meaning given them by logicians and mathematicians.

For M. Couturat it is not non-contradiction that proves existence, but existence that proves non-contradiction. To establish the existence of a class, it is necessary therefore to establish by an *example* that there is an individual belonging to this class:

> But, it will be said, how is the existence of this individual proved? Must not this existence be established, in order that the existence of the class of which it is a part may be deduced? Well, no; however paradoxical the assertion may appear, we never demonstrate the existence of an individual. Individuals, just because they are individuals, are always considered as existent. We never have to express that an individual exists, absolutely speaking, but only that it exists in a class.

M. Couturat finds his own assertion paradoxical, and he will certainly not be the only one. Yet it must have a meaning. It doubtless means that the existence of an individual, alone in the world, and of which nothing is affirmed, cannot involve contradiction; in so far as it is all alone it evidently will not embarrass anyone. Well, so let it be; we shall admit the existence of the individual, 'absolutely speaking', but nothing more. It remains to prove the existence of the individual 'in a class', and for that it will always be necessary to prove that the affirmation 'Such an individual belongs to such a class' is neither contradictory in itself, nor to the other postulates adopted.

[Pieri has not fallen into the same error. He, too, wishes to prove existence by example; but he has a better sense of the requirements for such a proof: 'If A, B, C are propositions belonging to the same deductive system Γ, one can say that one has proved their compatibility if, in some domain Δ, one can find an *interpretation* of the primitive ideas of Γ that exhibit *all* the properties expressed by the propositions A, B, C, provided such a domain Δ does not contain any of these propositions among its premisses, and that the consistency of its principles has already been *established* or *granted a priori*.' This is perfectly correct, and I have nothing to change in it; and Pieri is yet clearer: 'This second condition entails the impossibility of establishing deductively (by means of the

---

[b] |In the version of 1908 this paragraph was altered to read:
I made two principal objections to the definition of integer adopted in logistic. What does Couturat say to the first of these objections?|

criterion indicated) the consistency of the premisses of logic necessary to discourse.' 'It will never be possible to prove deductively the truth or the consistency of the entire system of logical premisses.'

[So the compatibility of the fundamental postulates of logic is itself a postulate that it is necessary to admit and that it is impossible to prove deductively. And so we cannot affirm that compatibility except by a synthetic judgement *a priori*. But let us return to Couturat:]

'It is then', continues M. Couturat, 'arbitrary and misleading to maintain that a definition is valid only if we first prove it is not contradictory.' One could not claim in prouder and more energetic terms the liberty of contradiction. 'In any case the *onus probandi* rests upon those who believe that these principles are contradictory.' Postulates are presumed to be compatible until the contrary is proved, just as the accused person is presumed innocent.

Needless to add, I do not assent to this claim. But, you say, the demonstration you require of us is impossible, and you cannot ask us to jump over the moon. Pardon me; that is impossible for you, but not for us, who admit the principle of induction as a synthetic judgement *a priori*. And that would be necessary for you, as for us.

To demonstrate that a system of postulates implies no contradiction, it is necessary to apply the principle of complete induction; this mode of reasoning not only has nothing 'bizarre' about it, but is the only correct one. It is not 'unlikely' that it has ever been employed; and it is not hard to find 'examples and precedents' of it. I have cited two such instances borrowed from Hilbert's article. He is not the only one to have used it, and those who have not done so have been wrong. What I have blamed Hilbert for is not his having recourse to it (a born mathematician such as he could not fail to see a demonstration was necessary and this the only one possible), but his having recourse without recognizing the reasoning by recurrence.

IV

[I am obliged to insist on this process of reasoning. And indeed, Couturat claims ⟦prétend⟧ that it is founded, not on mathematical induction, but on ordinary induction; this evidently proves that he has understood nothing. Evidently the fault is my own, and my exposition undoubtedly was lacking in clarity; so it is necessary that I start afresh. For simplicity, I shall take Hilbert's first argument and insist on the details.

[Hilbert lays down the following axioms to define equality:

$$x = x;$$

$$\text{If } x = y \text{ and } y = z,\ x = z;$$

$$\text{If } x = y,\ \phi(x) = \phi(y)$$

and he wishes to prove that they are not contradictory. To do so, he proves that no matter how far one pushes their consequences, one never obtains anything but identities. We suppose, indeed, that one has already derived a certain number of equations and that these equations are all identities; and we apply yet again to these equations one of the three rules deduced from the three axioms. I say that the new equations obtained in this manner are likewise all identities.

[For the first, it is unnecessary to insist; we apply the second to two arbitrary equations, previously obtained:

$$X = Y, \ Y = Z.$$

[If these equations are identities, as we are supposing, they reduce to:

$$X = X, \ X = X$$

from which we can only derive

$$X = X$$

which is a new identity.

[We apply the third to an equation previously obtained:

$$X = Y.$$

If that equation is an identity, as we are supposing, and if it reduces to

$$X = X$$

then we deduce from it

$$\phi(X) = \phi(X)$$

which is likewise an identity.

[Thus from identities we can deduce only identities, and step by step, that is to say by mathematical induction, one sees that no matter how far we push our chain of reasoning we never obtain anything but identities.

[The example is perhaps too simple, but it suffices to show what this manner of reasoning consists in.

[So it is pointless to insist on the objections of Couturat; it makes no difference that the natural order of a proof is not linear, but branching. One never enunciates theorems except one after another; the order of those theorems is not absolutely imposed on us, and we can modify it slightly. But this does not mean that reasoning by recurrence is not applicable; I do not even need to say that we can choose the order of the theorems so that it does apply, since it applies whatever our choice.

[Pieri has better understood the question; however, he raises an objection: 'Even if', he says, 'the principle of induction were accepted as one of the axioms of logic, we should not know how to decide whether we are dealing with a *denumerable* sequence—that is to say, one to which the principle can be applied.'

[A reasoning formed of a *non-denumerable* series of propositions and syllogisms—whatever could that be? How is one to imagine it? Perhaps we could content ourselves to know that we shall *never* meet a contradiction, where *never* means at the end of a finite time, however long it may be; but we cannot be equally certain about a time after the end of eternity.

V ⟦IV⟧

THE SECOND OBJECTION

[I reserve for later the questions treated by Couturat in his paragraph IV which demand deeper examination; and this brings me to my second objection, which Couturat seeks to refute in his §V.

[Where, he asks, has Poincaré seen the logisticians commit the error with which he reproaches them? I begin by stating that in writing the incriminating phrase I did not think of the confusion committed by Russell in two different statements of the principle of induction. That confusion is found in a polemical article, but not in his principal work, and I do not wish to hold it against him.

[It is in Hilbert's article that I found the error in question;][c] today Hilbert is excommunicated, and Couturat does not regard him as a logistician; so he asks me if I have found the same error among the orthodox. No, I have not seen it in the pages I have read; I know not whether I should find it in the three hundred pages they have written which I have no desire to read.

Only, they must commit it the day they wish to make any application of mathematics. This science has not as sole object the eternal contemplation of its own navel; it has to do with nature and some day it will touch it. Then it will be necessary to shake off purely verbal definitions and to stop paying oneself with words.

To go back to the example of Hilbert: always the point at issue is reasoning by recurrence and the question of knowing whether a system of postulates is not contradictory. M. Couturat will doubtless say that then this does not touch him, but it perhaps will interest those who do not claim, as he does, the liberty of contradiction.

We wish to establish, as above, that we shall never encounter contradiction after any number of deductions whatever, provided this number be finite. For that, it is necessary to apply the principle of induction. Should we here understand by finite number every number to which by definition the principle of induction applies? Evidently not, else we should be led to most strange consequences.

To have the right to lay down a system of postulates, we must be sure they are not contradictory. This is a truth admitted by *most* scientists; I should have written *by all* before reading M. Couturat's last article. But what does this signify? Does it mean that we must be sure of not meeting contradiction after a *finite* number of propositions, the *finite* number being by definition that

---

[c] ⟦In the version of 1908, Poincaré altered this clause to read: 'I pointed out a second error of logistic in Hilbert's article'.⟧

which has all properties of recurrent nature, so that if one of these properties fails—if, for instance, we come upon a contradiction—we shall *agree* to say that the number in question is not finite?

In other words, do we mean that we must be sure not to meet contradictions, on condition of agreeing to stop just when we are about to encounter one? To state such a proposition is enough to condemn it.

So, Hilbert's reasoning not only assumes the principle of induction, but it supposes that this principle is given us not as a simple definition, but as a synthetic judgement *a priori*.

To sum up:

A demonstration is necessary.

The only demonstration possible is the proof by recurrence.

This is legitimate only if we admit the principle of induction and if we regard it not as a definition but as a synthetic judgement.

VI

[Is it worth the trouble of returning to Couturat's last two pages? Is it necessary to say that I have not given two incompatible statements of the principle of induction, of which the first (page 815, first article) is merely less complete than that on page 32 (second article), since one there speaks of integers without defining them? That the second statement does not have the grotesque sense which Couturat attributes to it, and that I never wished to say:

$$N = x \ni (0 \varepsilon s\colon n \varepsilon s \,.\, \supset_n . \, n + 1 \varepsilon s\colon \supset_s . \, x \varepsilon s)?$$

[And what is the meaning of 'an integer is a number which can be defined by recurrence'? That is to say, an integer is a number which can be obtained by successive additions; or, if you prefer, an integer is a number from which one can return to zero by successive subtractions. And then you ask, 'How many subtractions?'—and I reply, 'It does not matter.' If we take an arbitrary infinite number (say, aleph zero) we cannot return to zero either by a finite number of subtractions or by an infinite number, since aleph-zero minus one equals aleph-zero.

[I cannot write all this in Peanian since I cannot speak that language with sufficient certainty; but I can put it into formulae which you could easily translate into Peanian.

[An integer is a cardinal number; zero is an integer—every integer has a successor different from itself which is also an integer—every integer but zero has a predecessor different from itself which is also an integer.

[There would perhaps be some advantage in modifying this definition, but it was the one I had in mind when I spoke of definition by recurrence; what matters to me is that it does not *analytically* imply the principle of induction. (See below, §10.)]

VII ⟦V⟧

The Cantor Antinomies

Now to examine Russell's important new memoir. This memoir was written with the view to conquer the difficulties raised by those Cantor antinomies to which frequent allusion has already been made. Cantor thought he could construct a science of the infinite; others went on in the way he opened, but they soon ran foul of strange contradictions. These antinomies are already numerous, but the most celebrated are:

1. The Burali-Forti antinomy;
2. The Zermelo–König antinomy;
3. The Richard antinomy.

Cantor proved that the ordinal numbers (the question is of transfinite ordinal numbers, a new notion introduced by him) can be ranged in a linear series; that is to say, that of two unequal ordinals one is always less than the other. Burali-Forti proves the contrary; and in fact he says in substance that if one could range *all* the ordinals in a linear series, this series would define an ordinal greater than *all* the others; we could afterwards adjoin 1 and would obtain again an ordinal which would be *still greater*, and this is contradictory.

[In consequence of my article, Burali-Forti has written to Couturat. There is no contradiction, he claims, because Cantor's result applies to *well*-ordered sets, while mine applies to *perfectly*-ordered sets.

[Burali-Forti's letter is cited by Couturat on page 229 of his last article; but it has been mangled to the point of becoming absurd. Is it he himself who has committed an error, is it Couturat who has translated badly, or is it the fault of the printer? I know not. Happily, the text is easy to reconstruct; it suffices to invert each sentence.

[He is made to say: A perfectly-ordered class is also a well-ordered class, but not conversely; while he certainly wished to say: A well-ordered class is also a perfectly-ordered class, but not conversely. And in fact if one goes back to the text cited one finds: *Ogni classe ben ordinata è anche perfettamente ordinata, ma non vice-versa*.

[Even after this correction his explanation is not satisfactory. For Burali-Forti's reasoning easily applies to well-ordered sets and to Cantor's ordinal numbers; in particular, it is easy to prove that the series of all ordinal numbers forms a *well*-ordered set.]

We shall return later to the Zermelo–König antinomy, which is of a slightly different nature. The Richard antinomy (*Revue générale des sciences*, June 30, 1905) is as follows: Consider all the decimal numbers definable by a finite number of words; these decimal numbers form an aggregate E, and it is easy to see that this aggregate is *countable*, that is, we can *number* the different decimal numbers of this assemblage from 1 to infinity. Suppose the numbering effected, and define a number N as follows: If the nth decimal of the nth number of the assemblage E is

$$0, 1, 2, 3, 4, 5, 6, 7, 8, 9$$

the nth decimal of N shall be:

$$1, 2, 3, 4, 5, 6, 7, 8, 1, 1.$$

As we see, N is not equal to the nth number of E, and as n is arbitrary, N does not appertain to E; and yet N should belong to this assemblage, since we have defined it with a finite number of words.

We shall see later that M. Richard has himself given with much sagacity the explanation of his paradox and that this extends, *mutatis mutandis*, to the other like paradoxes.[d]

VIII ⟦VI⟧

ZIGZAG THEORY AND NO-CLASS THEORY

What is Mr. Russell's attitude in presence of these contradictions? After having analysed those of which we have just spoken, and cited still others, after having given them a form recalling Epimenides, he does not hesitate to conclude: 'A propositional function of one variable does not always determine a class.' A propositional function (that is to say a definition) does not always determine a class. A 'propositional function' or 'norm' may be 'non-predicative'. And this does not mean that these non-predicative propositions determine an empty class, a null class; this does not mean that there is no value of x satisfying the definition and capable of being one of the elements of the class. The elements exist, but they have no right to unite in a syndicate to form a class.

But this is only the beginning, and it is needful to know how to recognize whether a definition is or is not predicative. To solve this problem Russell hesitates between three theories which he calls:

A. the zigzag theory;
B. the theory of limitation of size; and,
C. the no-class theory.

---

[d] ⟦In the 1908 version, Poincaré added the following sentences at this point:

Again, Russell cites another quite amusing paradox: *What is the least whole number which cannot be defined by a phrase composed of less than a hundred English words*?

This number exists; and in fact the numbers capable of being defined by a like phrase are evidently finite in number, since the words of the English language are not infinite in number. Therefore among them will be one less than all the others. And, on the other hand, this number does not exist, because its definition implies contradiction. This number, in fact, is defined by the phrase in italics, which is composed of less than a hundred English words; and by definition this number should not be capable of definition by a like phrase.⟧

According to the zigzag theory 'propositional functions determine a class when they are very simple, and cease to do so only when they are complicated and obscure.' Who, now, is to decide whether a definition may be regarded as simple enough to be acceptable? To this question there is no answer, if it be not the loyal avowal of a complete inability:

> The axioms as to what functions are predicative have to be exceedingly complicated and cannot be recommended by any intrinsic plausibility. This is a defect which might be remedied by greater ingenuity, or by the help of some hitherto unnoticed distinction. But, hitherto, in attempting to set up the axioms for this theory, I have found no guiding principle except the avoidance of contradictions; and this, by itself, is a very insufficient principle, since it leaves us always exposed to the risk that further deductions will elicit contradictions.

This theory therefore remains very obscure; in this night a single light—the word zigzag. What Russell calls the 'zigzaginess' is doubtless the particular characteristic which distinguishes the argument of Epimenides.

According to the theory of limitation of size, a class would cease to have the right to exist if it were too extended. Perhaps it might be infinite, but it should not be too much so.

But we always meet again the same difficulty; at what precise moment does it begin to be too much so? 'A great difficulty of this theory is that it does not tell us how far up the series of ordinals it is legitimate to go.' Of course this difficulty is not solved, and Russell passes on to the third theory.

In the no-classes theory it is forbidden to speak the word 'class', and this word must be replaced by various periphrases. What a change for logistic, which talks only of classes and classes of classes! It becomes necessary to remake the whole of logistic. Imagine how a page of logistic would look upon suppressing all the propositions where it is a question of class. There would only be some scattered survivors in the midst of a blank page. *Apparent rari nantes in gurgite vasto* ⟦'Scattered and few they appear, swimming the vasty deep'⟧.

Be that as it may, we see how Russell hesitates and the modifications to which he submits the fundamental principles he has hitherto adopted. Criteria are needed to decide whether a definition is too complex or too extended, and these criteria can only be justified by an appeal to intuition.

[We hasten to add that at the end of the memoir a telegram of the last hour informs us that these hesitations have ceased: 'From further investigation, I now feel hardly any doubt that the no-classes theory affords the complete solution of all the difficulties ....'

[Be that as it may, logistic is to be remade, and it is unclear how much can be saved: 'I hope in future to work out this theory to the point where it will appear exactly how much of mathematics it preserves, and how much it forces us to abandon.' Needless to add, Cantorism and logistic are alone under consideration; real mathematics, that which is good for something, may continue to develop in accordance with its own principles without bothering about the storms which rage outside it, and go on step by step with its usual conquests, which are final and which it never has to abandon.[e]]

IX ‖VII‖

The true solution

What choice ought we to make among these different theories? It seems to me that the solution is contained in a letter of M. Richard of which I have spoken above, to be found in the *Revue générale des sciences* of June 30, 1905. After having set forth the antinomy we have called Richard's antinomy, he gives its explanation.

Recall what has already been said of this antinomy. E is the aggregate of *all* the numbers definable by a finite number of words, *without introducing the notion of the aggregate E itself*. Else the definition of E would contain a vicious circle; we must not define E by the aggregate E itself.

Now we have defined N with a finite number of words, it is true, but with the aid of the notion of the aggregate E. And this is why N is not part of E.

In the example selected by M. Richard, the conclusion presents itself with complete evidence, and the evidence will appear still stronger on consulting the text of the letter itself.

[But the same explanation holds for the other antinomies, and in particular for that of Burali-Forti. There one introduces the set E of all ordinal numbers; that is to say, of all ordinal numbers that can be defined without introducing the notion of the set E itself; the ordinal number which corresponds to the order type defined by that set E is thus excluded.

[Thus *the definitions which ought to be regarded as non-predicative are those which contain a vicious circle*. And the preceding examples sufficiently show what I understand by this. Is this what Russell calls the 'zigzaginess?' I put the question without answering it.

[For example, the definition of aleph-one is non-predicative; the reasoning by which Cantor seeks to establish the existence of this number seems to me precisely the same as that of Burali-Forti. So I am not sure that aleph-one exists.[f]

---

[e] ‖In the 1908 version, Poincaré replaced these paragraphs with the following:

It is toward the no-class theory that Russell finally inclines. Be that as it may, logistic is to be remade, and it is not clear how much of it can be saved. Needless to add that Cantorism and logistic are alone under consideration; real mathematics, that which is good for something, may continue to develop in accordance with its own principles without bothering about the storms which rage outside it, and go on step by step with its usual conquests which are final and which it never has to abandon.‖

[f] ‖In the 1908 version, Poincaré replaced these paragraphs with the following:

But the same explanation holds good for the other antinomies, as is easily verified. Thus *the definitions which should be regarded as not predicative are those which contain a vicious circle*. And the preceding examples show what I mean by that. Is this what Russell calls the 'zigzaginess'? I put the question without answering it.‖

X ⟦VIII⟧

The proofs of the principle of induction[g]

[I am now ready to return to the questions treated by Couturat in his §IV and which I set aside. Can one prove the principle of induction, and, if one regards it as a definition, can one prove that the definition is not contradictory? Let us examine the proofs that have been proposed and which I reduce to three: that of Whitehead–Russell; that of Burali-Forti recalled by Pieri in his last article; and that of Zermelo, which I shall expound later.

[And first, to make the nature of the question more comprehensible, let us profit from several felicitous new terms introduced by Russell in his recent memoir.

[We call every class containing zero and containing $n + 1$ if it contains n a *recurrent class*.

[We call every number which is a part of *all* the recurrent classes an *inductive number*.

[We call the cardinal number of a class that is not equivalent to any of its parts a *finite number*.

[To avoid any sort of confusion, it will be convenient to supplement this nomenclature; for there exists another definition of finite number, namely, a number n is finite if it is not equal to $n - 1$. So we shall say that a *finite integer* is a cardinal number n not equal to $n - 1$.

[It is clear that every finite number is a finite integer, but the converse is not evident; to prove it, it is necessary to invoke Bernstein's theorem, which we shall discuss later.

[One also sees at once that the class of finite integers is recurrent, and thus that every inductive number is a finite integer.

[It remains to be seen whether every finite number is an inductive number; and the same for every finite integer. To establish this point would be to prove the principle of induction in the sense which I gave in §VI, as can be easily seen by consulting that paragraph. But it does not seem that this has yet been achieved. Russell, who is deemed to have proved it,[1] is extremely dubious, as

---

[g] ⟦In the version of 1908, Poincaré begins §VIII with the following paragraphs:

Let us now examine the pretended demonstrations of the principle of induction and in particular those of Whitehead and Burali-Forti.

We shall speak of Whitehead's first, and take advantage of certain felicitous new terms introduced by Russell in his recent memoir.

Call every class containing zero, and containing $n + 1$ if it contains n, a *recurrent class*.

Call every number which is a part of *all* the recurrent classes an *inductive number*.

Upon what condition will this latter definition, which plays an essential role in Whitehead's proof, be 'predicative' and consequently acceptable?

In accordance with what has been said, it is necessary to understand by *all* the recurrent classes, all those in whose definition the notion of inductive number does not enter. Else we fall again upon the vicious circle which has engendered the antinomies.

Now Whitehead has not taken this precaution. Whitehead's reasoning is therefore fallacious; it is the same which led to the antinomies.

—Poincaré then skips to the middle of §XI of the original version.⟧

[1] Couturat attributes the proof to Whitehead, but Whitehead attributes it to Russell (*American Journal of Mathematics*, Vol. XXIV). (All of section 3, he says, is due to Russell.)

he says in his last article: 'But, so far as I know, we cannot prove that the number of classes contained in a finite class is always finite, or *that every finite number is an inductive number.*'

XI

[I shall expound as best I can Whitehead's proof; readers who find my exposition lacking in clarity can turn to the original text (*American Journal of Mathematics*, Vol. XXIV).

[The inductive numbers exist because zero belongs by definition to all recurrent classes; we see at once that every inductive number is a finite integer. The problem is to establish the converse, that every non-inductive number is not a finite integer, and that every class whose cardinal number is not inductive contains a class whose cardinal number is aleph-zero. To show this, we establish the following proposition:

[If n is not inductive then $n - m$ is not inductive for any m.

[And in fact the class of numbers m such that, if n is any non-inductive number, then $n - m$ is non-inductive—this class, I say, is recurrent; but if m is inductive it ought to belong to it.

[It is easy to verify that this class (which we call K) is recurrent. For zero belongs to it (because $n - 0$ is not inductive if n is not); moreover, if m belongs to it then so does $m + 1$ (for if $n - m$ is not inductive, the same is true for $n - m - 1$).

[I pursue the proof no further, since the error occurs here:

[*The definition of inductive number is not predicative*, if one admits the criterion of §IX. An inductive number is one which belongs to all the recurrent classes; if we wish to avoid a vicious circle we must understand: *to all the recurrent classes in whose definition the notion of inductive number does not already intervene.*

[Now, the class K defined above does not satisfy this condition. The notion of inductive number figures in its definition. It is the class of numbers m such that, if n is a given non-*inductive* number, then $n - m$ is not *inductive*. The word inductive is repeated twice. We pass over the first, because it is a question of a *given* non-inductive number; but the second time admits no excuse.

[Another example. We wish to prove that the sum of two inductive numbers is an inductive number. For (we say) the class K of numbers which, if added to a given inductive number n, yield an *inductive* number is evidently recurrent. But this is invalid: the class K is indeed recurrent, but the word *inductive* figures in its definition.]

Whitehead's reasoning is therefore fallacious; it is the same which led to the antinomies. It was illegitimate when it gave false results; it remains illegitimate when by chance it leads to a true result.

A definition containing a vicious circle defines nothing. It is of no use to say, we are sure, whatever meaning we may give to our definition, zero at least belongs to the class of inductive numbers; it is not a question of knowing whether this class is void, but whether it can be rigorously delimited. A 'non-predicative' class is not an empty class; it is a class whose boundary is undetermined.

Needless to add, this particular objection leaves in force the general objections applicable to all the demonstrations.

XII ‖IX‖

Burali-Forti has given another proof in his article, 'Le classi finite' (*Atti di Torino*, Vol. XXXII). But he is obliged to assume two postulates: First, there always exists at least one infinite class. The second is thus expressed:

$$u \in K(K - \Lambda) . \supset . u < v'u.$$

The first postulate is not more evident than the principle to be proved. The second not only is not evident, but it is false, as Whitehead has shown; as moreover any recruit would see at the first glance, if the axiom had been stated in intelligible language, since it means that the number of combinations which can be formed with several objects is less than the number of these objects.

[But Pieri observes that, without changing the proof, one can replace this false axiom by another according to which if the number of objects is finite so is the number of combinations. This new axiom is true, but it is no more evident than the proposition to be proved. So it does not resolve the question, and I have nothing further to say about Burali-Forti's proof; I limit myself to saying that in spite of his oversight and the difficulty of reading it, this memoir contains very interesting things.

[Zermelo was good enough to write me a letter where he proposes a proof of the principle of induction. He calls a well-ordered set where every element (except the first) has an immediate predecessor a simple series; and this simple series is finite if it has a last element. One easily sees that there are finite simple series.

[It is clear that the principle of induction applies to these series, since in one of its forms it says that in a class of integers there always exists one less than all the others, i.e. that the series of integers is 'well-ordered'. This proof does not differ essentially from the others, and most of the objections remain. Moreover, what needs to be proved is that there exists at least one *infinite* simple series.]

XIII ‖X‖

Zermelo's axiom

A famous proof by Zermelo rests on the following axiom: In any aggregate (or in each aggregate of an assemblage of aggregates) we can always choose an element *at random* (even if this assemblage of aggregates should contain an infinity of aggregates). This axiom had been applied a thousand times without being stated, but, once stated, it aroused doubts. Some mathematicians, for instance M. Borel, resolutely reject it; others admire it. Let us see what, according to his last article, Russell thinks of it. [He makes no pronouncement: 'Whether Zermelo's axiom is true or false is a question which, while more fundamental matters are in doubt, is very likely to remain unanswered.' He contents himself with exhibiting several new forms that can be given to the question; but the considerations which he raises are very suggestive.[h]]

And first a picturesque example: Suppose we have as many pairs of shoes as there are whole numbers, so that we can number the *pairs* from one to infinity, how many shoes shall we have? Will the number of shoes be equal to the number of pairs? Yes, if in each pair the right shoe is distinguishable from the left; it will in fact suffice to give the number $2n - 1$ to the right shoe of the nth pair, and the number $2n$ to the left shoe of the nth pair. No, if the right shoe is just like the left, because a similar operation would become impossible—unless we admit Zermelo's axiom, since then we could choose *at random* in each pair the shoe to be regarded as the right.

[In the end, Russell shows that if one abandons Zermelo's axiom one is led to abandon what he calls the multiplicative axiom, on which rests the definition of multiplication of two transfinite cardinal numbers. So everything that Couturat calls the theory of cardinal number collapses at a stroke.

[To be sure, Russell does not abandon all hope of rebuilding:

['The complete solution of our difficulties, we may surmise, is more likely to come from clearer notions in logic than from the technical advance of mathematics; but until the solution is found we cannot be sure how much of mathematics it will leave intact.'

[Again, *true* mathematics, mathematics where one does not wallow in the actual infinite, is not in question. But it is interesting to investigate what Russell understands by 'clearer notions in logic'. To understand this, it is necessary to re-read what precedes page 49:

['The multiplicative axiom has been employed constantly in proofs of theorems concerning transfinite numbers. It is open to everybody, as yet, to accept it as a self-evident truth, but it remains possible that it may turn out to be capable of disproof by *reductio ad absurdum*. It may also, of course, be capable of proof, but that is far less probable.'

[Thus Russell still hopes that, proceeding from other postulates, one can *deductively* prove that Zermelo's axiom is false, or that it is true. Needless to say, this hope seems to me illusory. There are no 'clearer notions in logic' that will extricate us from the difficulty; nor will 'the technical advance of mathematics'. The axioms in question will never be anything other than propositions admitted by some as 'self-evident' and doubted by others. Each will always believe his own intuition. There will always be one point on which everybody will be in agreement. The axiom is 'self evident' for finite classes; but if it is unprovable for infinite classes, it is undoubtedly also unprovable for finite

[h] ‖In the version of 1908, Poincaré abridged this passage to read: 'He makes no pronouncement, but the considerations which he raises are very suggestive.'‖

classes (which have not been distinguished at this stage of the theory); so it is a synthetic judgement *a priori* without which the 'cardinal theory' would be impossible, both for finite numbers and for infinite.

XIV

[BERNSTEIN'S THEOREM

[I return to this theorem not to reply to Couturat but to clarify several points in light of the preceding considerations. The theorem can be stated in the following manner:

[If a set A_0 can be decomposed into three parts, H_0, Q_0, and A_1 such that

$$A_0 = H_0 + Q_0 + A_1$$

and if A_1 is equivalent to A_0 so that

$$A_1 \sim A_0$$

then $A_1 + Q_0$ will be equivalent to A_0.

[For if A_1 is equivalent to A_0, then to each element of A_0 there corresponds an element of A_1 which one can call its image. If B is a set contained in A_0, the images of the elements of B form a set which we can call the image of B and which we shall designate by $\phi(B)$. So one has $\phi(A_0) = A_1$. We set:

$$\phi(A_1) = A_2,\ \phi(H_0) = H_1,\ \phi(Q_0) = Q_1$$

$$\phi(A_2) = A_3,\ \phi(H_1) = H_2,\ \phi(Q_1) = Q_2$$

. . . .

and so on. One has

$$A_n = H_n + Q_n + A_{n+1}$$

when the index n is any inductive number.

[Now let C be the set of *all* elements that belong to *all* A_n where the index n is an inductive number. One proves without difficulty that $\phi(C) = C$.

[Now I say that

$$A_0 = H_0 + Q_0 + H_1 + Q_1 + \ldots + C \tag{1}$$

i.e. that every element of A_0 which belongs to no H_n or to no Q_n where the index n is an inductive number must belong to *every* A_n with inductive index and thus to C.

[For the class K formed by the A_n containing an element which does not belong to any of the $A_k - A_{k+1}$ (where the index k is an *inductive* number) is a recurrent class.

[Clearly, the equality (1) implies the stated proposition; but since the notion of *inductive* number figures in the definition of class K, we find the same blemish noted in §XI. What are we to say? The proof of Bernstein's theorem remains legitimate, but on condition that one regards the principle of induction as a synthetic judgement and not as a definition, because that definition would be 'non-predicative'.

[Zermelo sends me another proof of Bernstein's theorem. We consider all sets B that contain Q_0 and that contain their proper image $\phi(B)$. Let R be the set formed by the common elements of *all* the sets B; one sees that the image of R, i.e. $\phi(R)$, is formed of the elements common to all the $\phi(B)$; these elements belong to every B, and consequently to R. It follows that R, which contains Q_0, also contains $\phi(R)$.

[One then shows that

$$R = Q_0 + \phi(R)$$

for otherwise $Q_0 + \phi(R)$ would be a set B not containing R, since it would only be a part of it.

[This point having been established, one sees that $Q_0 + A_1$, where

$$\mathrm{Q}_0 + [\mathrm{A}_1 - \phi(\mathrm{R})] + \phi(\mathrm{R}) = [\mathrm{A}_1 - \phi(\mathrm{R})] + \mathrm{R}$$

is equivalent to

$$[A_1 - \phi(R)] + \phi(R) = A_1$$

since $\phi(R)$ is equivalent to R; and finally to A_0, since A_1 is equivalent to A_0.

[The error is again the same: R belongs to *all* the sets B; on penalty of a vicious circle, this must be taken to say, to all the sets B whose definition does not involve the notion of R. This excludes the set $Q_0 + \phi(R)$ which depends on R. So the definition of the set R is not predicative.

[And now one could deduce Bernstein's theorem from the celebrated theorem of Zermelo; but we would always run up against the same obstacle.

[What does Zermelo do? He considers a set E and its subsets. He chooses at random in each of these subsets an element which he calls the distinguished element of this subset. This is possible if one admits 'Zermelo's axiom' mentioned above. Then he defines what he calls the M_γ. He calls Γ the logical sum of *all* the M_γ, such that any element of an M_γ is an element of Γ. Now it is a matter of showing that Γ is nothing other than the entire set E. For if it were not, then the set $E - \Gamma$ would contain a distinguished element A; the set $\Gamma + A$ would be a M_γ such that A would belong to a M_γ without belonging to Γ, which is contrary to the hypothesis.

[It is always the same thing: the definition of Γ is not predicative. The logical sum of *all* the M_γ must be taken to be the logical sum of all M_γ in whose definition the notion of Γ does not figure; and then the M_γ formed by Γ and the distinguished element of $E - \Gamma$ ought to be excluded. Moreover, although I am rather disposed to admit Zermelo's axiom, I reject his proof, which for an instant made me believe that aleph-one could indeed exist.]

XV ⟦XI⟧

Conclusions

A demonstration truly founded upon the principles of analytic logic will be composed of a series of propositions. Some, serving as premisses, will be identities or definitions; the others will be deduced from the premisses step by step. But though the bond between each proposition and the following is immediately evident, it will not at first sight appear how we get from the first to the last, which we may be tempted to regard as a new truth. But if we replace successively the different expressions therein by their definition, and if this operation be carried as far as possible, there will finally remain only identities, so that all will reduce to an immense tautology. Logic therefore remains sterile unless made fruitful by intuition.

This I wrote long ago; logistic professes the contrary and thinks it has proved it by actually proving new truths. By what mechanism?

Why in applying to their reasonings the procedure just described—namely, replacing the terms defined by their definitions—do we not see them dissolve into identities like ordinary reasonings? It is because this procedure is not applicable to them. And why? Because their definitions are not predicative and present this sort of hidden vicious circle which I have pointed out above; non-predicative definitions cannot be substituted for the terms defined. Under these conditions *logistic is not sterile, it engenders antinomies*.

It is the belief in the existence of the actual infinite which has given birth to those non-predicative definitions. Let me explain. In these definitions the word 'all' figures, as is seen in the examples cited above. The word 'all' has a very precise meaning when it is a question of a finite number of objects; to have another one, when the objects are infinite in number, would require there being an actual (given complete) infinity. Otherwise *all* these objects could not be conceived as postulated anteriorly to their definition, and then if the definition of a notion N depends upon *all* the objects A, it may be infected with a vicious circle, if among the objects A are some indefinable without the intervention of the notion N itself.[i]

There is no actual (given complete) infinity. The Cantorians have forgotten this, and they have fallen into contradiction. It is true that Cantorism has been of service, but this was when applied to a real problem whose terms were precisely defined, and when we could advance without fear.

---

[i] ⟦In the version of 1908, Poincaré inserted the following paragraph at this point: The rules of formal logic express simply the properties of all possible classifications. But for them to be applicable it is necessary that these classifications be immutable and that we have no need to modify them in the course of the reasoning. If we have to classify only a finite number of objects, it is easy to keep our classifications without change. If the objects are *indefinite* in number, that is to say if one is constantly exposed to seeing new and unforeseen objects arise, it may happen that the appearance of a new object may require the classification to be modified, and thus it is we are exposed to antinomies.⟧

Logistic also forgot it, like the Cantorians, and encountered the same difficulties. But the question is to know whether they went this way by accident or whether it was a necessity for them.

For me, the question is not doubtful; belief in an actual infinity is essential in the Russell logic. It is just this which distinguishes it from the Hilbert logic. Hilbert takes the viewpoint of extension, precisely in order to avoid the Cantorian antinomies. Russell takes the viewpoint of comprehension. Consequently for him the genus is anterior to the species, and the *summum genus* is anterior to all. That would not be inconvenient if the *summum genus* was finite; but if it is infinite, it is necessary to postulate the infinite, that is to say to regard the infinite as actual (given complete).

And we have not only infinite classes; when we pass from the genus to the species in restricting the concept by new conditions, these conditions are still infinite in number. Because they express generally that the envisaged object presents such or such a relation with all the objects of an infinite class.

But that is ancient history. Russell has perceived the peril and takes counsel. He is about to change everything, and, what is easily understood, he is preparing not only to introduce new principles which shall allow of operations formerly forbidden, but he is preparing to forbid operations he formerly thought legitimate. Not content to adore what he burned, he is about to burn what he adored, which is more serious. He does not add a new wing to the building, he undermines its foundation.

The old logistic is dead, so much so that already the zigzag theory and the no-classes theory are disputing over the succession. To judge the new logistic, we must wait until it exists.

G. ON TRANSFINITE NUMBERS (*POINCARÉ 1910*)

The following selection discusses Richard's paradox, impredicative definitions, and transfinite set theory; it was delivered as a lecture in Göttingen on 27 April 1909. It was translated from the German by William Ewald; references to *Poincaré 1910* should be to the paragraph numbers, which have been added in this edition.

‖1‖ Gentlemen! I wish to speak to you today about the concept of transfinite cardinal number; in particular, I want to speak first of an *apparent* contradiction that this concept contains. About that I say the following in advance: in my view an object is only thinkable when it can be defined with a finite number of words. An object that is in this sense finitely definable, I shall for brevity call simply 'definable'. Accordingly, an undefinable object is also unthinkable. Similarly, I shall call a law 'expressible' if it can be expressed in a finite number of words.

⟦2⟧ Now, Richard has shown that the totality of definable objects is denumerable, that is, that the cardinal number of this totality is $\aleph_0$. The proof is quite simple: if α is the number of words in the dictionary, then with n words one can define at most α^n objects. If one now lets n grow beyond all limits, one sees that one never gets beyond a denumerable totality. The power of the set of thinkable objects would then be $\aleph_0$. Schoenflies has objected against this proof that one can define several objects, indeed even infinitely many, with a single definition. As an example he cites the definition of constant functions, of which there are obviously infinitely many. But this objection is inadmissible, because by such definitions it is not at all the individual objects that are defined, but their totality, which is a single object: in our example the *set* of constant functions. The objection of Schoenflies is therefore not conclusive.

⟦3⟧ Now, as is well known, Cantor proved that the continuum is not denumerable; this contradicts the proof of Richard. The question therefore arises which of the two proofs is correct. I maintain that they are both correct and the contradiction only apparent. To support this contention I shall give a new proof of the Cantorian theorem: we therefore assume that an interval AB is given, and a law by which every point of the interval is correlated with a whole number. For the sake of simplicity we shall designate points by the numbers correlated to them. We now divide our interval by two arbitrarily chosen points A_1 and A_2 into three parts, which we designate as sub-intervals of level 1; these again we divide into three parts and obtain sub-intervals of level 2; we imagine this process continued into the infinite, whereby the length of the sub-intervals decreases beneath every bound. Now point 1 belongs to one or, if it coincides with A_1 or A_2, at most two of the sub-intervals of level 1; there is therefore certainly one to which it does not belong. Here we look for the point with the lowest number, which now must be at least 2. Among the three sub-intervals of the second level which belong to the interval of the first level in which we find ourselves, there is again at least one to which the last-considered point does not belong. We continue the process with this interval, and so obtain a sequence of intervals which has the following properties: each of them is contained in all the preceeding intervals, and an interval of the nth level contains none of the points 1 to $n - 1$. From the first property it follows that there must be at least one point which is common to them all; but from the second property it follows that the number of this point must be greater than any finite number—that is, no number can be correlated with it.

⟦4⟧ Now what have we presupposed for this proof? We have presupposed a law that correlates a whole number to every point of the interval. Then we were able to define a point to which no whole number is correlated. In this regard the different proofs of this theorem do not differ. But for that it was necessary that the law first be determinate. According to Richard, such a law would seemingly have to exist, but Cantor has proved the opposite. How do we get out of this dilemma? To start, we ask about the meaning of the word 'definable'. We take the table of all finite sentences and strike out all those which define no point. We correlate the remaining sentences with the whole numbers. If we now undertake the scrutiny of the table anew, it will in general turn out that we must now let several sentences stand that we had earlier struck out. For earlier the sentences in which one spoke of the law of correlation itself had no meaning, since the points were not yet correlated to the whole numbers. These sentences now have a meaning and must remain in our table. Should we now set up a new law of correlation the same difficulty would repeat itself, and so on *ad infinitum*. But herein lies the solution of the apparent contradiction between Cantor and Richard. Let M_0 be the set of whole numbers, M_1 the set of all points of our interval definable after the first scrutiny of the table of all finite sentences, G_1 the law of correlation between the two sets. A new set M_2 of definable points arises through this law. But a new law G_2 belongs to $M_1 + M_2$, through which there arises a new set M_3, etc. Now, Richard's proof shows that, wherever I break the process off, a law always exists, while Cantor proves that the process can be continued arbitrarily far. Therefore there exists no contradiction between the two.

⟦5⟧ The appearance of contradiction comes from the fact that Richard's law of correlation lacks a property which I designate, with an expression borrowed from the English philosophers, as 'predicative'. (In Russell, from whom I borrow the word, a definition of two concepts A and A' is not predicative if A occurs in the definition of A' and conversely.) I understand by this the following: every law of correlation presupposes a determinate classification. I now call a correlation predicative, if the corresponding classification is predicative. And I call a classification predicative if it is not changed by the introduction of new elements. But this is not the case for Richard's classification; rather the introduction of the law of correlation alters the division of the sentences into those which have a meaning and those which have none. What is here meant by the word 'predicative' is best illustrated by an example. If I am to deposit a set of objects into a number of boxes two things can occur: either the objects already deposited are conclusively in their places, or, when I deposit a new object, I must always take the others out again (or at any rate some of them). In the first case I call the classification predicative, in the second not. Russell has given a good example of a non-predicative definition: let A be the smallest whole number whose definition requires more than a hundred German words. A must exist, since one can define only a finite number of numbers with a hundred words. But the definition which we have just given of this number contains less than a hundred words. And the number A is thus *defined* as *undefinable*.

⟦6⟧ Now, Zermelo has objected against the rejection of non-predicative definitions that a great part of mathematics would become invalid as well, for instance, the proof of the existence of a root of an algebraic equation.

[7] This proof, as is well known, runs as follows:

[8] An equation $F(x) = 0$ is given. One now proves that $|F(x)|$ must have a minimum; let x_0 be one of the arguments for which the minimum occurs, so that

$$|F(x)| \geq |F(x_0)|.$$

From this it then follows that $F(x_0) = 0$. Now here the definition of $F(x_0)$ is not predicative, for this value depends upon the totality of the values of $F(x)$, to which it itself belongs.

[9] I cannot admit the legitimacy of this objection. One can reshape the proof so that the non-predicative definition disappears. To this end, I consider the totality of arguments of the form $(m + ni)/p$ where m, n, and p are whole numbers. Then I can draw the same conclusions as before, but the value of the argument for which the minimum of $|F(x)|$ occurs does not in general belong to the arguments considered. In this way we avoid the circle in the proof. One can demand of every mathematical proof that the definitions, etc., occurring therein be predicative; otherwise the proof would not be rigorous.

[10] How do things now stand with the classical proof of the Bernstein theorem? Is it unobjectionable? The theorem states, as is well known, that if three sets A, B, and C are given, where A is contained in B and B in C, and if A is equivalent to C, then A must also be equivalent to B. So here too it is a question of a law of correlation. If the first law of correlation (between A and C) is predicative, then the proof shows that there must also be a predicative law of correlation between A and B.

[11] Now, as far as the second transfinite cardinal $\aleph_1$ is concerned, I am not entirely convinced that it exists. One reaches it by considering the totality of ordinal numbers of the power $\aleph_0$; it is clear that this totality must be of a higher power. But the question arises whether it is self-contained, and therefore of whether we may speak of its power without contradiction. There is not in any case an actual infinite.

[12] What, then, are we to think of the famous *problem of the continuum*? Can one well-order the points of space? What do we mean thereby? There are two cases possible here: either one asserts that the law of well-ordering is finitely statable, and then this assertion is unproven; even Zermelo does not claim to have proved such an assertion. Or we grant the possibility that the law is not finitely statable. Then I can no longer attach any sense to this statement; it is for me merely empty words. Here lies the difficulty. And that is indeed the cause of the conflict over the theorem of Zermelo, a theorem that is nearly a stroke of genius. This conflict is very peculiar: one side rejects the postulate of choice but holds the proof to be correct; the other admits the postulate of choice, but does not acknowledge the proof.

[13] However I could speak about this for many more hours without solving the question.

23
The French analysts

The publication of Zermelo's first proof that every set can be well-ordered (*Zermelo 1904*, translated in *van Heijenoort 1967*) touched off an international controversy in 1905. In France Zermelo's proof provoked the exchange of letters translated below among Emile Borel, René Baire, Jacques Hadamard, and Henri Lebesgue. In Germany *Mathematische Annalen* carried six articles on the proof by Borel, Julius König, Philip Jourdain, Felix Bernstein, Arthur Schoenflies, and Georg Hamel; all but Hamel were critical of the proof. So, too, were Russell and Poincaré, and many other mathematicians in Europe and America who wrote on the subject. (Poincaré's criticisms are translated above in *Poincaré 1906b*, §§XII–XIII.)

The variety of the responses will give some idea of the confusion produced by Zermelo's paper. Some mathematicians (for example Hamel, Hadamard, and Hausdorff) accepted Zermelo's article more or less in its entirety. Others, notably the German writers in *Mathematische Annalen*, accepted the axiom of choice, but argued that the proof of the well-ordering theorem ran foul of the Burali-Forti paradox. Poincaré, like the Germans, accepted the axiom while rejecting the proof, but on the subtler ground that Zermelo's argument relied on impredicative definitions. The French analysts, in contrast, accepted the proof while rejecting the axiom. And Russell expressed uncertainty: 'Whether, in particular, Zermelo's axiom is true or false is a question which, while more fundamental matters are in doubt, is very likely to remain unanswered' (*Russell 1906a*, conclusion). For a detailed account of the debate, see *Moore 1982*, especially Chapter 2; and see also the discussion of Zermelo in *Hallett 1984*.

The debate among the French analysts was initiated by Borel's note in the *Annalen*, written in late 1904 at Hilbert's request, and translated below. Borel (like Baire and Lebesgue) made constant use of set-theoretic techniques in his studies; indeed, concepts like *derived set*, *denumerable sequence*, *accumulation point*, and *cardinality* were central tools for the French analysts, and grew directly out of Cantor's set-theoretic researches. But as early as his *Leçons sur la théorie des fonctions* (*Borel 1898*)—and thus well before the public dissemination of the set-theoretic paradoxes or of Zermelo's proof—Borel had expressed his unease about transfinite set theory and about sets with cardinality greater than that of the continuum. His objections were in part pragmatic (since the higher cardinals had little use in analysis), but also in part based on philosophical principles; and the philosophical objections come to the fore in the following review of Zermelo's proof. Borel is willing to countenance finitely many choices, and even denumerably many choices; but when Zermelo postulates uncountably many choices, Borel declares them to be 'outside the realm of mathematics'.

The translation of *Borel 1905* is by William Ewald.

A. SOME REMARKS ON THE PRINCIPLES OF THE THEORY OF SETS (*BOREL 1905*)

At the request of the editors of this journal, I shall briefly summarize some thoughts suggested to me by the interesting Note of Zermelo.[1]

One of the most important problems one can pose regarding an arbitrary set M is the following:

A.—*To put M in the form of a well-ordered set.*

Zermelo's remarkable result can be described as follows: to know how to resolve problem **A**, it is sufficient to know how to resolve the following problem **B**:

B.—*Given an arbitrary subset M' of M, to choose in a* determinate *(but otherwise arbitrary) manner an element m' of M' to which one gives the name of* distinguished element of M'; *this choice is to be made for all subsets M' of M.*

It is evident that every solution of problem **A** yields a particular solution to problem **B**; but the converse is not evident; and thanks to Zermelo we know that *problems* **A** *and* **B** *are equivalent.*

But this result, despite its interest, cannot be considered a general solution to problem **A**. In effect, if problem **B** is to be regarded as solved relative to a given set M, it is necessary to give (at least in theory) a means for determining the distinguished element m' of an arbitrary subset M'; and this problem appears exceptionally difficult if one supposes, to fix ideas, that M coincides with the continuum.

So one cannot regard as valid the following reasoning, to which Zermelo alludes: 'In a given set M' it is possible to choose *ad libitum* the distinguished element m'; *this choice having been made for each of the sets M', it can be made for the set of these sets.*'

Such reasoning seems to me no better grounded than the following: 'To well-order a set M, it suffices to choose arbitrarily an element to which one assigns the rank 1, then another to which one assigns rank 2, and so on *transfinitely*, that is, until one has exhausted all the elements of M by the sequence of transfinite numbers.' Now, no mathematician could regard this latter reasoning

---

[1] *Math. Annalen*, Vol. 59 (1904), 514–16.

as valid. It seems to me that the objections that one can raise against it apply equally well against any reasoning where one supposes an *arbitrary choice* to be made a non-denumerable infinity of times; such reasonings are outside the realm of mathematics.[2]

Paris, 1 December 1904.

B. FIVE LETTERS ON SET THEORY (*BAIRE* et alii *1905*)

In the exchange of letters that followed the publication of Borel's note, Borel soon found himself waging a war on two fronts. On the one hand, Jacques Hadamard (in Letter 1) came to the defence of Zermelo's axiom. Hadamard drew a pair of distinctions: (i) between *dependent* and *independent* choices; and, (ii) between the *existence* of a mathematical correspondence and our *ability to describe* it. Hadamard takes a strongly Platonist line: our abilities are a matter of empirical psychology, and 'outside mathematics'; what matters is *existence*. Zermelo's axiom, because it involves independent choices, can legitimately be applied in the transfinite; Borel's criticisms would hold water only if Zermelo's choices were applied in a succession, with each choice depending on those that had been made earlier. Borel forwarded Hadamard's letter to Baire. Baire responded (Letter 2) by rejecting altogether the concept of a completed infinity. In his view, the only infinite is a potential infinite, and in the end all mathematics must reduce to the finite. In particular, Baire criticized Borel for his acceptance of denumerable sets.

Lebesgue, in a long and careful analysis of the two preceding letters, takes a similar position. For him, the central question is Hadamard's distinction (ii); but, in direct opposition to Hadamard, Lebesgue declares that an existence proof requires *definability*. On this basis, Lebesgue regards much of Zermelo's argument as being 'too little Kroneckerian to have any meaning'. Lebesgue also rejects Borel's special treatment of denumerable infinites, and of the theorem that every infinite set has a denumerable subset. Hadamard (Letter 4) responds by accusing Baire and Lebesgue of importing illegitimate empirical considerations into mathematics. In the fifth and final letter, Borel adopts a position that was to become central to Hilbert's thinking about the infinite. For Borel, transfinite arguments are void of significance in themselves, but they can be

[2] Permit me here to cite several lines of a letter from Baire (of Montpellier) which seem to me to sum up with considerable polish an opinion which I find very just and which is undoubtedly very widespread: 'Personally, I doubt that one can ever find a common measure between the continuum (or, what comes to the same thing, the set of all sequences of positive integers) and the well-ordered sets; as I see it, each of these two things is defined only potentially ‖virtuellement‖, and it may be that these two potentialities are irreducible.'

used as a heuristic aid in deriving legitimate mathematical results that can be expressed in a finite number of words.

For further discussion of the five letters, see *Monna 1975* and *Moore 1982*.

The translation of the five letters is by Gregory H. Moore, and originally appeared as an appendix to *Moore 1982*. References to *Baire* et alii *1905* should be to the original pagination, which is given in the margin; vertical lines in the text indicate the page breaks.

I. *Letter from Hadamard to Borel*

I have read with interest the arguments that you put forward (second issue of Volume LX of *Mathematische Annalen*) against Zermelo's proof, found in the previous volume. However, I do not share your opinion on this matter. I do not agree, first of all, with the comparison that you make between the fact which Zermelo uses as his starting-point and an argument which would enumerate the elements of the set one after another *transfinitely*. Indeed, there is a fundamental difference between the two cases: The latter argument requires a sequence of successive choices, *each of which depends on those made previously*; this is the reason why it is inadmissible to apply it transfinitely. I do not see how any analogy can be drawn, from the point of view which concerns us, between the choices in question and those used by Zermelo, which are *independent of each other*.

Moreover, you take exception to his procedure for a *non-denumerable* infinity of choices. But, for my part, I see no difference in this regard between the case of a non-denumerable infinity and that of a denumerable infinity. The
261| difference | would be evident if the choices in question depended on each other
262 in some way, because then it would be necessary to consider the order in which one made them. To me the difference appears, once again, to vanish completely in the case of independent choices.

What is certain is that Zermelo provides no method to carry out *effectively* the operation which he mentions, and it remains doubtful that anyone will be able to supply such a method in the future. Undoubtedly, it would have been more interesting to resolve the problem in this manner. But the question posed in this way (the effective determination of the desired correspondence) is none the less completely distinct from the one that we are examining (does such a correspondence exist?). Between them lies all the difference, and it is fundamental, separating what Tannery[1] calls a *correspondence* that can be *defined* from a correspondence that can be *described*. Several important mathematical ques-

[1] *Revue générale des Sciences*. Vol. VIII, 1897, pp. 133ff.

tions would completely change their meaning, and their solutions, if the first word were replaced by the second. You use correspondences, whose *existence* you establish without being able to *describe* them, in your important argument about ‖complex‖ series which can be continued along an arc across their circle of convergence. If only those entire series were considered whose law of formation can be described, the earlier view (i.e., that entire series which can be continued along an arc across their circle of convergence are the exception) ought, in my opinion, to be regarded as the true one. Furthermore, this is purely a matter of taste, since the notion of a correspondence 'which can be described' is, to borrow your expression, 'outside mathematics'. It belongs to the field of psychology and concerns a property of our minds. To discover whether the correspondence used by Zermelo can be specified *in fact* is a question of this sort.

To render the existence of this correspondence possible, it appears sufficient to take *one* element from any given set, just as the following proposition A suffices for B:

A. *Given a number x, there exists a number y which is not* | *a value obtained* 262|
from x in any algebraic equation with integer coefficients. 263

B. *There exists a function y of x such that, for every x, y is not an algebraic number and is not a value obtained from x in any algebraic equation with integer coefficients.*

Undoubtedly, one can form such functions. But what I claim is that this fact is in no way necessary in order to assert the correctness of theorem B. I believe that many mathematicians would not take any more trouble than I do to verify this fact before using the theorem in question.

J. HADAMARD

II. *Letter from Baire to Hadamard*

Borel has communicated to me the letter in which you express your viewpoint in the great debate resulting from Zermelo's note. I beg your indulgence in presenting some thoughts that it suggested to me.

As you know, I share Borel's opinion in general, and if I depart from it, it is to go further than he does.

Let us suppose that one tries to apply Zermelo's method to the set M of sequences of positive integers. One takes from M a distinguished element m_1; there remains the set $M - \{m_1\}$, from which one takes a distinguished element m_2; and so on. Each of these successive choices, indeed, depends on those that precede it. But, so you say along with Zermelo, the choices are independent of each other because he permits as a starting-point *some choice of a distinguished element in EVERY subset of M*. I do not find this satisfactory. To me it conceals the difficulty *by immersing it in a still greater difficulty.*

The expression *a given set* is used continually. Does it make sense? Not always, in my opinion. As soon as one speaks of the infinite (even the

denumerable, and it is here that I am tempted to be more radical than Borel), the comparison, *conscious or unconscious*, with a bag of marbles passed from hand to hand must disappear completely. We are then, I believe, in the realm
263| of *potentiality* ⟦dans le *virtuel*⟧. | That is to say, we establish conventions that
264 ultimately permit us, when an object is defined *by a new convention*, to assert certain properties of this object. But to hold that one can go farther than this does not seem legitimate to me. In particular, when a set is given (we agree to say, for example, that we are given the set of sequences of positive integers), *I consider it false to regard the subsets of this set as given*. I refuse, *a fortiori*, to attach any meaning to the act of supposing that a choice has been made in every subset of a set.

Zermelo says: 'Let us suppose that to each subset of M there corresponds one of its elements.' This supposition is, I grant, in no way contradictory. Hence all that it proves, as far as I am concerned, is that we do not perceive a contradiction in supposing that, in each set which is defined for us, the elements are positionally related to each other in exactly the same way as the elements of a well-ordered set. In order to say, then, that one has established that every set can be put in the form of a well-ordered set, the meaning of these words must be extended in an extraordinary way and, I would add, a fallacious one.

In the preceding paragraphs I have only managed to express my thinking very incompletely. I stated my viewpoint in the letter that Borel cited in his note. For me, progress in this matter would consist in delimiting the domain of the definable. And, despite appearances, in the last analysis everything must be reduced to the finite.

R. BAIRE

III. *Letter from Lebesgue to Borel*

You ask for my opinion about Zermelo's note (*Math. Annalen*, Vol. LIX), about your objections to it (*Math. Annalen*, Vol. LX), and about the letter from Hadamard that you communicated to me. Here is my reply. Forgive me for being so lengthy: I have tried to be clear.

First of all, I agree with you on the following point: Zermelo has very cleverly shown that we know how to resolve problem A:

264| | A. *To put a set M in the form of a well-ordered set*,
265 whenever we know how to resolve problem B:

B. *To assign to each set M′, formed from elements of M, a particular element m′ of M′*.

Unfortunately, problem B is not easy to resolve, so it seems, except for the sets that we know how to well-order. As a result, we do not have a general solution to problem A.

I strongly doubt that a general solution can be given to this problem, at least if one accepts (as Cantor does) that to define a set M is to name a property

P which is possessed by certain elements of a previously defined set *N* and which characterizes, by definition, the elements of *M*. In fact, with this definition, we know nothing about the elements of *M* other than this: They possess all the *unknown* properties of the elements of *N* and they are the only ones that possess the *unknown* property *P*. Nothing about this permits two elements of *M* to be distinguished from each other, still less to be arranged as they would need to be in order to resolve A.

This objection, made *a priori* to any attempt to resolve A, obviously disappears if we particularize *N* or *P*. The objection disappears, for example, if *N* is the set of numbers. In general, all that one can hope to do is to indicate certain problems, such as B, whose resolution would entail that of A and which are possible in certain particular cases that are rarely encountered. In my opinion, this is why Zermelo's argument is interesting.

I believe that Hadamard is more faithful than you are to Zermelo's thought when he interprets this author's note as an attempt, not to resolve A effectively, but to demonstrate the existence of a solution. The question comes down to this, which is hardly new: *Can one prove the existence of a mathematical object without defining it*?

This is obviously a matter of convention. Nevertheless, I believe that we can only build solidly *by granting that it is impossible to demonstrate the existence of an object without defining it*. From this perspective, closely related to Kronecker's and Drach's, there is nothing to distinguish between A and problem C:

C. *Can every set be well-ordered*?

| I would have nothing more to say if the convention that I mentioned were 265|
universally adopted. But I must admit that one often uses, and that I myself 266
have often used, the word *existence* in other senses. For example, when Cantor's well-known argument is interpreted as saying that *there exists a non-denumerable infinity of numbers*, no means is given to name such an infinity. It is only shown, as you have said before me, that whenever one has a denumerable infinity of numbers, one can define a number not belonging to this infinity. (Here the word *define* always means *to name a property characterizing what is defined*.) This sort of existence can be used in an argument in the following fashion: A property is true if negating it leads to the assertion that all numbers can be arranged in a denumerable sequence. I believe that this kind of existence can enter an argument only in such a fashion.

Zermelo utilizes the *existence* of a *correspondence* between the subsets of *M* and certain of their elements. You see, even if the existence of these correspondences were not questionable, owing to the way in which their existence had been proved, it would not be self-evident that one had the right to use their existence in the way that Zermelo did.

I come to the argument that you state in the following way: 'It is possible to choose *ad libitum* a distinguished element *m'* from a particular set *M*; since this choice can be made for each of the sets *M'*, it can be made for the set of these sets.' From this argument the existence of the correspondences seems to follow.

First of all, when M' is given, is it self-evident that one can choose m'? It would be self-evident if M' existed in the almost Kroneckerian sense that I mentioned earlier, since to say that M' exists would then be to assert that one knew how to name certain of its elements. But let us extend the meaning of the word *exist*. The set Γ of correspondences between the subsets M' and the distinguished elements m' certainly *exists* for Hadamard and Zermelo; the latter even represents the number of its elements by a transfinite product. However, do we know how to choose an element from Γ? Obviously not, since this would give a determinate solution to problem B in the case of M.

It is true that I use the word *to choose* in the sense of *to name* and that
266| perhaps it suffices for Zermelo's argument that | *to choose* mean *to think of*.
267 Yet it must be noted all the same that what one is thinking of is not stated and that Zermelo's argument still requires one to think *always of the same determinate correspondence*. Hadamard believes, it seems to me, that it is not necessary to prove that one can *determine* a unique element. In my opinion, this is the source of our differences of judgement.

So as to convey more clearly the difficulty that I see, I remind you that in my thesis I proved the existence (in a sense that is not Kroneckerian and is perhaps difficult to make precise) of sets that were measurable but were not Borel-measurable. Nevertheless, I continued to doubt that any such set could ever be named. Under these conditions, would I have the right to base an argument on this hypothesis—*I assume as chosen a set that is measurable but not Borel-measurable*—even though I doubt that anyone could ever name one?

Thus I already see a difficulty with the assertion that 'in a determinate M' I can choose a determinate m'', since there exist sets (the set C for example, which can be regarded as a set M' coming from a more general set) in which it is perhaps impossible to choose an element. Then there is the difficulty that you pointed out concerning the infinity of choices. As a result, if we wish to regard Zermelo's argument as completely general, it must be granted that we are speaking about an infinity of choices whose power may be very large. Moreover, no law is given either for this infinity or for the choices. We do not know if it is possible to name a rule defining a set of choices having the power of the set of the M'. We do not know if it is possible, given an M', to name an m'.

In sum, when I scrutinize Zermelo's argument, I find it, like a number of other general arguments about sets, too little Kroneckerian to have meaning (only as an existence theorem giving a solution to C, of course).

You allude to the following argument: 'To well-order a set, it suffices to choose one element from it, then another, and so on.' Certainly this argument presents enormous difficulties which are even greater, at least in appearance, than Zermelo's. And I am tempted to believe, as Hadamard does, that
267| progress | has been made by replacing an infinity of successive choices, which
268 depend on each other, with an unordered infinity of independent choices.

Perhaps this is only an illusion. Perhaps the apparent simplification resides only in the fact that an ordered infinity of choices must be replaced by an unordered infinity, but one of higher power. Consequently, the fact that one can reduce to the single difficulty, placed at the beginning of Zermelo's argument, all the difficulties of the simplistic argument that you cited merely shows that this single difficulty is very great. In any case, it does not seem to me that the difficulty vanishes just because it concerns an unordered set of independent choices. For example, if I believe that there exist functions $y(x)$ such that, whatever x may be, y is never a value obtained from x in any algebraic equation with integer coefficients, this is because I believe, as does Hadamard, that it is possible to construct such a function. But in my opinion this does not follow immediately from the existence, whatever x may be, of numbers y which are not a value obtained from x in any equation with integer coefficients.[2]

I agree completely with Hadamard when he states that to speak of an infinity of choices without giving a rule presents a difficulty that is just as great whether or not the infinity is denumerable. When one says, as in the argument that you criticize, 'since this choice can be made for each of the sets M', it can be made for the set of these sets', one says nothing unless the terms being used are explained. To make a choice can be to write down or to name the element chosen. To make an infinity of choices cannot be to write down or to name the elements chosen, one by one; life is too short. Hence, one must say what it means to make them. By this, we understand in general that a rule is given which defines the elements chosen. For me as for Hadamard, this rule is equally indispensable whether or not the infinity is denumerable.

All the same, perhaps I still agree with you on this point since, although I
find no theoretical difference between | the two kinds of infinity, from the prac- 268|
tical point of view I distinguish strongly between them. When I hear of a rule 269
defining a transfinite infinity of choices, I am very suspicious because I have never seen such a rule, whereas I know of rules defining a denumerable infinity of choices. Still, this is only a question of habit. Upon reflection, I see difficulties which, in my opinion, are sometimes just as great in arguments involving only a denumerable infinity of choices as in arguments involving a transfinite number. For example, if I do not regard the classical argument as establishing the proposition that every non-denumerable set contains a subset whose power is that of Cantor's second number-class, I do not grant any greater validity to the argument showing that a set which is not finite has a denumerable subset. Although I seriously doubt that a set will ever be named which is neither finite nor infinite, it has not been proved to my satisfaction that such a set is impossible. But I have already spoken to you about these questions.

H. Lebesgue

[2] While correcting the proofs, I will add that in fact the argument by which we ordinarily justify Hadamard's statement A (p. 262) justifies at the same time statement B. And, in my opinion, it is because it justifies B that it justifies A.

IV. *Letter from Hadamard to Borel*

The question appears quite clear to me now, after Lebesgue's letter. More and more plainly, it comes down to the distinction, made in Tannery's article, between what is *determined* and what can be *described*.

In this matter Lebesgue, Baire, and you have adopted Kronecker's viewpoint, which until now I believed to be peculiar to him. You answer in the negative the question posed by Lebesgue (above, p. 265): Can one prove the existence of a mathematical object without defining it? I answer it in the affirmative. I take as my own, in other words, the answer that Lebesgue himself gave regarding the set Γ (p. 266).

I grant that it is impossible for *us*, at least at present, to *name* an element in this set. That is the issue for you; it is not the issue for me.

There is only one point, it seems to me, where Lebesgue is inconsistent with
269| himself. That is when he does or | does not allow himself to use the existence
270 of an object, according to the way in which its existence was proved. For me, the *existence* about which he speaks is a fact like any other, or else it does not occur.

As for Baire, the question takes the same form. I would prefer not to base it as he does (p. 264), following Hilbert, on the *non-contradictory*, which still seems to me to depend on psychology and to take into account the properties of our brains. I do not understand very well how Zermelo could have *proved* that *we do not perceive* a contradiction, *etc*. This cannot be *proved* but only *ascertained*: One perceived it or one did not perceive it.

Leaving this point aside, it is clear that the principal question, that of knowing if a set can be well-ordered, does not mean the same thing to Baire (any more than to you or Lebesgue) that it does to me. I would say rather—Is a well-ordering possible?—and not even—Can *one* well-order a set?—for fear of having to think who this *one* might be. Baire would say: Can *we* well-order it? An altogether subjective question, to my way of thinking.

Consequently, there are two conceptions of mathematics, two mentalities, in evidence. After all that has been said up to this point, I do not see any reason for changing mine. I do not mean to impose it. At the most, I shall note in its favour the arguments that I stated in the *Revue générale des Sciences* (30 March 1905), to wit:

1. I believe that in essence the debate is the same as the one which arose between Riemann and his predecessors over the notion of function. The *rule* that Lebesgue demands appears to me to resemble closely the analytic expression on which Riemann's adversaries insisted so strongly.[3] And even an analytic

---

[3] I believe it necessary to reiterate this point, which, if I were to express myself fully, appears to form the essence of the debate. From the invention of the infinitesimal calculus to the present, it seems to me, the essential progress in mathematics has resulted from successively annexing notions which, for the Greeks or the Renaissance geometers or the predecessors of Riemann, were 'outside mathematics' because it was impossible to describe them.

expression that is not too unusual. Not only does the *cardinality* of the choices fail to alter the question, but, it seems to me, their *uniqueness* does not alter it either. I do not see | how we have the right to say, "For each value of x there exists a number satisfying. . . . Let y be this number . . .," whereas, since "the bride is too beautiful," we cannot say, "For each value of x there exists an infinity of numbers satisfying. . . . Let y be one of the number . . . ."

2. Tannery's arbitrary choices lead to numbers v which *we would be* incapable of defining. I do not think that these numbers fail to exist.

As for the arguments proposed by Bernstein (*Math. Annalen*, Vol. LX, p. 187) and, consequently, his objections to Zermelo's proof, I do not find them convincing. All the same, my opinion on this matter is independent of the question that we have been discussing.

Bernstein begins with Burali-Forti's paradox (*Circolo Matematico di Palermo*, 1897) concerning the set W of *all* ordinal numbers. To circumvent the contradiction obtained by Burali-Forti, he supposes the ordinal number of W to be such that it is impossible to add one to it. In my opinion this supposition, as well as the arguments that Bernstein adduces in its favour, is unacceptable. In Cantor's theory the order established between the elements of W and the additional element (it is this order which Bernstein attacks) is merely a *convention* that one is always free to make and that the properties of W, whatever they may be, cannot alter.

The solution is different. It is the very existence of the set W that leads to a contradiction. In the definition of W, the general definition of the word *set* is incorrectly applied. We have the right to form a set only from previously existing objects, and it is easily seen that the definition of W supposes the contrary.

Same observation for *the set of all sets* (Hilbert, Heidelberg Congress ‖*Hilbert 1904*‖).

Let us return to the original question. I submit in this regard, not an argument (since I believe that we shall rest eternally on our respective positions) but a consequence of your principles.

Cantor considered the set of all those functions on the interval (0, 1) that assume only the values zero and one. To my mind this set has | a clear meaning 271|
and its power is $2^{\aleph}$, as Cantor stated. Likewise, the set of all functions of x 272
makes sense to me, and I see clearly that its power is $\aleph^{\aleph}$.

What meaning does all of this have for you? It appears obvious to me that it cannot have any. For on each function you impose an additional condition which has no mathematical meaning—that of being *describable to us.*

Or rather, this is what it means: From your point of view, one should consider only those functions definable in a finite number of words. But, for this reason, the two sets formed above are *countable* and, indeed, so is every other possible set.

J. HADAMARD

V. *Letter from Borel to Hadamard*

. . . First of all, I would like to call your attention to an interesting remark that Lebesgue made at the meeting of the Society on 4 May: How can Zermelo be certain that in the different parts of his argument he is always speaking of *the same* choice of distinguished elements, since he characterizes them in no way *for himself*. (Here it is not a question that someone may contradict him but rather of his being intelligible to himself.)

As for your new objection, here is my response.

I prefer not to write alephs. Nevertheless, I willingly state arguments equivalent to those which you mention, without many illusions about their intrinsic value, *but intending them to suggest other more serious arguments*. To give you a practical example, I refer to Note III which I inserted at the end of my recent little book (*Leçons sur les fonctions de variables réelles*, edited by Maurice Fréchet). The argument used there was obviously suggested by Cantor's argument, which I reported in my first *Leçons sur la théorie des fonctions*, page 107.[4]

272| | The form that I adopted in Note III is not absolutely satisfactory, as I
273 indicated at the bottom of the last page of my book. But the analogous argument which Lebesgue gives in his article in the *Journal de Jordan* (1905) is, I believe, completely irreproachable, in the sense that it leads to a precise result expressible in a finite number of words. Nevertheless, it originated from that of Cantor.

One may wonder what is the real value of these arguments that I do not regard as absolutely valid but that still lead ultimately to effective results. In fact, it seems that if they were completely devoid of value, they could not lead to anything, since they would be meaningless collections of words. This, I believe, would be too harsh. They have a value analogous to certain theories in mathematical physics, through which we do not claim to express reality but rather to have a guide that aids us, by analogy, in predicting new phenomena, which must then be verified. It would require considerable research to learn what is the real and precise sense that can be attributed to arguments of this sort. Such research would be useless, or at least it would require more effort than it would be worth. How these overly abstract arguments are related to the concrete becomes clear when the need is felt.

I would agree with you that it is self-contradictory to speak of the set of all sets, for, by the argument from page 107 cited above, we can form a set whose power is still greater. But I believe that this contradiction arises because sets that are not really defined have been introduced.

E. Borel

---

[4] In Notes I and II of this little book, I continually used arguments of the sort that you deny me the right to make. So I was constantly filled with scruples, and each of these two Notes ended with a very restrictive remark.

24
David Hilbert (1862–1943)

Hilbert's research into the foundations of mathematics was undertaken during two distinct periods at the beginning and end of his career; the two periods were separated by an interval of about a dozen years. To the first period belong the first three editions of the *Grundlagen der Geometrie*, two articles on the foundations of arithmetic (*Hilbert 1900a* and *1904*), and parts of his celebrated address on mathematical problems (*Hilbert 1900b*). To the second period belong the books he co-authored with Wilhelm Ackermann (*Grundzüge der theoretischen Logik*, *1928*) and Paul Bernays (*Grundlagen der Mathematik*, two volumes, *1934* and *1939*), as well as eight articles on proof theory, logic, and the philosophy of mathematics.

The selections that follow contain all of Hilbert's published articles on the foundations of mathematics, with the following exceptions: 'On the foundations of logic and arithmetic' (*Hilbert 1904*), 'On the infinite' (*Hilbert 1926*), 'The foundations of mathematics' (*Hilbert 1928a*), 'Problems in the foundation of mathematics' (*Hilbert 1928c*), and 'Proof of the *Tertium non datur*' (*Hilbert 1931b*). The first three of these articles are of cardinal importance for the foundations of mathematics; they have been translated by Stefan Bauer-Mengelberg, and are readily available in *van Heijenoort 1967*. The last two articles are today of little interest save to Hilbert specialists, and have therefore been omitted from the present collection.

In addition to this published material, Hilbert's *Nachlass* in Göttingen contains, besides an extensive collection of letters and manuscripts in the Niedersächsische Staats- und Universitätsbibliothek, a collection of some eighty volumes of official notes of his lectures. These notes were made by his assistants (among them, Ackermann, Courant, Hellinger, Schönfinkel, and Bernays), corrected by Hilbert, bound, and deposited in the library of the Mathematical Institute in Göttingen. A list of the subjects covered by Hilbert in his Göttingen lectures can be found in the appendix to the third volume of Hilbert's *Gesammelte Abhandlungen* (*Hilbert 1932–5*). Aproximately twenty of these volumes deal with topics in logic, set theory, proof theory, the foundations of geometry, and the philosophy of mathematics and physics; these volumes are at present being prepared for publication.

Hilbert was born in 1862 in Kant's city of Königsberg; he attended university in Königsberg, and was appointed Ordinary Professor there in 1893. Among his teachers were Adolf Hurwitz and Ferdinand von Lindemann; among his close friends, Hermann Minkowski, who later joined Hilbert as a professor in

Göttingen in 1902. Most of Hilbert's work during his Königsberg period was on the then-flourishing theory of algebraic invariants; by proving his finite basis theorem (discussed below in *Hilbert 1918*, §§45–53) Hilbert solved the central problem in the field, and put the capstone on the theory of invariants, whose origins go back to George Boole. During this time in Königsberg, Hilbert also simplified the existing proofs of the transcendence of e and π.

In 1895, at the invitation of Felix Klein, Hilbert left Königsberg to accept a professorship in Göttingen; he remained there for the rest of his life. Hilbert's subsequent career can be more or less tidily divided into five periods, each concerned with a specific cluster of mathematical problems. (As Weyl noted in *1944b*, Hilbert liked to concentrate his energies, and, having moved on to a new area, he devoted to it his undivided attention.)

In the first period, lasting roughly from 1893 to 1898, Hilbert worked intensively on algebraic number theory. He published a series of papers on the subject, including his magisterial *Theorie der algebraischen Zahlkörper* of 1897; this so-called *Zahlbericht*, described by Weyl as 'a jewel of mathematical literature', systematized algebraic number theory, and exerted a powerful influence on its development in the twentieth century. The *Zahlbericht* provided the starting-point for the work of Artin, Hecke, Hasse, Chevalley, and Weyl, among others. A full discussion of its influence and of Hilbert's contributions to number theory can be found in *Weyl 1944b* and in *Hasse 1932*.

The second segment of Hilbert's career was devoted to the foundations of geometry and to the axiomatic method, with some excursions into the foundations of arithmetic. This period lasted roughly from 1898 to 1903; the first two selections below date from this period.

The third period lasted from 1903 until about 1910. During this time Hilbert worked on the theory of integral equations and on the calculus of variations. After Minkowski's death in 1909 Hilbert entered a fourth period, and until the early 1920s he turned his energies to mathematical physics; he worked in particular on the kinetic theory of gases, on general relatively, and on the theory of radiation.

Much of Hilbert's work thus lies outside the scope of this collection; but useful specialist essays surveying various aspects of Hilbert's contributions to mainstream mathematics were included by the editors of his *Gesammelte Abhandlungen* (*Hilbert 1932–5*). In addition to the essay by Hasse on number theory, see the essays *van der Waerden 1933* on algebra, *Schmidt 1933* on geometry, *Hellinger 1933* on integral equations, and *Bernays 1935* on the foundations of mathematics.

The fifth and final phase of Hilbert's career begins with the delivery in 1917 of 'Axiomatic thought', in which, after a thirteen-year hiatus, he turned again to the foundations of mathematics. In the winter semester of 1917–18 he delivered a course of lectures on *Prinzipien der Mathematik*; the lectures were written up by Paul Bernays, and contain the germ of many of the ideas that were to appear in *Hilbert and Ackermann 1928*. But Hilbert did not devote his full attention to technical work in foundations until the early 1920s. Most of the selections below date from this final phase of Hilbert's career (when he was in his sixties). Hilbert's ideas on logic and proof theory dominated research in the foundations of mathematics in the 1920s, inspiring such younger logicians as Ackermann, Bernays, von Neumann, Schönfinkel, Herbrand, and Gentzen, and paving the way for Kurt Gödel. The influence of Hilbert during this period is amply documented in the articles collected in *van Heijenoort 1967*, a work which should therefore be read in parallel with the present selections.

The two obituary notices of Hilbert by Hermann Weyl (*1944a* and *1944b*) contain valuable discussions of the entirety of Hilbert's work. The three essays Bernays *1922b*, *1935*, and *1948* discuss Hilbert's contributions to the foundations of arithmetic and of geometry; *Kreisel 1958a* and *1976* and *Sieg 1988* discuss subsequent developments in Hilbert's programme. For a biography of Hilbert, see *Reid 1970*.

A. ON THE CONCEPT OF NUMBER (*HILBERT 1900a*)

This article is Hilbert's first essay on the foundations of arithmetic; it builds directly upon his investigations into the axiomatic foundations of geometry. According to Hermann Weyl, Hilbert had already stated the essence of his conception of axiomatics in conversation as early as 1891: 'It must be possible to replace in all geometric statements the words *point*, *line*, *plane*, by *table*, *chair*, *mug*.'[a]

This insight in itself was not original to Hilbert. Many earlier mathematicians had pointed out that the terms used in a mathematical theory need not have a unique interpretation; indeed, this issue has been a central topic, implicitly or explicitly, in many of the previous readings—perhaps most strikingly in Lambert's remarks on the axiom of parallels, but also in Berkeley's *Analyst* (*Berkeley 1734*), in Gauss's remarks on the complex integers (*Gauss 1831*), in Duncan Gregory's 'On the real nature of symbolical algebra' (*Gregory 1840*), in Boole's introduction to the *Mathematical analysis of logic* (*Boole 1847*), in De Morgan's work on 'double algebra' (*De Morgan 1849b*), in Dedekind's *Was sind und was sollen die Zahlen?* (*Dedekind 1888*), and in many other authors not represented in this collection (such as Hermann Grassmann, Giuseppe Peano, and Moritz Pasch).

But these writers did not perceive the full implications of their remarks. The

---

[a] *Weyl 1944b*, p. 153. Weyl names Otto Blumenthal as his source. This anecdote is also recounted by Blumenthal in his biographical sketch of Hilbert in *Hilbert 1932–5*, Vol. iii, p. 403.

syntactic theory of algebra developed by the British was for the most part used in an *ad hoc* way to defend the legitimacy of various branches of mathematics—the complex numbers, the theory of differential operators, algebraic logic—but was not itself thought of as a tool for making new mathematical discoveries. In practice, having paid lip-service to the syntactic theory of algebra, the mathematicians of the late nineteenth century proceeded to explore a wide array of new algebraic structures; but they saw no need to construct formal systems on the model of *De Morgan 1849b*, or even to provide a rigorous axiomatic framework for the systems they were studying.

Hilbert was the first to blend together into a single system the ideas of a purely syntactic calculus, capable of multiple interpretations, and generating its theorems by finite, gapless deductions, and explicitly to show how this idea of a formal system could be made to yield powerful mathematical results. Specifically, in the *Grundlagen der Geometrie* (*Hilbert 1899*) he used what are now called model-theoretic techniques to investigate the logical relationships that obtain between the various axioms of geometry; his study of arithmetical models of the geometric axioms yielded not only independence results and a relative consistency proof, but also new non-Archimedean geometries, a new topological characterization of the plane, new insights into the nature of continuity, and deep new theorems about the relationships between geometry and arithmetic. These metageometric investigations were to have a considerable influence not only on geometry, but also on logic: for proof theory and model theory grew out of Hilbert's adept exploitation of the insight contained in his remark about *table*, *chair*, and *mug*.

In the present essay, Hilbert extends to the foundations of real analysis the axiomatic method that had done such signal service in his geometric studies. He contrasts the *axiomatic method* with the *genetic method* that had previously been the standard approach in arithmetical investigations, and is well exemplified by Dedekind's *Habilitation* address, 'On the introduction of new functions in mathematics' (*Dedekind 1854*). Hilbert remarks—a theme that was to recur constantly in his later writings—that his axiomatic method is supposed to furnish a tool for investigating not only the foundations of mathematics, but also the foundations of the physical sciences—an aspect of his thought that should be borne in mind by those who are tempted to see Hilbert as a 'formalist', i.e. as one who held that mathematics is merely a meaningless game played with formal symbols.

Hilbert characterizes the real numbers axiomatically as an ordered, Archimedean field that cannot be embedded in any larger such field; apart from the last condition (Hilbert's 'Axiom of Completeness') the style of the axiomatization is thoroughly modern, and was the first of its kind.[b] In §3, Hilbert men-

[b] The 'Axiom of Completeness' was criticized at the time both for its logical complexity (the axiom quantifies over all models of the other axioms, rather than simply stating properties of the real numbers) and for not obviously being the statement of a continuity condition for the real line. Suggestions were made for replacing it with the so-called Cantorian axiom of continuity, which

tions the need to prove the consistency and completeness of the given axioms; and he rather confidently remarks that the proof of consistency 'needs only a suitable modification of well-known methods of inference'. Hilbert was, in fact, to spend the rest of his career searching for such a proof.

It is not entirely clear what 'well-known methods' Hilbert had in mind when he wrote these words in 1899. In his *Grundlagen der Geometrie* he had proved the independence results using techniques that are today called *model-theoretic*: To show that a given axiom *A* is independent of the remaining axioms, Hilbert constructed a model in which *A* fails but the remaining axioms are satisfied. Similarly, he proved the consistency of the entire stock of axioms by endowing them with an arithmetical model (specifically, the model furnished by analytic geometry). This gave him a consistency proof *relative* to the theory of the real numbers; for a contradiction in the geometrical axioms would, by his arithmetical model, also yield a contradiction in the axioms for the real numbers themselves. The problem now was to prove the consistency of the theory of the real numbers. It seems from his remarks about this problem in his 'Mathematical problems' address (*Hilbert 1900b*) that he somehow planned to modify the techniques used by Weierstrass, Dedekind, and Cantor in the theory of irrational numbers; but the details are murky. And despite the confident tone in his *1900a*, Hilbert placed the problem second on his famous list of problems; so perhaps he was not as confident as he sounded.

In 1904 in his Heidelberg lecture 'On the foundations of logic and arithmetic' (translated in *van Heijenoort 1967*) Hilbert described an essentially different strategy for proving consistency. He observed that, while a consistency proof could be furnished for geometry by providing an arithmetical interpretation of the geometric axioms, the same strategy could not be applied to arithmetic: 'Recourse to another fundamental discipline does not seem to be allowed when the foundations of arithmetic are at issue.' Instead of proceeding semantically, Hilbert proposed to prove consistency directly with a syntactic consistency proof. Mathematical proofs were to be translated into a special formal language; this language was then itself to be the object of a mathematical investigation, which would culminate in a proof that a formal contradiction could never be derived within the system.

Although *Hilbert 1904* contains the germ of Hilbert's later proof theory, he is still not fully clear about the details, nor about the power of the modes of inference he is willing to countenance in the consistency proof itself. Indeed, Hilbert's presentation gave some mathematicians the impression that he was

stipulates that, for every nested sequence of intervals, there exists at least one point belonging to every interval in the sequence. For a discussion of the relative merits of the Hilbert and Cantor axioms, see the articles *Enriques 1907–10*; *Baldus 1928a*, *1928b*, and 1930; *Schmidt 1933*; *Hertz 1934*.

It should be observed that Hilbert's axioms for the real numbers are an adaptation of seventeen *propositions* about real numbers that Hilbert had stated in §13 of the first edition of his *Grundlagen der Geometrie* (*Hilbert 1899*). The present article *1900a* adds the Completeness Axiom and turns the resulting list of eighteen *axioms* into a characterization of the reals. The eighteen axioms were then incorporated into the second *1903* edition of the *Grundlagen*. (I owe this observation to Michael Hallett.)

reasoning in a circle. Poincaré was quick to leap on the point, and to accuse Hilbert of presupposing the truth of mathematical induction in order to prove its consistency. (See Poincaré's three articles on *Les mathématiques et la logique*, especially the second article in the series, *Poincaré 1906a*.) This objection was not fully met until Hilbert's proof-theoretic papers of the 1920s, which for the first time drew a clear distinction between the strong induction principles expressed in the formal language, and the weaker one used in the metalanguage.

After publishing *1904*, Hilbert suspended his research in the foundations of mathematics; although he presented new technical results in his lecture course of 1917–18, he did not begin to publish his research again until 1922.[c]

The translation of *Hilbert 1900a* is by William Ewald; references should be to the paragraph numbers, which have been added in this edition.

⟦1⟧ If we cast an eye over the numerous works that exist in the literature on the principles of *arithmetic* and on the axioms of *geometry*, and if we compare them with one another, then, in addition to many analogies and relationships between these two subjects, we nevertheless notice a difference in the *method* of investigation.

⟦2⟧ Let us first recall the manner of introducing the concept of number. Starting from the concept of the number 1, one usually imagines the further rational positive integers 2, 3, 4 . . . as arising through the process of counting, and one develops their laws of calculation; then, by requiring that subtraction be universally applicable, one attains the negative numbers; next one defines fractions, say as a pair of numbers—so that every linear function possesses a zero; and finally one defines the real number as a cut or a fundamental sequence, thereby achieving the result that every entire rational indefinite (and indeed every continuous indefinite) function possesses a zero. We can call this method of introducing the concept of number the *genetic method*, because the most general concept of real number is *engendered* ⟦*erzeugt*⟧ by the successive extension of the simple concept of number.

⟦3⟧ One proceeds essentially differently in the construction of geometry. Here one customarily begins by assuming the existence of all the elements, i.e. one postulates at the outset three systems of things (namely, the points, lines, and planes) and then—essentially on the pattern of Euclid—brings these elements into relationship with one another by means of certain axioms—namely, the axioms of linking ⟦Verknüpfung⟧, of ordering, of congruence, and of continuity.[1] The necessary task then arises of showing the *consistency* and

---

[c] *Hilbert 1904* did, however, have a strong impact on the book *Neue Grundlagen der Logik, Arithmetik, und Mengenlehre* by Julius König (*König 1914*). Hilbert appears not to have been aware of König's book, but it influenced von Neumann in his work on consistency proofs (*von Neumann 1927*).

[1] See Hilbert, *Grundlagen der Geometrie, Festschrift zur Enthüllung des Gauss-Weber-Denkmals in Göttingen*. Leipzig 1899.

the *completeness* of these axioms, i.e. it must be proved that the application of the given axioms can never lead to contradictions, and, further, that the system of axioms is adequate to prove all geometrical propositions. We shall call this procedure of investigation the *axiomatic method.*

⟦4⟧ We raise the question whether the genetic method is in fact the only suitable one for the study of the concept of number, and the axiomatic method for the foundations of geometry; it also seems of interest to compare the two methods and to investigate which method is more advantageous for a logical investigation of the foundations of mechanics or other physical disciplines.

⟦5⟧ My opinion is this: *Despite the high pedagogic and heuristic value of the genetic method, for the final presentation and the complete logical grounding ⟦Sicherung⟧ of our knowledge the axiomatic method deserves the first rank.*

⟦6⟧ In the theory of the number concept, the axiomatic method takes the following form:

⟦7⟧ We think a system of things ⟦denken ein System von Dingen⟧; we call these things numbers and designate ⟦bezeichnen⟧ them by a, b, c, . . . We think these numbers in certain reciprocal relationships ⟦Beziehungen⟧ whose exact and complete description occurs through the following axioms:

1. Axioms of linking

I 1. From the number a and the number b there arises through 'addition' a determinate number c; in symbols:

$$a + b = c \text{ or } c = a + b.$$

I 2. If a and b are given numbers, then there always exists one and only one number x and also one and only one number y such that

$$a + x = b \text{ and } y + a = b$$

respectively.

I 3. There is a determinate number—it is called 0—such that for every a

$$a + 0 = a \text{ and } 0 + a = a.$$

I 4. From the number a and the number b there arises in yet another way, through 'multiplication', a determinate number c; in symbols:

$$ab = c \text{ or } c = ab.$$

I 5. If a and b are arbitrary given numbers and a is not 0, then there always exists one and only one number x, and also one and only one number y, such that

$$ax = b \text{ and } ya = b.$$

I 6. There is a determinate number—it is called 1—such that for every a

$$a \,.\, 1 = a \text{ and } 1 \,.\, a = a.$$

II. Axioms of calculation

If a, b, c are arbitrary numbers, then the following formulae always hold:

II 1. $$a + (b + c) = (a + b) + c$$

II 2. $$a + b = b + a$$

II 3. $$a(bc) = (ab)c$$

II 4. $$a(b + c) = ab + ac$$

II 5. $$(a + b)c = ac + bc$$

II 6. $$ab = ba.$$

III. Axioms of ordering

III 1. If a, b are any two different numbers, then a determinate one of them (say a) is always greater ($>$) than the other; the latter is then called the smaller. In symbols:

$$a > b \text{ and } b < a.$$

III 2. If $a > b$ and $b > c$, then $a > c$.

III 3. If $a > b$, then we always have

$$a + c > b + c \text{ and } c + a > c + b.$$

III 4. If $a > b$ and $c > 0$, then we always have

$$ac > bc \text{ and } ca > cb.$$

IV. Axioms of continuity

IV 1. (*Archimedean axiom.) If* $a > 0$ and $b > 0$ are two arbitrary numbers, then it is always possible to add a to itself so often that the resulting sum has the property that

$$a + a + \ldots + a > b.$$

IV 2. (*Axiom of Completeness.*) It is not possible to add to the system of numbers another system of things so that the axioms I, II, III, and IV 1 are also all satisfied in the combined system; in short, the numbers form a system of things which is incapable of being extended while continuing to satisfy all the axioms.

⟦8⟧ In axiom IV 1 we have presupposed the concept of finite number ⟦Anzahl⟧.

⟦9⟧ Several of the axioms I 1–6, II 1–6, III 1–4, IV 1–2 are consequences of the others, so we are faced with the task of discussing the logical dependencies of the above axioms. This task furnishes many new and fruitful ideas for the investigation of the principles of arithmetic. For example, we recognize the following facts:

⟦10⟧ The existence of the number 0 (axiom I 3) is a consequence of axioms I 1, 2 and II 1; so it rests essentially on the associative law of addition.

⟦11⟧ The existence of the number 1 (axiom I 6) is a consequence of axioms I 4, 5 and II 3; so it rests essentially on the associative law of multiplication.

⟦12⟧ The commutative law of addition (axiom II 2) is a consequence of axioms I, II 1, 4, 5; so it appears essentially as a consequence of the associative law of addition and of the two distributive laws.

⟦13⟧ Proof. We have

$$(a + b)(1 + 1) = (a + b)1 + (a + b)1 = a + b + a + b,$$
$$= a(1 + 1) + b(1 + 1) = a + a + b + b;$$

consequently,

$$a + b + a + b = a + a + b + b,$$

and therefore, by I 2,

$$b + a = a + b.$$

⟦14⟧ The commutative law of multiplication (axiom II 6) is a consequence of axioms I, II 1–5, III, IV 1, but is not a consequence of axioms I, II 1–5, III; thus, that law can be deduced from the remaining axioms if and only if one adjoins the Archimedian axiom (axiom IV 1). This fact has special significance for the foundations of geometry.[2]

⟦15⟧ Axioms IV 1 and IV 2 are independent of each other; they make no statement about the concept of convergence or about the existence of limits, but nevertheless they imply (as one can show) Bolzano's theorem about the existence of a point of condensation ⟦Verdichtungsstelle⟧. We therefore recognize the agreement of our number-system with the usual system of real numbers.

⟦16⟧ To prove the consistency of the above axioms, on needs only a suitable modification of familar methods of inference. In this proof I also see the proof of the existence of the totality of real numbers, or—in the terminology of G. Cantor—the proof that the system of real numbers is a consistent (finished) set ⟦consistente (fertige) Menge⟧.

⟦17⟧ Under the conception described above, the doubts which have been raised against the existence of the totality of all real numbers (and against the existence of infinite sets generally) lose all justification; for by the set of real numbers we do not have to imagine, say, the totality of all possible laws according to which the elements of a fundamental sequence can proceed, but rather—as just described—a system of things whose mutual relations are given by the *finite and closed* system of axioms I–IV, and about which new statements are valid only if one can derive them from the axioms by means of a finite number of logical inferences.

⟦18⟧ If we should wish to prove in a similar manner the existence of a totality of all powers (or of all Cantorian alephs), this attempt would fail; for in fact the totality of all powers does not exist, or—in Cantor's terminology—the system of all powers is an inconsistent (unfinished) set.

Göttingen, 12 October 1899.

[2] See D. Hilbert, l.c., Ch. VI.

B. *FROM* MATHEMATICAL PROBLEMS (*HILBERT 1900b*)

The following selection contains the introductory material from Hilbert's address to the International Congress of Mathematicians in Paris in 1900, the address in which Hilbert presented to the mathematical community his famous list of 23 unsolved problems. (For an account of the subsequent fate of the Hilbert problems see *Browder 1976*.) The address states some of the characteristic themes of his philosophy of mathematics—the solvability in principle of every mathematical problem; the connection between rigour and simplicity; the nature of mathematical existence; the connection between physics and mathematics, experience and thought. (For a late treatment of similar themes, see *Hilbert 1930b*.) The present selection also contains his discussion of the first two Hilbert problems: the Continuum Hypothesis, and the Consistency of Arithmetic.

The translation is by Mary Winston Newson, and originally appeared in 1902; it has been lightly revised by William Ewald to bring its terminology into conformity with the other translations of Hilbert in the present work, and to restore the paragraphing to that of Hilbert's original text. References to *Hilbert 1900b* should be to the paragraph numbers, which have been added in this edition.

⟦1⟧ Who of us would not be glad to lift the veil behind which the future lies hidden; to cast a glance at the next advances of our science and at the secrets of its development during future centuries? What particular goals will there be toward which the leading mathematical spirits of coming generations will strive? What new methods and new facts in the wide and rich field of mathematical thought will the new centuries disclose?

⟦2⟧ History teaches the continuity of the development of science. We know that every age has its own problems, which the following age either solves or casts aside as profitless and replaces by new ones. If we would obtain an idea of the probable development of mathematical knowledge in the immediate future, we must let the unsettled questions pass before our minds and look over the problems which the science of today sets and whose solution we expect from the future. To such a review of problems the present day, lying at the meeting of the centuries, seems to me well adapted. For the close of a great epoch not only invites us to look back into the past but also directs our thoughts to the unknown future.

⟦3⟧ The deep significance of certain problems for the advance of mathematical science in general and the important role which they play in the work of the individual investigator are not to be denied. As long as a branch of science offers an abundance of problems, so long is it alive ; a lack of problems foreshadows extinction or the cessation of independent development. Just as every human undertaking pursues certain objects, so also mathematical research requires its problems. It is by the solution of problems that the investigator tests the temper of his steel; he finds new methods and new outlooks, and gains a wider and freer horizon.

⟦4⟧ It is difficult and often impossible to judge the value of a problem correctly in advance; for the final award depends upon the gain which science obtains from the problem. Nevertheless we can ask whether there are general criteria which mark a good mathematical problem.

⟦5⟧ An old French mathematician said: 'A mathematical theory is not to be considered complete until you have made it so clear that you can explain it to the first man whom you meet on the street.' This clearness and ease of comprehension, here insisted on for a mathematical theory, I should still more demand for a mathematical problem if it is to be perfect; for what is clear and easily comprehended attracts, the complicated repels us.

⟦6⟧ Moreover, a mathematical problem should be difficult in order to entice us, yet not completely inaccessible, lest it mock at our efforts. It should be to us a guide post on the mazy paths to hidden truths, and ultimately a reminder of our pleasure in the successful solution.

⟦7⟧ The mathematicians of past centuries were accustomed to devote themselves to the solution of difficult particular problems with passionate zeal. They knew the value of difficult problems. I remind you only of the 'problem of the line of quickest descent', proposed by John Bernoulli. Experience teaches, explains Bernoulli in the public announcement of this problem, that lofty minds are led to strive for the advance of science by nothing more than by laying before them difficult and at the same time useful problems, and he therefore hopes to earn the thanks of the mathematical world by following the example of men like Mersenne, Pascal, Fermat, Viviani, and others and laying before the distinguished analysts of his time a problem by which, as a touchstone, they may test the value of their methods and measure their strength. The calculus of variations owes its origin to this problem of Bernoulli and to similar problems.

⟦8⟧ Fermat had asserted, as is well known, that the diophantine equation

$$x^n + y^n = z^n$$

(x, y, and z integers) is unsolvable—except in certain self-evident cases. The attempt to prove this impossibility offers a striking example of the inspiring effect which such a very special and apparently unimportant problem may have upon science. For Kummer, incited by Fermat's problem, was led to the introduction of ideal numbers and to the discovery of the law of the unique decomposition of the numbers of a circular field into ideal prime factors—a law which today, in its generalization to any algebraic field by Dedekind and Kronecker, stands at the centre of the modern theory of numbers and whose significance extends far beyond the boundaries of number theory into the realm of algebra and the theory of functions.

⟦9⟧ To speak of a very different region of research, I remind you of the problem of three bodies. The fruitful methods and the far-reaching principles which Poincaré has brought into celestial mechanics and which are today recognized and applied in practical astronomy are due to the circumstance that he undertook to treat anew that difficult problem and to approach nearer a solution.

⟦10⟧ The two last-mentioned problems—that of Fermat and the problem of the three bodies—seem to us almost like opposite poles—the former a free invention of pure reason, belonging to the region of abstract number theory, the latter forced upon us by astronomy and necessary to an understanding of the simplest fundamental phenomena of nature.

⟦11⟧ But it often happens also that the same special problem finds application in the most unlike branches of mathematical knowledge. So, for example, the problem of the shortest line plays a chief and historically important part in the foundations of geometry, in the theory of curved lines and surfaces, in mechanics, and in the calculus of variations. And how convincingly has F. Klein, in his works on the icosahedron, pictured the significance which attaches to the problem of the regular polyhedra in elementary geometry, in group theory, in the theory of equations, and in that of linear differential equations.

⟦12⟧ In order to throw light on the importance of certain problems, I may also refer to Weierstrass, who spoke of it as his happy fortune that he found at the outset of his scientific career a problem so important as Jacobi's problem of inversion on which to work.

⟦13⟧ Having now recalled to mind the general importance of problems in mathematics, let us turn to the question from what sources this science derives it problems. Surely the first and oldest problems in every branch of mathematics spring from experience and are suggested by the world of external phenomena. Even the rules of calculation with integers must have been discovered in this fashion in a lower stage of human civilization, just as the child of today learns the application of these laws by empirical methods. The same is true of the first problems of geometry, the problems bequeathed us by antiquity, such as the duplication of the cube, the squaring of the circle; also the oldest problems in the theory of the solution of numerical equations, in the theory of curves and the differential and integral calculus, in the calculus of variations, the theory of Fourier series and the theory of potential—to say nothing of the further abundance of problems properly belonging to mechanics, astronomy, and physics.

⟦14⟧ But, in the further development of a branch of mathematics, the human mind, encouraged by the success of its solutions, becomes conscious of its independence. It evolves from itself alone, often without appreciable influence from without, by means of logical combination, generalization, specialization, by separating and collecting ideas in fortunate ways, new and fruitful problems, and appears then itself as the real questioner. Thus arose the problem of prime numbers and the other problems of number theory, Galois's theory of equations, the theory of algebraic invariants, the theory of abelian and automorphic functions; indeed almost all the finer questions of modern arthmetic and function theory arise in this way.

⟦15⟧ In the mean time, while the creative power of pure reason is at work, the outer world again comes into play, forces upon us new questions from actual experience, opens up new branches of mathematics, and while we seek to conquer these new fields of knowledge for the realm of pure thought, we often find the answers to old unsolved problems and thus at the same time advance most successfully the old theories. And it seems to me that the numerous and surprising analogies and that apparently harmony which the mathematician so often perceives in the questions, methods, and ideas of the various branches of his science, have their origin in this ever-recurring interplay between thought and experience.

⟦16⟧ It remains to discuss briefly what general requirements may be justly laid down for the solution of a mathematical problem. I should say first of all, this: that it shall be possible to establish the correctness of the solution by means of a finite number of steps based upon a finite number of hypotheses which are implied in the statement of the problem and which must always be exactly formulated. This requirement of logical deduction by means of a finite number of processes is simply the requirement of rigour in reasoning. Indeed the requirement of rigour, which has become proverbial in mathematics, corresponds to a universal philosophical necessity of our understanding; and, on the other hand, only by satisfying this requirement do the thought content and the suggestiveness of the problem attain their full effect. A new problem, especially when it comes from the world of outer experience, is like a young twig, which thrives and bears fruit only when it is grafted carefully and in accordance with strict horticultural rules upon the old stem, the established achievements of our mathematical science.

⟦17⟧ Besides it is an error to believe that rigour in the proof is the enemy of simplicity. On the countrary, we find it confirmed by numerous examples that the rigorous method is at the same time the simpler and the more easily comprehended. The very effort for rigour forces us to find out simpler methods of proof. It also frequently leads the way to methods which are more capable of development than the old methods of less rigour. Thus the theory of algebraic curves experienced a considerable simplification and attained greater unity by means of the more rigorous function-theoretical methods and the consistent introduction of transcendental devices. Further, the proof that the power series permits the application of the four elementary arithmetical operations as well as the term by term differentiation and integration, and the recognition of the utility of the power series depending upon this proof contributed materially to the simplification of all analysis, particularly of the theory of elimination and the theory of differential equations, and also of the existence proofs demanded in those theories. But the most striking example for my statement is the calculus of variations. The treatment of the first and second variations of definite integrals required in part extremely complicated calculations, and the processes applied by the old mathematicians had not the needful rigour. Weierstrass showed us the way to a new and sure foundation of the calculus of variations. By the examples of the simple and double integral I will show briefly, at the close of my lecture, how this way leads at once to a surprising simplification of the calculus of variations. For in the demonstration of the necessary and sufficient criteria for the occurrence of a maximum and minimum, the calculation of the second variation, and in part, indeed, the wearisome reasoning connected with the first variation may be completely dispensed with—to say nothing of the advance which is involved in the removal of the restriction to variations for which the differential coefficients of the function vary but slightly.

⟦18⟧ While insisting on rigour in the proof as a requirement for a perfect solution of a problem, I should like, on the other hand, to oppose the opinion that only the concepts of analysis, or even those of arithmetic alone, are susceptible of a fully rigorous treatment. This opinion, occasionally advocated by eminent men, I consider entirely erroneous. Such a one-sided interpretation of the requirement of rigour would soon lead to the ignoring of all concepts arising from geometry, mechanics, and physics, to a stoppage of the flow of new material from the outside world, and finally, indeed, as a last consequence, to the rejection of the ideas of the continuum and of the irrational number. But what an important nerve, vital to mathematical science, would be cut by the extirpation of geometry and mathematical physics! On the contrary I think that wherever, from the side of the theory of knowledge or in geometry, or from the theories of natural or physical science, mathematical ideas come up, the problem arises for mathematical science to investigate the principles underlying these ideas and so to establish them upon a simple and complete system of axioms, that the exactness of the new ideas and their applicability to deduction shall be in no respect inferior to those of the old arithmetical concepts.

⟦19⟧ To new concepts correspond, necessarily, new signs. These we choose in such a way that they remind us of the phenomena which were the occasion for the formation of the new concepts. So the geometrical figures are signs or mnemonic symbols of space intuition and are used as such by all mathematicians. Who does not always use along with the double inequality $a > b > c$ the picture of three points following one another on a straight line as the geometrical picture of the idea 'between'? Who does not make use of drawings of segments and rectangles enclosed in one another, when it is required to prove with perfect rigour a difficult theorem on the continuity of functions or the existence of points of condensation? Who could dispense with the figure of the triangle, the circle with its centre, or with the cross of three perpendicular axes? Or who would give up the representation of the vector field, or the picture of a family of curves or surfaces with its envelope which plays so important a part in differential geometry, in the theory of differential equations, in the foundation of the calculus of variations and in other purely mathematical sciences?

⟦20⟧ The arithmetical symbols are written diagrams and the geometrical figures are graphic formulae; and no mathematician could spare these graphic formulae, any more than in calculation the insertion and removal of parentheses or the use of other analytical signs.

⟦21⟧ The use of geometrical signs as a means of strict proof presupposes the exact knowledge and complete mastery of the axioms which underlie those figures; and in order that these geometrical figures may be incorporated in the general treasure of mathematical signs, there is necessary a rigorous axiomatic investigation of their conceptual content. Just as in adding two numbers, one must place the digits under each other in the right order, so that only the rules of calculation, i.e., the axioms of arithmetic, determine the correct use of the digits, so the use of geometrical signs is determined by the axioms of geometrical concepts and their combinations.

⟦22⟧ The agreement between geometrical and arithmetical thought is shown also in that we do not habitually follow the chain of reasoning back to the axioms in arithmetical, any more than in geometrical, discussions. On the contrary we apply, especially in first attacking a problem, a rapid, unconscious, not absolutely sure combination, trusting to a certain arithmetical feeling for the behaviour of the arithmetical symbols, which we could dispense with as little in arithmetic as with the geometrical imagination in geometry. As an example of an arithmetical theory operating rigorously with geometrical ideas and signs, I may mention Minkowski's work, *Die Geometrie der Zahlen.**

⟦23⟧ Some remarks upon the difficulties which mathematical problems may offer, and the means of surmounting them, may be in place here.

⟦24⟧ If we do not succeed in solving a mathematical problem, the reason frequently consists in our failure to recognize the more general standpoint from which the problem before us appears only as a single link in a chain of related problems. After finding this standpoint, not only is this problem frequently more accessible to our investigation, but at the same time we come into possession of a method which is applicable also to related problems. The introduction of complex paths of integration by Cauchy and of the notion of the *ideals* in number theory by Kummer may serve as examples. This way of finding general methods is certainly the most practicable and the most certain; for he who seeks for methods without having a definite problem in mind seeks for the most part in vain.

⟦25⟧ In dealing with mathematical problems, specialization plays, as I believe, a still more important part than generalization. Perhaps in most cases where we seek in vain the answer to a question, the cause of the failure lies in the fact that problems simpler and easier than the one in hand have been either not at all or incompletely solved. All depends, then, on finding out these easier problems, and on solving them by means of devices as perfect as possible and of concepts capable of generalization. This rule is one of the most important levers for overcoming mathematical difficulties and it seems to me that it is used almost always, though perhaps unconsciously.

⟦26⟧ Occasionally it happens that we seek the solution under insufficient hypotheses or in an incorrect sense, and for this reason do not succeed. The problem then arises: to show the impossibility of the solution under the given hypotheses, or in the sense contemplated. Such proofs of impossibility were

---

* Leipzig, 1896.

effected by the ancients, for instance when they showed that the ratio of the hypotenuse to the side of an isosceles right triangle is irrational. In later mathematics, the question as to the impossibility of certain solutions plays a preëminent part, and we perceive in this way that old and difficult problems, such as the proof of the axiom of parallels, the squaring of the circle, or the solution of equations of the fifth degree by radicals have finally found fully satisfactory and rigorous solutions, although in another sense than that originally intended.

⟦27⟧ It is probably this important fact along with other philosophical reasons that gives rise to the conviction (which every mathematician shares, but which no one has as yet supported by a proof) that every definite mathematical problem must necessarily be susceptible of an exact settlement, either in the form of an actual answer to the question asked, or by the proof of the impossibility of its solution and therewith the necessary failure of all attempts. Take any definite unsolved problem, such as the question as to the irrationality of the Euler–Mascheroni constant C, or the existence of an infinite number of prime numbers of the form $2^n + 1$. However unapproachable these problems may seem to us and however helpless we stand before them, we have, nevertheless, the firm conviction that their solution must follow by a finite number of purely logical processes.

⟦28⟧ Is this axiom of the solvability of every problem a peculiarity characteristic of mathematical thought alone, or is it possibly a general law inherent in the nature of the mind, that all questions which it asks must be answerable? For in other sciences also one meets old problems which have been settled in a manner most satisfactory and most useful to science by the proof of their impossibility. I instance the problem of perpetual motion. After seeking in vain for the construction of a perpetual motion machine, the relations were investigated which must subsist between the forces of nature if such a machine is to be impossible;* and this inverted question led to the discovery of the law of the conservation of energy, which, again, explained the impossibility of perpetual motion in the sense originally intended.

⟦29⟧ This conviction of the solvability of every mathematical problem is a powerful incentive to the worker. We hear within us the perpetual call: There is the problem. Seek its solution. You can find it by pure reason, for in mathematics there is no *ignorabimus*.

⟦30⟧ The supply of problems in mathematics is inexhaustible, and as soon as one problem is solved numerous others come forth in its place. Permit me in the following, tentatively as it were, to mention particular definite problems, drawn from various branches of mathematics, from the discussion of which an advancement of science may be expected.

⟦31⟧ Let us look at the principles of analysis and geometry. The most sug-

* See Helmholtz, 'Über die Wechselwirkung der Naturkräfte und die darauf bezüglichen neuesten Ermittelungen der Physik'; Vortrag, gehalten in Königsberg, 1854.

gestive and notable achievements of the last century in this field are, as it seems to me, the arithmetical formulation of the concept of the continuum in the works of Cauchy, Bolzano, and Cantor, and the discovery of non-Euclidean geometry by Gauss, Bolyai, and Lobatchevsky. I therefore first direct your attention to some problems belonging to these fields.

1. Cantor's problem of the cardinal number of the continuum

⟦32⟧ Two systems, i.e., two sets of ordinary real numbers or points, are said to be (according to Cantor) equivalent or of equal *cardinal number*, if they can be brought into a relation to one another such that to every number of the one set corresponds one and only one definite number of the other. The investigations of Cantor on such sets of points suggest a very plausible theorem, which nevertheless, in spite of the most strenuous efforts, no one has succeeded in proving. This is the theorem:

⟦33⟧ Every system of infinitely many real numbers, i.e., every set of numbers (or points), is either equivalent to the set of natural integers, 1, 2, 3, . . . or to the set of all real numbers and therefore to the continuum, that is, to the points of a line; *as regards equivalence there are, therefore, only two sets of numbers, the countable set and the continuum.*

⟦34⟧ From this theorem it would follow at once that the continuum has the next cardinal number beyond that of the countable set; the proof of this theorem would, therefore, form a new bridge between the countable set and the continuum.

⟦35⟧ Let me mention another very remarkable statement of Cantor's which stands in the closest connection with the theorem mentioned and which, perhaps, offers the key to its proof. Any system of real numbers is said to be ordered, if for every two numbers of the system it is determined which one is the earlier and which the later, and if at the same time this determination is of such a kind that, if a is before b and b is before c, then a always comes before c. The natural arrangement of numbers of a system is defined to be that in which the smaller precedes the larger. But there are, as is easily seen, infinitely many other ways in which the numbers of a system may be arranged.

⟦36⟧ If we think of a definite arrangement of numbers and select from them a particular system of these numbers, a so-called partial system or set, this partial system will also prove to be ordered. Now Cantor considers a particular kind of ordered set which he designates as a well-ordered set and which is characterized in this way, that not only in the set itself but also in every partial set there exists a first number. The system of integers 1, 2, 3, . . . in their natural order is evidently a well-ordered set. On the other hand the system of all real numbers, i.e., the continuum in its natural order, is evidently not well ordered. For if we think of the points of a segment of a straight line, with its initial point excluded, as our partial set, it will have no first element. The question now arises whether the totality of all numbers may not be arranged in another manner so that every partial set may have a first element, i.e., whether the continuum cannot be considered as a well-ordered set—a question which Cantor thinks must be answered in the affirmative. It appears to me most desirable to obtain a direct proof of this remarkable statement of Cantor's, perhaps by actually giving an arrangement of numbers such that in every partial system a first number can be pointed out.

2. The consistency of the arithmetical axioms

⟦37⟧ When we are engaged in investigating the foundations of a science, we must set up a system of axioms which contains an exact and complete description of the relations subsisting between the elementary ideas of that science. The axioms so set up are at the same time the definitions of those elementary ideas; and no statement within the realm of the science whose foundation we are testing is held to be correct unless it can be derived from those axioms by means of a finite number of logical steps. Upon closer consideration the question arises: *Whether, in any way, certain statements of single axioms depend upon one another, and whether the axioms may not therefore contain certain parts in common, which must be isolated if one wishes to arrive at a system of axioms that shall be altogether independent of one another.*

⟦38⟧ But above all I wish to designate the following as the most important among the numerous questions which can be asked with regard to the axioms: *To prove that they are not contradictory, that is, that a finite number of logical steps based upon them can never lead to contradictory results.*

⟦39⟧ In geometry, the proof of the consistency of the axioms can be effected by constructing a suitable field of numbers, such that analogous relations between the numbers of this field correspond to the geometrical axioms. Any contradiction in the deductions from the geometrical axioms must thereupon be recognizable in the arithmetic of this field of numbers. In this way the desired proof for the consistency of the geometrical axioms is made to depend upon the theorem of the consistency of the arithmetical axioms.

⟦40⟧ On the other hand a direct method is needed for the proof of the consistency of the arithmetical axioms.

⟦41⟧ The axioms of arithmetic are essentially nothing else than the known rules of calculation, with the addition of the axiom of continuity. I recently collected them* and in so doing replaced the axiom of continuity by two simpler axioms, namely, the well-known axiom of Archimedes, and a new axiom essentially as follows: that numbers from a system of things which is capable of no further extension, as long as all the other axioms hold (axiom of completeness). I am convinced that it must be possible to find a direct proof for the consistency of the arithmetical axioms, by means of a careful study and suitable modification of the known methods of reasoning in the theory of irrational numbers.

* *Jahresbericht der Deutschen Mathematiker-Vereinigung*, Vol. 8 (1900), p. 180.

⟦42⟧ To show the significance of the problem from another point of view, I add the following observation: If contradictory attributes be assigned to a concept, I say, that *mathematically the concept does not exist*. So, for example, a real number whose square is -1 does not exist mathematically. But if it can be proved that the attributes assigned to the concept can never lead to a contradiction by the application of a finite number of logical inferences, I say that the mathematical existence of the concept (for example, of a number or a function which satisfies certain conditions) is thereby proved. In the case before us, where we are concerned with the axioms of real numbers in arithmetic, the proof of the consistency of the axioms is at the same time the proof of the mathematical existence of the complete system of real numbers or of the continuum. Indeed, when the proof for the consistency of the axioms shall be fully accomplished, the doubts which have been expressed occasionally as to the existence of the complete system of real numbers will become totally groundless. The totality of real numbers, i.e., the continuum according to the point of view just indicated, is not the totality of all possible series in decimal fractions, or of all possible laws according to which the elements of a fundamental sequence may proceed. It is rather a system of things whose mutual relations are governed by the axioms set up and for which all propositions, and only those, are true which can be derived from the axioms by a finite number of logical inferences. In my opinion, the concept of the continuum is strictly logically tenable in this sense only. It seems to me, indeed, that this corresponds best also to what experience and intuition tell us. The concept of the continuum or even that of the system of all functions exists, then, in exactly the same sense as the system of integral, rational numbers, for example, or as Cantor's higher classes of numbers and cardinal numbers. For I am convinced that the existence of the latter, just as that of the continuum, can be proved in the sense I have described; unlike the system of *all* cardinal numbers or of *all* Cantor's alephs, for which, as may be shown, a system of axioms, consistent in my sense, cannot be set up. Either of these systems is, therefore, according to my terminology, mathematically non-existent.

C. AXIOMATIC THOUGHT (*HILBERT 1918*)

The following address, delivered before the Swiss Mathematical Society in neutral Zurich on 11 September 1917, marks the (published) start of Hilbert's second period of research into the foundations of mathematics. In the interval since *Hilbert 1904*, numerous important developments had taken place in the foundations of mathematics. Hilbert's younger Göttingen colleague Ernst Zermelo had proved his well-ordering theorem (*Zermelo 1904* and *1908a*). The paradoxes of set theory had become widely known, and had led to spirited discussions between Russell, Poincaré, Richard, König, Zermelo, and Peano. Hilbert's axiomatic method, first employed in his *Grundlagen der Geometrie* (*1899*), had been applied in numerous investigations in geometry, algebra, and mathematical physics. Poincaré published his criticisms of *Hilbert 1904* and of the logicists (*Poincaré 1905b*, *1906a*, and *1906b*); Zermelo supplied axioms for the set theory of Dedekind and Cantor (*Zermelo 1908b*); Whitehead and Russell published *Principia mathematica*; Brouwer began to develop his intuitionistic mathematics; Hilbert's gifted former pupil, Hermann Weyl, was drawn to Brouwer's ideas, and was soon to publish *Das Kontinuum* (*Weyl 1917*).

Hilbert in the following lecture does not directly grapple with technical problems raised by these developments; nor was he to do so in print until his proof-theory papers of 1922. Brouwer and Weyl, who were to be prominent targets in his later articles, are not mentioned; Cantor, Poincaré, and the paradoxes are touched on only briefly. Instead, Hilbert surveys the role of axiomatization in mathematics and the physical sciences; he restates and refines his account of the axiomatic method and, in the wake of the *Principia* and of Zermelo's axiomatization, points again to the need for a direct proof of the consistency of number theory and set theory. He also seeks to turn the attention of mathematicians to the study of *proofs*, and in addition to the consistency and independence of the axioms proposes the problems of the *decidability* in a finite number of steps of mathematical problems, the criteria for the *simplicity* of a proof, the relationship between the *content* and the *formalism* of mathematics, and the *solvability in principle* of every mathematical problem. As Hilbert admits, his suggestions here are programmatic, and all the details remain to be worked out.

The essay is noteworthy for the numerous illustrations of the axiomatic method drawn from various branches of mathematics and physics, and for the relative lack of space devoted to the set-theoretic paradoxes. Hilbert is persistently misconstrued as a 'formalist', i.e. as somebody who was so shaken by the paradoxes that he took up the theory that mathematics is merely a game played with meaningless symbols. But the intellectual background to Hilbert's proof theory was richer than this. To be sure, the paradoxes were an important goad; but he had more positive ambitions than the mere avoidance of paradox. (Hilbert, in fact, in his unpublished Göttingen lectures, repeatedly indicated that in his opinion Zermelo's work had successfully resolved the known paradoxes.) His study of the axiomatic foundations of geometry had yielded a rich harvest of mathematical results—non-Archimedean geometries, a new topological characterization of the plane, new theorems on the nature of continuity—and there was every reason to hope that the same powerful tool would prove equally useful in the other branches of mathematics and physics mentioned in this article.

As for the term 'formalist', it is so misleading that it should be abandoned altogether as a label for Hilbert's philosophy of mathematics. On the face of it, 'formalism' is not a felicitous description for the style of reasoning one finds in *Anschauliche Geometrie*—'*Intuitive geometry*' (*Hilbert and Cohn-Vossen*

1932), a work which was written at the high-point of Hilbert's work in proof theory; nor does it accurately convey the convictions of the man who, in his Göttingen lectures, derided those who saw mathematics as a mere heaping-up of consequences mechanically derived from a given stock of axioms; nor the man who ended his essay, 'On the infinite', by saying, 'We gain a conviction that runs counter to the earlier endeavours of Frege and Dedekind, the conviction that, if scientific knowledge is to be possible, certain intuitive conceptions ⟦Vorstellungen⟧ and insights are indispensable; logic alone does not suffice' (*Hilbert 1926*). As the present selection makes clear, Hilbert viewed formal axiom systems instrumentally, as a powerful tool for mathematical research, a tool to be employed when a field had reached a point of sufficient ripeness. But he nowhere suggests that the whole of mathematics can simply be *identified* with the study of formal systems; and indeed in his proof-theoretical writings he took considerable pains to point out that the genuine mathematics—*inhaltliche Mathematik*—takes place, not in the formalism, but in the metalanguage. For all these reasons, Hilbert himself rejected the label 'formalist' (see, for example, *Hilbert 1931a*, §35), and students of his thought would do well to follow his example.

The translation is by William Ewald; references to *Hilbert 1918* should be to the paragraph numbers, which have been added in this edition.

⟦1⟧ Just as in the life of nations the individual nation can only thrive when all neighbouring nations are in good health; and just as the interest of states demands, not only that order prevail within every individual state, but also that the relationships of the states among themselves be in good order; so it is in the life of the sciences. In due recognition of this fact the most important bearers of mathematical thought have always evinced great interest in the laws and the structure of the neighbouring sciences; above all for the benefit of mathematics itself they have always cultivated the relations to the neighbouring sciences, especially to the great empires of physics and epistemology. I believe that the essence of these relations, and the reason for their fruitfulness, will appear most clearly if I describe for you the general method of research which seems to be coming more and more into its own in modern mathematics: I mean the *axiomatic method*.

⟦2⟧ When we assemble the facts of a definite, more-or-less comprehensive field of knowledge, we soon notice that these facts are capable of being ordered. This ordering always comes about with the help of a certain *framework of concepts* ⟦*Fachwerk von Begriffen*⟧ in the following way: a concept of this framework corresponds to each individual object of the field of knowledge, and a logical relation between concepts corresponds to every fact within the field of knowledge. The framework of concepts is nothing other than the *theory* of the field of knowledge.

⟦3⟧ Thus the facts of geometry order themselves into a geometry, the facts of arithmetic into a theory of numbers, the facts of statics, mechanics, electrodynamics into a theory of statics, mechanics, electrodynamics, or the facts from the physics of gases into a theory of gases. It is precisely the same with the fields of knowledge of thermodynamics, geometrical optics, elementary radiation-theory, the conduction of heat, or also with the calculus of probabilities or the theory of sets. It even holds of special fields of knowledge in pure mathematics, such as the theory of surfaces, the theory of Galois equations, and the theory of prime numbers, no less than for several fields of knowledge that lie far from mathematics, such as certain parts of psychophysics or the theory of money.

⟦4⟧ If we consider a particular theory more closely, we always see that a few distinguished propositions of the field of knowledge underlie the construction of the framework of concepts, and these propositions then suffice by themselves for the construction, in accordance with logical principles, of the entire framework.

⟦5⟧ Thus in geometry the proposition of the linearity of the equation of the plane and of the orthogonal transformation of point-coordinates is completely adequate to produce the whole broad science of spatial Euclidean geometry purely by means of analysis. Moreover, the laws of calculation and the rules for integers suffice for the construction of number theory. In statics the same role is played by the proposition of the parallelogram of forces; in mechanics, say, by the Lagrangian differential equations of motion; and in electrodynamics by the Maxwell equations together with the requirement of the rigidity and charge of the electron. Thermodynamics can be completely built up from the concept of energy function and the definition of temperature and pressure as derivatives of its variables, entropy and volume. At the heart of the elementary theory of radiation is Kirchhoff's theorem on the relationships between emission and absorption; in the calculus of probabilities the Gaussian law of errors is the fundamental proposition; in the theory of gases, the proposition that entropy is the negative logarithm of the probability of the state; in the theory of surfaces, the representation of the element of arc by the quadratic differential form; in the theory of equations, the proposition concerning the existence of roots; in the theory of prime numbers, the proposition concerning the reality and frequency of Rieman's function $\zeta(s)$.

⟦6⟧ These fundamental propositions can be regarded from an initial standpoint as the *axioms of the individual fields of knowledge*: the progressive development of the individual field of knowledge then lies solely in the further logical construction of the already mentioned framework of concepts. This standpoint is especially predominant in pure mathematics, and to the corresponding manner of working we owe the mighty development of geometry, of arithmetic, of the theory of functions, and of the whole of analysis.

⟦7⟧ Thus in the cases mentioned above the problem of grounding the individual field of knowledge had found a solution; but this solution was only temporary. In fact, in the individual fields of knowledge the need arose to ground the fundamental axiomatic propositions themselves. So one acquired 'proofs' of the linearity of the equation of the plane and the orthogonality of the transformation expressing a movement, of the laws of arithmetical calculation, of the parallelogram of forces, of the Lagrangian equations of motion, of Kirchhoff's law regarding emission and absorption, of the law of entropy, and of the proposition concerning the existence of roots of an equation.

⟦8⟧ But critical examination of these 'proofs' shows that they are not in themselves proofs, but basically only make it possible to trace things back to certain deeper propositions, which in turn are now to be regarded as new axioms instead of the propositions to be proved. The actual so-called *axioms* of geometry, arithmetic, statics, mechanics, radiation theory, or thermodynamics arose in this way. These axioms form a layer of axioms which lies deeper than the axiom-layer given by the recently-mentioned fundamental theorems of the individual field of knowledge. The procedure of the axiomatic method, as it is expressed here, amounts to a *deepening of the foundations* of the individual domains of knowledge—a deepening that is necessary for every edifice that one wishes to expand and to build higher while preserving its stability.

⟦9⟧ If the theory of a field of knowledge—that is, the framework of concepts that represents it—is to serve its purpose of orienting and ordering, then it must satisfy two requirements above all: *first* it should give us an overview of the *independence* and *dependence* of the propositions of the theory; *second*, it should give us a guarantee of the *consistency* of all the propositions of the theory. In particular, the axioms of each theory are to be examined from these two points of view.

⟦10⟧ Let us first consider the independence or dependence of the axioms.

⟦11⟧ The *axiom of parallels* in geometry is the classical example of the independence of an axiom. When he placed the parallel postulate among the axioms, Euclid thereby denied that the proposition of parallels is implied by the other axioms. Euclid's method of investigation became the paradigm for axiomatic research, and since Euclid geometry has been the prime example of an axiomatic science.

⟦12⟧ Classical mechanics furnishes another example of an investigation of the independence of axioms. The Lagrangian equations of motion were temporarily able to count as axioms of mechanics—for mechanics can of course be entirely based on these equations when they are generally formulated for arbitrary forces and arbitrary side-constraints. But further investigation shows that it is not necessary in the construction of mechanics to presuppose arbitrary forces or arbitrary side-constraints; thus the system of presuppositions can be reduced. This piece of knowledge leads, on the one hand, to the axiom system of Boltzmann, who assumes only forces (and indeed special central forces) but no side-constraints, and the axiom system of Hertz, who discards forces and makes do with side-constraints (and indeed special side-constraints with rigid connections). These two axiom systems form a deeper layer in the progressive axiomatization of mechanics.

⟦13⟧ If in establishing the theory of Galois equations we assume as an axiom the existence of roots of an equation, then this is certainly a dependent axiom; for, as Gauss was the first to show, that existence theorem can be proved from the axioms of arithmetic.

⟦14⟧ Something similar would happen if we were to assume as an axiom in the theory of prime numbers the proposition about the reality of the zeroes of the Riemann $\zeta(s)$-function: as we progress to a deeper layer of purely arithmetical axioms the proof of this reality-proposition would become necessary, and only this proof would guarantee the reliability of the important conclusions which we have already achieved for the theory of prime numbers by taking it as a postulate.

⟦15⟧ A particularly interesting question for axiomatics concerns the independence of the propositions of a field of knowledge from the axiom of *continuity*.

⟦16⟧ In the theory of real numbers it is shown that the axiom of measurement—the so-called Archimedean axiom—is independent of all the other arithmetical axioms. As everybody knows, this information is of great significance for geometry; but it seems to me to be of capital interest for physics as well, for it leads to the following result: the fact that by adjoining terrestrial distances to one another we can achieve the dimensions and distances of bodies in outer space (that is, that we can measure heavenly distances with an earthly yardstick) and the fact that the distances within an atom can all be expressed in terms of metres—these facts are not at all a mere logical consequence of propositions about the congruence of triangles or about geometric configurations, but are a result of empirical research. The validity of the Archimedean axiom in nature stands in just as much need of confirmation by experiment as does the familiar proposition about the sum of the angles of a triangle.

⟦17⟧ In general, I should like to formulate the axiom of continuity in physics as follows: 'If for the validity of a proposition of physics we prescribe any degree of accuracy whatsoever, then it is possible to indicate small regions within which the presuppositions that have been made for the proposition may vary freely, without the deviation of the proposition exceeding the prescribed degree of accuracy.' This axiom basically does nothing more than express something that already lies in the essence of experiment; it is constantly presupposed by the physicists, although it has not previously been formulated.

⟦18⟧ For example, if one follows Planck and derives the second law of thermodynamics from the axiom of the impossibility of a *perpetuum mobile of the second sort*, then this axiom of continuity must be used in the derivation.

⟦19⟧ By invoking the theorem that the continuum can be well-ordered, Hamel has shown in a most interesting manner that, in the foundations of statics, the axiom of continuity is necessary for the proof of the theorem concerning the *parallelogram of forces*—at any rate, given the most obvious choice of other axioms.

‖20‖ The axioms of classical mechanics can be deepened if, using the axiom of continuity, one imagines continuous motion to be decomposed into small straight-line movements caused by discrete impulses and following one another in rapid succession. One then applies Bertrand's maximum principle as the essential axiom of mechanics, according to which the motion that actually occurs after each impulse is that which maximizes the kinetic energy of the system with respect to all motions that are compatible with the law of the conservation of energy.

‖21‖ The most recent ways of laying the foundations of physics—of electrodynamics in particular—are all theories of the continuum, and therefore raise the demand for continuity in the most extreme fashion. But I should prefer not to discuss them because the investigations are not yet completed.

‖22‖ We shall now examine the second of the two points of view mentioned above, namely, the question concerning the *consistency* of the axioms. This question is obviously of the greatest importance, for the presence of a contradiction in a theory manifestly threatens the contents of the entire theory.

‖23‖ Even for successful theories that have long been accepted, it is difficult to know that they are internally consistent: I remind you of the reversibility and recurrence paradox in the kinetic theory of gases.

‖24‖ It often happens that the internal consistency of a theory is regarded as obvious, while in reality the proof requires deep mathematical developments. For example, consider a problem from the elementary theory of the *conduction of heat*—namely, the distribution of temperatures within a homogeneous body whose surfaces are maintained at a definite temperature that varies from place to place: then in fact the requirement that there be an equilibrium of temperatures involves no internal theoretical contradiction. But to know this it is necessary to prove that the familiar boundary-value problem of potential theory is always solvable; for only this proof shows that a temperature distribution satisfying the equations of the conduction of heat is at all possible.

‖25‖ But particularly in physics it is not sufficient that the propositions of a theory be in harmony with each other; there remains the requirement that they not contradict the propositions of a neighbouring field of knowledge.

‖26‖ Thus, as I showed earlier, the axioms of the elementary theory of radiation can be used to prove not only *Kirchhoff's law* of emission and absorption, but also a special law about the reflection and refraction of individual beams of light, namely, the law: If two beams of natural light and of the same energy each fall on the surface separating two media from different sides in such a way that one beam after its reflection, and the other after its passage, each have the same direction, then the beam that arises from uniting the two is also of natural light and of the same energy. This theorem is, as the facts show, not at all in contradiction with optics, but can be derived as a conclusion from the electromagnetic theory of light.

‖27‖ As is well known, the results of the *kinetic theory of gases* are in full harmony with *thermodynamics*.

‖28‖ Similarly, *electrodynamic inertia* and *Einsteinian gravitation* are compatible with the corresponding concepts of the classical theories, since the classical concepts can be conceived as limiting cases of the more general concepts in the new theories.

⟦29⟧ In contrast, *modern quantum theory* and our developing knowledge of the internal structure of the atom have led to laws which virtually contradict the earlier electrodynamics, which was essentially built on the Maxwell equations; modern electrodynamics therefore needs—as everybody acknowledges—a new foundation and essential reformulation.

⟦30⟧ As one can see from what has already been said, the contradictions that arise in physical theories are always eliminated by changing the selection of the axioms; the difficulty is to make the selection so that all the observed physical laws are logical consequences of the chosen axioms.

⟦31⟧ But matters are different when contradictions appear in purely theoretical fields of knowledge. Set theory contains the classic example of such an occurrence, namely, in the *paradox of the set of all sets*, which goes back to Cantor. This paradox is so serious that distinguished mathematicians, for example, Kronecker and Poincaré, felt compelled by it to deny that set theory—one of the most fruitful and powerful branches of knowledge anywhere in mathematics—has any justification for existing.

⟦32⟧ But in this precarious state of affairs as well, the axiomatic method came to the rescue. By setting up appropriate axioms which in a precise way restricted both the arbitrariness of the definitions of sets and the admissibility of statements about their elements, Zermelo succeeded in developing set theory in such a way that the contradictions disappear, but the scope and applicability of set theory remain the same.

⟦33⟧ In all previous cases it was a matter of contradictions that had emerged in the course of the development of a theory and that needed to be eliminated by a reformulation of the axiom system. But if we wish to restore the reputation of mathematics as the exemplar of the most rigorous science it is not enough merely to avoid the existing contradictions. The chief requirement of the theory of axioms must go farther, namely, to show that within every field of knowledge contradictions based on the underlying axiom-system are *absolutely impossible*.

⟦34⟧ In accordance with this requirement I have proved the consistency of the axioms laid down in the *Grundlagen der Geometrie* by showing that any contradiction in the consequences of the geometrical axioms must necessarily apppear in the arithmetic of the system of real numbers as well.

⟦35⟧ For the fields of physical knowledge too, it is clearly sufficient to reduce the problem of *internal consistency* to the consistency of the arithmetical axioms. Thus I showed the consistency of the axioms of the *elementary theory of radiation* by constructing its axiom system out of analytically independent pieces—presupposing in the process the consistency of analysis.

⟦36⟧ One may and should in some circumstances proceed similarly in the construction of a mathematical theory. For example, if in the development of the theory of Galois groups we have taken the proposition of the *existence of roots* as an axiom, or if in the theory of prime numbers we have taken the hypothesis concerning the *reality of the zeros* of the Riemann $\zeta(s)$-function as an axiom, then in each case the proof of the consistency of the axiom system comes down to a proof, using the means of analysis, of the proposition of the existence of roots or of the Riemann hypothesis concerning $\zeta(s)$—and only then has the theory been securely completed.

⟦37⟧ The problem of the consistency of the axiom system for the *real numbers* can likewise be reduced by the use of set-theoretic concepts to the same problem for the integers: this is the merit of the theories of the irrational numbers developed by Weierstrass and Dedekind.

⟦38⟧ In only two cases is this method of reduction to another special domain of knowledge clearly not available, namely, when it is a matter of the axioms for the *integers* themselves, and when it is a matter of the foundation of *set theory*; for here there is no other discipline besides logic which it would be possible to invoke.

⟦39⟧ But since the examination of consistency is a task that cannot be avoided, it appears necessary to axiomatize logic itself and to prove that number theory and set theory are only parts of logic.

⟦40⟧ This method was prepared long ago (not least by Frege's profound investigations); it has been most successfully explained by the acute mathematician and logician Russell. One could regard the completion of this magnificent Russellian enterprise of the *axiomatization of logic* as the crowning achievement of the work of axiomatization as a whole.

⟦41⟧ But this completion will require further work. When we consider the matter more closely we soon recognize that the question of the consistency of the integers and of sets is not one that stands alone, but that it belongs to a vast domain of difficult epistemological questions which have a specifically mathematical tint: for example (to characterize this domain of questions briefly) the problem of the *solvability in principle of every mathematical question*, the problem of the subsequent *checkability* of the results of a mathematical investigation, the question of a *criterion of simplicity* for mathematical proofs, the question of the relationship between *content and formalism* in mathematics and logic, and finally the problem of the *decidability* of a mathematical question in a finite number of operations.

⟦42⟧ We cannot rest content with the axiomatization of logic until all questions of this sort and their interconnections have been understood and cleared up.

⟦43⟧ Among the mentioned questions, the last—namely, the one concerning decidability in a finite number of operations—is the best-known and the most discussed; for it goes to the essence of mathematical thought.

⟦44⟧ I should like to increase the interest in this question by indicating several particular mathematical problems in which it plays a role.

⟦45⟧ In the theory of *algebraic invariants* we have the fundamental theorem that there is always a finite number of whole rational invariants by means of which all other such invariants can be represented. In my opinion, my first general proof of this theorem completely satisfied our requirements of simplicity and perspicuity; but it is impossible to reformulate this proof so that we can obtain from it a statable bound for the number of the finitely many invariants of the full system, let alone obtain an actual listing of them. Instead, new principles and considerations of a completely different sort were necessary in order to show that the construction of the full system of invariants requires only a finite number of operations, and that this number is less than a bound that can be stated before the calculation.

⟦46⟧ We see the same thing happening in an example from the *theory of surfaces*. It is a fundamental question in the geometry of surfaces of the fourth order to determine the maximum number of separate sheets it takes to make up such a surface.

⟦47⟧ The first step towards an answer to this question is the proof that the number of sheets of a curved surface must be finite. This can easily be shown function-theoretically as follows. One assumes the existence of infinitely many sheets, and selects a point inside each spatial region bounded by a sheet. A point of accumulation for these infinitely many chosen points would then be a point of a singularity that is excluded for an algebraic surface.

⟦48⟧ This function-theoretic path does not at all lead to an upper bound for the number of surface-sheets. For that, we need instead certain observations on the number of cut-points, which then show that the number of sheets certainly cannot be greater than 12.

⟦49⟧ The second method, entirely different from the first, in turn cannot be applied or transformed to decide whether a surface of the fourth order with twelve sheets actually exists.

⟦50⟧ Since a quaternary form of the fourth order possesses 35 homogeneous coefficients, we can conceive of a given surface of the fourth order as a point in 34-dimensional space. The discriminant of the quaternary form of fourth order is of degree 108 in its coefficients; if it is set equal to zero, it accordingly represents in 34-dimensional space a surface of order 108. Since the coefficients of the discriminant are themselves determinate integers, the topological character of the discriminant surface can be precisely determined by the rules that are familiar to us from 2- and 3-dimensional space; so we can obtain precise information about the nature and significance of the individual subdomains into which the discriminant surface partitions the 34-dimensional space. Now, the surfaces of fourth order represented by points of these subdomains all certainly possess the same sheet-number; and it is accordingly possible to establish, by a long and wearying but finite calculation, whether we have a surface of fourth order with $n \leq 12$ sheets or not.

⟦51⟧ The geometric method just described is thus a third way of treating our question about the maximum number of sheets of a surface of the fourth order. It proves the decidability of this question in a finite number of operations. So in principle an important demand of our problem has been satisfied: it has been reduced to a problem of the level of difficulty of determining the $10^{(10^{10})}$th numeral in the decimal expansion for π—a task which is clearly solvable, but which remains unsolved.

⟦52⟧ Rather, it took a profound and difficult algebraic-geometric investigation by Rohn to show that 11 sheets are not possible in a surface of the fourth order, while 10 sheets actually occur. Only this fourth method delivered the full solution of the problem.

⟦53⟧ These particular discussions show how a variety of methods of proof can be applied to the same problem, and they ought to suggest how necessary it is to study the essence of mathematical proof itself if one wishes to answer such questions as the one about decidability in a finite number of operations.

⟦54⟧ All such questions of principle, which I characterized above and of which the question just discussed—that is, the question about decidability in a finite number of operations—was only the last, seem to me to form an important new field of research which remains to be developed. To conquer this field we must, I am persuaded, make the concept of specifically mathematical proof itself into an object of investigation, just as the astronomer considers the movement of his position, the physicist studies the theory of his apparatus, and the philosopher criticizes reason itself.

⟦55⟧ To be sure, the execution of this programme is at present still an unsolved task.

⟦56⟧ In conclusion, I should like to sum up in a few sentences my general conception of the essence of the axiomatic method. I believe: anything at all that can be the object of scientific thought becomes dependent on the axiomatic method, and thereby indirectly on mathematics, as soon as it is ripe for the formation of a theory. By pushing ahead to ever deeper layers of axioms in the sense explained above we also win ever-deeper insights into the essence of scientific thought itself, and we become ever more conscious of the unity of our knowledge. In the sign of the axiomatic method, mathematics is summoned to a leading role in science.

D. THE NEW GROUNDING OF MATHEMATICS. FIRST REPORT.[1] (*HILBERT 1922a*)

The next two selections show Hilbert's proof theory growing to maturity; they also mark the start of his debates with the intuitionists. In the early days of his career, Hilbert had championed the new set-theoretic ideas of Cantor, a stand which had pitted him against the proto-intuitionist position of Kronecker. (Hilbert's views on Kronecker are discussed above in the first two Kronecker

[1] This report is essentially the contents of the lectures I delivered in the spring of this year in Copenhagen at the invitation of their Mathematical Society, and in the summer in Hamburg at the invitation of the University's Mathematical Seminar.

selections and in the accompanying Introductory Notes.) After Zermelo's axiomatization of set theory (*Zermelo 1908b*) Cantor's ideas seemed to win broad acceptance, and Kronecker's radical criticisms receded from view. But in the years following the First World War, Brouwer and then Weyl published influential critiques of classical mathematics that seemed to bring Kronecker back to life (notably *Brouwer 1913*, *Weyl 1917*, *Brouwer 1919*, and *Weyl 1920*); in response, Hilbert added to the other tasks of his foundational research that of rebutting these critiques and of clearing away for ever the doubts engendered by the paradoxes.

Despite Hilbert's fiery polemics against Kronecker, Weyl, and Brouwer, it should be observed that the entire controversy is an internal feud among constructivists.[a] At the level of metamathematics, Hilbert adopted the intuitionistic criticisms of infinitary mathematics, and sought to use only reasoning that was intuitionistically acceptable; indeed, at this level, Hilbert's finitism went further than that of Brouwer himself.[b] The disagreements stemmed rather from a difference of opinion about what constitutes a foundation for mathematics, and concerned, first, the desirability of formalized mathematics *überhaupt*; second, the usefulness, legitimacy, and mathematical interest of the classical, infinitary modes of inference expressed in Hilbert's formal system.

In contrast to *Hilbert 1904*, the present essay now draws a clear distinction between the logico-mathematical formalism and the *inhaltliche*[c] metamathematical reasonings about it; see, for example, §§33, 59. This distinction allows Hilbert to answer the charge of circularity raised in *Poincaré 1905b* against *Hilbert 1904*. Poincaré had charged that Hilbert needed to presuppose the truth of mathematical induction in order to prove its consistency; but Hilbert can now distinguish (as he does in §31) between the strong principle of complete induction expressed in the formal language and the weaker principle used in the metalanguage.

Hilbert sets out the basic ideas of his proof theory, describes a simple formal axiom system for a fragment of arithmetic, and proves its consistency; he also lays the groundwork for his later investigations of the foundations of set theory and real analysis. In particular, at the end of the article he mentions that the *tertium non datur* is to be formalized by means of a transfinite functional $\chi(f)$ such that $\chi(f) = 0$ iff $f(a) = 1$ for every positive integer a; otherwise, $\chi(f)$ is the least a such that $f(a) \neq 1$. But these matters are here only mentioned, and Hilbert makes no attempt to supply a consistency proof for the transfinite part of his theory.

---

[a] Hilbert was later explicitly to acknowledge the similarities between his constructivist views and those of Kronecker; see *Hilbert 1930b*, §8. This point is also made in *Weyl 1944b*, and by Kreisel in *Kreisel and Newman 1969*.

[b] This fact, not universally appreciated in the 1920s, became mathematically important when, after the proof of the incompleteness theorems (and specifically of the unprovability of consistency), Gödel and Gentzen independently observed that the formal system of classical number theory can be proved consistent relative to Heyting's 1930 system of intuitionistic number theory. Hence intuitionistic reasoning diverges from finitistic. See *Gödel 1933a*. (Gentzen withdrew his paper from publication on the appearance of Gödel's paper.)

[c] This word, following the example set by Stefan Bauer-Mengelberg in *van Heijenoort 1967*, has here been translated with the neologism 'contentual'.

The translation of *Hilbert 1922a* is by William Ewald; references should be to the paragraph numbers, which have been added in this edition.

⟦1⟧ The foundations of mathematics have long been studied by the most diverse authors and in the most manifold ways. In consequence, splendid sequences of ideas have been developed; important and enduring results have been achieved. If I now believe a deeper treatment of the problem to be requisite, and if I attempt such a deeper treatment, this is done not so much to fortify individual mathematical theories as because, in my opinion, all previous investigations into the foundations of mathematics fail to show us a way of formulating the questions concerning foundations so that an unambiguous answer must result. But that is what I require: in mathematical matters there should be in principle no doubt; it should not be possible for half-truths or truths of fundamentally different sorts to exist. Thus—to give as an example a difficult and remote item on the agenda—it must be possible to formulate Zermelo's postulate of choice in such a way that, in the same sense of 'valid', it becomes just as valid and reliable as the arithmetical proposition that $2 + 2 = 4$. I am of the opinion that the foundations of mathematics are capable of full clarity and knowledge, and that the problem of grounding our science is difficult but nevertheless conclusively solvable. The purpose of this interim report is to characterize briefly the sense in which I believe I can attain the solution, and to explain the means.

⟦2⟧ At present there is moreover a topical interest in this subject. Distinguished and highly accomplished mathematicians, Weyl and Brouwer, are seeking the solution to these problems by following what I believe to be a false path.

⟦3⟧ In his critique of previous ways of grounding the number-concept, Weyl asserts that the usual procedure contains a circle (*circulus vitiosus*). He finds this circle in the fact that, in the definition of real numbers, partitions ⟦Einteilungen⟧ are used that are determined by whether there exist real numbers with a given property. But in my opinion matters are as follows: If one takes the usual definitions of real number by Dedekind cuts, sequences of numbers, or fundamental series, one sees that for the mathematicians various methodological standpoints exist side by side. The standpoint that Weyl chooses and from which he exhibits his vicious circle is not at all one of these standpoints; instead, it seems to me to be artificially concocted. Weyl justifies his peculiar standpoint by saying that it preserves the principle of constructivity ⟦das konstruktive Prinzip⟧; but in my opinion precisely because he ends with a circle he should have realized that his standpoint (and therefore the principle of constructivity as he conceives of it and applies it) is not usable, that it blocks the path into analysis.

⟦4⟧ The standpoints usually taken by mathematicians do not rest on the principle of constructivity at all; nor do they exhibit Weyl's circle. Here there are essentially two standpoints:

⟦5⟧ First, one says roughly: a real number is a partition of the rational numbers which possesses the Dedekind-cut property. In this definition the concept of a partition of the rational numbers is precisely delimited with respect to its scope and content. The familiar objection to this standpoint is that the concept of a partition of the rational numbers amounts to the same thing as the concept of a set; but the general concept of set has in fact given rise to paradoxes. If Weyl is making some version of this objection, then the immediate reply is that it is not compelling. The concept of set in the most general sense is not admissible without qualification; but this does not in any way mean that there is anything amiss with the concept of a set of integers. And the paradoxes of set theory cannot be regarded as proving that the concept of a set of integers leads to contradictions. On the contrary: all our mathematical experience speaks for the correctness and consistency of this concept.

⟦6⟧ But if someone were to assert that it is a violation of mathematical rigour to make such a tacit presupposition in the construction of mathematical science, then I point to the second standpoint for grounding the concept of number. This standpoint—the method of axiomatic grounding—is not vulnerable to this objection. It is characterized as follows. The continuum of real numbers is a system of things which are linked to one another by determinate relations, the so-called axioms. In particular, in place of the definition of real number by Dedekind cut, we have the two axioms of continuity, namely, the Archimedean axiom and the so-called completeness axiom. To be sure, the Dedekind cuts can then also be used to specify individual real numbers, but they do not provide the definition of the concept of real number. Rather, a real number is conceptually just a thing belonging to our system.

⟦7⟧ This grounding of the theory of the continuum is not at all opposed to intuition ⟦Anschauung⟧. The concept of extensive magnitude, as we derive it from intuition, is independent of the concept of number ⟦Anzahl⟧; and it is therefore thoroughly in keeping with intuition if we make a fundamental distinction between number and measuring-number ⟦Maßzahl⟧ or quantity.

⟦8⟧ This standpoint is logically completely unobjectionable, and the only thing that remains to be decided is, whether a system of the requisite sort is thinkable, that is, whether the axioms do not, say, lead to a contradiction. Now there is scarcely any subject, either within or outside the mathematical sciences, that has been so thoroughly studied as real analysis. Mathematicians have pursued to the uttermost the modes of inference that rest on the concept of sets of numbers, and not even the shadow of an inconsistency has appeared. If Weyl here sees an 'inner instability of the foundations on which the empire is constructed', and if he worries about 'the impending dissolution of the commonwealth of analysis', then he is seeing ghosts. Rather, despite the application of the boldest and most manifold combinations of the subtlest techniques, a complete security of inference and a clear unanimity of results reigns in analysis. We are therefore justified in assuming those axioms which are the basis of this security and agreement; to dispute this justification would mean to take away in advance from all science the possibility of its functioning: here, if anywhere, axiomatics is called for.

⟦9⟧ To be sure, the problem arises of proving the consistency of the axioms; this is a well-known problem, and for decades I have never lost sight of it. This report concerns the solution of this problem.

⟦10⟧ What Weyl and Brouwer do amounts in principle to following the erstwhile path of Kronecker: they seek to ground mathematics by throwing overboard all phenomena that make them uneasy and by establishing a dictatorship of prohibitions *à la* Kronecker. But this means to dismember and mutilate our science, and if we follow such reformers, we run the danger of losing a large number of our most valuable treasures. Weyl and Brouwer calumniate the general concept of irrational number, of function, even of number-theoretic function, the Cantorian numbers of the higher number-classes, etc.; the proposition that among infinitely many integers there is always a smallest, and even the logical *tertium non datur* (for example, in the assertion: either there is only a finite number of prime numbers, or there are infinitely many)—all of these are examples of forbidden propositions or modes of inference. I believe that, just as Kronecker in his day was unable to get rid of the irrational numbers (Weyl and Brouwer, incidentally, allow the preservation of a torso) so today Weyl and Brouwer will be unable to push their programme through. No: Brouwer is not, as Weyl believes, the revolution, but only a repetition, with the old tools, of an attempted coup that, in its day, was undertaken with more dash, but nevertheless failed completely; and now that the power of the state has been armed and strengthened by Frege, Dedekind, and Cantor, this coup is doomed to fail.

⟦11⟧ To sum up, I should like to say: if one speaks of a mathematical crisis, in any case one may not speak, as Weyl does, of a new crisis. He has artificially imported the vicious circle into analysis. His account of the uncertainty of the results of modern analysis does not correspond to the actual state of affairs. And as for the constructive tendencies that he and Brouwer emphasize so strongly, in my opinion it is precisely Weyl who has failed to see the path to the fulfilment of these tendencies. In my opinion, only the path taken here in pursuit of axiomatics will do full justice to the constructive tendencies, to the extent that they are natural.

⟦12⟧ The goal of finding a secure foundation for mathematics is also my own. I should like to regain for mathematics the old reputation for incontestable truth, which it appears to have lost as a result of the paradoxes of set theory; but I believe that this can be done while fully preserving its accomplishments. The method that I follow is none other than the axiomatic. Its essence is as follows.

⟦13⟧ In order to investigate a subfield of a science, one bases it on the smallest possible number of principles, which are to be as simple, intuitive, and comprehensible as possible, and which one collects together and sets up as axioms. Nothing prevents us from taking as axioms propositions which are provable, or which we believe are provable. Indeed, as history shows, this procedure is perfectly in order: examples are Legendre's prime-number postulate in the theory of quadratic residues, Riemann's conjecture about the zeroes of $\zeta(s)$, the theorem about the existence of roots in algebra, and finally the so-called ergodic hypothesis, a mathematical proposition from whose proof we are today far removed, but which nevertheless has become the foundation of statistical mechanics.

⟦14⟧ The axiomatic method is and remains the indispensable tool, appropriate to our minds, for all exact research in any field whatsoever: it is logically incontestable and at the same time fruitful; it thereby guarantees the maximum flexibility in research. To proceed axiomatically means in this sense nothing else than to think with consciousness: although in earlier times without the axiomatic method it happened that one naïvely believed in certain interconnections as though they were dogmas, the axiomatic method of eliminates this *naïveté*, but leaves us the advantage of belief.

⟦15⟧ But now it is a question of something even more important. Precisely by means of the formulation which I believe I am able to give of the axiomatic method, we shall see how it leads us to full clarity about the principles of inference in mathematics. As I have already said, we can never be certain in advance of the consistency of our axioms if we do not have a special proof of it. Axiomatics therefore compels us to take a stand on this difficult epistemological problem. The proof of the consistency of the axioms succeeds in many cases—for instance, in geometry, in thermodynamics, in the theory of radiation, and in other physical disciplines—in that one reduces the proof to the question of the consistency of the axioms of analysis; this question is in turn a hitherto unsolved problem.

⟦16⟧ Until now there has scarcely been a serious attempt to represent the consistency of the axioms, whether in number theory or analysis or set theory.

⟦17⟧ Kronecker coined the slogan: God created the integer, everything else is the work of man ⟦Die ganze Zahl schuf der liebe Gott, alles andere ist Menschenwerk⟧. Accordingly he despised—the classical prohibiting dictator—everything that did not seem to him to be integer; on the other hand, it was also far from his practice and that of his school to think further about the integer itself.

⟦18⟧ Poincaré was from the start convinced of the impossibility of a proof of the consistency of the axioms of arithmetic. According to him, the principle of complete induction is a property of our mind—i.e. (in the language of Kronecker) it was created by God. His objection that this principle could never be proved except by the use of complete induction itself is unjustified and will be refuted by my theory.

⟦19⟧ The importance of our question about the consistency of the axioms is well recognized by the philosophers; but in this literature as well I do not find anywhere a clear demand for the solution of the problem in the mathematical sense.

⟦20⟧ In contrast, our question is influenced in its essence by the old attempts to ground number theory and analysis on set theory, and set theory on pure logic.

⟦21⟧ Frege tried to ground number theory on pure logic; Dedekind tried to ground it on set theory as a chapter of pure logic: both failed to reach their goal. Frege did not treat carefully enough the usual concept-formations of logic in their application to mathematics: so he held the scope of a concept to be something immediately given, and he believed he was entitled to take these scopes themselves unrestrictedly as things. He thus fell to some extent into an extreme realism of concepts. Something similar happened to Dedekind; his classic error consists in the fact that he took the system of all things as a starting-point. However dazzling and captivating Dedekind's idea of grounding the finite numbers on the infinite appears, today we know beyond doubt—not least because of the following report—that this path cannot be travelled.

⟦22⟧ Nevertheless, the acute investigations of Frege and Dedekind have brought forth the most valuable fruit; Frege and Dedekind have inaugurated the modern critique of analysis, and this—carried forward by men like Cantor, Zermelo, and Russell—does not 'culminate', as Weyl maintains, 'in chaos and senselessness': rather we are indebted to it on the one hand for profound theories resting on an axiomatic foundation (especially the theories of Zermelo and Russell); and on the other hand for the proper development of the so-called logical calculus, whose basic ideas prove more and more to be an indispensable tool in logico-mathematical investigations.

⟦23⟧ This is in my opinion roughly the present state of the question concerning the foundations of mathematics. Accordingly, a satisfactory conclusion to the research into these foundations can only be attained by the solution of the problem of the consistency of the axioms of analysis. If we can produce this proof, then we can say that mathematical statements are in fact incontestable and ultimate truths—a piece of knowledge that (also because of its general philosophical character) is of the greatest significance for us.

⟦24⟧ We turn to the solution of this problem.

⟦25⟧ As we saw, abstract operation with general concept-scopes and contents has proved to be inadequate and uncertain. Instead, as a precondition for the application of logical inferences and for the activation of logical operations, something must already be given in representation ⟦in der Vorstellung⟧: certain extra-logical discrete objects, which exist intuitively as immediate experience before all thought. If logical inference is to be certain, then these objects must be capable of being completely surveyed in all their parts, and their presentation, their difference, their succession (like the objects themselves) must exist for us immediately, intuitively, as something that cannot be reduced to something else. Because I take this standpoint, the objects ⟦Gegenstände⟧ of number theory are for me—in direct contrast to Dedekind and Frege—the signs themselves, whose shape ⟦Gestalt⟧ can be generally and certainly recognized by us—independently of space and time, of the special conditions of the production of the sign, and of insignificant differences in the finished product.[1] The

[1] In this sense, I call signs of the same shape 'the same sign' for short.

solid philosophical attitude that I think is required for the grounding of pure mathematics—as well as for all scientific thought, understanding, and communication—is this: *In the beginning was the sign.*

⟦26⟧ With this philosophical attitude we turn first to the theory of elementary arithmetic, and ask ourselves whether and to what extent, on this purely intuitive basis of concrete signs, the science of number theory would come into existence. We therefore begin with the following explanation of the numbers.

⟦27⟧ The sign 1 is a number.

⟦28⟧ A sign that begins with 1 and ends with 1, and such that in between + always follows 1 and 1 always follows +, is likewise a number; for example, the number-signs

$$1 + 1$$

$$1 + 1 + 1.$$

⟦29⟧ These number-signs ⟦Zahlzeichen⟧, which are numbers and which completely make up the numbers, are themselves the object of our consideration, but otherwise they have no *meaning* ⟦*Bedeutung*⟧ of any sort.[a] In addition to these signs, we make use of yet other signs that *mean* something and serve for communication, for instance, the sign 2 as an abbreviation for the number-sign 1 + 1 or the sign 3 as an abbreviation for the number-sign 1 + 1 + 1; moreover, we use the signs =, >, which serve for the communication of assertions. Thus $2 + 3 = 3 + 2$ is not to be a formula,[b] but is merely to serve to communicate the fact that $2 + 3$ and $3 + 2$, with respect to the abbreviations we are using, are the same number-sign, namely, the number-sign $1 + 1 + 1 + 1 + 1$. Nor is $3 > 2$ a formula; rather, it serves to communicate the fact that the sign 3 (that is, $1 + 1 + 1$) extends beyond the sign 2 (that is, $1 + 1$), or that the latter sign is a part of the former.

⟦30⟧ For purposes of communication we also use letters $\mathfrak{a}$, $\mathfrak{b}$, $\mathfrak{c}$ for number-signs. Then $\mathfrak{b} > \mathfrak{a}$ is also not a formula, but only the communication that the number-sign $\mathfrak{b}$ extends beyond the number-sign $\mathfrak{a}$. And in the same way, from our present standpoint $\mathfrak{a} + \mathfrak{b} = \mathfrak{b} + \mathfrak{a}$ would be only the communication of the fact that the number-sign $\mathfrak{a} + \mathfrak{b}$ is the same as $\mathfrak{b} + \mathfrak{a}$. And then, as regards content, the point of this communication would be seen in the following manner. If, as we are entitled to assume, $\mathfrak{b} > \mathfrak{a}$ (that is, the number-sign $\mathfrak{b}$ extends beyond $\mathfrak{a}$), then $\mathfrak{b}$ can be decomposed in the form $\mathfrak{a} + \mathfrak{c}$, where $\mathfrak{c}$ serves to communicate a number; then one need only show that $\mathfrak{a} + \mathfrak{a} + \mathfrak{c} = \mathfrak{a} + \mathfrak{c} + \mathfrak{a}$, that is, that $\mathfrak{a} + \mathfrak{a} + \mathfrak{c}$ is the same number-sign as $\mathfrak{a} + \mathfrak{c} + \mathfrak{a}$. But this is the case as long as

---

[a] ⟦The expression 'sign without meaning' caused offense to the philosphers. (See, for example, the note of Aloys Müller, 'Über Zahlen als Zeichen', and the reply by P. Bernays, both in *Math. Ann.* vol. 90 (1923).) In Hilbert's later writings on the foundations of mathematics, the term 'number-sign' was replaced by 'numeral' ⟦'Ziffer'⟧.—Note of Bernays.⟧

[b] ⟦Hilbert here uses the word 'formula' in the narrow sense, i.e. for the formulae of formalized mathematics. But one could of course equally well speak here of formulae with meaning, just as one speaks of signs with meaning.—Note of Bernays.⟧

$\mathfrak{a} + \mathfrak{c}$ is the same sign as $\mathfrak{c} + \mathfrak{a}$, i.e. $\mathfrak{a} + \mathfrak{c} = \mathfrak{c} + \mathfrak{a}$. But here, in contrast to the original communication, at least one 1 has been removed by the decomposition of $\mathfrak{a}$, and this procedure of decomposition can be continued until the summands that are to be exchanged agree with each other. For every number-sign $\mathfrak{a}$ is built up in the manner described from the signs 1 and +; it can therefore also be decomposed by the splitting and cancellation of the individual signs.

⟦31⟧ When we develop number theory in this way, there are no axioms, and no contradictions of any sort are possible. We simply have concrete signs as objects, we operate with them, and we make contentual ⟦inhaltliche⟧ statements about them. And in particular, regarding the proof just given that $\mathfrak{a} + \mathfrak{b} = \mathfrak{b} + \mathfrak{a}$, I should like to stress that this proof is merely a procedure that rests on the construction and deconstruction of number-signs and that it is essentially different from the principle that plays such a prominent role in higher arithmetic, namely, the principle of complete induction or of inference from n to $n + 1$. This principle is rather, as we shall see, a formal principle that carries us farther and that belongs to a higher level; it needs proof, and the proof can be given.

⟦32⟧ We can of course make considerable further progress in number theory using the intuitive and contentual manner of treatment which we have depicted and applied. But we cannot conceive the whole of mathematics in such a way. Already when we cross over into higher arithmetic and algebra—for example, if we wish to make assertions about infinitely many numbers or functions—the contentual procedure breaks down. For we cannot write down number-signs or introduce abbreviations for infinitely many numbers; if we were not to heed this difficulty, we should immediately fall into the sort of nonsense that Frege quite correctly reprehends in his critical remarks about the traditional definitions of irrational numbers. And analysis cannot be constructed by a concrete procedure of the sort we have just given for elementary number theory. For we cannot come close to exhausting the essence of analysis merely by using that sort of contentual communication; rather, we need real, actual formulae for its construction.

⟦33⟧ But we can achieve an analogous point of view if we move to a higher level of contemplation, from which the axioms, formulae, and proofs of the mathematical theory are themselves the objects of a contentual investigation. But for this purpose the usual contentual ideas of the mathematical theory must be replaced by formulae and rules, and imitated by formalisms. In other words, we need to have a strict formalization of the entire mathematical theory, inclusive of its proofs, so that—following the example of the logical calculus—the mathematical inferences and definitions become a formal part of the edifice of mathematics. The axioms, formulae, and proofs that make up this formal edifice are precisely what the number-signs were in the construction of elementary number theory which I described earlier; and with them alone, as with the number-signs in number theory, contentual thought takes place—i.e. only with them is actual thought practised. In this way the contentual thoughts (which of course we can never wholly do without or eliminate) are removed elsewhere—

to a higher plane, as it were; and at the same time it becomes possible to draw a sharp and systematic distinction in mathematics between the formulae and formal proofs on the one hand, and the contentual ideas on the other.

⟦34⟧ In the present paper my task is to show how this basic idea can be carried out in a rigorous and unobjectionable manner, and to show that our problem of proving the consistency of the axioms of arithmetic and analysis is thereby solved.

⟦35⟧ As we have just seen, the signs 1 and + suffice for concrete-contentual number theory. For the construction of the whole of mathematics we shall introduce additional signs of different sorts and explain how to operate with them. We distinguish:

I. *Individual signs* (mostly Greek letters):
- **1.** 1, + (parts of the number-signs),
- **2.** $\phi(*)$, $\psi(*)$, $\sigma(*, *)$, $\delta(*, *)$, $\mu(*, *)$ (individual functions with empty positions, individual functions of functions),
- **3.** $=$ (equality), $\neq$ (inequality), $>$ (greater than) (mathematical signs),
- **4.** Z (is a number), Φ (is a function),
- **5.** $\rightarrow$ ('implies', a logical sign),
- **6.** ( ) (all-sign).

II. *Variables* (Latin letters):
- **1.** a, b, c, d, p, q, r, s (basic variables),
- **2.** f(*), g(*) (variable functions, variable functions of functions),
- **3.** A, B, C, D, S, T, U, V, W (variable formulae).

III. *Signs for communication* (German letters):
- **1.** $\mathfrak{a}$, $\mathfrak{b}$, $\mathfrak{c}$, $\mathfrak{k}$ (functionals),
- **2.** $\mathfrak{A}$, $\mathfrak{B}$, $\mathfrak{C}$, $\mathfrak{R}$, $\mathfrak{S}$, $\mathfrak{T}$ (formulae).

⟦36⟧ Next it is necessary to make some remarks about the operation of these signs.

⟦37⟧ Signs standing next to one another are called a line; lines standing on top of one another are called a figure.

⟦38⟧ Individual signs (I) and variables (II) are the only signs that occur in the calculus, and they make up the formal edifice; the final species of signs (III) serves merely for communication in the contentual reflections. We shall in general use Greek letters for individual signs (I), Latin letters for variables (II), and German letters for the communication-signs (III). These last signs (III) shall also occasionally and provisionally serve as *abbreviatory signs*; an abbreviatory sign is a sign which exists only to shorten the length of the printed expression, and which *refers to* another definite sign. I stress however that the introduction of abbreviatory signs is not necessary to the construction of mathematics, and that we need these signs (III) only for communication in the proper sense of the term—i.e. in the contentual operation on the formal proofs.

⟦39⟧ A number-sign; a basic variable; an individual or a variable function whose empty positions are filled by number-signs, basic variables, or functions; or an individual or variable function of functions whose empty positions are filled, is called a *functional*. A functional can always itself be placed into an appropriate empty position; if in the process the empty positions of a function or a function of functions are all filled, then the resulting line is also called a functional. A functional is thus a sign that is made up of signs from the classes I 1, 2; II 1, 2 but not from I 3, 4, 5, 6; II 3.

⟦40⟧ If one places a functional on either side of the sign = or of the sign ≠, the resulting line is called a *prime formula*; likewise a prime formula arises if one fills the empty position of the logical sign Z with a functional. Thus, if $\mathfrak{a}$ and $\mathfrak{b}$ designate functionals, then

$$\mathfrak{a} = \mathfrak{b}$$

$$\mathfrak{a} \neq \mathfrak{b}$$

$$Z(\mathfrak{a})$$

are prime formulae.

⟦41⟧ If one places a prime formula or a variable formula (II 3)[2] on both sides of an implication sign, then an *implication formula* arises. If one places a prime formula or a variable formula or an implication formula on both sides of an implication sign, then the resulting line is also called a *formula*. And in general

$$\mathfrak{A} \rightarrow \mathfrak{B}$$

is to be a formula, if $\mathfrak{A}$ and $\mathfrak{B}$ are variables or already constructed formulae.

⟦42⟧ Certain formulae, which serve as the building-blocks of the edifice of mathematics, are called *axioms*.

⟦43⟧ In the treatment of the axioms, and in operating with them, the following general rules are to be observed:

⟦44⟧ Individual signs are irreplaceable; basic variables can be replaced at will by functionals.

⟦45⟧ Parentheses are used in the ordinary way to separate out parts of signs; they serve to mark empty positions and they lend certainty and precision to the insertion of lines.

⟦46⟧ The all-sign (I 6) is a logical sign: a parenthesis with a variable inside; the following subformula, which in general contains this variable, is marked off by a special parenthesis and is thereby made recognizable as the scope of the all-sign. The following special rules hold for the all-sign:

⟦47⟧ A variable in a formula is called 'free' if it does not occur in an all-sign of this formula; an all-sign containing a free variable may be prefixed to any formula, so that the entire formula is the scope of the all-sign. Conversely, an all-sign whose scope is the rest of the formula can always be omitted.

⟦48⟧ A variable occurring in an all-sign may be replaced there and simul-

[2] The variable in question can still have one or more functionals as arguments. Thus, for example, $C(1, a)$ is a variable formula.

taneously in the corresponding scope by any other variable that does not occur in that scope.

⟦49⟧ Two all-signs may be interchanged if they occur in immediate succession and if their scopes extend equally far.

⟦50⟧ If a part of a formula takes the form

$$(b)(\mathfrak{A} \rightarrow \mathfrak{B}(b))$$

where $\mathfrak{A}$ does not contain the variable b, then (b) may be placed after the sign $\rightarrow$ so that we obtain the formula:

$$\mathfrak{A} \rightarrow (b)\mathfrak{B}(b).$$

⟦51⟧ Using our new formal standpoint, we shall now show how we acquire the theorems of elementary calculation. For this we need a table of axioms, which begins as follows:

1. $a = a$,

2. $1 + (a + 1) = (1 + a) + 1$,

3. $a = b \rightarrow a + 1 = b + 1$,

4. $a + 1 = b + 1 \rightarrow a = b$

5. $a = c \rightarrow (b = c \rightarrow a = b)$.

Moreover, we make use of the following inference schema:

$$\frac{\begin{array}{c}\mathfrak{S}\\ \mathfrak{S} \rightarrow \mathfrak{T}\end{array}}{\mathfrak{T}}.$$

Then the formal proofs for the number equations can be given in the manner shown by the following special example:

⟦52⟧ From axiom 1 by substitution we get

$$1 = 1;$$

moreover, using the abbreviatory sign 2 for $1 + 1$ and the abbreviatory sign 3 for $2 + 1$

$$2 = 2 \tag{1}$$

and

$$3 = 3. \tag{2}$$

⟦53⟧ From axiom 2 we also get by substitution

$$1 + (1 + 1) = (1 + 1) + 1$$

or

$$1 + 2 = 2 + 1$$

or

$$1 + 2 = 3. \tag{3}$$

⟦54⟧ From axiom 5, we obtain by substitution

$$3 = 3 \rightarrow (1 + 2 = 3 \rightarrow 3 = 1 + 2),$$

and because of (2) we get, by an application of the inference schema, the formula

$$1 + 2 = 3 \rightarrow 3 = 1 + 2$$

and finally by (3) we get, by an application of the inference schema, the formula

$$3 = 1 + 2.$$

⟦55⟧ This is therefore a formula that can be proved from the axioms we have already introduced.

⟦56⟧ Since we do not yet get all the formulae we need from the axioms we have hitherto introduced, the path is open to us to introduce additional axioms. But first we need a stipulation of what a proof is, and a precise description of the use of the axioms.

⟦57⟧ A *proof* is a figure, which we must be able to view as such; it consists of inferences according to the schema

$$\frac{\begin{array}{c}\mathfrak{S}\\ \mathfrak{S} \rightarrow \mathfrak{T}\end{array}}{\mathfrak{T}}$$

where at each stage each of the premisses—that is, each of the formulae $\mathfrak{S}$ and $\mathfrak{S} \rightarrow \mathfrak{T}$—is either an axiom, or results directly from an axiom by substitution, or agrees with the *end-formula* $\mathfrak{T}$ of an inference that occurs earlier in the proof, or results from such an end-formula by substitution.

⟦58⟧ A formula is said to be *provable* if it is an axiom or results from an axiom by substitution or is the end-formula of a proof or results from such an end-formula by substitution. Thus the concept 'provable' is to be understood relative to the underlying axiom-system. This relativism is natural and necessary; it causes no harm, since the axiom system is constantly being extended, and the formal structure ⟦Aufbau⟧, in keeping with our constructive tendency, is always becoming more complete.

⟦59⟧ To reach our goal, we must make the proofs as such the object of our investigation; we are thus compelled to a sort of *proof theory* which studies operations with the proofs themselves. For concrete-intuitive number theory, which we treated first, the numbers were the objectual and the displayable, and the proofs of theorems about the numbers fell into the domain of the thinkable. In our present investigation, proof itself is something concrete and displayable; the contentual reflections follow the proofs themselves. Just as the physicist investigates his apparatus and the astronomer investigates his location; just as the philosopher practises the critique of reason; so, in my opinion, the mathematician has to secure his theorems by a critique of his proofs, and for this he needs proof theory.

⟦60⟧ Recall now in particular our intention to prove the consistency of the axioms. From the present standpoint this problem seems to be meaningless, since at present the only 'provable' formulae that arise are formulae that are as it were equivalent to purely positive assertions, and which can accordingly produce no contradiction: we could allow $1 = 1 + 1$ to count as a formula along with $1 = 1$, provided it were a provable formula yielded by our rules of inference. But if our formalism is to offer a full replacement for the earlier, real theory consisting of inferences and assertions then a contentual contradiction must have its formal equivalent. In order that this should be so, we must take inequality to be a positive expression like equality, and we must introduce it as a new sign $\neq$ with new axioms; this sign is then operated in a manner that accords with our earlier rules. And then we declare an axiom system to be *consistent* if the formulae

$$\mathfrak{a} = \mathfrak{b} \text{ and } \mathfrak{a} \neq \mathfrak{b}$$

are never simultaneously provable formulas, where $\mathfrak{a}$ and $\mathfrak{b}$ designate functionals.

⟦61⟧ In accordance with this general plan, we introduce the new axiom

6. $$\mathrm{a} + 1 \neq 1;$$

for the sake of simplicity, we now delete axiom 2. Then the first test of a genuine proof of consistency in our new proof theory lies in the proof of the following theorem:

The axiom system that consists of the following five axioms:

1. $a = a$,

3. $a = b \rightarrow a + 1 = b + 1$,

4. $a + 1 = b + 1 \rightarrow a = b$,

5. $a = c \rightarrow (b = c \rightarrow a = b)$,

6. $a + 1 \neq 1$

is consistent.

⟦62⟧ The proof of this theorem takes several steps; first we prove:

⟦63⟧ Lemma. A provable formula can contain at most two occurrences of the sign $\rightarrow$.

⟦64⟧ For suppose we had a proof for a formula with more than two $\rightarrow$ signs. Then we proceed through this proof until we find the first formula that has this property, i.e. such that no previous formula in the proof of this formula contains $\rightarrow$ more than twice. This formula cannot result directly from an axiom by substitution, for the letters a, b, c appearing in the axioms can be replaced only by functionals, and these do not introduce any new $\rightarrow$ sign. But neither can that formula appear as the end-formula $\mathfrak{T}$ of an inference; for then the second premiss $\mathfrak{S} \rightarrow \mathfrak{T}$ of this inference would be an earlier formula with more than two $\rightarrow$ signs; and therefore $\mathfrak{T}$ would not be the first formula with this property.

⟦65⟧ Next we prove:

⟦66⟧ Lemma. A formula $\mathfrak{a} = \mathfrak{b}$ is provable only if $\mathfrak{a}$ and $\mathfrak{b}$ are the same sign.

⟦67⟧ To prove this, once again we distinguish the two cases. First, suppose the formula is the direct result of substituting into an axiom. Then only axiom 1 comes into consideration, and in this case our theorem obviously holds. Second, we assume we have a proof with the end-formula $\mathfrak{a} = \mathfrak{b}$, such that $\mathfrak{a}$ and $\mathfrak{b}$ are not the same sign and such that no such formula occurs earlier in the proof. Then in our inference schema $\mathfrak{T}$ must agree with $\mathfrak{a} = \mathfrak{b}$ and $\mathfrak{S}$ must be a provable formula; so the second premiss would have had the form

$$\mathfrak{S} \rightarrow \mathfrak{a} = \mathfrak{b}. \tag{4}$$

This formula in turn must either be the result of substitution into an axiom, or be the end-formula of a proof. In the first case, only axioms 3 and 4 come into consideration; if axiom 3, then $\mathfrak{a}$ would have to be of the form $\mathfrak{a}' + 1$ and $\mathfrak{b}$ of the form $\mathfrak{b}' + 1$, and $\mathfrak{S}$ would have to be the formula $\mathfrak{a}' = \mathfrak{b}'$. But if $\mathfrak{a}'$ and $\mathfrak{b}'$ were the same sign, then $\mathfrak{a}$ and $\mathfrak{b}$ would have to be as well, contrary to our hypothesis. But if, on the other hand, $\mathfrak{a}'$ and $\mathfrak{b}'$ were not the same sign, then $\mathfrak{S}$ (that is, $\mathfrak{a}' = \mathfrak{b}'$) would be a formula of the sort under discussion and would appear in the proof before $\mathfrak{T}$; and this too cannot be. But if axiom 4 is involved, then $\mathfrak{S}$ must be the formula $\mathfrak{a} + 1 = \mathfrak{b} + 1$, in which we do not have the same sign on both sides of the equality sign; but this too is impossible, as $\mathfrak{S}$ occurs earlier in the proof. So the only remaining possibility is that (4) is the end-formula of a proof whose last inference would have to have the form

$$\frac{\begin{array}{c}\mathfrak{A}\\ \mathfrak{A} \rightarrow (\mathfrak{S} \rightarrow \mathfrak{a} = \mathfrak{b})\end{array}}{\mathfrak{S} \rightarrow \mathfrak{a} = \mathfrak{b}}.$$

Accordingly we investigate the origins of the second premiss

$$\mathfrak{A} \rightarrow (\mathfrak{S} \rightarrow \mathfrak{a} = \mathfrak{b}).$$

⟦68⟧ If this premiss resulted directly from substitution into an axiom, then only axiom 5 would come into consideration, in which case $\mathfrak{S}$ would have to be of the form $\mathfrak{b} = \mathfrak{c}$ and $\mathfrak{A}$ would have to be of the form $\mathfrak{a} = \mathfrak{c}$. Now, if $\mathfrak{c}$ were the same as $\mathfrak{b}$, then $\mathfrak{A}$ would just be $\mathfrak{a} = \mathfrak{b}$, and this formula would therefore already have appeared at an earlier place in the proof. But if $\mathfrak{c}$ were not the same as $\mathfrak{b}$, then the formula $\mathfrak{b} = \mathfrak{c}$ is a formula occurring earlier in the proof having the property originally required of $\mathfrak{T}$. Accordingly, the only remaining possibility is that (5) is the end-formula of an inference; but then the second premiss of this inference must be a formula with at least three → signs, and by our previous lemma, this could not be a provable formula.

⟦69⟧ With this, our second lemma has been proved.

⟦70⟧ We earlier declared an axiom-system to be consistent if

$$\mathfrak{a} = \mathfrak{b} \text{ and } \mathfrak{a} \neq \mathfrak{b}$$

are never simultaneously provable formulae. Now since according to the lemma just proved $\mathfrak{a} = \mathfrak{b}$ is a provable formula only if $\mathfrak{a}$ and $\mathfrak{b}$ are the same sign, our proof of the consistency of our axioms reduces to the problem of showing that our axiom system can never give rise to a provable formula with the form

$$\mathfrak{a} \neq \mathfrak{a}. \tag{6}$$

We prove this as follows.

⟦71⟧ To obtain a formula of the form of (6) and containing the sign ≠ directly from an axiom by substitution it would be necessary to use axiom 6; but a formula arising from axiom 6 by substitution always has the form

$$\mathfrak{a}' + 1 \neq 1$$

and here $\mathfrak{a}' + 1$ is certainly not the same sign as 1. If, on the other hand, (6) should arise as the end-formula of an inference, then the second premiss of this inference must have the form

$$\mathfrak{S} \to \mathfrak{a} \neq \mathfrak{a} \tag{7}$$

and since such a formula cannot possibly arise from substitution into an axiom, this formula (7) must itself be the result of an inference. The second premiss of this inference would then have to be

$$\mathfrak{T} \to (\mathfrak{S} \to \mathfrak{a} \neq \mathfrak{a})$$

and this formula too would for the same reason have to come from an inference whose second premiss would necessarily have the form

$$\mathfrak{A} \to (\mathfrak{T} \to (\mathfrak{S} \to \mathfrak{a} \neq \mathfrak{a})).$$

But by our first lemma, such a formula is not provable because it certainly contains more than three → signs. This excludes the possibility that (6) is a provable formula, and this completes our proof of the consistency of our axiom-system.

⟦72⟧ A new goal would be to carry out the corresponding investigation after having reintroduced the previously excluded Axiom 2. And in fact it is possible in this way to demonstrate the consistency of the axiom-system

1. $a = a,$

2. $1 + (a + 1) = (1 + a) + 1,$

3. $a = b \to a + 1 = b + 1,$

4. $a + 1 = b + 1 \to a = b,$

5. $$a = c \rightarrow (b = c \rightarrow a = b),$$

6. $$a + 1 \neq 1.$$

⟦73⟧ We have hitherto introduced no logical sign apart from the → sign and the all-sign; in particular, we have avoided the formalization of the logical operation '*not*'. This way of treating negation is characteristic for our proof theory: a formal equivalent for the missing negation lies solely in the sign ≠; by introducing this sign, inequality is expressed just as positively and treated in the same way as its counterpart equality. Contentually we use negation only in the proof of consistency and only in so far as it corresponds to our basic point of view. In the light of this circumstance, it seems to me that our proof theory also yields us an epistemologically important insight into the meaning and the essence of negation.

⟦74⟧ The logical concept '*all*' is exhibited in our theory by the variables that appear there and by the rules we have laid down for operating with them and with the all-sign.

⟦75⟧ The only logical concept still to be formalized is the concept 'there exists'—a concept which, as is well known, can be expressed in formal logic by negation and the concept 'all'. But since negation cannot have any direct representation in our proof theory, the formalization of 'there exists' is achieved here by introducing individual function-signs through a kind of implicit definition; so that 'that which exists' is as it were actually produced by a function. The simplest example of this is the following:

⟦76⟧ To express: 'If $\mathfrak{a}$ is not 1, then "there exists" a number which precedes $\mathfrak{a}$', we introduce the function-sign $\delta(*)$ with one empty position as an individual sign, and we lay down as an axiom the formula

7. $$a \neq 1 \rightarrow a = \delta(a) + 1.$$

⟦77⟧ I shall here merely mention that it can then be proved by contentual considerations that the axiom-system consisting of axioms 1–7 is consistent.

⟦78⟧ Although these explanations contain only the very beginning of my proof theory, we can nevertheless perceive in them the general tendency and direction in which the new grounding of mathematics ought to proceed. Two points emerge in particular.

⟦79⟧ *First*: everything that hitherto made up mathematics proper is now to be strictly formalized, so that *mathematics proper*, or mathematics in the strict sense, becomes a stock of provable formulae. The formulae of this stock are distinguished from the usual formulae of mathematics only by the fact that, besides the mathematical signs, they also contain the → sign, the all-sign, and the sign for statements. This circumstance corresponds to a conviction I have long maintained,[3] namely, that a simultaneous construction of arithmetic and

[3] See my lecture, 'Über den Zahlbegriff', *Jber. dtsch. Math.-Ver.*, Vol. 8, 1900, pp. 180–184, reprinted as Appendix VI to my 'Grundlagen der Geometrie'.

formal logic is necessary because of the close connection and inseparability of arithmetical and logical truths.

⟦80⟧ *Secondly*: in addition to this proper mathematics, there appears a mathematics that is to some extent new, a *metamathematics* which serves to safeguard it by protecting it from the terror of unnecessary prohibitions as well as from the difficulty of paradoxes. In this metamathematics—in contrast to the purely formal modes of inference in mathematics proper—we apply contentual inference; in particular, to the proof of the consistency of the axioms.

⟦81⟧ The development of mathematical science accordingly takes place in two ways that constantly alternate: (i) the derivation of new 'provable' formulae from the axioms by means of formal inference; and, (ii) the adjunction of new axioms together with a proof of their consistency by means of contentual inference.

⟦82⟧ In keeping with the principles and tendencies we have just described, let us now carry out the new grounding of mathematics.

⟦83⟧ Our previous stock of axioms is merely the axioms 1–7 that we have already mentioned. These axioms are of a purely arithmetical character; the provable formulae that follow from them do not in the least supply a foundation for the theory of real numbers, and they make up only a small part even of arithmetic. A glance at these axioms 1–7 shows that the only variables that appear (small Latin letters without empty positions) are basic variables. But even for the grounding of arithmetic axioms of such a sort are utterly inadequate. Rather, we need a sequence of axioms that contain variable formulae (large Latin letters); in particular, we lay down the following arithmetical axioms, each with a variable formula:

Axiom of mathematical equality

8. $$a = b \rightarrow (A(a) \rightarrow A(b)).$$

Axiom of complete induction

9. $$(a)(A(a) \rightarrow A(a+1)) \rightarrow \{A(1) \rightarrow (Z(b) \rightarrow A(b))\}.$$

⟦84⟧ We furthermore need a stock of such axioms corresponding to the usual patterns of logical inference; they are the following four axioms with variable formulae:

Axioms of logical inference

10. $$A \rightarrow (B \rightarrow A),$$

11. $$\{A \rightarrow (A \rightarrow B)\} \rightarrow (A \rightarrow B),$$

12. $$\{A \rightarrow (B \rightarrow C)\} \rightarrow \{B \rightarrow (A \rightarrow C)\},$$

13. $$(B \rightarrow C) \rightarrow \{(A \rightarrow B) \rightarrow (B \rightarrow C)\}.$$

⟦85⟧ Moreover, we also need two axioms for mathematical inequality; these axioms serve as an equivalent for certain modes of inference that are indispensable in contentual reflections, namely, the following axioms:

Axioms of mathematical inequality

14. $$a \neq a \rightarrow A,$$

15. $$(a = b \rightarrow A) \rightarrow \{(a \neq b \rightarrow A) \rightarrow A\}.$$

⟦86⟧ As I have already mentioned, the axioms 1–7 are only some of the arithmetical axioms that are necessary. To complete them, we need above all to introduce the logical function-sign Z ('to be a positive integer'). On the other hand, it is necessary to restrict axiom 6. If at the same time, for the sake of orthographic uniformity, we use the sign $*-1$ instead of the function-sign $\delta(*)$; and if we generalize and complete axioms 2 and 7; and if we discard axioms 3, 4, and 5 (because they are now provable formulae), we finally obtain the following axioms in place of 1–7:

Arithmetical axioms

16. $$Z(1),$$

17. $$Z(a) \rightarrow Z(a+1),$$

18. $$Z(a) \rightarrow (a \neq 1 \rightarrow Z(a-1)),$$

19. $$Z(a) \rightarrow (a+1 \neq 1),$$

20. $$(a+1)-1=a,$$

21. $$(a-1)+1=a,$$

22. $$a+(b+1)=(a+b)+1,$$

23. $$a-(b+1)=(a-b)-1.$$

⟦87⟧ If we take this axiom-system 1, 8–23 as a basis,[c] then we can obtain the entire stock of formulae and theorems of arithmetic purely by application of our rules, i.e. in a formal manner.

⟦88⟧ The first important goal is to prove the consistency of this axiom system 1, 8–23. This proof can in fact be carried out, and the mode of inference of complete induction (axiom 9), which is characteristic of arithmetic, is thereby secured.[d]

⟦89⟧ But the most essential step still remains to be taken, namely, the proof of the applicability of the logical principle *tertium non datur* in the sense of the admissibility of the inference, for infinitely many numbers, functions, or functions of functions, that a statement either holds for all these numbers, func-

[c] ⟦A schema for the introduction of functions by *recursion equations* must also be added.—Note of Bernays.⟧

[d] ⟦As matters have turned out, the mentioned proof is valid only if one excludes the all-sign and replaces Axiom 9 by the induction *schema*.—Note of Bernays.⟧

tions, and functions of functions, or that there necessarily exists one among them for which the statement does not hold. Only with the proof of the applicability of this principle will the grounding of the theory of real numbers have been accomplished, and the bridges be erected to analysis and set theory.

⟦90⟧ This proof can be carried out on the basis of the fundamental ideas I have described. I introduce certain functions of functions τ and α by laying down axiom systems, and I prove the consistency of these axiom-systems.[e]

⟦91⟧ The simplest example of a function of functions serving this end is the function of functions $\kappa(f)$, where the argument f is a variable number-theoretic function of the basic variable a, so that

$$Z(a) \rightarrow \{f(a) \neq 1 - 1 \rightarrow Z(f(a))\}$$

holds, where $\kappa(f) = 1 - 1$ if f has the value 1 for all a; otherwise, $\kappa(f)$ is the least argument for which f is not 1. The axiom system for this $\kappa(f)$ is:

24. $$(\kappa(f) = 1 - 1) \rightarrow (Z(a) \rightarrow f(a) = 1),$$

25. $$(\kappa(f) \neq 1 - 1) \rightarrow Z\kappa(f),$$

26. $$(\kappa(f) \neq 1 - 1) \rightarrow (f(\kappa(f)) \neq 1),$$

27. $$Z(a) \rightarrow \{Z(\kappa(f) - a) \rightarrow f(\kappa(f) - a) = 1\}.$$

⟦92⟧ In a similar manner, a certain pair of functions of functions τ, α that belong together can be introduced, by means of which the complete grounding of the theory of real numbers—and in particular the proof of the existence of upper bounds for any arbitrary set of real numbers—becomes possible.

⟦93⟧ To conclude this first report, I should like to remark that P. Bernays has been of the greatest assistance to me in working out the ideas presented here.

E. THE LOGICAL FOUNDATIONS OF MATHEMATICS (*HILBERT 1923a*)

This article, delivered as a lecture to the Deutsche Naturforscher Gesellschaft in Leipzig, September 1922, is a sequel to *Hilbert 1922a*, and brings Hilbert's proof theory to maturity. Hilbert here introduces several technical refinements and clarifications to his theory. Specifically: (i) He improves the formal system by adding a special sign for formal negation. (Previously he had allowed only the primitive sign $\neq$ for numerical inequality.) (ii) He refines his account of the distinction between the formal language and the metalanguage; in par-

[e] ⟦Hilbert here refers to his attempt to treat the transfinite functions in his consistency proof. But it is still uncertain whether one can attain the desired goal by these means.—Note of Bernays.⟧

ticular, he now distinguishes clearly between the modes of inference that are permissible in each. In the metalanguage, one operates with a *finite logic* (§11), that is, a logic dealing with finite totalities; but in the formal language the modes of inference are more powerful. In §§13–15 Hilbert identifies the application of the quantifiers to infinite totalities as the precise point where traditional mathematics departs from finite logic. (iii) He outlines a consistency proof for an elementary, quantifier-free formal system of number theory. (iv) He begins to extend his proof theory to analysis and set theory, and at the end of the paper he sketches a strategy for proving the consistency of a version of Zermelo's axiom of choice for real numbers. (v) To handle the transfinite parts of mathematics, he introduces a special operator τ on predicates; τ is governed by the single *transfinite axiom*

$$A(\tau A) \rightarrow A(a).$$

τ is to be thought of as a counterexample operator. That is, for any predicate A, τA is supposed to choose a counterexample to A if such a counterexample exists. Intuitively, the axiom says that if A holds even of the purported counterexample, then it holds generally. Hilbert can now define the quantifiers in terms of τ. For $(\forall x)A(x)$ holds just in case $A(x)$ holds even for $\tau(A)$; thus, we can define,

$$(\forall x)A(x) \equiv A(\tau A).$$

By similar reasoning,

$$(\exists x)A(x) \equiv A(\tau \bar{A}).$$

So the operator τ ties together the quantifiers, a principle of transfinite choice, and the *tertium non datur* for infinite totalities. For by the transfinite axiom and the definition of the quantifiers, it follows (§20) that

$$(\overline{\forall a})A(a) \equiv (\exists a) - \bar{A}(a);$$

and this formula Hilbert identifies as the *tertium non datur* for infinite totalities (§15).

The τ-operator (which Hilbert and Bernays soon replaced by its better-known dual, ε) also allowed Hilbert to eliminate quantifiers from his formal language: they are replaced by the ε-operator, which is governed by the single transfinite axiom. This greatly simplifies the proof theory. For a formalized proof then consists solely of substitutions and of applications of the sentential calculus. Hilbert's strategy for proving consistency was this: to replace the (finitely many) ε-terms in any proof by successively assigning numerical values to them in an effective manner. The hope was that this procedure could be shown to terminate in a finite number of steps, leaving behind only unobjectionable, quantifier-free formulae.

Ackermann, in his dissertation (*Ackermann 1924*) believed that he had carried out a Hilbert-style consistency proof for analysis; but he soon realized that his argument was incomplete. Von Neumann (*1927*) gave a consistency proof for first-order number theory with quantifier-free induction; and it seemed to the Hilbert school that the realization of Hilbert's programme was just around the corner. Instead, what lay around the corner was Gödel's second incompleteness theorem, which came as an unwelcome surprise and complicated the prospects for Hilbert's programme greatly.[a]

A survey of the further technical development of Hilbert's proof theory after 1922 is beyond the scope of this book. Hilbert's own important papers *1926* and *1928a* are translated in *van Heijenoort 1967*, which also contains many important proof-theoretic papers by other authors. Other important contemporary works on Hilbert's proof theory are *Hilbert and Ackermann 1928*, *Hilbert and Bernays 1934* and *1939*, *Gentzen 1969*, *Herbrand 1971*, and *Ackermann 1940*; see also the accounts by *Kreisel 1958a* and *1976* and by *Sieg 1988*.

⟦1⟧ My investigations in the new grounding of mathematics[1] have as their goal nothing less than this: to eliminate, once and for all, the general doubt about the reliability of mathematical inference. We can see how necessary such an investigation is, if we think of how changeable and imprecise the intuitions of even the most distinguished mathematicians have been in this area, or if we remember that the inferences that were previously regarded as the most certain in mathematics have been challenged by some of the most renowned mathematicians of modern times.

⟦2⟧ I believe that the difficulties of principle under discussion cannot be fully solved without a theory of mathematical proof itself. With the assistance of Paul Bernays I have developed this proof theory so far that it can in fact be used to give an unobjectionable foundation to analysis and set theory; indeed, I believe that I have come so far that one can now successfully confront the great classical problems of set theory such as the continuum problem, and the equally important open problems of mathematical logic.

⟦3⟧ It is not possible to explain here this entire theory with its long and difficult ramifications. But in the course of the investigation an array of new

---

[a] The Gödel results did not, however, in any straightforward manner *refute* Hilbert's programme. Indeed, Gödel himself, in his *1931*, immediately after proving his theorem on the unprovability of consistency (Theorem XI), observed: 'The entire proof of Theorem XI carries over word for word to the axiom system of set theory, *M*, and to that of classical mathematics, *A*, and here, too, it yields the result: There is no consistency proof for *M*, or for *A*, that could be formalized in *M*, or *A*, respectively, provided *M*, or *A*, is consistent. I wish to note expressly that Theorem XI (and the corresponding results for *M* and *A*) do not contradict Hilbert's formalistic viewpoint. For this viewpoint presupposes only the existence of a consistency proof in which nothing but finitary means of proof are used, and it is conceivable that there exist finitary proofs that *cannot* be expressed in the formalism of *P* (or of *M* or *A*).'

[1] See my lectures held in Hamburg and Copenhagen. *Abhandlungen aus dem mathematischen Seminar der Hamburgischen Universität*, 1922. ⟦*Hilbert 1922*, translated above.⟧

insights and interconnections have come to light, each of which is interesting for its own sake, independently of the others. I should like to discuss here what I believe to be such a new insight—an insight that goes to the core of my proof theory in the most profound way.

‖4‖ Let us recall the axiom of choice in set theory, which Zermelo was the first to formulate, and on the basis of which he gave his ingenious proof of the well-ordering of the continuum. The objections that were raised against this proof—and against the developments in set theory that are bound up with it—were essentially directed against the axiom of choice. And even today, most people probably believe that the admissibility of the axiom of choice is dubious, but that the other modes of inference that occur in set theory in general (and in Zermelo's proof in particular) are not objectionable to the same extent. I believe this opinion is false. Instead, logical analysis of the sort studied in my proof theory shows that the essential thought underlying the principle of choice is a general logical principle which is necessary and indispensable even for the most elementary rudiments of mathematical inference. If we make these rudiments secure, we simultaneously establish the principle of choice: my proof theory does both.

‖5‖ The fundamental idea of my proof theory is as follows:

‖6‖ Everything that previously made up mathematics is to be rigorously formalized, so that mathematics proper or mathematics in the strict sense becomes a stock of formulae. These formulae are distinguished from the ordinary formulae of mathematics only by the fact that they contain logical signs in addition to the ordinary signs—in particular, the logical signs for 'implies' ($\rightarrow$) and for 'not' ($\overline{\ \ }$).[2] Certain formulae that serve as building-blocks for the formal edifice of mathematics are called axioms. A proof is a figure that must intuitively ‖anschaulich‖ appear to us as such; it consists of inferences using the inference-schema

$$\frac{\begin{array}{c}\mathfrak{S}\\ \mathfrak{S}\rightarrow\mathfrak{T}\end{array}}{\mathfrak{T}}$$

where in every case each of the premisses—that is, the formulas $\mathfrak{S}$ and $\mathfrak{S}\rightarrow\mathfrak{T}$—either is an axiom, or results directly from an axiom by substitution, or agrees with the end-formula $\mathfrak{T}$ of an inference that appears earlier in the proof, or results from such an end-formula by substitution. A formula shall be called provable if it either is an axiom, or results from an axiom by substitution, or is the end-formula of a proof.

‖7‖ In addition to this formalized mathematics proper, we have a mathematics that is to some extent new: a metamathematics that is necessary for securing mathematics, and in which—in contrast to the purely formal modes of

[2] In the article cited above I avoided using this sign; but it has turned out that the sign for 'not' can appear without danger in the present slightly altered version of my theory.

inference in mathematics proper—one applies contentual inference, but only to prove the consistency of the axioms. In this metamethematics we operate with the proofs of mathematics proper, and these proofs are themselves the object of the contentual investigation. Thus the development of mathematical science as a whole takes place in two ways that constantly alternate: on the one hand we derive new provable formulae from the axioms by formal inference; on the other, we adjoin new axioms and prove their consistency by contentual inference.

⟦8⟧ The axioms and provable theorems (i.e. the formulae that arise in this interplay) are the images of the thoughts that make up the usual procedure of traditional mathematics; but they are not themselves the truths in an absolute sense. Rather, the absolute truths are the insights that my proof theory furnishes into the provability and the consistency of these formal systems.

⟦9⟧ This programme already affects the choice of axioms for our proof theory. We begin the sequence of axioms as follows:

I. Axioms of implication

1. $$A \rightarrow (B \rightarrow A)$$

(Adjunction of a presupposition)

2. $$\{A \rightarrow (A \rightarrow B)\} \rightarrow (A \rightarrow B)$$

(Deletion of a presupposition)

3. $$\{A \rightarrow (B \rightarrow C)\} \rightarrow \{B \rightarrow (A \rightarrow C)\}$$

(Exchange of presuppositions)

4. $$(B \rightarrow C) \rightarrow \{(A \rightarrow B) \rightarrow (A \rightarrow C)\}$$

(Elimination of a statement)

II. Axioms of negation

5. $$A \rightarrow (\bar{A} \rightarrow B)$$

(Law of contradiction)

6. $$(A \rightarrow B) \rightarrow \{(\bar{A} \rightarrow B) \rightarrow B\}$$

(Principle of *tertium non datur*)

III. Axioms of equality

7. $$a = a$$

8. $$a = b \rightarrow (A(a) \rightarrow A(b))$$

IV. Axioms of number

9. $$a + 1 \neq 0$$

10. $$\delta(a + 1) = a$$

⟦10⟧ Concerning 9, I note that the formal negation of $a = b$ (i.e. $\overline{a = b}$) is also written $a \neq b$, and therefore $a + 1 \neq 0$ is the formal negation of $a + 1 = 0$.

⟦11⟧ From these axioms 1–10 we easily obtain the positive integers and the numerical equations that hold for them. The elementary theory of numbers can also be obtained from these beginnings by means of 'finite' logic and purely intuitive thought ⟦durch rein anschauliche Überlegungen⟧ (which includes recursion and intuitive induction for finite existing totalities); here it is not necessary to apply any dubious or problematical mode of inference.[3]

⟦12⟧ The provable formulae that we acquire in this way all have the character of the finite; that is, the thoughts whose images ⟦Abbilder⟧ they are can also be obtained contentually and immediately, without resort to any axioms, from the examination of finite totalities.

⟦13⟧ But in our proof theory we wish to go beyond this domain of finite logic, and we wish to obtain provable formulae that are the images of the transfinite theorems of ordinary mathematics. And the true demonstration of the power of our proof theory will come when, after having added certain transfinite axioms, we are able to give the consistency proof. Now, where do we first depart from the concrete intuitive and from the finite? Obviously, already with the application of the concepts '*all*' and '*there exists*'. The facts about these concepts are as follows. The assertion that *all* the objects of a finite existing surveyable totality possess a particular property is logically equivalent to a conjunction of several individual assertions; for instance 'all the benches in this auditorium are wooden' means: 'this bench is wooden and that bench is wooden and . . . and that bench over there is wooden'. Similarly, the assertion that there *exists* an object with a property in a finite totality is equivalent to a disjunction of individual assertions; for example, 'there exists among these pieces of chalk one that is red' means: 'this piece of chalk is red or that piece of chalk is red or . . . or that piece of chalk over there is red'.

⟦14⟧ From this we deduce the *tertium non datur* for finite totalities in the following form: either all the objects have a fixed property or there exists an object that does not have this property; and simultaneously we obtain (using the usual signs—'for all a': (a); 'not for all a': $(\bar{a})$; 'there exists an a': $(\mathrm{E}a)$; 'there exists no a': $(\overline{\mathrm{E}a})$) the strict validity of the equivalences

$$(\bar{a})A(a) \text{ eq. } (\mathrm{E}a)\bar{A}(a)$$

and

$$(\overline{\mathrm{E}a})A(a) \text{ eq. } (a)\bar{A}(a);$$

where $A(a)$ denotes a statement with a variable a, i.e. a predicate.

⟦15⟧ But in mathematics these equivalences are customarily assumed,

[3] In the definitive presentation of my theory, the grounding of elementary number theory also takes place by means of axioms; but here, merely for the sake of brevity, I appeal to the direct intuitive grounding.

without further proof, to be valid for infinitely many individuals as well; and with this step we leave the domain of the finite and enter the domain of transfinite modes of inference. If we were constantly and blithely to apply to infinite totalities procedures that are admissible in the finite case, then we would open the floodgates of error. This is the same source of mistakes that we are familiar with from analysis. In analysis, we are allowed to extend theorems that are valid for finite sums and products to infinite sums and products only if a special investigation of convergence guarantees the inference; similarly here we may not treat the infinite sums and products

$$A_1 \& A_2 \& A_3 \& \dots$$
$$A_1 \vee A_2 \vee A_3 \vee \dots$$

as though they were finite, unless the proof theory we are about to discuss permits such a treatment.

‖16‖ Let us consider the equivalences we have just set up. For an infinite number of things the negation of the universal judgement $(a)Aa$ has no precise content whatever; nor does the negation of the existential judgement $(\mathrm{E}a)Aa$. To be sure, these negations can on occasion make sense, namely, if the assertion $(a)Aa$ is refuted by a counterexample or if a contradiction is derived from the assumption that $(a)Aa$ or $(\mathrm{E}a)Aa$. But these cases do not contradict each other; for if $A(a)$ does not hold for all a, we do not yet know that there actually exists an object with the property not-A; nor can we simply say: either $(a)Aa$ (or: $(\mathrm{E}a)Aa$) holds, or else these assertions actually lead to a contradiction. For finite totalities '*there exists*' ‖es gibt‖ and '*there is available*' ‖es liegt vor‖ are synonymous; for infinite totalities, only the latter concept is clear as it stands.

‖17‖ We therefore see that, if we wish to give a rigorous grounding of mathematics, we are not entitled to adopt as logically unproblematic the usual modes of inference that we find in analysis. Rather, our task is precisely to discover why and to what extent we always obtain correct results from the application of transfinite modes of inference of the sort that occur in analysis and set theory. The free use and the full mastery of the transfinite is to be achieved on the territory of the finite! How is the solution of this task possible?

‖18‖ In accordance with our plan we shall adjoin to the four previous axiom groups new groups that express transfinite modes of inference. I use the idea that underlies the principle of choice by introducing a logical function

$$\tau(A) \text{ or } \tau_a(A(a))$$

which assigns a definite object $\tau(A)$ to each predicate $A(a)$—that is, to every statement with a variable a. This function τ is to satisfy the following axiom:

V. Transfinite axiom

11. $$A(\tau A) \rightarrow A(a).$$

‖19‖ In ordinary language, this axiom says: if a predicate A applies to the object τA, then it applies to all objects a. The function τ is a fixed individual function of a single predicate-variable A; we shall call it the *transfinite function* and axiom 11 the *transfinite axiom*. In order to get an intuitive sense of its content, let us take for A the predicate '*bribable*'. Then τA would be a definite man with such a firm sense of justice that, if he were to turn out to be bribable, then in fact all men whatsoever are bribable.

⟦20⟧ The transfinite axiom V is to be regarded as the original source of all transfinite concepts, principles, and axioms. That is, if we add the following axioms:

VI. Axioms defining the all- and existence-signs

$$A(\tau A) \rightarrow (a)A(a),$$

$$(a)A(a) \rightarrow A(\tau A),$$

$$A(\tau \bar{A}) \rightarrow (\mathrm{E}a)A(a) \;,$$

$$(\mathrm{E}a)A(a) \rightarrow A(\tau \bar{A}),$$

then all the purely logical transfinite priricíples turn out to be provable formulae, namely:

$$(a)A(a) \rightarrow Aa$$

(Aristotelian principle),

$$A(a) \rightarrow (\mathrm{E}a)Aa$$

(Existential principle),

$$(\bar{a})Aa \rightarrow (\mathrm{E}a)\bar{A}a,$$

$$(\mathrm{E}a)\bar{A}a \rightarrow (\bar{a})Aa,$$

$$(\overline{\mathrm{E}a})Aa \rightarrow (a)\bar{A}a,$$

$$(a)\bar{A}a \rightarrow (\overline{\mathrm{E}a})Aa.$$

⟦21⟧ In these last four formulae, the equivalences we earlier laid down for finite totalities, and also the *tertium non datur*, are recognized as valid for infinite totalities.[4]

⟦22⟧ After these explanations we see that everything comes down to the proof of the consistency of axioms I to V (1 to 11).

⟦23⟧ The basic idea of such a proof is always as follows: we assume that we are presented with a concrete proof having the end-formula $0 \neq 0$; the presence of a contradiction can in fact be reduced to this case. Then, by considering the matter in a finite and contentual way, we show that this cannot be a proof satisfying our requirements.

---

[4] I owe to P. Bernays the observation that the *single* formula 11 suffices for the derivation of all these formulae.

⟦24⟧ We must first give the proof of the consistency of axioms I to IV (1 to 10). We proceed by gradually altering the supposed proof that $0 \neq 0$. In particular:

⟦25⟧ 1. By repeating and omitting formulae, the proof can be turned into a proof in which each formula serves to justify one and only one 'descendant' formula. In this way the proof is decomposed into threads which begin in the axioms and end in the end-formula.

⟦26⟧ 2. The variables appearing in the proof can be eliminated.

⟦27⟧ 3. We can rearrange the proof so that every formula contains only logical signs and number-signs

$$0,\ 0+1,\ 0+1+1,\ \ldots$$

so that every formula of the proof becomes a 'numerical' formula.

⟦28⟧ 4. Every formula is brought into a certain logical 'normal form'.

⟦29⟧ After these operations have been carried out, it is possible to check each formula of the proof directly, that is, to determine whether it is 'correct' or 'false' in a certain sense that can be precisely stated. Now, if the supposed proof were to satisfy all our requirements, then clearly each formula of the proof would have to pass this test in turn. Thus the end-formula $0 \neq 0$ would also have to be 'correct'; but it is not correct.

⟦30⟧ In this way a proof of the consistency of axiom-groups I to IV (1 to 10) can be given—although, to be sure, the details would take more time than I have available in this lecture.

⟦31⟧ But for the moment it is precisely the consistency of axiom V (11) that commands our interest, for this axiom is to furnish the justification for the transfinite mode of inference in mathematics.

⟦32⟧ I should like to develop the essential core of this proof in somewhat greater detail, using the first and simplest case as an example. This first case appears as soon as we extend our number theory, which until now has remained strictly finite. This happens when in axiom V (11) we let the objects a be the number-signs (i.e. the positive integers, including 0) and the predicates $A(a)$ be the equations $f(a) = 0$, where f is an ordinary function of whole numbers. The logical function τ assigns an object to every predicate; that is, it assigns a number to every mathematical function f. Thus τ becomes an ordinary whole-numbered function of functions—i.e. if f is a fixed function, τ is a fixed number. We call it $\tau(f)$, so that

$$\tau(f) = \tau_a(f(a) = 0);$$

axiom V then turns into the axiom

12. $$f(\tau(f)) = 0 \rightarrow f(a) = 0.$$

⟦33⟧ We can most easily acquire a sense of the functional $\tau(f)$ if we think of $\tau(f)$ as being the number 0 provided $f(a) = 0$ for every a; otherwise $\tau(f)$ is the least a for which $f(a) \neq 0$.[a] The function $\tau(f)$ is a transfinite function

[a] ⟦In later articles Hilbert used the symbol $\mu(f)$ instead of $\tau(f)$. See also footnote b below.⟧

and is one of the functions prohibited by Brouwer and Weyl. Everything depends on proving that, if axiom 12 is adjoined to axioms 1 to 10, no contradiction results.

[34] To this end, let us take the proof of the consistency of axioms 1 to 10 and try to extend it to the present case. We now have to consider a new difficulty, namely, that in the given proof the sign $\tau(f)$ appears, and that arbitrarily many particular functions ϕ, ϕ', ... can be substituted for the function-variable f. But for the moment let us make the simplifying assumption that only a single such special function ϕ appears as a substitute for f, so that the given proof can ultimately be turned into a proof that contains only logical-signs, number-signs, and $\tau(\phi)$, where ϕ designates a special function whose definition does not use τ.

[35] We now perform the following operations on this proof:

[36] 1. We substitute for all occurrences of $\tau(\phi)$ the number-sign 0; we do this as it were provisionally and experimentally. Our proof then becomes a succession of 'numerical' formulae; all these formulae are in our earlier sense 'correct', with the possible exception of those that come from axiom 12. But if in 12 we take ϕ for f and make the proper substitutions for a and substitute the number-sign 0 for $\tau(\phi)$, then the only formulae that arise are of the form

$$\phi(0) = 0 \to \phi(\mathfrak{z}) = 0.$$

[37] Since $\mathfrak{z}$ here designates a number-sign and ϕ is a function defined by recursion—definition by recursion is easily handled in our formalism—then $\phi(\mathfrak{z})$ reduces to a number-sign as well. It will essentially be a question of whether in these formulae the reduction of $\phi(\mathfrak{z})$ to a number-sign always produces the number-sign 0, or at some point $\phi(\mathfrak{z})$ gives rise to a number-sign that is different from 0. In the former case, we have already attained our proof of consistency. For then all the formulae that come from axiom 12 are already correct in themselves. The sequence of formulae which we obtain from the proof becomes in turn a proof in which we can check all the formulae one by one and see that they are correct; so the false formula $0 \neq 0$ cannot appear as the end-formula.

[38] 2. Now suppose the second alternative holds. Then we have a $\mathfrak{z}$ such that

$$\phi(\mathfrak{z}) = 0$$

is a false formula. We then undertake a different operation with the given proof: we substitute for every occurrence of $\tau(\phi)$ not 0 but the number sign $\mathfrak{z}$. The formulae that come from axiom 12 then all have the form

$$\phi(\mathfrak{z}) = 0 \to \phi(\mathfrak{s}) = 0,$$

and these formulae are correct in themselves, since the formula preceding the implication-sign is false. The proof once again becomes a proof with purely numerical formulae which are correct, so that the end-formula cannot be $0 \neq 0$.

[39] With this, the proof of the consistency of the transfinite function $\tau(f)$ has been completed; at the same time we have secured the *tertium non datur*

for the concept of the infinite sequence of numbers which is represented by the whole-numbered variable in f: that is, on the basis of the axioms of negation II (5 and 6) formal negation amounts to the contradictory opposite; but $f\tau(f) \neq 0$ is the formal negation of $f\tau(f) = 0$, and, on the other hand, according to VI, $f\tau(f) \neq 0$ is equivalent to $(\mathrm{E}a)(f(a) \neq 0)$ and $f\tau(f) = 0$ is equivalent to $(a)(f(a) = 0)$.

⟦40⟧ One can understand the solution that my proof theory gives of the difficulty as follows. Our thought is finite; when we think, a finite process takes place. This self-activating truth ⟦sich von selbst betätigende Wahrheit⟧ is as it were used in my proof theory in such a manner that, if a contradiction were to emerge at any point, then, as soon as we recognize this contradiction, the relevant choice from the infinitely many things would also have to have been made. Accordingly, it is not asserted in my proof theory that we can always find an object from among the infinitely many objects, but rather that one can always act as though the choice had been made without risking an error. We can concede to Weyl the presence of a circle, but this circle is not vicious. Rather, the application of *tertium non datur* can never lead to danger.

⟦41⟧ In my proof theory, the transfinite axioms and formulae are adjoined to the finite axioms, just as in the theory of complex variables the imaginary elements are adjoined to the real, and just as in geometry the ideal constructions are adjoined to the actual. The motivation and the success of the procedure is the same in my proof theory as it is there: that is, the adjoining of the transfinite axioms results in the simplification and completion of the theory.

⟦42⟧ In consequence of the previous discussion, the transfinite function $\tau(f)$ may be applied throughout mathematics—in mathematical proofs as well as in the definition of new functions or the formation of new concepts.

⟦43⟧ The function

$$\phi(a) = \left[a^{\sqrt{a}}\right]$$

may serve as an example of the definition of a function, where the right-hand expression is either 0 or 1 according as $a^{\sqrt{a}}$ is a rational or an irrational number.

⟦44⟧ As for the application in proofs, one can usually tell quite easily whether a proof one finds in the literature makes an essential use of a transfinite function. A good example is provided by my two utterly different proofs of the finiteness of the full system of invariants. In the first I apply a transfinite mode of inference; but not in the second. My first proof of the finiteness of the full system of invariants is of such a sort that the transfinite mode of inference is essential and cannot be eradicated. To be sure, a finite theorem can probably always be proved without the application of a transfinite mode of inference—as my second proof of the theorem of the finiteness of the full system of invariants shows—but this contention is of the same sort as the contention that every mathematical proposition can either be proved or refuted. P. Gordan had an unclear sense of the transfinite mode of inference in my first invariant proof; he expressed it by calling the proof 'theology'. He then modified the presentation of my proof by bringing in his symbolism, and believed that he had in this way stripped the proof of its 'theological' character. But in fact the transfinite mode of inference was merely hidden behind the formalism.

‖45‖ In the same way that we earlier proved the consistency of the transfinite function of functions $\tau(f)$ we can also prove the consistency of the function of functions $\mu(f)$ which—like $\tau(f)$—has the property that it is 0 if $f(a)$ vanishes for all values of a, but otherwise assumes the smallest value for which $f(a)$ is not zero.[b] The principle of complete induction

$$A(0) \to (a)(A(a) \to A(a+1)) \to A(a)$$

then turns out to be a provable formula.[c]

‖46‖ In order to ground analysis we define a real number z lying between 0 and 1 by a dyadic fraction, and the dyadic fraction in turn by a function $f(n)$ which is only capable of assuming the values 0 or 1:

$$z = 0.a_1 a_2 a_3 \ldots \quad (a_n = f(n)).$$

‖47‖ An example of a transfinitely defined dyadic fraction is:

$$0.[2^{\sqrt{2}}]\,[3^{\sqrt{3}}]\,[4^{\sqrt{4}}] \ \ldots;$$

this expression represents a well defined real number, even though in the present state of science not even the first position can be calculated.

‖48‖ The foundation of analysis is the theorem about upper bounds. The transfinite function τ makes possible the proof of the theorem that the upper bound of a sequence of real numbers always exists.

‖49‖ To see this it is advisable first to introduce the logical signs & for 'and' and $\vee$ for 'or'. We do this by using our previous logical signs $\to$ and $\overline{}$ as follows:

$$\mathfrak{A} \,\&\, \mathfrak{B} \text{ and } \mathfrak{A} \vee \mathfrak{B}$$

shall be the same as

$$\overline{\mathfrak{A} \to \overline{\mathfrak{B}}} \text{ and } \overline{\mathfrak{A}} \to \mathfrak{B}$$

respectively. Then we abbreviate the formula

---

[b] ‖The axioms for $\mu(f)$ are:

$$f(\mu(f)) = 0 \to f(a) = 0, \quad (a)(f(a) = 0) \to \mu(f) = 0,$$
$$f(a) \neq 0 \to \mu(f) \leq a.$$

—Note of Bernays.‖

[c] ‖The axiom

$$a \neq 0 \to a = \delta(a) + 1$$

must be added to the axioms IV.—Note of Bernays.‖

$$(a)(fa = 0 \vee fa = 1) \,\&\, (a)(\mathrm{E}b)(f(a+b) = 1)$$

by $\mathfrak{R}f$, i.e. $\mathfrak{R}f$ means that the function fa represents a real number lying between 0 (exclusive) and 1 (inclusive) as the (always infinite) dyadic fraction

$$0.f(1)f(2)f(3)\ \ldots$$

A sequence $\zeta_1, \zeta_2, \zeta_3, \ldots$ of real numbers is then represented by a function $\phi(a, n)$ for which the formula $\mathfrak{R}\phi(a, n)$ is provable for arbitrary integral n. The further course of the proof rests on the following thought. In the schema

$$\zeta_1 = 0.\phi(1, 1)\phi(2, 1)\phi(3, 1)\ \ldots$$
$$\zeta_2 = 0.\phi(1, 2)\phi(2, 2)\phi(3, 2)\ \ldots$$
$$\zeta_3 = 0.\phi(1, 3)\phi(2, 3)\phi(3, 3)\ \ldots$$
$$\ldots\ldots\ldots\ldots\ldots$$

we consider first the numerals in the first vertical column after the decimal point. If these are all 0 (i.e. if $\phi(1, n) = 0$ for all n) then one takes $\psi(1) = 0$; otherwise, $\psi(1) = 1$. Now if in the second vertical column all those numerals are 0 for which the corresponding numeral in the same horizontal row in the first column is $\psi(1)$, then one takes $\psi(2) = 0$; otherwise, $\psi(2) = 1$. If in the third vertical column all those numerals are 0 for which both of the numerals in the first and second columns of the same horizontal row are $\psi(1)$ and $\psi(2)$ respectively, then $\psi(3) = 0$; otherwise, $\psi(3) = 1$, etc. On the basis of this observation we can define the upper bound $\psi(a)$ of the sequence $\phi(a, n)$ of real numbers by the following simultaneous recursion:

$$\chi(0, n) = 0$$
$$\psi(a+1) = \pi_n\{\chi(a, n) = 0 \rightarrow \phi(a+1, n) = 0\}$$
$$\chi(a+1, n) = \chi(a, n) + \iota(\psi(a+1), \phi(a+1, n)).$$

Here $\iota(a, b)$ is the function of a, b which yields 0 or 1 according as $a = b$ or $a \neq b$, and π_n is the transfinite function defined by the following axioms

$$(n)\mathfrak{A}(n) \rightarrow \pi_n(\mathfrak{A}n) = 0,$$
$$(\bar{n})\mathfrak{A}(n) \rightarrow \pi_n(\mathfrak{A}n) = 1;$$

in words: $\pi_n(\mathfrak{A}n)$ is 0 or 1 according as the statement $\mathfrak{A}$ holds for all n or not.

⟦50⟧ One can now in the spirit of my proof theory rigorously prove that $\mathfrak{R}\psi$ holds and that moreover the real number $\psi(n)$ has the upper-bound property, where the concept 'smaller' is defined for any two real numbers f, g by the formula

$$(\mathrm{E}a)\{(b)(b < a \rightarrow fb = gb) \,\&\, fa = 0 \,\&\, ga = 1\}.$$

⟦51⟧ Now suppose that instead of a sequence of real numbers we are given an arbitrary set of real numbers—say, in that for the function-variable f a fixed statement $\mathfrak{R}(f)$ is given which both characterizes f as a function representing a real number and also distinguishes precisely the real numbers of the set. The upper bound $\psi(a)$ of this set $\mathfrak{R}(f)$ of real numbers is then obtained by the following simultaneous recursion:

$$\chi(0, f) = 0$$

$$\psi(a+1) \qquad = \pi_f\{\mathfrak{R}f \rightarrow (\chi(a, f) = 0 \rightarrow f(a+1) = 0)\}$$

$$\chi(a+1, f) \quad = \chi(a, f) + \iota(\psi(a+1), f(a+1));$$

where π_f denotes the transfinite function defined by the axioms:

$$(f)\mathfrak{A}(f) \rightarrow \pi_f(\mathfrak{A}f) = 0,$$

$$(\overline{f})\mathfrak{A}(f) \rightarrow \pi_f(\mathfrak{A}f) = 1.$$

⟦52⟧ In conclusion I should like to make an application of these ideas to Zermelo's principle of choice for sets of sets of real numbers. Previously a set of real numbers f was given by a definite statement $\mathfrak{R}(f)$ with f as a function-variable; we now add the axioms

$$\mathfrak{R}f \rightarrow \nu(f) = 1$$

$$\overline{\mathfrak{R}f} \rightarrow \nu(f) = 0$$

whose consistency is easily recognized. In this way the set is defined by the function of functions $\nu(f)$, which has the value 1 for the real numbers f of the set and the value 0 for all other real numbers f. The formula, valid for $\mathfrak{R}$,

$$\mathfrak{R}f \rightarrow \mathfrak{R}f$$

yields

$$\nu(f) = 1 \rightarrow \mathfrak{R}f.$$

ν is a special function of functions; let r be the corresponding variable, i.e. a variable for functions of functions whose argument is an ordinary function of one argument.

⟦53⟧ A special set of sets of real numbers is then represented by a special statement $\mathfrak{M}(r)$ containing r, for which the formula

$$\mathfrak{M}(r) \,\&\, (rf = 1) \rightarrow \mathfrak{R}f$$

is valid. We assume that this set of sets has the property that every set of real numbers which is an element of it contains at least one real number, or, in formulae,

$$\mathfrak{M}(r) \rightarrow (\mathrm{E}f)(r(f) = 1).$$

⟦54⟧ Now we define a transfinite function τ_f like the earlier τ_a, only with the difference that in place of the number-variable a we take a function-variable f—i.e. τ_f is defined by the axiom

$$r(\tau_f(r)) = 0 \rightarrow r(f) = 0 \qquad \textbf{(12*)}$$

which corresponds to our axiom 12 for τ_a and can also be obtained directly from the logical axiom V (11), if one there takes the objects to be the functions f and the predicates to be the equations $r(f) = 0$; then τ_f has the property of alway representing a function, while the argument is a function of functions r.

⟦55⟧ We now have the following provable formulae:

$$(\mathrm{E}f)(r(f) = 1) \rightarrow (\bar{f})(rf = 0),$$
$$(\bar{f})(rf = 0) \rightarrow r(\tau_f(r)) \neq 0,$$
$$r(\tau_f(r)) \neq 0 \rightarrow r(\tau_f(r)) = 1,$$

and thus

$$\mathfrak{M}(r) \rightarrow r(\tau_f(r)) = 1;$$

i.e. to every element r of the set $\mathfrak{M}(r)$ there corresponds a whole-numbered function, namely, $\tau_f(r)$. This function represents a real number; for from an earlier formula it follows at once that $\mathfrak{R}(\tau_f(r))$. The functions $\tau_f(r)$ form a set; for in order to obtain a statement defining the totality of these functions (we shall call them $g(a)$) we need only state that each of them agrees with the representative $\tau_f(r)$ of a set r belonging to $\mathfrak{M}$; this is done by the formula

$$(\mathrm{E}r)\{\mathfrak{M}(r) \,\&\, (a)(g(a) = \tau_f r(a))\}$$

so that according to the original method of representation a set really exists.

⟦56⟧ This yields the proof of Zermelo's principle of choice for sets of sets of real numbers.

⟦57⟧ Because of the appearance of the existence-sign (E r) it is still necessary to prove the consistency of the transfinite function $\tau_f(r)$, which belongs to the new variable sort r. This proof has to follow the pattern of the proof for the transfinite function τ_a, as do the proofs for π_n and π_f.

⟦58⟧ The task remains to give a precise version of the basic ideas just sketched; when this problem has been solved the grounding of analysis will be completed and a path will have been broken to the grounding of set theory.

F. THE GROUNDING OF ELEMENTARY NUMBER THEORY (*HILBERT 1931a*)

In the years following the previous selection (*Hilbert 1923a*) Hilbert published an influential pair of technical articles on proof theory: 'On the infinite' (*Hilbert 1926*) and 'The foundations of mathematics' (*Hilbert 1928a*); both are translated by Stefan Bauer-Mengelberg in *van Heijenoort 1967*, and are therefore omitted from this collection. In 1928 he also published with Wilhelm Ackermann the *Grundzüge der theoretischen Logik* (*Hilbert and Ackermann 1928*). (The core ideas of this book go back to Hilbert's lecture course for 1917–18.) The following lecture was read to the Philosophische Gesellschaft in Hamburg in December, 1930. It contains some interesting remarks about the philosophical motivations underlying Hilbert's proof theory, as well as a polemical response to his critics. The editors of *Hilbert 1932–5* reproduced only the technical section (§§16–30) of *Hilbert 1931a*; the present translation has been made from the full version, which appeared in the *Mathematische Annalen*.

The translation of *Hilbert 1931a* is by William Ewald; references should be to the paragraph numbers, which have been added in this edition.

⟦1⟧ If in the realm of mathematics we investigate the two sources of our knowledge, experience and pure thought, we come across a number of ideas that are perhaps also of philosophical interest. In particular, all these ideas point to similarities between these two sources of knowledge which, in themselves, are constituted so differently. For instance, we observe the unity of material in matter;[a] on the other hand, the unity of foundations certainly crops up in our thought as a demand which we seek to fulfil, and often also achieve. The unity of the laws of nature, which we so often encounter in such surprising ways, can serve as an example for both sources of knowledge. But even more striking than this point of view of unity is a phenomenon that we call pre-established harmony, and that clearly attests a connection between nature and thought. The most wonderful and magnificent example of pre-established harmony is Einstein's famous theory of relativity. Here the quite complicated differential equations for the gravitational potential are uniquely derived from the general requirement of invariance alone. And this derivation would not have been possible without the profound and difficult mathematical investigations of Riemann, which existed long before. Even in mathematical analysis it is an isolated occurrence that such a complicated special formal system with numerical coefficients arises from a general thought. My proof theory, which I shall discuss later in this article, is also an example of pre-established harmony. For it uses the so-called logical calculus, which was devised earlier and for quite different purposes, namely, solely for the shortening and communication of statements.

⟦2⟧ However, attentive reflection leads us to see that, besides experience and thought, there is yet a third source of knowledge. Even if today we can no longer agree with Kant in the details, nevertheless the most general and

[a] ⟦die Einheit des Stoffes in der Materie⟧

fundamental idea of the Kantian epistemology retains its significance: to ascertain the *a priori* intuitive mode of thought ⟦Einstellung⟧, and thereby to investigate the condition of the possibility of all knowledge. In my opinion, this is essentially what happens in my investigations of the principles of mathematics. The *a priori* is here nothing more and nothing less than a fundamental mode of thought ⟦Grundeinstellung⟧, which I also call the finite mode of thought: something is already given to us in advance in our faculty of representation: certain extra-logical concrete objects that exist intuitively as an immediate experience before all thought. If logical inference is to be certain, then these objects must be completely surveyable in all their parts, and their presentation, their differences, their succeeding one another or their being arrayed next to one another is immediately and intuitively given to us, along with the objects, as something that neither can be further reduced to anything else, nor needs such a reduction. This is the fundamental mode of thought that I hold to be necessary for mathematics and for all scientific thought, understanding, and communication, and without which mental activity is not possible at all.

⟦3⟧ In this way, I believe myself to have recognized and characterized the third source of knowledge that accompanies experience and logic.

⟦4⟧ The *a priori* insights ⟦Einsichten⟧ are those intuitive insights as well as the logical insights that are achieved within the frame of the finite mode of thought. In particular, we see:

⟦5⟧ There are propositions that Kant regarded as *a priori*, and that we ascribe to experience. For example, all the basic facts of geometry, as well as the elementary properties of space and matter. But there are also propositions that are generally held to be *a priori*, but which cannot be achieved within the frame of the finite mode of thought—for example the principle of the *tertium non datur*, as well as the so-called transfinite statements generally.

⟦6⟧ The most obvious application and the first appearance of transfinite statements occurs in number theory, and with this we come to the principal topic of the present lecture. It is already remarkable and philosophically significant that the first and simplest questions about the numbers 1, 2, 3, . . . present such profound difficulties. These difficulties must be overcome; for how can knowledge be possible at all if not even number theory can be given a firm foundation, and if full unity and absolute correctness cannot be demanded even here!

⟦7⟧ It would be too great a digression and also superfluous to discuss the many false paths that are today recognized as such: some tried to define the numbers purely logically; others simply took the usual number-theoretical modes of inference to be self-evident. On both paths they encountered obstacles that proved to be insuperable. One path was not yet trodden, which seemed the most obvious to a mathematician. Before I describe this path, which in fact leads to the goal, I should like to make some remarks about the most important dates in the pre-history of this problem.

⟦8⟧ In the year 1888 as a young *Privatdozent* from Königsberg I made a tour of the German universities. At my first stop, in Berlin, I heard people in all mathematical circles, both young and old, discuss Dedekind's *Was sind und was sollen die Zahlen?*, which had then just appeared; their remarks were mostly critical. This essay is, along with the investigation of Frege, the most important and profound early attempt to ground elementary number theory. At roughly the same time, and so more than a generation ago, Kronecker clearly expressed a conception which he illustrated with numerous examples; this conception today essentially coincides with our finite mode of thought.

⟦9⟧ In those days we young mathematicians, *Privatdozenten* and students, played the game of transforming transfinite proofs of mathematical theorems into finite terms, in accordance with Kronecker's paradigm. Kronecker only made the mistake of declaring the transfinite mode of inference to be inadmissible. He issued prohibitions against the transfinite mode of inference; in particular, according to him, one was not allowed to infer that, if a statement $\mathfrak{A}(n)$ does not hold for every integer n, then there must exist an integer n for which that statement is false. At the time, the whole of mathematics unanimously rejected his prohibitions and went on to the business of the day.

⟦10⟧ How in fact do matters stand with the use of the transfinite modes of inference?

⟦11⟧ The theory of number fields, for example, is a finely articulated edifice which has been built up into the heavens, and which is bound up with the most fully developed theories of analysis. In beauty and perfection, it towers far above all other products of the human mind, and every step of the way it uses the *tertium non datur* and transfinite modes of inference of the sort forbidden by Kronecker. All the heroes of the mind prior to Gauss—just like those from Gauss, Hermite, Jacobi, until Poincaré—have used the transfinite mode of inference in the boldest and most manifold ways, and at no point did even the slightest discrepancy appear. Finally, if we just think of all the applications and make it clear to ourselves what a multitude of transfinite inferences of the most difficult and arduous sort are contained in for example the theory of relativity and quantum theory, and how nature nevertheless precisely conforms to these results—the beam of the fixed star, the planet Mercury, and the most complicated spectra here on earth and at a distance of hundreds of thousands of light years—how, in this situation, could we even for a moment doubt the legitimacy of applying the *tertium non datur*, just because of Kronecker's pretty eyes and just because a few philosophers disguised as mathematicians have put forward reasons that are utterly arbitrary and not even precisely formulable?

⟦12⟧ Every piece of scientific knowledge whatsoever rests on a reasonable estimation of probability, invoking agreement and reciprocal relationships: think of theories in physics or astronomy (for example, the construction of the world of stars), or, in biology, of the laws of heredity or of the idea of evolution—all results that we today view as firmly settled truths. It would be the death of all science and the end of all progress if we could not even allow such laws as those of elementary arithmetic to count as truths. Nevertheless, even today Kronecker still has his followers who do not believe in the admissibility of the *tertium non datur*: this is probably the crassest lack of faith that can be met with in the history of mankind.

⟦13⟧ However, a science like mathematics must not rely upon faith, however strong that faith might be; it has rather the duty to provide complete clarity. Now, since the applicability of the *tertium non datur* to finitely many statements is self-evident, our entire attention immediately turns to the concept 'infinite'. I have already begun an extensive investigation into the infinite, but can here give only the upshot of this investigation.

⟦14⟧ Physics teaches that a homogeneous continuum that would allow continued divisibility and would thus realize the infinitely small is nowhere encountered in reality. The infinite divisibility of a continuum is an operation that only exists in thought—is only an idea, which is refuted by our observations of nature and the experiences of physics and chemistry. On the other hand, in astronomy there are grave doubts about the existence of infinite space, and thus of the infinitely large. And all of our action is finite; the infinite has no place in it. The infinite is realized nowhere; it does not exist in nature, nor is it admissible as a foundation of our rational thought. And yet we cannot dispense with the unconditional application of the *tertium non datur* and of negation, since otherwise the gapless and unified construction of our science would be impossible. So operation with the infinite must be secured in the finite; and precisely this occurs in my proof theory.

⟦15⟧ I pursue an important goal with this new grounding of mathematics. I should like to rid the world of the question of the foundations of mathematics once and for all by making every mathematical statement into a formula that can be concretely exhibited and rigorously derived, and thereby bring mathematical concept-formations and inferences into such a form that they are irrefutable and yet furnish a model ⟦Bild⟧ of the entire science.

⟦16⟧ The fundamental idea of my proof theory is as follows:

⟦17⟧ Everything that makes up mathematics in the traditional sense is rigorously formalized, so that mathematics proper (or mathematics in the narrow sense) becomes a stock of formulae. These are distinguished from the usual formulae of mathematics only in the following way: that besides the usual signs, the logical signs appear as well—in particular, the signs for 'implies' ($\rightarrow$) and for 'not' ($\overline{\ \ }$). Certain formulae that serve as a foundation for the formal edifice of mathematics are called axioms. A proof is a figure, which must be intuitively presented to us as such; it consists of inferences, where each of the premisses is either an axiom, or agrees with the end-formula of an inference that comes earlier in the proof, or results from such a formula by substitution. Instead of contentual inference, in proof theory we have an external action according to rules, namely, the use of the inference schemata and of substitution. A formula shall be called provable if it is either an axiom or the end-formula of a proof.

⟦18⟧ This proper, formalized mathematics is accompanied by a mathematics that is to a certain extent new—a metamathematics that is necessary to secure formalized mathematics. In this metamathematics—in contrast to the purely formal modes of inference of mathematics proper—contentual inference is applied, but only to prove the consistency of the axioms.

⟦19⟧ The axioms and provable theorems, i.e. the formulae that arise in this interplay ⟦Wechselspiel⟧, are the images of the thoughts that make up the usual procedure of traditional mathematics.

⟦20⟧ The choice of axioms for our proof theory is already indicated by this programme. In selecting the axioms, we distinguish between qualitatively distinct groups, just as we do in geometry.

I. Axioms of implication:

$$A \rightarrow (B \rightarrow A)$$

(Adjunction of a presupposition);

$$(A \rightarrow B) \rightarrow \{(B \rightarrow C) \rightarrow (A \rightarrow C)\}$$

(Elimination of a statement);

$$\{A \rightarrow (A \rightarrow B)\} \rightarrow (A \rightarrow B).$$

II. Axioms concerning 'and' (&) as well as 'or' (∨).

III. Axioms of negation:

$$\{A \rightarrow (B \,\&\, \bar{B})\} \rightarrow \bar{A}$$

(Principle of contradiction);

$$\bar{\bar{A}} \rightarrow A$$

(Principle of double negation).

These axioms of groups I, II, and III are none other than the axioms of the propositional calculus.

IV. Transfinite axioms:

$$(x)A(x) \rightarrow A(b)$$

(Inference from the general to the particular, Aristotelian axiom);

Converse *via* the schema:

$$\frac{\mathfrak{A} \rightarrow \mathfrak{B}(a)}{\mathfrak{A} \rightarrow (x)\mathfrak{B}(x)};$$

$$A(a) \rightarrow (\mathrm{E}x)A(x)$$

converse again *via* schema. Further formulae are derivable, for example:

$$\overline{(x)}A(x) \rightleftarrows (\mathrm{E}x)\bar{A}(x)$$

(if a predicate does not hold for all arguments then there is a counterexample, and conversely);

$$(\overline{Ex})A(x) \rightleftarrows (x)\bar{A}(x)$$

(if there is no example for a statement, then the statement is false for all arguments, and conversely).
The axioms of this group IV are those of the predicate calculus.

Next come the special mathematical axioms:

V. Axioms of equality:

$$a = a;$$

$$a = b \rightarrow (A(a) \rightarrow A(b))$$

and

VI. Axioms of number:

$$a + 1 \neq 0;$$

as well as the axiom of complete induction and the schema of recursion.

⟦21⟧ The proof of consistency has recently been carried so far by Ackermann and von Neumann that the axioms just listed for elementary number theory have been shown to be consistent, and therefore the transfinite modes of inference—in particular the mode of inference of *tertium non datur*—can be seen to be admissible in the domain of elementary number theory. Our most important further task is to show the following (compare *Mathematische Annalen* **102**, p. 6):[b]

1. If a statement can be shown to be consistent, then it is also provable; moreover,

2. If a proposition $\mathfrak{S}$ can be proved consistent with the axioms of number theory, then $\bar{\mathfrak{S}}$ cannot be proved consistent with those axioms as well.

⟦22⟧ I have succeeded in proving these theorems at least for certain simple cases. I obtained this result by adding to the already given rules of inference (substitution and inference schema) the following equally finite new rule of inference:

⟦23⟧ If it has been proved, for any given numeral $\mathfrak{z}$, that the formula

$$\mathfrak{A}(\mathfrak{z})$$

is always a correct numerical formula, then the formula

$$(x)\mathfrak{A}(x)$$

can be laid down as a starting formula ⟦Ausgangsformel⟧.

⟦24⟧ Recall that the statement $(x)\mathfrak{A}(x)$ extends far wider than the formula $\mathfrak{A}(\mathfrak{z})$, where $\mathfrak{z}$ is an arbitrary given numeral. For in the former case not merely a numeral, but any expression of our formalism having a numerical character

[b] ⟦*Hilbert 1928c.*⟧

can be substituted for x in $\mathfrak{A}(x)$; moreover, the negation can be formed in accordance with the logical calculus.

⟦25⟧ First we observe that even with the addition of the new rule the axiom system remains consistent.

⟦26⟧ For suppose we are given a proof-figure which ends in a contradiction.

⟦27⟧ The previous proof of consistency consists in transforming all formulae of the given proof into numerical formulae according to a determinate procedure; then it is a matter of checking that all starting formulae are correct. Now, under our procedure, the formulae that are written down in accordance with our new rule also are transformed into numerical formulae; in particular, from $(x)\mathfrak{A}(x)$ we obtain $\mathfrak{A}(\mathfrak{z})$, where $\mathfrak{z}$ is a determinate numeral. But by the presupposition of the new rule, this formula is correct as well. So as before our procedure converts all starting formulae of the proof-figure into correct formulae. The proof of consistency has thus been given.

⟦28⟧ Now let $\mathfrak{S}$ be a formula of the form

$$(x)\mathfrak{A}(x)$$

that contains no variables other than x, and let it be consistent with the axioms. Then $\mathfrak{A}(\mathfrak{z})$ is certainly correct as soon as a numeral is substituted for $\mathfrak{z}$; for otherwise $\overline{\mathfrak{A}}(\mathfrak{z})$ would be correct and therefore provable, and this would contradict $(x)\mathfrak{A}(x)$, contrary to our hypothesis.

⟦29⟧ Therefore our formula $\mathfrak{S}$ is proved by the new rule of inference. So Theorem 1 holds for every statement $\mathfrak{S}$ of the form $(x)\mathfrak{A}(x)$ that contains no variable other than x. And for these statements of the form $\mathfrak{S}$ the validity of Theorem 2 follows from that of Theorem 1, which has just been proved.

⟦30⟧ If we now consider a statement $\mathfrak{T}$ of the form

$$\mathfrak{T}: (\mathrm{E}x)\mathfrak{A}(x)$$

then clearly the negation of this statement

$$\overline{\mathfrak{T}}: (x)\overline{\mathfrak{A}}(x)$$

is of the previously considered form $\mathfrak{S}$. So by Theorem 2 it is not possible to give a proof of the consistency of both statements $\mathfrak{T}$ and $\overline{\mathfrak{T}}$. So let us suppose that the proof of the consistency of $\mathfrak{T}$ has been given; it follows that the proof of the consistency of $\overline{\mathfrak{T}}$ cannot be given as well. Thus Theorem 2 has also been proved for every statement of the form $\mathfrak{T}$. To be sure, we cannot infer from this that $\mathfrak{T}$ is provable.—

⟦31⟧ Objections of various sorts have been raised against my proof theory; they are all unjustified. Let me make the following remarks on this matter:

⟦32⟧ 1. The critics of my theory should indicate the precise spot in my proof where my alleged error is supposed to occur. Otherwise, I shall refuse to examine their arguments.

⟦33⟧ 2. My theory has been subjected to the reproach that, although the theorems are indeed consistent, they are not for that reason proved. To be sure, they are provable, as I have shown here in simple cases. More generally, it turns out (as I was convinced from the outset) that the attainment of consistency is the essential thing in proof theory, and the question of provability (possibly with a suitable extension of the conditions that preserves the finite character) is settled at the same time. However, it cannot be demanded of a theory that all the relevant questions which it poses be fully solved at the outset; it suffices if the path to that goal has been indicated.

⟦34⟧ 3. Critics of my theory should use concepts such as 'consistent' the way I use them, and not the way other authors imagine them to be defined. On these points, my interpretation is authoritative because it is the only one that comes into consideration for my theory.

⟦35⟧ 4. The objections to my theory often fasten on incidental and wholly indifferent matters; for example, when they are directed against the term 'ideal'—which I use and which, despite all objections, I consider perfectly apt and an aid to understanding. And in many other instances one-sided prejudices and slogans are cheerfully introduced into the fray. I have already discussed the reproach of formalism in earlier essays. Formulae are a necessary aid to logical investigation. To be sure, their use demands precise mental labour, and makes empty twaddle impossible.

⟦36⟧ 5. Until now there has existed no other theory (and indeed, in my opinion no other theory is conceivable) that has been equally successful. For my proof theory does nothing other than to imitate the intimate activity of our understanding, and to make a protocol of the rules whereby our thinking actually proceeds. Thought takes place parallel to speaking and writing: by the formation and placing together of sentences ⟦Sätzen⟧. And for justification I need neither God, like Kronecker, nor the assumption of a special capacity of our understanding directed towards the principle of complete induction, like Poincaré, nor some ur-intuition like Brouwer, nor, like Whitehead and Russell, the axioms of infinity and reducibility, which are real, contentual presuppositions, not compensated for by proofs of consistency, and of which the latter is not even plausible.

⟦37⟧ In a recent philosophical lecture I find the sentence:

'The nothing is the absolute negation of the allness of being.'

⟦38⟧ This sentence is instructive for the following reason: in spite of its brevity, it illustrates all of the principal offences against the principles that are laid down in my proof theory. Concepts like 'the allness of being' contain a contradiction in themselves and already endanger the sense of every statement. But apart from this, negation is now applied to the problematic concept of the allness of being. It is precisely one of the most important tasks of proof theory to present clearly the sense and admissibility of negation: negation is a formal process, by means of which, from a statement $\mathfrak{S}$, another arises, which is bound to $\mathfrak{S}$ by the axioms of negation mentioned above (essentially, the principle of contradiction and *tertium non datur*). The process of negation is a necessary means of theoretical investigation; its unconditional application first makes possible the completeness and closure of logic. But in general the statement that arises through negation is an ideal statement, and to take this ideal statement as being in itself a real statement would be to misunderstand the nature and essence of thought.—

⟦39⟧ I believe that in my proof theory I have fully attained what I desired and promised: The world has thereby been rid, once and for all, of the question of the foundations of mathematics as such.

⟦40⟧ The philosophers will be interested that a science like mathematics exists at all. For us mathematicians, the task is to guard it like a relic, so that one day *all* human knowledge whatsoever will partake of the same precision and clarity. That this must and will occur is my firm conviction.

G. LOGIC AND THE KNOWLEDGE OF NATURE (*HILBERT 1930b*)

The following address was delivered in Königsberg, Kant's and Hilbert's native city, in the autumn of 1930 on the occasion of Hilbert's being awarded the honorary citizenship of the city. Hilbert had retired in the spring; the Königsberg address was his last major public appearance, and sums up his conception of the practice of mathematics. A few years later, the Hilbert circle in Göttingen was purged by the Nazis. Many of his friends and colleagues were driven into exile. Hilbert, now in his seventies, remained behind; he led a lonely retirement in Göttingen, where he died in 1943. The concluding words of the Königsberg address are engraved on his tombstone.

The translation is by William Ewald; references to *Hilbert 1930b* should be to the paragraph numbers, which have been added in this edition.

⟦1⟧ The knowledge of nature and of life is our noblest task. All human striving and willing flows into it, and in it we have met with ever-increasing success. In the last decades, we have achieved a richer and deeper knowledge of nature than previously in so many centuries. Today we shall, in keeping with our subject, use this fortunate situation to treat an old philosophical problem, namely, the vexed question about the share which thought, on the one hand, and experience, on the other, have in our knowledge. This old question is legitimate, for to answer it is at bottom to ascertain the general character of our knowledge in the natural sciences, and the sense in which the knowledge which we collect in the scientific enterprise is true.

⟦2⟧ Without being disrespectful towards the old philosophers and inquirers, we can today count on a correct answer to this question with more confidence than they could—for two reasons. The first has already been mentioned. It is the rapid tempo with which our sciences are developing today.

⟦3⟧ The most important discoveries of the older period, from Copernicus, Kepler, Galileo, and Newton to Maxwell, are distributed at large intervals over almost four centuries. The modern period begins with the discovery of Hertzian waves. And now events follow thick and fast: Röntgen discovers his X-rays, Curie discovers radioactivity, Planck establishes quantum theory. And in the most recent period, the discoveries of new phenomena and surprising connections come tumbling over each other, so that the abundance of visions becomes almost alarming: Rutherford's theory of radioactivity, Einstein's *hv*-law, Bohr's explanation of the spectrum, Moseley's numbering of the elements, Einstein's theory of relativity, Rutherford's decomposition of nitrogen, Bohr's construction of the elements, Aston's theory of isotopes.

⟦4⟧ So in physics alone we have witnessed an unbroken sequence of discoveries—and what discoveries! In power, not a single one is inferior to the achievements of the older period; furthermore they are temporally more compressed, and yet they are internally just as many-faceted. And in them theory and practice, thought and experience constantly prove to be most intimately intertwined. Now theory hurries ahead, now experiment—always mutually confirming, completing, and stimulating each other. A similar story holds for chemistry, astronomy, and the biological disciplines.

⟦5⟧ We therefore have two advantages over the old philosophers: we have lived through a great number of such discoveries, and we have become acquainted with the resulting new points of view during their formation. Moreover, among the new discoveries there were many which modified old, firmly-rooted conceptions and ideas, or completely eradicated them. We need only think, for example, of the new conception of time in relativity theory, or of the decomposition of chemical elements, and of how prejudices have thereby been eradicated that earlier it would not even have occurred to anybody to touch.

⟦6⟧ But yet a second circumstance helps us today towards a solution of that old philosophical problem. Not only have the technique of experimentation and the art of erecting theoretical edifices in physics attained new heights, but their counterpart, the science of logic, has also made fundamental progress. Today there is a general method for the theoretical treatment of questions in the natural sciences, which in every case facilitates the precise formulation of the problem and helps prepare its solution—namely, the axiomatic method.

⟦7⟧ How do matters stand with this axiomatics, which is today on everybody's lips? Now, the basic idea rests on the fact that generally even in comprehensive fields of knowledge ⟦Wissensgebieten⟧ a few propositions—called axioms—suffice for the purely logical construction of the entire edifice of theory. But their significance is not fully explained by this remark. Examples will be the easiest way to elucidate the axiomatic method. The oldest and best-known example of the axiomatic method is Euclid's geometry. But I should prefer to illustrate the axiomatic method quite briefly with a striking example from modern biology.

‖8‖ *Drosophila* is a small fly, but our interest in it is great; it has been the object of the most extensive, the most careful, and the most successful breeding experiments. This fly is usually grey, red-eyed, unspeckled, round-winged, and long-winged. But there are also flies with deviant special characteristics: they are yellow rather than grey, they are white-eyed rather than red-eyed, etc. These five special characteristics are often coupled. That is, when a fly is yellow, it is also white-eyed and speckled, split-winged and stump-winged. And when it is stump-winged then it is also yellow and white-eyed, and so on. But now with suitable cross-breeding later generations exhibit a smaller number of deviations from these commonly occurring couplings—indeed, the percentage is a definite constant. The numbers which one thus finds experimentally tally with the linear Euclidean axioms of congruence and with the axioms for the geometrical concept 'between', and so the laws of heredity result as an application of the axioms of linear congruence, that is, of the marking-off of intervals; so simple and exact, and at the same time so wonderful, that probably no fantasy, no matter how bold, could have devised it.

‖9‖ The following is a further example of the axiomatic method in an entirely different field:

‖10‖ In our theoretical sciences we are used to the application of formal thought processes and abstract methods. The axiomatic method belongs to logic. By the word logic, many understand something very boring and difficult. Today the science of logic has become easy to understand and very interesting. For example, one has recognized that already in daily life methods and concept-formations are used that demand a great measure of abstraction and that can be understood only by the unconscious application of axiomatic methods. For example, the general process of negation, and in particular the concept 'infinite'. As far as the concept 'infinite' is concerned, we must be clear to ourselves that 'infinite' has no intuitive meaning ‖anschauliche Bedeutung‖ and that without more detailed investigation it has absolutely no sense. For everywhere there are only finite things. There is no infinite speed, and no force or effect that propagates itself infinitely fast. Moreover the effect itself is of a discrete nature and exists only in quanta. There is absolutely nothing continuous that can be divided infinitely often. Even light has atomic structure, just like the quanta of action. I firmly believe that even space is only of finite extent, and one day the astronomers will be able to tell us how many kilometres long, high and broad it is. And although there are in reality often cases of very large numbers (for instance, the distance of the stars in kilometres, or the number of essentially different games of chess) nevertheless endlessness or infinity, because it is the negation of a condition that prevails everywhere, is a gigantic abstraction—practicable only through the conscious or unconscious application of the axiomatic method. This conception of the infinite, which I have grounded through detailed investigations, answers a number of important questions; in particular, it shows the baselessness of the Kantian antinomies of space and of the unlimited possibility of division, and thus of the difficulties that crop up thereby.

⟦11⟧ If we now turn to our problem itself, the problem of how nature and thought are interconnected, we shall need here to state three principal points of view. The first connects with the problem we have just discussed, that of infinity. We saw: the infinite is nowhere realized; it neither occurs in nature nor is it admissible as a foundation in our thought without special precautions. Here I already see an important parallelism between nature and thought, a fundamental agreement between experience and theory.

⟦12⟧ We perceive yet another parallelism: our thought proceeds from unity and seeks to form unity; we observe the unity of material ⟦Stoffes⟧ in matter ⟦Materie⟧ and we everywhere detect the unity of the laws of nature. Thus nature in reality greatly accommodates us in our research, as though she were prepared and pleased to unveil her secrets. The sparse distribution of mass in outer space made possible the discovery and the more precise confirmation of Newton's law. Despite the great speed of light, Michelson was able to establish with certainty the invalidity of the law of addition of speeds, because our earth moves just quickly enough in its orbit about the sun. Mercury just did us the service of executing its perihelion motion so that we could thereby check Einstein's theory. And the rays from the stars pass by the sun in such a way that we can observe their deflection.

⟦13⟧ But even more striking is an occurrence which is virtually an embodiment and realization of mathematical thoughts, and which we shall call, in a sense different from that in Leibniz, pre-established harmony. The older examples for this are conic sections, which one studied long before one suspected that our planets or even electrons move in such a course. But the most magnificent and wonderful example of pre-established harmony is Einstein's famous theory of relativity. Here solely through the general demand for invariance together with the principle of greatest simplicity the differential equations for the gravitational potential are constructed mathematically and uniquely. This construction would not have been possible without the profound and difficult mathematical investigations of Riemann, which existed long before. Recently cases have been piling up in which precisely the most important mathematical theorems, the ones that stand at the centre of mathematical attention, are at the same time the ones that are needed in physics. I had developed the theory of infinitely many variables from pure mathematical interest, and had even used the term spectral analysis, without any inkling that it would one day be realized in the actual spectrum of physics.

⟦14⟧ We can understand this agreement between nature and thought, between experience and theory, only if we take into account both the formal element and the mechanism that is connected with it; and we must do this both for nature and for our understanding. The mathematical process of elimination furnishes, it seems, the resting-points and stations where both the bodies in the real world and the thoughts in the mental world linger and, in the process, present themselves for control and comparison.

⟦15⟧ But this pre-established harmony does not yet exhaust the relations between nature and thought, and does not yet uncover the deepest secrets of our problem. To come to this, let us turn our gaze to the entire complex of physical and astronomical knowledge. In modern science we find a point of view that goes far beyond our science's old goals and formulations of questions: modern science does not merely, like classical mechanics, teach how from present data we can determine in advance the future movements and expected appearances, but it also shows that the present actual state of physical matter on earth and in space is not accidental or arbitrary, but follows from the laws of physics.

⟦16⟧ The most important proofs of this are Bohr's atomic models, the construction of the cosmos, and finally the entire evolutionary history of organic life. It appears that the pursuit of these methods would actually have to lead to a system of laws of nature that applies to the whole of reality, and we should then in fact need only thought (that is, conceptual deduction) to acquire all physical knowledge. Hegel would then have been right to contend that it is possible to deduce all the happenings of nature from concepts. But this inference does not hold true. For what is the origin of the laws of the world? How do we acquire them? And who teaches us that they fit reality? The answer is, that experience alone makes this possible. In contrast to Hegel, we recognize that the laws of the world can be acquired in no other way than through experience. In the construction of the framework of physical concepts, various speculative points of view may cooperate: but whether the laws that have been postulated and the logical framework of concepts that has been constructed out of them are correct is something that only experience is competent to decide. Sometimes an idea had its origin in pure thought, such as, for example, the atomism of Democritus, while the existence of atoms was proved by experimental physics only two thousand years later. Sometimes experience leads the way and forces the speculative point of view on the mind. Thus the powerful impetus of the Michelson experiment cleared away the deeply rooted prejudice of absolute time and made it possible for Einstein to grasp the idea of general relativity.

⟦17⟧ Whoever wishes nevertheless to deny that the laws of the world come from experience must maintain that besides deduction and experience there is a third source of knowledge.

⟦18⟧ Philosophers have in fact maintained—and Kant is the classical representative of this standpoint—that besides logic and experience we have a certain *a priori* knowledge of reality. Now I admit that already for the construction of the theoretical framework certain *a priori* insights are necessary and that they always underlie the genesis of our knowledge. I also believe that mathematical knowledge in the end rests on a kind of intuitive insight ⟦anschaulicher Einsicht⟧ of this sort, and even that we need a certain intuitive *a priori* outlook for the construction of number theory. Thus the most general and fundamental idea of the Kantian epistemology retains its significance: namely the philosophical problem of determining that intuitive, *a priori* outlook and thereby of investigating the condition of the possibility of all conceptual knowledge and of every experience. I believe that in essence this has occurred in my investigations into the principles of mathematics. The *a priori* is nothing more and nothing less than a fundamental outlook, or the expression of certain indispensable preconditions of thought and experience. But we must draw the boundary between what we possess *a priori* and what requires experience differently than Kant: Kant greatly overestimated the role and the extent of the *a priori*.

⟦19⟧ In the days of Kant one could think that the representations ⟦Vorstellungen⟧ that one had of space and time were just as immediately and generally applicable to reality as, for example, our representations of number, sequence, and quantity—representations which we constantly use in the manner familiar to us from mathematical and physical theory. Then in fact the theory of space and time (and in particular geometry) would be something that, like arithmetic, precedes all knowledge of nature. But even before the development of physics compelled it, this Kantian standpoint had already been abandoned by Riemann and Helmholtz—and quite rightly, for geometry is nothing other than that part of the total framework of physical concepts which models the possible positional relations between rigid bodies in the world of real things. That there are movable rigid bodies at all, and what their positional relations are, is solely a matter of experience. As Gauss already recognized, the proposition that the sum of the angles in a triangle equals two right angles and that the axiom of parallels is valid is in the end to be determined or refuted solely by experiment. If, for example, all the facts expressed by the theorems on congruence were to agree with experience, but if, in contrast, the sum of the angles in a triangle constructed out of rigid bars should turn out to be less than two right angles, then nobody would hit on the idea that the axiom of parallels is valid in the space of actual bodies.

⟦20⟧ The inclusion of something in the domain of the *a priori* demands the greatest caution; however, much of the knowledge that earlier was counted as *a priori* is today recognized as being not even correct. The most striking example of this is the idea ⟦Vorstellung⟧ of the absolute present. There is no absolute present, however accustomed we are from childhood to assume that there is; for in daily life it is always a matter of short distances and slow movements. If things were otherwise, it would have occurred to nobody to introduce absolute time. But even such deep thinkers as Newton and Kant did not dream of doubting the absoluteness of time. The cautious Newton formulated it as crassly as possible: absolute true time, in virtue of its nature, flows by itself, uniformly and without relation to any object.[a] Newton thereby cut off every possibility of retreat or compromise. Kant, the critical philosopher, is here not critical at all, for he simply accepted Newton. Only Einstein definitively

---

[a] ⟦The above quotation is translated from Hilbert's German. The standard English version is: 'Absolute true and mathematical time, of itself and from its own nature, flows equably and without relation to anything external.' *Newton 1934*, Scholium to Definition 8.⟧

liberated us from this prejudice—in what will always remain one of the most powerful acts of the human mind—and the *a priori* theory, which went altogether too far, could not have been reduced to absurdity more strikingly than by this advance of physical science. For the assumption of absolute time has as a consequence, *inter alia*, the theorem that the composition of speeds is given by the addition of the speeds (a proposition which, by the way, can, it seems, hardly be surpassed in evidence and popular intelligibility), and yet the most varied sorts of experiments in optics, astronomy, and the theory of electricity compelled the conclusion that this theorem of the addition of speeds is not correct. In fact, another quite complicated law holds for the composition of two speeds. We can say: in recent times the conception of the empirical nature of geometry, as represented by Gauss and Helmholtz, has become a secure result of science. It must today serve as a firm point of support for all philosophical speculations that concern space and time. The Einsteinian theory of gravitation makes it manifest: geometry is nothing more than a branch of physics; geometric truths are in no principled way whatsoever different from physical truths. Thus, for example, the Pythagorean theorem and the Newtonian laws of attraction are essentially related in that they are both governed by the same fundamental physical concept, that of the potential. But even more is certain for anybody who knows the Einsteinian theory of gravitation: these two laws, which are so different and which until now were apparently widely separated, the one a theorem of elementary geometry, known since antiquity and taught in schools everywhere, the other a law about the action of masses on each other—these two laws are not only of the same character, but are just parts of one and the same general law.

⟦21⟧ That the facts of geometry and physics are in principle of the same kind could hardly have emerged more drastically. To be sure, in the usual logical construction and in our ordinary daily experiences which have been familiar since childhood, the geometric and kinematic propositions precede the dynamic; this circumstance explains how one could forget that they are experiences at all. We therefore see: the Kantian theory of the *a priori* still contains anthropological dross from which it must be liberated; afterwards only the *a priori* attitude ⟦Einstellung⟧ is left over which also underlies pure mathematical knowledge: essentially it is the finite attitude which I have characterized in several works.[1]

⟦22⟧ The instrument which mediates between theory and practice, between thought and observation, is mathematics; it builds the connecting bridges, and makes them ever sounder. Thus it happens that our entire modern culture, in so far as it rests on the penetration and utilization of nature, has its foundation in mathematics. Galileo already said: Only that man can understand nature who has become acquainted with her language and with the signs in which she speaks

[1] See 'Über das Unendliche', *Math. Ann.* Vol. 95 (1926), p. 161. 'Die Grundlagen der Mathematik', *Abh. a. d. math. Sem. d. Hamburgischen Universität*, Vol. 6 (1928), p. 65. Reprinted as Appendices VIII and IX of *Grundlagen der Geometrie*, 7th edn, 1930.

to us; this language is mathematics, and her signs are mathematical figures. Kant made the remark: 'I maintain that in every particular natural science, the amount of of mathematics it contains is equal to the amount of genuine science we can find in it.' And in fact, we have not mastered a theory in the natural sciences until we have extracted and fully revealed its mathematical core. Without mathematics, modern astronomy and physics would not be possible; these sciences, in their theoretical parts, almost dissolve into mathematics. It is to these applications and numerous others that mathematics owes the prestige which it enjoys in the general public.

⟦23⟧ Nevertheless, mathematicians have refused to allow the applications to serve as the standard of value for mathematics. The prince of mathematicians, Gauss (who was to be sure at the same time an applied mathematician *par excellence*)—Gauss, who created anew entire sciences, like the theory of errors, or geodesy, in order to allow mathematics to play the leading role; who, when the astronomers had lost and could not find the newly discovered planet Ceres—an especially important and interesting planet—devised a new mathematical theory which he was able to use to predict its location; who invented the telegraph, and many other practical things—Gauss was nevertheless of the same opinion. The pure theory of numbers is the domain of mathematics that until now has found no application. But it is precisely number theory that Gauss called the queen of mathematics, and that is glorified by him and almost all great mathematicians. Gauss speaks of the magical allure which has made number theory the favourite science of the foremost mathematicians—to say nothing of its inexhaustible wealth, in which it outstrips all other parts of mathematics. Gauss recounts how already in early youth the fascination of number-theoretic investigations ensnared him, so that he could never leave them. He praises Fermat, Euler, Lagrange, and Legendre as men of incomparable glory, because they had found the pathway to the inner sanctuary of this heavenly science, and had shown the riches with which it is filled. And the mathematicians before Gauss and after Gauss speak with equal enthusiasm—mathematicians like Lejeune-Dirichlet, Kummer, Hermite, Kronecker, and Minkowski. Kronecker compares the number-theorists to the Lotos Eaters, who, once they have tasted something of this fare, can never again desist.

⟦24⟧ Poincaré too, the most brilliant mathematician of his generation, who was also a profound physicist and astronomer, is of the same opinion. With remarkable asperity Poincaré once attacked Tolstoy, who had said that it was foolish to demand 'science for the sake of science'. 'As we choose our pursuits', Tolstoy asked, 'should we allow ourselves to be led by the moods of our greed for knowledge? Would it not be better to make the decision according to its usefulness, that is, according to our practical and moral needs?' Strange, that it should be Tolstoy whom we mathematicians must reject as a dull realist and a narrow-hearted utilitarian. Poincaré argued against Tolstoy that, if one had followed Tolstoy's recipe, science would never have arisen at all. One need only open one's eyes, Poincaré concludes, to see that, for example, the achievements of industry would never have seen the light of day if only practical people had existed, and if these achievements had not been promoted by disinterested fools who never thought about practical applications. All of us are of the same opinion.

⟦25⟧ Our great Königsberg mathematician Jacobi also thought so—Jacobi, whose name stands next to that of Gauss, and is pronounced with reverence by every student of our subject. When the famous Fourier once said that the chief purpose of mathematics is the explanation of natural phenomena, it was Jacobi who scolded him with all the passion of his temperament. A philosopher like Fourier ought to have known, Jacobi cried, that the sole aim of all science is the honour of the human spirit, and that from this point of view a problem of pure number theory is every bit as valuable as a problem with practical applications.

⟦26⟧ Whoever feels the truth of the magnificent manner of thinking and of the world-view that shines forth in these words of Jacobi will not fall into retrogressive and fruitless scepticism; he will not believe those who today, with a philosophical air and a superior tone, prophesy the downfall of culture and fall into an *ignorabimus*. For the mathematician there is no *ignorabimus*, nor, in my opinion, for any part of natural science. The philosopher Comte once said—with the intention of mentioning a problem that was certainly unsolvable—that science would never be able to fathom the secret of the chemical composition of the heavenly bodies. A few years later this problem was solved by the spectral analysis of Bunsen and Kirchhoff, and today we can say that we claim the most distant stars as the most important physical and chemical laboratories—laboratories such as we find nowhere on earth. The real reason why Comte was unable to find an unsolvable problem is, in my opinion, that there are absolutely no unsolvable problems. Instead of the foolish *ignorabimus*, our answer is on the contrary:

We must know,
We shall know.

25
Luitzen Egbertus Jean Brouwer (1881–1966)

Brouwer, the father of intuitionism, was the most eminent Dutch mathematician of the twentieth century; he lived in Holland all his life. He was educated at the University of Amsterdam, where he studied mathematics and philosophy, and where he obtained the degree of Doctor in Mathematics and Physics; his dissertation, *On the foundations of mathematics*, was defended on 19 February 1907.

Although constructive mathematics and the critique of logic were to be lifelong preoccupations for Brouwer (they were already central topics in the dissertation), his first major contributions to mathematics were not to intuitionism but to topology. He was drawn to topology by his study of Schoenflies's work on the theory of point-manifolds (*Schoenflies 1908*), and in the years 1909–13 he produced a remarkable outpouring of papers—on average, ten *per annum*—on such matters as the transformation of surfaces, *n*-dimensional topology, Jordan's theorem, polyhedral approximations, fixed-points of continuous mappings, the concept of dimension, homotopy classes of maps, and topological group theory. These pioneering works, written at the start of his career when he was still a *privaat-docent* (unsalaried lecturer), laid the groundwork for much of twentieth-century topology, and form the bulk of the second volume of his *Collected works*. The details of his contributions to topology lie beyond the scope of the present volume; but a survey discussion can be found in the important biographical notice of Brouwer for the Royal Society, *Kreisel and Newman 1969*.[a]

[a] In general, Brouwer employed classical logic in his topological works; and when in later years he gave intuitionistically valid versions of his results, he presented them under titles like 'An intuitionist correction of the fixed-point theorem on the sphere' (*Brouwer 1975*, Vol. i, pp. 506–7, originally published in 1952). For this reason, it has often been asserted that his work in topology and his work in intuitionism are entirely disjoint. However, Brouwer's 1907 dissertation shows him struggling simultaneously with the theory of the continuum and with the logical foundations of mathematics, trying to work out the implications of his philosophical views for both subjects; and Kreisel points out an important connection between Brouwer's work in topology and his constructivism:

'Brouwer, like most of his contemporaries, used as a tool approximations by geometric figures which are determined by a finite number of points: polygons in the plane, polyhedra in space. But unlike most of them, except for example Poincaré, another constructivist, his use of those figures was very elementary; or "algebraic" as we should say now. In his style of work one never thinks of a line or surface as made up of an infinite number of points in contrast to point set topology.

In 1912 Brouwer was appointed to a professorship at the University of Amsterdam; a few years later he was given the title, Professor of Set Theory, Function Theory, and Axiomatics, a name which reflected the new direction of his research. Despite offers of professorships at other institutions (in particular, Groningen, Leiden, Göttingen, and Berlin), Brouwer stayed in Amsterdam until his retirement in 1951. Almost immediately his research interests began to shift away from topology and back towards the foundations of mathematics; indeed, he wrote almost nothing on topology after his initial four-year sprint of 1909–13. Brouwer's most important intuitionistic writings have been gathered in the first volume of his *Collected works*; some important supplementary material is contained in *van Dalen 1981* and the appendices to *van Stigt 1990*.

Already in his dissertation Brouwer had expressed strong opinions on constructive mathematics. At the time, the legitimacy of transfinite set theory was being hotly debated—in particular, Zermelo's first proof of the well-ordering theorem (*Zermelo 1904*). The interrelated problems of the *continuum* and of *completed infinities* were once again creating conceptual difficulties at the heart of mathematics, just as they had done in classical antiquity with the discovery of incommensurable quantities and in the seventeenth century with the debates over the foundations of the calculus. In previous selections we have seen how such mathematicians as Kronecker, Borel, Lebesgue, and Poincaré, the chief forerunners of Brouwer's intuitionism, reacted to the work of Cantor and Zermelo.[b] However, Brouwer's intuitionism was to emerge as the most thorough and radical critique of classical mathematics; and where thinkers like Poincaré wavered in their constructivism (for example by giving, at the end of his career, a new proof of the non-denumerability of the continuum (*Poincaré*

'Inevitably one wonders about the heuristic value of Brouwer's (or Poincaré's) general logical ideas for his topology and, on a higher level, about the value of his general philosophical ideas for his logic. (In a few philosophical papers on knowledge and its origins Brouwer elaborated on the primacy of the will over the intellect.) Now his logical ideas, which he published several years before his topological work, were not only novel, but almost detailed enough to deduce rigorously some of his topological innovations from them, though he himself had not done so explicitly. In contrast his philosophical views, though quite consistent with his logic (his stress on the will corresponding to his interest in constructions, in what we do ourselves) were undeveloped and, literally, commonplace: the observer, language and linguistic conventions had a big role in the epistemology of the time' (*Kreisel and Newman 1969*, p. 41).

[b] It is important to observe that the set-theoretic paradoxes were not the primary spur to these constructivist debates: the methodology of Cantor and the theorems of Zermelo were radical departures from traditional conceptions of mathematical technique, and would have given rise to controversy with or without the paradoxes—just as Newton's 'method of fluxions' attracted Berkeley's criticisms despite the absence of any formal contradictions in the foundations of the calculus. Brouwer, like Berkeley, was driven to his intuitionistic position in large measure by epistemology; and he was propelled less by the fear that classical mathematics might turn out to be false than by the conviction that it was meaningless. Although Brouwer discussed the paradoxes in his dissertation, it is clear that they were never at the centre of his thought, and that he regarded them as but a surface symptom of a more deeply-rooted disease. Consider the structure of his dissertation. The work is divided into three parts. The first two parts (*The construction of mathematics* and *Mathematics and experience*) present his general philosophical position; the paradoxes appear only in the third part (*Mathematics and logic*), where they are presented as a consequence of the more general issues.

1910)), Brouwer, who never altered his position in its fundamentals, developed a full-blown constructivist philosophy of mathematics, and showed how to erect a fruitful mathematical research programme on its foundations.

Brouwer's programme emerged gradually over two decades. The dissertation of 1907 is fundamental for understanding the evolution of his thought. It already contains his *Grundgedanke*, the conception of mathematics as a *construction*; it also contains characteristically zealous polemics against the logico-linguistic approach to foundations. Despite flashes of technical brilliance (such as his prescient distinction between—in effect—formal mathematics and meta-mathematics) the dissertation was in many respects an immature work, filled with criticisms of other thinkers rather than with positive mathematical ideas for a new, constructive mathematics. In the dissertation Brouwer was still using the law of the excluded middle in his reasoning about real numbers; not until *Brouwer 1908* did he explicitly present the illegitimate use of the law of the excluded middle as the focal point of his disagreement with classical mathematics.

In his inaugural lecture for the Amsterdam professorship, 'Intuitionism and formalism' (*Brouwer 1913*), delivered at the end of his topological period, he began to fill in some of the mathematical consequences of his constructivist views; he coined the names 'intuitionism' and 'formalism', and gave the first tightly-reasoned exposition of intuitionism; though at this time he refrained from *explicitly* identifying himself as an intuitionist.[c]

---

[c] Brouwer's coinage of the term 'formalism' has not been entirely fortunate for philosophical nomenclature. In his usage, it often embraces a large number of divergent tendencies: as late as *Brouwer 1952* (reprinted below) he labelled Dedekind, Cantor, Peano, Hilbert, Russell, Zermelo, and Couturat 'Old Formalists'. (Brouwer never mentions Frege, and apparently was unaware of his work.) The use of the term 'New Formalist' (or simply 'formalist') to describe Hilbert is misleading if the term is taken (as it often is) to imply that Hilbert thought mathematics was a meaningless game played with symbols. For reasons discussed above in the Introductory Notes to the selections from Hilbert, this brand of 'formalism' is a caricature of Hilbert's position; and indeed Hilbert (in contrast to some of his followers) did not himself accept the label 'formalist' (see *Hilbert 1931a*, §35).

Incidentally, two points about Hilbert's relationship to Brouwer should be noticed here. First, although many of Hilbert's foundational papers in the 1920s were explicitly directed against the intuitionists, Hilbert's general foundational project, including his search for consistency proofs, antedated Brouwer's dissertation by at least seven years. (Brouwer himself, in *1928b* (translated below) notes that the Hilbert programme grew naturally out of Hilbert's studies of the axiomatic foundations of geometry.) Second, even though Hilbert sought to prove the consistency of classical mathematics, the *finitary* reasoning he permitted in his metamathematical consistency proofs was even more restricted than Brouwer's intuitionism; so that—seen from a certain angle—the controversy between Brouwer and Hilbert, for all its intensity of feeling, was less a struggle between classical mathematics and constructive mathematics than a struggle between two varieties of constructivism. Indeed, in his 'Intuitionistic reflections on formalism' (*Brouwer 1927c*) Brouwer points out somewhat peevishly that he had anticipated Hilbert's distinction between formal mathematics and metamathematics, and contends that Hilbert had adopted without attribution his criticisms of the unrestricted use of the law of the excluded middle in metamathematics. Indeed: 'formalism has received nothing but benefactions from intuitionism, and may expect further benefactions. The formalistic school should therefore accord some recognition to intuitionism, instead of polemicizing against it in sneering tones while not even observing proper mention of authorship' *van Heijenoort 1967*, p. 492).

Most of Brouwer's mature papers on intuitionism were written during the decade 1918–28. This was the period when, attempting to build a mathematics that did not rely on the law of the excluded middle, he developed intuitionistic set theory, topology, and real analysis; this decade was also the period of his well-known controversies with Hilbert, which are documented in *van Heijenoort 1967*. The central new concept was *choice sequences* (and the related ideas of *spreads* and *fans*), which first made their appearance in *Brouwer 1918*. Choice sequences were a major contribution to constructive mathematics. Pre-intuitionists like Poincaré and Weyl had never been able to produce a satisfactory theory of the continuum: although they were willing to countenance statements (arrived at by mathematical induction) about the totality of the integers, they were unwilling to allow arithmetical operations to be applied to completed infinite totalities; thus they were unable to accept the theory of the real numbers that had been developed by Dedekind, Weierstrass, and Cantor, and were left without a satisfactory account of the real line. Brouwer rescued geometry for intuitionistic mathematics, essentially by viewing the continuum as perpetually in the process of creation: in the new theory, points of the real line develop as choice-sequences and reasoning about them takes place on the basis of the finite amount of information that is available to date.[d]

Suddenly at the end of this period Brouwer lapsed into a fourteen-year silence: from 1928 until 1942 he published nothing on intuitionistic mathematics.[e] Between 1942 and his death in an automobile accident in 1966,

---

It is unclear whether Hilbert knew of Brouwer's distinction between formal mathematics and metamathematics. (The terms are not Brouwer's, who spoke of 'first-order' and 'second-order' mathematics.) The distinction is made in Brouwer's dissertation of 1907, a work that was available only in Dutch. But Brouwer may have communicated the substance of his idea to Hilbert in conversation.

[d] The concept of choice sequence emerged only gradually in Brouwer's work. In *Brouwer 1913* (§§29 and 31) he briefly mentions choice sequences. But although he admits their validity in *classical* mathematics, it is unclear from the context whether he means to allow them into intuitionism as well; and in any event he in 1913 puts them to no mathematical use. In his review of Schoenflies and Hahn (*Brouwer 1914*) he clearly allows choice sequences into intuitionistic mathematics, although he deals with this topic only briefly. And by *Brouwer 1918* (the paper that inaugurates his decade of research into intuitionistic mathematics) choice sequences have moved to centre stage. Brouwer was to put choice sequences to work in logic as well as in the theory of the continuum, using them (first in *1927c*) to provide intuitionistic counterexamples to the law of the excluded middle. For a survey of subsequent research on choice sequences, see *Troelstra 1977*.

[e] In 1927 Brouwer published five articles on intuitionistic mathematics; in March, 1928 he delivered his two Vienna lectures, and published nothing more on intuitionism until 1942. The reason for this silence is unknown. Brouwer's bitter quarrels with Hilbert over intuitionism (which culminated in Hilbert's curt dismissal of Brouwer from the editorial board of *Mathematische Annalen*), and the fact that his technical discoveries in intuitionistic mathematics seemed to have dried up, undoubtedly contributed to his depression and feeling of isolation at this time. Kreisel conjectures that Brouwer was intellectually shocked by Heyting's formal laws for elementary intuitionistic logic and by Gödel's incompleteness theorems (*Heyting 1930*; *Gödel 1931*); see pp. 43–5 of *Kreisel and Newman 1969* for a discussion of why Brouwer would have found these results disturbing. Although Brouwer's silence began two years before the appearance in print of the first of these articles, *Heyting 1930*, Heyting had communicated his axiomatization to Brouwer by July 1928. Brouwer did not approve of the project of formalization (which, according to Heyting, he regarded as 'a sterile exercise'); but otherwise his reaction was favourable, and he even offered to publish Heyting's article in the *Mathematische Annalen* (*van Stigt 1990*, pp. 289–90).

Brouwer published a number of short articles on intuitionism, totalling about one hundred pages; many of these late articles dealt with solipsism and the creative subject.

Before he lapsed into silence in 1928, Brouwer delivered a pair of lectures in Vienna on 10 and 14 March; the first dealt with intuitionistic philosophy, and the second with the continuum. These two lectures (translated below) are a landmark in Brouwer's writings: the conclusion of his decade of research into intuitionistic mathematics, and the most complete and mature statement of his philosophical views.

The literature on intuitionism, both as a philosophical doctrine and as a foundational research programme, is enormous. Brouwer's own writings are collected in *Brouwer 1975*; important articles by his followers are reprinted in *van Heijenoort 1967* and *Benacerraf and Putnam 1964*. The most extensive study of Brouwer's own thought is *van Stigt 1990*, which contains useful biographical information and translations of important unpublished manuscripts. *Heyting 1956* is still the best short introduction to intuitionistic mathematics. For more recent treatments, see *Dummett 1978* (which contains extensive discussions, from a point of view very different from Brouwer's own, of the philosophical underpinnings of intuitionistic logic) and *Troelstra and van Dalen 1988* (which treats constructivism in general, and which contains a bibliography of the technical literature). For a discussion of Brouwer's technical accomplishments, see *Kreisel and Newman 1969*, which treats his contributions to topology as well as to intuitionism. A complete bibliography of Brouwer's known writings, published and unpublished, is to be found in *van Stigt 1990*.

A. MATHEMATICS, SCIENCE, AND LANGUAGE (*BROUWER 1928a*)

The following lecture was delivered in Vienna on 10 March 1928.[a] It is primarily devoted to intuitionistic philosophy; a companion lecture, delivered four days later and translated in the next selection, is primarily devoted to intuitionistic mathematics.

This lecture is Brouwer's fullest statement of his mature philosophical position, and comes at the end of his most productive decade of research into intuitionistic mathematics. The tone is characteristically oracular: Brouwer had no hankering for the kind of analytical linguistic precision pursued by the Vienna Circle (and indeed thought that language was a most imperfect vehicle for expressing his ideas).

[a] Brouwer's lecture has a historical significance beyond the intrinsic interest of its ideas. In the Viennese audience was Ludwig Wittgenstein, who, after writing the *Tractatus logico-philosophicus*, had abandoned philosophy. He was persuaded by Herbert Feigl to attend Brouwer's lecture. According to Feigl, Wittgenstein was so stimulated that he ended his exile from philosophy and took up the researches that eventually led to the *Philosophical investigations* (*Hacker 1986*, p. 120).

Brouwer's philosophical pronouncements have often been dismissed as cranky mysticism. But whatever their intrinsic worth, they were central to his thought, and some brief hints of the way they influenced his technical work in foundations throughout his career may perhaps be in order.

The philosophy of mathematics in the 1907 dissertation had its roots in more general ideas about mind and language which Brouwer shared with the 'Significs' movement, a school of Dutch philosophy active in Amsterdam in the early decades of the century.[b] Significs (in contrast to the tradition of analytical philosophy) sharply distinguished between language and thought, and attempted to develop a theory of the mental elements (perceptual, emotive, and volitional) underlying human linguistic activity. Language was viewed principally as a tool for influencing the behaviour and the mental states of others: a thesis that the Significs philosophers applied, not only to the emotive language of literature and politics, but also to the seemingly more descriptive language of mathematics and science.

Brouwer added to these doctrines the central idea that mathematics is neither a linguistically expressed theory, nor a body of abstract truths about some objective reality, but rather an *activity*, a creation of the human mind: mathematics is constructed in the individual consciousness from what Brouwer, in *1913*, was to call the 'basic intuition of two-oneness' that is derived from the perception of a move of time.[c]

---

[b] The movement was not officially constituted until 1915, with Brouwer and Gerrit Mannoury as principal movers; but the general philosophical ideas were present in Dutch intellectual circles earlier. See Brouwer's description in *Brouwer 1975*, Vol. i, pp. 465–76.

[c] For a mature statement of this view, see the writings reproduced below: for example, *Brouwer 1952*, p. 141. The following extended quotation from the dissertation of 1907 will convey the flavour of his early views on the construction of mathematics:

'*Mathematics can deal with no other matter than that which it has itself constructed.* In the preceding pages it has been shown for the fundamental parts of mathematics how they can be *built up* from units of perception, by simple juxtaposition, by building sequence of type ω or η, or by building continua, while at every stage in the process complete systems which have been constructed before can be taken as new units.

'In the third chapter it will be explained why no mathematics can exist which has not been intuitively built up in this way, why consequently the only possible foundation of mathematics must be sought in this construction under the obligation carefully to watch which constructions intuition allows and which not, and why any other attempt at such a foundation is condemned to failure.

'Often it is quite simple to construct inside such a structure, independently of how it originated, new structures, as the elements of which we take elements of the original structure or systems of these, arranged in a new way, but bearing in mind their original arrangement. The so-called 'properties' of a system express the possibility of constructing such new systems having a certain connection with the given system.

'And it is exactly this *imbedding* of new systems in a given system that plays an important part in building up mathematics, often in the form of an inquiry into the possibility or impossibility of an imbedding satisfying certain conditions, and in the case of possibility into the ways in which it is possible.

'Examples have been treated above as inquiries into the possibility of imbedding transformation groups satisfying certain conditions *into given systems*. (There we may consider the continuum of group parameters as the system to be imbedded, and the character of that continuum as a continuum of group parameters, as the condition of the imbedding.) And in this form we have given, among other things, substrata of several recent investigations which were intended to shed light upon the foundations of mathematics, and which have mathematical meaning only in the sense of our interpretation—this last on the basis of ideas to be defended in the third chapter, for which the preceding development can be considered as an illustration' (*Brouwer 1907*, conclusion of Ch. 1; trans. A. Heyting).

These two ideas—mathematics as a *construction* and mathematics as an essentially *languageless* activity—were the taproot of Brouwer's intuitionism; a number of interrelated doctrines, both philosophical and mathematical, flow from these initial premisses about the relationship between mind, language, and mathematics.

First, for Brouwer there is a wide gulf between mathematics (a necessarily private mental activity) and its linguistic expression. The linguistic expression is not, in his view essential to mathematics, but is instead merely an adjunct—a crutch to support the memory of the individual during the mind's process of mathematical construction, and a tool for influencing others to carry out the constructions for themselves. With respect to this latter feature, Brouwer stresses that mathematical language is an imperfect medium of communication, and may not succeed in getting the hearer to perform the construction in the desired way.

This conclusion has important implications for his conception of a *mathematical proof*. On his view, a proof is identical with the carrying out of a mathematical construction; but because there is no absolutely secure language for mathematics, it follows that—in contrast to the view both of the logicists and of the Hilbert school—a proof cannot be identified with a fixed sequence of logical relationships among propositions (or among their linguistic representations). The words printed in mathematics texts are to be viewed as a kind of exhortation, instructions to enable the reader to carry out the intended construction for himself; they are not a rigid set of commands, a series of steps leading inexorably from premisses to conclusion.

By the same line of reasoning, the *paradoxes* become a purely linguistic matter, and thus a problem only for those mathematicians who fail to make Brouwer's sharp distinction between mathematics and language. Brouwer regarded the paradoxes as mere concatenations of words: words that do not correspond to any mathematical construction that can be executed by the creating subject. As he says, one simply sees that the construction no longer *goes*; logic and language are beside the point.[d] (For similar reasons, *Brouwer 1913* was critical of Zermelo's axioms: in particular, of the Axiom of Choice and of *Aussonderung*. The objection was not that they lead to—linguistic—contradiction, but that they are mere words detached from any actual mathematical construction.)

Theoretical logic for Brouwer becomes a mathematical activity, namely, the study, by means of mathematics, of certain regularities in the linguistic scaffolding of mathematics. Logic is thus a kind of applied mathematics (a subject

---

[d] Thus Brouwer, in *1907*, imagining a dialogue with a logician, writes:

> '"But", the logician will retort, "it might have happened that in the course of these reasonings a contradiction turned up between the newly deduced relations and those that had been kept in store. This contradiction, to be sure, will be observed as a logical figure, and this observation will be based upon the *principium contradictionis*." To this we can reply: "The words of your mathematical demonstration merely accompany a mathematical *construction* that is effected without words. At the point where you announce the contradiction, I simply perceive that the construction no longer *goes*, that the required structure cannot be imbedded in the given basic structure. And when I make this observation, I do not think of a *principium contradictionis*"' (*Brouwer 1975*, Vol. i, p. 73; trans. A. Heyting).

for which, incidentally, Brouwer had a certain distaste), and is dependent on mathematics both for its linguistic subject-matter, and for its mode of investigation. Brouwer's position is thus in direct contradiction to that of the logicists: for him, *logic* rests on a foundation of *mathematics*, rather than the other way around.[e] This view of logic underlies Brouwer's relative lack of interest in the subject; it is also the basis of his claim to have anticipated Hilbert's distinction between formal mathematics and metamathematics.

It follows from the foregoing doctrines that logic cannot be fully exhausted by the syntactic rules of any *formal system*. For the number of permissible mathematical constructions (and the regularities in their linguistic expression) is not eternally fixed, but is free to grow and to change with human experience;

---

[e] Indeed, the basic intuition of two-oneness underlies, not only pure mathematics and theoretical logic, but also many scarcely conscious everyday mental processes, including the mental organization of the objects of the external world; so that in this sense mathematics is broader than any of the special sciences, and does not rest upon any foundation more fundamental than itself.

Brouwer described the relationship between logic and mathematics in his dissertation of 1907 as follows:

'People try by means of sounds and symbols to originate in other people copies of mathematical constructions and reasonings which they have made themselves; by the same means they try to aid their own memory. In this way the *mathematical language* comes into being, and as its special case the *language of logical reasoning*.[1]

⟦Brouwer's footnote reads: 'Even in domains of mathematics where no relations of whole and part enter, *the relations which were in the mind are often transformed into relations of whole and part*, when they must be communicated verbally to other people; hereby the usual language of mathematics in general is imbued with that of logical reasoning. However, this fact is due only to the centuries-old tradition of logical terms in language, in connection with its limited vocabulary.'⟧

'With which mathematical notions a spoken or written symbol will be made to correspond, this choice will take into account as economically as possible the most common mathematical systems and methods of reasoning; therefore it will in general differ according to the milieu. In particular, the answer to the question which domains of mathematics will be accompanied by a language, not only among professional mathematicians, but also in daily life, will depend for every nation anew upon the question, which domains of mathematics have found most applications to the guidance of action or as a means of understanding about action.

'Therefore it is easily conceivable that, given the same organization of the human intellect and consequently the same mathematics, a different language would have been formed, into which the language of logical reasoning, well known to us, would not fit. Probably there are still peoples, living isolated from our culture, for whom this is actually the case. And no more is it excluded that in a later stage of development the logical reasonings will lose their present position in the languages of the cultured peoples.

'Man, inclined to take a mathematical view of everything, has also applied this bias to mathematical language, and in former centuries exclusively to the language of logical reasonings: the science arising from this activity is *theoretical logic*.

'It is only in the last twenty years (though the earliest traces go back to Leibniz) that people have started looking in the same way at *mathematical language in general*; this is the content of *logistic*, in so far as it is studied without overrating its value.[1]

⟦Brouwer's footnote reads: 'See p. 159 seq. ⟦in *Brouwer 1907*⟧. The systems of logistic developed until now consider a mathematical language which uses excessively the words of theoretical logic, and which sometimes, where excessive use leads to illicit use, gave rise to error.'⟧

'We infer that theoretical logic as well as logistic are *empirical sciences* and that they *apply* mathematics; consequently they can yield no information whatsoever on the organization of the human intellect; there would be better reasons to reckon them under *ethnography* than under *psychology*.

'And the language of logical reasonings is no more an *application of theoretical logic* (for that matter, if this were the case, of which science would the language of theoretical logic itself be an application?) than the human body is an application of anatomy, (*Brouwer 1975*, Vol. i, pp. 73–4; trans. A. Heyting).

formal systems, because they are static, are in principle unable to capture the dynamic and open-ended realm of creative mathematical activity.[f]

Most famously, these philosophical and mathematical doctrines led Brouwer to reject the law of the excluded middle. Since, for Brouwer, the truth or falsity of a mathematical proposition is determined, not by its correspondence to some objective mathematical reality, but by a construction (a proof), it becomes necessary to reinterpret the logical connectives in the well-known intuitionistic manner; the law of the excluded middle then in effect says that every mathematical problem can either be shown by a suitable construction to be true, or shown by a suitable construction to lead to a contradiction. Brouwer was thus led to identify the law of the excluded middle with the claim (promoted, for example, in *Hilbert 1900b*) that every mathematical problem is in principle solvable: a claim which Brouwer saw no reason to accept (although he was unable to find a convincing example of a mathematical problem that is in principle unsolvable).[g] In general, Brouwer paid little attention to the problem of developing an intuitionistic logic before *Brouwer 1925*, and even thereafter his attitude remained cool. Certainly he continued to regard logic as a weak tool for achieving mathematical results. He not only insisted (as one would expect) that a proof of *non-contradiction* does not allow one to infer *existence*; but further that the conclusion of *any* valid inference of intuitionistic logic must still be confirmed by a direct, mental construction, and may always turn out to be incorrect.

The lecture translated below presents a number of difficulties. The prose style

---

[f] Heyting shared Brouwer's view of the limitations of formal systems, and indeed began his *1930* by saying,

> 'Intuitionist mathematics is thought-activity, and all language, including formal language, is only a means of communication. It is in principle impossible to construct a system of formulae that would be equivalent to intuitionist mathematics, for the possibilities in human thinking cannot be reduced to a finite number of predetermined rules.'

Although Heyting did not regard his formal presentation of intuitionistic logic as either complete or final, he gradually came to acknowledge the 'great advantages' of formalization, and even its 'necessity' for a clear exposition of the theory of choice sequences (*van Stigt 1990*, p. 290). See *Heyting 1956* for further discussion of the evolving intuitionistic attitude towards formalization.

[g] Brouwer rejected the law of the excluded middle for the first time in *Brouwer 1908*. The relevant passage is the following (*Brouwer 1975*, Vol. i, p. 109; trans. A. Heyting):

> 'Now consider the *principium tertii exclusi*: It claims that every supposition is either true or false; in mathematics this means that for every supposed imbedding of a system into another, satisfying certain given conditions, we can either accomplish such an imbedding by a construction, or we can arrive by a construction at the arrestment of the process which would lead to the imbedding. It follows that the question of the validity of the *principium tertii exclusi* is equivalent to the question *whether unsolvable mathematical problems can exist*. There is not a shred of proof for the conviction, which has sometimes been put forward,[1] that there exist no unsolvable mathematical problems.
>
> ‖Brouwer's footnote 1 reads: 'See D. Hilbert, *Mathematische Probleme*, Göttinger Nachr. 1900, pp. 253–297. Schoenflies ‖*Schoenflies 1908*, Ch. 1, §7‖ also maintains unconditionally the method of indirect proof, which he erroneously considers as depending only upon the *principium contradictionis*.'‖
>
> 'In so far as only finite discrete systems are introduced, the investigation whether an imbedding is possible or not can always be carried out and admits a definite result, so in this case the *principium tertii exclusi* is reliable as a principle of reasoning.[2]
>
> ‖Brouwer's footnote 2 reads: 'This investigation can even in every case be made by a machine or by a trained animal; it does not require the basic intuition of mathematics, living in a human mind. But with respect to questions regarding infinite sets the basic intuition is indispensable; by disregarding this fact, Peano and Russell, Cantor and Bernstein have fallen into errors.'‖

is turgid, and at times obscure. Some of the obscurity may have been unavoidable (Brouwer was dealing with ideas that, as he stresses, are difficult to express in words). In addition, German was not his native language, and he sometimes expressed himself clumsily in it. Students of Brouwer's philosophy are urged to bear these facts in mind, and not to venture a novel interpretation until they have carefully consulted the original German text.

In a pamphlet published in 1933 and entitled 'Willen, weten, spreken' ('Volition, knowledge, language'), Brouwer reprinted sections I and II of the present lecture in Dutch translation; in the Dutch version he incorporated some significant additions. The additions, as translated by Arend Heyting (*Brouwer 1975*, Vol. i, pp. 598–9), are given in footnotes below. (The full text of the Dutch version is translated by Walter van Stigt in *van Stigt 1990*, pp. 418–31.)

Brouwer's technical terminology in Part III has been translated as follows: *fliehende Eigenschaft* = fleeing property; *Lösungszahl* = critical number; *Pendelzahl* = oscillatory shrinking number; *Näherungszahl* = kernel. These translations reflect Brouwer's own chosen terminology in his subsequent English writings: see (*Brouwer 1975*, Vol. i, pp. 622–7). The translations *Oberzahl* = up-number; *Unterzahl* = down-number are taken from Heyting's translation of an article originally written by Brouwer in Dutch: see (*Brouwer 1975*, Vol. i, p. 444). *Mathematische Betrachtung* has throughout been translated as 'mathematical contemplation'; but the German noun also carries the senses: reflection, viewing, observation, consideration.

The translation of *Brouwer 1928a* is by William Ewald; references should be to the section numbers, which appeared in the original edition.

I[h]

Mathematics, science, and language are the chief functions of the activity of mankind, by means of which it dominates nature and maintains order in its midst. These functions have their origin in three modes of operation of the will-to-life of the individual man: 1. mathematical contemplation ‖Betrachtung‖, 2. mathematical abstraction, and 3. the imposition of the will by means of sound.[i]

1. *Mathematical contemplation* arises in two phases as an act of the will in the service of the instinct for self-preservation of the individual man: the phase

Incidentally, Brouwer in 1907 submitted a list of twenty-one statements to be defended together with his doctoral thesis. The last statement on the list held that Hilbert's conviction, expressed in *Hilbert 1900b*, that every well-formed mathematical problem is in principle solvable, is unfounded (*Brouwer 1975*, Vol. i, p. 101). But Brouwer does not seem at the time to have made the connection to the law of the excluded middle.

[h] ‖In the 1933 Dutch version of this lecture, this section was given the heading, 'Reflection on mathematics, science, and language'.‖

[i] ‖The 1933 version continues: 'All three are subject to free volition with respect to extent and to modality.'‖

of the *temporal attitude* ‖*zeitlichen Einstellung*‖, and the phase of the *causal attitude* ‖*kausalen Einstellung*‖. The former is nothing other than the intellectual ur-phenomenon[j] of the falling apart of a life-moment into two qualitatively distinct things; one senses the one thing as yielding to the other while nevertheless being maintained in the act of memory. At the same time, the split life-moment is separated from the I and transferred to a world of its own to be designated as the world of perception ‖Anschauungswelt‖. The temporal twoness that has arisen through the temporal attitude—or the two-membered temporal appearance-sequence—can then itself be conceived as one of the members of a new twoness, thereby creating the temporal threeness, and so on. In this way the *temporal appearance-sequence of arbitrary multiplicity* arises by means of the self-unfolding of the intellectual ur-phenomenon. The *causal attitude* then consists in the act of the will of 'identifying' different temporal appearance-sequences that extend over the past and the future. In this way there arises a common substrate of these identified sequences, to be called a *causal sequence*. A special case of the causal attitude is the mental formation ‖gedankliche Bildung‖ of objects, i.e. of enduring (simple or compound) things of the world of perception; at the same time, the world of perception is itself stabilized in this way.[k] As I have said, the two stages of mathematical contemplation are not at all passive attitudes, but are on the contrary acts of the will; everybody can produce the inner experience that one can, as a matter of free choice, either dream oneself into a state without a temporal attitude and without a division between the I and the world of perception, or else bring about the latter division by exercising one's own powers, and thus call forth the condensation of individual things in the world of perception. And the equating of different temporal sequences (which is never absolutely unavoidable) is just as arbitrary, just as much a matter of free choice.

The sole justification of mathematical contemplation is the 'usefulness' ‖'Zweckmäßigkeit'‖ of the '*mathematical action*' that follows from it—by which I understand the following. The causal attitude puts men into a position to bring about *indirectly*, through cool calculation, from a given appearance-sequence, a later appearance, called a *goal*, that is instinctively desired, but that cannot be brought about by a direct impulse; men do this by calling forth, from the sequence, an earlier appearance (called a *means*), which perhaps is not desirable in itself, and which then carries along with it the desired appearance as a consequence.[l]

---

[j] ‖Urphänomen. The prefix 'ur' conveys the sense of 'first, original, primordial, most primitive, most elemental.'‖

[k] ‖The 1933 version reads: 'As a consequence of the appearance of objects (including one's own person and one's fellow-creatures) the world of perception is itself more-or-less stabilized. Objects differ widely in their degree of *egoicity*, i.e. the degree to which the desire for their stability is accepted as a guiding force for the free will.'‖

[l] ‖The 1933 version adds: 'The performance of mathematical actions and the choice of the ends they are to serve are also subject to free volition. In particular the degree of egoicity of objects is expressed in the choice of ends. It is even an essential condition for any initiative of the will that the order of this degree is clear; where it is blurred out, dreaming away can only be interrupted by automatic routine action.'‖

Of course, a causal sequence possesses no further existence than as the correlate of an attitude of the human will that calls forth a mathematical action, and there is no such thing as the existence of a causal connection of the world independent of men. On the contrary, the so-called causal connection of the world is a power of thought, directed outwards, in the service of a dark function of the will of man, to which the world more or less defencelessly submits itself, just as the serpent makes its prey defenceless through its hypnotizing glance, or the squid by squirting its ink.

The causal attitude has the consequence moreover that already at a low level of culture man, in order to stabilize the domain that he causally influences, seeks to create a *sphere of order* about himself which is subject to his control; in this sphere he first isolates the causal sequences that are serviceable to him (i.e. protects them from disturbing secondary appearances) and then brings about new causal sequences—doing this both through the material construction of new, enduring objects and instruments, and also through the more-or-less organized subjection of the will of his fellow men[m] to his own will.

2. However, the full development of the mechanism of mathematical activity is only made possible at higher levels of culture—specifically, by mathematical *abstraction*, by means of which one deprives twoness of its substantial content ‖dinglichen Inhaltes‖ and retains only the empty form, the common substrate of all twonesses. This common substrate of all twonesses is the *ur-intuition of mathematics*, and its self-unfolding introduces (*inter alia*) the infinite as a conceptual reality, specifically (in a manner that shall not here be discussed in greater detail) it furnishes, first, the totality of natural numbers, then the totality of the real numbers, and finally the whole of pure mathematics.

The effectiveness of mathematical abstraction rests on the fact that many causal sequences are considerably easier to master if one *projects* them on to *subsystems* of such pure-mathematical systems, i.e. embeds their contentless abstractions, as subsystems, into such more extended pure-mathematical systems. In this way the relationships existing within the extended system can also be used to obtain an overview of the narrower system—which often brings about a drastic simplification. In this manner *scientific theories* arise, in which, besides the elements of the causal sequences, the extended system of pure mathematics that plays a centralizing role in the overview appears as a *hypothesis*.[n] In particular, certain scientific theories are designated as *theories of the exact sciences*; first, these theories apply to especially stable causal sequences (whether they be exclusively *observed* as laws of nature, or whether they be technical facts that are *artificially produced*); second, one attempts to use their

[m] ‖In a footnote to the published version of the second lecture in this series, *Die Struktur des Kontinuums* (*1928b*), Brouwer stated that 'fellow men' should here be changed to 'fellow creatures'.‖

[n] ‖The 1933 version adds: 'For instance, it is a quite essential hypothesis for a mathematical theory about my fellow creatures that each of them possesses a mathematical-scientific mechanism of observation, action, and reflection which is analogous to mine.'‖

hypotheses to achieve a great simplification; and, third, the causal sequences correspond to special values of *numerical parameters*, whose full value-domain belongs to the overarching mathematical system. In particular, in the theories of the exact sciences there occurs the phenomenon of the *heuristic character of scientific hypotheses*; this consists in the fact that, for sequences that were originally introduced as hypothetical, actual causal sequences of the world of perception are later discovered which occupy the same place in the overarching mathematical system.

3. Mathematical contemplation and mathematical action, which at first functioned in the service of the will of the individual man, can now, just like any initially autonomous aggressive or defensive activity, be placed as *labour* in the service of a commanding will—whether it be the individual will of another man, or else the parallel will of a group of men or of the whole of mankind. This occurs either directly, by *suggestion*, (for example, by instilling anxiety or fear, by enticement, or by arousing the imagination) or indirectly by means of *training the reason* ‖*Vernunftdressur*‖, i.e. by influencing the experience of the individual who is to be made serviceable in such a way that a mathematical contemplation is evoked in him which triggers the hope of pleasure (or fear of displeasure) that determines his will-to-work.[o]

Among the mathematical contemplations that are imposed on each man by the parallel will of all mankind, the most important is the presupposition of the hypothetical 'spatio-temporal world' as the common bearer of all the temporal appearance-sequences of all individuals; and next the exact and technical sciences, to the extent that they do not—in the form of industrial secrets—serve special interests.

The principal example of a mathematical contemplation imposed by a limited group of men (who are bound together by for example the state or a profession) is the acknowledgement and preservation of the *organization* of the group, i.e. of the power-grid ‖Stromnetz‖ for the transmission of the will, by means of which, within the group, the individual mathematical contemplations and acts of labour are compelled to occur. This organization of individual human groups is therefore much less stable than are the exact and technical sciences because, first, the organization is never in control of all the external material circumstances that it needs to take into account, and so, to remain effective, it must constantly adapt to the changes in the external material circumstances; and, second, because its effectiveness does not depend just on its organizatorial effectiveness, but also on the *loyalty* and the *contentedness*[1] of the individuals that are subject to it—and these can be only imperfectly acquired and maintained. For especially in the higher ranks loyalty is endangered by the clash of

[o] ‖The 1933 version adds: 'However, such a mathematical theory often becomes unconscious once it has aroused the readiness to action, while nevertheless this readiness remains as an automatic routine.'‖

[1] The discontentedness of the single individuals is destructive to the organization of the group because it brings about the formation of sub-communities, which initiate mathematical contemplations aimed at altering the organization of the larger commmunity.

individual interests with group interests; and contentedness is imperfect in the lower ranks because those lower down the scale, although they in general recognize that certain organizatorial arrangements promote the common wishes and needs of the members of the group, do not recognize that the prevailing organization is the only right one, nor that they themselves have been assigned the right position within it.

Now, the method which the organization adopts for the training of reason is inadequate to maintain loyalty and contentedness in organized human groups, even imperfectly; rather, every organization is compelled to carry out the propaganda of *moral theories* as well—that is, of mathematical contemplations which base the necessity of the existing organization, not just on the egoistic perception of shared goals and needs, but also on moral values for the maintenance of life—i.e. values that prescind from egoistic contemplation. Examples that immediately spring to mind are the moral values of the religious commandments as well as the concepts of fatherland, property, and family—all of which are protected and propagated by the community.

The propaganda of moral values can hardly use the training of reason, so it must resort above all to suggestion, in particular to the arousal of the imagination. Furthermore, the power of the moral values does not rest exclusively on the organized propaganda of the corresponding human group, but also on the quiet action of the mathematical contemplations of single individuals, in whom the moral values enter as a rejection of the egoistical impulses of others.

In organized human groups, at primitive levels of culture and in primitive relationships, the transmission of the will is achieved by a simple gesture; and here the cry is particularly effective. But in matters belonging to the organization of a higher human community the tasks to be imposed are too various and too complicated to be brought about by simple cries. To make the regular causing of these works possible by sounds that request or command, one must subject to a mathematical contemplation the totality of the decrees, objects, and theories which play a role in the mathematical actions demanded from the subservient. *Elementary linguistic signals* are correlated to the elements of the system of pure mathematics which belongs to the scientific theory that grows out of this mathematical contemplation; organized *language* operates with them according to *grammatical rules* that are taken from the same scientific theory—and this allows the great majority of the necessary transmissions of the will in the cultural community to be carried out. Language is therefore absolutely a function of the activity of social man. Even if the individual man in total isolation uses language to prop up his memory, this is only because he must take the sciences and the organization of the community into consideration. And even if non-active transcendent processes are accompanied by language, this is because all human activity is subject to the transcendent influx of the free will.[p]

---

[p] [The 1933 version adds: 'Of course the scientific theories which are at the basis of languages are far from exact. On the contrary, the greater part of the stability and the formal exactness which a language seems to possess by its grammar and its dictionary gets lost in practice because ordinary

II[q]

However, for transmission of the will, and in particular for transmission of the will mediated by speech, there exists neither exactness nor certainty. And this state of affairs remains unaltered if the transmission of the will is concerned with the construction of systems of pure mathematics. *Thus for pure mathematics as well there exists no certain language*, i.e. no language that excludes misunderstandings in conversation and, when it is being used to prop up the memory, protects against errors (i.e. against the confounding of different mathematical entities). This circumstance cannot be remedied, as the *formalistic school* attempts, by submitting mathematical language itself (i.e. the system of signs which serves to evoke purely mathematical constructions in other men) to a mathematical contemplation, revising it so as to give it the preciseness and stability of a material instrument or of a phenomenon of exact science, and then coming to an understanding ‖sich verständigen‖ about it in a language of the second order, a metalanguage ‖Übersprache‖. For, first, in the use of mathematical language this metalanguage can indeed guard against misunderstandings and errors with great probability (because it applies to a surveyable finite set of enduring objects and to the pure mathematics of the finite system that is abstracted therefrom), but in keeping with the essence of language it cannot do so with absolute certainty; second, even if it could do so, the possibility of misunderstanding the constructions of pure mathematics that are indicated by such a precise language would by no means be removed.

The efforts of the formalist school (whose origin, according to what we have just said, can be traced back to the false belief in a magic reach ‖Tragweite‖ of language—or, at least, in a reach that exceeds its character as a means for transmission of the will) can be explained from this point of view as the natural

---

life needs many more elementary notions than the elementary words and associations of words which language offers. On the other hand stability and exactness of the language is not necessary in practice because people are drilled by a common will to an automatism of understanding incomplete sentences.

‘Everything said so far is *reasonable reflection*, i.e. mathematical speculation in which the *content* neither of ends, nor of objects in the world of perception is involved. It is an essential hypothesis for human understanding that the structure of this reasonable reflection is the same for all individuals. Therefore it represents an eminent social value as a means of avoiding confusion in fixing the principles of social organization and in consolidating social life.

‘*Moral reflection* is of a completely different and much more individual nature. It tests the objects of the world of perception and also the mathematical activity itself with respect to their egoicity, and consequently their right to exist as sources of guiding force for the free will. Moral reflection tries to approach the connection between ends to be chosen and the origin and design of our life, which are clear as well as mysterious, by dwelling with tense vigilance on the borderline between dreaming away and perception of time. Causality appears there only ephemerally, and there is no place for mathematical action. Its language is inexact and unstable, more suggestive than adequate, “it ought not to be taken literally”. It assumes a “prophetic” character in the rare cases where an inspiration is received which is transferable or which conveys a tendency to collective action.

‘Still, moral reflection is not without social importance, firstly because, though practised in complete solitude, it induces a feeling for social justice and readiness to struggle against evil, and besides because from its prophetical language there crystallize now and then the most useful moral theories.’‖

[q] ‖In the 1933 version this section received the heading: ‘Criticism of the attempts to purge mathematics by linguistic means’.‖

consequence of a much older, more primal, more consequential and more deeply rooted error, namely, the reckless trust in *classical logic*. This trust arose as follows: already in antiquity man possessed a very perfect ‖vollkommen‖ language (i.e. one that practically excluded misunderstandings) for the mathematical contemplation of finite groups of things of the objective spatio-temporal world (which things had always been conceived as *unitary* and *persisting*). For this language there are certain forms of transition from correct ‖zutreffenden‖ statements (i.e. statements that indicate actual mathematical contemplations) to other correct statements; these forms of transition were designated as the laws of *identity*, *contradiction*, the *excluded middle*, and the *syllogism*, and were gathered together under the name of *logical principles*. When one applied these principles purely linguistically, i.e. derived linguistic expressions from other linguistic expressions with their help, without thinking about the mathematical contemplations indicated by these statements, it turned out that *the principles proved themselves*, i.e. it was found that every statement obtained in this way was capable of triggering an actual mathematical contemplation which turned out to be practically 'identical' for all linguistically-raised men in the objective spatio-temporal world.

Furthermore, the logical principles held good also when one applied them, in a checkable fashion, quite generally to the language of science or to occurrences in other parts of practical life—at any rate, so long as one treated only such occurrences as were ruled by laws of nature in whose solidity one had learnt to believe. And then one came to trust statements that had been derived by means of the logical principles even when they were not susceptible to a direct check. In particular, this trust was given to the law of the excluded middle—and that in the extended form, according to which an earlier occurrence is presumed to have happened, not just on grounds of absurdity, but also on the grounds of the practical impossibility of finding another explanation for an established fact. And on this trust are based, not just theoretical sciences like paleontology and cosmogony, but also governmental institutions like the rules of procedure for a criminal trial. Nevertheless it happened that, by applying logical considerations, one arrived at false results in affairs of the world of perception; but an experience of this sort led only to a suitable revision of the underlying facts or natural laws, and never to a revocation of the trust in the logical principles.

But the practical reliability of the logical principles (i.e. of the law for combining sentences in the language of finite mathematics) is in matters concerning the world of perception merely a consequence of the following more general fact. Mankind successfully controls the great majority of observable objects and mechanisms of the world of perception as they occur within extended complexes of facts and occurrences because it considers and treats the system of the states of these objects and mechanisms in the spatio-temporal world as part of a finite, discrete system with finitely many linkage-relations between the elements. In other words, the practical reliability of the logical principles rests on the fact that much of the world of perception, in its finite organization, shows more loyalty and contentedness than mankind itself. Man has for ages been blind to this sober interpretation because he did not recognize that words are nothing but a means of transmission for the will, and regarded them, in virtue of a naïve superstition, as a means of indicating ‖Andeutungsmittel‖ fetish-like '*concepts*'. These 'concepts'—as well as the linkages existing between them—were supposed to possess an existence independent of the causal attitude of men, and the logical principles were supposed to represent *a priori* laws governing the concepts and their linkages. In consequence, the opinion prevailed that concept-linkages ‖Begriffsverknüpfungen‖ which were derived from undeniable axioms (i.e. from concept-linkages which correspond to the establishment ‖Konstatierung‖ of undeniable facts or laws of nature) with the help of the logical principles (possibly by means of reducing their opposite to absurdity), if they themselves yielded checkable statements about the world of perception, could then always survive this check; and if they did not, then they were to be regarded as '*ideal truths*', and were taken to be just as admissible. 'Ideal truths' of this sort were then derived by the philosophers with confident zeal for centuries. When contradictions cropped up now and then as uncomfortable side-effects, thereby provoking doubt about the correctness of these reasonings ‖Entwicklungen‖, this doubt was never directed against the reliability of the logical principles, but always against the undeniability of the axioms, i.e. the concept-linkages which underlay the reasonings. And many an axiom had to be rejected or modified precisely because of the contradictions that arose among the ideal truths that followed from the axiom.

Finally, in imitation of the philosophers, the mathematicians too took the logical principles of the language of finite mathematics and applied them without scruple in the pure-mathematical study of infinite systems. In this way, statements of 'ideal truths' were also derived for the mathematics of infinite systems (and of the sets that appear in set theory and that are created by means of the comprehension axiom), and mathematicians took these statements to be more than empty words. Until here too (namely, after the introduction of set theory) contradictions arose—and contradictions of such a sort, that they could not simply be eradicated by a suitable revision of the axioms. These contradictions (which were far more startling in mathematics than in philosophy) were at first attacked with the formalistic efforts mentioned above. In particular, the fundamental axiomatic linkages of mathematical concepts and the forms of transition between the various linkages of mathematical concepts (in particular to the extent that they are involved with the creation of sets and the admission of elements to sets) are here subjected to a thoroughgoing analysis and revision, into which revision, of course, the linguistic action ‖Wirkung‖ of the logical principles is also drawn—and all the while the belief in a sense of language independent of the transmission of the will is maintained. But the *sense* of the mathematical concepts and concept-linkages is not discussed any more closely; the ultimate goal of these efforts (to which one has not even come close) is a *consistent reformulation of the language of mathematics*, which shall moreover encompass the entire edifice ‖Lehrgebäude‖ of previous mathematics, with the exception of a few small amputations to take care of the contradictions.

III

In contrast, *intuitionism* brings the extra-linguistic existence of mathematics to consciousness, and, in order on this basis to examine the correctness of previous mathematics, investigates first the extent to which the logical principles that have played such a leading role in the construction of this mathematics can also function, in the mathematics of the infinite, as a practically reliable means of transition between constructions of pure mathematics. This investigation yields a positive result for the principles of identity, contradiction, and the syllogism, but a negative one for the principle of the excluded middle—i.e. it turns out that in general no mathematical reality corresponds to the statements of the latter principle and to the inferences that are based upon it.

In order to explain this with a few examples, let us designate as a *fleeing property*[r] a property such that, for any given natural number, either the existence or the absurdity of the property can be proved, while one can neither calculate a natural number that possesses the property, nor prove the absurdity of the property for all natural numbers. By the *critical-number* λ_f of a fleeing property f we shall understand the (hypothetical) smallest natural number that possesses the property; by an *up-number* (a *down-number*) of f, a number that is not less than (less than) the critical-number. One sees at once that for an arbitrary fleeing property every natural number is recognizably either an up-number or a down-number; in the former case, the fleeing property is no longer fleeing. We call the fleeing property f *parity-free* if one can prove its absurdity neither for the positive nor for the negative natural numbers.[s] We define the *binary oscillatory shrinking number* p_f *belonging to the parity-free fleeing property* f to be the real number that is the limit of the convergent sequence $a_1, a_2, \ldots$, where a_ν equals $(-1/2)^\nu$ if ν is a down-number of f, and $(-1/2)^{\lambda_f}$ if ν is an up-number of f. This binary oscillatory shrinking number is neither equal to zero, nor different from it—in violation of the principle of the excluded middle.[t] If we understand a *non-positive* real number to be a real number that cannot possibly be positive, then the binary oscillatory shrinking number is neither positive nor non-positive, in violation of the principle of the excluded middle. If moreover we call both the positive and the non-positive numbers *comparable with zero*, and the real numbers that cannot possibly be compared with zero *incomparable with zero*, then the binary oscillatory shrinking number is neither comparable nor incomparable with zero, in violation of the principle of the excluded middle. And if we call a real number *g rational*

---

[r] ‖See the Introductory Note for an explanation of the translation of Brouwer's technical terminology in the paragraphs that follow.‖

[s] ‖In a footnote to the published version of the second lecture in this series, *Die Struktur des Kontinuums* (*1928b*), Brouwer stated that 'positive' and 'negative' should here be changed to 'even' and 'odd'.‖

[t] ‖Brouwer is here and in the following examples using the *weak* sense of negation, so that the first clause of this sentence should be read as saying that 'We neither have a proof that the dual oscillatory shrinking number is equal to zero, nor a proof that it is different from it.' The strong, mathematical negation Brouwer expresses by words like 'impossible' or 'absurd'.‖

if it is either equal to zero, or if it is possible to determine two positive or negative whole numbers p and q such that $g = p/q$, and *irrational* if the assumption of the rationality of g can be reduced to absurdity, then the above-mentioned binary oscillatory shrinking number is neither rational nor irrational, in violation of the principle of the excluded middle.

If we define the *binary kernel n_f belonging to the parity-free fleeing property f* to be the real number that is the limit of the convergent sequence b_1, b_2, . . ., where b_ν equals $(1/2)^\nu$ if ν is a down-number of f, and equals $(1/2)^{\lambda_f}$ if ν is an up-number of f, and if in the Euclidean plane with a rectangular coordinate system we draw a straight line l through the points $(1, p_f)$ and $(-1, n_f)$, then, first, the x-axis and l are not parallel, although their being parallel is not absurd; second, they do not coincide, although their coincidence is not absurd; third, they do not intersect, although their intersection is not absurd.

The x-axis and l are neither parallel, nor coincident, nor intersecting; so the theorem resting on the principle of the excluded middle that two straight lines in the Euclidean plane are either parallel or coincide or intersect turns out to be invalid.

If the parity-free character of f should be lost, then the validity of the principle of the excluded middle would return either for intersection or for being parallel. But only when f completely loses its character as a fleeing property does the principle once again hold for the three properties of coincidence, intersection, and being parallel.

Let us in the Euclidean plane with a rectangular coordinate system consider the unit square q with the corner-points (0, 0), (0, 1), (1, 0), and (1, 1). Let us designate the square surface determined by q as Q, and the point (p_f, p_f) as P. Then P does not lie on q, although the coincidence of P with q is not absurd; moreover, P does not belong to Q, although the membership of P in Q is not absurd. Finally, P does not belong to q nor to the interior of q nor to the exterior of q; so the Jordan Curve Theorem, which rests on the law of the excluded middle, and which says that a simple closed curve divides the plane into two regions in such a way that every point of the plane belongs either to the curve or to one of the regions, also turns out to be invalid.

Let us consider the infinite series with positive terms $b_1 + b_2 + b_3 + \ldots$ where the b_ν have the same meaning as above. This series does not converge, although its convergence is not absurd; by the same token, it does not diverge, although its divergence is not absurd. So the theorem, resting on the law of the excluded middle, which says that every infinite series with positive terms is either convergent or divergent turns out to be invalid. But one of the most important convergence criteria in the theory of infinite series, Kummer's convergence criterion, rests on this theorem, or one essentially equivalent to it. And in fact counterexamples show that this criterion cannot be maintained in the face of the intuitionistic critique. However the existence of roots itself is guaranteed by new intuitionistic proofs.

Let us consider the algebraic equation $x^3 - 3x + 2b^3 = 0$ where $b = 1 + p_f$. The discriminant of this equation equals $-108(1 - b^6)$, and is therefore neither equal to zero, nor different from it. So the second Gaussian proof of the existence of a root is not applicable to this algebraic equation. All other classical proofs of the existence of roots become invalid in the light of the intuitionistic critique.

These examples will make it clear that intuitionism has far-reaching consequences for mathematics. In fact, if the intuitionistic insights prevail, then considerable portions of the previous mathematical edifice must collapse, and new portions must be erected in an utterly new style. And the parts that remain stand in need of thorough reconstruction.

But we will avoid making further excursions into higher mathematics, and instead merely make a few remarks about basic principles. The first is that, along with the principle of the excluded middle, indirect proof in its general form (i.e. the derivation of a property by the *reductio ad absurdum* of its opposite) becomes invalid. For the oscillatory shrinking number p_f mentioned above is not rational, although its irrationality is absurd, and is not comparable with zero, although its incomparability with zero is absurd. But it is interesting that for *negative properties* (i.e. properties which themselves express an absurdity) *the method of indirect proof remains in force undiminished.* For intuitionistic mathematics contains the theorem, that *absurdity of the absurdity of the absurdity is equivalent to absurdity*, so that an arbitrary non-disappearing finite sequence of absurdity-predicates, 'absurdity of the absurdity of . . . of the absurdity', which in previous mathematics expressed either correctness or absurdity, is in intuitionistic mathematics equivalent either with absurdity, or with the absurdity of the absurdity.

Finally we remark that the principle of the excluded middle in intuitionistic mathematics, although it is not *correct*, nevertheless, if one presupposes it exclusively for *finite species* of properties, is *consistent*—which, first, explains why the errors of previous mathematics could hold their ground for so long, and, second, can count as an encouraging circumstance for the formalistic efforts. For on the basis of the intuitionistic insights it is possible to derive, not just *correct theories* that can be developed independently of the principle of the excluded middle, but also (even using this principle, with the constraints mentioned above) *non-contradictory* theories, which encompass a much larger part of previous mathematics than correct theories do. So a suitable mechanization of the language of this intuitionistically non-contradictory mathematics would furnish precisely what the formalistic school has taken as its goal.

On the other hand, the simultaneous statement of the principle of the excluded middle for *arbitrary species* of properties can very well be contradictory. For instance, the following statement can be shown to be contradictory: All real numbers are either rational or irrational. In view of this fact, the construction of the consistent formalistic linguistic edifice will still demand the very greatest care and caution.

B. THE STRUCTURE OF THE CONTINUUM (*BROUWER 1928b*)

The following lecture, the companion to the preceding selection, was delivered in Vienna on 14 March 1928; Brouwer was to publish nothing further on intuitionism until 1942. The lecture is thus his last word at the end of a full decade of research into the intuitionistic continuum.[a] Brouwer here reviews the way in which the discovery of non-Euclidean geometry and of transfinite set theory had created epistemological difficulties in the theory of the continuum. He then proceeds to describe the intuitionistic continuum and the fundamental ways in which it differs from the traditional conception of the real line; in the process, he shows how classical concepts like *density*, *compactness*, and *order* split into several related but distinct intuitionistic concepts.

The translation of Brouwer's technical vocabulary is as described above in the Introductory Note to *Brouwer 1928a*. The German term *reduziertes Kontinuum*—the continuum of *lawlike* Cauchy sequences—has here been translated as 'reduced continuum'. In some of his later writings, Brouwer used the misleading phrase 'classical continuum' to refer to the same notion. (See for example *van Heijenoort 1967*, p. 342 or *Brouwer 1952*, below.) The term *überabzählbar* has been translated as 'more-than-denumerably many'; the translation 'non-denumerable' would clash with Brouwer's special sense of negation. (Incidentally, Brouwer, in his set-theoretical writings, drew his own idiosyncratic distinctions between the concepts *abzählbar*, *aufzählbar*, *auszählbar*, *durchzählbar*, *nachzählbar*, *zählbar*, and the like; see for example *Brouwer 1925*.)

The translation of *Brouwer 1928b* is by William Ewald; references should be to the section numbers, which appeared in the original edition.

I

Since antiquity, people have regarded the arithmetical or geometric continuum as something given: but they were nevertheless far from being clear or in agreement about the microscopic content of this continuum. Thus, until a few centuries ago, one did not quite believe in the existence of irrational numbers in the number-continuum, and the existence of transcendental numbers was corroborated only in the nineteenth century. This uncertainty about the precise content of the continuum did not prevent the educated layman of the nineteenth century from finding a certain peace in the conception of Kant and Schopenhauer, according to which the continuum (just like, moreover, the whole of mathematics) is regarded as a pure intuition *a priori*, and therefore as independent of experience, as having an exact and unambiguous existence. Only in the

[a] Choice sequences—the central concept in Brouwer's theory of the continuum—were first introduced in *Brouwer 1918*; see the discussion above, p. 1169, footnote d.

second half of the nineteenth century was this peace disturbed by three innovations in the epistemological situation:

1. The insight emerged that the continuous space in which one embedded the appearances of the objective world was subject to Euclidean laws only from practical considerations of simplicity, and that the non-Euclidean geometry discovered by Lobachevsky and Bolyai, and the non-Archimedean geometry introduced by Veronese, could provide just as consistent a description of the appearances in question as Euclidean geometry. The initial rejection of these geometries was completely overcome by their arithmetization at the hands respectively of Riemann, Beltrami, Cayley, and Klein, and of Levi-Cività and Hahn. Thus arose the peculiar situation that the non-Archimedean continuum, which had shown itself to be just as adequate as the Archimedean to the *a priori* demands to be placed on the continuum, could only plausibly be made actual ‖verwirklicht‖ with the help of the latter—so that the doubt about the *a priori* necessity of the Archimedean continuum had to be grounded in precisely the *a priori* consistency of this continuum itself.

2. In the set theory that had meanwhile been discovered by Cantor the continuum appeared in a form that rested upon logical operations, and that was therefore fundamentally different from the form that was given by pure intuition *a priori*.

3. The successes of the axiomatic method in studying the foundations of geometry (a method that was inspired precisely by the discovery of non-Euclidean and non-Archimedean geometry) led repeatedly to the expectation that the same method could also be successful in the analysis of the structure of the continuum. For the results of the axiomatic investigations in the realm of geometry were usually so formulated that the consistency of the geometric theories was reduced to that of the number-continuum;[1] with the result that the proof of the consistency of the latter was presented as a urgent task.

In order in this situation to regain certainty for the introduction of the continuum (and of mathematical figures in general) two methods were applied at first:

1. The *formalistic* (*Dedekind*, *Peano*, *Russell*, *Zermelo*, *Hilbert*): this method completely renounced any geometric or arithmetical intuition and restricted its object of investigation to mathematical language, which it attempted to regulate in such a way that the linguistic figure of contradiction would be excluded. It therefore seeks as a consistent sequence of natural numbers (or a consistent continuum) a linguistically consistent theory which exhibits a sufficient kinship with the language of the theory of the natural numbers (or of the continuum), which had previously counted as a reasonable theory. This 'sufficiency of kinship' is of course in the end a matter of taste.

---

[1] To be sure, here it was tacitly presupposed that, from the fact that the existence of A carries with it the existence of B, it follows that the consistency of A also encompasses the consistency of B. This presupposition is false from the intuitionistic standpoint. ‖Heyting interprets this opaque footnote as follows. 'Probably Brouwer means the following: Let A denote the conjunction of the axioms of the arithmetic of real numbers, B the conjunction of the axioms of Euclidean geometry. From any model of A we can construct a model of B. It does not follow that we can derive a formal consistency proof of B from any formal consistency proof of A' (*Brouwer 1975*, Vol. i, p. 599).‖

2. The *old-intuitionistic* (*Poincaré*, *Borel*): this method denies that the three mentioned innovations have any meaning for the theory of the natural numbers, and upholds the conception of the theory of the natural numbers as a collection of synthetic judgements *a priori*. Furthermore, it holds that the consistency proofs of the formalists presuppose the theory of complete induction, and therefore the core of the theory of the natural numbers.[2] The old-intuitionist school uses essentially classical logic for the development of the mathematics that goes beyond the natural numbers: indeed, the (Archimedean) continuum is constructed by means of the comprehension axiom in the first place as the species of the species of 'coinciding' ⟦'zusammengehöriger'⟧ subspecies of the rational numbers; or as the species of the species of 'coinciding' convergent fundamental sequences of rational numbers; or finally as the species of Dedekind cuts of the rational numbers. In these manufacturings ⟦Erzeugungen⟧ of the continuum (which are regarded as equivalent to one another) an extra-mathematical creative power (i.e. a power that goes beyond the constructive) must already be ascribed to logic, because, of the three named species that represent the continuum, in each case only a 'denumerably-unfinished' ⟦'abzählbar-unfertige'⟧ subspecies can be constructively produced by mathematical thought-operations; but such a subspecies is inadequate for all those mathematical theories that employ a concept of measure ⟦Inhalt⟧, for the denumerably-unfinished subspecies of the continuum (or of n-dimensional space) possess the linear (or n-dimensional) measure zero.

The faith in theoretical logic thus almost plays an even greater role in the old-intuitionist method than in the formalist, for the latter draws all the logical laws into the sphere of its linguistic investigations, while the old intuitionists observe a certain caution (mistrust against 'impredicative' definitions) at most in the *application* of these laws (which in themselves are regarded as given).

Of course, in the old-intuitionist method logic is also formally applied in the language of the theory of the natural numbers; but here it is only conceived as an aid to prop up the memory in the non-linguistic, step-by-step coming into being ⟦gliedweisen Zustandekommen⟧ of a complicated inner intuition or of a complicated synthetic judgement *a priori*.

Intuitionism now raises the following objections to the above introductions of the (Archimedean) continuum:

1. All three definitions are untenable because they furnish, not more-than-denumerably many, but only denumerably-unfinished many elements of the continuum (the extra-mathematical creative power of logic is rejected).

2. The definition of the continuum as a species of the species of 'coinciding' sub-species of the rational numbers is untenable because with the demise of the law of the excluded middle the existence of the 'upper bound' also becomes

[2] This insight has moreover in recent years also broken through among the formalists themselves.

invalid. For example, let λ_f be the critical-number of a fleeing-property f,[3] and consider the infinite sum $a_1 + a_2 + \ldots$ with the property that $a_1 = 1$; $a_{\nu+1} = \frac{1}{2} a_\nu$ if $\nu \neq \lambda_f$; and $a_{\nu+1} = 1$ if $\nu = \lambda_f$. And consider further those rational numbers r for which an n can be determined such that $r < a_1 + a_2 + \ldots + a_n$. The species of these rational numbers clearly possesses no 'upper bound.'

3. We can see that the definitions by means of Dedekind cuts and by means of species of coinciding convergent fundamental sequences are not equivalent on intuitionistic principles by considering the binary oscillatory shrinking number p_f of the parity-free fleeing property f, which is representable as a species of coinciding convergent fundamental sequences of rational numbers, but not as a Dedekind cut of the rational numbers.

In the intuitionistic theory the species of species of coinciding convergent fundamental sequences of rational numbers forms only a part of the continuum; it is called the *reduced continuum* ‖*reduziertes Kontinuum*‖ that overlaps the system of rational numbers (or, what comes to the same thing, the system of finite binary fractions, or of finite decimal fractions). If we consider in particular the unit-continuum, then an element l of the corresponding *reduced unit-continuum* has the property, in relation to the rational numbers $r_1, r_2, \ldots$ that are denumerated by a fundamental sequence and are unitarily restricted ‖unitär beschränkt‖ (i.e. are non-negative and do not exceed 1), that for any n there exists an index $i_n \leq n$ such that for every $\nu \leq n$ (with the possible exception of i_n) either the relation $l \geq r_\nu$ or the relation $l \leq r_\nu$ can be established, while i_{n+1} equals either i_n or $n + 1$. Among the mentioned elements of the reduced unit-continuum the Dedekind cuts are then distinguished by the fact that the exceptional index i_n falls away.

We also say that the Dedekind cuts are distinguished among the elements l by the fact that they possess a *degree of precision* ‖*Präzisionslage*‖ *of the first order* in relation to the unitarily restricted rational numbers; we can consequently define further degrees of precision, for example as follows. For an element l of the above-mentioned sort there exists the *degree of precision of the second order and of the first* (or *second*) *sort* with respect to the unitarily restricted rational numbers if, for any n, either the relation $l \geq r_n$ or the relation $l < r_n$ holds (either the relation $l \leq r_n$ or the relation $l > r_n$ holds). One sees the scope of the degree of precision of the second order most readily if one considers the degrees of precision, not with respect to the rational numbers, but with respect to the finite decimal fractions. Then the degree of precision of the first order means that the element l allows a decimal fraction development; if one has a degree of precision of the second order and the first sort,

[3] For this terminology, introduced in my lecture 'Mathematics, Science, and Language' belonging to the same series, see *Monatshefte f. Math. u. Phys.* XXXVI, p. 161. On that page, lines 21–22, should be 'neither for even nor for odd' instead of 'neither for positive nor for negative'. Moreover, on p. 154, line 9, the words 'fellow men' should be replaced by 'fellow creatures'. ‖The changes have been introduced in the translation of *Brouwer 1928a* above.‖

then the existence of a final digit different from 9 is excluded; and if one has a degree of precision of the second order and the second sort, then the existence of a final digit different from 0 is excluded—so that in both cases a decimal fraction development is *unambiguously determined*. As the *degree of precision of the third order* of the element l with respect to the unitarily restricted rational numbers we can define the property that, for every n, one of the relations $l \gg r_n$, $l \ll r_n$, and $l = r_n$ holds. (Here $l \gg r_n$ means that an r_m $(r_m > r_n)$ can be given such that $l > r_m$.) The degree of precision of the third order with respect to the unitarily restricted rational numbers holds for those elements, and only those elements, l that can be developed as regular (possibly terminating) continued fractions (and here one does not need to know in advance whether the termination actually takes place or not).

In the intuitionistic theory, to obtain the *full unit-continuum* ‖*volle Einheitskontinuum*‖ that overlaps the unitarily restricted rational numbers (or the finite decimal fractions) it is necessary to introduce 'unfinished elements' in addition to the 'finished elements' of the reduced continuum; we do this by allowing, in addition to the convergent *fundamental sequences* of unitarily restricted rational numbers, convergent *sequences* (produced by free *choice*) of such rational numbers. In order to express the finished more-than-denumerable multiplicity of this full unit continuum, we first extract from it a suitable *representing species of sequences*, that is, a species of *special* convergent sequences such that every convergent sequence of unitarily restricted rational numbers is coincident ‖zusammengehörig‖ with one of the sequences belonging to the representing species. And then from the latter we extract a suitable *locally individualized* subspecies, i.e. a subspecies such that only equal elements of it belong to equal elements of the full unit continuum. In this way we produce each of these two successive extracts in the form which, in intuitionistic mathematics, is the primary way of creating finished more-than-denumerable species, namely in the form of a *spread* ‖*Menge*‖, and indeed a spread of the simplest sort, a *pure finite spread*, which is defined as follows:

A *pure n-finite spread* is a *law* on the basis of which, whenever one of the numbers 1, 2, . . ., n is chosen again, each of these choices produces a determinate symbol-sequence. Any sequence of symbol-sequences produced in this manner from an unrestricted sequence of choices ‖Wahlfolge‖ is called an *element of the spread* (and because the sequence of choices is unrestricted, it will in general not be representable as finished).

In accordance with this definition we obtain in the pure 3-finite spread a species of sequences representing the unit continuum on the basis of which every mth choice produces a unitarily restricted finite binary fraction $c_m = a_m \cdot 2^{-m-1}$ (a_m a non-negative integer) such that $c_1 = 1/4$, $1/2$, or $3/4$, and further $c_m = c_{m-1} - 2^{-m-1}$, c_{m-1}, or $c_{m-1} + 2^{-m-1}$ according as the number 1, 2, or 3 is chosen. Those elements of this spread for which, at every even stage, number 2 is chosen form in turn a pure 3-finite spread, but this time one that is locally individualized and that therefore exhibits a subspecies of the full unit continuum with the degree of multiplicity of a pure 3-finite spread. Only with this degree of multiplicity—which is essential to pure n-finite spreads

($n \geq 2$), and which possesses 'spreading character' ⟦'Ausdehnungscharakter'⟧ because it admits unfinished elements—is it possible for a subspecies of the continuum to have a measure different from zero.

The introduction of the spread-construction, on which the finished more-than-denumerable multiplicity of the continuum rests, stands in need of no further reflection ⟦Besinnung⟧ after the reflection that has already taken place on the mathematical ur-intuition of twoity[4] and that underlies the whole of intuitionism. And the spread construction contains no *petitio principii* (so that the Kant–Schopenhauer view mentioned above—the view which regards the continuum as a pure intuition *a priori*—is in its essentials maintained in intuitionism). For the ur-intuition contains the possibility of an *interpolation between two elements* (namely the contemplation of the *combination* ⟦*Bindung*⟧ as a new element), and therefore also contains the possibility of the construction in the intuitive continuum of a spread of closed intervals that do not touch one another, which arises in the following way. Intervals are successively fixed with the indices 0, 1, 1/2, 1/4, 3/4, 1/8, 3/8, 5/8, 7/8, 1/16, . . ., with order-relations determined by the natural ordering of these indices. But now, if we choose first the interval 0 and 1 and a point p in between, we can then choose the intervals 1/2, 1/4, 3/4, 1/8, . . . in such a manner that they all omit the point p and so that the position of p with respect to the interval-spread corresponds to an arbitrary infinite binary fraction, specified in advance and generated by an unrestricted sequence of *free choices* between the numbers 0 and 1. Conversely, therefore, in the intuitive continuum, in which an ordered interval-spread M of the above sort is constructed, there exists for every infinite binary fraction generated by an unrestricted sequence of free choices between the numbers 0 and 1 a point whose position with respect to M is unambiguously characterized by the relevant infinite binary fraction. This existence-justification of pure 2-finite spreads can be extended to arbitrary n-finite spreads, and also to more general arbitrary spreads (which do not come into consideration here).

II

We shall now investigate the extent to which the fundamental properties of the formalist and old-intuitionist continuum (properties that until recently were generally assumed to hold) remain true for the intuitionist continuum. We begin by listing the properties that are here under consideration:

1. *Discreteness*. A species is *discrete* if, for any two of its elements, it is determined either that they are equal or that they are different.

2. *Ordering*. A species is *ordered* if, for every pair of its elements (a, b) an *ordering relation* $a < b$ (which is equivalent to $b > a$) is defined such that the equality of a and b is equivalent to the simultaneous absurdity of $a < b$ and $a > b$; such that $a > b$ and $a < b$ reciprocally exclude each other; such that if

[4] *Ibid.*, p. 154. ⟦*Brouwer 1928a*, §I.⟧

$a \neq b$, then either $a < b$ or $a > b$; such that $a < b$ and $b < c$ implies $a < c$; and such that $a < b$, $a = h$, and $b = k$ imply $h < k$.

3. *Density in itself.* The concept of this property rests on the definition of a boundary-element. The element a of an ordered species S is a *boundary-element* of the increasing fundamental sequence $a_1, a_2, \ldots$ $(a_1 < a_2 \ldots < a)$ if, for every $b < a$, there exists an $a_n > b$. Boundary elements of decreasing fundamental sequences are defined analogously. Both sorts of boundary element are called principal elements ‖Hauptelemente‖. An ordered species in which every element is principal is *dense in itself.*

4. *Separability in itself.* An ordered species S is *separable in itself* if one can indicate a fundamental sequence F in it such that between any two different elements of S there is an element of F.

5. *Connectedness.* To define this property for ordered species, let us call two subspecies α and β of the ordered species *S orderwise-separated* if every element of α precedes every element of β. The ordered species S is *connected* if for every division of S into two orderwise-separated subspecies α and β, either α has a last element and β no first, or β has a first element and α no last.

In the classical theory this property of connectedness can be decomposed into properties 6 and 7:

6. *Everywhere-denseness.* An ordered species is *everywhere dense* if between any two distinct elements a and b of the species there lies an element c—i.e., for any two distinct elements a and b of the species there exists an element c such that either $a < c < b$ or $a > c > b$.

7. *Compactness.* This says for ordered species that for every *nested-sequence of intervals*—i.e. for every unlimited sequence of closed intervals $I_1, I_2, \ldots$ where each $I_{\nu+1}$ is a sub-species of I_ν—there exists an element common to all the I_ν.

We shall now examine these seven properties in sequence and see whether they apply to the intuitionistic continuum.

1. That the intuitionistic continuum (and likewise the reduced continuum) *is not discrete* follows, for example, from the fact that the number $1/2 + p_f$ is neither equal to $1/2$ nor different from $1/2$ (where p_f is the binary oscillatory shrinking number[5] of the fleeing property f).

2. That the continuum *is not ordered* by the sequence of its elements that is taken from intuition is shown by the element p determined by the convergent sequence $c_1, c_2, \ldots$. In this sequence c_1 is chosen to be the zero-point, and every $c_{\nu+1}$ is chosen to equal c_ν, with the single exception that, as soon as I become aware of a critical number λ_f of a given fleeing property f, I choose the next c_ν to equal $-2^{-\nu-1}$; and that, as soon as I become aware of a proof of the absurdity of this critical number, I choose the next c_ν to equal $2^{-\nu-1}$. This element p is different from zero, but is also neither smaller than zero, nor greater than zero.

To be sure, the preceding counterexample shows only that the indistinctly

[5] *ibid.*, p. 161 ‖§III‖.

experienced ‖unscharf empfundene‖ ‘natural ordering’ of the continuum yields no ordering of the intuitionistic continuum. But one can show that any other ordering of the intuitionistic continuum (and even of the reduced continuum) is hopeless;[6] a result that stands in sharp contrast to the opinion (until recently widespread) that the continuum can not only be ordered in many different ways, but can even be well-ordered.

In place of order, the intuitionistic continuum has the somewhat weaker property of *virtual order*, which is defined as follows: A species S is virtually ordered if the ordering relation $<$ is determined, not for the full species of pairs of different elements of S, but only for a subspecies of S, and such that the following five axioms are satisfied:

1. The relations $r = s$, $r < s$, $r > s$ are mutually exclusive.
2. If $r = u$, $s = v$, and $r < s$, then $u < v$.
3. If $r < s$ and $s < t$, then $r < t$.
4. If the relations $r > s$ and $r = s$ are both absurd, then $r < s$.
5. If the relations $r > s$ and $r < s$ are both absurd, then $r = s$.

These axioms guarantee the existence of any such relation $a = b$ or $c < d$ between elements of S that can be consistently added to the existing relations of this sort, so that a virtual order is seen to be *saturated* ‖*unerweiterbar*‖.[a]

A virtual order of this sort that broadly agrees with naïve intuition can be constructed for the intuitionistic continuum in the following manner. Let π' and π'' be two elements of the continuum. We write $\pi' <\!\circ\, \pi''$ if an end-segment of a sequence p' of π' is separated from an end-segment of a sequence p'' of π'' by two different rational numbers in the ‘natural ordering’; $\pi' \leq\!\circ\, \pi''$ if $\pi' \circ\!> \pi''$ is impossible; $\pi' < \pi''$ if $\pi' \leq\!\circ\, \pi''$ and moreover $\pi' \neq \pi''$.[b] For this

[6] Briefly put, to order the full continuum one would have to possess a method for solving all mathematical problems; to order the reduced continuum, one would have to possess a method for solving all mathematical problems that belong to a definite, very general category (which we shall not describe in further detail here).

[a] ‖Brouwer first introduced the concept of a virtual order in §1 of *Zur Begründung der intuitionistischen Mathematik II* (*Mathematische Annalen* **95**, 453–72 (1926)); and the concept of saturated order in *Virtuelle Ordnung und unerweiterbare Ordnung* (*Journal für die reine und angewandte Mathematik*, **157**, 255–7 (1927)). In the latter paper he offered a proof that the two concepts are intuitionistically equivalent. However, he subsequently realized that half of his proof—the proof that every virtual order is saturated—contained a fallacy. In a handwritten note, dated 25 March 1933 and never published by Brouwer, he pointed out the fallacy, and tried to correct it by changing the definition of virtual order. The note is printed in *Brouwer 1975*, Vol. i, p. 596.‖

[b] ‖Elsewhere Brouwer defines the virtual order of the continuum more precisely as follows:

‘1. If a is a rational number, and b a real number defined by the convergent infinite sequence of rational numbers $b_1, b_2, \ldots$, we say that b is *measurably greater* than a and a *measurably smaller* than b (expressed by $b \circ\!> a$ and $a <\!\circ\, b$), if for two suitable natural numbers m and p the inequality $b_\nu - a > 2^{-p}$ holds for each $\nu > m$.

‘2. If a and b are real numbers, we say that b is *measurably greater* than a, and a *measurably smaller* than b (expressed by $b \circ\!> a$ and $a <\!\circ\, b$), if for some suitable natural number n the inequality $b - a > 2^{-n}$ holds.

‘3. Indicating for two real numbers a and b the absurdity of $a <\!\circ\, b$ by $a \circ\!\geq b$ and $b \leq\!\circ\, a$, we say that b is *greater* than a and a *smaller* than b (expressed by $b > a$ and $a < b$), if $a \neq b$ as well as $a \leq\!\circ\, b$.’ (*Brouwer 1975*, Vol. i, p. 504.)‖

ordering relation the five axioms of a virtual ordering are indeed fulfilled.

3. *Density in itself* in the above sense (and extended to virtual orderings) does *not* hold for the intuitionistic continuum, for the above characterization of its elements as principal elements fails. For example, consider a characterization of the element 1/2 as a principal element on the basis of a convergent sequence $a_1 < a_2 \ldots < 1/2$. Then construct a sequence $d_1, d_2, \ldots$ in the following way: We choose successively $d_1 = a_1$, $d_2 = a_2$, and in this way set $d_\nu = a_\nu$ so long as we remain unaware either of a critical number or of the absurdity of a critical number for a given fleeing property; but if, between the determinations of d_ν and $d_{\nu+1}$ one of these two events occurs, then we set $d_\mu = d_\nu = a_\nu$ for $\mu > \nu$. The element d of the continuum belonging to this sequence $d_1, d_2, \ldots$ is $< 1/2$; nevertheless, no a_ν can be indicated such that $a_\nu > d$. Therefore, for the full intuitionistic continuum density-in-itself is at least hopeless (in the sense of the preceding footnote); and the same can be shown for the reduced continuum.

To restore the property of density-in-itself to the intuitionistic continuum (or to the reduced continuum) we subject the definition to a logical transformation—that is, we give it another form (one that in the classical conception is equivalent to the earlier definition, but that in the intuitionistic conception is not equivalent). To that end, let us first explain the concept of *interval* for virtually ordered species. For two arbitrary elements a and b of the virtually ordered species S, the *closed interval ab* is the species of those elements c of S for which neither the relations $c > a$ and $c > b$, nor the relations $c < a$ and $c < b$ can both hold. The *open interval ab* is the species of those elements c of S that lie *between a* and b, i.e. that are, first, different from both a and b, and, second, belong to the closed interval ab. The elements a and b are called *end-elements* of both the closed and the open interval ab. If $a < b$, then this 'between' is clearly equivalent with the classical 'between'. An element e of the virtually ordered species S is called a *principal element* if there exists an unlimited sequence of different closed intervals such that each succeeding interval is contained in its predecessor, such that each interval contains the element e and such that any element belonging to every interval is identical with e. On the basis of this definition of a principal element, the intuitionistic continuum (and the reduced continuum) is once again dense in itself.

4. The untenability for the intuitionistic continuum of *separability in itself* (as extended to virtual orderings) can be shown as follows. Let F be the discrete and ordered fundamental sequence on which the separability in itself of the continuum K is to rest. Let p_1 be the first element of F. We can assume that $p_1 > 2^{-n}$ for a suitable natural number n. Let p_2 be the first element of F following p_1 that lies between p_1 and the zero-point; p_3 be the first element of F following p_2 that lies between p_1 and p_2; p_4 be the first element of F following p_3 that lies between p_1 and p_3; and so on. We construct a convergent sequence $m_1, m_2, \ldots$ of elements of F as follows. We set $m_\nu = p_\nu$ so long as we know neither of a critical number nor of the absurdity of a critical number for a given fleeing property; but if between the determinations of m_k and m_{k+1} a critical number is found or the absurdity of such a critical number is

proven, then we set $m_\nu = p_k$ for $\nu > k$. The element p of K that belongs to this convergent sequence is different from p_1; nevertheless, no element of K can be given that lies between p and p_1. Therefore, for the full intuitionistic continuum separability in itself (as defined above) is at least hopeless (in the sense of the preceding footnote); the same can be shown for the reduced continuum.

The property of separability in itself can also be restored for the intuitionistic continuum by a logical transformation of the definition. We call two different elements a and b of a virtually ordered species *strictly different* ‖*scharf verschieden*‖ and the interval *ab extended* if the conjugate subspecies ‖Komplementärspezies‖ $k(ab)$ of the open interval ab is *split*[7] (in a sense explained below, under 5) into a subspecies $k_1(ab)$ whose elements are $\leq a$ and $\leq b$, and a subspecies $k_2(ab)$ whose elements are $\geq a$ and $\geq b$. And for the separability in itself of the virtually ordered species S we require only that there exist in S a discrete and ordered fundamental sequence F such that between any two strictly different elements of S there lies an element of F.

5. To examine the intuitionistic continuum with regard to the property of *connectedness*, we must first consider the concept of 'division' more closely. For in intuitionism this concept needs to be made more precise through the concepts of *compounding* ‖*Zusammensetzung*‖ and of *splitting*. We say that the species P is compounded from its disjoint subspecies Q and R if the existence of an element of P not in $Q \cup R$ can be shown to be absurd. And we say that the species P *splits* into its disjoint subspecies Q and R if $P = Q \cup R$. We shall use the terms *strongly connected* or *weakly connected* according as the underlying concept of division is a compounding-concept or a splitting-concept. Then in the full intuitionistic continuum the concept of *weak connectedness* is *empty* ‖*inhaltslos*‖ because of the theorem of the *unsplittability of the continuum*, which says that, given any splitting of the continuum into discrete species of subspecies, one of these species is identical with the continuum. In contrast, *strong connectedness* is false for the full intuitionistic continuum, as the following counterexample shows: We consider the fundamental sequence $a_1, a_2, \ldots$ where, if λ_f is the critical number of a given fleeing property, $a_\nu = 1 - 2^{-\nu}$ for $\nu < \lambda_f$ and $a_\nu - 2 - 2^{-\nu}$ for $\nu \geq \lambda_f$. Let K_1 be the species of those elements e of K for which a ν exists such that $a_\nu > e$; and K_2 and let be the species of those elements e of K for which no ν can exist such that $a_\nu > e$. These species K_1 and K_2 are orderwise-separated, and K is compounded from K_1 and K_2; nevertheless, K_1 does not possess a last element, nor K_2 a first.

We see in the following way that weak connectedness cannot simultaneously have a meaning and be applicable to the reduced continuum. Let e be the last element of the 'front' subspecies α, where we are entitled to assume that $e \geq \frac{1}{4}$ and $e \leq \frac{3}{4}$. Then on the one hand every element $b > e$ of the reduced continuum must belong to β; on the other hand, $b' > e$ for every element b' of β. The subspecies β is therefore identical with the subspecies of elements $>e$,

[7] Then certainly either $a < b$ or $b < a$.

and the subspecies α is identical with the subspecies of elements $\leq e$, so that every element of the reduced continuum would have to be either $\leq e$ or $>e$, which is not the case. The falsity of strong connectedness for the reduced continuum follows from the same counterexample as for the full continuum.

To restore the property of connectedness to the intuitionistic continuum by means of a logical transformation of the definition, we say that the virtually ordered species S is *exhaustively divided* into the orderwise-separated subspecies α and β from which it is compounded, if for any two strongly different elements a and $b (a < b)$ either all elements $\leq a$ belong to α or all elements $\geq b$ belong to β. And we call the virtually ordered species S *freely connected* (or *restrictedly connected*) if, for any exhaustive division of S (or any exhaustive division of S determined by a law) into two orderwise-separated subspecies α and β, there exists an element e of S such that every element $<e$ belongs to α and every element $>e$ belongs to β. Then the full intuitionistic continuum is freely connected, and the reduced continuum is restrictedly connected.

6. It is out of the question that the intuitionistic continuum (or the reduced continuum) is *everywhere dense* in the sense of the above definition. For this property requires in the first place that for any two different elements a and b either the relation $a < b$ or the relation $a > b$ holds; i.e. it presupposes the ordering of the relevant species. But if we conceive the word 'between' for virtually ordered species in the sense explained under (3) above, then between any two elements of the full continuum K a further element of K can be determined; so that in this way the property of being everywhere dense is restored for the full continuum. The same thing can be done in the same way for the reduced continuum.

7. *Compactness* in the above sense (as extended to virtual orderings) holds *neither* for the intuitionistic continuum *nor* for the reduced continuum. We can see this from the following example. Let λ_f be the critical number of the parity-free fleeing property f, and let $I_\nu = \left(-\frac{1}{2}, +\frac{1}{2}\right)$ if $\nu < \lambda_f$; $I_\nu = \left(-\frac{1}{2}, -\frac{1}{4}\right)$ if $\nu \geq \lambda_f$ and λ_f is odd; $I_\nu = \left(+\frac{1}{4}, +\frac{1}{2}\right)$ if $\nu \geq \lambda_f$ and λ_f is even. Then the sequence of nested intervals I_1, $I_2, \ldots$ does not possess an element common to all I_ν.

To restore the property of compactness to the intuitionistic continuum (or to the reduced continuum) by a logical transformation of the definition, we define a *predeterminate sequence of nested intervals* to be a *fundamental sequence* of closed intervals $I_1, I_2, \ldots$ where every $I_{\nu+1}$ is a subspecies of I_ν. We call a sequence of nested intervals $I_1, I_2, \ldots$ *hollow* ‖*hohl*‖ if for every element π of the relevant virtualy ordered species there exists a definite ν_π such that π cannot belong to I_{ν_π}. And we define *free* (or *restricted*) *compactness* as the impossibility of the existence of a hollow sequence of nested intervals (or of a hollow predeterminate sequence of nested intervals). Then the full intuitionistic continuum is freely compact, and the reduced continuum restrictedly compact.

We obtain another suitable logical transformation of the definition if we call a sequence of nested intervals $I_1, I_2, \ldots$ *unrestrictedly contracting* if for every extended interval I a ν can be determined such that I_ν cannot possibly be a subspecies of I.[c] And we define a virtually ordered species S to be *freely compact* (or *restrictedly compact*) if for every unrestrictedly contracting sequence of nested intervals (or for every unrestrictedly contracting predeterminate sequence of nested intervals) $I_1, I_2, \ldots$ an element of S exists that is common to all I_ν. In this sense as well the full intuitionistic continuum is freely compact and the reduced continuum restrictedly compact.

C. HISTORICAL BACKGROUND, PRINCIPLES, AND METHODS OF INTUITIONISM (*BROUWER 1952*)

The following lecture was read to Section A of the South African Association for the Advancement of Science, Cape Town, in July 1952. Brouwer's principal concern is to give a short introduction to intuitionism, recounting its origins, describing its philosophical underpinnings, and briefly surveying some of its principal theorems. Although the technical details are presented more fully elsewhere (notably in *Brouwer 1954*), this article is the best introduction to Brouwer's views in the late phase of his career.

After the initial historical and philosophical remarks, Brouwer introduces the notion of infinitely proceeding sequences; he states without proof the fan theorem, and describes the construction of the intuitionistic two-dimensional Cartesian plane. He ends the article with a proof of perhaps the most startling theorem of intuitionistic analysis: every real-valued function defined on the closed interval [0,1] is uniformly continuous.

This theorem first appeared in *Brouwer 1923*, with a proof that Brouwer soon realized was unsatisfactory. (In effect, he had assumed without proof 'König's Lemma', which says that a finitely-branching tree, all of whose branches are finite, is itself finite. Brouwer had tacitly been using this lemma since *Brouwer 1918*.) In *Brouwer 1924*, in the process of re-proving the uniform continuity theorem, he stated and proved for the first time the theorems he was later to call the *fan theorem* and the *bar theorem*. In *1927a* he gave a polished exposition of both theorems and of the uniform continuity theorem; this article is partially translated in *van Heijenoort 1967*, pp. 457–63, with a valuable Introductory Note by Charles Parsons analysing the difficulties with Brouwer's argument, and giving a summary of Brouwer's technical terminology. Readers who wish to pursue the technical aspects of these theorems are referred to that article and to the monograph *Troelstra 1977*.

---

[c] ⟦Brouwer here evidently means 'such that I cannot possibly be a subspecies of I_ν'.⟧

In the article that follows, Brouwer uses the phrase 'classical Cartesian plane' to refer to the species of points defined by predeterminate (i.e. lawlike) convergent infinite sequences. Earlier, in *1928b*, Brouwer had used the less misleading phrase 'reduced continuum' to refer to the same concept.

References to *Brouwer 1952* should be to the page numbers of the original article, as given in the margins.

---

139 The historical development of the mental mechanism of mathematical thought is naturally closely connected with the modifications which, in the course of history, have come about in the prevailing philosophical ideas *firstly* concerning the origin of mathematical certainty, *secondly* concerning the delimitation of the object of mathematical science. And that the mental mechanism of mathematical thought during so many centuries has undergone so little fundamental change is due to the circumstance that, in spite of all revolutions undergone by philosophy in general, the belief in the existence of properties of time and space, immutable and independent of language and experience, remained well-nigh intact until far into the nineteenth century. Exact knowledge of these properties was called mathematics, and was generally pursued in the following way: for some familiar regularities of (outer or inner) experience which, with any attainable degree of approximation, *seemed invariable*, *complete invariability was postulated*. These regularities were called *axioms* and were put into language. Thereupon extensive systems of properties were developed from the linguistic substratum of the axioms by means of *reasoning* guided by experience but linguistically following and using the principles of *classical logic*.

We will call the standpoint governing this mode of thinking and working the *observational standpoint*, and the long period characterized by this standpoint the *observational period*.

During the observational period mathematics was considered functionally, if not existentially, dependent on logic, and logic itself was considered autonomous.

For space the observational standpoint became untenable when, in the course of the nineteenth and the beginning of the twentieth century, as a consequence of a series of discoveries with which the names of Lobatchevsky, Bolyai, Riemann, Cayley, Klein, Hilbert, Einstein, Levi-Cività, and Hahn are associated, mathematics was gradually transformed into a mere science of numbers. Simultaneously, besides observational space, a great number of other spaces, sometimes exclusively originating from logical speculations, with properties distinct from the traditional but no less beautiful, gradually found an arithmetical representation. Consequently the science of classical (Euclidean) three-dimensional space had to continue its existence as a chapter without priority, on the one hand, of (exact) science of numbers, on the other hand, as applied mathematics, of (naturally only approximative) descriptive natural science.

Encouraged by the important part which, in this process of extending the domain of conceivable geometry, had been played by the *logico-linguistic method*, which, without any guidance by experience, operated on words by means of logical rules, the *Old Formalist school* (Dedekind, Cantor, Peano, Hilbert, Russell, Zermelo, Couturat) finally, for the purpose of a rigorous treatment of mathematics *and logic* (though not for the purpose of choosing the subjects of investigation of these sciences) rejected any element extraneous to language and logic. Thus logic and mathematics were divested by this school both of their essential difference in character and of their autonomy. However, the hope originally fostered by the Old Formalists that mathematical science erected according to their principles would be crowned one day with a proof of non-contradictority, was never fulfilled, and, nowadays, in view of the results of certain investigations of the last few decades, has, I think, been relinquished. | 139|

Of a totally different orientation was the *Pre-intuitionist school*, led mainly 140
by Poincaré, Borel, and Lebesque. These thinkers seem to have maintained a modified observational standpoint for the introduction of natural numbers, of the principle of complete induction, and of all mathematical entities and theories springing from this source without the intervention of axioms of existence, hence for what might be called the 'separable' parts of arithmetic and algebra. For these parts of mathematics, even for such theorems as were deduced by means of classical logic, they postulated an existence and exactness independent of language and logic, and regarded their non-contradictority as certain, even without logical proof. For the continuum, however, they seem not to have sought an origin extraneous to language and logic. On some occasions they seem to have contented themselves with an ever-unfinished and ever-denumerable system of 'real numbers', generated by an ever-unfinished and ever-denumerable system of laws defining convergent infinite sequences of rational numbers. In doing so they seem to have overlooked that such an ever-unfinished and ever-denumerable system of 'real numbers' is incapable of fulfilling the mathematical functions of the continuum, for the simple reason that it *cannot have a measure positively differing from zero*. On other occasions they seem to have introduced the continuum by having recourse to some logical axiom of existence lacking sensory as well as epistemological evidence, such as the 'axiom of ordinal connectedness', or the 'axiom of completeness'. But in both cases, in their further development of mathematics, they unreservedly continued to apply classical logic, including the principle of the excluded third. They did so regardless of the fact that the non-contradictority of systems thus constructed had become very doubtful after the discovery of the logico-mathematical antinomies.

Thus, in point of fact, Pre-intuitionism re-established on the one hand the essential difference in character between logic and mathematics, and on the other hand the autonomy of logic and of a part of mathematics. On these two autonomous domains of thought the rest of mathematics remained dependent.

When the Old Formalist standpoint had been badly shaken, mainly by Pre-intuitionist criticism, Hilbert founded the *New Formalist school*, which postulated existence and exactness independent of language—it is true not for mathematics proper, but for *meta-mathematics* or *mathematics of the second order*, i.e. the scientific consideration of the symbols occurring in purified mathematical language, and of the rules of manipulation of these symbols. Thus New Formalism, in contrast with Old Formalism, consciously and *in confesso*, made use of the intuition of natural numbers and of complete induction. It is true that autonomy was postulated here for a much smaller part of mathematics than in the case of Pre-intuitionism.

But no attention was paid by New Formalism to the circumstance that, between the perfection of mathematical language and the perfection of mathematics proper, no clear connection can be seen.

The situation left by Formalism and Pre-intuitionism can be summarized as follows: for the elementary theory of natural numbers, the principle of complete induction, and more or less considerable parts of algebra and theory of numbers, exact existence, absolute reliability, and non-contradictority were universally acknowledged, independently of language and without proof. There was little concern over the existence of the continuum. Introduction of a set of predeterminate real numbers with a positive measure was attempted by logico-linguistic means, but a proof of the non-contradictory existence of such a set was lacking. For the whole of mathematics the rules of classical logic were accepted as reliable aids in the search for exact truths.

In this situation intuitionism intervened with two acts, of which the first seems necessarily to lead to destructive and sterilizing consequences; then, however, the second yields ample possibilities for recovery and new developments. To begin with, the

FIRST ACT OF INTUITIONISM

completely separates mathematics from mathematical language, in particular
140| *from | the phenomena of language which are described by theoretical logic, and*
141 *recognizes that intuitionist mathematics is an essentially languageless activity of the mind having its origin in the perception of a* ***move of time****, i.e. of the falling apart of a life moment into two distinct things, one of which gives way to the other, but is retained by memory. If the two-ity thus born is divested of all quality, there remains the* ***empty form of the common substratum of all two-ities****. It is this common substratum, this empty form, which is the* ***basic intuition of mathematics****.*

How much of 'separable' mathematics can be rebuilt in a slightly modified form, by unlimited self-unfolding of the basic intuition, is introspectively realized.

In the edifice of mathematical thought thus erected, language plays no other part than that of an efficient, but never infallible or exact, technique for memorizing mathematical constructions, and for suggesting them to others; so that mathematical language by itself can never create new mathematical systems. But on account of the highly logical character of usual mathematical language the following question naturally presents itself:

Suppose that an intuitionist mathematical construction has been carefully described by means of words, and then, the introspective character of the mathematical construction being ignored for a moment, its linguistic description is considered by itself and submitted to a linguistic application of a principle of classical logic. Is it then always possible to perform a languageless mathematical construction finding its expression in the logico-linguistic figure in question?

After a careful examination one answers this question in the *affirmative* (if one allows for the inevitable inadequacy of language as a mode of description) as far as the principles of contradiction and syllogism are concerned; but in the *negative* (except in special cases) with regard to the principle of the excluded third, so that the latter principle, as an instrument for discovering new mathematical truths, must be rejected.

Indeed, if each linguistic application of the principle of the excluded third in a mathematical argument were to accompany some actual intuitionist-mathematical construction, this would mean that each intuitionist-mathematical assertion (i.e. each assignment of a property to an intuitionist-mathematical entity) can be *judged*, i.e. can either be proved or be reduced to absurdity.

Now every construction of a bounded finite character in a finite mathematical system can be attempted only in a finite number of ways, and each attempt can either be carried through to completion, or be continued until further progress is impossible. It follows that every assertion of possibility of a construction of a bounded finite character in a finite mathematical system can be judged. So, in this exceptional case, application of the principle of the excluded third is permissible.

In order to show that this is not so for infinite systems, we shall call a hypothetical property f of natural numbers a *fleeing property*, if it satisfies the following conditions:

1. for each natural number it can be decided either that it possesses the property f, or that it cannot possibly possess the property f;

2. no method is known for calculating a natural number possessing the property f;

3. the assumption of existence of a natural number possessing the property f is not known to lead to an absurdity.

In particular, a fleeing property is called *opaque*, if the assumption of existence of a natural number possessing f is not known to be non-contradictory either.

Obvious examples of fleeing properties can easily be given.

Now should we assert of a fleeing property f, on the grounds of the principle of the excluded third, that a natural number possessing the property f either exists or cannot exist, then this assertion, precisely because of the nature of fleeing properties, would be an utter falsehood; which shows conclusively that, in the language of intuitionist mathematics, blind applications of the said principle are not permissible.

141| From the intuitionist standpoint the dogma of the universal validity of the |
142 principle of the excluded third in mathematics can only be considered as a phenomenon of history of civilization, of the same order as the former belief in the rationality of π or in the rotation of the firmament about the earth. That the dogma was nevertheless able to retain its currency for so long, may perhaps be explained by the following two circumstances: firstly that (as is easily recognized) within a given domain of mathematical entities previously obtained, for a single assertion the principle is non-contradictory; secondly that the principle stands the test of application to an extensive group of everyday phenomena of the exterior world.

We have seen in the preceding how the first act of intuitionism affected classical mathematics in two ways: in the first place, owing to the disappearance of the logical basis for the continuum, so large a part becomes illusory that essentially only the separable parts of algebra and theory of numbers remain; in the second place, even in this remaining portion, several chapters based on the principle of the excluded third have to be rejected. Under these circumstances one might fear that intuitionist mathematics must necessarily be poor and anaemic, and in particular would have no place for analysis. But this fear would have presupposed that infinite sequences generated by the intuitionist self-unfolding of the basic intuition would have to be fundamental sequences, i.e. predeterminate infinite sequences which, like classical ones, proceed in such a way that, from the beginning, the mth term is fixed for each m. Such, however, is not the case; on the contrary, a much wider field of development, which includes analysis, and in several places far exceeds the frontiers of classical mathematics, is opened by the

SECOND ACT OF INTUITIONISM

which recognizes the possibility of generating new mathematical entities:

firstly, in the form of ***infinitely proceeding sequences*** $p_1, p_2, \ldots,$ *whose terms are choosen more or less freely from mathematical entities previously acquired; in such a way that the freedom of choice existing perhaps for the first element* p_1 *may be subjected to a lasting restriction at some following* p_ν*, and again and again to sharper lasting restrictions or even abolition at further subsequent* p_ν*'s,* while all these restricting interventions, as well as the choices of*

* In former publications I have sometimes admitted restrictions of freedom with regard also to future restrictions of freedom. However this admission is not justified by close introspection, and moreover would endanger the simplicity and rigour of further developments.

the p_ν's themselves, at any stage may be made to depend on possible future mathematical experiences of the creating subject;

secondly, in the form of mathematical ***species, i.e. properties supposable for mathematical entities previously acquired,*** *and satisfying the condition that, if they hold for a certain mathematical entity, they also hold for all mathematical entities which have been defined to be equal to it, relations of equality having to be symmetric, reflexive, and transitive; mathematical entities previously acquired for which the property holds are called* ***elements*** *of the species.*

With regard to this definition of species we have to remark firstly that, during the development of intuitionist mathematics, some species will have to be considered as being re-defined time and again in the same way, secondly that a species can very well be an element of another species, but never an element of itself.

Two mathematical entities are called *different*, if their equality has been proved to be absurd.

Two infinitely proceeding sequences of mathematical entities a_1, a_2, ..., and b_1, b_2, ... are called *equal* or *identical*, if $a_\nu = b_\nu$ for each ν, and *distinct*, if a natural number s can be indicated such that a_s and b_s are different.

The second act of intuitionism creates the possibility of introducing the *intuitionist continuum* as the species of the *more or less freely proceeding* convergent
infinite sequences of rational numbers,[†] and more | generally the *intuitionist n-* 142|
dimensional Cartesian space as the species of the *more or less freely proceeding* 143
convergent infinite sequences of the 'n-dimensional rational grid', which expression may be considered self-explanatory. These species will prove to be susceptible of a standard representation making them considerably more surveyable and manageable than the classical species of the predeterminate real numbers and of the predeterminate real points of Cartesian n-dimensional spaces.

The development of this standard representation must be preceded by the introduction of some new concepts.

By a *node of order n* we understand a sequence of n natural numbers ($n \geq 1$), called the *indices* of the node.

A node p' of order $n + m$, ($m \geq 1$), will be called an *mth descendant* of a node p of order n, and p will be called the *mth ascendant of p'*, if p is an initial segment of p'. For $m = 1$, p' is also called an *immediate descendant of p*, and p the *immediate ascendant* of p'.

A *finite sequence of nodes* consisting of a node p_1 of order 1, an immediate

[†] As the common notion of a rational number and the common notion of a convergent infinite sequence are both imbued with images of measure, the method followed in the text for the introduction of the continuum might suggest that the intuitionist continuum depends on the concept of measure. This, however, is by no means the case. The inuitionist closed continuum can be spread over an arbitrary fundamental sequence which has been completely ordered as an everywhere dense species with a first and a last element, and has been provided with a definition of convergence based exclusively on the relations constituting its everywhere dense order. The metrical method of introducing the continuum which is given in the text was chosen to abbreviate the approach to some applications of the fan theorem.

descendant p_2 of p_1, an immediate descendant p_3 of p_2, ..., up to an immediate descendant p_n of p_{n-1}, will be called a *rod of order n*.

An *infinite* (not necessarily predeterminate) *sequence of nodes* consisting of a node p_1 of order 1, an immediate descendant p_2 of p_1, an immediate descendant p_3 of p_2, and so on *ad infinitum*, will be called an *arrow*.

Naturally an arrow may grow in complete freedom , i.e. in the passage from p_ν to $p_{\nu+1}$, the choice of a new index for $p_{\nu+1}$ to be joined to those of p_ν may be completely free for each ν, for as long as the creating subject may desire. On the other hand this freedom in the generation of the arrow may at any stage be completely abolished, at the beginning or at any p_ν, by means of a law fixing all further nodes in advance. From this moment the arrow concerned will be called a *sharp arrow*. Furthermore, the freedom in the generation of the arrow, without being completely abolished, may, at any p_ν, undergo some restriction, and this restriction may be intensified at further p_ν's. Finally all these interventions, by virtue of the second act of intuitionism, may, at any stage, be made to depend on possible future mathematical experiences of the creating subject.

We will consider a species of nodes σ to which a law $W(\sigma)$ assigns the following nodes: of order 1 the natural numbers which do not exceed a certain definite natural number m_o, and of each order $n + 1$ the immediate descendants of each node p of order n belonging to σ whose $(n + 1)$th index, joined to those of p, does not exceed a certain definite natural number m_p. Then this law $W(\sigma)$ at the same time defines the species $w(\sigma)$ of the arrows consisting exclusively of nodes of σ. This species of arrows $w(\sigma)$ is called a *fan*, and the law $W(\sigma)$ is called a *fan key*.

For fans can be proved the

FAN THEOREM: *If to each arrow α of a fan F has been assigned a natural number $\mu(\alpha)$, then a natural number s can be indicated such that, for any α, $\mu(\alpha)$ is completely determined by the sth node of α*. If follows that, moreover, for $\mu(\alpha)$ a finite maximum can be indicated.

Passing now to the development of the standard representation of the intuitionist continuum and the intuitionist n-dimensional Cartesian space, we will treat explicity only the case of the intuitionist two-dimensional space, also called the *intuitionist plane*. The same reasoning, with little modification, applies to other values of n.

Calling the two-dimensional rational grid simply the 'rational grid', we shall understand by a *limiting point* an element of the intuitionist plane, i.e. a (not necessarily predeterminate) convergent infinite sequence of elements of the rational grid. A predeterminate limiting point will also be called a *sharp limiting point*. Again, regarding as self-explanatory the meaning of *coincidence* of two limiting points, we shall call the species of the limiting points coinciding with a given limiting point a *limiting point core*, and a limiting point core containing a sharp limiting point, a *sharp limiting point core*.

Denoting by a and n arbitrary integers, we will consider the species of the

finite binary fractions $a.2^{-n}$ in their natural order, and we will call a pair of these | fractions a *grid interval*. In particular the grid intervals consisting, for 143|
a certain n and a certain a, of $a.2^{-(n+1)}$ and $(a+2).2^{-(n+1)}$, will be called 144
$\lambda^{(n)}$-*grid intervals*. They will be $\kappa^{(n)}$-*grid intervals* if a is even. All $\lambda^{(n)}$-grid intervals, for all n, will be λ-*grid intervals*, and all $\kappa^{(n)}$-grid intervals, for all n, will be κ-*grid intervals*, By a *grid square* we shall understand an ordered pair (i.e. a pair consisting of a 'first element' and a 'second element') of grid intervals, by a $\lambda^{(n)}$-*grid square* a similar pair of $\lambda^{(n)}$-grid intervals, by a $\boldsymbol{\kappa}^{(n)}$-*grid square* a similar pair of $\kappa^{(n)}$-grid intervals. All $\lambda^{(n)}$-grid squares, for all n, are λ-*grid squares*, and all $\kappa^{(n)}$-grid squares, for all n, are κ-*grid squares*. With regard to the mutual position of two λ-grid squares a and b, the meaning of the following expressions may be supposed self-explanatory: *a lies inside b*, *a lies outside b*, *a touches b internally*, *a touches b externally*. Furthermore we shall say that a and b *overlap* if a λ-square lying inside both can be indicated; that *a lies within b* if a lies inside b, and does not touch b; and that a and b *lie apart* if they lie outside each other and do not touch each other.

The union g of a fundamental sequence $\kappa_1(g)$, $\kappa_2(g)$, ... of κ-grid squares lying outside each other will be called a *grid area*, if for each $\kappa_\nu(g)$ we can indicate a finite number of elements $\kappa_{\nu 1}(g)$, $\kappa_{\nu 2}(g)$, ..., $\kappa_{\nu m_\nu}(g)$ of the same fundamental sequence lying outside each other, and together *enclosing* $\kappa_\nu(g)$, i.e. all touching $\kappa_\nu(g)$ externally in such a way that no place is left for any further κ-grid square touching $\kappa_\nu(g)$ externally, and lying outside $\kappa_{\nu 1}(g)$, ..., $\kappa_{\nu m_\nu}(g)$.

The union of an arbitrary finite number of κ-grid squares lying outside each other will be called a *grid portion*. With regard to the mutual position of two λ-grid squares or grid portions a and b, the meaning of the following expressions may be considered self-explanatory: *a lies inside b*, *a lies outside b*, *a and b touch each other externally*, *a and b lie apart*, *a and b overlap*, while a will be said to lie *within* b, if it lies inside b, and cannot possibly touch any κ-grid square lying outside b. A λ-grid square or grid portion a will be said to *lie inside* or *within the grid area* g if for an s suitably chosen it lies inside the union of $\kappa_1(g)$, $\kappa_2(g)$, ..., $\kappa_{s-1}(g)$, $\kappa_s(g)$. The grid area g will be said to *lie within the λ-grid square*, *grid portion* or *grid area b* if $\kappa_\nu(g)$ lies within b for each ν.

What is meant by the *measure of a λ-grid square* and, in this connection, by *measurability of a grid area*, and by the *measure of a measurable grid area*, may be considered self-explanatory.

A (not necessarily predeterminate) infinite sequence of λ-grid squares k_1, k_2, ..., such that $k_{\nu+1}$ lies within k_ν for each ν, will be called a *binary point* or simply a *point*.

Two binary points k'_1, k'_2, ... and k''_1, k''_2, ... will be said to *coincide* if it is certain that k'_μ and k''_ν overlap for each μ and each ν. Obviously coincidence is a transitive relation. The species of the binary points coinciding with a given binary point will be called a *binary point core* or simply a *point core*.

A *limiting point* will be said to *lie inside the λ-grid square* or *grid portion* b if it coincides with a limiting point r_1, r_2, ... possessing a tail segment inside

b; it will be said to *lie within the λ-grid square* or *grid portion b* if it lies inside a λ-grid square lying within b.

A *point* $k_1, k_2, \ldots$ will be said to *lie inside the λ-grid square* or *grid portion b* if k_ν and b overlap for each ν; it will be said to *lie within the λ-grid square* or *grid portion b* if it lies inside a λ-grid square lying within b.

A point or limiting point will be said to *lie inside* or *within the grid area g* or to *be surrounded by the grid area g* if it lies inside a grid portion lying within g.

By a $k^{(\nu)}$ $(\upsilon > 0)$ we shall understand a $\lambda^{(4\nu+1)}$-grid square, and by a *standard point* a point $k_1, k_2, \ldots$ for which each k_ν is a $k^{(\nu)}$. It can be proved that each limiting point p *coincides* with a standard point q, i.e. to each limiting point p can be assigned a standard point q, in such a way that within each $k^{(\nu)}$ of q lies a tail segment of p. Furthermore, coinciding limiting points coincide within coinciding standard points.

If by the 'unity grid square' L we understand the $\kappa^{(0)}$-grid square consisting
144| of two equal $\kappa^{(0)}$-grid intervals (0, 1), it | can be proved in particular that each
145 limiting point p lying inside L coincides with a standard point q lying inside L. The species of the unitary limiting points, i.e. the limiting points lying inside L, will be called the *unitary intuitionist plane* or simply the *unitary plane*, and the species of the unitary standard points, i.e. the standard points lying inside L, will be called the *unitary standard plane*.

By *counting* the finite species of the k' overlapping L, and for each $k^{(\nu)}$ overlapping L *counting* the finite species of the $k^{(\nu+1)}$ lying inside this $k^{(\nu)}$, and overlapping L, we bring about a (1, 1) correspondence between the unitary standard plane and a fan w. This correspondence has far-reaching consequences.

If, for example, we attempt to surround the unitary plane with a grid area ψ, we shall in particular have to surround the unitary standard plane with ψ. So, by virtue of the fan theorem, a natural number m can be indicated such that all unitary standard points corresponding to arrows of w containing the same rod K of order m, must lie inside one and the same grid portion $\rho(K)$ lying within ψ. Now indicating by G the species of the unitary standard points containing K, by H the $k^{(m)}$ corresponding to K, and by H' the grid portion consisting of all $\kappa^{(4m+5)}$-grid squares lying inside L and within H (so covering a grid square concentric and homothetic with H and with side length $\frac{7}{8}$ of the side of H), we remark that if there existed a λ-square lying inside H' and outside $\rho(K)$, a square of a point of G, so a point of G, could be indicated lying outside $\rho(K)$. Consequently no λ-square lying inside H' can lie outside $\rho(K)$, i.e. H' must lie inside $\rho(K)$, hence within ψ. This being the case, independently of the choice of K from the rods of w of order m, finally also L proves to lie within ψ. It follows that *a measurable grid area surrounding all unitary limiting points of the intuitionist plane must have a measure* ≥ 1.

How different the plight of the classical Cartesian plane appears if we suppose a procedure which, after the choice of a fixed natural number n, at the end of the mth century from today, will surround the species of all predetermined limiting points defined until then, with a measurable grid area g_m whose measure does not exceed 2^{-n-m}. Then the union of $g_1, g_2, g_3, \ldots$ would in the course of centuries constitute a grid area g whose measure would never exceed 2^{-n}, and which would in due time surround all present and future limiting points of the classical Cartesian plane. Hence, as n could be chosen arbitrarily large, there can be no question of any positive measure for the classical Cartesian plane.

Defining limiting numbers, numbers, standard numbers, limiting number cores, and number cores analogously with limiting points, points, standard points, limiting point cores, and point cores respectively, and considering the notion of a '*distance*' of two limiting number cores self-explanatory, we finally will prove, by means of the fan theorem, that *each full unitary function of the unitary continuum* (i.e. each assignment of a unitary limiting number core $f(z)$ to *each* unitary limiting number core z) *is uniformly continuous.*

For, such a full function implies an assignment of a unitary standard number $\phi(x)$ to *each* unitary standard number x, in such a way that, to coinciding x, coinciding $\phi(x)$ are assigned. It is with regard to this assignment $\phi(x)$ that we make the following successive statements:

first, to *each* natural number p_1, a natural number p_2 can be assigned, such that each two standard numbers coinciding with standard numbers whose arrows contain the same rod of order p_2, have a distance smaller than 2^{-p_1};

second (by virtue of the fan theorem), to each natural number p_2 a natural number p_3 can be assigned such that the first p_2 squares of $\phi(x)$ are everywhere completely defined by the first p_3 squares of x, so that to all standard numbers x whose arrows contain the same rod of order p_3, are assigned the same $k^{(p_2)}$ of $\phi(x)$, and to all standard numbers x coinciding with standard numbers whose arrows contain the same rod of order p_3, are assigned standard numbers $\phi(x)$ coinciding with standard numbers whose arrows contain the same rod of order p_2; | 145|

third to each natural number p_3 a natural number p_4 can be assigned, such 146
that two arbitrary standard numbers x with a distance $< 2^{-p_4}$ have overlapping $k^{(p_3+1)}$'s, so that both these standard numbers lie within one and the same $k^{(p)}$, hence coincide with two standard numbers x whose arrows contain the same rod of order p_3.

Consequently to each natural number p_1 has been assigned a natural number p_4 such that to each two standard numbers x with a distance $< 2^{-p_4}$ have been assigned standard numbers $\phi(x)$ with a distance $< 2^{-p_1}$, so that also to each two limiting number cores z with a distance $< 2^{-p_4}$ have been assigned limiting number cores $f(z)$ with a distance $< 2^{-p_1}$. *Precisely this is the meaning of saying that $f(z)$ is uniformly continuous.*

26
Ernst Friedrich Ferdinand Zermelo (1871–1953)

After studying at the universities of Berlin, Halle, and Freiburg, Zermelo was granted a doctorate from the University of Berlin in 1894 with a dissertation on the calculus of variations. In 1896 he published a famous paper in thermodynamics which applied the Poincaré recurrence theorem to show that the kinetic theory of gases can allow no irreversible processes (*Zermelo 1896*), and he then engaged in a celebrated discussion with Boltzmann on the nature of these processes. He became a *Privatdozent* at Göttingen in 1899 with a *Habilitationsschrift* on hydrodynamics. In 1910 he was appointed ordinary professor in Zürich, resigning in 1916 because of ill health. He lived without position in the Black Forest until 1926, when he was appointed honorary professor at the University of Freiburg im Breisgau. He resigned from this position in 1935 in anticipation of expulsion for his remarks against Hitler and the Third Reich and for refusing to give the Hitler salute (*Pinl 1969*, p. 221). He requested his position back after the war, a request which was granted in 1946.

Despite the importance of his work in mathematical physics, Zermelo is most famous for his contributions to set-theory, on which he lectured from at least as early as 1900–1. He discovered 'Russell's paradox' independently, very likely as early as 1899 or 1900.[a] In 1904 he published the first of two proofs of the well-ordering theorem (*1904*), following this in 1908 with the second proof (*1908a*),[b] and with the first axiomatization of set-theory (*1908b*). The work of these two 1908 papers is connected. The first prefaces the new proof with a discussion replying to the numerous attacks on the proof of *1904*, in particular to the use of the *axiom of choice*, formulated there for the first time.[c] Following this discursive part, Zermelo then sets out to show that a limited number of specific principles (including a version of choice) is sufficient to deduce the well-ordering theorem. These principles are very close to the system of axioms set out in *1908b* for the first time, a system which avoids the known antinomies, and which was to become the basis of modern set-theory.

Despite the enormous impact of this body of work, there are apparently no publications on set-theory between the two 1909 papers on the treatment of

[a] This calculation relies on a letter of Hilbert to Frege of 7 November 1903. (See *Frege 1976* or *1980*. See also *Husserl 1902* or *Rang and Thomas 1981*.)

[b] The mathematical relation between the two proofs is explained in *Hallett 1984*, §7.3.

[c] For the history of this, see *Moore 1982*.

arithmetic in set-theory (*1909a*, *1909b*) and those of the late 1920s and early 1930s. In 1932, Zermelo published a collected edition of the mathematical and philosophical works of Cantor, useful, apart from making available Cantor's papers, as a limited source of information for some of Zermelo's own views.[d] The *1908b* paper was called 'Investigations in the foundations of set-theory. I'. 'II' was never published, although the title of the 1908 paper explains the subtitle of the paper translated here—'Über Grenzzahlen und Mengenbereiche: Neue Untersuchungen über die Grundlagen der Mengenlehre' (*Zermelo 1930*).

The paper sets out to show that the standard set-theoretic axioms (what Zermelo calls the 'constitutive axioms', thus the *ZF* axioms minus the axiom of infinity), have an unending sequence of different models, thus that they are non-categorical. In his *1908b*, Zermelo proceeds by axiomatizing 'domains' of sets, imposing conditions on these domains by insisting that, if a domain contains some sets at all, then it must contain other sets of the kind the axioms specify.[e] Clearly the same approach lies behind this later treatment, except that here a good deal more is said about the structure of domains which can satisfy the basic axioms. In fact, he argues that these 'normal' domains can be uniquely described by specifying two numbers, the cardinality of the set of all urelements of the domain (the 'basis'), i.e. the set of all elements (apart from the empty set) which have no members, and the 'boundary number' of the domain, which he shows must satisfy the conditions (I) and (II) explained below. In doing this, Zermelo concentrates essentially on what have become known as 'natural models' of set-theory.[f] These are models given by certain levels in the so-called cumulative hierarchy, whose specification (urelements aside) depends only on specifying the ordinal indexing the level reached in this hierarchy, in effect what Zermelo calls the 'boundary number'. Thus, the 'division' Zermelo uses in the 'First Development Theorem' (p. 1225, below) is in effect the division of the normal domain P into the cumulative levels P_γ defined by $P_{\alpha+1}=P_\alpha \cup \mathcal{P}(P_\alpha)$, for successor ordinals, and $P_\alpha = \bigcup_{\beta<\alpha} P_\beta$ for ordinals, where $\mathcal{P}(x)$ stands for the 'power set of x'.[g] If P is a domain satisfying the axioms, then we arrive at the 'boundary number' of P by looking at the upper bound π of the set Z of all 'ordinals' in P, or rather the order type π of the (well-ordered) set G_u of all sets in P with a common initial urelement u which 'represent' ordinals (see below, p. 1222). Zermelo shows that the stage P_π is itself a normal domain and must be identical to P. He then proves that such 'boundary numbers' must satisfy his conditions (I) and (II), which amounts to showing that numbers indexing models of *ZF* in the cumulative hierarchy must be 'strongly inaccessible ordinals (cardinals)'—

[d] This is *Cantor 1932*.

[e] Zermelo just assumes that these domains exist. (See *1908b*, and below, p. 1219.) For a discussion of the relation between this assumption and Hilbert's views on mathematical existence, see *Hallett 1994*, pp. 59–65.

[f] See *Tarski 1956*, and *Montague and Vaught 1959*.

[g] The cumulative hierarchy was introduced in *von Neumann 1929*, and implicitly in *Mirimanoff 1917a*, for which see *Hallett 1984*, §4.4. Zermelo makes no reference to either work.

ordinals which are easily defined in ordinary set-theoretic terms.[h] Hence we see that the resulting models themselves can be specified completely in ordinary set-theoretic terms, providing only that one assumes the existence of the requisite ordinal numbers, the existence of which, however, cannot be guaranteed by the constitutive axioms for set theory, even with the usual axiom of infinity.

The conditions (I) and (II), and the notions surrounding them, can be explained as follows. A limit ordinal α is said to be *cofinal* with a limit number β if there is an increasing sequence of numbers ξ_γ indexed by β, with each ξ_γ less than α, such that $\lim \xi_\gamma = \alpha$.[i] (As an example, ω_ω is cofinal with ω.) A *singular* ordinal is one that is cofinal with a strictly smaller ordinal (for example ω_ω, which is cofinal with ω); a *regular* ordinal is one that is not.[j] Hausdorff (*1908*, p. 442; *1914*, p. 130) shows that an infinite regular ordinal must be an *initial* number, i.e. it must be one of the numbers ω_α, and thus the first number of a number class, and hence one such that any number smaller than it is of a smaller cardinality.[k] Zermelo's condition (I) is just that π must be an infinite and regular, and thus an initial, number.

But Zermelo points out that π must be much more than this, that it must be at least what he calls an 'exorbitant number', or what is now known as a *weakly inaccessible* number. The term 'exorbitant', and its origin, is explained as follows. A function f from ordinals to ordinals is called a 'normal function' (initially by *Hausdorff 1914*, p. 114) if it is monotone and continuous, i.e. if $\beta < \alpha$ implies that $f(\beta) < f(\alpha)$, and if $f(\alpha) = \lim f(\xi)$ whenever $\alpha = \lim \xi$. A 'critical number' (or a 'fixed point') of such a function is a number α such that $\alpha = f(\alpha)$. The function $f(a) = \omega(\alpha) = \omega_\alpha$ is clearly a normal function of α, and consideration of its fixed points is what gives rise to the characterization of 'exorbitant' numbers. Hausdorff notes that an infinite initial number ω_a whose index α is a limit number is, of course, cofinal with that index. Thus, since any number is certainly cofinal with itself, and since we obviously have $\alpha \leq \omega_\alpha$, such a number can only be regular if the index α actually *equals* ω_α, and thus if the number α is a fixed point of the function $\omega(\alpha)$. As Hausdorff points out, the smallest of these numbers is still singular. Thus:

> If there are regular initial numbers with a limit index, and so far it has not been possible to find a contradiction in this assumption, then the smallest of them is of such an exorbitant magnitude that it would scarcely come into consideration for the usual purposes of set-theory (*Hausdorff 1914*, p. 131).

---

[h] The term 'inaccessible' seems to be due to Sierpinski and Tarski (*1930*). They mean this to apply to ordinals (or cardinals) which would now be called 'strongly inaccessible'. See below.

[i] See *Hausdorff 1906*, p. 124; *1908*, p. 440; and *1914*, p. 86.

[j] See *Hausdorff 1908*, p. 442.

[k] In the current conception, due to von Neumann in his *1928b*, the cardinals $\aleph_\alpha$ are actually identified with the initial numbers ω_a. In the less formal Hausdorff tradition, on which Zermelo seems to be relying here, $\aleph_\alpha$ is the cardinality of ω_α, and thus, in Zermelo's notation, $\aleph_\alpha = \bar{\omega}_\alpha$.

'Exorbitant numbers' α are thus those which are infinite initial numbers ω_β, with β a limit ordinal, and which are also regular. This is exactly the definition of weakly inaccessible numbers. The existence of such numbers bigger than ω cannot be proved from the usual axioms of set theory (including the axiom of infinity).

As Zermelo observes, if 'Cantor's conjecture' were true generally, then 'exorbitant' or weakly inaccessible numbers would also be what have become known as *strongly inaccessible* numbers, i.e. would be (in the presence of the axiom of choice) infinite, regular, and such that if a cardinal number m is smaller than $\bar{\pi}$ (the cardinality of π), then 2^m is also smaller than $\bar{\pi}$. ('Cantor's conjecture', or course, is the generalized Cantor Continuum Hypothesis, which, when stated in terms of aleph numbers, says that $2^{\aleph_\alpha} = \aleph_{\alpha+1}$.) Thus, if π were weakly inaccessible, we would have a guarantee that a 'natural model' indexed by it would contain enough sets to satisfy the power set axiom. But, in the absence of any proof of the generalized Cantor conjecture, we appear to have no such guarantee, for there is no guarantee that if a cardinal number m is less than a weakly inaccessible π, then 2^m is also less than $\bar{\pi}$. Zermelo's second condition (II) on a boundary number π not surprisingly states that π must be of this kind. The first step is to define a normal function ψ from ordinals to ordinals in such a way that $\psi(\beta + 1)$ has the power of $2^{\overline{\psi(\beta)}}$. The condition is then that π must also be a 'fixed point' or a 'critical number' of this 'normal function', and this is equivalent to demanding that π must be strongly inaccessible.[l]

The claim that indices of 'natural models' of *ZF* must be strongly inaccessible is interesting. Tarski was the first to point out that if *Z* is Zermelo set-theory, i.e. the set-theory we get from Zermelo–Fraenkel (with the axiom of foundation) by adopting the axiom of separation in place of the the axiom of replacement, and if ZF^- is Zermelo–Fraenkel theory *without* the axiom of infinity, then the first ordinals for which natural models of that ordinal index exist for these theories are $\omega + \omega$ and ω respectively.[m] If π is the first infinite strongly inaccessible number $> \omega$, then there are certainly 'natural' models of *ZF* (including the axiom of infinity) which have rank π. However, Montague and Vaught (*1959*) indicate that having inaccessible rank is not, as Zermelo seems to establish, a *necessary* condition for being a model of *ZF*, providing *ZF* is taken to be a *first-order* theory. In other words, they show that there must be ordinals $\rho < \pi$ such that there are natural models of *ZF* with index ρ. However, Zermelo's strong inaccessibility condition does turn out to be both sufficient *and* necessary if *ZF* is taken to be a *second-order* theory, and all the evidence suggests that Zermelo *did* construe the axioms in this way.[n]

It is important to note that Zermelo does not just accept agnostically the possibility of inaccessible numbers, as others did before him: he actually makes

---

[l] For a treatment of inaccessible numbers, see *Levy 1979*, pp. 137–41.
[m] See *Tarski 1956*.
[n] See *Hallett 1984*, pp. 266–9.

use of them. Indeed, he makes clear his assumption that there is an *unbounded* sequence of such numbers, taking ω to be the smallest such, and thus that there is a corresponding range of models of the basic axioms, models with the same basis being ordered by inclusion according to the size of their 'boundary numbers'. These chains of models have the property that the non-sets of one model exist as sets in those succeeding.[o] Zermelo sees in this a 'satisfactory' clarification of what he calls the 'ultrafinite antinomies', presumably because a collection like Russell's can be thought of as both not a set (in one model) and a set (in all the succeeding models). According to Zermelo, the antinomies arise from mistaking a particular model (or domain, to use Zermelo's other term) $\mathfrak{B}$ for the complete or absolute universe of sets. The temptation to think that $\mathfrak{B}$ is itself actually a set will then lead to contradiction, for if $\mathfrak{B}$ is the complete universe of sets, and $\mathfrak{B}$ is a set, then $\mathfrak{B}$ must belong to $\mathfrak{B}$, from which contradictions will follow. But there is no contradiction if we simply allow (as Zermelo does) that $\mathfrak{B}$ is a set, not in $\mathfrak{B}$, but in a different, bigger domain $\mathfrak{B}^*$, in other words, that $\mathfrak{B}$ is *not* the completed universe of sets.

The temptation to think of $\mathfrak{B}$ as a set is revealed in Zermelo's plausibility argument for the existence of the unbounded sequence of 'boundary numbers', this being based, in effect, on the following. Consider the sequence of all the ordinals whose existence can be proved from Zermelo's basic axioms, thus all the ordinals in the smallest model $\mathfrak{B}$. This sequence must have an ordinal limit, whose existence certainly cannot be proved on the basis of those axioms, and this limit will be a boundary number. (For the argument, see below, p. 1231.) This is reminiscent, both of Russell's 'self-reproductive processes' and of Dummett's 'indefinitely extensible concepts', for whenever it seems as if we have all the ordinals, the very collection of these appears to give rise to an ordinal which cannot be in the original collection (see *Russell 1906a = 1973*, pp. 152–4, and *Dummett 1991*, p. 316). Plausible as Zermelo's argument sounds in the case of ordinals, these are themselves just sets—or, rather, mirror certain sets. Hence, the argument seems not far away from the far less plausible claim that, since all the objects in $\mathfrak{B}$ are sets (apart from the *Urelemente*), $\mathfrak{B}$ must be a set, too. Thus, what we have is a return to something like a principle of arbitrary set formation, a principle which Zermelo repudiates in his axiomatization of 1908 (see *1908b*, p. 261) in favour of the *Aussonderungsaxiom*. However, in the context of the 1930 paper, the principle gives rise, not to contradiction, but to the indefinite extensibility of the various natural models.

The term 'ultrafinite' to refer to both the antinomies and the putative sets involved in them was first introduced in Hessenberg's 1906 monograph (*Hessenberg 1906*, p. 628), in the introduction to which, incidentally, Hessenberg expresses his thanks to Zermelo for reading the proofs and for making available 'his own unpublished investigations' (p. 483).

Zermelo's mention of the Kantian antinomies in this context (p. 1233, below)

---

[o] Note that this will only be the case if one assumes a *second-order* formulation of the *ZF* axioms.

is worth pointing out. In his remarks in *Cantor 1932* (p. 377), Zermelo suggests that there is at least a 'formal analogy' between these and the set-theoretic antinomies. Such an analogy was also suggested by Hessenberg (*1906*, p. 633) and by Fraenkel (*1928*, pp. 212–13), and indeed by Hilbert in some lectures in 1905.[p]

As Zermelo stresses, the existence of non-isomorphic models for the *basic* axioms must mean that these axioms are non-categorical. One should not see in this a confirmation of the considerable body of work in logic from which it follows that the first-order Zermelo–Fraenkel axiomatization is highly non-categorical. For one thing, as was mentioned above, there is good reason to think that Zermelo had in mind, if anything like a modern formal system at all, a genuinely second-order formulation, where separation (*Aussonderung*), replacement (*Ersetzung*), and foundation (*Fundierung*) are regarded as single axioms containing a second-order quantifier over properties, and not as schemata.[q] For another, Zermelo does not claim that the axiom system for *ZF* has *elementarily equivalent* non-isomorphic models, i.e. non-isomorphic models which make exactly the same sentences of the language true. Indeed, it is clear that, although Zermelo's non-isomorphic models agree on the basic axioms, they disagree on those sentences which assert the existence of their own 'boundary' number. Furthermore, the same theorem that shows this (the First Isomorphism Theorem, p. 1228–9 below) also shows that domains whose 'basis' number (the cardinality of the set of urelements) is the same, and that have the same 'boundary number', *are* isomorphic. This is *not* so in the first-order case.

Zermelo briefly considers various attempts (he mentions those of Finsler, Fraenkel, and von Neumann) to reduce the multiplicity of models by adding new axioms. He dismisses these attempts because he sees them as imposing unnecessary restrictions by focusing on certain types of model, as opposed to 'set-theory itself'. This, says Zermelo, threatens the 'applicability of set-theory', which he regards as directly connected to the non-categoricity of the axioms, and thus to the 'length' of models, and also to their 'breadth', i.e., the variability in the number of urelements in the basis. (See pp. 1227 and 1232, below.) In Zermelo's view, 'set-theory as a science' must first be developed 'in the fullest generality', and only then ought attention to be directed to 'the comparative investigation of individual *models*' (below, p. 1232).

Zermelo mentions two axioms explicitly, that employed by von Neumann, to which we will come in a minute, and Fraenkel's 'Axiom of Restrictedness ⟦Beschränktheit⟧' (see *Fraenkel 1928*, p. 355), which says: *There are no sets other than those shown to exist by the basic axioms of the system.* In all probability, the axiom of Finsler referred to is his 'Axiom of Completeness',

---

[p] For a discussion of this analogy, see *Hallett 1984*, §6.2.

[q] The formulation of these key axioms turns partly on the resolution of the 'definiteness issue' mentioned by Zermelo in his note to p. 1220, below. For a discussion of this, see *Hallett 1984*, §§7.4 and 8.2.

consciously modelled after Hilbert's *Vollständigkeitsaxiom* for the real numbers and for Euclidean geometry. The axiom can be paraphrased (for this context) as follows: *The sets form a system of things incapable of extension without violation of the basic axioms* (see *Finsler 1926*, p. 691). Finsler's axiom system for set-theory is quite different from Zermelo's, however, and cannot be discussed here. For Hilbert's *Vollständigkeitsaxiom*, see *Hilbert 1900a*, Axiom Group IV, this volume, p. 1094.) Fraenkel's axiom is intended to restrict attention to the *smallest* model (without urelements) of the basic axioms, and in this respect is similar in intention to the induction postulate in the second-order formulation of the Dedekind–Peano axioms. Finsler, on the other hand, states that his axiom (in the spirit of Hilbert's *Vollständigkeitsaxiom*) is intended to pick out 'the greatest possible system' of sets (*1926*, p. 697.) Both axioms, therefore, deliberately seek to restrict the very generality which Zermelo sees as a desirable feature. Indeed, Fraenkel's axiom, when formalized as a schema in the language of ordinary set-theory (without urelements), enables one to prove that inaccessible ordinals (and thus Zermelo's uncountable 'boundary numbers') do not exist, thus confirming Zermelo's doubts. (This work is reported in *Fraenkel et alii 1973*, §6.4, where the philosophical import of Fraenkel's axiom is discussed at length.) As for Finsler's axiom, the maximality of the domain again rules out the variation in the size of the class of urelements that would seem to be necessary for the greatest possible flexibility in applications. And, quite apart from the fact that it rules out 'small' models of set theory, how much can actually be proved from it about the existence of large cardinals?

The axiom of von Neumann in question is one of his proper-class axioms, namely that which states: *All classes which are not also sets can be put into one-to-one correspondence with each other (respectively, with the class of all sets).* This principle is Axiom IV 2 in both *von Neumann 1925* and *von Neumann 1928b*. It implies the Global Axiom of Choice *and* the Replacement Axiom. See *Hallett 1984*, §8.3.) This axiom looks very different from the other two, for it is not metamathematical in formulation. However, as Gödel pointed out (see *Hallett 1984*, p. 291), it too might be regarded as a maximal principle (as Hilbert's *Vollständigkeitsaxiom* is), for it says that all collections which do not lead to inconsistencies when assumed to be sets really *are* sets, i.e., no proper-classes intermediate in size between sets and the whole universe can be squeezed in. However, in this respect, it is somewhat ambiguous, for it also says that there are the same number of ordinals as there are sets, which can be regarded either as a maximal principle (for ordinals) or a minimal principle (for sets). In any case, Zermelo's reason for disapproval is tied to his claim that, in his framework, while von Neumann's axiom holds for 'unit domains' (those domains satisfying his basic axioms but with only a single urelement in their 'basis'), it does not hold in general, i.e., when there is no such restriction on the 'basis'. Once again, Zermelo sees the axiom as a threat to the generality he values (see the Second Development Theorem and the subsequent discussion, below, pp. 1226–7).

The non-categoricity of the axioms which Zermelo demonstrates reveals an incompleteness of a sort, highlighted by the connection Zermelo draws between the 'ultrafinite antinomies' and indefinite extensibility, for this seems to show that there will always be sets (indeed, an unending sequence of them) that the basic axioms are incapable of revealing to be sets. Moreover, Zermelo's argument (adverted to above) is to the effect that the very fixing of a model reveals an ordinal that cannot be *in* that model. This incompleteness is worth a brief discussion.

One can see in this part of Zermelo's paper an implicit attack on Hilbert's axiomatic method, based as this is on the principle that a finite number of axioms, together with a logical system allowing only finite proofs according to certain rules, fully determines its subject-matter. Part of what Zermelo appears to argue for is that consideration of the (standard) axioms alone cannot suffice, for there is an unbounded sequence of very different models among which the axioms are incapable of distinguishing, and the mathematical characterization of these models (which Zermelo himself gives) involves existence assumptions which themselves go beyond anything that can be settled by the axioms. Indeed, the incompleteness touched on above can be given the sharp form: there are questions concerning the existence of ordinals and powers which cannot be settled on the basis of the axiom system alone. Once a formal link between consistency and existence is established, a connection emerges between the incompleteness revealed by Zermelo for set-theory and that for all first-order theories (of a certain character) demonstrated shortly afterwards by Gödel in 1931. (See *Hallett 1994* for a preliminary discussion.) In any case, Zermelo's paper indicates a reduction in importance of axiomatized set-theory in Hilbert's sense. (That Zermelo regarded the finitistic stress of Hilbert's approach to axiom systems as inadequate is confirmed by some of Zermelo's other views at around this time. See, for example, *Zermelo 1932*, and his letter to Gödel of 29.x.1931, published in *Grattan-Guinness 1979*.)

The comparison with Gödel is worth pursuing, although a full treatment is out of the question here. In the first place, Gödel, unlike Zermelo, retains Hilbert's conception of the centrality to established mathematics of circumscribed axiom systems, for he holds that a mathematically established result must be something stated in a specific, precisely circumscribed language and proved in an appropriate axiom system. (Gödel retains Hilbert's conception of an axiom system as something that is 'finitely presented', though he weakens 'finitely presented' to 'recursively presented', this again being a condition on the language, the deductive framework, and the list of axioms.) Secondly, Gödel was aware that the incompleteness of mathematics is far more pervasive than is indicated by the limited incompleteness of the basic axioms for set-theory that concerned Zermelo. The incompleteness results of 1931 (*Gödel 1931*) show that all axiom systems precise enough to satisfy the Hilbert conception are *necessarily* incomplete. (In his *1951*, Gödel says that this shows that what he calls 'proper mathematics' is inexhaustible or incompletable, proper mathematics being what is codified in an acceptable way in an axiom system, which is incomplete provided only that the system is of a certain minimal strength.) The incompleteness which concerned Zermelo is essentially that surrounding the (transfinitely) indefinite iterability of the power set operation. That discovered by Gödel has a less direct source, for it involves the formalization (via coding) of the concept of provability from the axioms. None the less, both have to do with the *consistency* of the axioms (in Zermelo's case, with the existence of set models for the axioms, in Gödel's case, with the existence of finitary, syntactic consistency proofs). Gödel assimilated them to each other in the following way. He observed that adding to the standard axioms new axioms of infinity allowing us to prove the existence of ordinals like Zermelo's uncountable 'boundary numbers' will render it *provable* in the new axiom system that the original, unextended axiom system has a model, and thus will render possible a proof of what the unextended system itself could not prove, namely the consistency of the (unextended) system. While such a piecemeal extension will not achieve completeness, it simultaneously closes some of the incompletenesses Zermelo reveals, and also renders decidable the Gödel sentence left undecidable by the original system, all this, of course, without deviating from Hilbert's conception of axiom systems, for the recursiveness of the system is preserved. Thus, for Gödel, as for Zermelo, there is an appeal to something over and beyond the axioms; for Gödel, this is a source of new *axioms*, and is not used as a direct attack on the inadequacy of the axiomatic method. And the new axioms considered by Gödel are quite unlike the extra axioms considered by Zermelo. For one thing, they involve no new conceptual machinery such as quantification over classes or models, just as Zermelo's characterization of boundary numbers uses no new machinery. For another, they do not attempt to put upper limits on the 'boundary numbers'.

This appeal to axioms yielding new ordinals as a way of tackling incompleteness is in keeping with Gödel's remark in the 1931 paper (*Gödel 1931*, n. 48a) which suggests that the incompletenesses revealed there can be closed by adding 'higher types' (even into the transfinite), since adding a new (limit) ordinal allows (via the cumulative hierarchy) the addition of (transfinitely many) power sets, each of which might be construed as being an increase in type (see also the lecture *Gödel 1933b*). It is also the origin of Gödel's 'large cardinals programme' in set-theory, which goes further, for it is based on the idea that *all* incompletenesses in set-theory (not just those of metamathematical origin) can be eventually overcome by the addition of strong enough axioms of infinity, such as (but not solely) the assertion of the existence of Zermelo's boundary numbers. The programme is based on the following conjecture put forward by Gödel in his short lecture from 1946 (*Gödel 1946*): For any true set-theoretic statement, we can find an axiom of infinity that renders possible the proof of this statement when this axiom is added to the basic system. Thus, Zermelo's unending sequence of models of the basic ('constitutive') axioms is mirrored in Gödel by an unending sequence of *axiom systems* extending the basic axioms. (Gödel's views on this are best represented in the two lectures *1933b* and *1946*, the two papers *1947* and *1964*, and the lecture *1951*.)

The most striking of the undecided propositions of set-theory of purely mathematical content is Cantor's Continuum Hypothesis (*CH*), the generalized version of which (*GCH*) is mentioned several times by Zermelo in his article. Zermelo implicitly refers to the fact that *GCH* has not yet been decided. By 1940 at the latest (see *Wang 1987*, p. 109), Gödel was already convinced that this proposition is undecidable by the standard axioms, a fact finally shown by Cohen's work of 1963. (In 1938, Gödel showed that $ZF + GCH$ is consistent if *ZF* is. Cohen succeeded in showing that $ZF + \neg CH$ is also consistent if *ZF* is, which finally shows that neither *GCH* nor $\neg GCH$ can be proved from *ZF* itself.) Unfortunately, despite the striking consequences of certain large cardinal axioms, no plausible such axiom has ever been proposed which even hints at a decision of the continuum problem. (For more on this, see *Fraenkel et alii 1973*, §6, or *Hallett 1984*, §2.3(*b*) and (*c*).)

Certain other features of Zermelo's paper are worth pointing out and clarifying.

Firstly, unlike other earlier writers, such as Fraenkel, Zermelo clearly allows that there might be urelements, that is, objects, other than the empty set, which themselves have no members. Indeed, he sees in this the possibility of wide-spread application of set-theory (see below, pp. 1227, 1232).

Secondly, Zermelo adopts what he calls the 'Zermelo–Fraenkel system', which is the core of what we now know by that name. (Zermelo denotes the axioms by the first letters of their German names. The one exception is the power set axiom, the axiom of the *Potenzmenge*. This is denoted by a capital 'U', which obviously stands for *Untermengen*—subsets.) The central differences between this and the 'Zermelo system' going back to *1908b* are, as Zermelo points out, the omissions of the axioms of choice and infinity, and the additions of the axioms of replacement and foundation as axioms of 'general' set-theory.[r] The replacement axiom was first proposed in *Fraenkel 1922a* and in *Skolem 1922*, though its potential was not shown until von Neumann's work in his *1925*, *1928a*, and *1928b*, all stemming from 1922.[s] The axiom of foundation appears in various forms before Zermelo in the work of von Neumann, and might even be suggested by various reflections of Mirimanoff and Skolem.[t] Zermelo insists that its adoption constitutes no *genuine* restriction (see below, p. 1220).

Thirdly, Zermelo introduces (in §2), and defines by transfinite induction, what he calls 'basic sequences'. These, apart from the fact that they can start

[r] The term 'general' was introduced by Fraenkel in his *1922a* (see p. 233), although he was not constant as to whether to allow the axiom of infinity as part of 'general set theory'. (See, for example, *Fraenkel 1927*, pp. 139–41, and *Fraenkel and Bar-Hillel 1958*, p. 81.) However, it seems that he never accepted replacement as a 'general' axiom.

[s] See *Hallett 1984*, §8.2.

[t] See *Skolem 1922* and *Mirimanoff 1917a*, and, for an analysis of Mirimanoff's work, *Hallett 1984*, §4.4. See also *Fraenkel et alii 1973*, Ch. II, §5.1, for a discussion of the various versions of foundation and for further references. The desire to exclude what Mirimanoff calls 'extraordinary' (i.e. non-well-founded) sets is part of what lies behind Fraenkel's 'Axiom of Restriction', the principle criticized by Zermelo on p. 1232, below. See *Fraenkel 1923*, pp. 218–19.

with any urelement and not just with the empty set, are really the von Neumann ordinals, which stem, in published form, from *von Neumann 1923*. The use of 'basic sequences' to 'represent' ordinals is quite different from the way Zermelo represents the finite ordinals (the natural numbers) in his earlier *1908b*, where 0 is identified with the empty set, and $n + 1$ is defined as $\{n\}$. However, according to Bernays (*1941*, p. 6), Zermelo was in possession of what has become known as the 'von Neumann conception' as early as 1915, and the ability to represent the ordinals by sets of this kind is already exhibited in *Mirimanoff 1917a* and *1917b*.[u] Bernays's remark ought to be taken seriously, not least because it is supported by comments Zermelo made to Hilbert in a postcard from 1913, by remarks of Bernays himself to Zermelo in 1920, and by von Neumann's statement (in his *1928a*, p. 321 of the reprinting in *von Neumann 1961*) that Zermelo informed him that he had such a conception already in 1916. (The direct evidence is rather mixed. Zermelo made somewhat sketchy notes on this conception of the ordinals, notes which may go back to 1915. See *Hallett 1984*, pp. 277–80.) It seems clear that Zermelo, Mirimanoff, and von Neumann all arrived at the von Neumann conception independently. Nevertheless, it should be noted that Zermelo does not *identify* the sets so defined with the ordinals, and neither does Mirimanoff: these sets merely 'represent' ordinals.

In the light of this, it is worth pointing out that Zermelo does not acknowledge von Neumann's public priority in any of the possession or use of the axiom of foundation, the ordinal numbers, or the cumulative hierarchy. In particular, there is no reference to *von Neumann 1929*, where the cumulative hierarchy is used to prove the relative consistency of the von Neumann axiom IV 2, despite the fact that Zermelo points out, as mentioned above, that von Neumann's axiom must hold in any cumulative hierarchy (or in any 'unit domain') indexed by one of his 'boundary numbers'. The fact that he acknowledges *other* contributions of von Neumann, as well as those of Fraenkel, Baer, Hausdorff, and Tarski, leads one to think that he discovered these things *independently*, which in any case, we have good reason to think in the case of the ordinals.

Valuable works dealing with Zermelo and his contributions are: *Fraenkel 1953*, *Fraenkel 1967*, *Fraenkel and Bar-Hillel 1958*, *Fraenkel et alii 1973*, *Hallett 1984*, *Moore 1980*, *Moore 1982*, and *van Rootselaar 1976*.

In his article, Zermelo uses both the mathematical sign '$\in$' and the lower-case Greek 'ε' to stand for the 'basic relation' of membership. It is not clear what difference this is meant to reflect, but the translation respects both uses.

References should be to the section numbers, which appeared in the original edition.

Michael Hallett

[u] See *Hallett 1984*, pp. 273–5.

A. ON BOUNDARY NUMBERS AND DOMAINS OF SETS: NEW INVESTIGATIONS IN THE FOUNDATIONS OF SET THEORY (*ZERMELO 1930*)

The following work investigates 'domains' composed of sets and urelements in which are satisfied the 'general' axioms of set-theory, the 'Zermelo–Fraenkel axioms' with one addition. It presents a demonstration that a 'normal domain' of this kind is determined up to isomorphic mappings by two numbers, by the power of its 'basis', that is, by the cardinality of the totality of the domain's 'urelements' (which are not proper sets), and by its 'characteristic', the ordinal type of all the 'basic sequences' contained in it or, alternatively, of all ordinal numbers which can be represented in it by sets. It is shown that both of these numbers can be chosen arbitrarily and independently of one another, providing the 'characteristic' is a 'boundary number', namely, that it is simultaneously a 'core number ‖Kernzahl‖', or a 'regular initial number', and an 'eigenvalue' or 'critical number' of a certain 'normal function'. In this way, the boundless ‖schrankenlose‖ extendability of the transfinite number series permits the presentation of set-theory itself as an unlimited sequence of well-differentiated 'models'. The sharp differentiation between these models of the (non-categorical!) axiom system yields a satisfactory clarification of the 'ultrafinite antinomies', for it is always the case that the 'non-sets' of one model appear as 'sets' both in the the next model and in all subsequent ones.

The 'basic sequences' are one important element in the investigation; these are the simplest representatives of the ordinal numbers present in each normal domain. A second important feature is the 'development' of a normal domain, its division into a well-ordered sequence of separated 'layers'. According to this, the sets of a given layer are always 'grounded ‖würzelnd‖' in the preceding layers in the sense that their elements are in these and even serve as material for following layers.

§*1* *The constitutive axioms*

The axiom system for set-theory which underlies our investigation is, in essentials, that of 'Zermelo–Fraenkel', namely my axioms of 1908[1] with Fraenkel's 'replacement axiom' added. But there are two alterations. Firstly, my 'axiom of infinity' is omitted, since it does not belong to 'general' set-theory; and secondly the 'axiom of foundation' is added in order to exclude 'circular' or 'non-grounded' ‖abgründige‖ sets. Accordingly, we denote by the term 'extended *ZF* system' or '*ZF'* system' the following axioms:

(B) *Axiom of definiteness ‖Bestimmheit‖*: Every set, in so far as it possesses elements, is completely determined by them.

[1] *Mathematische Annalen*, **65**, pp. 261–81.

(A) *Axiom of separation* ‖*Aussonderung*‖: Given any propositional function $\mathfrak{f}(x)$, it is possible to separate from any set m the subset $m_{\mathfrak{f}}$ consisting of just those elements x for which $\mathfrak{f}(x)$ is true. Or: to each part of a set there corresponds a set which contains all the elements of that part.[2]

(P) *Axiom of pairing* ‖*Paarung*‖: If a and b are any two elements, then there exists a set which contains both of them as members.

(U) *Axiom of power set* ‖*Potenzmenge*‖: For every set m there is a set $\mathfrak{U}m$ containing all subsets ‖Untermengen‖ of m as elements, including the null set and the set m itself. An arbitrarily chosen 'urelement' u_0 replaces the 'null set' here.

(V) *Axiom of union* ‖*Vereinigung*‖: For every set m, there is a set $\mathfrak{S}m$ which contains all elements of the elements of m.

(E) *Axiom of replacement* ‖*Ersetzung*‖: If one replaces the elements x of a set m in a unique way by any arbitrary elements x' of the domain, then the domain also contains a set m' which has all the objects x' as elements.

(F) *Axiom of foundation* ‖*Fundierung*‖: Every (decreasing ‖rückschreitende‖) chain of elements, in which each term is an element of the preceding term, breaks off with finite index at an urelement. Or equally: Every partial domain T contains at least one element t_0 none of whose elements are in T.

This last axiom, which excludes all 'circular' sets, and thus all 'sets that contain themselves', and in general all 'groundless' sets, has always been satisfied in all practical applications of set-theory. Thus, for the time being, it presents no essential restriction of the theory.

We shall not be concerned with the 'independence' of the axioms here. When appropriately interpreted, axiom (A) can be derived from (E), and also (P) from (U) and (E). The 'axiom of choice' is not expressly formulated here, since it has a different character from that of the other axioms and cannot serve in the delimitation of the domains. However, our whole investigation assumes it as a general logical principle; consequently it is taken for granted in what follows that every set is capable of being *well-ordered*.

We take as starting-point the axiom system BAPUVEF, the '*ZF' system*', and we call a 'normal domain' a domain consisting of 'sets' and 'urelements' which satisfies the *ZF'* system with regard to the 'basic relation' $a \in b$. We will treat 'domains' of this kind, their 'elements', their 'subdomains', their 'sums' and 'intersections', exactly like *sets*, and thus according to general set-theoretical concepts and axioms, for there is no means of distinguishing them from sets in any way which essentially matters. However, we will always denote

---

[2] Here the propositional function $\mathfrak{f}(x)$ can be quite *arbitrary*, as can the replacement function in (E). Thus none of the consequences of restricting these functions to a particular class are relevant for the point of view taken here. I withhold a thoroughgoing exposition of the issue of "definiteness", except to cite my last note in this Journal (*Fundamenta Mathematicae*, **14**, pp. 339–44) and the critical "Remarks" thereon of T. Skolem (*ibid.*, **15**, pp. 337–41).

them as 'domains' and not as 'sets' in order to distinguish them from the 'sets' which are the elements of the domain in question.

§2 *The basic sequences of a normal domain and its characteristic*

I call a '*basic sequence*' a *well-ordered set in which each element* (*with the exception of the first, which must be an 'urelement'*) *is identical with the set of all elements preceding it.*

Thus the following basic sequences arise from the urelement u:

$$g_0 = u,\ g_1 = \{u\},\ g_2 = \{u, \{u\}\},\ g_3 = \{u, \{u\},\ \{u, \{u\}\}\},$$

and so on according to the rule

$$g_{\alpha+1} = g_\alpha + \{g_\alpha\}, \text{ and } g_\alpha = \sum_{\beta < \alpha} g_\beta$$

whenever α is a limit number.

Generally put, a basic sequence is a *set which is ordered by the ε-relation*, and this ordering, by (F), must be a *well-ordering*. As is easily verified, the following theorems hold, among others:

(1) Each element of a basic sequence is an element of all succeeding sequences, and it contains all preceding sequences as elements.

(2) Each element of a basic sequence is itself a basic sequence, as is each segment ‖Abschnitt‖ of a basic sequence.

(3) If we add to each basic sequence ‖g‖ the set itself as the last element, then we get a new basic sequence $g' = g + \{g\}$. In this way, the ordinal type is increased by precisely 1.

(4) If T is a set of basic sequences with the same initial element u, then the union $\mathfrak{S}T$ is also a basic sequence. $\mathfrak{S}T$ contains all the elements of T both as segments and, apart from itself, as elements. The ordinal type of the new sequence is also the one which follows next after all those given ‖by T‖.[a]

(5) Given two distinct basic sequences with identical initial elements, then the one with smaller ordinal type is always a segment, and a member, of the other. For this reason we call this sequence the 'smaller' of the two.

(6) If u is an urelement in a normal domain and r is a well-ordered set of type ρ in that domain, then the domain also contains a basic sequence g_ρ similar to r and with u as initial element.

For, assuming that the theorem is correct for all ordinal numbers $\rho < \alpha$, then it also holds for $\rho = \alpha$. If $\alpha = \beta + 1$ and g_β has the type β, then, according

[a] ‖This is not generally correct. If T is a set of ordinals ('basic sequences'), then $\bigcup T$ is greater than or *equal to* all the ordinals ('basic sequences') in T. Zermelo's statement is only correct if T contains no greatest element.‖

to (3), g' has the type $\beta + 1 = \alpha$. If, on the other hand, α is a limit number, then the union $\Sigma\, g_\beta$ of all g_β for $\beta < \alpha$ is itself a basic sequence, according to (4), and indeed of type α, since each of its proper segments itself is a $g_\beta < g_\alpha$.

(7) The totality of *all* the basic sequences g_α in a normal domain P starting from a common initial element u forms a well-defined subdomain G_u of P, and the corresponding ordinal numbers α form a well-defined segment Z_π of the number series with ordinal type π. But the domain P contains no 'set' w which has all these basic sequences as elements. In fact it can contain no well-ordered set of ordinal type π. Rather π is only the *upper-limit* of all those ordinal numbers which are represented in P by sets, for otherwise the 'Burali-Forti antinomy' would arise.

We shall call the ordinal number π so defined the 'boundary number' or the 'characteristic' of the normal domain ‖P‖. π is by no means arbitrary, however; it must satisfy certain conditions in order to be a 'boundary number'. These are the following:

(I) Every boundary number is a '*core number* ‖*Kernzahl*‖', i.e. a 'regular initial number', which means that it is *not cofinal with any smaller number.*[3]

If π were cofinal with any $\rho < \pi$, then the segment Z_π of the number series would contain a partial sequence of ordinal type ρ, consisting of numbers $\alpha_\nu < \pi$ which would belong to no proper segment $Z_\alpha < Z_\pi$. For each of these numbers α_ν, there would then be in P a basic sequence g_{α_ν} of the same ordinal type; thus, according to (4), the union of all these g_{α_ν} would again be a basic sequence of the normal domain. But, according to the assumption, the ordinal number of this sequence must be $\alpha = \lim \alpha_\nu = \pi$ ‖which is impossible‖. Thus π is a 'core number' or a 'regular initial number'. Indeed, as we can see, ‖being a 'boundary number',‖ it must actually be a regular initial number of the 'second kind',[b] and thus an 'exorbitant' number (Hausdorff, *op. cit.*, p. 131[c]). Suppose that π were equal to some $\omega_{\nu+1}$, then $\omega_\nu < \pi$, and the domain would contain a basic sequence g_{ω_ν} of this type. It would also contain its power set $m = \mathfrak{U} g_{\omega_\nu}$, which would have the cardinal number $\mathfrak{m} > \bar{\omega}_\nu$, in which case we would have $m \geq \bar{\omega}_{\nu+1} = \bar{\pi}$, which contradicts the definition of π.

If Cantor's conjecture that the power set $\mathfrak{U}m$ always has the next highest power ‖to that of m‖ were demonstrated, then ‖, if π were an 'exorbitant number',‖ $2^{\mathfrak{m}} < \bar{\pi}$ would always follow from $\mathfrak{m} < \pi$, and thus every 'exorbitant' number π would thus be the 'boundary number' of a normal domain.[4]

---

[3] Compare F. Hausdorff, *Grundzüge der Mengenlehre*, first edition, Ch. IV, §4. ‖'Cofinal' is defined in §4 (p. 86) of Hausdorff's book, though the application of this to the definition of 'regular' and 'singular' numbers does not come until Ch. V, §6, p. 130.‖

[b] ‖By 'second kind' here Zermelo means that it must have a limit number as index. Following Cantor's use, successor ordinals were known as 'ordinals of the first kind', and limit ordinals 'ordinals of the second kind'. See *Cantor 1883d*, translated above.‖

[c] ‖See the introductory comments.‖

[4] Compare R. Baer, 'Zur Axiomatik der Kardinalzahlarithmetik', *Mathematische Zeitschrift*, **29**, p. 382f., including footnote 8 on p. 382. ‖Footnote 8 is actually on p. 383. Zermelo might have meant footnote 3, which *is* on p. 382.‖

However, since this matter is still undecided, we need a further condition ‖beyond just being an exorbitant number‖ for the characterization of the 'boundary numbers', a condition which can be provided by making use of a certain 'normal function'.[5]

Suppose ξ is an arbitrary ordinal represented in a normal domain. Then, as well as the basic sequence g_ξ, the domain also contains, by (U), a basic sequence with the index $\xi^* = \phi(\xi)$, the initial number of the number-class with the cardinal number $2^{\bar{\xi}}$.[d] This function $\phi(\xi)$ is not itself a normal function, since the same functional values can correspond to different arguments ξ. However, we can define such a function by iterating ϕ as follows:

$$(1)\quad \psi(0) = 0,\ (2)\ \psi(\xi + 1) = \psi(\xi)^* = \phi\psi(\xi),\ (3)\ \psi(\alpha) = \lim_{\xi < \alpha} \psi(\xi)$$

if a is a limit number. In this way, the function ψ is uniquely determined for arbitrary arguments ξ, and the conditions of being a normal function are fulfilled. For, from $a < \beta$ it follows that $a + 1 \leq \beta$, and therefore, by transfinite induction,

$$\psi(\alpha) < \psi(\alpha)^* = \psi(\alpha + 1) \leq \psi(\beta)$$

And from (3), it follows generally that $\lim \psi(\alpha_\nu) = \psi(\lim \alpha_\nu)$, and thus that the function is 'continuous'. It is not just the case that the *whole* number series is mapped on to a similar part of itself by the function $\psi(\xi)$, but that in addition each *segment* Z_π corresponding to a normal domain is also mapped onto a part of itself. In other words, if $\alpha < \pi$, then $\psi(\alpha) < \pi$, as can be shown by induction. Assume that it is always the case that $\psi(\xi) < \pi$ for all $\xi < \alpha$. Then, if α is of the first kind ‖and $= \xi + 1$‖, ‖$\psi(\alpha) =$‖ $\psi(\xi + 1) = \psi(\xi)^* < \pi$, because the normal domain contains the power set $\mathfrak{U}m$ whenever it contains m. Thus $\psi(\alpha) < \pi$. On the other hand, let α be a limit ordinal. The elements of the basic sequence g_α are themselves basic sequences g_ξ of smaller type. To these there correspond uniquely the basic sequences $g_{\psi(\xi)} < g_\pi$ ‖since, by induction hypothesis, $\xi < \pi$ entails $\psi(\xi) < \pi$‖. Thus, according to (E), these latter are themselves elements of a set in P, and their union $\Sigma\, g_{\psi(\xi)}$, according to (4), is itself a basic sequence g_ρ of the normal domain. However, $\rho = \lim\ \psi(\xi) = \psi(a)$, and therefore, as asserted, $\psi(a) < \pi$. ‖Finally, $\psi(\pi) = \pi$.‖ For $\psi(\pi) = \lim_{\alpha < \pi} \psi(\alpha)$; thus, if it were the case that $\pi < \psi(\pi)$, then there would be an $\alpha < \pi$ for which $\psi(\alpha) > \pi$, and this would contradict what has already been proved. ‖Thus, since we must have $\pi \leq \psi(\pi)$, $\pi = \psi(\pi)$‖ in this way we arrive at the *second condition*:

---

[5] See Hausdorff *op. cit.*, Ch. V, §3, p. 130. With respect to the normal function $\psi(\xi)$ used here, compare also A. Tarski, *Fundamenta Mathematicae*, 7, pp. 1–5. ‖It should be noted that p. 130 of Hausdorff's book does not fall in §3 but rather in §5. However, although p. 130 is important for the definitions of regularity and singularity, etc., Zermelo probably means to direct attention to p. 116 of §3. Here Hausdorff shows how to turn an ordinal function $\phi(\alpha)$ such that $\phi(\alpha) > \alpha$ into a *normal* function in just the way that Zermelo does to obtain his ψ.‖

[d] ‖Zermelo has $2^{\bar{\xi}*}$ here, which makes little sense.‖.

(II) *Every 'boundary number' or 'characteristic' of a normal domain is also an 'eigenvalue' or a 'critical number' of the normal function $\psi(\xi)$ defined above.*

These two conditions on 'boundary numbers' are essentially independent of each other, in so far as (I) just postulates the core number property. That there can be no core number of the *first* kind then follows immediately from the second condition. For two successive (transfinite) initial numbers ω_ν and $\omega_{\nu+1}$ we have namely $\omega_\nu < \omega_\nu + 1 < \omega_{\nu+1}$, and therefore

$$\omega_{\nu+1} \leq \omega_\nu^* \leq \psi(\omega_\nu)^* = \psi(\omega_\nu + 1) < \psi(\omega_{\nu+1})$$

Thus, $\omega_{\nu+1}$ is certainly not an eigenvalue of the normal function ‖ψ‖. However, if Cantor's conjecture were correct, then each 'exorbitant' number, each 'core number of the second kind', automatically would fulfil the second condition.[6] For in this case, if $\psi(\xi) = \omega_\xi$ for all transfinite ξ, and if $\xi < \pi$ always implies $\omega_\xi < \pi$, then the normal function ω_ξ would have eigenvalues $< \pi$, and π, as the limit of all these eigenvalues, would itself be an eigenvalue, i.e. $\pi = \omega_\pi$. However, the conjecture remains undecided for the moment, and it will be shown in the following that the two conditions on 'boundary numbers' as set out are indeed *sufficient* in the sense that every number π which satisfies both can be the characteristic of a normal domain.

§3. The development of a normal domain

We denote by 'normal domain' any domain of sets and urelements which satisfies the *ZF′* axioms. Such a normal domain can also possess partial domains whose ε-relation, holding between the elements of this partial domain, itself satisfies the axioms. In fact, the following lemma holds:

Lemma. *A partial domain M of a normal domain P is itself a normal domain if* (1) *along with each of its sets it also contains the elements of this set, and conversely* (2) *if M contains all the elements of a set m belonging to P, then it also contains m itself. If in addition M also includes ‖umfaßt‖ the whole 'basis' of the total domain P, then it is identical with P.*

Assume that these conditions are satisfied for M, and that the *ZF′* axioms hold for P; then these axioms also hold in M. The axioms (U) and (V) hold because of (2). As for the replacement axiom, naturally this must be understood as saying that the elements x' which replace the elements x again belong to the partial domain M. In the particular case where M includes the whole basis, then the remainder domain $R = P - M$ does not contain any urelements, and each of the elements of R would then be a set r whose elements, by assumption, would not all be in M. Thus some of them must be in R, and this then con-

[6] Compare R. Baer, *op. cit.* as in n. 4, p. 33 ‖1222‖.

tradicts the axiom (F).[e]

On the other hand, it can easily happen that a normal domain P ‖properly‖ contains another normal domain with smaller basis $Q' \subset Q$. Such a case arises from the domain P ‖with basis $Q \supset Q'$‖ if we restrict ourselves to just those sets ‖m‖ from which all descending element chains $m \ni m_1 \ni m_2 \ni m_3 \ldots$ end, not only in urelements, as they must according to (F), but in urelements from Q'.

First development theorem. *Every normal domain P with characteristic π can be divided into a well-ordered sequence of non-empty and pairwise disjoint 'layers' Q_α of type π, in such a way that each layer Q_α includes all the elements of P which do not appear in an earlier layer and whose elements belong to the associated 'segment' P_α, i.e. to the sum of the preceding layers* ‖$= \sum_{\beta<\alpha} Q_\beta$‖. *The first layer Q_0 includes all the urelements.*

The partial domains *or 'segments'* P_α are defined by transfinite induction as follows:

(1) $P_1 = Q_0 = Q$ contains the whole basis, the totality of urelements.

(2) $P_{\alpha+1} = P_\alpha + Q_\alpha$ contains all those elements of P 'grounded' in P_α, i.e. all those elements whose members fall in P_α.

(3) If α is a limit number, then P_α is the sum or union of all preceding P_β with smaller indices $\beta < \alpha$.

By means of these conditions, every P_α, and hence also $P_\pi = \sum_{\alpha<\pi} P_\alpha$, is uniquely determined by the preceding ‖P_β‖, and satisfies the conditions of the theorem, since we can prove that P_π is identical with P. Every layer $Q_\alpha = P_{\alpha+1} - P_\alpha$ contains the basic sequence g_α with the same index, as can be shown by induction. Assume $g_0 = u$, then this lies in $Q_0 = P_1$. Suppose now that the statement holds for all numbers $\beta < \alpha$. According to this, all the elements of g_α, which are themselves basic sequences g_β, are in preceding layers Q_β, and thus also in P_α. Hence g_α itself is in $P_{\alpha+1}$. But g_α is not in P_α, since otherwise it would belong to a level Q_β, a layer which already contains g_β, an element of g_α, contradicting the conditions of the construction.[f] Thus g_α lies in Q_α, and this layer is non-empty.

We can now apply the above 'lemma' to the sub-domain P_π of P. We consider a set r of the normal domain ‖P‖, all of whose elements r_ν are taken to be in P_π. We assume each r_ν is in the layer Q_{α_ν}. The ordinal numbers α_ν, not

---

[e] ‖Assume that r is in R. Since r does not belong to M, then there must be at least one element a_1 of r that does not belong to M, and which thus also belongs to R. Similarly, a_1 must contain an element a_2 which does not belong to M, and so on. In this way we can construct an infinite descending membership chain of sets which all belong to P, thus contradicting the fact that P is supposed to satisfy axiom (F).‖

[f] ‖If g_α were actually in Q_β, then the conditions on the construction of Q_β would demand that $g_\beta \notin Q_\beta$.‖

all necessarily different, form a well-ordered set of type $\rho < \pi$, since the power of ρ cannot be greater than that of r. However, since π is a 'boundary number', then by (I) it cannot be cofinal with any smaller number. Hence, the a_ν possess an upper limit $\alpha < \pi$, and all the r_ν, which are elements of r, must already be contained in P_π. Thus r itself is in $P_{\alpha+1}$, and consequently also in P_π. The sub-domain ⟦P_π⟧ thus contains all sets of P which are 'grounded' in it ⟦condition (2) of the lemma⟧, as well as all elements of its elements ⟦condition (1) of the lemma⟧. Thus, since it also contains the whole basis, ⟦by the lemma⟧ it is identical with the normal domain being developed. The proof of our theorem is thus complete.

Let us call a 'unit domain' a normal domain with 'basis number 1', i.e. a domain which arises from a *single* urelement. Then we have the following theorem concerning the development of such a domain:

Second development theorem. *In the development of a unit domain, every segment P_α has the power of $\psi(\alpha)$. P_α contains only sets of ⟦strictly⟧ smaller cardinal number ⟦than that of $\psi(\alpha)$⟧, while the corresponding layer Q_α already contains sets of this power. Every segment of the first kind $P_{\beta+1}$ contains as sets all sub-domains of the immediately preceding P_β, and every segment of the second kind contains all preceding segments and their sub-domains. The unit domain itse has the power of its characteristic π, and contains as sets all its sub-domains of smaller power.*

The proof of this is again carried out by transfinite induction. Assume that the assertion for P_α is correct when stated for all smaller indices $\beta < \alpha$. It certainly holds for $\beta = 1$, $P_1 = Q$ and $\psi(1) = 1$. Now let $\alpha = \beta + 1$—in other words, α is of the first kind. According to the assumption, P_β has the power of $\psi(\beta) < \psi(\pi) = \pi$, and contains only sets of smaller cardinal number than $\psi(\beta)$, namely all smaller segments and their subsets. Since $\alpha = \beta + 1$, $P_\alpha = P_\beta + Q_\beta$, and P_α certainly contains *all* subsets of P_β, and thus only such subsets, since each subset of a smaller segment is also a subset of the larger segment. Thus $P_\alpha = P_{\beta+1}$ is of the power $\mathfrak{p}_\alpha = 2^{\overline{\psi(\beta)}} = \overline{\psi(\beta+1)} = \overline{\psi(\alpha)}$, but contains only sets with cardinality $\leq \overline{\psi(\beta)} < \overline{\psi(\alpha)}$. On the other hand, the associated layer Q_α contains a set with cardinal number $\overline{\psi(\alpha)} < \bar{\pi}$, namely the set P_α itself, which must be in P according to the replacement axiom. However, Q_α contains no larger set, since it is only formed from subsets of P_α. If α is a limit number $< \pi$, then $P_\alpha = \sum_{\beta<\alpha} P_\beta$ is the sum of all smaller segments P_β, which, according to assumption, possess the required property. P_α thus contains as elements only subsets of these domains, and indeed *all* such subsets, since each subset of P_β is contained in the following segment $P_{\beta+1}$ as an element. Each of these sets has a cardinal number not larger than $\overline{\psi(\beta)} < \overline{\psi(\alpha)}$, and the power of the segment P_α itself is therefore given by

$$\mathfrak{p}_\alpha = \lim_{\beta<\alpha} \mathfrak{p}_\beta = \lim_{\beta<\alpha} \overline{\psi(\alpha)} = \overline{\psi(\alpha)} \leq \bar{\pi}$$

Thus, for every $\alpha < \pi$, the layer Q_α has the property claimed, namely that it contains sets of cardinal number $\overline{\psi(\alpha)}$ but of no larger cardinality, since it contains just P_α itself and all of its subsets. For $\alpha = \pi$, however, the power of P is $\lim_{\alpha<\pi} \overline{\psi(\alpha)} = \overline{\psi(\pi)} = \bar{\pi}$. To each subdomain of smaller power than π there corresponds in P an equivalent basic sequence, and therefore, according to (E), also a set which includes all its elements. For *unit domains*, therefore, the *von Neumann* 'axiom' holds, according to which *only* those partial domains which are of the same power as the total domain are 'too big' to be 'sets'.[7] However, this by no means holds for *all* normal domains; and restricting set theory just to 'unit domains' would rob it for the most part of its applicability.

On the basis of this, we can now modify the development of an *arbitrary* normal domain in such a way that every 'layer' Q_α contains *only* those sets from P whose cardinal number is not greater than would be the case for unit domains, namely those which are $\leq \overline{\psi(\alpha)}$. Nevertheless, all the sets of P are eventually included, since for $\alpha < \pi$ we always have $\psi(\alpha) < \psi(\pi) = \pi$, and because $\pi = \lim_{\beta<\alpha} \psi(\alpha)$, each $\rho < \pi$ is eventually overtaken by a $\psi(\alpha)$. The '*canonical*' development which arises in this way has the advantage that any single layer Q_α can be completely determined, independently of the total domain and its characteristic, solely by its index and by the 'basis' of the normal domain. Thus, for a given basis, the different developments corresponding to the different boundary numbers always coincide at first.

Third development theorem. (Theorem of "canonical" development.) *Every normal domain with basis Q can be divided into a well-ordered sequence of separated 'layers' Q_α, beginning with Q, such that again each Q_α has the sum of the preceding layers as a 'segment', and such that each Q_α contains as sets all those sub-domains of the associated segment P_α which are not themselves in that segment and which do not have a power greater than $\psi(\alpha)$. This last restriction is not needed in the case of 'unit domains', where 'free' and 'canonical' developments coincide. With a 'canonical' development, every segment P_τ, whose index τ satisfies the 'boundary number' conditions (I) and (II), is itself a normal domain.*

The proof is analogous to that of the First Development Theorem. As we did there, we begin by defining the segments P_α and the layers Q_α inductively by the conditions:

(1) $P_1 = Q_0 = Q$.

(2) $P_{\alpha+1} = P_\alpha + Q_\alpha$ contains all sets of the normal domain P whose elements lie in P_α and whose cardinal numbers are $\leq \overline{\psi(\alpha)}$.

(3) For limit numbers α, $P_\alpha = \sum_{\beta<\alpha} P_\beta$, i.e. the sum of all smaller P_β.

---

[7] See J. von Neumann, 'Die Axiomatisierung der Mengenlehre', *Mathematische Zeitschrift*, **26**, pp. 669–752 (1928). The axiom mentioned is Axiom IV 2, and is explained on pp. 677ff. ⟦in particular on pp. 679–80⟧.

In this way, the segments P_α and the corresponding layers $Q_\alpha = P_{\alpha+1} - P_\alpha$ are uniquely determined for all indices $\alpha \leq \pi$, and P_π is a well-determined subdomain of P which, according to the 'lemma', is identical with P, since P_π contains the whole basis, the elements of its elements, as well as all the sets which are 'grounded' in it. This last condition is fulfilled, since every set r with cardinal number $\bar{\rho} < \bar{\pi}$ grounded in P_α appears at the latest in the layer Q_ρ because of the fact that $\rho \leq \psi(\rho)$.

Now let $\tau \leq \pi$ be a boundary number, and let P_τ be the segment of the canonical development of P which corresponds to it. Then, according to the construction, P_τ contains only sets with cardinal number $\bar{\rho} \leq \overline{\psi(\alpha)} < \overline{\psi(\tau)} = \bar{\tau}$, since each such set must belong to a layer Q_α for some $\alpha < \tau$. Conversely, each set r with cardinality $< \tau$ which is grounded in P_τ must be grounded already in some smaller segment P_α, since τ is a 'core number' ⟦and not cofinal with any smaller α⟧, and hence it must belong to a larger ⟦segment⟧ P_γ, where γ needs to be no bigger than the ordinal type ρ of r, since $\rho \leq \psi(\rho)$ ⟦which is $< \psi(\tau) = \tau$⟧. Thus P_τ actually contains all sets of P which are grounded in it and whose cardinal numbers are smaller than $\bar{\tau}$, in particular all sets formed by 'replacement' inside P. If r is an arbitrary set in P, well-ordered with type $\rho < \tau$, then $\psi(\rho) < \psi(\tau) = \tau$ and $2^{\bar{\rho}} \leq 2^{\overline{\psi(\rho)}} = \overline{\psi(\rho+1)} < \bar{\tau}$, and thus $\mathfrak{U}r$ is also an element of P_τ. If, finally, $\mathfrak{r}_\nu$ is a sequence of cardinal numbers $< \bar{\tau}$, well-ordered with type $\sigma < \tau$, then the upper bound $\mathfrak{r}'$ of these numbers is $< \bar{\tau}$, because otherwise τ would be cofinal with σ, and thus not a core number. Hence $\sum_\nu \mathfrak{r}_\nu \leq \bar{\sigma}\mathfrak{r}' < \bar{\tau}$. This means that axiom (V) is satisfied in the segment P_τ, and hence that this latter is a normal domain. In this way, we have also shown that conditions (I) and (II) on the characteristics of normal domains set out in §2 are indeed *sufficient*, in other words that every ordinal number τ which satisfies these conditions can indeed be a 'boundary number' of a normal domain. It is, of course, assumed that the number τ belongs to a domain which fulfils the ZF' axioms.

§4. *Isomorphisms and automorphisms of normal domains*

Two normal domains are called 'isomorphic' if the elements ⟦x⟧ of one of them can be mapped one-to-one on to the elements x' of the other in such a way that each basic relation $a \in b$ in the one goes over into the relation $a' \in b'$ in the other and conversely. Thus, in the case of isomorphic domains, to each urelement u of one of them there corresponds an urelement u' of the other, to each set m there corresponds an equivalent set m', to each basic sequence g_α a basic sequence g'_α with the same index, and to the 'basis' Q an equivalent basis Q'; and the 'boundary number' or 'characteristic' π corresponds to itself. That the last two conditions are alone sufficient for isomorphism is shown by the following theorem:

First isomorphism theorem. *Two normal domains with the same charac-*

teristic and with equivalent bases are isomorphic, and indeed the isomorphic mapping of the two domains on to one another is uniquely *determined by the mapping of their bases.*

For the proof, we avail ourselves of the 'development theorems', and we can suppose here either the 'free' or the 'canonical' development. We imagine the bases Q and Q' of the normal domains mapped on to each other in a one-to-one way, in other words, so that to each urelement u there corresponds a definite u'. We then show by induction that we can associate to each *segment* P_α of the development of P an isomorphic segment P'_α in the development of the other domain ⟦P'⟧, and indeed *uniquely* for each $\alpha \leq \pi$. Two segments P'_α, P'_β with different indices cannot possibly be isomorphic, since the larger always contains basic sequences for which there are no similar sequences in the smaller. Now assume that P_α can be mapped isomorphically on to P'_α. This certainly is the case for $P_1 = Q$, $P'_1 = Q'$. Then all smaller segments ⟦P_β⟧ are simultaneously mapped on to smaller segments ⟦P'_β⟧ of P', and in all these mappings a given element x is always mapped on to the *same* element x' of P'. Now let r be a set in Q_α. All the elements r_ν of r are in P_α ⟦by definition of Q_α⟧; thus the elements corresponding to the r_ν lie in P'_α, and and according to the axiom (E) they are elements of an equivalent set r', for, since $\pi = \pi'$, the normal domain P' certainly contains sets of this power. The set r' must appear in $P'_{\alpha+1}$, even in the case of a 'canonical development'. But it is not in P'_α, since otherwise the corresponding set r would lie in P_α (because of the isomorphism), and thus not in Q_α. It follows that to each element r of Q_α there corresponds uniquely an element r' of Q'_α and conversely; thus the segment $P_{\alpha+1} = P_\alpha + Q_\alpha$ is univocally and isomorphically mapped on to the segment $P'_{\alpha+1}$ of the other normal domain. Now let α be a limit number, and assume that for every smaller $\beta < a$ the segment P_β is uniquely isomorphic to the segment P'_β. Since each element x of P_α certainly belongs to some P_β, in all the mappings from P_β to P'_β there corresponds to x a quite determinate element x' of P'_α, and whenever we have $a \in b$, we also have $a' \in b'$, since there is always a segment P_β which contains both elements ⟦a and b⟧. Thus, P_α is uniquely isomorphic to P'_α for arbitrary limit numbers $\alpha \leq \pi$. Because $P_\pi = P$ and $P'_\pi = P'$, both normal domains are isomorphic to each other as claimed.

Second isomorphism theorem. *Given two normal domains with equivalent bases and different boundary numbers π and π', it is always the case that one is isomorphic to a canonical segment of the other.*

Suppose ⟦we have two normal domains P and P',⟧ $Q \sim Q'$, and that ⟦these domains have characteristics π and π' with⟧ $\pi < \pi'$. Then, according to the Third 'Development Theorem' on p. 39 ⟦1227⟧, the 'canonical segment' P'_π is a normal domain, since π is a boundary number. This P'_π has both the same characteristic as P and an equivalent basis, and so according to the previous theorem it is isomorphic to P.

Third isomorphism theorem. *Given two normal domains with the same characteristic, one is always isomorphic to a (proper or improper) sub-domain of the other.*

Let P be a normal domain and let $Q' \subset Q$ be a part of its basis. Then we consider the totality of all those elements ⟦m⟧ of P from which each descending chain of elements $m \ni m_1 \ni m_2 \ni m_3 \ldots$ ends, not just in an urelement, as it must according to axiom (F), but in an urelement in Q'. The partial domain P' of P defined in this way fulfils all the conditions of the 'lemma' in §3. Thus it is itself a normal domain with the same characteristic π ⟦as P⟧, for it contains all the basic sequences arising from Q', and is isomorphic to each normal domain P'' with the same characteristic π whose basis Q'' is equivalent to Q'. From the assumed comparability of the arbitrary sets Q and Q'' the claim follows immediately ⟦by the First Isomorphism Theorem⟧.

The 'structure' of a normal domain, i.e. what it has in common with all isomorphic domains, or, in other words, its 'model type', is, by what has been proved here, determined by *two* numbers, by the power of its basis $\mathfrak{q}$ and by its characteristic π. The first can be considered as the 'breadth' of the normal domain, and can be chosen completely *arbitrarily*, while the second, the 'height' of the domain, must have the properties of 'boundary numbers' set out in §2. These 'model types' ⟦μ⟧ thus form a *twofold well-ordered manifold* with the property that one model type is always isomorphic to a component of another, thus $\mu \leq \mu'$, whenever $\mathfrak{q} \leq \mathfrak{q}'$ and $\pi \leq \pi'$, i.e. when *both* determining numbers of the one are smaller than or equal to those of the other.

According to the 'first isomorphism theorem', the isomorphic mapping of two normal domains on to one another, if such exists, is *uniquely* determined by the mapping of their bases. It thus follows that an isomorphic mapping of a normal domain on to *itself*, thus an 'automorphism', is only possible by a 'permutation' of its basis, and is therefore *impossible* for 'unit domains', which only contain a single urelement. Likewise, if the basis Q is *infinite*, isomorphic mappings of a normal domain P on to a *part* P' of itself arise from each equivalent mapping of Q on to one of its parts Q'. Indeed, as we saw from the proof of the last theorem, to each sub-basis there corresponds a normal partial domain P' of the same characteristic π. In particular, to each individual urelement u there corresponds the associated 'unit domain', and to equivalent partial bases there correspond isomorphic partial domains which, in analogy with field theory, we can call 'conjugated'. Thus we have the following:

Automorphism theorem. *Automorphisms, i.e. isomorphic mappings of a normal domain on to itself, correspond in a one-one way to equivalent mappings of the basis on to itself. It follows that automorphisms are only possible when the basis number* $\mathfrak{q} > 1$. *All unit domains are "monomorphic". The group of all automorphisms is isomorphic to the permutation group of the basis. In addition, "meromorphisms", i.e. isomorphic mappings of the normal domain on to a part of itself, correspond to the one-one mappings of the (infinite) basis on to equivalent parts.*

§5. *Existence, consistency, and categoricity*

Our considerations hitherto have assumed the existence of different 'normal domains', and in any case assume the consistency of the set-theoretic axioms.

We shall not attempt here to *prove* this consistency logically and formally. On the contrary, under the general *assumption* of the freedom from contradiction of set theory, we shall examine the (mathematical, i.e. ideal) *existence* of the different model types considered here. Thus, we assume the existence of set domains with an arbitrary basis which satisfy the *ZF* axioms. Given this, then there are certainly domains which in addition satisfy the axiom of 'foundation' (F). For if M is a set domain with the assumed property, then all the elements of M which also satisfy the axiom (F), among them naturally the urelements, form a well-defined sub-domain N of M which satisfies all the ZF' axioms and thus forms a normal domain with given basis Q.

By making the *basis* smaller, we arrive at normal domains which are partial domains of the first; this has already been shown in the proof of the 'Third Isomorphism Theorem' of §4. On the other hand, it is not clear without further explanation whether decreasing or increasing the *characteristic* leads to new types of normal domains. Each 'boundary number' must fulfil the conditions (I) and (II) of §2, and one may question whether there are such numbers at all, and indeed how many of them there are. Certainly ω, the initial number of the second number-class, is such a number, for it is an 'eigenvalue' of the function and $\psi(\xi)$ and not cofinal with any smaller number. Indeed, ω is the characteristic of the least normal domain, a domain which arises in the following way. Starting from a normal domain M, we leave out all those sets $\|m\|$ whose 'descending element chains' $m \ni m_1 \ni m_2 \ni m_3 \ldots$, formed according to (F), contain an infinite set.[g] The domain arrived at in this way, a domain which contains only *finite* sets, satisfies all the conditions of a normal domain, and at the same time is the segment P_ω, with index ω, of the 'canonical development' of the original normal domain $\|M\|$. By its mere existence, this 'finitistic' domain, to which, despite its own infiniteness, the 'intuitionists' can scarcely object, can serve at least to demonstrate the consistency of the ZF' axioms. On the other hand, it can *not* claim to be a true 'model' of Cantorian set theory, because it contains no infinite sets. From this circumstance stems my earlier 'axiom of infinity', which postulates the existence of at least one 'infinite' set. The *smallest* normal domain which satisfies this latter condition as well, and which I call the 'Cantorian' domain, would then have the characteristic π_1, that is, the smallest eigenvalue of the ψ-function which has the core number property. It is thus a 'regular initial number of the second kind', even if it is not the *smallest* 'exorbitant' number—at least as long as Cantor's hypothesis is not proved.

But do there exist, after ω, any such 'boundary numbers'? Certainly, in so far as there is an 'infinitistic' set theory at all, that is, in so far as there are normal domains with infinite sets. For the *totality* of all the 'basic sequences' of such a domain itself has an ordinal type π, even if there is no set of this type π *inside* the domain. And if there are 'boundary numbers' $\pi > \omega$ at all, then among them is a *smallest* π_1. Certainly neither their existence nor their

[g] |The m here are essentially the *hereditarily finite* sets of M.|

non-existence can be 'proved', i.e. be derived from the general ZF' axioms, because, for example, the boundary number ω exists in the 'Cantorian' domain, but not in the 'finitistic' domain. In other words, the question can be answered *differently* in different 'models' of set theory, and thus is not yet settled by the axioms alone. Our axiom system is *non-categorical*, which in this case is not a disadvantage but rather an *advantage*, for on this very fact rests the enormous importance and unlimited applicability of set theory. Naturally one can always *force* categoricity artificially by the addition of further 'axioms', but always at the cost of generality. Such new postulates, like those proposed by Fraenkel,[8] Finsler,[9] and von Neumann,[10] among others, do not concern set theory *as such*, but rather only characterize a quite special model chosen by the author concerned. As a rule, the 'unit domains' are the ones preferred, and hence, as already remarked on p. 38 ⟦1227⟧, the *applicability of set theory has to be given up. In addition, it is customary to confine attention to the smallest* infinite domain, the 'Cantorian', and I see little advantage in doing this. Rather, set theory as a *science* ⟦*Wissenschaft*⟧ must be developed in the fullest generality, and then the comparative investigation of individual *models* can be undertaken as a particular problem.

How are distinct models with a common basis differentiated in set theory, in particular the different 'unit domains'? As we saw, this can be done by their 'characteristics', i.e. by the totality of ordinal numbers represented in them by 'sets', and thus by the totality of the 'basic sequences' contained in them and starting with the single given urelement. Since only 'boundary numbers' can serve as 'characteristics', then every 'unit model' is determined uniquely by the totality of the *basic sequences with boundary number type* present (or not present) in it. By specifying this number, the presence of which can be guaranteed in the various different cases by suitable postulates, the model type is 'categorically' fixed. At the same time, the power $\bar{\pi}$ of the corresponding unit domain is fixed via the characteristic π, as indeed is stated by the Second Development Theorem of §3. If we now put forward the general hypothesis that *every categorically determined domain can also be interpreted* ⟦*aufgefaßt*⟧ *as a set in some way*, i.e. can appear as an element of a (properly chosen) normal domain, then it follows that to each normal domain there is a higher domain with the same basis, to each unit domain there is a higher unit domain, and finally, therefore, that to each 'boundary number' π there is a greater boundary number π'. In the same way, from each infinite sequence of different normal domains with a common basis, which are such that of any two one always con-

---

[8] See A. Fraenkel, *Einleitung in die Mengenlehre*, third edition, §18, 5, p. 355, the "Axiom of Restrictedness ⟦Beschränktheit⟧".

[9] See P. Finsler, 'Über die Grundlegung der Mengenlehre', *Mathematische Zeitschrift*, **25**, pp. 683–713.

On this, see also R. Baer, 'Über ein Vollständigkeitsaxiom in der Mengenlehre', *Mathematische Zeitschrift*, **27** (1928), pp. 536–9.

[10] See J. von Neumann, as above, n. 7.

tains the other as canonical segment, there arises, through union and fusion ‖Verschmelzung‖, a categorically determined domain of sets which again can be extended ‖ergänzt‖ to a normal domain of higher characteristic. Thus, to each categorically determined totality of 'boundary numbers' there follows a greater such number, and the series of 'all' boundary numbers is unbounded in the same way as the number series itself. Thus to each transfinite index there corresponds in a one-to-one fashion a determinate boundary number. Naturally, again this *cannot* be 'proved' from the ZF' axioms, since the behaviour claimed goes beyond any individual normal domain. Rather, the *existence of an unbounded sequence of boundary numbers* must be postulated as a new *axiom* of 'meta-set theory', and in so doing the 'consistency' question must be looked at more closely. However, restricting myself just to the present sketch and to an indication of how the ideas are to be worked out subsequently, then the following, which might be regarded as the essential result of the present investigation, ought to be clear already:

Scientific reactionaries and anti-mathematicians have so eagerly and lovingly appealed to the 'ultrafinite antinomies' in their struggle against set theory. But these are only apparent 'contradictions', and depend solely on confusing *set theory itself*, which is not categorically determined by its axioms, with individual *models* representing it. What appears as an 'ultrafinite non- or super-set' in one model is, in the succeeding model, a perfectly good, valid set with both a cardinal number and an ordinal type, and is itself a foundation stone for the construction of a new domain. To the unbounded series of Cantor ordinals there corresponds a similarly unbounded double-series of essentially different set-theoretic models, in each of which the whole classical theory is expressed. The two polar opposite tendencies of the thinking spirit, the idea of creative *advance* and that of collection and *completion* ‖*Abschluß*‖, ideas which also lie behind the Kantian 'antinomies', find their symbolic representation and their symbolic reconciliation in the transfinite number series based on the concept of well-ordering. This series reaches no true completion in its unrestricted advance, but possesses only relative stopping-points, just those 'boundary numbers' which separate the higher model types from the lower. Thus the set-theoretic 'antinomies', when correctly understood, do not lead to a cramping and mutilation of mathematical science, but rather to an, as yet, unsurveyable unfolding and enriching of that science.

In concluding this article, I must offer heartfelt thanks to my colleague Dr Arnold Scholz for his most gracious support. He offered me much valuable advice, both with the working out of this investigation and in the proof-reading.

Freiburg im Breisgau, 13 April 1930.

27
Godfrey Harold Hardy (1877–1947)

G.H. Hardy, the most eminent British mathematician of his generation, was educated at Trinity College, Cambridge, where he was elected to a fellowship in 1900; he remained at Trinity as a lecturer in mathematics until 1919, when he was appointed Savilian Professor of Geometry at Oxford. He returned to Cambridge in 1931 as Sadlerian Professor of Pure Mathematics, a post he held until his retirement in 1942.

Hardy wrote some 350 papers on analysis and number theory, treating such topics as infinite series, the theory of integration, Riemann's zeta function, inequalities, and trigonometric series; nearly a hundred of these papers were written jointly with J.E. Littlewood. Hardy also wrote seven books on mathematics. Their titles display the range of his research interests: *The integration of functions of a single variable* (1905), *A course of pure mathematics* (1908), *Orders of infinity* (1910), *The general theory of Dirichlet's series* (with M. Riesz, 1915), *Inequalities* (with Littlewood and Polya, 1934), *The theory of numbers* (with E.M. Wright, 1938), *Fourier series* (with W.W. Rogosinski, 1944), and *Divergent series* (1948). The *Course of pure mathematics* was especially influential, and introduced modern techniques of analysis into British mathematical education. Hardy also wrote an elegant autobiography, *A mathematician's apology* (1940), in which, as well as giving his views on the nature and purpose of pure mathematics, he described his celebrated collaboration with the Indian number-theorist Ramanujan.

A. SIR GEORGE STOKES AND THE CONCEPT OF UNIFORM CONVERGENCE (*HARDY 1918*)

Many of the earlier readings have treated the arithmetization of mathematics in the nineteenth century—the introduction of analytic definitions of *continuity* and *limit* and *derivative*; the rigorous definition of *real number*; the proofs of important theorems of analysis such as the intermediate value theorem or Cantor's proof of the existence of transcendental numbers. Like most history, the history of mathematics is written from the point of view of the victors, and presents itself as a sequence of successful discoveries. Hardy's article, written long after the theory of real analysis had reached its maturity, is a meticulous examination, with the clear vision of hindsight, of an important but neglected aspect of the history of ideas: how a great mathematician *failed* to make an important discovery, despite having had all the necessary tools at hand.

Hardy's footnote symbols, originally daggers and asterisks, have here been changed to numerals. References to *Hardy 1918* should be to the section numbers, which appeared in the original edition.

---

1. The discovery of the notion of uniform convergence is generally and rightly attributed to Weierstrass, Stokes, and Seidel. The idea is present implicitly in Abel's proof of his celebrated theorem on the continuity of power series; but the three mathematicians mentioned were the first to recognise it explicitly and formulate it in general terms.[1] Their work was quite independent, and it would be generally agreed that the debt which mathematics owes to each of them is in no way diminished by any anticipation on the part of the others. Each, as it happens, has some special claim to recognition. Weierstrass's discovery was the earliest, and he alone fully realised its far-reaching importance as one of the fundamental ideas of analysis. Stokes has the actual priority of publication; and Seidel's work is but a year later and, while narrower in its scope than that of Stokes, is even sharper and clearer.

My object in writing this note is to call attention to and, so far as I can, explain two puzzling features in the justly famous memoir[2] in which Stokes announces his discovery. The memoir is remarkable in many respects, containing a general discussion of the possible modes of convergence, both of series and of integrals, far in advance of the current ideas of the time. It contains also two serious mistakes, mistakes which seem at first sight almost inexplicable on the part of a mathematician of so much originality and penetration.

The first mistake is one of omission. It does not seem to have occurred to Stokes that his discovery had any bearing whatever on the question of term by term integration of an infinite series. The same criticism, it is true, may be made of Seidel's paper. But Seidel is merely silent on the subject. Stokes, on the other hand, quotes the false theorem that a convergent series may always be integrated term by term, and refers, apparently with approval, to the erroneous proof offered by Cauchy and Moigno.[3]

---

[1] The idea was rediscovered by Cauchy, five or six years after the publication of the work of Stokes and Seidel. See Pringsheim, 'Grundlagen der allgemeinen Funktionenlehre', *Encykl. der Math. Wiss.*, II A 1, §17, p. 35.

[2] 'On the critical values of the sums of periodic series', *Trans. Camb. Phil. Soc.*, Vol. 8, 1847, pp. 533–583 (*Mathematical and physical papers*, Vol. 1, pp. 236–313.)

[3] See p. 242 of Stokes's memoir (as printed in the collected papers).

Of this there is, I think, a fairly simple and indeed a double explanation. In the first place it must be remembered that Stokes was primarily a mathematical physicist. He was also a most acute pure mathematician; but he approached pure mathematics in the spirit in which a physicist approaches natural phenomena, not looking for difficulties, but trying to explain those which forced themselves upon his attention. The difficulties connected with continuity and discontinuity are of this character. The theorem that a convergent series of continuous functions has necessarily a continuous sum is one whose falsity is open and aggressive: examples to the contrary obtrude themselves on analyst and physicist alike. The falsity of this theorem Stokes therefore observed and corrected. The falsity of the corresponding theorem concerning integration lies somewhat deeper. It is easy enough, when one's attention has been called to it, to see that the proof of Cauchy and Moigno is invalid. But there are no particularly obvious examples to the contrary: simple and natural examples are indeed somewhat difficult to construct.[4] And Stokes, his suspicions never having been excited, seems to have accepted the false theorem without examination or reflection.

This is half the explanation. The second half, I think, lies in the distinctions between different modes of uniform convergence which I shall consider in a moment.

Stokes's second mistake is more obvious and striking. He proves, quite accurately, that uniform convergence implies continuity.[5] He then enunciates and offers a proof[6] of the converse theorem, which is false. The error is not one merely of haste or inattention. The argument is as explicit and as clearly stated in one case as in the other; and, up to the last sentence, it is perfectly correct. He proves that continuity involves *something*, and then states, without further argument, that this something is what he has just defined as uniform convergence. It is merely this last statement that is false.

Stokes's mistake seems at first sight so palpable that I was for some time quite at a loss to imagine how he could have made it. A closer examination of his memoir, and a comparison of his work with other work of a very much later date, has made the lapse a good deal more intelligible to me; and my attempts to understand it have led me to a number of remarks which, although they contain very little that is really novel, are, I think, of some historical and intrinsic interest.

2. There are no less than *seven* different senses, all important, in which a series may be said to be uniformly convergent.

I shall write the series in the form

$$\sum_{1}^{\infty} u_n(x);$$

[4] See Bromwich, *Infinite series*. pp. 116–118; Hardy, 'Notes on some points in the integral calculus', XL, *Messenger of Mathematics*, Vol. 44, 1915, pp. 145–149.

[5] p. 282. I use 'uniform' instead of Stokes's 'not infinitely slow'.

[6] p. 283.

and I shall suppose, for simplicity, that every term of the series is continuous, and the series convergent, for every x of the interval $a \leqslant x \leqslant b$. I shall denote the sum of the series by $s(x)$; and I shall write

$$s_n(x) = u_1(x) + u_2(x) + \ldots + u_n(x), \; s(x) = s_n(x) + r_n(x).$$

The fundamental inequality in all my definitions will be of the type

$$|r_n(x)| \leqslant \varepsilon \qquad \text{(A)}.$$

I shall refer to this inequality simply as (A).

When we define uniform convergence, in one sense or another, we have to choose various numbers in a definite logical order, those which are chosen later being, in general, functions of those which are chosen before. I shall write each number in a form in which all the arguments of which it is a function appear explicitly: thus $n_0(\xi, \varepsilon)$ is a function of ξ and ε, $n_0(\varepsilon)$ one of ε alone.

It will sometimes happen that one of the later numbers depends upon several earlier numbers *already connected by functional relations*, so that it is really a function of a selection of these numbers only. Thus δ may have been determined as a function of ε; and n_0 may have to be determined as a function of ξ, ε, and δ, so that it is in reality a function of ξ and ε only. I shall express this by writing

$$n_0 = n_0(\xi, \varepsilon, \delta) = n_0(\xi, \varepsilon);$$

and I shall use a similar notation in other cases of the same kind.

3. The first three senses of uniform convergence are as follows.

A1: Uniform convergence throughout an interval. *The series is said to be uniformly convergent throughout the interval* (a, b) *if to every positive* ε *corresponds an* $n_0(\varepsilon)$ *such that* (A) *is true for* $n \geqslant n_0(\varepsilon)$ *and* $a \leqslant x \leqslant b$.

This is the ordinary or 'classical', and most important, sense, the sense in which uniform convergence is defined in every treatise on the theory of series.

A2: Uniform convergence in the neighbourhood of a point. *The series is said to be uniformly convergent in the neighbourhood of the point* ξ *of the interval* (a, b) *if an interval* $(\xi - \delta(\xi), \xi + \delta(\xi))$[7] *can be found throughout which it is uniformly convergent; that is to say if a positive* $\delta(\xi)$ *exists such that* (A) *is true for every positive* ε, *for* $n \geqslant n_0(\xi, \delta, \varepsilon) = n_0(\xi, \varepsilon)$, *and for* $\xi - \delta(\xi) \leqslant x \leqslant \xi + \delta(\xi)$.

A3: Uniform convergence at a point. *The series is said to be uniformly convergent at the point* $x = \xi$ *(or for* $x = \xi$*) if to every positive* ε *correspond a positive* $\delta(\xi, \varepsilon)$ *and an* $n_0(\xi, \varepsilon, \delta) = n_0(\xi, \varepsilon)$ *such that* (A) *is true for* $n \geqslant n_0(\xi, \varepsilon)$ *and for* $\xi - \delta(\xi, \varepsilon) \leqslant x \leqslant \xi + \delta(\xi, \varepsilon)$.

4. Before proceeding further it will be well to make a few remarks concerning these definitions and their relations to one another.

---

[7] A trivial change is of course required in the definition if $\xi = a$ or $\xi = b$. The same point naturally arises in the later definitions.

The idea of uniform convergence *in the neighbourhood of a particular point* (Definition **A2**) is substantially that defined by Seidel in 1848.[8] It is clear, however, that definitions **A1** and **A2** were both familiar to Weierstrass as early as 1841 or 1842.[9] It is obvious that a series uniformly convergent throughout an interval is uniformly convergent in the neighbourhood of every point of the interval. The converse theorem is important and by no means obvious, and was first proved by Weierstrass[10] in a memoir published in 1880. This theorem would now be proved by a simple application of the 'Heine–Borel Theorem', and is a particular case of a theorem which will be referred to in a moment.

Definition **A3** appears first, in the form in which I state it, in a paper of W.H. Young published in 1903;[11] but the idea is present in an earlier paper of Osgood.[12] The essential difference between definitions **A2** and **A3** is that in the latter δ is chosen *after* ε and is a function of ξ and ε, while in the former it is chosen before ε and is a function of ξ alone. In each case n_0 is a function of two independent variables, ξ and ε. It is plain that uniform convergence in the neighbourhood of ξ involves uniform convergence at ξ, and at (and indeed in the neighbourhood of) all points sufficiently near to ξ. But uniform convergence at ξ does not involve uniform convergence in the neighbourhood of ξ.

It is important, however, to observe that *uniform convergence at every point of an interval involves uniform convergence throughout the interval.* This important theorem is proved very simply by Young, in his paper already quoted, by means of the Heine–Borel Theorem;[3] and it plainly includes, as a particular case, Weierstrass's theorem referred to above.

5. It seems to me that the definition given by Stokes is not any one of **A1**, **A2**, **A3**; and that, if we are to understand him rightly, we must consider another parallel group of definitions. These definitions differ from those given above

---

[8] 'Note über eine Eigenschaft der Reihen, welche discontinuirliche Functionen darstellen', *Münchener Abhandlungen*, Vol. 7, 1848, pp. 381–394. This memoir has been reprinted in Ostwald's *Klassiker der exakten Wissenschaften*, no. 116. The reference there given to Vol. 5, 1847, is incorrect.

[9] For detailed references bearing on this and similar historical points, see Pringsheim's article already quoted.

[10] See the memoir 'Zur Functionenlehre' (*Abhandlungen aus der Funktionenlehre*, pp. 69–104 (pp. 71–2)).

[11] 'On non-uniform convergence and term-by-term integration of series', *Proc. London Math. Soc.*, Ser. 2, Vol. 1, pp. 89–102.

[12] 'Non-uniform convergence and the integration of series', *American Journal of Math.*, Vol. 19, 1897, pp. 155–190. See Prof. Young's remarks on this point at the beginning of his later paper 'On uniform and non-uniform convergence of a series of continuous functions and the distinction of right and left', *Proc. London Math. Soc.*, Ser. 2, Vol. 6, 1907, pp. 29–51.

[13] Choose ε and determine $\delta(\xi, \varepsilon)$ and $n_0(\xi, \varepsilon)$, as in definition **A3**, for every ξ of the interval. Every point of (a, b) is included in an interval $(\xi - \delta, \xi + \delta)$. By the Heine–Borel Theorem, every point of (a, b) is included in one or other of a finite sub-set of these intervals. If $N(\varepsilon)$ is the largest of the n_0's corresponding to each of the intervals of this finite sub-set, then (A) is true for $n \geqslant N$ and $a \leqslant x \leqslant b$.

This is the essence of the proof, though, like all proofs of the same character, it requires a somewhat more careful statement if all appearance of dependence upon Zermelo's *Auswahlsprinzip* is to be avoided.

in that (A) is supposed to be satisfied, not for *all* sufficiently large values of n, but only for *an infinity* of values.

B1: Quasi-uniform convergence throughout an interval. *The series is said to be quasi-uniformly convergent throughout* (a, b) *if to every positive* ε *and every* N *corresponds an* $n_0(\varepsilon, N)$ *greater than* N *and such that* (A) *is true for* $n = n_0(\varepsilon, N)$ *and* $a \leqslant x \leqslant b$.

B2: Quasi-uniform convergence in the neighbourhood of a point. *The series is said to be quasi-uniformly convergent in the neighbourhood of* ξ *if an interval* $(\xi - \delta(\xi), \xi + \delta(\xi))$ *can be found throughout which it is quasi-uniformly convergent; i.e., if a positive* $\delta(\xi)$ *exists such that* (A) *is true for every positive* ε, *every* N, *an* $n_0(\xi, \delta, \varepsilon, N) = n_0(\xi, \varepsilon, N)$ *greater than* N, *and* $\xi - \delta(\xi) \leqslant x \leqslant \xi + \delta(\xi)$.

B3: Quasi-uniform convergence at a point. *The series is said to be quasi-uniformly convergent for* $x = \xi$ *if to every positive* ε *and every* N *correspond a positive* $\delta(\xi, \varepsilon, N)$ *and an*

$$n_0(\xi, \varepsilon, \delta, N) = n_0(\xi, \varepsilon, N),$$

greater than N, *such that* (A) *is true for* $n = n_0(\xi, \varepsilon, N)$ *and for*

$$\xi - \delta(\xi, \varepsilon, N) \leqslant x \leqslant \xi + \delta(\xi, \varepsilon, N).$$

Definition **B1** is to be attributed to Dini or to Darboux.[14] Another form of it has been given by Hobson.[15] As Arzelà and Hobson[16] have pointed out, a series is quasi-uniformly convergent throughout an interval if, and only if, it can be made uniformly convergent by an appropriate bracketing of its terms.

Definition **B2** is for us at the moment of peculiar interest, for (as I shall show in a moment) it is really *this* definition that is given by Stokes.

Definition **B3** is also of great interest, both in itself and in relation to Stokes's memoir. For *the necessary and sufficient condition that* $s(x)$ *should be continuous for* $x = \xi$ *is that the series should be quasi-uniformly convergent for* $x = \xi$. This theorem is in substance due to Dini.[17] I give the proof, as it is essential for the criticism of Stokes's memoir.

(1) *The condition is sufficient*. For

$$|s(x) - s(\xi)| \leqslant |s_n(x) - s_n(\xi)| + |r_n(x)| + |r_n(\xi)|.$$

Choose ε, N, $\delta(\xi, \varepsilon, N)$, and $n = n_0(\xi, \varepsilon, N)$ as in definition **B3**. Then $|r_n(x)| < \varepsilon$ for $\xi - \delta \leqslant x \leqslant \xi + \delta$. Now that n is fixed we can choose δ_1 less than δ and such that $|s_n(x) - s_n(\xi)| < \varepsilon$ for $\xi - \delta_1 \leqslant x \leqslant \xi + \delta_1$. And thus

$$|s(x) - s(\xi)| < 3\varepsilon$$

---

[14] See Pringsheim, *l.c.*

[15] 'On modes of convergence of an infinite series of functions of a real variable', *Proc. London Math. Soc.*, Ser. 2, Vol. 1, 1903, pp. 373–387. Hobson (following Dini) uses the expression 'simply uniformly'.

[16] *L.c.*, p. 375.

[17] *Fondamenti* . . ., p. 107 (German translation, *Grundlagen* . . ., pp. 143–145).

for $\xi - \delta_1 \leqslant x \leqslant \xi + \delta_1$, so that $s(x)$ is continuous for $x = \xi$.

It is plain that this argument proves, *a fortiori*, that **A2**, **A3**, and **B2** all furnish sufficient conditions for continuity at a point, and **A1** and **B1** sufficient conditions for continuity throughout an interval.

(2) *The condition is necessary.* For

$$|r_n(x)| \leqslant |s(x) - s(\xi)| + |r_n(\xi)| + |s_n(x) - s_n(\xi)|.$$

Suppose that ε and N are given. Then we can choose $\delta(\xi, \varepsilon)$ so that $|s(x) - s(\xi)| < \varepsilon$ for $\xi - \delta \leqslant x \leqslant \xi + \delta$, and $n_0(\xi, \varepsilon, N)$ so that $n_0 > N$ and $|r_{n_0}(\xi)| < \varepsilon$. And, when n_0 has thus been fixed, we can choose $\delta_1(\xi, \varepsilon, n_0) = \delta_1(\xi, \varepsilon, N)$ so that $\delta_1 < \delta$ and $|s_{n_0}(x) - s_{n_0}(\xi)| < \varepsilon$ for $\xi - \delta_1 \leqslant x \leqslant \xi + \delta_1$. Thus $|r_n(x)| < 3\varepsilon$ for $n = n_0 > N$ and $\xi - \delta_1 \leqslant x \leqslant \xi + \delta_1$, so that the series is quasi-uniformly convergent for $x = \xi$.

6. If a series is uniformly convergent at every point ξ of an interval, it is (as we saw in §4) uniformly convergent throughout the interval: definition **A3** (and *a fortiori* definition **A2**) passes over, in virtue of the Heine–Borel Theorem, into definition **A1**. It is important to observe that this relation does not hold between **B3** (or **B2**) and **B1**: a series quasi-uniformly convergent at every point of an interval (or in the neighbourhood of every such point) is not necessarily quasi-uniformly convergent throughout the interval. We can apply the Heine–Borel Theorem in the manner indicated in the first sentences of footnote 13; but the last stage of the argument, in which every one of a finite number of different integers is replaced by the largest of them, fails. What we obtain is *the necessary and sufficient condition that* $s(x)$ *should be continuous throughout the interval*; and this is not the condition **B1** but a condition first formulated by Arzelà,[18] viz.:

C: Quasi-uniform convergence by intervals *(convergenza uniforme a tratti). The series is said to be quasi-uniformly convergent by intervals if to every positive* ε *and every* N *correspond a division of* (a, b) *into a finite number* $\nu(\varepsilon, N)$ *of intervals* $\delta_r(\varepsilon, N)$, *and a corresponding number of numbers* $n_r(\varepsilon, N)$, *all greater than* N, *and such that* (A) *is true for* $n = n_r (r = 1, 2, \ldots, \nu)$ *and all values of* x *which belong to* δ_r.

The deduction of Arzelà's criterion from **B3**, in the manner sketched above, was first made by Hobson.[19]

There is one further point which seems worth noticing here, although it is not directly connected with Stokes's memoir. Dini[20] proved that *if* $u_n(x) \geqslant 0$ *for all values of* n *and* x, *and* $s(x)$ *is continuous throughout* (a, b), *then the series is uniformly convergent throughout* (a, b). This theorem is now almost intuitive. For it is obvious that, for series of positive terms, quasi-uniform convergence in any one of the senses **B1**, **B2**, or **B3** involves uniform convergence

---

[18] 'Sulle serie di funzioni', *Memorie di Bologna*, Ser. 5, Vol. 8, 1900, pp. 131–186, 701–744.
[19] *L.c.*, pp. 380–382.
[20] *L.c.* (German edition), pp. 148–149. See also Bromwich, *Infinite series*, p. 125 (Ex. 3).

in the corresponding sense **A1**, **A2**, or **A3**. If then $s(x)$ is continuous throughout (a, b) it is continuous for every ξ of (a, b); and therefore the series is quasi-uniformly convergent for every ξ; and therefore uniformly convergent for every ξ; and therefore uniformly convergent throughout (a, b).

7. Let us now consider Stokes's definitions and proofs in the light of the preceding discussion.

It is clear, in the first place, that Stokes has in his mind some phenomenon characteristic of *a small, but fixed, neighbourhood of a point.*

'Let $u_1 + u_2 + \ldots$ (66)', he says,[21] 'be a convergent infinite series having U for its sum. Let $v_1 + v_2 + \ldots$ (67) be another infinite series of which the general term v_n is a function of the positive variable h and becomes equal to u_n when h vanishes. Suppose that *for a sufficiently small value of h and all inferior values* the series (67) is convergent, and has V for its sum. It might at first sight be supposed that the limit of V for $h = 0$ was necessarily equal to U. This however is not true. . . .

'THEOREM. The limit of V can never differ from U unless the convergency of the series (67) becomes infinitely slow when h vanishes.

'The convergency of the series is here said to become infinitely slow when, if n be the number of terms which must be taken in order to render the sum of the neglected series numerically less than a given quantity e, which may be as small as we please, n increases beyond all limit as h decreases beyond all limit.

'DEMONSTRATION. If the convergency do not become infinitely slow it will be possible to find *a number n, so great that for the value of h we begin with and for all inferior values greater than zero the sum of the neglected terms shall be numerically less than e. . . .*'

Stokes's words, and in particular those which I have italicised, seem to me to make two things perfectly clear.

(1) Stokes is considering neither a property of an interval (a, b) *im Grossen* (such as is contemplated in **A1** or **B1**), nor a property of a single point which (as in **A3** or **B3**) need not be shared by any neighbouring point, but a property of an interval *im Kleinen*, that is to say *a small but fixed interval chosen to include a particular point*. His definition is therefore one of the type of **A2** or **B2**.

Stokes's failure to perceive the bearing of his discovery on problems of integration is made much more natural when we realise that he is considering throughout a neighbourhood of a point and not an interval *im Grossen*. And this remark applies to Seidel as well.

(2) Stokes is considering an inequality satisfied for a special value of n, or at most an infinite sequence of values of n, and *not* necessarily for all values of n from a certain point onwards. In this respect there is a quite sharp distinction between Stokes's work and Seidel's. What Stokes defines is (to use the language of this note) a mode of *quasi-uniform* convergence and not one of strictly uniform convergence.

---

[21] p. 279.

It seems to me, then, that what Stokes defines is what I have called *quasi-uniform convergence in the neighbourhood of a point* (**B2**).

8. If we adopt this view, Stokes's mistake becomes very much more intelligible. He proves, quite correctly, that uniform convergence in his sense implies continuity: his proof, stated quite formally and by means of inequalities, is substantially that given in §5, under (1). He then continues[22] as follows.

'Conversely, if (66) is convergent, and if $U = V_0$,[23] the convergency of the series (67) cannot become infinitely slow when h vanishes. For if U_n', V_n' represent the sums of the terms after the nth in the series (66), (67) respectively, we have

$$V = V_n + V_n', \; U = U_n + U_n';$$

whence

$$V_n' = V - U - (V_n - U_n) + U_n'.$$

Now $V - U$, $V_n - U_n$ vanish with h, and U_n' vanishes when n becomes infinite. Hence *for a sufficiently small value of h and all inferior values, together with a value of n sufficiently large and independent of h*, the value of V_n' may be made numerically less than any given quantity e however small; *and therefore, by definition, the convergency of the series* (67) *does not become infinitely slow when h vanishes.*'

Now this argument is, until we reach the last sentence, perfectly accurate, and indeed, if we translate it into inequalities, substantially identical with that given in §5, under (2). Stokes proves, in fact, that continuity at ξ involves quasi-uniform convergence at ξ. Where he falls into error is simply in his final assertion that this property is that which he has previously defined, the mistake being due to a failure to observe that his intervals of values of h depend upon a prior choice of ε. In a word, he confuses, momentarily, **B2** and **B3**. The ordinary view that Stokes defined uniform convergence in the same sense as Weierstrass compels us to suppose that he confused **B3** with **A1**, or at any rate with **A2**: and this is hardly credible.

I add one final remark. If we could identify Stokes's idea with **B3**, instead of with **B2**, we could acquit him of having made any mistake at all, since **B3** really is a necessary and sufficient condition for continuity. We could then regard Stokes as having anticipated Dini's theorem. This view, however, does not seem to me to be tenable.

---

[22] p. 282. The italics are mine.

[23] V_0 is what Stokes calls 'the value of V for $h = 0$', by which he means, of course, its limit when h tends to 0.

B. MATHEMATICAL PROOF (*HARDY 1929a*)

The following article, originally delivered as the Rouse Ball lecture in Cambridge University, 1928, expresses Hardy's opinions on the three leading foundational schools of the 1920s: logicism, intuitionism, and Hilbert's proof-theoretical programme. Apart from some early contributions to set-theory, Hardy did little technical work in the foundations of mathematics; but he knew Russell well,[a] and he kept abreast of the current controversies.

Hardy begins by expressing his dissatisfaction with Russell's version of logicism; in particular, with the axiom of reducibility, which he finds contrived and unintuitive. After a hasty dismissal of intuitionism for its willingness to sacrifice large tracts of classical mathematics, Hardy turns to a careful examination of Hilbert's proof theory, dissenting from its constructivist foundations, but attempting to evaluate the technical merits of Hilbert's research programme, to defend him from the charge that consistency proofs must necessarily reason in a circle, and to explore the implications of his ideas about mathematical proof.

This article contains a bluff statement of Hardy's Platonist *credo*, a *credo* that virtually brushes aside the foundational debates about the infinite in mathematics:

> It is true, as Hilbert says, that no mathematician has completed an infinity of syllogisms. It is also true that there is no mathematician who has never drunk a glass of water, and, so far as I can see, one of these facts has neither more nor less logical importance than the other. There is no more *logical* reason why a mathematician should not prove an infinity of theorems in this world than why he should not (as he has been so often encouraged to hope) emit an infinite sequence of musical notes in the next.

(A constructivist might find this a convenient place to point out that Hardy was a fervent atheist.)

Hardy's views on the nature of mathematical reality do not constitute a fully-developed philosophical theory (he describes them as the prejudices of a working mathematician); and he does not come to grips with the essentially *epistemological* doubts about the *completed* infinite that had been voiced by Hilbert and Brouwer. His Platonism contains within itself a subtle tension between two opposing conceptions of mathematical proof, one objective and the other subjective. On the one hand (§§3, 16), proofs are patterns of mind-independent propositions about an objective reality: their truth or falsity does not depend upon human knowledge. On the other hand (§13) proofs have a subjective, psychological function, and are intended to persuade; in this sense, they are a kind of rhetorical 'gas'.

---

[a] Russell was one of Hardy's contemporaries at Trinity in the early years of the century. When the College dismissed Russell for opposing the First World War, Hardy took his side and later wrote a short book on the incident, *Bertrand Russell and Trinity* (1942).

Hardy is powerfully attracted to *both* conceptions of mathematical proof—evidently a difficult position to sustain. (Indeed, the philosophies of mathematics of both Hilbert and Brouwer can be regarded as attempts to work out an intermediate position, and in particular to avoid depicting the mathematician as a person who, with a mysterious inner eye, gazes over completed non-denumerable totalities, and reports on what he sees.) Hardy does not in the end show how the two conceptions are to be reconciled, and he lacks a fully-developed theory of the relationship between: (i) the objective propositions of mathematics; (ii) the language of mathematics; and, (iii) subjective mental states. But despite these shortcomings, this article, as an expression of Hardy's carefully considered and strongly held intuitions, brings out the depth and difficulty of the philosophical problems underlying the foundational disputes of the twenties.

References to *Hardy 1929a* should be to the section numbers, which appeared in the original edition.

---

1. I have chosen a subject for this lecture, after much hesitation, not from technical mathematics but from the doubtful ground disputed by mathematics, logic and philosophy; and I have done this deliberately, knowing that I shall be setting myself a task for which I have no sufficient qualifications. I have been influenced by three different motives. In the first place, the exercise will be good for me, since it will force me to think seriously about questions which a professional mathematician like myself is apt to neglect. Secondly, it is difficult to find a branch of pure mathematics suitable for popular exposition in an hour. Finally if, in a desperate attempt to be interesting, I lose myself in discussions where I am admittedly an amateur, then, whoever I may offend, I should certainly not have offended the founder of this lectureship and the Rouse Ball chair.

I do not regret my choice, but I am bound in self-defence to begin with a double apology. The first is to any real mathematical logicians who may be present. I am myself a professional pure mathematician in the narrow sense, and, in my own subject, quite as intolerant of amateurs as a self-respecting professional should be. I have therefore no difficulty in understanding that mathematical logic also is a subject for professionals; that it demands a detailed knowledge which I do not possess and, so long as I am active in my proper sphere, have hardly leisure to acquire; and that I am certain to be guilty of all sorts of confusions which would be impossible to a properly qualified logician. Indeed there is only one thought which gives me courage to proceed, and that is that I may be concerned less with strictly logical questions than with questions of general philosophy. However treacherous a ground mathematical logic, strictly interpreted, may be for an amateur, philosophy proper is a subject, on the one hand so hopelessly obscure, on the other so astonishingly elementary, that there knowledge hardly counts. If only a question be sufficiently fundamental, the arguments for any answer must be correspondingly crude and simple, and all men may meet to discuss it on more or less equal terms.

My second apology must be addressed to those mathematicians who dislike all discussions savouring of philosophy. But if I apologise to them, it is perhaps with less sincerity. I feel that this distaste is usually based on no better foundation than an unreasoning shrinking from anything unfamiliar, the distaste of the pragmatist for truth, of the engineer for mathematics, of the pavilion critic at Lords for the in-swinger and the two-eyed stance. It is reasonable to ask an audience like this to put aside this dislike of the fundamental for its own sake.

You must also remember that ordinary mathematics has a good deal at stake in some of these recent controversies. These controversies have seemed to threaten methods which we have used with confidence for nearly one hundred years. There are familiar elementary theorems—that any aggregate of real numbers has an upper bound, that any infinite aggregate has a point of condensation—the truth of which is simply denied by the 'intuitionist' school of logicians. There are also theorems of an apparently much less abstract or suspicious type, theorems for example in the theory of numbers, the only known proofs of which depend, in appearance at any rate, on principles which they reject.

2. It may not be possible to distinguish precisely between mathematics, mathematical logic, and philosophy, as the words are currently used. We can, however, by considering a few typical problems, recognise roughly the disputed tracts across which the boundaries must be drawn.

(i) *Is Goldbach's Theorem true*? Is any even number the sum of two primes? This is a strictly mathematical question to which all question of logic or philosophy seem irrelevant.

(ii) *Is the cardinal number of the continuum the same as that of Cantor's second number class*? This again appears to be a mathematical question; one would suppose that, if a proof were found, its kernel would lie in some sharp and characteristically mathematical idea. But the question lies much nearer to the borderline of logic, and a mathematician interested in the problem is likely to hold logical and even philosophical views of his own.

(iii) *What is the best system of primitives for the logic of propositions*? This is a question of mathematical logic in the strict professional sense. A logician qualified to discuss it will probably belong to some more or less definite philosophical school, but it is hardly likely that his philosophical views will have any very noticeable influence on his choice.

(iv) *What is a proposition, and what is meant by saying that it is true*? This, finally, is a problem of simple philosophy.

It is often said that mathematics can be fitted on to any philosophy, and up to a point it is obviously true. Relativity does not (whatever Eddington may say) compel us to be idealists. The theory of numbers does not commit us to any particular view of the nature of truth. However that may be, there is no doubt that mathematics does create very strong philosophical *prejudices*, and that the tests which a philosophy must satisfy before a mathematician will look at it are likely to be very different from those imposed by a biologist or a theologian. I am sure that my own philosophical prejudices are as strong as my philosophical knowledge is scanty.

One may divide philosophies into *sympathetic* and *unsympathetic*, those in which we should like to believe and those which we instinctively hate, and into *tenable* and *untenable*, those in which it is possible to believe and those in which it is not. To me, for example, and I imagine to most mathematicians, Behaviourism and Pragmatism are both unsympathetic and untenable. The philosophy of Mr. Bradley may be just tenable, but it is highly unsympathetic. The Cambridge New Realism, in its cruder forms, is very sympathetic, but I am afraid that, in the forms in which I like it best, it may be hardly tenable. 'Thin' philosophies, if I may adopt the expressive classification of William James, are generally sympathetic to me, and 'thick' ones unsympathetic. The problem is to find a philosophy which is both sympathetic and tenable; it is not reasonable to hope for any higher degree of assurance.

3. The crucial test of a philosophy, for a mathematician, is that it should give some sort of rational account of *propositions* and of *proof*. A mathematical theorem is a proposition; a mathematical proof is clearly in some sense a collection or pattern of propositions. It is plain then that if I ask what are, to a mathematician, the most obvious characteristics of a mathematical theorem or a mathematical proof, I am inviting philosophical discussion of the most fundamental kind. I wish to begin, however, by being as unsophisticated as I can, and I will therefore try to sketch what seems to be the view of mathematical common sense, the sort of view natural to a man who does not profess to be a logician but has spent his life in the search for mathematical truth. It is after all the misapprehensions of such a man that a logician may find the least fundamentally unreasonable and the least hopeless to remove.

I will begin then by enumerating some rough criteria which I think that a philosophy must satisfy if it is to be at all sympathetic to a working mathematician. I know too well how probable it is that just the most sympathetic philosophies will prove untenable.

(1) It seems to me that no philosophy can possibly be sympathetic to a mathematician which does not admit, in one manner or another, the immutable and unconditional validity of mathematical truth. Mathematical theorems are true or false; their truth or falsity is absolute and independent of our knowledge of them. In *some* sense, mathematical truth is part of objective reality.

'Any number is the sum of 4 squares'; 'any number is the sum of 3 squares'; 'any even number is the sum of 2 primes'. These are not convenient working hypotheses, or half-truths about the Absolute, or collections of marks on paper, or classes of noises summarising reactions of laryngeal glands. They are, in one sense or another, however elusive and sophisticated that sense may be, theorems concerning reality, of which the first is true, the second is false, and the third is either true or false, though which we do not know. They are not creations of our minds; Lagrange discovered the first in 1774; when he discovered it he discovered *something*: and to that something Lagrange, and the year 1774, are equally indifferent.

(2) When we know a mathematical theorem, there is something, some object, which we know; when we believe one, there is something which we believe; and this is so equally whether what we believe is true or false.

It is obvious that by this time we have escaped only too successfully from the domain of platitude and triviality. We have done no more than to make explicit a few of the instinctive prejudices of the 'mathematician in the street'. Yet with our first demand we have antagonised at least two-thirds of the philosophers in the world; and with the second we have reduced our first indiscretion to entire insignificance, since we have committed ourselves, in one form or another, to the objective reality of propositions, a doctrine rejected, I believe, not only by all philosophers, but also by all three of the current schools of mathematical logic.

(3) In spite of this I am going farther, and in a direction relevant to the recent controversies concerning 'transfinite' mathematics to which I shall return later. Mathematicians have always resented attempts by philosophers or logicians to lay down dogmas imposing limitations on mathematical truth or thought. And I am sure that the vast majority of mathematicians will rebel against the doctrine—even if it is supported by some of themselves, including mathematicians so celebrated as Hilbert and Weyl—that it is only the so-called 'finite' theorems of mathematics which possess a real significance. That 'the finite cannot understand the infinite' should surely be a theological and not a mathematical war-cry.

No one disputes that there are infinite processes which appear to be prohibited to us by the facts of the physical world. It is true, as Hilbert says, that no mathematician has completed an infinity of syllogisms. It is equally true that there is no mathematician who has never drunk a glass of water, and, so far as I can see, one of these facts has neither more nor less logical importance than the other. There is no more *logical* reason why a mathematician should not prove an infinity of theorems in this world than why he should not (as he has been so often encouraged to hope) emit an infinite sequence of musical notes in the next.

The history of mathematics shows conclusively that mathematicians do not evacuate permanently ground which they have conquered once. There have been many temporary retirements and shortenings of the line, but never a general retreat on a broad front. We may be confident that, whatever the precise issue of current controversies, there will be no general surrender of the ground which Weierstrass and his followers have won. 'No one', as Hilbert says himself, 'shall chase us from the paradise that Cantor has created': the worst that can happen to us is that we shall have to be a little more particular about our clothes.

4. Such then are the presuppositions and prejudices with which a working mathematician is likely to approach philosophical or logical systems. How far are they satisfied by the existing schools of mathematical logic? There are three such schools, the logisticians (represented at present by Whitehead, Russell,

Wittgenstein, and Ramsey), the finitists or intuitionists (Brouwer and Weyl), and the formalists (Hilbert and his pupils). I am primarily interested at the moment in the formalist school, first because it is perhaps the natural instinct of a mathematician (when it does not conflict with stronger desires) to be as formalistic as he can, secondly because I am sure that much too little attention has been paid to formalism in England, and finally because of the title of my lecture and because Hilbert's logic is above everything an explicit theory of mathematical proof. I must begin by a rapid summary of the most striking differences between these schools and of the difficulties which have brought them into existence. It is not my object to discuss these difficulties in detail, but what I have to say later can hardly be intelligible unless I give some sort of general explanation of their character. I can fortunately base this explanation on the extremely clear account of the situation given recently by Ramsey.

5. (1) I shall refer to the logisticians generally under the short title of 'Russell'. It is necessary to say that by 'Russell' I mean the Russell of *Principia Mathematica. Principia Mathematica* is not a treatise on philosophy, but it has a philosophical background, with which I am in general sympathy. I think that I can understand, in broad outline, how the logical edifice can support itself on *that* foundation. The problem of erecting it on the foundation of Russell's latest philosophical writings is one which I prefer to leave to bolder minds.

To Russell, then, logic and mathematics are substantial sciences which in some way give us information concerning the form and structure of reality. Mathematical theorems have *meanings*, which we can understand directly, and this is just what is important about them. In this, I may observe, Russell and the so-called 'intuitionists' are in complete agreement; and (since it is something of this sort which seems to me the natural implication of the word) I should prefer to avoid the use of 'intuitionism' as distinguishing one school from the other.

Mathematics is to Russell, up to a certain point at any rate, a branch of logic. It is concerned with particular kinds of assertions about reality, with particular logical concepts, propositions, classes, relations, and so forth. The propositions of logic and mathematics share certain general characteristics, in particular complete generality, though this is not an adequate description of them. There is no particular reason that I can see why any of this should be distasteful to us as mathematicians. It does not seem to conflict with the criteria which I suggested a moment ago; it seems likely at first sight even to indulge our desire for real propositions, though here we are ultimately disappointed.

There are certain definite points at which Russell's attempted reduction of mathematics to logic fails. In this, of course, there is nothing likely to astonish an unsophisticated mathematician. That mathematics should follow naturally, up to a point, from purely logical premisses, premisses to whose simplicity and 'self-evidence' no one can reasonably take exception, when proper allowance is made for the element of sophistication inevitable in a highly complex structure; but that it should then prove necessary to import fresh raw material and add new assumptions—all this is only what a mathematician might expect. In particular, I think that this is true of two of the three 'non-logical' axioms necessary in Russell's scheme; the Axiom of Infinity, that the universe contains an infinity of individuals, and the Multiplicative Axiom or Axiom of Zermelo, which is very famous but required only in particular theorems which might conceivably be discarded, and which I need not stop to explain, since I shall not refer to it further.

6. The situation is quite different with the third axiom, the notorious Axiom of Reducibility. The point here is much more important and also much more difficult. It is essential that I should say something about it, impossible that I should explain it fully. I cannot hope to find popular language clearer than Ramsey's, and I shall follow him very closely.

The theory of aggregates, in the classical form of Cantor and Dedekind, leads to certain antinomies, of which the most famous is Russell's paradox of the class of classes which are not members of themselves, a concept which may be shown to lead at once to flat contradiction. Russell met the difficulty by his Theory of Types.

Suppose that we are given a set S of properties, defined as being all properties of a certain kind K. Given an object x, we can ask whether x possesses *any* property of the kind K. If x has *any* such property, this is another property of x, say Σ; and we can then ask whether Σ can be itself a property of the set S, that is to say of the kind K. It is natural to suppose that the answer must be negative, since the idea of Σ already presupposes the totality S; and this is in fact Russell's answer. The property Σ is, he says, a property 'of higher order' than any property belonging to S; and so generally we must classify properties according to their orders, and any property defined by reference to all properties of a certain order must be a property of higher order. It is impossible to make any statement which is significant for properties of all orders simultaneously. Further, since, in Russell's logic, statements about classes are merely disguised statements about their defining properties, classes also must be divided into orders, and any statement about 'all' classes must really be confined to all classes of a certain order. This doctrine seems inherently plausible, and leads to an easy solution of Russell's and similar antinomies.

The theory of types has, however, very unfortunate mathematical consequences, since it appears to destroy some of the most fundamental theorems of analysis. The typical theorem is the theorem that any aggregate of numbers has an upper bound, a theorem which is substantially the same as what, in my *Pure Mathematics*, is called 'Dedekind's Theorem'. A real number is defined as a class of rationals. Suppose now that we are given a set S of real numbers x, i.e. a set of classes of rationals. The upper bound U of S is defined as the class of rationals which is the logical sum of the classes defining the various members of X, and it is taken for granted that this class stands on the same footing as the classes of which it is the sum. But a moment's consideration shows that this is not so. The classes which are the members of X are defined by certain properties of rationals, and the class which is U is defined by the property of belonging to *some one or other* of these classes, that is to say of possessing some one or other of these properties. Thus the defining characteristic of U involves a reference to *all* the defining characteristics of members of X, and is therefore a characteristic of higher order. It follows that, if we were to attempt to develop analysis without further assumptions, we should have to distinguish real numbers of different orders. We should have to say that the upper bound of an aggregate of real numbers of order n was a real number of order $n + 1$, and so on; and this, whether practicable or not (a point about which I express no opinion) would certainly be extremely inconvenient and probably intolerable.

Russell meets this difficulty by the Axiom of Reducibility, which asserts roughly that there is a property of the lowest order equivalent to any property of any order, not of course equivalent in *meaning*, but equivalent in *extension*, so that any object which possesses the one possesses the other, and they define the same class. The upper bound U may then be defined, not only by the property used to define it above, but also by the equivalent property of lower order, and it is thus a real number in the same sense as each of the numbers of which it is the upper bound. It is not disputed by anybody, so far as I know, that the axiom does yield a solution of the problem. Analysis can be developed in the classical manner and without further difficulty when once the truth of the axiom is granted; and there seems to be no ground for supposing that the axiom will lead to contradiction.

There are, however, objections to the axiom, about the force of which opinions may perhaps differ, but which have proved sufficient to prevent all other logicians from accepting it. It is complicated and (what is more important) very *unconvincing*. It has none of the 'self-evidence' of the properly logical assumptions; and it is obvious that Russell himself dislikes it very heartily and regards its presence in his system as a most regrettable necessity. Finally, an argument suggested in the rough by Ramsey, and developed in a more precise form by Waismann, appears to show conclusively that the axiom is definitely *not* a 'truth of logic' in the same sense as the other primitive propositions of *Principia Mathematica*. It is therefore impossible to regard Russell's solution as satisfactory, and this is about the only point on which the logicians, Russell himself included, are unanimous.

7. (2) I pass to the finitists, Brouwer and Weyl, and I shall dismiss them very shortly. Much as I admire the contributions of Brouwer and Weyl to constructive mathematics, I find their contribution to logic singularly unsympathetic. Finitism rejects, first, all attempts to push the analysis of mathematics beyond a certain point, and for this I see no sort of justification. I have no particular desire to be committed to the extreme Russellian doctrine, that all mathematics is logic and that mathematics has no fundamentals of its own. If it should turn out that there are parts of mathematics irreducible to logic, I do not see why I should be particularly distressed. On the other hand I see no reason for denying that, up to a point, the reduction has actually been made, and the arguments for denying in principle the possibility of a further reduction seem to me entirely inconclusive. That there is some particular sanctity about the notion of an

integer which should protect it against the humiliation of further analysis, that general existential propositions have no real significance, that there is some peculiar certainty in knowledge based, in some sense, in immediate perception of a finite number of sensible things—all these are dogmas to which the finitists seem to be comititted; and all of them seem to be founded on philosophical doctrines with which I have no sympathy, which indeed I find it extremely difficult to understand, and which seem to me, so far as I can understand them, to rest on all sorts of questionable assumptions, and in particular on an impossibly naïve attitude towards our knowledge of the physical world.

This, however, is a minor point for a mathematician. What is much more serious to a mathematician is that the mathematical consequences of finitism involve rejection not (like those of denying the Multiplicative Axiom) of particular isolated outworks of mathematics but of integral regions of ordinary analysis. It is no use trying to deny that the finitists have the better of the argument up to a point; the parts of analysis which they admit are unquestionably, at present, in a more secure position than the rest; and so long as finitism merely insists on this its position is unassailable. I cannot believe that mathematicians generally will be so ready to accept a check as final, so anxious to find metaphysical reasons for supposing that the prettiest path is that which passes on the side of the hedge away from the bull.

8. (3) I go on then to consider the logic of Hilbert and his school; and here I find it very necessary to distinguish between Hilbert the philosopher and Hilbert the mathematician. I dislike Hilbert's philosophy quite as much as I dislike that of Brouwer and Weyl, but I see no reason for supposing that the importance of his logic depends in any way on his philosophy.

I am sure that the Hilbert logic has been unreasonably neglected by English logicians. 'The formal school', says Ramsey, 'have concentrated on the propositions of mathematics, which they have pronounced to be meaningless formulæ to be manipulated according to certain rules, and mathematical knowledge they hold to consist in knowing what formulæ can be derived from what others consistently with the rules. Such being the propositions of mathematics, the account of its concepts, for example the number 2, immediately follows: "2" is a meaningless mark occurring in these meaningless formulæ. But, whatever may be thought of this as an account of mathematical propositions, it is obviously hopeless as a theory of mathematical concepts; for these occur not only in mathematical propositions, but also in those of everyday life. Thus "2" occurs not merely in "2 + 2 = 4", but also in "it is 2 miles to the station", which is not a meaningless formula but a significant proposition, in which "2" cannot conceivably be a meaningless mark. Nor can there be any doubt that "2" is used in the same sense in the two cases, for we can use "2 + 2 = 4" to infer from "it is 2 miles to the station and 2 miles on to the Gogs" to "it is 4 miles to the Gogs *via* the station", so that these ordinary meanings of "2" and "4" are clearly involved in "2 + 2 = 4".'

Let me say at once that this argument seems to me to be unanswerable and that, if I thought that this really was the beginning and the end of formalism,

I should agree with Ramsey's rather contemptuous rejection of it. But is it really credible that this is a fair account of Hilbert's view, the view of the man who has probably added to the structure of significant mathematics a richer and more beautiful aggregate of theorems than any other mathematician of his time? I can believe that Hilbert's philosophy is as inadequate as you please, but not that an ambitious mathematical theory which he has elaborated is trivial or ridiculous. It is impossible to suppose that Hilbert denies the significance and reality of mathematical concepts, and we have the best of reasons for refusing to believe it: 'the axioms and demonstrable theorems,' he says himself, 'which arise in our formalistic game, are the images of the ideas which form the subject-matter of the ordinary mathematics'.

I must, however, begin with a few remarks about the philosophical background which seems to lie behind Hilbert's views; and here of course I need not be alarmed if I find myself disagreeing with him as hopelessly as with the finitists. Hilbert's philosophy appears indeed to be in broad outline much the same as Weyl's, as Weyl himself has very fairly pointed out. There is the same rejection of the possibility of any purely logical analysis of mathematics: 'mathematics is occupied with a content given independently of all logic, and cannot in any way be founded on logic alone.' There is the same insistence on some sort of concrete, perceptible basis, for which Hilbert (with what justice I have no idea) claims the support of 'the philosophers and especially Kant': 'in order that we should be able to apply logical forms of reasoning, it is necessary that there should first be something given in presentation, some concrete, extra-logical object, immediately present to intuition and perceived independently of all thought. . . . In particular, in mathematics, the objects of our study are the concrete signs themselves.' There is, I think, no doubt at all that Hilbert does assert, quite unambiguously, that the subject matter of mathematics proper is the actual physical mark, not general formal relations between the marks, properties which one system of marks may share with another, but the black dots on paper which we see.

I had better state at once what is to me a fatal objection to this view. If Hilbert has made the Hilbert mathematics with a particular series of marks on a particular sheet of paper, and I copy them on another sheet, have I made a *new* mathematics? Surely it is *the same* mathematics, and that even if he writes in pencil and I in ink, and his marks are black while mine are red. Surely the Hilbert mathematics must be in some sense something which is common to all such sets of marks. I make this point here, because there are two questions which suggest themselves at once about Hilbert's marks. The first is whether we are studying the physical signs themselves or general formal relations in which they stand, and the second is whether these signs or relations have 'meaning' in the sense in which the symbols of mathematics are usually supposed to have meaning. It seems to me that the two questions are quite distinct.

9. It is no doubt this philosophical outlook, and this consequent insistence on the importance of the physical mark or sign, that inspire Hilbert's finitism, which appears at first sight as extreme as that of Brouwer and Weyl themselves.

I naturally find this attitude very disappointing; it seems to me that formalism is bound to die for want of air within the narrow confines of a finitistic system. But on the face of it Hilbert is entirely uncompromising: 'there is no infinite anywhere in reality', he says, and again 'is it not clear that, when we think we can recognise the reality of the infinite in any sense, we are merely allowing ourselves to be deceived by the enormity of the largeness or smallness which confronts us everywhere . . .?'.

Hilbert says that 'infinite theorems', theorems such as 'there are infinitely many primes', are not genuine propositions but 'ideal' propositions. I am not at all sure what he means by an 'ideal proposition', but I suppose that one thing at any rate that he would say (if he used Russell's language) is that the infinite is essentially *incomplete*. We know that mathematics is full of 'incomplete symbols', symbols which have no meaning in themselves, though larger collections of symbols of which they are parts have perfectly definite meanings. There are, for example, the ordinary 'operational' symbols; $\frac{\text{`}d\text{'}}{dx}$, '$\nabla^2$', '$\int_a^b \ldots dx$'. The most striking example is the '∞' of elementary analysis; we define '$\sum\limits_0^\infty$' and '$f(x) \to \infty$', but (at any rate in the ordinary presentations of the subject) we never define '∞' standing by itself. There is, in the classical analysis, no number ∞ standing on all fours with e or π; there is a sharp contrast here between the infinite of analysis and the infinite of geometry, in which 'the line at infinity', say $z = 0$, is on just the same footing as any other line.

It is one of Russell's admitted achievements to have recognised in a precise and explicit manner the immense importance of 'incomplete symbolism' in logic and philosophy also, and so to have shown how widely the correct analysis of a proposition may diverge from the analysis of unreflecting common sense. The standard example is that of propositions containing denoting phrases or *descriptions*, 'the so-and-so', 'the murderer', 'the author of Waverley'. The 'Waverley' argument applies to all propositions of the form '*a* is the *b*', and shows that the proposition cannot be analysed, as the words expressing it suggest, into an assertion of identity between '*a*' and 'the *b*'. I wish to know whether *a* is the *b*, whether Dr. Sheppard was the murderer of Roger Ackroyd; and in fact he was. If '*a*' and 'the *b*' are the same object, I can substitute one for the other in any proposition without destroying its sense or its truth; and therefore it appears that what I really wanted to know was whether Dr. Sheppard was Dr. Sheppard, which is obviously false. It follows that the analysis was wrong, and that there is no such object in reality as 'the *b*'; '*a* is the *b*' must be analysed in an entirely different manner.

I am not suggesting that Hilbert would accept the statement that the infinite is incomplete as an adequate account of his attitude towards it. No doubt he would want to go very much further. I have inserted this explanation merely (1) because I shall need it later and (2) because rival views about the infinite are apt to differ more violently in expression than reality, and the notion of an incomplete symbol might in some cases be a basis for a reconciliation between them. I have the less hope that it would do so in this case because Hilbert uses, as instances in support of his thesis that all 'infinite theorems' are in some sense 'ideal theorems', such divergent illustrations as (*a*) the infinite of analysis, (*b*) the infinite of geometry, and (*c*) the ideal numbers of higher arithmetic, and it seems to me quite impossible to regard all these as inspired by the same logical motive, the first representing a *purification* of mathematics by an agreement to regard certain notions as 'incomplete', the others an *enlargement* of it by the introduction of new elements as 'complete' as those which they generalise.

10. It is time, however, to proceed to some description of Hilbert's system, and I do this in language based upon that of v. Neumann, a pupil of Hilbert's whose statement I find sharper and more sympathetic than Hilbert's own.

(1) Hilbert's logic is a theory of proof. Its object is to provide a system of formal axioms for logic and mathematics, and a formal theory of logical and mathematical proof, which (*a*) is sufficiently comprehensive to generate the whole of recognised mathematics, and (*b*) can be proved to be consistent. The system of *Principia Mathematica* fulfils the first but not the second criterion.

(2) If we can do this, we shall be troubled by antinomies no more. But for this end the whole existing apparatus of axioms, proofs, and theorems must first be formalised strictly, so that to every mathematical theorem a formula will correspond. The structure of the formal system will of course be *suggested* by the current logic and mathematics. Every formula will *seem* to have a meaning, a meaning which we must afterwards forget.

(3) For example, we have the 'logical' formula

$$a \to (b \to a).$$

This is suggested by an obvious 'logical truth', the truth that (in Russell's symbolism) $a \,.\supset .\, b \supset a$, that a true proposition is a consequence of any hypothesis. This formula is an 'axiom', which means *simply* that it is one of the formulæ with which we start.

Similarly we have the formula (again an axiom)

$$Za \to Z(a+1),$$

which is suggested by the 'mathematical truth' that $a+1$ is an integer if a is one. We thus start with a finite system of axioms or 'given formulæ'. They are, so to say, the chessmen, the bat, ball, and stumps, *the material with which we play*.

(4) We also need *rules for the game*, of which there are two. Rule (1) is that we may substitute one formula inside another, in the first instance inside an axiom, while Rule (2) is embodied in the 'scheme of demonstration'

$$\begin{array}{c} a \\ a \to b \\ b \end{array} \qquad (A)$$

(which corresponds to the 'non-formal principle of inference' in *Principia Mathematica*). Such a scheme is called a *demonstration*, *a* the *hypothesis*, *b* the *conclusion*. A formula is said to be *demonstrable* (1) if it is an axiom, or (2) it is b, a and $a \rightarrow b$ being axioms, or (3) it is b, a and $a \rightarrow b$ being demonstrable, or (4) it is derivable from an axiom or a demonstrable formula by substitution. We have thus a quite precise concept of 'demonstration'. To use Weyl's illustration, we are playing chess. The *axioms* correspond to the given position of the pieces; the *process of proof* to the rules for moving them; and the *demonstrable formulæ* to all possible positions which can occur in the game.

(5) Let us observe in passing that there are far more axioms in Hilbert's scheme than in such a scheme as that of *Principia Mathematica*, and *no definitions* in the sense of *Principia Mathematica*. This is inevitable, since it is cardinal in Hilbert's logic that, however the formulæ of the system may have been suggested, the 'meanings' which suggested them lie entirely outside the system, so that the 'meaning' of a formula is to be forgotten immediately it is written down. The definitions of *Principia Mathematica* are the most important elements of the system, and embody 'philosophical' analyses of the meanings of the symbols used. The definition of a cardinal number, for example, presents to us at any rate one possible meaning of number, and tells us that that is the meaning with which Russell proposes to use the word. Hilbert is not concerned with that, or any, 'meaning' of 'number', and the only conceivable sense of a definition in his system is that of a symbolic convention which instructs us to replace a prolix formula by a more concise one.

11. (6) Mathematics proper, then, is reduced to a game like chess. We can, however, regard a game like chess from two quite different standpoints. In the first place we can *inspect*, or *construct*, chess, by reading the games whose aggregate constitutes chess, or playing new ones. Secondly, we can think and theorise *about* chess; we make judgements about it, and these judgements contain theorems which are in no sense part of the game. To take a definite illustration, which is in one form or another essential to the understanding of the Hilbert logic, we can judge, and in a sense *prove*, that *certain positions cannot occur*. There cannot be more than eighteen queens on the board; two knights cannot mate; these are true and provable theorems, not theorems *of* chess—the theorems *of* chess are the actual positions—but theorems *about* chess.

Similarly there is the Hilbert mathematics on the one hand, and what Hilbert calls 'metamathematics' on the other, the metamathematics being the aggregate of theorems *about* the mathematics; and of course it is the metamathematics which is the exciting subject and affords the real justification for our interest in this particular sort of mathematics. Suppose, for example, that we could find a finite system of rules which enabled us to say whether any given formula was demonstrable or not. This system would embody a theorem of metamathematics. There is of course no such theorem, and this is very fortunate, since if there were we should have a mechanical set of rules for the solution of all mathematical problems, and our activities as mathematicians would come to an end.

Such a theorem is not to be expected or desired, but there are metamathematical theorems of a different kind which it is entirely reasonable to expect and which it is in fact Hilbert's dominating aim to prove. These are the negative theorems of the kind which I illustrated a moment ago; they assert, for example, in chess, that two knights cannot mate, or that some other combination of the pieces is impossible, in mathematics that certain theorems cannot be demonstrated, that certain combinations of symbols cannot occur. In particular we may hope (and it is this hope that has inspired the whole construction of the logic) to show the impossibility of the combination

$$a . - a,$$

where $-$ is the symbol corresponding to the 'negation' of *Principia Mathematica*.

Let us suppose that our analysis of the game has established this, and then recur to the 'meanings' which suggested the game but were afterwards discarded. We may think about meanings *now*, because we are engaged in metamathematics, *outside* the game. It will plainly follow that the concepts and propositions which we symbolised cannot lead to contradiction. If this has been done, and for a formal system rich enough to be correlated with the whole of mathematics, the purpose of the Hilbert logic will have been achieved.

12. It is now time for me to interpolate a remark which gives the justification for the title of my lecture. It is obvious that to Hilbert *proof* means two quite different things. I have tried to anticipate the point in my choice of words: we fortunately have two words, *proof* and *demonstration*.

'Proof' has always meant at least two different things, even in ordinary mathematics. *We* distinguish vaguely and half-heartedly; in the Hilbert logic the distinction becomes absolutely sharp and clear. First, there is the *formal*, *mathematical*, *official* proof, the proof inside the system, the pattern (A), what I called the *demonstration*. These inside official proofs are, in the mathematics, the actual formulæ or patterns, in the metamathematics, the subject matter for discussion.

Secondly there are the proofs of the theorems of the metamathematics, the proof that two knights cannot mate. These are *informal*, *unofficial*, *significant* proofs, in which we reflect on the meaning of every step. The structure of these proofs is not dictated by our formal rules; in making them we are guided, as in ordinary life, by 'intuition' and common sense. 'Prof. Hardy will lecture at 12.00 today, because it says so in the *Reporter*, and because statements in the *Reporter* are always true.'

You must not imagine that the unofficial, metamathematical, non-formal, intuitionist proof is in any sense slacker or less 'rigorous' than the formal mathematical proof. The subject matter is abstract and complicated, and every step has to be scrutinised with the utmost care. We may even find it necessary to guide our thoughts by the introduction of new formalism, and it is quite likely that, if we do, we shall use over again the same symbols that we have used already. And here, of course, lies a danger; for we may be tempted to forget that we are using the same symbols in different contexts and with different aims; even Russell has been accused of making this mistake by logicians of the more formal schools. In the Hilbert logic at any rate the distinction is quite precise; the unofficial proof lies entirely outside the official system, and its object is simply *to produce conviction*, unofficial conviction of the absence of official contradiction—which is what we want.

13. At this point I should like to leave the Hilbert logic for a moment, and make a few general remarks about mathematical proof as we working mathematicians are familiar with it. It is generally held that mathematicians differ from other people in *proving* things, and that their proofs are in some sense *grounds* for their beliefs. Dedekind said that 'what is provable, ought not to be believed without proof'; and it is undeniable that a decent touch of scepticism has generally (and no doubt rightly) been regarded as some indication of a superior mind.

But if we ask ourselves why we believe particular mathematical theorems, it becomes obvious at once that there are very great differences. I believe the Prime Number Theorem because of de la Vallée-Poussin's proof of it, but I do not believe that $2 + 2 = 4$ because of the proof in *Principia Mathematica*. It is a truism to any mathematician that the 'obviousness' of a conclusion need not necessarily affect the interest of a proof.

I have myself always thought of a mathematician as in the first instance an *observer*, a man who gazes at a distant range of mountains and notes down his observations. His object is simply to distinguish clearly and notify to others as many different peaks as he can. There are some peaks which he can distinguish easily, while others are less clear. He sees A sharply, while of B he can obtain only transitory glimpses. At last he makes out a ridge which leads from A, and following it to its end he discovers that it culminates in B. B is now fixed in his vision, and from this point he can proceed to further discoveries. In other cases perhaps he can distinguish a ridge which vanishes in the distance, and conjectures that it leads to a peak in the clouds or below the horizon. But when he sees a peak he believes that it is there simply because he sees it. If he wishes someone else to see it, he *points to it*, either directly or through the chain of summits which led him to recognise it himself. When his pupil also sees it, the research, the argument, the *proof* is finished.

The analogy is a rough one, but I am sure that it is not altogether misleading. If we were to push it to its extreme we should be led to a rather paradoxical conclusion; that there is, strictly, no such thing as mathematical proof; that we can, in the last analysis, do nothing but *point*; that proofs are what Littlewood and I call *gas*, rhetorical flourishes designed to affect psychology, pictures on the board in the lecture, devices to stimulate the imagination of pupils. This is plainly not the whole truth, but there is a good deal in it. The image gives us a genuine approximation to the processes of mathematical pedagogy on the one hand and of mathematical discovery on the other; it is only the very unsophisticated outsider who imagines that mathematicians make discoveries by turning the handle of some miraculous machine. Finally the image gives us at any rate a crude picture of Hilbert's metamathematical proof, the sort of proof which is a *ground* for its conclusion and whose object is to *convince*.

On the other hand it is not disputed that mathematics is full of proofs, of undeniable interest and importance, whose purpose is not in the least to secure conviction. Our interest in these proofs depends on their formal and æsthetic properties. This is almost always so with *logical* proofs, Theorem 3.24 of *Principia Mathematica* is the law of contradiction, and it is certainly not because we require to be convinced of its truth that we are prepared to study its elaborate deduction from equally 'self-evident' premisses. Here we are interested in the pattern of proof *only*. In our practice as mathematicians, of course, we cannot distinguish so sharply, and our proofs are neither the one thing nor the other, but a more or less rational compromise between the two. Our object is *both* to exhibit the pattern and to obtain assent. We cannot exhibit the pattern completely, since it is far too elaborate; and we cannot be content with mere assent from a hearer blind to its beauty.

14. Let us return to the Hilbert logic. The very structure of the logic, its mere existence, are enough, I think, to prove two propositions of great importance. The first is that it is possible to establish the consistency of a system of axioms *internally*, that is to say by direct examination of its structure; and the second is that it is possible to prove a system consistent even when the axioms embody logical principles such as the law of contradiction itself. Each of these propositions has been disputed.

Consider for a moment the ordinary procedure of axiomatic geometry. In abstract geometry we consider unspecified systems of things, a class S of objects A, B, C, . . . which we call *points*, and sub-classes of these objects which we call *lines*. We make certain assumptions about these points and lines, which we call *axioms*, such as that there is a line which contains any given pair of points, that there is only one such line, and so on. To lay down a system of axioms in geometry is simply to limit the subject matter, to say that we propose to consider only objects of certain kinds. Thus, in a geometry which contains the two axioms I have mentioned, our 'points' might be the players in a tournament, and our 'lines' the opponents in a game, but the points and lines could not be undergraduates and colleges, because then the axioms would be untrue.

In a geometry we are not concerned with any *particular* meaning of 'point' or 'line'. We may say, if we like, that we are concerned with *all possible* meanings, or that we are not concerned with meanings at all; we might accept Hilbert's language, and say that we are concerned simply with *marks*, or we might say (what would, I think, be at any rate one stage nearer to the truth) that we are concerned with what Wittgenstein calls *forms*. It is possible that the question is mainly one of words. We assume merely that our unspecified subject matter possesses the properties stated in our axioms, and we set out to investigate its other properties, the theorems of our geometry, by the usual processes of logical inference.

Every geometry demands a *consistency theorem*, which is naturally not a theorem of the geometry. We have to prove that the axioms do not contradict one another. We produce an example, an 'interpretation', of the geometry, a set of objects which actually have the properties attributed by the axioms to our points and lines. In general in these discussions we take arithmetic or analysis for granted, and our example is one in which points and lines are sets of numbers. Thus our points might be the numbers 1, 2, and 3, and our lines the classes 23, 31, 12: these objects do in fact satisfy the particular axioms which I mentioned. It was by this process, for example, that the old difficulties about the possibility of non-Euclidean Geometry were ultimately settled. It has always been held, and no doubt correctly, that in Geometry, where only the 'subject-matter' is symbolised, and there is no attempt to symbolise the process of inference itself, there is no other possible method.

If we try to apply a similar process to arithmetic, we are met by a difficulty. It is natural that a mathematician should wish to treat arithmetic axiomatically, to say not (with *Principia Mathematica*) that a number is such or such a particular object, but that numbers are any set of objects which have certain properties: there are so many plausible definitions of a number, and the reasons for selecting one rather than another seem so purely technical. There is, however, an obvious difficulty about the inevitable proof of consistency. When we wanted such a proof for a geometry, we could appeal to arithmetic; but there is nothing in ordinary mathematics which comes before arithmetic, and it is not easy at first to see where any 'example' is to be found. There seems only one possibility, if we are to pursue the established method, and that is to find an example in which the role of number is played by some logical construct, such as the Frege–Russell class of similar classes, which can be shown to have the properties required. If we approach the subject from a standpoint different from that of *Principia Mathematica*, we may say that this is what the authors of that work have actually done.

Finally, if we have established consistency in geometry and arithmetic, can we do so in logic, or in a subject which includes logic? It has been held, and I think by Russell, that we cannot, because our formulæ symbolise, among other things, the logical processes which we use in examining it, because the rules of the game are required in forming the judgement that what purports to be an instance of the game really is one. Other logicians, with whom here I agree, have held that this is a misunderstanding, due to a failure to distinguish between the use of our symbolism inside and outside the formal system.

My own view is that even here the classical method, the method of instances, is available in principle, and that, in restricted subjects such as the logic of classes or of propositions, it can be and has been successfully carried through. If, however, we are as ambitious as Hilbert, so that our system is to cover the whole field of abstract thought, I imagine that the attempt to do what we want on these lines is hopeless. I cannot imagine where we could find an adequate image of so comprehensive a symbolism, except in the whole field of thought which it was actually constructed to symbolise. There remains only the 'internal' method followed by Hilbert, based on study of the formal properties of the rules themselves. Whatever we may think about the philosophical basis on which Hilbert has erected his system, and with whatever success he or his followers may pursue it, it seems to me unquestionable that this method is valid in principle, in mathematics in exactly the same sense as in chess. And in this case Hilbert is entirely justified in his claim that he has found a necessary condition for all systems of mathematical logic, and that 'even the assertions of intuitionism, however modest they may be, require first a certificate of authorisation from this tribunal'.

15. My remarks up to this point have been mainly explanations of things which I think I understand. The rest of what I have to say amounts to little more than a confession of a series of perplexities.

The first question which you will naturally ask is this: granted that Hilbert's method is valid in principle, what has it *done*? How far has the proof of consistency progressed? *Does* it establish freedom from contradiction in a domain co-extensive with mathematics? So far as I know the answer is, up to the present, *No*. There has been very substantial progress, and consistency has been proved up to a point beyond the point up to which success might be expected to be easy. The region accounted for includes the mathematics of the finitists, and that part of *Principia Mathematica* which is independent of the Axiom of Reducibility; but this region does not cover analysis.

It would be very reasonable to ask me, as an analyst, to explain my own attitude towards this hiatus in the foundations of analysis, and I do not profess to be able to give any satisfactory answer. I could only say this: in the first place, I am no finitist; I believe that the analysis of the text-books is true. Secondly, Ramsey has advanced a solution, which he does not profess to regard as entirely satisfactory, but in which I can find a good deal of encouragement. Ramsey makes a distinction, which seems to me obviously valid, between the properly mathematical antinomies, those which (like Russell's) would appear, unless precautions against them were taken, in the structure of mathematics itself, and those which appear to arise from some epistemological or psychological confusion concerning 'meaning' or 'definition'. He observes that Russell's theory of types can be divided into two parts, of which only the first, which is harmless, is required in order to dispose of the first category of antinomies, the second, from which all the trouble arises, being needed only for the antinomies of the second kind. He then puts forward a new theory, which might be described roughly as a revival, with appropriate safeguards, of the old-fashioned theory of classes in extension. In this theory there is no need for any axiom of reducibility; and this is at any rate the *sort* of solution that I should like to see. I cannot really doubt that there is a class which is the logical sum of any given set of classes, and this, or something like it, is all that is required by the Dedekind theory.

16. I will return for a moment in conclusion to the properly 'philosophical' question to which I referred at the beginning, about the reality or 'completensss' of propositions. I am entirely unable to exorcise my craving for real propositions, a weakness which is after all only natural in a mathematician, to whom mathematical theorems ought to be the first basic reality of life. But I can find no sort of encouragement wherever I turn.

Our first instinct is to suppose that a judgement, whether true or false, must be analysable into a mind and an object in relation. In a sense this is admitted to be true by everybody; it is undisputed that there is something objective, what Russell and Wittgenstein call the 'proposition as fact', which enters into any judgement. When we judge, we form a picture of the reality about which we are judging, a form of words, a set of marks or noises, which we suppose, rightly or wrongly, to afford an image of the facts. This is the 'proposition as fact'; the question is, what, if anything, is there more?

It can hardly be questioned that there is *something* more, something which is common to a whole class of factual propositions. If I say that 'George is the father of Edward', I create a factual proposition. If I and all other men say it, in all languages printed, written, or spoken, and formalise it in every conceivable symbolism, we create a class of facts, and there will plainly be something common to all these facts. This also is admitted; all such factual propositions have something in common, something which may be called their *form*. This, however, is by no means enough to satisfy me, since 'Edward is the father of George' has *the same* form as 'George is the father of Edward', while the propositions, if such there be, are plainly different.

In Russell's 'multiple relation' theory, the theory of truth accepted provisionally in the first edition of *Principia Mathematica*, no such entity as the proposition is recognised. A judgement is a complex of objects, of which a mind is one, my mind and 'George' and 'Edward' and 'fatherhood', if we treat all these for simplicity as simple objects. If George *is* the father of Edward (so that the judgement is true) then there is a smaller complex, the 'fact' that George is the father of Edward, which is a part of the larger complex which is the judgement. If the judgement is false, there is no such subordinate complex. In neither case is there anything which can be called the 'proposition'. First descriptions, then classes, then propositions have been washed away into the ocean of the incomplete.

I have myself always detested this theory of truth. Apart from my bewilderment about how a structure such as that of *Principia Mathematica* could possibly be built up on so bottomless a foundation, Russell's theory has always seemed to me to banish entirely the element of correspondence which I have felt to be essential in any theory. My own difficulty has always been this, that I find it impossible not to believe in false or uncertain propositions and almost equally difficult to believe in true ones. When we judge truly, there is something which is admitted, namely the fact; and it seems unreasonable to insist on the independent existence of the proposition as something distinct from either the judgement or the fact. When we judge falsely, there is no fact, and, unless we admit the proposition, there seems to be no foundation for our judgement. It seems, therefore, that there must be some subsidiary complex present in *any* judgement, and this is just what Russell's theory denies.

It was therefore with great relief that I found that Wittgenstein rejects Russell's theory, for a variety of reasons of which the most convincing seems to be that Russell's theory leaves it entirely unexplained why it should be impossible to judge a *nonsense*. It would seem, on Russell's theory, that if you can judge that Edward is the father of George, you should be equally capable of judging that Edward is the father of blue.

Wittgenstein's own theory, if I understand it correctly, is something like this. We begin with reality, the facts. Of these facts we construct pictures, the factual propositions. A factual proposition consists of objects, words, noises, chairs or tables, arranged in a certain *form*. This form is *the same* form as that of reality; it is only because the picture and the facts have the same form that they can be compared with one another. If the fact is that George is the father of Edward, then the picture 'Edward is the father of George' has the same form, and it is just because of this that we can say that the picture is a bad one, the proposition is false. 'The picture can represent every reality whose form it has.... The picture, however, cannot represent its form of representation; it shows it forth.... The picture has the logical form of representation in common with what it pictures.... It agrees with reality or not; it is right or wrong, true or false....'

There is, however, something beside the picture or factual proposition, namely the proposition in the sense which is relevant to logic. What is relevant to logic is not the factual proposition but what is common to all the factual propositions that can be pictures of a given state of affairs. A proposition is thus, in some sense, a *form*. The propositions of Hilbert's logic are also forms, but Wittgenstein's forms are more substantial than Hilbert's, since they contain what Russell and Wittgenstein call the 'logical constants', 'and', 'or', 'not', and so forth, whereas Hilbert's can hardly be said to 'contain' anything at all. These logical constants do not represent and are not represented, but are present in the proposition (that is to say the factual proposition) as in the fact. The proposition (that is to say here the logical proposition) is thus a form of logical constants, whereas Hilbert's propositions are so to say *pure* form.

I ask then, finally, whether there is anything in the proposition, as relevant to logic and as Wittgenstein seems to conceive it, which affords any justification for my belief in 'real' propositions, my invincible feeling that, if Littlewood and I both believe Goldbach's theorem, then there is something, and that the same something, in which we both believe, and that that same something will remain the same something when each of us is dead and when succeeding generations of more skilful mathematicians have proved our belief to be right or wrong. I hoped to find support for such a view, when I read that 'the essential in a proposition is that which is common to all propositions which can express the same sense' and that 'the proposition is the propositional sign in its projective relation to the world'. When I read further, both in the book itself and in what Russell says about it, I concluded that I had been deceived. I can find nothing, in Wittgenstein's theory, that is common to all the ways in which I can say that something is true and is not common also to many of the ways in which I can say that it is false. So here I can find no support for my belief; and if not here, where am I likely to find it? Yet my last remark must be that I am still convinced that it is true.

POSTSCRIPT

I have left this lecture as it was delivered, but I should like to add two remarks.

(1) My quotation from Mr. Ramsey at the beginning of §8 may lead to a misinterpretation of his general view of formalism. I understand from what Mr. Ramsey has written later, and from conversation with him, that his attitude towards Hilbert's logic is, up to a point at any rate, somewhat like my own, that is to say that he accepts the logic without accepting its philosophical foundation. In saying this, of course, I must not be interpreted as claiming Mr. Ramsey's approval for anything in particular that I say in the lecture.

(2) Prof. J.W. Alexander of Princeton has made the following remark to me concerning Hilbert's 'ideal theorems'. The fact that a great part of a formalism has been suggested by 'significant' concepts and propositions does not show that *all* its theorems must be capable of interpretation; there will generally be formulæ to which 'no meaning' can be attributed, and the study of these 'meaningless' formulæ may well advance our understanding of the relations of those which can be interpreted. Indeed (as v. Neumann has pointed out) the formalism *must* contain formulæ of this kind, since (e.g.) we can substitute a 'numerical' symbol inside a 'logical' formula, 2 for a and b in '$a \rightarrow (b \rightarrow a)$': no one has suggested any 'meaning' for '$2 \rightarrow (2 \rightarrow 2)$'.

It is natural to interpret Hilbert as meaning that his 'ideal theorems' are all of this kind; and that his logic does contain theorems 'ideal' in this sense is obvious after what I have just said. It is one thing to admit this, and another to admit that a particular proposition such as 'there are infinitely many primes' is 'ideal'. If I cannot admit that 'there are infinitely many primes' has no 'meaning', it is simply because it seems evident to me what the 'meaning' is.

28

Nicholas Bourbaki

The following article describes the methodology of Bourbaki's celebrated attempt to provide a unified exposition of all the basic branches of mathematics.

The Bourbaki programme began in the early 1930s. Having in the 1920s been educated almost exclusively in the French tradition of complex analysis, the young Bourbaki was unaware of the mathematics that had been developed outside France since 1914. Not until the publication of van der Waerden's *Moderne Algebra* did he learn of the approach to mathematics that had been developed in Göttingen by Hilbert, Noether, and their followers—an approach that combined abstract algebra with Hilbert's axiomatic method:

> This treatise made a great impression. I remember it—I was working on my thesis at that time; it was 1930 and I was in Berlin. I still remember the day that van der Waerden came out on sale. My ignorance in algebra was such that nowadays I would be refused admittance to a university. I rushed to those volumes and was stupefied to see the new world which opened before me. At that time my knowledge of algebra went no further than *mathématiques spéciales*, determinants, and a little on the solvability of equations and unicursal curves. I had graduated from the Ecole Normale and I did not know what an ideal was, and only just knew what a group was! This gives you an idea of what a young French mathematician knew in 1930. (*Dieudonné 1970*, pp. 136–7).

Having recognized the gaps in his mathematical training, the young Bourbaki, in large part as a matter of self-education, set out in 1934 to write a treatise in the style of van der Waerden that would describe the principal ideas of modern mathematics. The initial expectation was that the project would be completed in three years. But he quickly learned that modern mathematics was too vast to be described in a single work—as the acknowledged failure of the vast German *Enzyclopädie der mathematischen Wissenschaften* had indeed already made clear.

As Bourbaki and his project matured, he gradually developed the methodological approach described below. Consciously following in the footsteps of Dedekind and Hilbert, Bourbaki sought to present mathematics in terms of a small number of basic, axiomatized mathematical *structures*; the guiding idea was to find relatively few structures that would be adequate to handle relatively many mathematical situations, and in this manner to present a rational and coherent overview of large parts of mathematics. (Bourbaki, it should be observed, did not intend to denigrate those parts of mathematics that cannot be treated in this way: in his view, important domains like analytical number theory or general topology do not yet lend themselves to an axiomatic development, and are for that reason alone omitted from the treatise.)

The following article illustrates the extent to which the abstract approach of the Hilbert school permeated even French mathematics in the twentieth century, displacing the relatively more concrete and analytic, nineteenth-century style of Poincaré or Borel or Lebesgue. Bourbaki's article can be read as a comment upon, and a development of, the ideas presented in Hilbert's 'Axiomatic thought' (*Hilbert 1918*, translated above). Hilbert's inaptly-named 'formalism' is often regarded as an arid linguistic device, contrived principally to circumvent the set-theoretic paradoxes and the criticisms of the intuitionists—a mechanism to cut short philosophical argument.[a] But Bourbaki here makes evident the importance of Hilbert's axiomatic method for mainstream mathematics, presenting it as a powerful abstract tool that can unify large areas of mathematics and display the deep, hidden relationships between seemingly unrelated discoveries.

The translation of *Bourbaki 1948* is the authorized translation by Arnold Dresden, and was first published as *Bourbaki 1950*; references should be to the section numbers, which appeared in the original edition.

A. THE ARCHITECTURE OF MATHEMATICS (*BOURBAKI 1948*)

1. Mathematic or mathematics? To present a view of the entire field of mathematical science as it exists—this is an enterprise which presents, at first sight, almost insurmountable difficulties, on account of the extent and the varied character of the subject. As is the case in all other sciences, the number of mathematicians and the number of works devoted to mathematics have greatly increased since the end of the nineteenth century. The memoirs in pure

---

[a] Bourbaki himself, in some of his manifestations, has contributed to this way of looking at the philosophy of mathematics:

> On foundations we believe in the reality of mathematics, but of course when philosophers attack us with their paradoxes we rush to hide behind formalism and say: "Mathematics is just a combination of meaningless symbols", and then we bring out Chapters 1 and 2 on set theory. Finally we are left in peace to go back to our mathematics and do it as we have always done, with the feeling each mathematician has that he is working with something real. This sensation is probably an illusion, but it is very convenient. That is Bourbaki's attitude towards foundations (*Dieudonné 1970*, p. 145).

Although Hilbert had a pervasive influence over Bourbaki's mathematics, the influence did not extend to philosophy. In contrast to Bourbaki, Hilbert had a deep interest in questions of mathematical epistemology, and his proof-theoretical investigations were intended to explore and solve the problems of the foundations of mathematics, rather than merely to silence the carping of philosophers. Certainly Hilbert would never have dismissed the set-theoretic paradoxes as principally the affair of philosophers; nor would he have declared that "Mathematics is just a combination of meaningless symbols", a slogan that is a caricature of his actual position. For further discussion of these points, see the Introductory Notes to the selections above from Hilbert.

mathematics published in the world during a normal year cover several thousands of pages. Of course, not all of this material is of equal value; but, after full allowance has been made for the unavoidable tares, it remains true nevertheless that mathematical science is enriched each year by a mass of new results, that it spreads and branches out steadily into theories, which are subjected to modifications based on new foundations, compared and combined with one another. No mathematician, even were he to devote all his time to the task, would be able to follow all the details of this development. Many mathematicians take up quarters in a corner of the domain of mathematics, which they do not intend to leave; not only do they ignore almost completely what does not concern their special field, but they are unable to understand the language and the terminology used by colleagues who are working in a corner remote from their own. Even among those who have the widest training, there are none who do not feel lost in certain regions of the immense world of mathematics; those who, like Poincaré or Hilbert, put the seal of their genius on almost every domain, constitute a very great exception even among the men of greatest accomplishment.

It must therefore be out of the question to give to the uninitiated an exact picture of that which the mathematicians themselves cannot conceive in its totality. Nevertheless it is legitimate to ask whether this exuberant proliferation makes for the development of a strongly constructed organism, acquiring ever greater cohesion and unity with its new growths, or whether it is the external manifestation of a tendency towards a progressive splintering, inherent in the very nature of mathematics; whether the domain of mathematics is not becoming a tower of Babel, in which autonomous disciplines are being more and more widely separated from one another, not only in their aims, but also in their methods and even in their language. In other words, do we have today a mathematic or do we have several mathematics?

Although this question is perhaps of greater urgency now than ever before, it is by no means a new one; it has been asked almost from the very beginning of mathematical science. Indeed, quite apart from applied mathematics, there has always existed a dualism between the origins of geometry and of arithmetic (certainly in their elementary aspects), since the latter was at the start a science of discrete magnitude, while the former has always been a science of continuous extent; these two aspects have brought about two points of view which have been in opposition to each other since the discovery of irrationals. Indeed, it is exactly this discovery which defeated the first attempt to unify the science, *viz.*, the arithmetization of the Pythagoreans ('everything is number').

It would carry us too far if we were to attempt to follow the vicissitudes of the unitary conception of mathematics from the period of Pythagoras to the present time. Moreover this task would suit a philosopher better than a mathematician; for it is a common characteristic of the various attempts to integrate the whole of mathematics into a coherent whole—whether we think of Plato, of Descartes, or of Leibnitz, of arithmetization, or of the logistics of the nineteenth century—that they have all been made in connection with a philosophical system, more or less wide in scope; always starting from *a priori* views concerning the relations of mathematics with the twofold universe of the external world and the world of thought. We can do no better on this point than to refer the reader to the historical and critical study of L. Brunschvicg. [1][a] Our task is a more modest and a less extensive one; we shall not undertake to examine the relations of mathematics to reality or to the great categories of thought; we intend to remain within the field of mathematics and we shall look for an answer to the question which we have raised, by analysing the procedures of mathematics themselves.

2. Logical formalism and axiomatic method. After the more or less evident bankruptcy of the different systems, to which we have referred above, it looked at the beginning of the present century as if the attempt had just about been abandoned to conceive of mathematics as a science characterized by a definitely specified purpose and method; instead there was a tendency to look upon mathematics as 'a collection of disciplines based on particular, exactly specified concepts', interrelated by 'a thousand roads of communication', allowing the methods of any one of these disciplines to fertilize one or more of the others [1, page 447]. Today, we believe however that the internal evolution of mathematical science has, in spite of appearance, brought about a closer unity among its different parts, so as to create something like a central nucleus that is more coherent than it has ever been. The essential aspect of this evolution has been the systematic study of the relations existing between different mathematical theories, and which has led to what is generally known as the 'axiomatic method'.

The words 'formalism' and 'formalistic method' are also often used; but it is important to be on one's guard from the start against the confusion which may be caused by the use of these ill-defined words, and which is but too frequently made use of by the opponents of the axiomatic method. Everyone knows that superficially mathematics appears as this 'long chain of reasons' of which Descartes spoke; every mathematical theory is a concatenation of propositions, each one derived from the preceding ones in conformity with the rules of a logical system, which is essentially the one codified, since the time of Aristotle, under the name of 'formal logic', conveniently adapted to the particular aims of the mathematician. It is therefore a meaningless truism to say that this 'deductive reasoning' is a unifying principle for mathematics. So superficial a remark can certainly not account for the evident complexity of different mathematical theories, not any more than one could, for example, unite physics and biology into a single science on the ground that both use the experimental method. The method of reasoning by means of chains of syllogisms is nothing but a transforming mechanism, applicable just as well to one set of

[a] ‖Square-bracketed numerical note indices refer to Bourbaki's endnotes, which will be found at the end of the article.‖

premisses as to another; it could not serve therefore to characterize these premisses. In other words, it is the external form which the mathematician gives to his thought, the vehicle which makes it accessible to others,[1] in short, the language suited to mathematics; this is all, no further significance should be attached to it. To lay down the rules of this language, to set up its vocabulary and to clarify its syntax, all that is indeed extremely useful; indeed this constitutes one aspect of the axiomatic method, the one that can properly be called logical formalism (or 'logistics' as it is sometimes called). But we emphasize that it is but one aspect of this method, indeed the least interesting one.

What the axiomatic method sets as its essential aim, is exactly that which logical formalism by itself cannot supply, namely the profound intelligibility of mathematics. Just as the experimental method starts from the *a priori* belief in the permanence of natural laws, so the axiomatic method has its cornerstone in the conviction that not only is mathematics not a randomly developing concatenation of syllogisms, but neither is it a collection of more or less 'astute' tricks, arrived at by lucky combinations, in which purely technical cleverness wins the day. Where the superficial observer sees only two, or several, quite distinct theories, lending one another 'unexpected support' [1, page 446] through the intervention of a mathematician of genius, the axiomatic method teaches us to look for the deep-lying reasons for such a discovery, to find the common ideas of these theories, buried under the accumulation of details properly belonging to each of them, to bring these ideas forward and to put them in their proper light.

3. The notion of structure. In what form can this be done? It is here that the axiomatic method comes closest to the experimental method. Like the latter drawing its strength from the source of Cartesianism, it will 'divide the difficulties in order to overcome them better'. It will try, in the demonstrations of a theory, to separate out the principal mainsprings of its arguments; then, taking each of these separately and formulating it in abstract form, it will develop the consequences which follow from it alone. Returning after that to the theory under consideration, it will recombine the component elements, which had previously been separated out, and it will inquire how these different components influence one another. There is indeed nothing new in this classical going to-and-fro between analysis and synthesis; the originality of the method lies entirely in the way in which it is applied.

In order to illustrate the procedure which we have just sketched, by an example, we shall take one of the oldest (and also one of the simplest) of axiomatic theories, *viz.* that of the 'abstract groups'. Let us consider for example, the three following operations: 1. the addition of real numbers, their sum (positive,

---

[1] Indeed every mathematician knows that a proof has not really been 'understood' if one has done nothing more than verifying step by step the correctness of the deductions of which it is composed, and has not tried to gain a clear insight into the ideas which have led to the construction of this particular chain of deductions in preference to every other one.

negative, or zero) being defined in the usual manner; 2. the multiplication of integers '*modulo* a prime number p', (where the elements under consideration are the whole numbers $1, 2, \ldots, p-1$) and the 'product' of two of these numbers is, by agreement, defined as the remainder of the division of their usual product by p; 3. the 'composition' of displacements in three-dimensional Euclidean space, the 'resultant' (or 'product') of two displacements S, T (taken in this order) being defined as the displacement obtained by carrying out first the displacement T and then the displacement S. In each of these three theories, one makes correspond, by means of a procedure defined for each theory, to two elements x, y (taken in that order) of the set under consideration (in the first case the set of real numbers, in the second the set of numbers $1, 2, \ldots, p-1$, in the third the set of all displacements) a well-determined third element; we shall agree to designate this third element in all three cases by $x\tau y$ (this will be the sum of x and y if x and y are real numbers, their product '*modulo* p' if they are integers $\leqq p-1$, their resultant if they are displacements). If we now examine the various properties of this 'operation' in each of the three theories, we discover a remarkable parallelism; but, in each of the separate theories, the properties are interconnected, and an analysis of their logical connections leads us to select a small number of them which are independent (i.e., none of them is a logical consequence of all the others). For example,[2] one can take the three following, which we shall express by means of our symbolic notation, common to the three theories, but which it would be very easy to translate into the particular language of each of them:

(a) For all elements x, y, z, one has $x\tau(y\tau z) = (x\tau y)\tau z$ ('associativity' of the operation $x\tau y$);

(b) There exists an element e, such that for every element x, one has $e\tau x = x\tau e = x$ (for the addition of real numbers, it is the number 0; for multiplication '*modulo* p', it is the number 1; for the composition of displacements, it is the 'identical' displacement, which leaves every point of space fixed);

(c) Corresponding to every element x, there exists an element x' such that $x\tau x' = x'\tau x = e$ (for the addition of real numbers x' is the number $-x$; for the composition of displacements, x' is the 'inverse' displacement of x, i.e. the displacement which replaces each point that had been displaced by x to its original position; for multiplication '*modulo* p', the existence of x' follows from a very simple arithmetic argument.[3]

It follows then that the properties which can be expressed in the same way

[2] There is nothing absolute in this choice; several systems of axioms are known which are 'equivalent' to the one which we are stating explicitly, the axioms of each of these systems being logical consequences of the axioms of any other one.

[3] We observe that the remainders left when the numbers $x, x^2, \ldots x^n, \ldots$ are divided by p, cannot all be distinct; by expressing the fact that the two of these remainders are equal, one shows easily that a power x^m of x exists which has a remainder equal to 1; if now x' is the remainder of the division of x^{m-1} by p, we conclude that the product '*modulo* p' of x and x' is equal to 1.

in the three theories, by means of the common notation, are consequences of the three preceding ones. Let us try to show, for example that from $x\tau y = x\tau z$ follows $y = z$; one could do this in each of the theories by a reasoning peculiar to it. But, we can proceed as follows by a method that is applicable in all cases: from the relation $x\tau y = x\tau z$ we derive (x' having the meaning which was defined above) $x'\tau(x\tau y) = x'\tau(x\tau z)$; thence by applying (a), $(x'\tau x)\tau y = (x'\tau x)\tau z$; by means of (c), this relation takes the form $e\tau y = e\tau z$, and finally, by applying (b), $y = z$, which was to be proved. In this reasoning the nature of the elements x, y, z under consideration has been left completely out of account; we have not been concerned to know whether they are real numbers, or integers $\leqq p - 1$, or displacements; the only premiss that was of importance was that the operation $x\tau y$ on these elements has the properties (a), (b), and (c). Even if it were only to avoid irksome repetitions, it is readily seen that it would be convenient to develop once and for all the logical consequences of the three properties (a), (b), (c) only. For linguistic convenience, it is of course desirable to adopt a common terminology for the three sets. One says that a set in which an operation $x\tau y$ has been defined which has the three properties (a), (b), (c) is provided with a group structure (or, briefly, that it is a group); the properties (a), (b), (c) are called the axioms of[4] the group structures, and the development of their consequences constitutes setting up the axiomatic theory of groups.

It can now be made clear what is to be understood, in general, by a mathematical structure. The common character of the different concepts designated by this generic name is that they can be applied to sets of elements whose nature[5] has not been specified; to define a structure, one takes as given one or several relations, into which these elements enter[6] (in the case of groups, this was the relation $z = x\tau y$ between three arbitrary elements); then one postulates that the given relation, or relations, satisfy certain conditions (which are

[4] It goes without saying that there is no longer any connection between this interpretation of the word 'axiom' and its traditional meaning of 'evident truth'.

[5] We take here a naïve point of view and do not deal with the thorny questions, half philosophical, half mathematical, raised by the problem of the 'nature' of the mathematical 'beings' or 'objects'. Suffice it to say that the axiomatic studies of the nineteenth and twentieth centuries have gradually replaced the initial pluralism of the mental representation of these 'beings'—thought of at first as ideal 'abstractions' of sense experiences and retaining all their heterogeneity—by a unitary concept, gradually reducing all the mathematical notions, first to the concept of the natural number and then, in a second stage, to the notion of set. This latter concept, considered for a long time as 'primitive' and 'undefinable', has been the object of endless polemics, as a result of its extremely general character and on account of the very vague type of mental representation which it calls forth; the difficulties did not disappear until the notion of set itself disappeared (and with it all the metaphysical pseudo-problems concerning mathematical 'beings') in the light of the recent work on logical formalism. From this new point of view, mathematical structures become, properly speaking, the only 'objects' of mathematics. The reader will find fuller developments of this point in articles by J. Dieudonné [2] and H. Cartan [3].

[6] In effect, this definition of structures is not sufficiently general for the needs of mathematics; it is also necessary to consider the case in which the relations which define a structure hold not between elements of the set under consideration, but also between parts of this set and even, more generally, between elements of sets of still higher 'degree' in the terminology of the 'hierarchy of types'. For further details on this point, see [4].

explicity stated and which are the axioms of the structure under consideration).[7] To set up the axiomatic theory of a given structure, amounts to the deduction of the logical consequences of the axioms of the structure, excluding every other hypothesis on the elements under consideration (in particular, every hypothesis as to their own nature).

4. The great types of structures. The relations which form the starting point for the definition of a structure can be of very different characters. The one which occurs in the group structure is what one calls a 'law of composition,' i.e., a relation between three elements which determines the third uniquely as a function of the first two. When the relations which enter the definition of a structure are 'laws of composition', the corresponding structure is called an algebraic structure (for example, a field structure is defined by two laws of composition, with suitable axioms: the addition and multiplication of real numbers define a field structure on the set of these numbers).

Another important type is furnished by the structures defined by an order relation; this is a relation between two elements x, y which is expressed most frequently in the form 'x is at most equal to y,' and which we shall represent in general by $x\mathrm{R}y$. It is not at all supposed here that it determines one of the two elements x, y uniquely as a function of the other; the axioms to which it is subjected are the following: (a) for every x we have $x\mathrm{R}x$; (b) from the relations $x\mathrm{R}y$ and $y\mathrm{R}x$ follows $x = y$; (c) the relations $x\mathrm{R}y$ and $y\mathrm{R}z$ have as a consequence $x\mathrm{R}z$. An obvious example of a set with a structure of this kind is the set of integers (or that of real numbers), when the symbol R is replaced by the symbol $\leqq$. But it must be observed that we have not included among the axioms the following property, which seems to be inseparable from the popular notion of 'order', 'for every pair of elements x and y, either $x\mathrm{R}y$ or $y\mathrm{R}x$ holds'. In other words, the case in which x and y are incomparable is not excluded. This may seem paradoxical at first sight, but it is easy to give examples of very important order structures, in which such a phenomenon appears. This is what happens when X and Y denote parts of the same set and the relation $X\mathrm{R}Y$ is interpreted to mean 'X is contained in Y'; again when x and y are positive integers and $x\mathrm{R}y$ means 'x divides y'; also if $f(x)$ and $g(x)$ are real-valued functions defined on an interval $a \leqq x \leqq b$, while $f(x)\mathrm{R}g(x)$ is interpreted to mean 'for every x, $f(x) \leqq g(x)$'. These examples also give an indication of the great variety of domains in which order structures appear and thus point to the interest attached to their study.

We want to say a few words about a third large type of structures, *viz.* topological structures (or topologies): they furnish an abstract mathematical formulation of the intuitive concepts of neighbourhood, limit, and continuity, to which we are led by our idea of space. The degree of abstraction required

[7] Strictly speaking, one should, in the case of groups, count among the axioms, besides properties (a), (b), (c) stated above, the fact that the relation $z = x\tau y$ determines one and only one z when x and y are given; one usually considers this property as tacitly implied by the form in which the relation is written.

for the formulation of the axioms of such a structure is decidedly greater than it was in the preceding examples; the character of the present article makes it necessary to refer interested readers to special treatises. See, for example, **[5]**.

5. The standardization of mathematical technique. We have probably said enough to enable to reader to form a fairly accurate idea of the axiomatic method. It should be clear from what precedes that its most striking feature is to effect a considerable economy of thought. The 'structures' are tools for the mathematician; as soon as he has recognized, among the elements which he is studying, relations which satisfy the axioms of a known type, he has at his disposal immediately the entire arsenal of general theorems which belong to the structures of that type. Previously, on the other hand, he was obliged to forge for himself the means of attack on his problems; their power depended on his personal talents, and they were often loaded down with restrictive hypotheses, resulting from the peculiarities of the problem that was being studied. One could say that the axiomatic method is nothing but the 'Taylor system' for mathematics.

This is however, a very poor analogy; the mathematician does not work like a machine, nor as the workingman on a moving belt; we cannot over-emphasize the fundamental role played in his research by a special intuition,[8] which is not the popular sense-intuition, but rather a kind of direct divination (ahead of all reasoning) of the normal behaviour which he seems to have the right to expect of mathematical beings, with whom a long acquaintance has made him as familiar as with the beings of the real world. Now, each structure carries with it its own language, freighted with special intuitive references derived from the theories from which the axiomatic analysis described above has derived the structure. And, for the research worker who suddenly discovers this structure in the phenomena which he is studying, it is like a sudden modulation which orients at one stroke in an unexpected direction the intuitive course of his thought and which illumines with a new light the mathematical landscape in which he is moving about. Let us think—to take an old example—of the progress made at the beginning of the nineteenth century by the geometric representation of imaginaries. From our point of view, this amounted to discovering in the set of complex numbers a well-known topological structure, that of the Euclidean plane, with all the possibilities for applications which this involved; in the hands of Gauss, Abel, Cauchy, and Riemann, it gave new life to analysis in less than a century. Such examples have occurred repeatedly during the last fifty years; Hilbert space, and more generally, functional spaces, establishing topological structures in sets whose elements are no longer points, but functions; the theory of the Hensel p-adic numbers, where, in a still more astounding way, topology invades a region which had been until then the domain *par excellence* of the discrete, of the discontinuous, *viz.* the set of whole numbers; Haar measure, which enlarged enormously the field of application of the con-

[8] Like all intuition, this one also is frequently wrong.

cept of integral, and made possible a very profound analysis of the properties of continuous groups;—all of these are decisive instances of mathematical progress, of turning-points at which a stroke of genius brought a new orientation of a theory, by revealing the existence in it of a structure which did not *a priori* seem to play a part in it.

What all this amounts to is that mathematics has less than ever been reduced to a purely mechanical game of isolated formulae; more than ever does intuition dominate in the genesis of discoveries. But henceforth, it possesses the powerful tools furnished by the theory of the great types of structures; in a single view, it sweeps over immense domains, now unified by the axiomatic method, but which were formerly in a completely chaotic state.

6. A general survey. Let us now try, guided by the axiomatic concept, to look over the whole of the mathematical universe. It is clear that we shall no longer recognize the traditional order of things, which, just like the first nomenclatures of animal species, restricted itself to placing side by side the theories which showed greatest external similarity. In place of the sharply bounded compartments of algebra, of analysis, of the theory of numbers, and of geometry, we shall see, for example, that the theory of prime numbers is a close neighbour of the theory of algebraic curves, or that Euclidean geometry borders on the theory of integral equations. The organizing principle will be the concept of a hierarchy of structures, going from the simple to the complex, from the general to the particular.

At the centre of our universe are found the great types of structures, of which the principal ones were mentioned above; they might be called the mother-structures. A considerable diversity exists in each of these types; one has to distinguish between the most general structure of the type under consideration, with the smallest number of axioms, and those which are obtained by enriching the type with supplementary axioms, from each of which comes a harvest of new consequences. Thus, the theory of groups contains, beyond the general conclusions valid for all groups and depending only on the axioms enunciated above, a particular theory of finite groups (obtained by adding the axiom that the number of elements of the group is finite), a particular theory of abelian groups (in which $x\tau y = y\tau x$ for every x and y), as well as a theory of finite abelian groups (where these two axioms are supposed to hold simultaneously). Similarly, in the theory of ordered sets, one notices in particular those sets (as for example, the set of integers, or of real numbers) in which any two elements are comparable, and which are called totally ordered. Among the latter, further attention is given to the sets which are called well-ordered (in which, as in the set of integers greater than 0, every subset has a 'least element'). There is an analogous gradation among topological structures.

Beyond this first nucleus appear the structures which might be called multiple structures. They involve two or more of the great mother-structures simultaneously not in simple juxtaposition (which would not produce anything new), but combined organically by one or more axioms which set up a connection between them. Thus, one has topological algebra. This is a study of structures in which occur at the same time, one or more laws of composition and a topology, connected by the condition that the algebraic operations be (for the topology under consideration) continuous functions of the elements on which they operate. Not less important is algebraic topology, in which certain sets of points in space, defined by topological properties (simplexes, cycles, etc.) are themselves taken as elements on which laws of composition operate. The combination of order structures and algebraic structures is also fertile in results, leading, in one direction to the theory of divisibility and of ideals, and in another to integration and to the 'spectral theory' or operators, in which topology also joins in.

Farther along we come finally to the theories properly called particular. In these the elements of the sets under consideration, which in the general structures have remained entirely indeterminate, obtain a more definitely characterized individuality. At this point we merge with the theories of classical mathematics, the analysis of functions of a real or complex variable, differential geometry, algebraic geometry, theory of numbers. But they have no longer their former autonomy; they have become crossroads, where several more general mathematical structures meet and react upon one another.

To maintain a correct perspective, we must at once add to this rapid sketch the remark that it has to be looked upon as only a very rough approximation of the actual state of mathematics, as it exists; the sketch is schematic, and idealized as well as frozen.

Schematic—because in the actual procedures, things do not happen in as simple and as systematic a manner as has been described above. There occur, among other things, unexpected reverse movements, in which a specialized theory, such as the theory of real numbers, lends indispensable aid in the construction of a general theory like topology or integration.

Idealized—because it is far from true that in all fields of mathematics, the role of each of the great structures is clearly recognized and marked off; in certain theories (for example in the theory of numbers), there remain numerous isolated results, which it has thus far not been possible to classify, nor to connect in a satisfactory way with known structures.

Finally *frozen*,—for nothing is farther from the axiomatic method than a static conception of the science. We do not want to lead the reader to think that we claim to have traced out a definitive state of the science. The structures are not immutable, neither in number nor in their essential contents. It is quite possible that the future development of mathematics may increase the number of fundamental structures, revealing the fruitfulness of new axioms, or of new combinations of axioms. We can look forward to important progress from the invention of structures, by considering the progress which has resulted from actually known structures. On the other hand, these are by no means finished edifices; it would indeed be very surprising if all the essence had already been extracted from their principles. Thus, with these indispensable qualifications, we can become better aware of the internal life of mathematics, of its unity as well as of its diversity. It is like a big city, whose outlying districts and suburbs encroach incessantly, and in a somewhat chaotic manner, on the surrounding country, while the centre is rebuilt from time to time, each time in accordance with a more clearly conceived plan and a more majestic order, tearing down the old sections with their labyrinths of alleys, and projecting towards the periphery new avenues, more direct, broader, and more commodious.

7. Return to the past and conclusion. The concept which we have tried to present in the above paragraphs was not formed all at once; rather is it a stage in an evolution, which has been in progress for more than a half-century, and which has not escaped serious opposition, among philosophers as well as among mathematicians themselves. Many of the latter have been unwilling for a long time to see in axiomatics anything else than futile logical hairsplitting not capable of fructifying any theory whatever. This critical attitude can probably be accounted for by a purely historical accident. The first axiomatic treatments and those which caused the greatest stir (those of arithmetic by Dedekind and Peano, those of Euclidean geometry by Hilbert) dealt with univalent theories, i.e., theories which are entirely determined by their complete system of axioms; for this reason they could not be applied to any theory except the one from which they had been extracted (quite contrary to what we have seen, for instance, for the theory of groups). If the same had been true for all other structures, the reproach of sterility brought against the axiomatic method would have been fully justified.[9] But the further development of the method has revealed its power; and the repugnance which it still meets here and there, can only be explained by the natural difficulty of the mind to admit, in dealing with a concrete problem, that a form of intuition, which is not suggested directly by the given elements (and which often can be arrived at only by a higher and frequently difficult stage of abstraction), can turn out to be equally fruitful.

As concerns the objections of the philosophers, they are related to a domain, on which for reasons of inadequate competence we must guard ourselves from entering; the great problem of the relations between the empirical world and the mathematical world.[10] That there is an intimate connection between experimental phenomena and mathematical structures seems to be fully confirmed in the most unexpected manner by the recent discoveries of contemporary physics. But we are completely ignorant as to the underlying reasons for this fact (supposing that one could indeed attribute a meaning to these words) and we shall perhaps always remain ignorant of them. There certainly is one observation which might lead the philosophers to greater circumspection on this point in

[9] There also occurred, especially at the beginning of axiomatics, a whole crop of monster-structures, entirely without applications; their sole merit was that of showing the exact bearing of each axiom, by observing what happened if one omitted or changed it. There was of course a temptation to conclude that these were the only results that could be expected from the axiomatic method.

[10] We do not consider here the objections which have arisen from the application of the rules of formal logic to the reasoning in axiomatic theories; these are connected with logical difficulties encountered in the theory of sets. Suffice it to point out that these difficulties can be overcome in a way which leaves neither the slightest qualms nor any doubt as to the correctness of the reasoning; [2] and [3] are valuable references for this point.

the future: before the revolutionary developments of modern physics, a great deal of effort was spent on trying to derive mathematics from experimental truths, especially from immediate space intuitions. But, on the one hand, quantum physics has shown that this macroscopic intuition of reality covered microscopic phenomena of a totally different nature, connected with fields of mathematics which had certainly not been thought of for the purpose of applications to experimental science. And, on the other hand, the axiomatic method has shown that the 'truths' from which it was hoped to develop mathematics were but special aspects of general concepts, whose significance was not limited to these domains. Hence it turned out, after all was said and done, that this intimate connection, of which we were asked to admire the harmonious inner necessity, was nothing more than a fortuitous contact of two disciplines whose real connections are much more deeply hidden than could have been supposed *a priori*.

From the axiomatic point of view, mathematics appears thus as a storehouse of abstract forms—the mathematical structures; and it so happens—without our knowing why—that certain aspects of empirical reality fit themselves into these forms, as if through a kind of preadaptation. Of course, it cannot be denied that most of these forms had originally a very definite intuitive content; but, it is exactly by deliberately throwing out this content, that it has been possible to give these forms all the power which they were capable of displaying and to prepare them for new interpretations and for the development of their full power.

It is only in this sense of the word 'form' that one can call the axiomatic method a 'formalism'. The unity which it gives to mathematics is not the armour of formal logic, the unity of a lifeless skeleton; it is the nutritive fluid of an organism at the height of its development, the supple and fertile research instrument to which all the great mathematical thinkers since Gauss have contributed, all those who, in the words of Lejeune-Dirichlet, have always labored to 'substitute ideas for calculations'.

References

1. L. Brunschvicg, *Les étapes de la philosophie mathématique*, Paris, Alcan, 1912.

2. J. Dieudonné, Les méthodes axiomatiques modernes et les fondements des mathématiques, *Revue Scientifique*, LXXVII, 1939, pp. 224–232.

3. H. Cartan, Sur le fondement logique des mathématiques, *Revue Scientifique*, LXXXI, 1943, pp. 3–11.

4. N. Bourbaki, *Eléments de mathématique*, book I (fasc. de résultats), Actual. Scient. et Industr., no. 846.

5. ——, *Eléments de mathématique*, book III, introduction and Chapter I, Actual. Scient. et Industr., no. 858.

Bibliography

The following bibliography incorporates all references from both volumes in this collection; works marked with an asterisk are contained, either in whole or in part, in the present collection. The principles of dating the selections have been explained in the Introduction in Volume I. To make the bibliography as useful for historical reference as possible, an effort was made to supply full names of authors whenever they could be determined. Reprints or translations are listed only when they were considered likely to be of genuine assistance to the reader. Unless otherwise noted, all works listed as translations are into English. In addition to the titles listed here, the reader will find additional specialist bibliographical information in *Benacerraf and Putnam 1983*, *M. Cantor 1894–1908*, *Hallett 1984*, *van Heijenoort 1967*, *Hofmann 1963*, *Moore 1982*, and *Stäckel and Engel 1895*, as well as in the collected works of the mathematicians represented in this collection.

Ackermann, Wilhelm

1924 Begründung des "Tertium non datur" mittels der Hilbertschen Theorie der Widerspruchsfreiheit. *Mathematische Annalen*, **93**, 1–36.

1928a Zum Hilbertschen Aufbau der reellen Zahlen. *Mathematische Annalen*, **99**, 118–33. (Translated by Stefan Bauer-Mengelberg in *van Heijenoort 1967*, pp. 493–507.)

1928b Über die Erfüllbarkeit gewisser Zahlausdrücke. *Mathematische Annalen*, **100**, 638–49.

1940 Zur Widerspruchsfreiheit der Zahlentheorie. *Mathematische Annalen*, **117**, 162–94.

1951 Konstruktiver Aufbau eines Abschnitts der zweiten Cantorschen Zahlenklasse. *Mathematische Zeitschrift*, **53**, 403–13.

1954 *Solvable cases of the decision problem*. North-Holland, Amsterdam.

See Hilbert, David and Ackermann, Wilhelm.

Addison, John W.

1958 Separation principles in the hierarchies of classical and effective descriptive set theory. *Fundamenta Mathematicae*, **46**, 123–35.

Alexandroff, Paul

1916 Sur la puissance des ensembles mesurables, B. *Comptes rendues hebdomadaires des séances de l'académie des sciences, Paris*, **162**, 323–5.

Alexandroff, Paul and Urysohn, Paul
1929 Mémoire sur les éspaces topologiques compacts. *Koninklijke Nederlandse Akademie van Wetenschappen te Amsterdam, Proceedings of the Section of Mathematical Sciences*, **14**, 1-96.

Aristotle
1984 *The complete works of Aristotle*, (ed. Jonathan Barnes). Princeton University Press.

Artin, Emil and Schreier, Otto
1926 Algebraische Konstruktion reeller Körper. *Abhandlungen aus dem mathematischen Seminar der Hamburgischen Universität*, **5**, 85-99.

Arzelà, Cesare
1900 Sulle serie di funzioni. *Memorie de Bologna* (ser. 5), **8**, 131-86; 701-44.

Aspray, William and Kitcher, Philip (eds.)
1988 *History and philosophy of modern mathematics*. Minnesota Studies in the Philosophy of Science, Vol. 11. University of Minnesota, Minneapolis.

Bachmann, Heinz
1955 *Transfinite Zahlen*. Ergebnisse der Mathematik und ihrer Grenzgebiete, Vol. 1. Springer Verlag, Berlin.

Baer, Reinhold
1928 Über ein Vollständigkeitsaxiom in der Mengenlehre. *Mathematische Zeitschrift*, **27**, 536-9.

Baire, René
1898 Sur les fonctions discontinues qui se rattachent aux fonctions continues. *Comptes rendus hebdomadaires des séances de l'Académie des Sciences, Paris*, **129**, 1621-3.
1899 Sur les fonctions de variables réelles. *Annali di matematica pura ed applicata*, **3**, 1-123.
1905 *Leçons sur les fonctions discontinues*. Gauthier-Villars, Paris.

Baire, René; Borel, Emile; Hadamard, Jacques; and Lebesgue, Henri
1905* Cinq lettres sur la théorie des ensembles. *Bulletin de la Société Mathématique de France*, **33**, 261-73. (Translated by Gregory H. Moore, this collection, Vol. 2, pp. 1077-86.)

Baldus, Richard
1928a Zur Axiomatik der Geometrie I. Über Hilberts Vollständigkeitsaxiom. *Mathematische Annalen*, **100**, 321-33.
1928b Zur Axiomatik der Geometrie II. Vereinfachungen des Archimedischen und Cantorschen Axioms. *Atti del Congresso Internazionale dei Mathematici, Bologna, 3-10 Settembre 1928*, **4**, 271-5.
1930 Zur Axiomatik der Geometrie III. Über das Archimedische und das Cantorsche Axiom. *Sitzungsberichte der Heidelberger Akademie der Wissenschaften*, **5**, 3-12.

Baltzer, Richard
1866–7 *Die Elemente der Mathematik*. Hirzel, Leipzig.

Banach, Stefan
1923 Sur le problème de la mesure. *Fundamenta Mathematicae*, **4**, 7–33.
1930 Über additive Massenfunktionen in abstrakten Mengen. *Fundamenta Mathematicae*, **15**, 97–101.

Bar-Hillel, Yehoshua (ed.)
1961 *Essays on the foundations of mathematics*. Magnes Press, Hebrew University, Jerusalem.

Baron, Margaret Elanor
1969 *The origins of the infinitesimal calculus*. Pergamon, Oxford.

Barrow, Isaac
1670 *Lectiones geometricae*. In *The mathematical works of Isaac Barrow* (ed. William Whewell, Cambridge University Press, 1860).

Barwise, Jon (ed.)
1977 *Handbook of mathematical logic*. North-Holland, Amsterdam.

Baumann, Julius
1868–9 *Die Lehre von Raum, Zeit und Mathematik*, 2 vols. Reimer, Berlin.
1908 Dedekind und Bolzano. *Annalen der Naturphilosophie*, **7**, 444–9.

Baumgarten, Alexander Gottlieb
1739 *Metaphysica*. Hemmerde, Halle.

Baxter, Andrew
c.1730 *An enquiry into the nature of the human soul*. J. Bettenham, London.

Becker, Oskar
1923 Phänomenologische Begründung der Geometrie. *Jahrbuch für Philosophie und phänomenologische Forschung*, **6**, 385–560.
1927 Mathematische Existenz. *Jahrbuch für Philosophie und phänomenologische Forschung*, **7**, 439–809.
1933–6 Eudoxus Studien. *Quellen und Studien zur Geschichte der Mathematik, Astronomie und Physik*, Abteilung B: Studien, **2** (1933), 311–33, 369–87; **3** (1936), 236–44, 370–410.
1964 (ed.) *Grundlagen der Mathematik in geschichtlicher Entwicklung*, (2nd extended edn, 1974). Suhrkamp, Frankfurt.

Benacerraf, Paul and Putnam, Hilary (eds.)
1964 *Philosophy of mathematics: Selected readings*. Prentice-Hall, Englewood Cliffs, NJ.
1983 Second edition of *Benacerraf and Putnam 1964*.

Bendixson, Ivar
1883 Quelques théorèmes de la théorie des ensembles de points. *Acta Mathematica*, **2**, 415–29.

Berkeley, George

1707 *Arithmetica et miscellanea mathematica*. Churchill, London and Pepyat, Dublin.

1707-8* *Philosophical commentaries*. (First published in *Berkeley 1871*; reprinted in *Berkeley 1948-57*, Vol. 1, pp. 7-139, and partially reprinted in this collection, Vol. 1, pp. 13-16.)

1710* *A treatise concerning the principles of human knowledge*. Rhames, Dublin. (2nd edn: Jacob Tonson, London, 1734. Partially reprinted in this collection, Vol. 1, pp. 21-37.)

1721* *De motu*. Jacob Tonson, London. (Reprinted, with a translation by the editors, in *Berkeley 1948-57*, Vol. 4, pp. 11-52. Translation reprinted in this collection, Vol. 1, pp. 37-54.)

1732 *Alciphron: or, the minute philosopher. In seven dialogues. Containing an apology for the Christian religion, against those who are called free-thinkers*. Jacob Tonson, London and G. Risk, Dublin.

1734* *The analyst, or, A discourse addressed to an infidel mathematician*. Jacob Tonson, London. (Reprinted in *Berkeley 1948-57*, Vol. 4, pp. 65-102, and in this collection, Vol. 1, pp. 60-92.)

1735a *A defence of free-thinking in mathematics*. Rhames, Dublin. (Reprinted in *Berkeley 1948-57*, Vol. 4, pp. 109-41.)

1735b *Reasons for not replying to Mr. Walton's* Full answer. (Reprinted in *Berkeley 1948-57*, Vol. 4, 147-56.)

1871 *The works of George Berkeley, D.D., formerly Bishop of Cloyne: including many of his writings hitherto unpublished*, 4 vols, (ed. Alexander Campbell Fraser). Clarendon Press, Oxford.

1901* Of infinites. (First printed in 1901 in *Hermathena*, Vol. 11 (ed. Swift Payne Johnston). Reprinted in *Berkeley 1948-57*, Vol. 4, pp. 235-8, and in this collection, Vol. 1, pp. 16-19. Probably written in 1707.)

1948-57 *The works of George Berkeley Bishop of Cloyne*, 9 vols, (ed. Arthur Aston Luce and Thomas Edmund Jessop). Thomas Nelson, London.

Bernays, Paul

(A bibliography of the writings of Bernays can be found in *Müller 1976*.)

1922a Über Hilberts Gedanken zur Grundlegung der Arithmetik. *Jahresbericht der Deutschen Mathematiker-Vereinigung*, **31**, 1st section, 10-19.

1922b Die Bedeutung Hilberts für die Philosophie der Mathematik. *Die Naturwissenschaften*, **10**, 93-9.

1928 Zusatz zu Hilberts Vortrag über 'Die Grundlagen der Mathematik', *Abhandlungen aus dem mathematischen Seminar der Hamburgischen Universität*, **6**, 89-92. (Translated by Stefan Bauer-Mengelberg and Dagfinn Føllesdal in *van Heijenoort 1967*, pp. 485-9.)

1930 Die Philosophie der Mathematik und die Hilbertsche Beweistheorie. *Blätter für deutsche Philosophie*, **4**, 326-67. (Reprinted in *Bernays 1976*.)

1932 Methoden des Nachweises von Widerspruchsfreiheit und ihre Grenzen. In *Verhandlungen des Internationalen Mathematiker-Kongresses, Zürich, 1932*, Vol. 2, pp. 342-3. Orell Füssle, Zürich and Leipzig.

1934 Sur le platonisme dans les mathématiques. *Enseignement mathématique*, **34**, (1935), 52-69.

1935 Hilberts Untersuchungen über die Grundlagen der Arithmetik. In *Hilbert 1932–5*, Vol. 3, pp. 196–216.

1937 A system of axiomatic set theory, Part I. *Journal of Symbolic Logic*, **2**, 65–77.

1941 A system of axiomatic set theory, Part II. *Journal of Symbolic Logic*, **6**, 1–17.

1942 A system of axiomatic set theory, Part III. *Journal of Symbolic Logic*, **7**, 65–89.

1948 Bemerkungen zu den Grundlagen der Geometrie. In *Studies and essays presented to Richard Courant on his 60th birthday*, pp. 29–44. Interscience, New York.

1967 Hilbert, David. In *P. Edwards 1967*, Vol. 3, pp. 496–504.

1976 *Abhandlungen zur Philosophie der Mathematik*. Wissenschaftliche Buchgesellschaft, Darmstadt.

See Hilbert, David and Bernays, Paul.

Bernoulli, Jacob (= Jacques = James) (1654–1705)

1744 *Opera*, 2 vols. G. Cramer, Geneva.

Bernoulli, John (= Johann = Jean) (1667–1748)

1742 *Opera omnia*. Bousquet, Lausanne and Geneva.

Bernstein, Felix

1904 Bemerkungen zur Mengenlehre. *Nachrichten von der Königlichen Gesellschaft der Wissenschaften zu Göttingen, Mathematisch-physikalische Klasse*, 557–60.

1905a Über die Reihe der transfiniten Ordnungszahlen, *Mathematische Annalen*, **60**, 187–93.

1905b Zum Kontinuumproblem. *Mathematische Annalen*, **60**, 463–4.

1905c Zur Mengenlehre. *Jahresbericht der Deutschen Mathematiker-Vereinigung*, **14**, 198–9.

1905d Untersuchungen aus der Mengenlehre. *Mathematische Annalen*, **61**, 117–55.

1905e Die Theorie der reellen Zahlen. *Jahresbericht der Deutschen Mathematiker-Vereinigung*, **14**, 447–9.

1919 Die Mengenlehre Georg Cantors und der Finitismus. *Jahresbericht der Deutschen Mathematiker-Vereinigung*, **28**, 63–78.

Bertrand, Joseph

1849 *Traité d'arithmétique*. Hachette, Paris.

Biermann, Kurt R.

1960 Vorschläge zur Wahl von Mathematikern in die Berliner Akademie. *Abhandlungen der Deutschen Akademie der Wissenschaften zu Berlin, Klasse für Mathematik, Physik und Technik,* **3**, 29–34.

1966 Richard Dedekind im Urteil der Berliner Akademie. *Forschungen und Fortschritte*, **40**, 301–2.

Birkhoff, Garrett

1973a (ed.) *A source book in classical analysis*. Harvard University Press, Cambridge, Mass.

1973b Current trends in algebra. *American Mathematical Monthly*, **80**, 760–82.

1976 The rise of modern algebra. In *Men and institutions in American mathematics*, Graduate Studies, Texas Tech University, No. 13, pp. 41–85.

Birkhoff, Garrett and Bennett, M.K.

1988 Felix Klein and his 'Erlanger Programm'. In *Aspray and Kitcher 1988*, pp. 145–76.

Bolzano, Bernard

1804* *Betrachtungen über einige Gegenstände der Elementargeometrie*. Karl Barth, Prague. (Translated by Stephen Russ, this collection, Vol. 1, pp. 172–4.)

1810* *Beiträge zu einer begründeteren Darstellung der Mathematik*. Caspar Widtmann, Prague. (Translated by Stephen Russ, this collection, Vol. 1, pp. 174–224.)

1816 *Der binomische Lehrsatz, und als Folgerung aus ihm der polynomische und die Reihen, die zur Berechnung der Logarithmen und Exponentialgrössen dienen, genauer als bisher erwiesen*. C.W. Enders, Prague.

1817a* *Rein analytischer Beweis des Lehrsatzes, dass zwischen je zwei Werten, die ein entgegengesetztes Resultat gewähren, wenigstens eine reelle Wurzel der Gleichung liege*. Gottlieb Haase, Prague. (Translated by Stephen Russ, this collection, Vol. 1, pp. 225–48.)

1817b *Die drey Probleme der Rectification, der Complanation und der Cubierung, ohne Betrachtung des unendlich Kleinen. ohne die Annahme des Archimedes und ohne irgend eine nicht streng erweisliche Voraussetzung gelöst; zugleich als Probe einer gänzlichen Umgestaltung der Raumwissenschaft allen Mathematikern zur Prüfung vorgelegt*. Gotthelf Kummer, Leipzig.

1836 *Lebensbeschreibung des Dr. B. Bolzano mit einigen seiner ungedrückten Aufsätze*. Seidel, Sulzbach.

1837 *Dr. Bolzanos Wissenschaftslehre. Versuch einer ausführlichen und grössenteils neuen Darstellung der Logik, mit steter Rücksicht auf deren bisherige Bearbeiter*. Seidel, Sulzbach.

1842 *Versuch einer objektiven Begründung der Lehre von der Zusammensetzung der Kräfte*. Kronberger and Rziwnas, Prague.

1843 *Versuch einer objektiven Begründung der Lehre von den drei Dimensionen des Raumes*. Kronberger and Rziwnas, Prague.

1849 *Über die Einteilung der schönen Künste. Eine aesthetische Abhandlung*. Gottlieb Haase, Prague.

1851 *Dr. Bernard Bolzanos Paradoxien des Unendlichen, herausgegeben aus dem schriftlichen Nachlasse des Verfassers von Dr. Fr. Prihonsky*. Reclam, Leipzig.

1930 *Funktionenlehre*, (ed. Karel Rychlik). Königliche Böhmische Gesellschaft der Wissenschaften, Prague. (Written 1834.)

1950* *Paradoxes of the infinite*. Routledge and Kegan Paul, London. (Translation by Donald A. Steele of *Bolzano 1851*. Partially reprinted in this collection, Vol. 1, pp. 249–92.)

1969- *Bernard Bolzano Gesamtausgabe*, (ed. Eduard Winter *et alii*). Fromann, Stuttgart.

1972 *Theory of Science*, (ed. and trans. Rolf George). Blackwell, Oxford. (Selections from *Bolzano 1837*.)

See Russ, Stephen.

Bonola, Roberto

1955 *Non-Euclidean geometry, A critical and historical study of its developments*, (trans. H.S. Carslaw). Dover, New York.

Boole, George

1841 On certain theorems in the calculus of variations. *Cambridge Mathematical Journal*, **2**, 97–102.

1842a Exposition of a general theory of linear transformations, Part I. *Cambridge Mathematical Journal*. **3**, 1–20.

1842b Exposition of a general theory of linear transformations, Part II. *Cambridge Mathematical Journal*, **3**, 106–19.

1844 On a general method in analysis. *Philosophical Transactions of the Royal Society of London*, **134**, 225–82.

1845 Notes on linear transformations. *Cambridge Mathematical Journal*. **4**, 167–71.

1847* *The mathematical analysis of logic*. Macmillan, Barclay and Macmillan, Cambridge. (Reprinted Blackwell, Oxford, 1948, and in this collection, Vol. 1, pp. 451–509.)

1848 The calculus of logic. *Cambridge and Dublin Mathematical Journal*, May 1848. (Reprinted in *Boole 1952*, pp. 125–40.)

1851 On the theory of probabilities, and in particular on Mitchell's problem of the distribution of the fixed stars. *Philosophical Magazine*, **1**, 521–30.

1854 *An investigation of the laws of thought, on which are founded the mathematical theories of logic and probabilities*. Macmillan, Cambridge.

1859 *A treatise on differential equations*. Macmillan, Cambridge.

1860 *A treatise on the calculus of finite differences*. Macmillan, Cambridge.

1952 *Studies in logic and probability*, (ed. Rush Rhees). Watts, London and Open Court, La Salle.

Boole, George and De Morgan, Augustus

1982 *The Boole-De Morgan correspondence, 1842–1864*, (ed. G.C. Smith). Oxford University Press.

Boole, Mary Everest

1878 The home side of a scientific mind. *The University Magazine*, **1**, 105–14; 173–83; 326–36; 454–60.

Boolos, George

1990 The standard of equality of numbers. In idem (ed.), *Meaning and method: Essays in honour of Hilary Putnam*. Cambridge University Press.

Borchardt, C.W.

For Borchardt's Journal, *see* Crelle, August Leopold.

Borel, Emile

1898 *Leçons sur la théorie des fonctions*. Gauthier-Villars, Paris.

1899 A propos de 'l'infini nouveau'. *Revue Philosophique*, **48**, 383-90. (Reprinted in *Borel 1914b*, *1950*, and *1972*, Vol. 4. pp. 2113-20.)

1905* Quelques remarques sur les principes de la théorie des ensembles. *Mathematische Annalen*, **60**, 194–5. (Translated by William Ewald, this collection, Vol. 2, pp. 1076–7.)

1908a Sur les principes de la théorie des ensembles. *Atti del IV Congresso Internazionale dei Matematici (Roma)*, **2**, 15–17. (Reprinted in *Borel 1914b*, *1950*, and *1972*, Vol. 3. pp. 1267–9.)

1908b Les 'paradoxes' de la théorie des ensembles. *Annales scientifiques de l'Ecole Normale Supérieure* (ser. 3), **25**, 443-8. (Reprinted in *Borel 1914b*, *1950*, and *1972*, Vol. 3. pp. 1271-6.)

1909 La théorie des ensembles et les progrès rècents de la théorie des fonctions. *Revue générale des sciences pures et appliquées*, **20**, 315-24. (Reprinted in *Borel 1972*, Vol. 3, pp. 1277-307.)

1912 La philosophie mathématique et l'infini. *Revue du mois*, **14**, 219-27. (Reprinted in *Borel 1914b*, *1950*, and *1972*, Vol. 4, pp. 2127-36.)

1914a L'infini mathématique et la réalité. *Revue du mois*, **18**, 71-83. (Reprinted in *Borel 1914b*, *1950*, and *1972*, Vol. 4, pp. 2137-50.)

1914b Second edition of *Borel 1898*.

1946 L'axiome du choix et la mesure des ensembles. *Comptes rendus hebdomadaires des séances de l'académie des sciences*, **222**, 309-10. (Reprinted in *Borel 1972*, Vol. 3, pp. 1373-4.)

1947 Les paradoxes de l'axiome du choix. *Comptes rendus hebdomadaires des séances de l'académie des sciences*, **224**, 1537-8. (Reprinted in *Borel 1950* and *1972*, Vol. 3, pp. 1377–8.)

1950 Fourth edition of *Borel 1898*.

1972 *Oeuvres*, 4 vols. Centre National de la Recherche Scientifique, Paris.

See Baire, René; Borel, Emile; Hadamard, Jacques, and Lebesgue, Henri.

Bourbaki, Nicolas (or Nicholas, or Nicolaus)

1939 *Eléments de mathématique. Première partie: Les structures fondamentales de l'analyse. Livre I. Théorie des ensembles*. Hermann, Paris.

1940 *Eléments de mathématique. Topologie générale*. Hermann, Paris.

1948 *L'architecture des mathématiques*. In *Le Lionnais 1948*, pp. 35–47. (Translated by Arnold Dresden as *Bourbaki 1950*.)

1949 Foundations of mathematics for the working mathematician. *The Journal of Symbolic Logic*, **14**, 1–8.

1950* The architecture of mathematics. *American Mathematical Monthly*, **57**, 221–32. (Authorized translation by Arnold Dresden of *Bourbaki 1948*. Reprinted in this collection, Vol. 2, pp. 1265–76.)

1968 *Elements of mathematics: Theory of sets*. Hermann, Paris.

1969 *Eléments d'histoire des mathématiques*, (2nd edn). Hermann, Paris.

Boyer, Carl B.

1939 *The concepts of the calculus*. Columbia University Press, New York. (Reprinted 1949 and 1959, with the title, *The history of the calculus and its conceptual development*, Dover, New York.)

Brewer, James W. and Smith, Martha K., (eds.)

1978 *Emmy Noether: A tribute to her life and work*. Marcel Dekker, New York.

Bromwich, Thomas John I'Anson

1908 *An introduction to the theory of infinite series.* Macmillan, London.

Brouwer, Luitzen Egbertus Jean

(A complete bibliography of Brouwer's writings, published and unpublished, is contained in *van Stigt 1990.* Partial bibliographies are in *Brouwer 1975*, *Kreisel and Newman 1969*, and *van Heijenoort 1967.* All the works listed below are reprinted in *Brouwer 1975*, Vol. 1.)

1907 *Over de grondslagen der wiskunde.* Doctoral dissertation, Amsterdam. (Defended 19 Feb. 1907.) (Translated by Arend Heyting as *On the foundations of mathematics* in *Brouwer 1975*, Vol. 1, pp. 11-101.)

1908 De onbetrouwbaarheid der logische principes. *Tijdschrift voor wijsbegeerte*, **2**, 152–8. (Translated by Arend Heyting as 'The unreliability of the logical principles' in *Brouwer 1975*, Vol. 1, pp. 107–11.)

1913 Intuitionism and formalism. *Bulletin of the American Mathematical Society*, **20**, 81-96. (Reprinted in *Benacerraf and Putnam 1964*, pp. 66–77.)

1914 Review of: A. Schoenflies and H. Hahn, *Die Entwicklung der Mengenlehre und ihrer Anwendungen* (Leipzig and Berlin, 1913). *Jahresbericht der Deutschen Mathematiker-Vereinigung*, **23**, 78-83.

1918 Begründung der Mengenlehre unabhängig vom logischen Satz vom ausgeschlossenen Dritten. *Verhandeling Koninklijke Nederlandse Akademie van Wetenschappen te Amsterdam*, **12**, no. 5 (43 pp.); **12**, no. 7 (33 pp.).

1919 Intuitionistische Mengenlehre. *Jahresbericht der Deutschen Mathematiker-Vereinigung*, **28**, 203-8.

1923 Begründung der funktionenlehre unabhängig vom logischen Satz vom ausgeschlossenen Dritten. Erster Teil, Stetigkeit, Messbarkeit, Derivierbarkeit. *Koninklijke Nederlandse Akademie van Wetenschappen te Amsterdam*, Verhandelingen, 1st section, **13**, no. 2 (24 pp.).

1924 Beweis, dass jede volle Funktion gleichmässig stetig ist. *Koninklijke Nederlandse Akademie van Wetenschappen te Amsterdam*, Proc. **27**, 189-93.

1925 Intuitionistische Zerlegung intuitionistischer Grundbegriffe. *Jahresbericht der Deutschen Mathematiker-Vereinigung*, **33**, 251-6.

1926 Zur Begründung der intuitionistischen Mathematik, II. *Mathematische Annalen*, **95**, 453–72.

1927a Über Definitionsbereiche von Funktionen. *Mathematische Annalen*, **97**, 60–75. (Partial translation by Stefan Bauer-Mengelberg in *van Heijenoort 1967*, pp. 446-63.)

1927b Virtuelle Ordnung und unerweiterbare Ordnung. *Journal für die reine und angewandte Mathematik*, **157**, 255–7.

1927c Intuitionistische Betrachtungen über den Formalismus. *Koninklijke Akademie van Wetenschappen te Amsterdam, Proceedings of the Section of Sciences*, **31**, 374-9; also in *Sitzungsberichte der Preussischen Akademie der Wissenschaften, Physikalisch-mathematische Klasse*, 48–52. (Translated in part by Stefan Bauer-Mengelberg in *van Heijenoort 1967*, pp. 490–2.)

1928a* Mathematik, Wissenschaft und Sprache. (Lecture, delivered in Vienna on 10 March 1928.) *Monatshefte für Mathematik und Physik*, **36**,

153–64. (Translated by William Ewald, this collection, Vol. 2, pp. 1170–85.)

1928b* *Die Struktur des Kontinuums.* (Lecture, delivered in Vienna on 14 March 1928.) Gistel, Vienna. (Translated by William Ewald, this collection, Vol. 2, pp. 1186–97.)

1948 Consciousness, philosophy, and mathematics. In *Proceedings of the Tenth International Congress of Philosophy (Amsterdam, August 11–18, 1948)*, pp. 1235–49. North-Holland, Amsterdam.

1952* Historical background, principles, and methods of intuitionism. *South African Journal of Science*, **49**, 139–46. (Reprinted in this collection, Vol. 2, pp. 1197–1207.)

1954 Points and spaces. *Canadian Journal of Mathematics*, **6**, 1–17.

1975 *Collected works*, 2 vols, (ed. Arend Heyting and Hans Freudenthal). North-Holland, Amsterdam.

Browder, Felix E. (ed.)

1976 *Mathematical developments arising from Hilbert's problems.* Proceedings of symposia in pure mathematics, Vol. 28, parts 1 and 2. American Mathematical Society, Providence.

Brunschvicg, Léon

1912 *Les étapes de la philosophie mathématique.* Alcan, Paris.

Burali-Forti, Cesare

1894 *Logica mathematica.* Hoepli, Milan.

1897a Una questione sui numeri transfiniti. *Rendiconti del Circolo matematico di Palermo*, **11**, 154–64. (Translated by Jean van Heijenoort in *van Heijenoort 1967*, pp. 104–11.)

1897b Sulle classe ben ordinate. *Rendiconti del Circolo matematico di Palermo*, **11**, 260. (Translated by Jean van Heijenoort in *van Heijenoort 1967*, pp. 111–12.)

Cajori, Florian

1919 *A history of the conceptions of limits and fluxions in Great Britain from Newton to Woodhouse.* Open Court, Chicago.

1924 *A history of mathematics.* Macmillan, New York.

1925 Indivisibles and 'ghosts of departed quantities' in the history of mathematics. *Scientia*, **37**, 303–6.

1928 *A history of mathematical notations, Vol. 1: Notations in elementary mathematics.* Open Court, La Salle.

Cantor, Georg

1870 Beweis, daß ein für jeden reellen Wert von x durch eine trigonometrische Reihe gegebene Funktion f(x) sich nur auf eine einzige Weise in dieser Form darstellen läßt. *Journal für die reine und angewandte Mathematik*, **72**, 139–42. (Reprinted in *Cantor 1932*, pp. 80–3.)

1871 Über trigonometrische Reihen. *Mathematische Annalen*, **4**, 139–43. (Reprinted in *Cantor 1932*, pp. 87–91.)

1872 Über die Ausdehnung eines Satzes aus der theorie der trigonometrischen Reihen. *Mathematische Annalen*, **5**, 123–32. (Reprinted in *Cantor 1932*, pp. 92–101.)

1874* Über eine Eigenschaft des Inbegriffs aller reellen algebraischen Zahlen. *Journal für die reine und angewandte Mathematik*, **77**, 258–62. (Reprinted in *Cantor 1932*, pp. 115–18. Translated by William Ewald, this collection, Vol. 2, pp. 839–43.)

1878 Ein Beitrag zur Mannigfaltigkeitslehre. *Journal für die reine und angewandte Mathematik*, **84**, 242-58. (Reprinted in *Cantor 1932*, pp. 119-33.)

1879a Über einen Satz aus der Theorie der stetigen Mannigfaltigkeiten. *Nachrichten der Königlichen Gesellschaft der Wissenschaften zu Göttingen*, 127-35. (Reprinted in *Cantor 1932*, pp. 134-8.)

1879b Über unendliche, lineare Punktmannigfaltigkeiten, 1. *Mathematische Annalen*, **15**, 1-7. (Reprinted in *Cantor 1932*, pp. 139-45.)

1880 Über unendliche, lineare Punktmannigfaltigkeiten, 2. *Mathematische Annalen*, **17**, 355-8. (Reprinted in *Cantor 1932*, pp. 145-8.)

1882 Über unendliche, lineare Punktmannigfaltigkeiten, 3. *Mathematische Annalen*, **20**, 113-21. (Reprinted in *Cantor 1932*, pp. 149-57.)

1883a Über unendliche, lineare Punktmannigfaltigkeiten, 4. *Mathematische Annalen*, **21**, 51-8. (Reprinted in *Cantor 1932*, pp. 157-64.)

1883b Über unendliche, lineare Punktmannigfaltigkeiten, 5. *Mathematische Annalen*, **21**, 545-86. (Reprinted in *Cantor 1932*, pp. 165-209.)

1883c Sur divers théorèmes de la théorie des ensembles de pointes situés dans un éspace continu à *n*-dimensions. *Acta Mathematica*, **2**, 409-14.

1883d* *Grundlagen einer allgemeinen Mannigfaltigkeitslehre. Ein mathematisch-philosophischer Versuch in der Lehre des Unendlichen*. Teubner, Leipzig. (Separate printing of *Cantor 1883b*, with an additional preface and footnotes not reproduced in *Cantor 1932*. Translated by William Ewald, this collection, Vol. 2, pp. 878–920.)

1884a Über unendliche, lineare Punktmannigfaltigkeiten, 6. *Mathematische Annalen*, **23**, 453-88. (Reprinted in *Cantor 1932*, pp. 210-46.)

1884b De la puissance des ensembles parfaits des points. *Acta Mathematica*, **4**, 381-92. (Reprinted in *Cantor 1932*, pp. 252-60.)

1885 Rezension der Schrift von G. Frege *Die Grundlagen der Arithmetik*. *Deutsche Literaturzeitung*, **6**, 728-9. (Reprinted in *Cantor 1932*, pp. 440-1.)

1886a Über die verschiedenen Ansichten in Bezug auf die actualunendlichen Zahlen. *Bihang Till Koniglen Svenska Vetenskaps Akademiens Handligar*, **11**, 1-10. (Not reprinted in *Cantor 1932*.)

1886b Über die verschiedenen Standpunkte in Bezug auf das aktuelle Unendliche. *Zeitschrift für Philosophie und philosophische Kritik*, **88**, 224-33. (Reprinted in *Cantor 1932*, pp. 370-7.)

1887-8 Mitteilungen zur Lehre vom Transfiniten I, II. *Zeitschrift für Philosophie und philosophische Kritik*, **91**, 81-125, 252-70; **92**, 250-65. (Reprinted in *Cantor 1932*, pp. 378-439.)

1891* Über eine elementare Frage der Mannigfaltigkeitslehre. *Jahresbericht der Deutschen Mathematiker-Vereinigung*, **1**, 75-8. (Reprinted in *Cantor 1932*, pp. 278–80. Translated by William Ewald, this collection, Vol. 2, pp. 920–2.)

1895 Beiträge zur Begründung der transfiniten Mengenlehre, 1. *Mathematische Annalen*, **46**, 481-512. (Reprinted in *Cantor 1932*, pp. 282-311.)

1897 Beiträge zur Begründung der transfiniten Mengenlehre, 2. *Mathematische Annalen*, **49**, 207–46. (Reprinted in *Cantor 1932*, pp. 312–56.)

1932 *Gesammelte Abhandlungen mathematischen und philosophischen Inhalts*, (ed. Ernst Zermelo). Springer Verlag, Berlin.

1991 *Georg Cantor: Briefe*, (ed. H. Meschkowski and W. Nilson). Springer Verlag, Berlin, Heidelberg, and New York.

Cantor, Georg and Dedekind, Richard
See Noether, Emmy and Cavaillès, Jean.

Cantor, Moritz
1894–1908 *Vorlesungen über die Geschichte der Mathematik*, 4 vols, (2nd edn). Teubner, Leipzig.

Carnot, Lazare Nicolas Marguerite
1813 *Réflexions sur la métaphysique du calcul infinitésimal*, (2nd edn). Courcier, Paris. (1st edn: Duprat, Paris, 1787.)

Cartan, Henri Paul
1943 Sur le fondament logique des mathématiques. *Revue scientifique*, **81**, 3–11.

Cauchy, Augustin-Louis
1821 *Cours d'analyse algébrique*. Imprimerie Royale, Paris. (Reprinted in *Cauchy 1882–1958*, Vol. 3 (1897).)

1882–1958 *Œuvres complètes d'Augustin Cauchy*, 12 + 15 vols. Gauthier-Villars, Paris.

Cavaillès, Jean
1947 *Transfini et continu*. Hermann, Paris. (Reprinted in *Cavaillès 1962*.)

1962 *Philosophie mathématique*. Hermann, Paris.

See Noether, Emmy and Cavaillès, Jean.

Cavalieri, Bonaventura
1635 *Geometrica indivisibilibus continuorum nova quadam ratione promota*. C. Ferroni, Bologna. (Also published 1653.)

Cayley, Arthur
1854 On the theory of groups. *Philosophical Magazine*, **7**, 40–7. (Reprinted in *Cayley 1889–98*, Vol. 2, pp. 123–30.)

1857 Note on the recent progress of theoretical dynamics, *Reports of the British Association*, **1857**, 1–42. (Reprinted in *Cayley 1889–98*, Vol. 3, pp. 156–204.)

1883* Presidential address to the British Association, September 1883. *Report of the British Association for the Advancement of Science*, **1883**, 3–37. (Reprinted in *Cayley 1889–98*, Vol. 9, pp. 429–59, and in this collection, Vol. 1, pp. 542–73.)

1889–98 *Collected mathematical papers*, 13 vols. Cambridge University Press.

Chihara, Charles
1973 *Ontology and the vicious circle principle*. Cornell University Press, Ithaca.

Clairault, Alexis Claude
1797 *Elémens d'algèbre*, 2 vols, (5th edn). Duprat, Paris.

Clifford, William Kingdon
1872* On the aims and instruments of scientific thought. (Lecture delivered before the members of the British Association, Brighton, 19 August, 1872; first published in *Clifford 1901*, Vol. 1, pp. 139–80. Reprinted in this collection, Vol. 1, pp. 524–41.)
1873* On the hypotheses which lie at the bases of geometry. *Nature*, **7**, 14–17, 36, 37, 183–4. (Translation of *Riemann 1868*). (Reprinted in *Clifford 1882*, pp. 55–71, and in this collection, Vol. 2, pp. 652–61.)
1876* On the space theory of matter. *Proceedings of the Cambridge Philosophical Society*, **2**, 157–8. (Read 21 Feb. 1870; reprinted in *Clifford 1882*, pp. 21–2, and in this collection, Vol. 1, pp. 523–4.)
1882 *Mathematical papers*. Macmillan, London.
1885 *The common sense of the exact sciences*. Kegan, Paul, London. (Reprinted, with an introduction by James R. Newman, Knopf, New York, 1946.)
1901 *Lectures and essays*, 2 vols, (ed. Leslie Stephen and Sir Frederick Pollock). Macmillan, London.

Cohn, Jonas
1896 *Geschichte des Unendlichkeitsproblems im abendländischen Denken bis Kant*. Engelmann, Leipzig.

Courant, Richard and Robbins, Herbert
1948 *What is mathematics?* Oxford University Press.

Couturat, Louis
1896 *De l'infini mathématique*. Alcan, Paris. (Reprinted Franklin, New York, 1969; and Blanchard, Paris, 1973.)
1901 *La logique de Leibniz d'après des documents inédits*. Alcan, Paris.
1904 Kant et la mathématique moderne. *Bulletin de la société française de philosophie*, **4**, 125–234.
1905a *Les principes des mathématiques*. Alcan, Paris.
1905b *L'algèbre de la logique*. Gauthier-Villars, Paris.
1905c Les définitions mathématiques. *L'enseignement mathématique*, **7**, 27–40.
1906 Pour la logistique (réponse à M. Poincaré). *Revue de métaphysique et de morale*, **14**, 208–50.

Crelle, August Leopold (ed.)
1826–55 *Journal für die reine und angewandte Mathematik*. Berlin. (Often cited as '*Crelle's Journal*', or, subsequently, as '*Borchardt's Journal*'.)

Crowe, Michael J.
1967 *A history of vector analysis: The evolution of the idea of a vectorial system*. University of Notre Dame Press, South Bend, Indiana.

D'Alembert, Jean le Rond
1751 *Discours préliminaire de l'Encyclopédie*. In *Diderot* et alii *1751–72*, Vol. 1, pp. i–xlv.

1754* Différentiel. In *Diderot* et alii *1751–72*, Vol. 4. (Translated by William Ewald and Dirk J. Struik, this collection, Vol. 1, pp. 123–8.)

1765a* Infini. In *Diderot* et alii *1751–72*, Vol. 8. (Translated by William Ewald, this collection, Vol. 1, pp. 128–30.)

1765b* Limite. In *Diderot* et alii *1751–72*, Vol. 9. (Translated by William Ewald, this collection, Vol. 1, pp. 130–1.)

Dauben, Joseph W.

1971a *Georg Cantor: His mathematics and philosophy of the infinite.* Harvard University Press, Boston.

1971b The trigonometric background to Cantor's theory of sets. *Archive for History of Exact Sciences*, **7**, 181–216.

1982 Peirce's place in mathematics. *Historia Mathematica*, **9**, 311–25.

Davenport, Harold

1970 *The higher arithmetic*. Hutchinson, London.

Dawson, John W.

1984 Discussion on the foundation of mathematics. *History and Philosophy of Logic*, **5**, 111–29. (Translation of *Hahn* et alii *1931*.)

Dedekind, Richard

1854* Über die Einführung neuer Funktionen in der Mathematik. First published in *Dedekind 1930–2*, Vol. 3, pp. 428–38. (Delivered as a *Habilitationsvorlesung* in Göttingen on 30 June 1854. Translated by William Ewald, this collection, Vol. 2, pp. 754–62.)

1871* Über die Komposition der binären quadratischen Formen. Supplement X to Dirichlet's *Vorlesungen über Zahlentheorie*, 2nd edn, pp. 423–62. (Reprinted in *Dedekind 1930–2*, Vol. 3, pp. 223–61. Translated by William Ewald, this collection, Vol. 2, pp. 762–5.)

1872* *Stetigkeit und irrationale Zahlen*. Vieweg, Braunschweig. (Translated by Wooster W. Beman as 'Continuity and irrational numbers' in *Dedekind 1901*. Reprinted, with corrections by William Ewald, this collection, Vol. 2, pp. 765–79.)

1877* Sur la théorie des nombres entiers algébriques. *Belletin des sciences mathématiques et astronomiques*, **11**, 278–88. (Reprinted in *Dedekind 1930–2*, Vol. 3, pp. 262–98. Translated by David Reed, this collection, Vol. 2, pp. 779–87.)

1888* *Was sind und was sollen die Zahlen?* Vieweg, Braunschweig.

1890 Über den Begriff des Unendlichen. Manuscript in the Niedersächsische Staats- und Universitätsbibliothek, Göttingen (Cod. Ms. Dedekind XIII). Published, with a French translation, in *Sinaceur 1974*.

1893* Second edition of *Dedekind 1888*, with a new preface. (Translated by Wooster W. Beman as 'The nature and meaning of numbers' in *Dedekind 1901*. Reprinted, with corrections by William Ewald, this collection, Vol. 2, pp. 787–833.)

1894* Über die Theorie der ganzen algebraischen Zahlen. Supplement XI to Dirichlet's *Vorlesungen über Zahlentheorie*, 4th edn, pp. 434–657. (Reprinted in *Dedekind 1930–2*, Vol. 3, pp. 1–222. Partially translated by William Ewald, this collection, Vol. 2, pp. 833–4.)

1901 *Essays on the theory of numbers*, (trans. Wooster W. Beman). Open Court, Chicago. (Reprinted Dover, New York, 1963.)

1930–2 *Gesammelte mathematische Werke*, 3 vols, (ed. Robert Fricke, Emmy Noether, and Öystein Ore). Vieweg, Braunschweig.

1985 *Vorlesung über Differential- und Integralrechnung 1861/62 (in einer Mitschrift von Heinrich Bechthold)*, (ed. Max-Albert Knus and Winfried Scharlan). Dokumente zur Geschichte der Mathematik, Band 1. Deutsche Mathematiker Vereinigung-Vieweg, Braunschweig-Wiesbaden.

See Dirichlet, Peter Gustav Lejeune-; *and see* Noether, Emmy and Cavaillès, Jean.

De Morgan, Augustus

1831 *On the study and difficulties of mathematics*. Society for the Diffusion of Useful Knowledge, London.

1836 *A treatise of the calculus of functions*. Encyclopedia Metropolitana, London.

1837 *The elements of algebra preliminary to the differential calculus and fit for the higher classes of schools in which the principles of arithmetic are taught*. Taylor and Walton, London.

1839 *First notions of logic, preparatory to the study of geometry*. Taylor and Walton, London. (Reprinted as Chapter 1 of *De Morgan 1847*.)

1842a* On the foundation of algebra. *Transactions of the Cambridge Philosophical Society*, **7**, 173–87. (Reprinted in this collection, Vol. 1, pp. 336–48.) (Read 9 Dec. 1839.)

1842b On the foundation of algebra, no. II. *Transactions of the Cambridge Philosophical Society*, **7**, 287–300. (Read 29 Nov. 1841.)

1846 On the syllogism, no. I. On the structure of the syllogism, and on the application of the theory of probabilities to questions of argument and authority. *Transactions of the Cambridge Philosophical Society*, **8**, 379–408. (Read 9 Nov. 1846.)

1847 *Formal logic: or, The calculus of inference, necessary and probable*. Taylor and. Walton, London.

1849a On the foundation of algebra, nos. III and IV. *Transactions of the Cambridge Philosophical Society*, **8**, 139–42 and 241–54. (Read 27 Nov. 1843 and 28 Oct. 1844.)

1849b* *Trigonometry and double algebra*. Taylor, Walton, and Maberly, London. (Partially reprinted in this collection, Vol. 1, pp. 349–61.)

1850 On the syllogism, no. II. On the symbols of logic, the theory of the syllogism, and in particular of the copula. *Transactions of the Cambridge Philosophical Society*, **9**, 79–127. (Read 25 Feb. 1850.)

1852 On the early history of infinitesimals in England. *Philosophical Magazine* (ser. 4), **4**, 321–30.

1860a *Syllabus of a proposed system of logic*. Walton and Maberly, London.

1860b Logic. Entry in the *English Cyclopaedia*, Vol. 5. (Reprinted, with omissions, in *De Morgan 1966*, pp. 247–70). Bradbury and Evans, London.

1864a On the syllogism, no. III, and on logic in general. *Transactions of the Cambridge Philosophical Society*, **10**, 173–230. (Read 8 Feb. 1858.)

1864b On the syllogism, no. IV, and on the logic of relations. *Transactions of the Cambridge Philosophical Society*, **10**, 331–58. (Read 23 Apr. 1860).

1864c On the syllogism, no. V. and on various points of the onymatic system. *Transactions of the Cambridge Philosophical Society*, **10**, 428-87. (Read 4 May 1863.)

1872 *A budget of paradoxes*. Longmans, Green, London.

1966 On the syllogism and other logical writings, (ed. Peter Heath). Yale University Press, New Haven.

See Boole, George and De Morgan, Augustus.

Dick, Auguste

1970 *Emmy Noether, 1881-1935*. Birkhäuser, Basle.

Diderot, Denis *et alii* (eds.)

1751-72 *Encyclopédie, ou Dictionnaire raisonné des sciences, des arts et des métiers, par une société de gens de lettres*, 28 vols. Samuel Faulche, Neufchastel [Neuchâtel].

Dieudonné Jean A.

1939 Les méthodes axiomatiques et les fondements des mathématiques. *Revue scientifique*, **77**, 224-32.

1951 L'axiomatique dans les mathématiques modernes. *Actualités scientifiques et industrielles*, **1137**, 47-53.

1970 The work of Nicholas Bourbaki. *American Mathematical Monthly*, **77**, 134-45.

1971 Modern axiomatic methods and the foundations of mathematics. In *Great currents of mathematical thought*, Vol. 2, (ed. Le Lionnais), (trans. R.A. Hall), pp. 251-66. Dover, New York.

1972 The historical development of algebraic geometry. *American Mathematical Monthly*, **79**, 827-66.

1975 Poincaré, Henri. In *Gillispie* et alii *1970-6*, Vol. 11, pp. 51-61.

1978 (ed.) *Abrégé d'histoire des mathématiques, 1700-1900*, 2 vols. Hermann, Paris.

Dini, Ulisse

1878 *Fondamenti per la teorica delle funzioni di variabili reali*. Nistri, Pisa.

1892 *Grundlagen für eine Theorie der Functionen einer veränderlichen reelen-Grösse*, (trans. Jacob Lüroth and Adolf Schepp). Teubner, Leipzig. (German translation of *Dini 1878*.)

Dirichlet, Peter Gustav Lejeune-

1837 Über die Darstellung ganz willkürlicher Functionen durch Sinus- und Cosinusreihen. *Reportorium der Physik*, **1**, 152-74.

1863 *Vorlesungen über Zahlentheorie*, (ed. Richard Dedekind). Vieweg, Braunschweig. (Subsequent editions in 1871, 1879, 1894. These contain important 'Supplements' written by Dedekind, of which portions are translated by William Ewald in this collection, Vol. 2, pp. 762-5 and 833-4.)

Dryer, Douglas Poole

1966 *Kant's solution for verification in metaphysics*. Allen and Unwin, London.

du Bois-Reymond, Paul

1875 Sur la grandeur relative des infinis des fonctions. *Annali di matematica pura ed applicata* (ser. 2), **4**, 338–53.

1880a *Zur Geschichte der trigonometrischen Reihen. Eine Entgegnung.* Laupp, Tübingen.

1880b Der Beweis des Fundamentalsatzes der Integralrechnung. *Mathematische Annalen*, **16**, 115–28.

1882 *Die allgemeine Funktionenlehre.* Laupp, Tübingen.

Dugac, Pierre

1976 *Richard Dedekind et les fondaments des mathématiques.* Vrin, Paris.

Dummett, Michael

1975 The philosophical basis of intuitionistic logic. In *Logic colloquium '73* (ed. H.E. Rose and J.C. Shepherdson), pp. 5–40. North-Holland, Amsterdam. (Reprinted in *Dummett 1978*, pp. 215–47, and in *Benacerraf and Putnam 1983*, pp. 97–129.)

1976 Frege on the consistency of mathematical theories. In *Studien zu Frege* (ed. M. Schirn) Vol. 1, pp. 229–42. Fromann-Holzboog, Stuttgart.

1977 *Elements of intuitionism.* Oxford University Press.

1978 *Truth and other enigmas.* Duckworth, London.

1991a *Frege and other philosophers.* Oxford University Press.

1991b *Frege: Philosophy of mathematics.* Harvard University Press, Cambridge, Mass.

Edwards, Charles Henry

1979 *The historical development of the calculus.* Springer Verlag, New York.

Edwards, Harold M.

1977 *Fermat's last theorem.* Springer Verlag, Berlin.

1980 The genesis of ideal theory. *Archive for History of Exact Sciences*, **23**, 321–78.

Edwards, Paul

1967 *The encyclopedia of philosophy.* Macmillan, New York.

Encyklopädie

1898–1935 *Encyklopädie der mathematischen Wissenschaften, mit Einschluss ihrer Anwendungen.* Herausgegeben im Auftrag der Akademien der Wissenschaften zu Göttingen, Leipzig, München, und Wien. Teubner, Leipzig. (Numerous volumes; appeared in parts.)

Engel, Friedrich

See Stäckel, Paul and Engel, Friedrich.

Enriques, Federigo

1907–10 Prinzipien der Geometrie. In *Enzyclopädie der mathematischen Wissenschaften*, (ed. W. Fr. Meyer and H. Mohrmann), Vol. III, part 1, first half, pp. 6–129. Teubner, Leipzig.

Euclid

1896 *Data, cum commentario Marini et scholiis antiquis*, (ed. H. Mengel). Leipzig.

1925 *The thirteen books of the elements*, 2 vols, (2nd edn; trans. and ed. Thomas L. Heath, with extensive commentary). Cambridge University Press. (Reprinted Dover, New York, 1956.)

Fang, Joong

1970 *Bourbaki: Towards a philosophy of modern mathematics*. Paideia, Hauppauge, New York.

Feferman, Solomon

1964 Systems of predicative analysis. *Journal of Symbolic Logic*, **29**, 1–30.

1968 Systems of predicative analysis II: Representations of ordinals. *Journal of Symbolic Logic*, **33**, 193–219.

Ferreiros, José

1993 On the relations between Georg Cantor and Richard Dedekind. *Historia mathematica*, **20**, 343–63.

Feuer, Lewis S.

1988 Sylvester in Virginia. *The Mathematical Intelligencer*, **9**, 13–19.

Finsler, Paul

1926 Über die Grundlegung der Mengenlehre. Erster Teil. Die Mengen und ihre Axiome. *Mathematische Zeitschrift*, **25**, 683–712. (Reprinted in *Finsler 1975*.)

1975 *Aufsätze zur Mengenlehre*. Wissenschaftliche Buchgesellschaft, Darmstadt.

Fisch, Max F.

1980 The range of Peirce's relevance. *The Monist*, **63**, 269–76.

Fischer, Ernst Gottfried

1808 *Untersuchung über den eigentlichen Sinn der Höheren Analysis nebst einer idealen Übersicht der Mathematik und Naturkunde nach ihrem ganzen Umfange*. S.F. Weiss, Berlin.

Folina, Janet M.

1992 *Poincaré and the philosophy of mathematics*. St Martin's Press, New York.

Fontenelle, Bernard

1727 *Elémens de la géométrie de l'infini*, Suite des Mémoires de l'Académie royale des sciences. De l'Imprimerie royale, Paris.

Fraenkel, Abraham

(A partial bibliography of Fraenkel's writings can be found in *Bar-Hillel 1961*.)

1915 Über die Teiler der Null und die Zerlegung von Ringen. *Journal für die reine und angewandte Mathematik*, **145**, 1–43.

1916 *Über gewisse Teilbereiche und Erweiterungen von Ringen*. Teubner, Leipzig and Berlin.

1919 *Einleitung in die Mengenlehre*. Springer Verlag, Berlin.

1921 Über die Zermelosche Begründung der Mengenlehre. *Jahresbericht der Deutschen Mathematiker-Vereinigung* (*Angelegenheiten*), **30**, 97–8.

1922a Zu den Grundlagen der Cantor-Zermeloschen Mengenlehre. *Mathematische Annalen*, **86**, 230–7.

1922b Über den Begriff 'definit' und die Unabhängigkeit des Auswahlaxioms. *Sitzungsberichte der Preussischen Akademie der Wissenschaften, physikalische-mathematische Klasse*, 253–7. (Translated by Beverly Woodward in *van Heijenoort 1967*, pp. 284–9.)

1922c Axiomatische Begründung der transfiniten Kardinalzahlen. I. *Mathematische Zeitschrift*, **13**, 153–88.

1922d Zu den Grundlagen der Mengenlehre. *Jahresbericht der Deutschen Mathematiker-Vereinigung* (*Angelegenheiten*), **31**, 101–2.

1923 Second expanded edition of *Fraenkel 1919*.

1924 Die neueren Ideen zur Grundlegung der Analysis und Mengenlehre. *Jahresbericht der Deutschen Mathematiker-Vereinigung*, **33**, 97–103.

1925 Untersuchungen über die Grundlagen der Mengenlehre. *Mathematische Zeitschrift*, **22**, 250–73.

1927 *Zehn Vorlesungen über die Grundlegung der Mengenlehre*. Teubner, Leipzig.

1928 Third edition of *Fraenkel 1919*.

1930 Georg Cantor. *Jahresbericht der Deutschen Mathematiker-Vereinigung*, **39**, 189–266. (An abridged version of this article appears in *Cantor 1932*.)

1935a Sur l'axiome du choix. *Enseignement mathématique*, **34**, 32–51.

1935b Sur la notion d'éxistence dans les mathématiques. *Enseignement mathématique*, **34**, 18–32.

1937 Über eine abgeschwächte Fassung des Auswahlaxioms. *Journal of Symbolic Logic*, **2**, 1–25.

1953 *Abstract set theory*. North-Holland, Amsterdam.

1961 Revised second edition of *Fraenkel 1953*.

1967 *Lebenskreise: aus den Errinerungen eines jüdischen Mathematikers*. Deutsche Verlags-Anstalt, Stuttgart.

Fraenkel, Abraham and Bar-Hillel, Yehoshua

1958 *Foundations of set theory*. North-Holland, Amsterdam.

Fraenkel, Abraham; Bar-Hillel, Yehoshua; and Levy, Azriel

1973 Second edition of *Fraenkel and Bar-Hillel 1958*.

Fréchet, Maurice

1934 *L'arithmétique de l'infini*. Actualités scientifiques et industrielles, Vol. 144. Hermann, Paris.

Frege, Gottlob

(Bibliographies of Frege's writings can be found in *van Heijenoort 1967* and in *Frege 1972*.)

1879 *Begriffschrift, eine der arithmetischen nachgebildete Formelsprache des reinen Denkens*. Nebert, Halle. (Translated by Stefan Bauer-Mengelberg in *van Heijenoort 1967*, pp. 1–82.)

1882 Über den Zweck der Begriffschrift. In *Sitzungsberichte der Jenaischen Gesellschaft für Medicin und Naturwissenschaft für das Jahr 1882*, 1–10.

1884 *Die Grundlagen der Arithmetik, eine logisch-mathematische Untersuchung über den Begriff der Zahl.* Koebner, Breslau. (Translated by John Langshaw Austin as *The foundations of arithmetic, A logico-mathematical enquiry into the concept of number*, Basil Blackwell, Oxford, 1950; rev. 2nd edn, 1953.)

1891 *Function und Begriff.* H. Pohle, Jena.

1893 *Grundgesetze der Arithmetik, begriffschriftlich abgeleitet* (Vol. 1). H. Pohle, Jena.

1894 Review of Edmund Husserl, *Philosophie der Arithmetik. Zeitschrift für Philosophie und philosophische Kritik*, **103**, 313–32.

1903a Volume 2 of *Frege 1893*.

1903b Über die Grundlagen der Geometrie. *Jahresbericht der Deutschen Mathematiker-Vereinigung*, **12**, 319–24. (Reprinted in *Frege 1967*, pp. 262–6.)

1903c Über die Grundlagen der Geometrie. II. *Jahresbericht der Deutschen Mathematiker-Vereinigung*, **12**, 368–75. (Reprinted in *Frege 1967*, pp. 267–72.)

1906 Über die Grundlagen der Geometrie. I–III. *Jahresbericht der Deutschen Mathematiker-Vereinigung*, **15**, 293–309; 377–403; 423–30. (Reprinted in *Frege 1967*, pp. 281–323.)

1967 *Kleine Schriften*, (ed. Ignacio Angelelli). Georg Olms, Hildesheim.

1972 *Conceptual notation and related articles*, (ed. and trans. Terrell Bynum). Clarendon Press, Oxford. (Translation of *Frege 1879*.)

1976 *Nachgelassene Schriften und wissenschaftlicher Briefwechsel*, 2 vols, (ed. Gottfried Gabriel *et alii*). Felix Meiner, Hamburg.

1980 *Philosophical and mathematical correspondence*, (ed. Gottfried Gabriel *et alii*). Basil Blackwell, Oxford.

1983 *Nachgelassene Schriften*, 2 vols, (2nd edn, ed. Hans Hermes *et alii*). Felix Meiner, Hamburg.

1984 *Collected papers on mathematics, logic, and philosophy*, (ed. Brian F. McGuinness). Basil Blackwell, Oxford.

Frend, William

1796–9 *The principles of algebra; or, the true theory of equations established on mathematical demonstration*, 2 vols. J. Davis, London.

Freudenthal, Hans

1955–6 Neure Fassung des Riemann–Helmholtz–Lieschen Raumproblems. *Mathematische Zeitschrift*, **63**, 374–405.

1957 Zur Geschichte der Grundlagen der Geometrie. *Nieue Archief voor Wiskunde*, **4**, 105–42.

1960 Die Grundlagen der Geometrie um die Wende des 19. Jahrhunderts. *Mathematische–Physikalische Semesterberichte*, *Göttingen*, **7**, 2–25.

1961 Im Umkreis der sogennanten Raumprobleme. In *Essays on the foundations of mathematics* (ed. Yehoshua Bar-Hillel). Magnes Press, Hebrew University, Jerusalem.

1973 Hilbert, David. In *Gillispie* et alii *1970–6*, Vol. 6, pp. 388–95.

Friedman, Michael

1992 *Kant and the exact sciences*. Harvard University Press, Cambridge, Mass.

Gauss, Carl Friedrich

1801 *Disquisitiones arithmeticae*. Fleischer, Leipzig.

1816* Review of J.C. Schwab, *Commentatio in primum elementorum Euclidis librum, qua veritatem geometriae principiis ontologicis niti evincitur, omnesque propositiones, axiomatum geometricorum loco habitae, demonstrantur* (Stuttgart, 1814), and of Matthias Metternich, *Vollständige Theorie der Parallel-Linien* (Mainz, 1815). *Göttingische gelehrte Anzeigen*, 20 April 1816. (Reprinted in *Stäckel and Engel 1895*, 220–3. Translated by William Ewald, this collection, Vol. 1, pp. 299–300.)

1822* Review of Carl Reinhard Müller, *Theorie der Parallelen* (Marburg, 1822). *Göttingische gelehrte Anzeigen*, 28 October 1822. (Reprinted in *Stäckel and Engel 1895*, pp. 223–6. Translated by William Ewald, this collection, Vol. 1, p. 300.)

1827 Disquisitiones generales circa superficies curvas. *Commentationes societatis regiae scientiarum Gottingensis recentiores*, Vol. 6. (Reprinted in *Gauss 1863–1929*, Vol. 4, (1873), pp. 217–58.)

1828 Theoria residuorum biquadraticorum. Commentatio prima. *Commentationes societatis regiae scientiarum Gottingensis recentiores*, Vol. 6. (Presented 5 Apr. 1825. Reprinted in *Gauss 1863–1929*, Vol. 2, pp. 65–92.)

1831* Anzeige der Theoria residuorum biquadraticorum, Commentatio secunda. *Göttingische gelehrte Anzeigen*, 23 April 1831. (Reprinted in *Gauss 1863–1929*, Vol. 2, (1876), pp. 169–78. Translated by William Ewald, this collection, Vol. 1, pp. 306–13.)

1832 Theoria residuorum biquadraticorum. Commentatio secunda. *Commentationes societatis regiae scientiarum Gottingensis recentiores*, Vol. 7. (Presented 15 April 1831. Reprinted in *Gauss 1863–1929*, Vol. 2, pp. 93–148.)

1860–5 *Briefwechsel zwischen C.F. Gauss und H.C. Schumacher*. Esch, Altona.

1863–1929 *Werke*, 12 vols. Königliche Gesellschaft der Wissenschaften, Göttingen, Leipzig, and Berlin.

1880 *Briefwechsel zwischen Gauss und Bessel*. Engelmann, Leipzig.

1929* Zur Metaphysik der Mathematik. In *Gauss 1863–1929*, Vol. 12, pp. 57–61 (1929). (Translated by William Ewald, this collection, Vol. 1, pp. 293–6.)

Gelbart, Stephen

1984 An elementary introduction to the Langlands program. *Bulletin of the American Mathematical Society*, NS, **10**, 177–219.

Gentzen, Gerhard

1969 *The collected papers of Gerhard Gentzen*, (ed. M.E. Szabo). North-Holland, Amsterdam.

Gergonne, Joseph-Diez

1818 Essai sur la théorie des définitions, *Annales de mathématiques pures et appliquées*, **9**, 1–35. (The *Annales de mathématiques pures et appliquées*—later renamed the *Journal de mathématiques pures et appliquées*—were edited by Gergonne, and are sometimes cited by his name.)

Gibson, George Alexander

1899 Berkeley's *Analyst* and its critics: An episode in the development of the doctrine of limits. *Bibliotheca Mathematica*, NS, **13**, 65–70.

Gillispie, Charles Coulston *et alii* (eds.)

1970–6 *Dictionary of scientific biography*, 14 vols. Scribner, New York.

Gödel, Kurt

1931 Über formal unentscheidbare Sätze der *Principia mathematica* und verwandter Systeme, I. *Monatshefte für Mathematik und Physik*, **38**, 173–98. (Reprinted, with English translation and a note added in 1965, in *Gödel 1986*, pp. 144–95.)

1933a Zur intuitionistischen Arithmetik und Zahlentheorie. (Reprinted in *Gödel 1986*, pp. 286–95.)

1933b The present situation in the foundations of mathematics. (Published in *Gödel 1994*, pp. 45–53.)

1946 Remarks before the Princeton bicentennial conference on problems in mathematics. In *The undecidable: basic papers on undecidable propositions, unsolvable problems and computable functions* (ed. Martin Davis), Raven Press, Hewlett, NY, pp. 84–8. (Reprinted in *Gödel 1990*, pp. 150–3.)

1947 What is Cantor's continuum problem? *American mathematical monthly*, **54**, 515–25. *Errata*, *ibid.*, **55**, 151. (Reprinted in *Gödel 1990*, pp. 176–87.)

1951 Some basic theorems on the foundations of mathematics and their implications. In *Gödel 1994*, pp. 304–23.

1964 What is Cantor's continuum problem? In *Benacerraf and Putnam 1964*, pp. 258–73. (Revised version of *Gödel 1947*. Reprinted in *Benacerraf and Putnam 1983*, pp. 470–85, and in *Gödel 1990*, pp. 254–70.)

1986 *Collected works*, Vol. 1, (ed. Solomon Feferman *et alii*). Clarendon Press, Oxford.

1990 *Collected works*. Vol. 2, (ed. Solomon Feferman *et alii*). Clarendon Press, Oxford.

1994 *Collected works*. Vol. 3, (ed. Solomon Feferman *et alii*). Clarendon Press, Oxford.

Goldfarb, Warren D.

1979 Logic in the twenties: the nature of the quantifier. *The Journal of Symbolic Logic*, **44**, 351–68.

1982 Logicism and logical truth. *The Journal of Philosophy*, **79**, 692–5.

1988 Poincaré against the Logicists. In *Aspray and Kitcher 1988*, pp. 61–81.

1989 Russell's reasons for ramification. In *Rereading Russell* (ed. C.W. Savage and C.A. Anderson) (Minnesota Studies in the Philosophy of Science Vol. XII), pp. 24–40. University of Minnesota, Minneapolis.

Gonseth, Ferdinand

1926 *Les fondaments des mathématiques*. Blanchard, Paris.

Grassmann, Hermann Günther

1844 *Die lineale Ausdehnungslehre, ein neuer Zweig der Mathematik*. Wigand, Leipzig. (2nd edn: 1878).

1861 *Lehrbuch der Arithmetik und Trigonometrie für höhere Lehranstalten*. Enslin, Berlin.

1865 Volume 2 of *Grassmann 1861*.

Grassmann, Robert

1872 *Die Formenlehre oder Mathematik in strenger Formentwicklung*. R. Grassmann, Stettin.

Grattan-Guinness, Ivor

1970 An unpublished paper by Georg Cantor: *Principien einer Theorie der Ordnungstypen. Erste Mittheilung*. *Acta Mathematica*, **124**, 65–107.

1971 The correspondence between George Cantor and Philip Jourdain. *Jahresbericht der Deutschen Mathematiker-Vereinigung*, **73**, 111–30.

1972a Bertrand Russell on his paradox and the multiplicative axiom. An unpublished letter to Philip Jourdain. *Journal of Philosophical Logic*, **1**, 103–10.

1972b Bolzano, Cauchy, and the 'new analysis' of the nineteenth century. *Archive for the History of Exact Sciences*, **6**, 372–400.

1974 The rediscovery of the Cantor–Dedekind correspondence. *Jahresbericht der Deutschen Mathematiker-Vereinigung*, **76**, 104–39.

1975 Preliminary notes on the historical significance of quantification and of the axioms of choice in the development of mathematical analysis. *Historia Mathematica*, **2**, 475–87.

1976 Review of *Dugac 1976*. *Annals of Science*, **33**, 589–93.

1977 *Dear Russell–Dear Jourdain*. Duckworth, London.

1978 How Bertrand Russell discovered his paradox. *Historia Mathematica*, **5**, 127–37.

1979 In memoriam Kurt Gödel: His 1931 correspondence with Zermelo on his incompletability theorem. *Historia Mathematica*, **6**, 294–304.

1980 (ed.) *From the calculus to set theory: 1630–1910. An introductory history*. Duckworth, London.

Graves, Robert Perceval

1882–9 *Life of Sir William Rowan Hamilton*, 3 vols. Longmans, London.

Gregory, Duncan Farquharson

1840* On the real nature of symbolical algebra. *Transactions of the Royal Society of Edinburgh*, **14**, 208–16. (Read 7 May 1838; reprinted in this collection, Vol. 1, pp. 323–30.)

1841 *Examples of the processes of the differential and integral calculus*. J. Deighton, Cambridge.

1845 *A treatise on the application of analysis to solid geometry*, (ed. William Walton). J. Deighton, Cambridge.

1865 *The mathematical writings of Duncan Farquharson Gregory*, (ed. William Walton). Deighton Bell, Cambridge.

Hacker, Peter, M.S.

1986 *Insight and illusion*, (2nd rev. edn). Oxford University Press.

Hadamard, Jacques

1897 Sur certaines applications possibles de la théorie des ensembles. *Verhandlungen des ersten Internationalen Mathematiker-Kongresses in Zürich vom 9 bis 11 August 1897*, 201–2.

1904 Le troisième Congrès international des Mathématiciens. *Revue générale des sciences pures et appliquées*, **15**, 961–2.

1905a La théorie des ensembles. *Revue générale des sciences pures et appliquées*, **16**, 241–2.

1905b Les principes des mathématiques et le problème des ensembles. *Revue générale des sciences pures et appliquées*, **16**, 541–3. (*See also Richard 1905.*)

1906 Les fondaments des mathématiques. *Bulletin des sciences mathématiques et astronomiques* (ser. 2), **51**, 66–73. (This is the preface to *Gonseth 1926.*)

1922 *The early scientific work of Henri Poincaré.* Rice Institute Pamphlet, Vol. 9, no. 3. The Institute, Houston.

1933 *The later scientific work of Henri Poincaré.* Rice Institute Pamphlet, Vol. 20, no. 1. The Institute, Houston.

1945 *The psychology of invention in the mathematical field.* Princeton University Press.

See also Baire, René; Borel, Emile; Hadamard, Jacques, and Lebesgue, Henri.

Hahn, Hans

1921 *Theorie der reellen Funktionen.* Springer Verlag, Berlin.

1980 *Empiricism, logic, and mathematics: philosophical papers*, (ed. Brian F. McGuinness). Reidel, Dordrecht.

Hahn, Hans; Carnap, Rudolf; Gödel, Kurt; Heyting, Arend; Reidermeister, Kurt; Scholz, Heinrich; and von Neumann, John

1931 Diskussion zur Grundlegung der Mathematik. *Erkenntnis*, **2**, 135–51. (Translated by John W. Dawson, Jr. in *Dawson 1984.*)

Hallett, Michael

1981 Russell, Jourdain, and 'limitation of size'. *British Journal for the Philosophy of Science*, **32**, 381–99.

1984 *Cantorian set theory and limitation of size.* Clarendon Press, Oxford.

1994 Logic and mathematical existence. In *Physik, Philosophie, und die Einheit der Wissenschaft. Für Erhard Scheibe* (ed. Lorenz Krüger and Brigitte Falkenburg), pp. 33–82. Bibliographisches Institut, Wissenschaftsverlag, Mannheim.

Hamilton, William

1852 *Discussions on philosophy and literature, education and university reform.* Longmans, London. (3rd edn: 1866.)

Hamilton, William Rowan

1834 On a general method in dynamics. *Philosophical Transactions of the Royal Society*, part 2 for 1834, 247–308. (Reprinted in *Hamilton 1931–67*, Vol. 2, pp. 103–61.)

1837* Theory of conjugate functions, or algebraic couples: with a preliminary and elementary essay on algebra as the science of pure time. *Transactions of the Royal Irish Academy*, **17**, 293–422. (Reprinted in *Hamilton 1931–67*, Vol. 3, pp. 3–96, and partially reprinted in this collection, Vol. 1, pp. 369–75.) (Read 4 Nov. 1833 and 1 June 1835.)

1853* *Lectures on quaternions.* Hodges and Smith, Dublin. (Preface reprinted in *Hamilton 1931–67*, Vol. 3, pp. 117–58, and in this collection, Vol. 1, pp. 375–425.)

1931-67 *The mathematical papers of Sir William Rowan Hamilton*, 3 vols, (ed. H. Halberstam and R.E. Ingram). Cambridge University Press.

Hankel, Hermann

1867 *Vorlesungen über die complexen Zahlen und ihren Functionen*. Voss, Leipzig.

Hankins, Thomas L.

1980 *Sir William Rowan Hamilton*. Johns Hopkins, Baltimore and London.

Hardy, Godfrey Harold

1904 A theorem concerning the infinite cardinal numbers. *The Quarterly Journal of Pure and Applied Mathematics*, **35**, 87-94.

1915 Notes on some points in the integral calculus. *Messenger of Mathematics*, **44**, 145-9.

1918* Sir George Stokes and the concept of uniform convergence. *Proceedings of the Cambridge Philosophical Society*, **19**, 148-56. (Reprinted in *Hardy 1979*, pp. 505-13, and in this collection, Vol. 2, pp. 1234-42.)

1925 What is geometry? *Mathematical Gazette*, **12**, 309-16. (Reprinted in *Hardy 1979*, pp. 519-26.)

1929a* Mathematical proof. *Mind*, **38**, 1-25. (Reprinted in *Hardy 1979*, pp. 581-606, and in this collection, Vol. 2, pp. 1243-63.)

1929b An introduction to the theory of numbers. *Bulletin of the American Mathematical Society*, **35**, 778-818. (Reprinted in *Hardy 1979*, pp. 540-80.)

1979 *Collected papers of G.H. Hardy*, Vol. 7, (ed. I.W. Busbridge and R.A. Rankin). Clarendon Press, Oxford.

Hasse, Helmut

1932 Zu Hilberts zahlentheoretischen Arbeiten. In *Hilbert 1932-5*, Vol. 1, pp. 528-35.

Hasse, Helmut and Scholz, Heinrich

1928 *Die Grundlagenkrise in der Griechischen Mathematik*. Metzner, Berlin.

Hatfield, Gary

1990 *The natural and the normative: Theories of spatial perception from Kant to Helmholtz*. MIT Press, Cambridge, Mass.

Haubrich, Ralf

1992 *Zur Entstehung der algebraischen Zahlentheorie Richard Dedekinds*. Ph.D. Dissertation, Georg-August Universität, Göttingen. (Expanded version forthcoming from Birkhäuser Verlag.)

Hausdorff, Felix

1904 Der Potenzbegriff in der Mengenlehre. *Jahresbericht der Deutschen Mathematiker-Vereinigung*, **13**, 569-71.

1906 Untersuchungen über Ordnungstypen. *Königlich Sächsischen Gesellschaft der Wissenschaften zu Leipzig, Mathematische-physikalische Klasse, Sitzungsberichte*, **58**, 106-69.

1907a Untersuchungen über Ordnungstypen. *Königlich Sächsischen Gesellschaft der Wissenschaften zu Leipzig, Mathematische-physikalische Klasse, Sitzungsberichte*, **59**, 84–159.

1907b Über dichte Ordnungstypen. *Jahresbericht der Deutschen Mathematiker-Vereinigung*, **16**, 541–6.

1908 Grundzüge einer Theorie der geordneten Mengen. *Mathematische Annalen*, **65**, 435–505.

1909 Die Graduierung nach dem Endverlauf. *Königlich Sächsischen Gesellschaft der Wissenschaften zu Leipzig, Mathematische-physikalische Klasse, Sitzungsberichte*, **61**, 297–334.

1914 *Grundzüge der Mengenlehre*. De Gruyter, Leipzig. (Reprinted Chelsea, New York, 1965.)

1916 Die Mächtigkeit der Borelschen Mengen. *Mathematische Annalen*, **77**, 430–7.

1927 Second, revised edition of *Hausdorff 1914*.

1937 Third, revised edition of *Hausdorff 1914*. (Translated as *Set theory*, Chelsea, New York, 1957.)

Hausen, Christian

1734 *Elementa matheseos*. Marcheana, Leipzig.

Heath, Thomas L.

1921 *A history of Greek mathematics*, 2 vols. Clarendon Press, Oxford.

Heine, Heinrich Eduard

1872 Die elemente der Funktionenlehre. *Journal für die reine und angewandte Mathematik*, **74**, 172–88.

Hellinger, Ernst

1933 Hilberts Arbeiten über Integralgleichungen und unendliche Gleichungssysteme. In *Hilbert 1932–5*, Vol. 3, pp. 94–145.

Helmholtz, Hermann von

1847 Über die Erhaltung der Kraft. *Vortrag in der physikalischen Gesellschaft zu Berlin am 23. Juli 1847*. G. Reimer, Berlin. (Reprinted in *Helmholtz 1882–95*, Vol. 1, pp. 12–75.)

1856–67 *Handbuch der Physiologischen Optik*, 3 vols. Leopold Voss, Leipzig.

1863 *Die Lehre von den Tonempfindungen als physiologische Grundlage für die Theorie der Musik*. Vieweg, Braunschweig.

1865–76 *Populäre wissenschaftliche Vorträge*, 3 vols. Vieweg, Braunschweig.

1868a Über die thatsächlichen Grundlagen der Geometrie. *Verhandlungen des naturhistorisch-medicinischen Vereins zu Heidelberg*, **4**, 197–202. (Additional material in Vol. **5**, 31–2 (April 1869).) (Reprinted in *Helmholtz 1882–95*, Vol. 2, pp. 610–17.)

1868b Über die Thatsachen, die der Geometrie zu Grunde liegen. *Nachrichten der Königlichen Gesellschaft der Wissenschaften zu Göttingen*, **9**, 193–221. (Reprinted in *Helmholtz 1882–95*, Vol. 2, pp. 618–39; and in *Helmholtz 1921*. Translated by Malcolm F. Lowe in *Helmholtz 1977*.)

1870 The axioms of geometry. *The Academy*, **1**, 128–31. (Later incorporated into *Helmholtz 1876*.)

1876* The origin and meaning of geometrical axioms. *Mind*, **1**, 301–21. (Reprinted in this collection, Vol. 2, pp. 663–85. German text in *Helmholtz 1865–76*, Vol. 3. Reprinted, with notes and commentary by Paul Hertz, in *Helmholtz 1921*; Hertz edition and notes translated, with title 'On the origin and significance of the axioms of geometry', by Malcolm F. Lowe in *Helmholtz 1977*.)

1878a* The origin and meaning of geometrical axioms (II). *Mind*, **3**, 212-25. (Reprinted in this collection, Vol. 2, pp. 685–9. German text in *Helmholtz 1882–95*, Vol. 2, pp. 640–62.)

1878b* Die Thatsachen in der Wahrnehmung. *Rede, gehalten zur Stiftungsfeier der Friedrich-Wilhelms-Universität zu Berlin am 3. August 1878*. Berlin. (Reprinted, with additions and corrections, A. Hirschwald, Berlin, 1879. Reprinted in *Helmholtz 1884* and *Helmholtz 1921*. Translations in *Helmholtz 1971* (by Russell Kahl) and in *Helmholtz 1977* (by Malcolm F. Lowe). Lowe translation reprinted in this collection, Vol. 2, pp. 689–727.)

1881 *Popular lectures on scientific subjects* (trans. E. Atkinson *et alii*). (2nd edn: 1893.) Longmans, Green, London. (Partial translation of *Helmholtz 1865–76*.)

1882-95 *Wissenschaftliche Abhandlungen von Hermann Helmholtz*, 3 vols. J.A. Barth, Leipzig. (Vol. 3 contains a complete bibliography of Helmholtz's published works.)

1884 *Vorträge und Reden*, 2 vols. Vieweg, Braunschweig. (3rd edn of *Helmholtz 1865–76*.)

1887* Zählen und Messen, erkenntnistheoretisch betrachtet. In *Philosophische Aufsätze, Eduard Zeller zu seinem fünfzigjährigen Doctorjubiläum gewidmet*, pp. 17-52. Fues, Leipzig. (Reprinted in *Helmholtz 1882-95*, Vol. 3, pp. 356-91. Translated by Malcolm F. Lowe in *Helmholtz 1977*; Lowe translation reprinted in this collection, Vol. II, pp. 727–52.)

1921 *Schriften zur Erkenntnistheorie*, (ed. Paul Hertz and Moritz Schlick). Springer Verlag, Berlin.

1971 *Selected writings of Hermann von Helmholtz*, (ed. Russell Kahl). Wesleyan University Press, Middletown, Conn.

1977 *Hermann von Helmholtz: Epistemological writings*, (ed. Robert Cohen, and Marx Wartofsky). Reidel, Dordrecht. (Translation of *Helmholtz 1921*, by Malcolm F. Lowe, with an introduction by the editors.)

Herbrand, Jacques

1971 *Logical writings*, (ed. and trans. Warren D. Goldfarb). Reidel, Dordrecht and Harvard University Press, Cambridge, Mass.

Hertz, P.

1934 Sur les axiomes d'Archimède et de Cantor. *Compte rendu des séances de la société de physique et d'histoire naturelle de Genève*, Vol. **51**.

Hessenberg, Gerhard

1905 Beweis des Desarguesschen Satzes aus dem Pascalschen. *Mathematische Annalen*, **61**, 161-72.

1906 *Grundbegriffe der Mengenlehre*. Vandenhoeck und Ruprecht, Göttingen.

1908 Willkürliche Schöpfungen des Verstandes? *Jahresbericht der Deutschen Mathematiker-Vereinigung*, **17**, 145–62.

1909 Kettentheorie und Wohlordnung. *Journal für die reine und angewandte Mathematik*, **135**, 81–133.

Heyting, Arend

1930 Die formalen Regeln der intuitionistischen Logik. *Sitzungsberichte der Preussischen Akademie der Wissenschaften, Physikalische-mathematische Klasse*, 42–56.

1956 *Intuitionism: An introduction*. North-Holland, Amsterdam.

Hilbert, David

1897 Die Theorie der algebraischen Zahlkörper. *Jahresbericht der Deutschen Mathematiker-Vereinigung*, **4**, 175–546. (Also known as the *Zahlbericht*. Reprinted in *Hilbert 1932–5*, Vol. 2. pp. 63–363.)

1899 Grundlagen der Geometrie. In *Festschrift zur Feier der Enthullung des Gauss-Weber Denkmals in Göttingen*. Teubner, Leipzig. (Translated by E.J. Townsend, Open Court, Chicago, 1902.) (Later editions published in 1903, 1909, 1911, 1922, 1923, and 1930; these contain supplements reprinting articles on the foundations of mathematics that do not appear in the *Gesammelte Abhandlungen*.)

1900a* Über den Zahlbegriff. *Jahresbericht der Deutschen Mathematiker-Vereinigung*, **8**, 180–94. (Reprinted in *Hilbert 1909*, pp. 256–62; *1913*, pp. 237–42; *1922b*, pp. 237–42; *1923b*, pp. 237–42; 1930a, pp. 241–6. Translated by William Ewald, this collection, Vol. 2, pp. 1089–95.)

1900b* Mathematische Probleme. Vortrag, gehalten auf dem internationalen Mathematiker-Kongress zu Paris. 1900. *Nachrichten der Königlichen Gesellschaft der Wissenschaften zu Göttingen*, 253–97. (Reprinted, with additions, in *Archiv der Mathematik und Physik* (ser. 3), **1**, (1901), 44–63, 213–37. French translation, with emendations and additions, in *Compte rendu du deuxième congrès international des mathématiciens tenu à Paris du 6 au 12 août 1900* (Gauthier-Villars, Paris, 1902), 58–114. Translated by Mary Winston Newson in *Bulletin of the American Mathematical Society* (ser. 2), **8** (1902), 437–79; partially reprinted in this collection, Vol. 2, pp. 1096–105.)

1903 Second edition of *Hilbert 1899*, Teubner, Leipzig.

1904 Über die Grundlagen der Logik und der Arithmetik. In *Verhandlungen des dritten internationalen Mathematiker-Kongresses in Heidelberg vom 8. bis 13. August 1904*, pp. 174–85. Teubner, Leipzig, 1905. (Reprinted in *Hilbert 1909*, pp. 263–79; *1913*, pp. 243–58; *1922b*, pp. 243–58; *1923b*, pp. 243–58; *1930a*, pp. 247–61. Translated by Beverly Woodward in *van Heijenoort 1967*, pp. 129–38.)

1905 On the foundations of logic and arithmetic. *The Monist*, **15**, 338–52. (Translation of *Hilbert 1904* by George Bruce Halsted.)

1909 Third edition of *Hilbert 1899*. Teubner, Leipzig and Berlin.

1913 Fourth edition of *Hilbert 1899*.

1918* Axiomatisches Denken. *Mathematische Annalen*, **78**, 405–15. (Reprinted in *Hilbert 1932–5*, Vol. 3, pp. 146–56. Translated by William Ewald, this collection, Vol. 2, pp. 1105–15.)

1922a* Neubegründung der Mathematik. Erste Mitteilung. *Abhandlungen aus dem mathematischen Seminar der Hamburgischen Universität*, **1**, 157–77. (Reprinted in *Hilbert 1932–5*, Vol. 3, pp. 157–77. Translated by William Ewald, this collection, Vol. 2, pp. 1115–34.)

1922b Fifth edition of *Hilbert 1899*.

1923a* Die logischen Grundlagen der Mathematik. *Mathematische Annalen*, **88**, 151–65. (Reprinted in *Hilbert 1932–5*, Vol. 3, pp. 178–91. Translated by William Ewald, this collection, Vol. 2, pp. 1134–48.)

1923b Sixth edition of *Hilbert 1899*.

1926 Über das Unendliche. *Mathematische Annalen*, **95**, 161–90. (Reprinted, in abridged form, in *Hilbert 1930a*, but with minor corrections and additions. Translated by Stefan Bauer-Mengelberg in *van Heijenoort 1967*, pp. 367–92.)

1927 Über das Unendliche. *Jahresbericht der Deutschen Mathematiker-Vereinigung*, **36**, 1st section, 201–15. (Abridged reprinting of *Hilbert 1926*.)

1928a Die Grundlagen der Mathematik. *Abhandlungen aus dem mathematischen Seminar der Hamburgischen Universität*, **6**, 65–85. (Reprinted in *Hilbert 1930a*, pp. 289–312. Translated by Stefan Bauer-Mengelberg and Dagfinn Føllesdal in *van Heijenoort 1967*, pp. 464–79.)

1928b *Die Grundlagen der Mathematik, mit Zusätzen von Hermann Weyl und Paul Bernays*. Hamburger mathematische Einzelschriften, Vol. 5. Teubner, Leipzig. (Contains reprints of *Hilbert 1928a*, *Weyl 1928*, and *Bernays 1928*.)

1928c Probleme der Grundlegung der Mathematik. *Mathematische Annalen*, **102**, 1–9. (Reprinted in *Hilbert 1930a*, pp. 313–23.)

1930a Seventh edition of *Hilbert 1899*.

1930b* Naturerkennen und Logik. *Die Naturwissenschaften*, **18**, 959–63. (Reprinted in *Hilbert 1932–5*, Vol. 3, pp. 378–87. Translated by William Ewald, this collection, Vol. 2, pp. 1157–65.)

1931a* Die Grundlegung der elementaren Zahlentheorie. *Mathematische Annalen*, **104**, 485–94. (Reprinted in part in *Hilbert 1932–5*, Vol. 3, pp. 192–5. Translation of the original version by William Ewald, this collection, Vol. 2, pp. 1148–57.)

1931b Beweis des *Tertium non Datur*. *Nachrichten von der Gesellschaft der Wissenschaften zu Göttingen, Mathematisch-physikalische Klasse*, 120–5.

1932–5 *Gesammelte Abhandlungen*, 3 vols. Springer Verlag, Berlin.

Hilbert, David and Ackermann, Wilhelm

1928 *Grundzüge der theoretischen Logik*. Springer Verlag, Berlin.

1938 Second edition of *Hilbert and Ackermann 1928*.

Hilbert, David and Bernays, Paul

1934 *Grundlagen der Mathematik. I*. Springer Verlag, Berlin.

1939 *Grundlagen der Mathematik. II*. Springer Verlag, Berlin.

1968 Second edition of *Hilbert and Bernays 1934*.

1970 Second edition of *Hilbert and Bernays 1939*.

Hilbert, David and Cohn-Vossen, Stefan

1932 *Anschauliche Geometrie*. Springer Verlag, Berlin.

Hobbes, Thomas

1845 *The Latin works of Thomas Hobbes*, 5 vols, (ed. William Molesworth). Green and Longman, London. (*De principiis et ratiocinatione geometrarum* is in Vol. 4, pp. 385–465.)

Hobson, Ernest William

1903 On modes of convergence of an infinite series of functions of a real variable. *Proceedings of the London Mathematical Society* (ser. 2), **1**, 373–87.

Hofmann, Joseph Ehrenfried

1963 *Geschichte der Mathematik*, 3 vols. De Gruyter, Berlin.

Hölder, Otto

1901 Die Axiome der Quantität und die Lehre von Mass. *Berichte über die Verhandlungen der Königlichen Sächsischen Gesellschaft der Wissenschaften zu Leipzig, Mathematisch-physikalische Klasse*, **53**, 1–64.

1914 *Die Arithmetik in strenger Begründung*. In Programmabhandlung der Philosophischen Fakultät zu Leipzig. Teubner, Leipzig. (2nd edn: Springer Verlag, Berlin 1929.)

1924 *Die mathematische Methode*. Springer Verlag, Berlin.

Hoppe, R.

1879 Review of *Frege 1879*. *Archiv der Mathematik und Physik*, (ser. 1), **63**, Litterarischer Bericht CCLII, 44–5. (Translated by Terrell Bynum in *Frege 1972*.)

Hospital, Guillaume Franççois Antoine de l'

See under L'Hospital.

Jacobi, Karl Gustav Jacob

1834 De binis quibuslibet functionibus homogenis. *Journal für die reine und angewandte Mathematik*, **12**, 1–69. (Reprinted in *Jacobi 1881–91*, Vol. 3, pp. 191–268).

1881–91 *Gesammelte Werke*, 7 vols. Reimer, Berlin.

Jesseph, Douglas M.

1993 *Berkeley's philosophy of mathematics*. University of Chicago Press.

Jevons, William Stanley

1879 *The principles of science, a treatise on logic and scientific method*, (3rd edn). Macmillan, London.

Johnson, Dale M.

1979 The problem of invariance of dimension in the growth of modern topology. *Archive for History of Exact Sciences*, **20**, 97–188.

1981 The problem of invariance of dimension in the growth of modern topology (Part Two). *Archive for History of Exact Sciences*, **25**, 85-267.

1987 L.E.J. Brouwer's coming of age as a topologist. In *Studies in the history of mathematics* (ed. E.R. Phillips), MAA Studies, Vol. 26, pp. 61-97. The Association, Providence.

Johnstone, Peter T.

1977 *Topos theory*. London Mathematical Society Mathematical Monographs, no. 10. Academic Press, London.

1983 The point of pointless topology. *Bulletin of the American Mathematical Society*, NS, **8**, 41-53.

Jordan, Camille

1893 *Cours d'analyse de l'Ecole polytechnique*, Vol. 1, (2nd edn). Gauthier-Villars, Paris.

Jourdain, Philip E.B.

1904 On the transfinite cardinal numbers of well-ordered aggregates. *Philosophical Magazine* (ser. 6), **7**, 61-75.

1905 On a proof that every aggregate can be well-ordered. *Mathematische Annalen*, **60**, 465-70.

1916 Richard Dedekind. *The Monist*, **26**, 415-27.

Kant, Immanuel

1747* *Gedanken von der wahren Schätzung der lebendingen Kräften und Beurteilung der Beweise, derer sich Herr von Leibniz und andere Mechaniker in dieser Streitsache bedient haben, nebst einigen vorhergehenden Betrachtungen, welche die Kraft der Körper überhaupt betreffen*. (Written and submitted 1746. Partial translation by William Ewald, this collection, Vol. 1, pp. 133–4.)

1755 *Allgemeine Naturgeschichte und Theorie des Himmels, oder Versuch von der Verfassung und dem mechanischen Ursprunge des ganzen Weltgebäudes nach Newtonischen Grundsätzen abgehandelt*. Johann Friedrich Petersen, Königsberg and Leipzig.

1781 *Critik der reinen Vernunft*. Hartknoch, Riga.

1787 *Kritik der reinen Vernunft*. Hartknoch, Riga. (Second edition of *Kant 1781*.)

1902-23 *Kants Werke*. Preussische Akademie der Wissenschaften, Berlin.

1933 *Critique of pure reason*. Macmillan, London. (Translation of *Kant 1781* and *1787* by Norman Kemp Smith.)

1972 *Briefwechsel*, (ed. Otto Schöndörfer). Meiner, Hamburg.

Kästner, Abraham Gotthelf

1758 *Anfangsgründe der Arithmetik, Geometrie, ebenen und sphärischen Trigonometric und Perspektiv*. Vandenhoeck, Königsberg.

1794 *Anfangsgründe der Analysis endlicher Grössen*, (3rd edn). Vandenhoeck and Ruprecht, Göttingen. (1st edn: 1760.)

Kennedy, Hubert

1979 James Mills Peirce and the cult of quaternions. *Historia Mathematica*, **6**, 423-9.

Klein, Felix

1872 Vergleichende Betrachtungen über neuere geometrische Forschungen. (Das Erlanger Programm). *Mathematische Annalen*, **43** (1893), 63–100. (Reprinted in *Klein 1921–3*, Vol. 1, pp. 460–97.) (Lecture read in Erlangen in Oct. 1872.)

1883 Über den allgemeinen Funktionsbegriff und dessen Darstellung durch eine willkürliche Kurve. *Mathematische Annalen*, **22**, 249–59. (Reprinted in *Klein 1921–3*, Vol. 2. pp. 214–24.) (Read in Erlangen on 8 Dec. 1873.)

1895 Über Arithmetisierung der Mathematik. *Nachrichten der Königlichen Gesellschaft der Wissenschaften zu Göttingen*, Geschäftliche Mitteilungen 1895, Heft 2. (Reprinted in *Klein 1921–3*, Vol. 2, pp. 232–40.)

1896* The arithmetizing of mathematics. *Bulletin of the American Mathematical Society*, **2**, 241–9. (Translation by Isabel Maddison of *Klein 1895*. Reprinted in this collection, Vol. 2, pp. 965–71.)

1897–1910 Über die Theorie des Kreisels, 4 vols. Teubner, Leipzig.

1906 Grenzfragen der Mathematik und Philosophie. *Wissenschaftliche Beilage zum 19. Jahresbericht der Philosophischen Gesellschaft an der Universität Wien, 1906*. (Reprinted in *Klein 1921–3*, Vol. 2, pp. 247–51.) (Lecture delivered on 14 Oct. 1905.)

1911* *The Evanston Colloquium lectures on mathematics*. Macmillan, New York. (Partially reprinted in this collection, Vol. 2, pp. 957–65.)

1921–3 *Gesammelte mathematische Abhandlungen*, 3 vols, (ed. R. Fricke, A. Ostrowski, H. Vermeil, E. Bessel-Hagen). Springer Verlag, Berlin.

1924–8 *Elementarmathematik vom höheren Standpunkt aus*, 3 vols, (3rd edn). Springer Verlag, Berlin.

1926–7 *Vorlesungen über die Entwicklung der Mathematik im 19. Jahrhundert*, 2 vols, (ed. Richard Courant, Otto Neugebauer, and Stephen Cohn-Vossen). Springer Verlag, Berlin.

Kline, Morris

1972 *Mathematical thought from ancient to modern times*. Oxford University Press, New York.

Klügel, Georg Simon

1763 *Conatuum praecipuorum theoriam parallelarum demonstrandi recensio*. F.A. Rosenbusch, Göttingen.

1803–31. *Mathematisches Wörterbuch*, 5 vols. Schwickert, Leipzig.

Kneale, William

1948 Boole and the revival of logic. *Mind*, **57**, 149–75.

Kneale, William and Kneale, Martha

1962 *The development of logic*. Clarendon Press, Oxford.

Koenigsberger, Leo

1902–3 *Hermann von Helmholtz*, 3 vols. Vieweg, Brunswick. (Abridged translation by Frances A. Welby, Clarendon Press, Oxford, 1906; reprinted New York, 1965.)

König, Julius

1904 Zum Kontinuum-Problem. In *Verhandlungen des dritten internationalen Mathematiker-Kongress in Heidelberg vom 8. bis 13 August 1904*, (Teubner, Leipzig, 1905), pp. 144-7.

1905a Über die Grundlagen der Mengenlehre und das Kontinuumproblem. *Mathematische Annalen*, **61**, 156-60. (Translated by Stefan Bauer-Mengelberg in *van Heijenoort 1967*, pp. 145-9.)

1905b Zum Kontinuumproblem. *Mathematische Annalen*, **60**, 177-80, 462.

1914 *Neue Grundlagen der Logik, Arithmetik und Mengenlehre*. Veit, Leipzig.

Kossak, Ernst

1872 *Die Elemente der Arithmetik*. Programm der Friedrichs-Werderschen Gymnasium, Berlin.

Kreisel, Georg

1958a Hilbert's programme. *Dialectica*, **12**, 346-72. (Revised version in *Benacerraf and Putnam 1964*, pp. 157-80.)

1958b Mathematical significance of consistency proofs. *Journal of Symbolic Logic*, **23**, 155-82.

1960 La predicativité. *Bulletin de la Société Mathématique de France*, **88**, 371-91.

1965 Mathematical logic. In *Lectures on modern mathematics*, Vol. 3, (ed. T. Saaty), pp. 95-195. J. Wiley, New York.

1967a Mathematical logic: what has it done for the philosophy of mathematics? In *Bertrand Russell: Philosopher of the century* (ed. R. Schoenmann), pp. 201-72 and 315-16. Allen and Unwin, London.

1967b Informal rigour and completeness proofs. In *Problems in the philosophy of mathematics. Proceedings of the International Colloquium in the Philosophy of Science* (ed. Imré Lakatos), pp. 138-86. North-Holland, Amsterdam.

1968 A survey of proof theory. *Journal of Symbolic Logic*, **33**, 321-88.

1971 A survey of proof theory II. In *Proceedings of the Second Scandinavian Logic Symposium* (ed. J. Fenstad), pp. 109-70. North-Holland, Amsterdam.

1976 What have we learned from Hilbert's second problem? In *Browder 1976*, pp. 93-130.

Kreisel, Georg and Newman, Maxwell Herman Alexander

1969 Luitzen Egbertus Jan Brouwer. *Biographical Memoirs of Fellows of the Royal Society*, **15**, 39-68.

Kronecker, Leopold

1881 Grundzüge einer arithmetischen Theorie der algebraischen Grössen. Festschrift zu Herrn Ernst Eduard Kummer's fünfzigjährigem Doctor-Jubiläum am 10 September 1881. *Journal für die reine und angewandte Mathematik*, **92**, 1-122. (Reprinted, with additions, as a monograph by G. Reimer, Berlin, 1882; and in *Kronecker 1895-1930*, Vol. 2, pp. 237-387.)

1886 Über einige anwendungen der Modulsysteme auf elementare algebraische Fragen. *Journal für die reine und angewandte*

Mathematik, **99**, 329–71. (Reprinted in *Kronecker 1895–1930*, Vol. 3, pp. 145–208.)

1887* Über den Zahlbegriff. In *Philosophische Aufsätze, Eduard Zeller zu seinem fünfzigjährigen Doctorjubiläum gewidmet*, pp. 261–74. Fues, Leipzig. (Also in *Journal für die reine und angewandte Mathematik*, **101**, 337–55. Translated by William Ewald, this collection, Vol. 2, pp. 947–55.) (Reprinted, with additions, in *Kronecker 1895–1930*, Vol. 3, pp. 251–74.)

1895–1930 *Werke*, 5 vols, (ed. Kurt Hensel). Teubner, Leipzig.

1901 *Vorlesungen über Zahlentheorie*, 2 vols, (ed. Kurt Hensel). Teubner, Leipzig.

Lacroix, Silvestre François

*c.*1805 *Elémens d'algèbre*, (5th edn). Courcier, Paris. (24 editions published altogether; this date for the fifth edition is only approximate.)

1811 *Anfangsgründe der Algebra*. Mainz. (German translation, by one Metternich, of the seventh edition of *Lacroix 1805*.)

See Peacock, George,

Lagrange, Joseph-Louis

1797 *Théorie des fonctions analytiques contenant les principes du calcul différentiel, dégagés de toute considération d'infiniment petits, d'évanouissans, de limites et de fluxions, et réduits à l'analyse algébrique des quantités finies*. Imprimerie de la Republique, Prairial, An 5 (= 1797), Paris.

1806 *Leçons sur le calcul des fonctions. Nouvelle edition*. Courcier, Paris. (1st edn: 1804.)

1808 *Traité de la résolution des équations numériques de tous les degrés*, (2nd edn). Courcier, Paris. (1st edn: Duprat, Paris, 1798.)

Lambert, Johann Heinrich

1761 *Cosmologische Briefe über die Einrichtung des Weltbaues*. E. Kletts, Augsburg.

1764 *Neues Organon, oder Gedanken über die Erforschung und Bezeichnung des Wahres und dessen Unterscheidung vom Irrthum und Schein*, 2 vols. Wendler, Leipzig.

1771 *Anlage zur Architectonic, oder Theorie des Einfachen und des Ersten in der philosophischen und mathematischen Erkenntnis*, 2 vols. Hartknoch, Riga.

1781–4 *Deutscher gelehrter Briefwechsel*, 4 vols, (ed. Joh. Bernoulli). Bernoulli, Berlin.

1786* Theorie der Parallellinien. *Magazin für reine und angewandte Mathematik für 1786*, 137–64, 325–58. (Reprinted in *Stäckel and Engel 1895*, pp. 152–208. Partial translation by William Ewald, this collection, Vol. 1, pp. 158–67.) (Written in 1766.)

1946–8 *Opera mathematica*, 2 vols. Orell Füssli Zurich.

1967– *Gesammelte philosophische Werke* (ed. Hans Werner Arndt). Olms, Hildesheim. (10 volumes planned.)

Landau, Edmund

1917 Richard Dedekind. *Nachrichten der Königlichen Gesellschaft der Wissenschaften zu Göttingen, Geschäftliche Mitteilungen*, 50–70.

Lasswitz, K.
1879 Review of *Frege 1879*. Jenaer Literaturzeitung, **6**, 248-9. (Translated by Terrell Bynum in *Frege 1972*.)

Lebesgue, Henri
1902 Intégrale, longeur, aire. *Annali di Matematica Pura ed Applicata* (ser. 3), **7**, 231-359.
1904 *Leçons sur l'intégration et la recherche des fonctions primitives*. Gauthier-Villars, Paris.
1905 Sur les fonctions représentables analytiquement. *Journal de mathématiques pures et appliquées*, **60**, 139-216.
1907 Contributions à l'étude des correspondances de M. Zermelo. *Bulletin de la Société Mathématique de France*, **35**, 202-12.
1917 Sur certaines démonstrations d'existence. *Bulletin de la Société Mathématique de France*. **45**, 132-44.
1918 Remarques sur les théories de la mesure et de l'intégration. *Annales scientifiques de l'Ecole Normale Supérieure* (ser. 3), **35**, 191-250.
1919 Sur les correspondances entre les points de deux espaces. *Fundamenta Mathematicae*, **2**, 256-85.
1922 *Notice sur les travaux scientifiques*. E. Privat, Toulouse.
1972-3 *Œuvres scientifiques*, 5 vols. Kundig, Geneva. *See* Baire, René; Borel, Emile; Hadamard, Jacques, and Lebesgue, Henri.

Legendre, Adrien Marie
1794 *Eléments de géométrie*. F. Didot, Paris. (12th edn: 1823.)
1833 Réflexions sur différentes manières de démontrer la théorie des parallèles ou le théorèm sur la somme des trois angles du triangle. *Mémoires de l'Académie royale des sciences*, **12**, 367-410.

Leibniz, Gottfried Wilhelm
1875-90 *Die philosophischen Schriften von Gottfried Wilhelm Leibniz*, 7 vols, (ed. Carl J. Gerhardt). Weidmann, Berlin. (Reprinted Georg Olms, Hildesheim, pp. 1960-1.)
1849-63 *Mathematische Schriften*, 7 vols, (ed. Carl J. Gerhardt). Asher, Berlin (Vols. 1-2) and Schmidt, Halle (Vols. 3-7).
1903 *Opuscules et fragments inédits de Leibniz*, (ed. Louis Couturat). Alcan, Paris. (Reprinted Georg Olms, Hildesheim, 1961.)

Le Lionnais, François
1948 *Les grands courants de la pensée mathématique*. Cahiers du Sud, Paris.

Levy, Azriel
1979 *Basic set theory*. Springer, Berlin.

Lewis, Clarence Irving
1918 *A survey of symbolic logic*. University of California, Berkeley. (Reprinted, omitting Chapters V and VI, Dover, New York, 1960.)

L'Hospital, Guillaume François Antoine de
1696 *Analyse des infiniment petits*. Imprimerie Royale, Paris.

Lipschitz, Rudolf Otto Sigismund

1877–80 *Lehrbuch der Analysis*, 2 vols. Cohen, Bonn.

1986 *Briefwechsel mit Cantor, Dedekind, Helmholtz, Kronecker, Weierstrass und anderen*, (ed. Winfried Scharlan). Dokumente zur Geschichte der Mathematik, Band 2. Deutsche Mathematiker Vereinigung-Vieweg, Braunschweig and Wiesbaden.

Lobatchevsky, Nikolai Ivanovich

1840 *Geometrische Untersuchungen zur Theorie der Parallellinien*. G. Funcke, Berlin.

Lorenzen, Paul

1955 *Einführung in die operative Logik und Mathematik*. Springer Verlag, Berlin.

1969 Second edition of *Lorenzen 1955*.

Löwenheim, Leopold

1940 Einkleidung der Mathematik in Schröderschen Relativkalkül. *The Journal of Symbolic Logic*, **5**, 1–15.

Luce, Arthur Aston

1949 *The Life of George Berkeley, Bishop of Cloyne*. Nelson, London.

Ludlam, William

1809 *The rudiments of mathematics: designed for the use of students at the universities*, (5th edn). John Evans, London.

MacHale, Desmond

1985 *George Boole: His life and work*. Boole Press, Dublin.

MacLane, Saunders

1981 Origin, rise, and decline of abstract algebra. In *American mathematical heritage: Algebra and applied mathematics*, Graduate Studies, Texas Tech University.

MacLaurin, Colin

1742* *A treatise of fluxions*, 2 vols. Ruddimans, Edinburgh. (Partially reprinted in this collection, Vol. 1, pp. 95–122.)

1748 *An account of Sir Isaac Newton's philosophical discoveries*. Murdoch, London.

Manning, Thomas

1796–8 *An introduction to arithmetic and algebra*, 2 vols. B. Flower, Cambridge.

Mansion, Paul

1908 Gauss contra Kant sur la géometrie non euclidienne. In *Bericht über den 3. internationalen Kongress für Philosphie, Heidelberg 1908* (ed. Theodor Elsenhans). Winter, Heidelberg.

Martin, Donald A.

1977 Descriptive set-theory: projective sets. In *Barwise 1977*, pp. 783–815.

Martin, Gottfried

1972 *Arithmetik und Kombinatorik bei Kant*. De Gruyter, Berlin.

Martin, Richard M.

1976 Some comments on De Morgan, Peirce, and the logic of relations. *Transactions of the Charles S. Peirce Society*, **12**, 223–30.

1978 Of servants, benefactors, and lovers: Peirce's algebra of relatives of 1870. *Journal of Philosophical Logic*, **7**, 27–48.

Maseres, Francis

1758 *A dissertation on the use of the negation sign in algebra*. S. Richardson, London. (An appendix contains 'Mr. Machin's quadrature of the circle'.)

Mercator, Nicolaus

c.1678 *Euclidis elementa geometriae*. London. (Approximate date.)

Meschkowski, Herbert

1961 *Denkweisen grosser Mathematiker*. Vieweg, Braunschweig. (Translated by John Dyer-Bennett as *Ways of thought of great mathematicians*, Holden-Day, San Francisco, 1964.)

1967 *Probleme des Unendlichen. Werk und Leben Georg Cantors*. Vieweg, Braunschweig.

Michael, Emily

1976 Peirce's earliest contact with scholastic logic. *Transactions of the Charles S. Peirce Society*, **12**, 46–55.

Michaelis, C. Th.

1880 Review of *Frege 1879*. *Zeitschrift für Völkerpsychologie und Sprachwissenschaft*, **12**, 232–40. (Translated by Terrell Bynum in *Frege 1972*.)

Michelsen, Johann Andreas

1789 *Gedanken über den gegenwärtigen Zustand der Mathematik und die Art der Vollkommenheit und Brauchbarkeit derselben zu vergrössern*. S.F. Hesse, Berlin.

1790 *Beiträge zur Beförderung des Studiums der Mathematik*. Berlin. (This work is cited by Bolzano; the editor has however been unable to locate a copy.)

Minnigerode, B.

1871 Bemerkungen über irrationale Zahlen. *Mathematische Annalen*, **4**, 497–8.

Mirimanoff, Dimitry

1917a Les antinomies de Russell et de Burali-Forti et le problème fondamental de la théorie des ensembles. *Enseignement mathématique*, **19**, 37–52.

1917b Remarques sur la théorie des ensembles et les antinomies cantoriennes. I. *Enseignement mathématique*, **19**, 209–17.

1921 Remarques sur la théorie des ensembles et les antinomies cantoriennes. II. *Enseignement mathématique*, **21**, 29–52.

Mitchell, Oscar Howard
1881 Some theorems in numbers. *American Journal of Mathematics*, **4**, 25–38.
1883 On a new algebra of logic. In *Studies in logic by members of the Johns Hopkins University* (ed. Charles S. Peirce). Little, Brown, Boston.

Möbius, August Ferdinand
1827 *Der barycentrische Calcul*. Barth, Leipzig.
1843 *Die Elemente der Mechanik des Himmels*. Weidmann, Leipzig.

Monna, A.F.
1972 The concept of function in the nineteenth and twentieth centuries, in particular with regard to the discussions between Baire, Borel, and Lebesgue. *Archive for History of Exact Sciences*, **9**, 57–84.
1975 *Dirichlet's principle*. Oosthoer, Scheltma, and Hoikema, Utrecht.

Montague, Richard and Vaught, Robert
1959 Natural models of set theories. *Fundamenta Mathematicae*, **47**, 219–42.

Mooij, Jan J.A.
1966 *La philosophie des mathématiques de Henri Poincaré*. Gauthier-Villars, Paris.

Moore, Gregory H.
1980 Beyond first-order logic: the historical interplay between mathematical logic and axiomatic set-theory. *History and Philosophy of Logic*, **1**, 95–137.
1982 *Zermelo's axiom of choice: Its origins, development, and influence*. Springer Verlag, New York.

Moore, Gregory H. and Garciadiego, Alejandro R.
1981 Burali-Forti's paradox: a reappraisal of its origins. *Historia Mathematica*, **8**, 319–50.

Müller, Gert H. (ed.)
1976 *Sets and classes: On the work by Paul Bernays*. North-Holland, Amsterdam.

Napier, John
1614 *Mirifici logarithmorum canonis descriptio*. Edinburgh. (Translated by Edward Wright as *A description of the admirable table of logarithms*, Waterson, London, 1618.)

Nelson, Leonard
1906 *Kant und die nichteuklidische Geometrie*. Schwetschke, Berlin.

Neugebauer, Otto
1969 *The exact sciences in antiquity*. Dover, New York.

Newman, Maxwell Herman Alexander
See Kreisel, Georg and Newman, Maxwell Herman Alexander.

Newton, Isaac
1686 *Philosophiae naturalis principia mathematica*. S. Pepys, London.
1700 *Tractatus de quadratura curvarum*. Paris. (Exact date and location uncertain; multiple later editions, e.g. London 1704 and 1706.)
1726 Third edition of *Newton 1686*.
1934 *Sir Isaac Newton's mathematical principles of natural philosophy and his system of the world*, (trans. Andrew Motte, 1729; rev. trans. Florian Cajori, 1934). University of California Press, Berkeley.
1972 *Philosophiae naturalis principia mathematica*, Vol. 1, (ed. Alexander Koyré and I. Bernard Cohen). Cambridge University Press. (A reprinting, with variant readings, of *Newton 1726*.)

Noether, Emmy
1921 Idealtheorie in Ringbereichen. *Mathematische Annalen*, **83**, 24–66.
1983 *Gesammelte Abhandlungen*, (ed. Nathan Jacobson). Springer Verlag, Berlin.

Noether, Emmy and Cavaillès, Jean
1937 *Briefwechsel Cantor-Dedekind*. Hermann, Paris. (Translated by William Ewald, this collection, Vol. 2, pp. 843–78.)

North, J.D.
1976 Sylvester, James Joseph. In *Gillispie* et alii *1970–6*, Vol. 13, pp. 216–22.

Nový, L.
1973 *Origins of modern algebra*. Academia, Prague.

Osgood, William Fogg
1897 Non-uniform convergence and the integration of series. *American Journal of Mathematics*, **19**, 155–90.

Pappus of Alexandria
1588 *Mathematicae collectiones*, (ed. Federigo Commandino). Pisa and Venice. (Multiple editions between 1588 and 1660.)

Paris Logic Group
1987 *Logic colloquium '85*, (ed. Anita B. Feferman and Solomon Feferman). North-Holland, Amsterdam.

Parsons, Charles D.
1965 Frege's theory of number. In *Philosophy in America* (ed. Max Black), pp. 180–203. Cornell University Press, Ithaca.
1990a The structuralist view of mathematical objects. *Synthese*, **84**, 303–46.
1990b The uniqueness of the natural numbers. *Iyyun*, **39**, 13–44.

Pasch, Moritz
1882a *Einleitung in die differential- und integral-Rechnung*. Teubner, Leipzig.

1882b *Vorlesungen über neuere Geometrie*. Teubner, Leipzig. (2nd edn: 1912.)

1927 *Mathematik am Ursprung. Gesammelte Abhandlungen über Grundfragen der Mathematik*. Meiner, Leipzig.

Peacock, George

1816 *A treatise on the differential and integral calculus. By Silvestre François Lacroix*, (trans. Charles Babbage, John Frederick William Herschel, and George Peacock). J. Smith, Cambridge.

1820 *A collection of examples of the application of the differential and integral calculus*. J. Smith, Cambridge.

1830 *A treatise on algebra*. Deighton, Cambridge.

1833 Report on the recent progress and present state of certain branches of analysis. *British Association for the Advancement of Science, Report 3*, 185–352.

1834 *Report on certain branches of analysis*. R. Taylor, London. (Separate printing of *Peacock 1833*.)

1842–5 *Treatise on algebra*, 2 vols. (2nd edn; 1st edn is *Peacock 1830*.) Deighton, Cambridge. (Vol. 1 (1842) entitled *Arithmetical algebra*; Vol. 2 (1845) entitled *Symbolical algebra*.)

Peano, Giuseppe

1889 *Arithmetices principia, novo methodo exposita*. Turin. (Reprinted in *Peano 1958*, pp. 20–55; trans. in *van Heijenoort 1967*, pp. 83–97.

1894 *Notations du logique mathématique: Introduction au formulaire de mathématique*. Charles Guadagnini, Turin.

1895–1908 *Formulaire de mathématiques*, 5 vols. Bocca, Turin.

1958 *Opere scelte*, Vol. 2. Editione Cremonese, Rome.

1973 *Selected works of Giuseppe Peano*, (ed. and trans. Hubert C. Kennedy). University of Toronto Press.

Peirce, Benjamin

1870* *Linear associative algebra*. Washington, DC. (Partially reprinted in this collection, Vol. 1, pp. 584–94.) (Lithograph.)

1881 —. *American Journal of Mathematics*, **4**, 97–229. (Reprinting of *B. Peirce 1870*, with notes and addenda by Charles Sanders Peirce.)

Peirce, Charles Sanders

1868 On a new list of categories. *Proceedings of the American Academy of Arts and Sciences*, **7**, 287–98. (Reprinted in *Peirce 1931–58*, Vol. 1, pp. 287–99; and in *Peirce 1982–*, Vol. 2, pp. 49–58.)

1869 The pairing of the elements. *Chemical News*, **4**, 339–40. (Reprinted in *Peirce 1982–*, Vol. 2, pp. 282–3.)

1870 Description of a notation for the logic of relatives, resulting from an amplification of the conceptions of Boole's calculus of logic. *Memoirs of the American Academy of Arts and Sciences*, NS, **9**, 317–78. (Reprinted in *Peirce 1931–58*, Vol. 3, pp. 27–98; and in *Peirce 1982–*, Vol. 2, pp. 359–429.)

1871 Review of Fraser's *The works of George Berkeley. North American Review*, **113**, 449–72. (Reprinted in *Peirce 1982–*, Vol. 2, pp. 462–87.)

1875 On the application of logical analysis to multiple algebra. *Proceedings of the American Academy of Arts and Sciences*, NS, **2**, 392–4. (Reprinted in *Peirce 1982–*, Vol. 3, pp. 177–9.)

1878 *Photometric researches: Made in the years 1872–1875*. Wilhelm Engelmann, Leipzig. (Reprinted in *Peirce 1982–*, Vol. 3, pp. 382–493.)

1880a On the algebra of logic. *American Journal of Mathematics*, **3**, 15–57. (Reprinted in *Peirce 1931–58*, Vol. 3, pp. 104–57; and in *Peirce 1982–*, Vol. 4, pp. 163–209.)

1880b On a Boolean algebra with one constant. (First published in 1933 in *Peirce 1931–58*, Vol. 4, pp. 13–18; title supplied by the editors. Also in *Peirce 1982–*, Vol. 4 pp. 218–21.)

1881* On the logic of number. *The American Journal of Mathematics*, **4**, 85–95. (Reprinted in *Peirce 1931–58*, Vol. 3, pp. 158–70; in *Peirce 1982–*, Vol. 4, pp. 299–309; and in this collection, Vol. 1, pp. 596–608.)

1883 A theory of probable inference. (With two Notes; Note B entitled 'The logic of relatives'.) In *Studies in logic by members of the Johns Hopkins University* (ed. Charles S. Peirce). Little, Brown, Boston. (Reprinted in *Peirce 1983*, and in *Peirce 1982–*, Vol. 4, pp. 409–466; Note B reprinted in *Peirce 1931–58*, Vol. 3, pp. 195–209.)

1885* On the algebra of logic: a contribution to the philosophy of notation. *The American Journal of Mathematics*, **7**, 180–202. (Reprinted in *Peirce 1931–58*, Vol. 3, pp. 210–38; in *Peirce 1982–*, Vol. 5, pp. 162–90; and in this collection, Vol. 1, pp. 608–32.)

1898* The logic of mathematics in relation to education. *Educational Review*, 1898, 209–16. (Reprinted in *Peirce 1931–58*, Vol. 3, pp. 346–59, and in this collection, Vol. 1, pp. 632–7.)

1902* The simplest mathematics. First published in 1933 in *Peirce 1931–58*, Vol. 4, pp. 189–262. (Reprinted in this collection, Vol. 1, pp. 637–48.)

1905 What pragmatism is. *The Monist*, **15**, 161–81.

1931–58 *Collected papers of Charles Sanders Peirce*, 8 vols, (ed. Charles Hartshorne, Paul Weiss, and Arthur Burks). Harvard University Press, Cambridge, Mass.

1976 *The new elements of mathematics*, 4 vols, (ed. Carolyn Eisele). Mouton, The Hague.

1982– *Writings of Charles Sanders Peirce: a chronological edition*, (ed. Max H. Fisch *et alii*). Indiana University Press, Bloomington.

1983 *Studies in logic by members of the Johns Hopkins University*, (ed. C.S. Peirce; with an introduction by Max H. Fisch). Benjamins, Amsterdam.

1985 *Historical perspectives on Peirce's logic of science*, 2 vols, (ed. Carolyn Eisele). Mouton, Berlin.

Peirce, Charles Sanders and Jastrow, Charles

1884 On small differences in sensation. *Memoirs of the National Academy of Sciences*, Vol. 3 (5th memoir), Part 1, 75–83.

Perry, Ralph Barton

1935 *The thought and character of William James*. Little, Brown, Boston.

Pinl, M.

1969 Kollegen in einer dunklen Zeit. *Jahresbericht der Deutschen Mathematiker-Vereinigung*, **71**, 167-228.

Pippard, Alfred Brian

1957 *Elements of classical thermodynamics*. Cambridge University Press.

Platner, Ernst

1793-1800 *Philosophische Aphorismen*, (2nd edn). Schwickert, Leipzig.

Poincaré, Henri

1894* Sur la Nature du Raisonnement mathématique. *Revue de métaphysique et de morale*, **2**, 371-84. (Reprinted with modifications in *Poincaré 1902*, pp. 29-44. Translated by George Bruce Halsted in *Poincaré 1913b*, pp. 31-42. Reprinted in this collection, Vol. 2, pp. 972-82.)

1898* On the foundations of geometry. *The Monist*, **9**, 1-43. (Reprinted in this collection, Vol. 2, pp. 982-1011.)

1900* Du rôle de l'intuition et de la logique en mathématiques. In *Compte rendu du Deuxième congrès international des mathématiciens tenu à Paris du 6 au 12 août 1900*, Gauthier-Villars, Paris, pp. 115-30. (Reprinted with modifications in *Poincaré 1905a*. Translated by George Bruce Halsted in *Poincaré 1913b*, pp. 210-22. Reprinted in this collection, Vol. 2, pp. 1012-20.)

1902 *La science et l'hypothèse*. Flammarion, Paris. (Translated by George Bruce Halsted in *Poincaré 1913b*.)

1905a *La valeur de la science*. Flammarion, Paris. (Translated by George Bruce Halsted in *Poincaré 1913b*.)

1905b* Les mathématiques et la logique. *Revue de métaphysique et de morale*, **13**, 815-35. (Page 813 contains a bibliography of articles in the *Revue de métaphysique* on foundations of mathematics.) (Reprinted with extensive deletions in *Poincaré 1908*, Chapter 3. Translated by George Bruce Halsted in *Poincaré 1913b*, pp. 448-59. Translation of original version (incorporating Halsted translation) by William Ewald, this collection, Vol. 2, pp. 1021-38.)

1906a* Les mathématiques et la logique. *Revue de métaphysigue et de morale*, **14**, 17-34. (Reprinted with extensive deletions in *Poincaré 1908*, Chapter 4, with the title 'Les logiques nouvelles'. Translated by George Bruce Halsted in *Poincaré 1913b*, pp. 460-71. Translation of original version (incorporating Halsted translation) by William Ewald, this collection, Vol. 2, pp. 1038-52.)

1906b* Les mathématiques et la logique. *Revue de métaphysique et de morale*, **14**, 294-317. (Reprinted with extensive deletions in *Poincaré 1908*, Chapter 5, with the title 'Les derniers efforts des Logisticiens'. Translated by George Bruce Halsted in *Poincaré 1913b*, pp. 472-85. Translation of original version (incorporating Halsted translation) by William Ewald, this collection, Vol. 2, pp. 1052-71.)

1906c A propos de la logistique. *Revue de métaphysique et de morale*, **14**, 866-8.

1908 *Science et méthode*. Flammarion, Paris. (Translated by George Bruce Halsted in *Poincaré 1913b*.)

1909a La Logique de l'infini. *Revue de métaphysique et de morale*, **17**, 461-82. (Reprinted in *Poincaré 1913a*.)

1909b Réflexions sur les deux notes précedents. *Acta mathematica*, **32**, 195-200.

1910* Über transfinite Zahlen. In Poincaré, *Sechs Vorträge über ausgewählte Gegenstände aus der reinen Mathematik und mathematischen Physik*. Teubner, Leipzig. (Reprinted in *Poincaré 1921–56*, Vol. 11, pp. 120–4. Translated by William Ewald, this collection, Vol. 2, pp. 1071–4.)

1913a *Dernières penséees*. Flammarion, Paris.

1913b *The foundations of science*. The Science Press, New York. (Contains translations by George Bruce Halsted of *Poincaré 1902*, *1905a*, and *1908*, a Preface by Poincaré, and an Introduction by Josiah Royce.)

1921-56 *Œuvres*, 11 Vols. Gauthier-Villars, Paris.

Popper, Karl

1953-4 Berkeley as a precursor of Mach. *British Journal for the Philosophy of Science*, **4**, 26-36.

Post, Emil Leon

1921 Introduction to a general theory of elementary propositions. *American Journal of Mathematics*, **43**, 163-85. (Reprinted in *van Heijenoort 1967*, pp. 264-83.)

Posy, Carl J.

1974 Brouwer's constructivism. *Synthese*, **27**, 125-59.

1975 Varieties of indeterminacy in the general theory of choice sequences. *Journal of Philosophical Logic*, **5**, 91-132.

Proclus

1970 *A commentary on the first book of Euclids 'Elements'*, (translated with introduction and notes by Glenn R. Morrow). Princeton University Press.

Purkert, Walter, and Ilgauds, Hans Joachim

1987 *Georg Cantor 1845-1918*. Birkhauser, Basle, Boston, and Stuttgart.

Quine, Willard van Orman

1940 *Mathematical logic*. Harvard University Press, Cambridge, Mass.

1955 On Frege's way out. *Mind*, **64**, 145-59.

1960 *Word and object*. MIT Press, Cambridge, Mass.

1963 *Set theory and its logic*. Harvard University Press, Cambridge, Mass.

Rados, Gustav

1906 Zur ersten Verteilung des Bolyai-Preises. *Mathematische Annalen*, **62**, 156-76.

Rang, B. and Thomas, W.

1981 Zermelo's discovery of the 'Russell paradox'. *Historia Mathematica*, **8**, 15-22.

Raphson, Joseph

1690 *Analysis aequationum universalis*. Abel Swalle, London.

1697 *De spatio reali, seu ente infinito conamen mathematico-metaphysicum*. Thomas Braddyll, London.

1715 *The history of fluxions, shewing in a compendious manner the first rise of, and various improvements made in, that incomparable method*. W. Pearson, London. (Also published in Latin as *Historia fluxionum*.)

Reid, Constance

1970 *Hilbert*. Springer Verlag, Berlin.

Riccardi, Pietro

1887–93 *Saggio di una bibliografia Euclidea*, 5 parts. *Istituto delle Scienze ed Arti Liberali, Memorie*, Ser. 4, Vols. 8 and 9; Ser. 5, Vol. 1. Istituto di Bologna.

Richard, Jules

1905 Lettre à Monsieur le rédacteur de la *Revue générale des Sciences. Acta Mathematica*, **30**, 295–6. (Translated by Jean van Heijenoort in *van Heijenoort 1967*, pp. 142–4.)

1907 Sur un paradoxe de la théorie des ensembles et sur l'axiome Zermelo. *Enseignement mathématique*, **9**, 94–8.

1929 Sur l'axiome de Zermelo. *Bulletin des sciences mathématiques et astronomiques* (ser. 2), **53**, 106–9.

Riemann, Bernhard

1861 Commentatio mathematica, qua respondere tentatur questioni ab Illma. Academia Parisiensi propositae. (First published in *Riemann 1876*, pp. 370–83; submitted unsuccessfully for a prize to the Paris Academy in 1861.)

1868* Über die Hypothesen, welche der Geometrie zu Grunde liegen. *Abhandlungen der Königlichen Gesellschaft der Wissenschaften zu Göttingen*, **13**, 133–52. (Read as a *Probevorlesung* in Göttingen, 10 June 1854. Reprinted in *Riemann 1876*, pp. 254–79. Translated by William Kingdon Clifford in *Clifford 1873*; translation reprinted in this collection, Vol. 2, pp. 652–61.)

1876 *Gesammelte mathematische Werke and wissenschaftlicher Nachlass*, (ed. Heinrich Weber and Richard Dedekind). Teubner, Leipzig.

1919 *Über die Hypothesen, welche der Geometrie zu Grunde liegen*, (ed. with commentary by Hermann Weyl). Springer Verlag, Berlin.

Rösling, Christian Lebrecht

1805 *Grundlehren von den Formen, Differenzen, Differentialen und Integralien der Functionen*. J.S. Palm, Erlangen.

Russ, Stephen B.

1980* A translation of Bolzano's paper on the intermediate value theorem. *Historia Mathematica*, **7**, 156–85. (Revised version reprinted in this collection, Vol. 1, pp. 225–48.)

Russell, Bertrand Arthur William

1901 Recent work in the philosophy of mathematics. *The International Monthly*, **4**, 83–101. (Reprinted under the title 'Mathematics and the metaphysicians' as Ch. 5 of *Mysticism and logic and other essays*, Longmans, Green, New York, 1918.)

1903 *The principles of mathematics*. Cambridge University Press.

1904 The axiom of infinity. *Hibbert Journal*, **2**, 809–12.

1906a On some difficulties in the theory of transfinite numbers and order types. *Proceedings of the London Mathematical Society* (ser. 2), **4**, 29–53. (Reprinted in *Russell 1973*, pp. 135–64.)

1906b Les paradoxes de la logique. *Revue de métaphysique et de morale*, **14**, 627–50. (Translated in *Russell 1973*, pp. 190–214.)

1907 The regressive method of discovering the principles of mathematics. (Read before the Cambridge Mathematical Club on 9 March 1907; first printed in *Russell 1973*, pp. 272–83.)

1908 Mathematical logic as based on the theory of types. *American Journal of Mathematics*, **30**, 222–62. (Reprinted in *van Heijenoort 1967*, pp. 150–82.)

1914 On scientific method in philosophy. Herbert Spencer Lecture, Oxford. (Reprinted in *Mysticism and logic and other essays*, Longmans, Green, New York, 1918.)

1937 Second edition of Russell 1903. Allen and Unwin, London.

1944 My mental development. In *Schilpp 1944*.

1956 *Portraits from memory and other essays*. Allen and Unwin, London.

1959 *My philosophical development*. Allen and Unwin, London.

1967 *The autobiography of Bertrand Russell, 1872–1914*. Allen and Unwin, London.

1973 *Essays in analysis*, (ed. Douglas Lackey). Braziller, New York.

See Whitehead, Alfred North and Russell, Bertrand.

Rychlik, Karel (ed.)

1962 *Theorie der reellen Zahlen in Bolzanos handschriftlichem Nachlass*. Verlag der Tschechoslowakischen Akademie der Wissenschaften, Prague.

Saccheri, Girolamo

1733 *Euclides ab omni naevo vindicatus*. (Translated by George Bruce Halsted in *American Mathematical Monthly*, 1–5, 1894–8. Reprinted, Open Court, Chicago, 1920; and Chelsea, New York, 1970. German translation in *Stäckel and Engel 1895*, pp. 41–136.)

Schilpp, Paul Arthur (ed.)

1941 *The philosophy of Alfred North Whitehead*. Northwestern University, Evanston.

1944 *The philosophy of Bertrand Russell*. Northwestern University, Evanston.

Schmidt, Arnold

1930 Die Stetigkeit in der absoluten Geometrie. *Sitzungsberichte der Heidelberger Akademie der Wissenschaften*, Vol. **5**.

1933 Zu Hilberts Grundlegung der Geometrie. In *Hilbert 1932–5*, Vol. 2, pp. 404–14.

Schoenflies, Arthur

1900 Die Entwicklung der Lehre von den Punktmannigfaltigkeiten. *Jahresbericht der Deutschen Mathematiker-Vereinigung*, **8**, 1–251.

1908 Die Entwicklung der Lehre von den Punktmannigfaltigkeiten. Zweiter Teil. *Jahresbericht der Deutschen Mathematiker-Vereinigung*, Ergänzungsband, **2**, 1–331.

1913 *Allgemeine Theorie der unendlichen Mengen und Theorie der Punktmengen*. Teubner, Leipzig. (Published as first half of *Entwicklung der Mengenlehre und ihrer Anwendungen*, ed. Arthur Schoenflies and Hans Hahn; second half never published.)

1922 Zur Erinnerung an Georg Cantor. *Jahresbericht der Deutschen Mathematiker-Vereinigung*, **31**, 90–106.

1927 Die Krisis in Cantors mathematischen Schaffen. *Acta Mathematica*, **50**, 1–23.

1928 Georg Cantor. *Mitteldeutsche Lebensbilder*, **3**, 548–63.

Scholz, Erhard

1982 Herbart's influence on Bernhard Riemann. *Historia mathematica*, **9**, 413–40.

Schröder, Ernst

1873 *Lehrbuch der Arithmetik und Algebra*. Teubner, Leipzig.

1880 Review of *Frege 1879*. *Zeitschrift für Mathematik und Physik*, **25**, 81–94. (Translated by Terrell Bynum in *Frege 1972*.)

1890 *Vorlesungen über die Algebra der Logik (exakte Logik)*, Vol. 1. Teubner, Leipzig. (Reprinted in *Schröder 1966*.)

1891— Vol. 2, part 1. (Reprinted in *Schröder 1966*.)

1895— Vol. 3, *Algebra und Logik der Relative*, part 1. (Reprinted in *Schröder 1966*.)

1966 *Vorlesungen über die Algebra der Logik*, 3 vols. Chelsea, New York.

Schrödinger, Erwin

1945 The Hamilton postage stamp: An announcement by the Irish minister of posts and telegraphs. In *A collection of papers in memory of Sir William Rowan Hamilton* (ed. David Eugene Smith), Scripta mathematica studies, no. 2. Scripta mathematica, New York.

Schultz, Johann

1784 *Entdeckte Theorie der Parallelen nebst einer Untersuchung über den Ursprung ihrer bisherigen Schwierigkeit*. Kanter, Königsberg.

1785 *Erläuterung über des Herrn Professor Kant Critik der reinen Vernunft*. Dengel, Königsberg. (2nd edn: Frankfurt and Leipzig, 1791.) [Same title can be found under the spelling Johann Schulze.] (Modernized reprint, (ed. R.C. Hafferberg), Rassmann, Jena and Leipzig, 1897.)

1788 *Versuch einer genauen Theorie des Unendlichen*. Hartung, Königsberg and Leipzig.

1789–92 *Prüfung der Kantischen Kritik der reinen Vernunft*, 2 vols. Hartung, Königsberg.

1790 *Anfangsgründe der reinen Mathesis.* Königsberg.
1797 *Kurzer Lehrbegriff der Mathematik.* Nicolovius, Königsberg.

Schweickart (or Schweikart), Ferdinand Karl
1807 *Theorie der Parallellinien, nebst dem Vorschlag ihrer Verbannung aus der Geometrie.* Jena.

Scott, Joseph F.
1938 *The mathematical work of John Wallis.* Oxford University Press.
1973 MacLaurin, Colin. In *Gillispie* et alii *1970–6*, Vol. 7, pp. 609–12.

Scriba, Christoph
1973 Lambert, Johann. In *Gillispie* et alii *1970–6*, Vol. 7, pp. 595–600.

Servois, François-Joseph
1814a Sur la théorie des quantités imaginaires. Lettre de M. Servois. *Annales de mathématiques pures et appliquées* (later renamed *Journal de mathématiques pures et appliquées*), **4**, 228–35. (The letter is dated 23 Nov. 1813.)
1814b Essai sur un nouveau mode d'exposition des principes du calcul différentiel. *Annales de mathématiques pures et appliquées* (later renamed *Journal de mathématiques pures et appliquées*), **5**, 93–140.

Shafarevitch, Igor R.
1973 Über einige Tendenzen in der Entwicklung der Mathematik. *Jahrbuch der Akademie der Wissenschaften in Göttingen*, 1973, 31–6.

Sheffer, Henry M.
1913 A set of five independent postulates for Boolean algebras, with application to logical constants. *Transactions of the American Mathematical Society*, **14**, 481–8.

Sieg, Wilfried
1988 Hilbert's program sixty years later. *Journal of Symbolic Logic*, **53**, 338–48.

Sierpinski, Waclaw
1928 *Leçons sur les nombres transfinis.* Gauthier-Villars, Paris.

Sierpinski, Waclaw and Tarski, Alfred
1930 Sur une propriété charactéristique des nombres inaccessibles. *Fundamenta Mathematicae*, **15**, 292–300.

Sigwart, Christoph
1904 *Logik*, (3rd edn). Mohr, Tübingen.

Sinaceur, Mohamed A.
1974 Über den Begriff des Unendlichens. *Revue d'histoire des sciences et de leurs applications*, **27**, 259–78. (Contains a German transcription and French translation of *Dedekind 1890*.)

Skolem, Thoralf
1922 Einige Bemerkungen zur axiomatischen Begründung der Mengenlehre. *Matematikerkongressen i Helsingfors den 4–7 Juli 1922*. (Translated in *van Heijenoort 1967*, 290–301.)

Sondheimer, Ernst and Rogerson, Alan
1981 *Numbers and infinity: An historical account of mathematical concepts*. Cambridge University Press.

Spivak, Michael
1970 *A comprehensive introduction to differential geometry*. Vol. 2. Privately published.

Stäckel, Paul and Engel, Friedrich (eds.)
1895 *Die Theorie der Parallellinien von Euklid bis auf Gauss*. Teubner, Leipzig.
1898 *Urkunden zur Geschichte der nichteuklidischen Geometrien*, 2 vols. Teubner, Leipzig.

Stammler, Gerhard
1922 *Berkeleys Philosophie der Mathematik*. Reuther and Reichard, Berlin.

Steck, Max
1970 *Bibliographia Lambertiana*, (2nd edn). Olms, Hildesheim.

Stein, Howard
1988 *Logos*, logic, and *logistikè*: some philosophical remarks on the nineteenth-century transformation of mathematics. In *Aspray and Kitcher 1988*, pp. 238–59.
1990 Eudoxus and Dedekind: on the ancient Greek theory of ratios and its relation to modern mathematics. *Synthese*, **84**, 163–211.

Stokes, (Sir) George
1880–1905 *Mathematical and physical papers*, 5 vols. Cambridge University Press.

Stolz, Otto
1881 B. Bolzano's Bedeutung in der Geschichte der Infinitesimalrechnung. *Mathematische Annalen*, **18**, 255–79.

Struik, Dirk J. (ed.)
1969 *A source book in mathematics, 1200–1800*. Harvard University Press, Cambridge, Mass.

Sylvester, James Joseph
1869* Presidential address to Section 'A' of the British Association. *Exeter British Association Report*, 1–9. (Reprinted in *Sylvester 1908*, pp. 650–61, and in this collection, Vol. 1, pp. 511–22.)
1870 *Laws of verse; or, Principles of versification exemplified in metrical translations, together with an annotated reprint of the inaugural Presidential address to the Mathematical and physical Section of the British Association at Exeter*. Longmans, London.

1908 *The collected mathematical papers of James Joseph Sylvester*, Vol. 2. Cambridge University Press.

Synge, John L.

1937 *Geometrical optics: An introduction to Hamilton's method*. Cambridge University Press.

Tannery, Jules

1886 *Introduction à la théorie des fonctions d'une variable*. Hermann, Paris.

1897 De l'infini mathématique. *Revue générale des sciences pures et appliquées*, **8**, 129–40.

Tannery, Paul

1879 Review of *Frege 1879*. *Revue philosophique*, **8**, 108–9. (Translated by Terrell Bynum in *Frege 1972*.)

1885 Le concept scientifique du continu. Zénon d'Elée et Georg Cantor. *Revue Philosophique*, **20**, 385–410.

Tarski, Alfred

1941 On the calculus of relations. *Journal of Symbolic Logic*, **6**, 73–89.

1956 Notions of proper models for set theory. *Bulletin of the American Mathematical Society*, **62**, 601.

Taurinus, Franz Adolph

1826 *Geometriae prima elementa*. Cologne.

Toepell, Michael-Markus

1986 *Über die Entstehung von David Hilberts 'Grundlagen der Geometrie'*. Vandenhoeck und Ruprecht, Göttingen.

Troelstra, Anne Sjerp

1977 *Choice sequences: A chapter of intuitionistic mathematics*. Clarendon Press, Oxford.

1982 The origin and development of Brouwer's concept of choice sequence. In *The L.E.J. Brouwer Centenary Symposium* (ed. A.S. Troelstra and D. van Dalen). North-Holland, Amsterdam.

1983 Analysing choice sequences. *Journal of Philosophical Logic*, **12**, 197–260.

1985 Choice sequences and informal rigour. *Synthese*, **62**, 217–27.

Troelstra, Anne Sjerp and van Dalen, Dirk

1988 *Constructivism in mathematics: An introduction*, 2 vols. North-Holland, Amsterdam.

Turnbull, Herbert Western

1947 Colin MacLaurin. *American Mathematical Monthly*, **54**, 318–22.

1951 *Bi-centenary of the death of Colin MacLaurin*. University Press, Aberdeen.

van Dalen, Dirk

1981 *Brouwer's Cambridge lectures on intuitionism*. Cambridge University Press.

van der Waerden, Bartel Leendert

1933 Nachwort zu Hilberts algebraischen Arbeiten. In *Hilbert 1932–5*, Vol. 2, pp. 401–3.

1975 On the sources of my book, *Moderne Algebra*. *Historia Mathematica*, **2**, 31–40.

1985 *A history of algebra*. Springer, Berlin.

van Heijenoort, Jean (ed.)

1967 *From Frege to Gödel: A source book in mathematical logic, 1879–1931*. Harvard University Press, Cambridge, Mass. (Second printing 1971.)

van Rootselaar, Bob

1962–6 Bolzano's theory of real numbers. *Archive for History of Exact Sciences*, **2**, 168–80.

1976 Zermelo, Ernst Friedrich Ferdinand. In *Gillispie* et alii *1970–6*, Vol. 14, pp. 613–16.

van Stigt, Walter

1990 *Brouwer's intuitionism*. North-Holland, Amsterdam.

Venn, John

1880 Review of *Frege 1879*. *Mind*, **5**, 297. (Reprinted in *Frege 1972*.)

Vilant, Nicholas

1798 *The elements of mathematical analysis, abridged, for the use of students*. F.W. Wingrave, London.

von Neumann, John

1923 Zur Einführung der transfiniten Zahlen. *Acta litterarum ac scientiarum Regiae Universitatis Hungaricae Francisco-Josephinae, Sectio scientiarum mathematicarum*, **1**, 199–208. (Translated by Stefan Bauer-Mengelberg in *van Heijenoort 1967*, pp. 346–54.)

1925 Eine Axiomatisierung der Mengenlehre. *Journal für die reine und angewandte Mathematik*, **154**, 219–40. Berichtigung, *ibid.*, **155**, 128. (Translated by Stefan Bauer-Mengelberg in *van Heijenoort 1967*, pp. 393–413.)

1927 Zur Hilbertschen Beweistheorie. *Mathematische Zeitschrift*, **26**, 1–46.

1928a Über die Definition durch transfinite Induktion und verwandte Fragen der allgemeinen Mengenlehre. *Mathematische Annalen*, **99**, 373–91.

1928b Die Axiomatisierung der Mengenlehre. *Mathematische Zeitschrift*, **27**, 669–752.

1929 Über eine Widerspruchsfreiheitsfrage in der axiomatischen Mengenlehre. *Journal für die reine und angewandte Mathematik*, **160**, 227–41.

1961 *John von Neumann: Collected works*, Vol. 1. Pergamon, Oxford.

von Waltershausen, W. Sartorius

1856 Metaphysik der Geometrie. In *Gauss 1863–1929*, Vol. 8, pp. 267–8 (1900). (This article is an excerpt from von Waltershausen's biographical recollections of Gauss in the memorial volume, *Gauss zum Gedächtnis*, pp. 80–1, Leipzig. The title 'Metaphysik der Geometrie' was supplied by the editors of Gauss's *Werke*.)

Vorländer, Karl

1977 *Immanuel Kant: der Mann und das Werk*, 2 vols, (2nd edn). Meiner, Hamburg.

Wallis, John

1685 *A treatise of algebra, both historical and practical*. J. Playford, London.

1693 *De postulato quinto et definitione quinta lib. 6. Euclidis disceptatio geometrica*. (Lecture on Euclid delivered in Oxford on the evening of 11 July 1663; reprinted in *Opera mathematica*, Vol. 2; German translation in *Stäckel and Engel 1895*, pp. 17-30.)

1695-9 *Opera mathematica*, 2 vols. Oxford University Press. (Reprinted 1968, Georg Olms, Hildesheim; Vol. 1 contains the *Arithmetica infinitarum*.)

Wang, Hao

1957 The axiomatization of arithmetic. *Journal of Symbolic Logic*, **22**, 145-58.

1974 *From mathematics to philosophy*. Routledge, London.

1987 *Reflections on Kurt Gödel*. MIT Press, Cambridge, Mass.

Warda, Arthur

1922 *Immanuel Kants Bücher*. Breslauer, Berlin.

Warnock, Geoffrey J.

1982 *Berkeley*, (2nd edn). Basil Blackwell, Oxford. (1st edn: 1953.)

Warren, John

1828 *A treatise on the geometrical representation of the square roots of negative quantities*. J. Smith, Cambridge.

Weber, Heinrich

1893 Leopold Kronecker. *Mathematische Annalen*, **43**, 1-25.

Weierstrass, Karl

1884 Zur Theorie der aus *n* Haupteinheiten gebildeten complexen Grössen. *Nachrichten der Königlichen Gesellschaft der Wissenschaften zu Göttingen*, 395-419.

1894-1927 *Mathematische Werke*, 7 vols. Herausgegeben unter Mitwirkung einer von der Königlichen Preussischen Akademie der Wissenschaften eingesetzten Kommission. Berlin.

Weyl, Hermann

1917 *Das Kontinuum. Kritische Untersuchungen über die Grundlagen der Analysis*. Veit, Leipzig.

1919 Der *circulus vitiosus* in der heutigen Begründung der Analysis. *Jahresbericht der Deutschen Mathematiker-Vereinigung*, **28**, 85–92.

1920 Über die neuen Grundlagenkrise der Mathematik. *Mathematische Zeitschrift*, **10**, 39–79.

1925 Die heutige Erkenntnislage in der Mathematik. *Symposion*, **1**, 1–32.

1928 Diskussionsbemerkungen zu dem zweiten Hilbertschen Vortrag über die Grundlagen der Mathematik. *Abhandlungen aus dem mathematischen Seminar der Hamburgischen Universität*, **6**, 86–8. (Translated by Stefan Bauer-Mengelberg in *van Heijenoort 1967*, pp. 480–4.)

1931 *Die Stufen des Unendlichen*. Fischer, Jena.

1932a *The open world*. Yale University Press, New Haven.

1932b Topologie und abstrakt Algebra als zwei Wege mathematischen Verständnises. *Unterrichtsblätter für Mathematik und Naturwissenschaften*, **38**, 177–88.

1944a Obituary: David Hilbert 1862–1943. *Obituary Notices of Fellows of the Royal Society*, **4**, 547–53.

1944b David Hilbert and his mathematical work. *Bulletin of the American Mathematical Society*, **50**, 612–54. (Reprinted in *Weyl 1968*, pp. 130–72.)

1949 *Philosophy of mathematics and natural science*. Princeton University Press.

1968 *Gesammelte Abhandlungen*, 4 vols, (ed. Komaravolu Chandrasekharan). Springer Verlag, Berlin.

Whately, Richard

1832 *Elements of logic*. William Jackson, New York.

Whitehead, Alfred North

1898 *A treatise on universal algebra, with applications*, Vol. 1. (No subsequent volumes published.) Cambridge University Press.

1902 On cardinal numbers. *American Journal of Mathematics*, **24**, 367–94.

1944 Autobiographical notes. In *Schilpp 1941*.

Whitehead, Alfred North and Russell, Bertrand

1910 *Principia mathematica* (Vol. 1). Cambridge University Press. (2nd edn: 1925.)

1912 — Vol. 2. (2nd edn: 1927.)

1913 — Vol. 3. (2nd edn: 1927.)

Wigner, Eugene P.

1960 The unreasonable effectiveness of mathematics in the natural sciences. *Communications in Pure and Applied Mathematics*, **13**, 1–14.

Young, William Henry

1903 On non-uniform convergence and term-by-term integration of series. *Proceedings of the London Mathematical Society* (ser. 2), **1**, 89–102.

1907 On uniform and non-uniform convergence of a series of continuous functions and the distinction of right and left. *Proceedings of the London Mathematical Society* (ser. 2), **6**, 29–51.

Zeller, Eduard

1875 *Geschichte der deutschen Philosophie seit Leibniz*, (2nd edn). Oldenbourg, München.

Zermelo, Ernst

1896 Über einen Satz der Dynamik und die mechanische Wäremetheorie. *Annalen der Physik und Chemie*, *Neue Folge*, **57**, 485–94.

1901 Addition transfiniter Cardinalzahlen. *Nachrichten der Königlichen Gesellschaft der Wissenschaften zu Göttingen*, 34–8.

1904 Beweis, daß jede Menge wohlgeordnet werden kann. (Aus einem an Herrn Hilbert gerichteten Brief). *Mathematische Annalen*, **59**, 514–16. (Translated by Stefan Bauer-Mengelberg in *van Heijenoort 1967*, pp. 139–41.)

1908a Neuer Beweis für die Möglichkeit einer Wohlordnung. *Mathematische Annalen*, **65**, 107–28. (Translated by Stefan Bauer-Mengelberg in *van Heijenoort 1967*, pp. 183–98.)

1908b Untersuchungen über die Grundlagen der Mengenlehre. I. *Mathematische Annalen*, **65**, 261–81. (Translated by Stefan Bauer-Mengelberg in *van Heijenoort 1967*, pp. 199–215.)

1909a Über die Grundlagen der Arithmetik. *Atti del IV Congresso Internazionale dei Matematici, Roma*, **2**, 8–11.

1909b Sur les ensembles finis et le principe de l'induction complète. *Acta Mathematica*, **32**, 185–93.

1914 Über ganze transcendente Zahlen. *Mathematische Annalen*, **75**, 434–42.

1929 Über den Begriff der Definitheit in der Axiomatik. *Fundamenta Mathematicae*, **14**, 339–44.

1930* Über Grenzzahlen und Mengenbereiche: Neue Untersuchungen über die Grundlagen der Mengenlehre. *Fundamenta Mathematicae*, **16**, 29–47. (Translated by Michael Hallett in this collection, Vol. 2, pp. 1219–33.)

1932 Über Stufen der Quantifikation und die Logik des Unendlichen. *Jahresbericht der Deutschen Mathematiker-Vereinigung* (*Angelegenheiten*), **41**, 85–8.

1935 Grundlagen einer allgemeinen Theorie der mathematischen Satzsysteme (erste Mitteilung). *Fundamenta Mathematicae*, **25**, 136–46.

Index

The following index includes references to both volumes in this collection. An attempt has been made to supply full names; when this was not possible, or when the attribution of a name is based on conjecture, the uncertain information is enclosed in brackets.